Computability

George Tourlakis

Computability

 Springer

George Tourlakis
Electrical Engineering and Computer
Science
York University
Toronto, ON, Canada

ISBN 978-3-030-83204-9   ISBN 978-3-030-83202-5   (eBook)
https://doi.org/10.1007/978-3-030-83202-5

This Springer imprint is published by the registered company Springer Nature Switzerland AG
The registered company address is: Gewerbestrasse 11, 6330 Cham, Switzerland

Για τον Νίκο την Ζωή και τον Ιάσονα

Preface

This volume is mostly about the *theoretical limitations of computing*: On one hand, we investigate what we *cannot* do at all through the application of *mechanical (computational) processes* —and why. This is the domain of *computability theory*. On the other hand, we study some problems that admit only extremely inefficient mechanical solutions —and explore why this is so. For example, is it always the enormous *size* of the output that makes a solution computationally inefficient? We will see (Chap. 14) that this is a deeper phenomenon and will discover that for any amount of "resources" that we might *a priori* decide to "spend" on any computation, we can build a *computable* function so that *every program* we might employ to compute it will *require*, for all but a finite number of inputs, *more* resources than we had *originally* set aside —and this remains true even if we restrict the outputs of our function to be only the numbers 0 or 1. The study of this phenomenon belongs to *(computational) complexity theory*.

Admittedly, anyone who has programmed a computer, or studied some discrete mathematics, or has done both, will normally have no trouble recognising a *mechanical (computational) process* (or *algorithm*) when they see one. For example, they ought to readily recognise *Euclid's algorithm* for finding the "greatest common divisor",[1] $gcd(a, b)$, as such a process. The same is true of the *binary search algorithm* used to search for a given value in a sorted (ascending or descending order) finite array of values.

Most of us will define a mechanical process as any *finitely described, precise, step-by-step* process, usually performed in terms of a menu of *permissible instructions*, which must be carried out in a certain order. Moreover, these instructions must be extremely *simple*, and such that they can be *mechanically* carried out *unambiguously*, say, by pencil and paper.[2] Here are two examples of instructions

---

[1] "Greatest common divisor" is established jargon, where "greatest" means "largest".

[2] Recipes describe neither mechanical, nor precise, processes. For example, a bit more or a bit less salt is not critical —unless the end-user is on a salt-free diet. And is it *really* necessary for all ingredients to be at room temperature in order to make mayonnaise?

that meet our informal requirements stated in this paragraph: (1) "change the i-th digit of a (natural) number written in decimal notation by adding 1 to it (where adding 1–9 makes it 0 but changes nothing else); another one: (2) "add 1 to a natural number n".

The simpler that these "mechanical" instructions are the easier it is for reasonable people to *agree* that they *can* be carried out mechanically and precisely.[3]

Hm. If all we want to achieve is to agree with reasonable people on the definition of an "algorithm" or "mechanical process", then what do we need a *theory* of computation for? The difficulty with algorithms —programs, as people who program computers call them— is not so much in recognising that a process *is* an algorithm, or to be able to *device* an algorithm to solve a straightforward problem; programmers do this for a living.

The difficulty manifests itself as soon as we want to study the inherent mathematical limitations of computing. That is, when we want to *prove* that a *particular problem*, or a *set of problems*, admits *no* solution via mechanical processes —admits no solution via a program. For example, consider the problems of the type "*check* whether *an arbitrary* given program P computes the constant function that, for every natural number input, outputs always the same number, 42".

We will see in this volume that such a "checker" —if we insist that it be implemented as a *mechanical* process— does *not* exist, regardless of whether the constant output is 42 or some other number.

How does one *prove* such a thing?

We need a *mathematical* theory of computation, whose *objects* of study are programs (mechanical processes), and *problems*. Then we can hope to prove a theorem of the type "there is no *program* that solves this *problem*".

 Usually, we prove this *non-existence* by contradiction, using tools such as *diagonalisation* or *reductions* (of one problem to another).

We will perform this kind of task for many problems in this book and we will develop powerful tools such as *Rice's theorem*, which, rather alarmingly, states: We *cannot have* a mechanical checker that decides whether an arbitrary program X has a given *property* or not, and this fact is true of *all conceivable mathematical* properties, *except two*: (1) The property that *no* program has, and (2) the property that *all* programs have.

On the other hand, in the mathematical domain, an instruction that is so complex as to require us *to prove a theorem in order to be allowed to apply it* is *certainly not* mechanical, because theorem proving in general is not. Nor is it *simple*.

[3] The "machines" of Turing (1936, 1937), the *Turing Machines*, or *TMs* of the literature, are used to define *mechanical processes on numbers* (equivalently, strings of symbols). Their instruction set is so primitive that adding 1 to a number is not allowed! (but this can be simulated, as a macro, of course, by using several Turing Machine instructions). Number manipulation at the primitive instruction level is restricted at the digit level on a TM.

By the way, a comment like this one, which is situated between two -signs is to be considered *worthy of attention*.[4]

The programs studied in a *mathematical* theory of computation cannot be technology-dependent nor can they be restricted by our physical world. In particular, the mathematisation of our "mechanical processes" will allow such processes to "compute" using arbitrary (natural) numbers, of *any length* and hence size, unlike real computers that have overall memory limitations, but also limitations *on the size of numbers that we can store in any program* variable.

Why do we disallow size limitations?

Consider the function which for any natural number x outputs the expression below

$$g(x) = 2^{2^{\cdot^{\cdot^{2}}}} \left.\right\} x \text{ 2's}$$

There is a very straightforward way to implement $g(x)$ recursively,[5] which I present in two versions, mathematically first, and then programmatically.

$$g(0) = 1$$

$$g(x+1) = 2^{g(x)}$$

or

procedure $g(x)$

  if $x = 0$ **then return** (1)

  else   **return** $(2^{g(x-1)})$

It would be a shame to disallow into our theory this very simple function — look at the trivial program!— for reasons of (output) size.[6] We *must* disallow any limitations on number size in our *theory* of *computability* lest we let our theory become the theory of *finite tables* (that represent functions of limited input and output size). Here is a tiny sample of *interesting* problems, both practically and theoretically speaking, that we will explore in this book to determine whether they are amenable to a programming solution, or not. We will find that *they are not*.

---

[4] This type of flagging an *important passage* of text originated in the writings of Bourbaki (1966), that some may find annoying. To be fair, *any exhortation to pay attention is likely to be annoying*, no matter what symbols you use to effect it: Boxes, red type, or road signs.

[5] *Primitive* recursively, but I do not want to be too far ahead of myself, so I am using a familiar term from elementary programming, say in the C language, with recursion.

[6] $2^{2^2} = 16$ but $2^{2^{2^2}} = 65{,}536$ while $2^{2^{2^{2^2}}}$ is astronomical.

- Do we have a mechanical checker for the *program correctness* problem? That is the problem "Is this (arbitrary) program (say, written in the C language) correct?"[7]
- Is there a mechanical process that can check whether a formula of a mathematical theory is a theorem? This is Hilbert's *Entscheidungsproblem*, or *decision problem*. Church proved that for the theory of natural numbers —the so called *Peano Arithmetic (PA)*— the answer is "no". Even the simpler theory obtained from PA by removing the *induction axiom* has a decision problem that is *unsolvable* by mechanical means.

There is an intimate connection between logic and computing that goes beyond the superficial, and rather obvious, analogy between a *proof* and a *computation* (both consist of a finite sequence of very elementary steps).

The most famous result on the *limitations of computing*[8] is perhaps the "unsolvability of the halting problem" (Theorem 5.8.4).

Even more famous are logic's *limitations of proofs* results, that is, Gödel's two *incompleteness theorems*, the first and the second (Gödel 1931, retold in Hilbert and Bernays 1968, Tourlakis 2003a).

We will visit Gödel's first incompleteness theorem several times in this volume, in Chap. 8 at first, and then we will discuss and prove several versions of it (including Rosser's) in Chap. 12.[9]

But this is noteworthy: Our most elementary proof of this result (in Chap. 8) is to show that if we assume that Gödel's theorem is *false*, then so is the unsolvability of the halting problem. In other words, *the unsolvability of the halting problem implies Gödel's first incompleteness theorem* supporting our claim that there is an intimate connection between logic and computation.[10]

There has been a lot of interest in logical circles in the early 1930s for the development of a mathematical theory of mechanical processes and the functions that they calculate —the *calculable* (or *computable*) functions. This was sparked by Hilbert's conjecture that his formalised versions of mathematical theories would have an *Entscheidungsproblem* (*decision problem*: "Is this sentence a theorem?") that would be *decidable* or *solvable* via *mechanical means*. Clearly, logicians were anxious to define (mathematically) *what "mechanical processes" or "mechanical*

---

[7] A "correct" program produces, *for every input*, the output that *is expected* by its *specification*.

[8] A "famous result" is not always hard to prove. For example, Russell's paradox or antinomy regarding Cantor's Set Theory is very easy to describe (it involves a proof that some "collection" is technically not a "set") even to a 1st year class on discrete mathematics. The "halting problem" is perhaps the easiest problem one can show not amenable to a computational solution. Even Gödel's first incompleteness theorem can be made "easy" to prove and understand after a bit of computability and cutting a few corners in the presentation —that is, omit the details of arithmetisation (cf. Chap. 8).

[9] For the statement and complete proof of the second theorem see Tourlakis (2003a) or the earlier (Hilbert and Bernays 1968).

[10] No wonder that computability is considered to be part of logic.

means" are in order to verify, or otherwise, Hilbert's conjecture! Towards this end, several logicians independently defined mathematisations of *mechanical process* or *algorithm* in the early 1930s (Turing 1936, 1937; Church 1936a; Kleene 1936; Post 1936) and there were some notable contributions in the 1960s as well (Markov 1960; Shepherdson and Sturgis 1963).

The first set of efforts (1930s) were quickly shown to be equivalent "extensionally":[11] The various very different mechanical processes introduced were, *provably*, all computing the members of the *same* set of *calculable* or *computable* functions! In fact the 1960s efforts also defined the very same set of functions as the ones produced in the 1930s (these proofs —of the equivalence of all these definitions— are retold in Tourlakis 1984).

There are so many mathematical models of computation out there! Which one should we use in this volume to *found* the theory of computation? The short (and correct) answer is *it does not matter; all models are extensionally the same*, that is why I emphasised "found" above: The model of computation is only meant to *found* the theory and lead (quickly, one hopes) to the pillars of the theory, the *universal* and *S-m-n* theorems along with the *normal form* theorems from which all else follows —no need to continue going back to the model after that.

Nevertheless, let me offer some comments on my choice of model of computation for the early elementary chapters of the present volume, the ones that deal with basics of computation —*without oracles*. Our choice is the URM (*unbounded register machine*) of Shepherdson and Sturgis (1963) through which we will define "computable function" and "computation".[12]

For our later chapters that are on computation *with* an oracle, and more generally computation of *functionals* —meaning functions that admit entire *functions on numbers as inputs*— we use the modern approach due to Kleene where the indexing (of programs) is built-in into the model.[13] Incidentally, an example of a functional, from calculus, that you may immediately relate to is the function $Int(a, b, f)$ where the inputs a and b are natural numbers and f is a function while the output is expected to be $\int_a^b f(x)dx$.

---

[11] A process is an *intentional* object. It *intends* to *finitely represent* a set, even build it —for example, an enumeration process. The built set is an *extensional* object. We know it by its extent; what it contains. For example, the *axiom of extensionality* in set theory, $(\forall y)(\forall z)((\forall x)(x \in y \equiv x \in z) \rightarrow y = z)$ says that two sets y and z will be equal if they have the same members —same extent. The axiom does not care how the sets came about.

[12] In some of our earlier works (Tourlakis 1984, 2003a) we have chosen instead the elegant but somewhat less intuitive Kleene approach of "μ-computable functions" (Kleene 1936, 1943) —one of the three approaches he developed.

[13] One can do computation with oracles using the URM model (e.g., Cutland 1980), as one can do so using the TM model. However the Kleene approach is more elegant and more direct and prepares the reader for further study of computation with even more complex inputs than functions on numbers.

Factors considered in choosing the URM model of computation include:

1. Quick access to the advanced results of the theory. To satisfy this parameter the *optimal* —in our opinion— approaches to the foundation of computability are

   - The completely number-theoretic approaches of Kleene, in particular the one via his μ-recursive functions formalism.[14]
   - The other, even quicker approach, where expressions such as $\{e\}(\vec{x}) \simeq y$ —interpreted as "the program (coded as the natural number) e, on input $\vec{x}$, produces output y" (Kleene 1959)— are taken as the starting point. Here the *index* or *program code*, or just "program" of each partial recursive function $\{e\}$ is already built-in in the inductive definition of $\{e\}(\vec{x}) \simeq y$.[15] In all other approaches one does a fair amount of "coding" work to show that computable functions can be so indexed.
   - Yet, the URM can reach the universal function, S-m-n and and normal form theorems almost as quickly, especially if, as we do in this volume, we by-pass arithmetisation in the elementary and foundational Chap. 5, relying instead on "Church's thesis".

2. Mathematical rigour. The first two approaches above are the most rigorous mathematically. However, one may want to trade some mathematical sophistication in order to reach a wider readership and still maintain a fairly rapid access to the major results of computability. Towards widening the readership scope authors invariably adopt "programming based" approaches to achieve the foundation of computability. Well chosen such programming foundations provide a fairly quick access to advanced results without unduly inhibiting the readers who are less mathematically advanced.

   - The "programming approach" is consistent with the oft stated belief that a pedagogically sound introduction to the theory of computation should take a student from the concrete to the abstract. This pedagogy supports audiences with diverse backgrounds, from undergraduate university computer science (or mathematics, logic, or philosophy) to graduate.
   - The choice of programming formalisms is not unlimited. Two prominent examples to choose from are TM-programming or URM-programming.
     But apropos pedagogy and the "from the concrete to the abstract" desideratum one must ask: What is the likely percentage of the readership who have programmed in *assembly language* ("concrete" TM-like programming) vs. those who have programmed in a high-level language like C or Python or JAVA or even FORTRAN? ("concrete" URM-like programming).

---

[14] We derive this as a *post-facto* characterisation in Sect. 5.6 and Chap. 7 but in Tourlakis (1984) this approach was our *starting point*; the foundation of computability.

[15] There is a tendency in the literature to say "induction" for proofs and say "recursive", rather than "inductive", for definitions. We just thought that the close proximity of terminology such as "each partial *recursive* function" and "*recursive* definition" might be confusing.

The answer to the question above tips the scale in favour of the register machine.

Besides, programs written in the TM language are not only hard for the students to understand, but are also hard for authors of books like this and similar ones to write precisely, *in detail*, and do so *within a rigorous mathematical discourse* —without hand-waving!

As far as I know, the only books in the entire computability literature that have used the Turing machine *mathematically* are the monograph of Davis (1958a) and the text by Boolos et al. (2003). Both works, quite appropriately, *abandon* "programming" (the TM) once they have proved rigorously what I have called above the "pillars of the theory".

All the other books (on the subject of computation) that I am aware of —and which have adopted the TM— on one hand, compromise mathematical rigour by extensively hand-waving in their proofs and constructions, *presumably because the TM model is so hard to* "program"[16]

On the other hand (which is probably worse than handwaving), these texts continue programming TMs until the very last theorem they present, thus misrepresenting computability as the practice of programming the Turing machine sprinkled with the *occasional* unsolvability result every now and then. Computability is not about programming in this or that language, but rather is about its *results* that detail what computation *is*, what it *does*, but, above all, *what it cannot do*.

This is a book on the theory of mechanical processes and computable functions, both treated as the objects of study in a mathematical theory, not as activities. So the amount of "programming" is held to a minimum. Yet, inevitably, whether founded via TMs or URMs, the theory of computation will *require* a little bit of "programming" *initially* (in the chosen model), to establish a few foundational results that bootstrap the theory. For example, we show by brute force programming (in the URM model) that

- The function that adds 1 to its input and returns the obtained value as output is computable.
- The function, which for every input returns *nothing* as output is computable.
- If two one variable functions are computable, then so is their composition.

Most of such programming lemmata are extremely tedious when proved for the TM language —unless one trades proof for hand-waving— *but are almost trivial for the URM language*. This is another reason for choosing URMs: Use programming as the foundation, but be quick, readable, and easily verifiable to be *correct* while going about it!

What is covered in this volume on computability is, partly, what is more or less "normal" to include in such a theory in the context of a reasonably sized book. It

---

[16] For example, try to program *primitive recursion* on a TM, or see (Boolos et al. 2003, p.30) where a TM for multiplication of two natural numbers is given.

is also, partly, influenced by the author's preferences (I admit that the first part, "normal", is influenced by the second).

The chapters, in outline are, as follows:

We start with a "Chapter 0" to provide a one-stop-shopping kind of reference on essential topics from discrete mathematics and elementary logic that are needed to make studying this volume fun.[17]

A *pessimist's* justification for a course in the theory of computation is that it serves to firm up the students' grasp of (discrete) mathematical techniques and mathematical reasoning, that weren't learnt in the prerequisite discrete math course. However I am rather *optimistic* that by the time the readers have studied the subject matter of this volume they will be equipped with substantially more than the mastery of the *prerequisite*.

For this to happen, it cannot be emphasised enough, the student of a theory of computation course *must* be already equipped with the knowledge expected to be acquired by the successful completion of a one-semester course on discrete mathematics that *does not* skimp on logic topics.

So why include a "Chap. 0" in this volume? This chapter is like the musical notes you see in front of professional Baroque players who hardly look at their notes while playing; they are just having fun! The notes are there to help *only if needed*!

The same holds with our "Chap. 0" and you: You ought to be familiar with the topics in said chapter, but I include it for your *reference*. To be consulted only if you *absolutely need it*, that is —and also, partly, to give you a guide as to which topics of discrete math you will have to call upon most frequently.

In order to be sure that readers have all the prerequisite tools available to them, if needed, our "Chap. 0" had to be longer than the *norm* especially on the topic of *formal logic*. This is due to our *need* of formal logic in order to present several (*complete*, mathematically) proofs of Gödel's first incompleteness theorem in Chap. 12.

I wanted, above all, to retell two stories in Chap. 0: *logic* and *induction*, that I often found are insufficiently present in the student's "toolbox", notwithstanding the earlier courses they may have taken. Now, (naïve) set theory topics included in discrete math courses provide not much more than a *vocabulary* of mathematical notation. I include this topic *mainly* to share with the readers early on that (1) *not all* functions of interest are *totally defined*, and (2) introduce them to Cantor's ingenious (and simple) *diagonalisation* argument that recurs in one or another shape and form, over and over, in the computability and complexity part of the theory of computation.

Enough about discrete math. Some *high-level programming* is also highly desirable (I should say, a prerequisite) to provide you with a context for all these *Un*computability results that are contained in this volume.

---

[17] What *is* essential is to some extent a subjective assessment.

Chapter 1 begins our story proper. It introduces the mathematisation of "mechanical process" by introducing the URM programs, which express such processes, *by definition*, in our mathematical foundation of computability. We call these processes computations and the functions that they compute *computable*.

How much mathematical rigour and how much intuition is a good mix? We favour both.

The main reason that compels us to teach (meta)theory in a computer science curriculum is not so much to prevent the innocent from trying in vain to program a solution for the halting problem (cf. Theorem 5.8.4), just as we do not teach courses in axiomatic Euclidean geometry in order to discourage "research" on circle squaring and angle trisecting.

Rather, we want the student to learn to mathematically *analyse* computational processes —not simply to *use* them— and (by doing the former) to become aware that not every problem is computationally solvable, and that, among the solvable ones, there are those that are computationally "*intractable*" in the sense that their computational solutions require prohibitive amounts of computational resources. It is our expectation that the reader will become proficient in the techniques that *establish* these limitations-results in the domain of computing.

The techniques the readers will learn to use in this volume are a significant extension of the formal mathematical methods they learnt in a discrete mathematics course.[18] These techniques, skills and results justify putting "science" in the term *computer science*.

The two premier tools, *diagonalisation* and *reductions*, that you will have learnt to use by studying this volume are ubiquitous in computability (undecidability, unprovability) and complexity (computational unfeasibility). Just consider the wide applicability of diagonalisation as evidence of the value and power of this tool: It was devised and used by Cantor to show that *the reals*, $\mathbb{R}$, is an *uncountable* set (cf. Sect. 0.5.3). It was then *used* by Russell, to show that Cantor's unrestricted use of "defining (mathematical) properties", $P(x)$, in building "sets" such as $\{x : P(x)$ is true$\}$ leads to a nasty *paradox*.[19] Then Gödel used it in his *diagonalisation lemma* to prove his incompleteness theorems (for example, cf. Lemma 9.1.1).

---

[18] The term "formal methods", when describing curriculum, uses the qualifier "formal" loosely to describe courses, such as *discrete mathematics* and *logic*. Also describes courses that are built upon the former two, where students are applying, *rigorously, mathematical techniques* in investigations of *program specification and correctness* (e.g., software engineering) as well as in investigations of the limitations of computing, computational complexity, logic, analysis of algorithms, AI, etc. The qualifier "formal" is, in this context, *not* that of Hilbert's; here it simply means "mathematically rigorous". Hilbert's "formal" mathematical theories are instead founded axiomatically via purely *syntactic* definitions, and the mathematical proofs carried out in said theories are also purely syntactic objects, devoid of semantics —based on *form* only.

[19] The word "paradox" is from the Greek παράδοξο, meaning inconsistent with one's intuition or belief; in short, a contradiction. In this connection we have the well known *Russell's paradox*, that the *collection* $\{x : x \notin x\}$ *cannot* be —technically— a *set*.

Then computer scientists used it over and over again, to prove the unsolvability of the halting problem (Theorem 5.8.4), the existence of total computable functions beyond the primitive recursive (Sect. 3.4), the existence of nontotal computable functions that *cannot be extended* to total computable functions (Example 6.2.1), and many other results in computability. Blum (1967) used diagonalisation (and the recursion theorem, which *is* a diagonalisation tool) in proving that there are total computable functions that have *no* best program; every program for them *can be sped up* (*Blum's speed-up theorem*) (Chap. 14).

It is also important to develop and strengthen the students' intuition that supports and discovers informal proofs! Students need to understand "what is going on" and help them see the essence of the argument rather than get lost in a maze of symbols. To this end we introduce Church's thesis in Chap. 5, and start using it in our proofs that *this or that construction leads to a computable function*, computable by a URM, that is.

This "thesis" is a *belief* —not a theorem or metatheorem— of Church, according to which any informal algorithm can be implemented as a Turing machine, or as any of the other equivalent mathematical computation models that were developed in the 1930s. It applies to Markov algorithms and URMs as well, as these two models are equivalent to all those proposed in the 1930s. Application of the "thesis" in proofs will strengthen intuition, an appreciation of "what's going on", and will offer *quick* and *user-friendly* progress in our study of the theory, especially in its *formative steps* in Chap. 5.

However, it is also *essential for the students to get a* generous *dose of* rigorous mathematical proofs in this study of computability, uncomputability and complexity, and we aim in this volume to also develop in the reader the ability to understand *and* apply such methods. Thus, we almost always[20] supplement a "proof by Church's thesis" with a rigorous mathematical proof. There are also instances where you cannot apply Church's thesis. For example, it is not applicable towards establishing that a function is *primitive* recursive, so rigorous mathematical methodology will be involved here out of necessity.

Apropos primitive recursive functions, these are introduced in Sect. 2.1. These functions provide us with a rich toolset, including coding/decoding tools and the ability to simulate URM computations mathematically (Sect. 2.4.4). This result — which uses the elementary tool of simultaneous (primitive) recursion— is at the heart of our proof that the Kleene predicate $T(x, y, z)$[21] is primitive recursive without doing a tedious arithmetisation of URM computations (the reader who absolutely wants to see such an arithmetisation is referred to Tourlakis (2012), or

---

[20] There are two (notable) instances where we did *not* do so supplement —in Chap. 5— namely, the proofs of the *universal function* and *S-m-n* theorems. This was so because we wanted to *not* do an arithmetisation of URM computations in this volume. However, in the chapter on oracle computation —where we have left the URM and switched to Kleene's indexing-based formalism— the S-m-n (and universal function) theorems *are* mathematically derived.

[21] $T(x, y, z)$ is true precisely when the URM at address x, when given as input y, has a computation of z steps (instruction-to-instruction steps).

simply wait until our Chap. 13 —but the latter arithmetisation is for a different foundation: of (oracle) computability, not for URM computability). The Kleene predicate is a pivotal tool in the development of computability.

Chapter 3 develops the *loop programs* of Meyer and Ritchie (1967). These programs are significantly less powerful than the URM programs. They *only* compute the primitive recursive functions; no more no less. In particular, *provably*, they compute only total functions, unlike the URMs. We will learn that while these loop programs can only compute a very small subset of "all the computable functions", nevertheless they are *significantly more than adequate* for programming solutions of any "practical", computationally solvable, problem. For example, even restricting the nesting of loop instructions to *as low as two*, we can compute — in principle— enormously large functions, which, for each input x, can produce astronomical outputs such as

$$\left.\begin{matrix} 2^{\cdot^{\cdot^{2^x}}} \\ 2^{\cdot} \end{matrix}\right\} 10^{350,000} \text{ 2's} \tag{1}$$

even with input $x = 0$.[22] The qualification above, "in principle", separates what theory can do from what real machines can do due to physical (as opposed to theoretical) limitations that impede the latter. This chapter ends by showing, informally by diagonalisation, that a total intuitively computable function exists that is *not* primitive recursive.

Chapter 4 is on the Ackermann function. Using a *majorisation argument* we prove that this function is *not* primitive recursive.[23] Rather startlingly, we also prove —via a simple arithmetisation of the "pencil-and-paper computation" of the Ackermann function— that the *graph* of the Ackermann function[24] is primitive recursive. An application of unbounded search to the graph shows that the Ackerman function is recursive.

Chapter 5 enlists Church's thesis (belief) that (essentially) states that an informally described program *can* be mathematised as a URM program. Here we present our first few unsolvability results, including the well known (at least by name) "halting problem" and offering a *diagonalisation proof* as to why it is unsolvable (by "mechanical means").

We also introduce here our first ad hoc *reduction argument* (these arguments are thoroughly developed and used in the following chapter) to prove that the *program*

---

[22] This result will have to wait until we study the complexity (definitional and dynamic) of primitive recursive functions, in Chap. 15.

[23] The majorisation argument consists in proving that the Ackermann function is too big to be primitive recursive. In a precise sense, it is strictly bigger (its outputs are) than any primitive recursive function.

[24] If f is a function of one argument, its graph is the relation $y = f(x)$.

correctness problem is also unsolvable by mechanical means. The chapter includes the definition of the Kleene T-predicate, the proof of its primitive recursiveness, the *normal form* theorems of Kleene, and a number theoretic (programming-independent) characterisation of the set of partial recursive functions. It also includes the basic foundational tools, the *universal* and *S-m-n* theorems (both proved invoking Church's thesis).

Chapter 6 thoroughly examines the technique of *reduction* used towards proving unsolvability and non-semi-recursiveness results. The closure properties of the semi-recursive relations are proved, and the recursively (computably) enumerable sets are shown to be just another way of looking at semi-recursive sets. The chapter concludes with the Rice's theorem and the two central Rice-like lemmata (Theorems 6.10.1 and 6.10.3).

In Chap. 7 we derive a characterisation for the set of recursive (and partial recursive) functions that does not require primitive recursion as a *given* operation. This result, as well as the *tool* used to obtain it —arithmetising primitive recursive computations using numbers expressed in "*2-adic* (pronounced "dyadic") as opposed to *binary* notation"— will be useful in Chap. 12.

Chapter 8 is a first visit to the incompleteness phenomenon in logic, presenting a reduction argument of the form "the unsolvability of the halting problem implies that Peano arithmetic is syntactically *incomplete* —that is, it cannot syntactically prove all arithmetical truths". This strongly connects unsolvability with unprovability.

The next chapter (Chap. 9) is on the second recursion Theorem (9.2.1) and a few applications, including another proof of Rice's theorem, and the technique of showing that recursive definitions have partial recursive solutions (*fixed points*, or *fixpoints*).

Apropos fixpoints, the preamble to Chap. 9 shows the connection between Kleene's version of the recursion theorem and the Carnap-Gödel version —the fixpoint or diagonalisation lemma— that was used by Gödel to construct a Peano arithmetic formula that says "I am not a theorem".

Chapter 10 introduces indices for primitive recursive derivations and produces a recursively defined universal function for $\mathcal{PR}$. This provides a mathematically complete and rigorous version of the hand-waving argument that we produced at the end of the loop-programs chapter, and shows here that the *universal function* for $\mathcal{PR}$ is *recursive* but *not* primitive recursive (in Sect. 3.4 we could only show that the universal function was "*intuitively* computable"). Some of the tools produced in this chapter have independent interest: An S-m-n theorem and a recursion theorem for $\mathcal{PR}$.

Chapter 11 has some advanced topics on the enumerations of recursive and semi-recursive sets while Chap. 12 is on creative and productive sets and revisits the Gödel incompleteness theorem, and proves Church's theorem on the unsolvability of Hilbert's *Entscheidungsproblem*, and the Rosser sharpening of the first incompleteness theorem. We prove the above two results for Robinson's arithmetic (that is, Peano arithmetic with the induction schema *removed*), with no recourse to any semantic help (the original statement and proof of Gödel's theorem includes no semantic ideas in either its statement or its proof, but in Chap. 8 we cheated and presented a semantic statement *and* proof). A mathematically rigorous deceptively short proof of Gödel's *second incompleteness theorem* —"*if Peano arithmetic is consistent, then it cannot syntactically prove this fact*"— is also given, but it *assumes* that the proofs of the so-called "*derivability conditions*" are provided to us by an oracle.[25] This chapter also contains proofs of *Tarski's theorem* on the *non definability of truth*.

Chapter 13 introduces "oracle computations", otherwise viewed as computing with arguments that are entire functions on the natural numbers —*extensionally*[26] passed as arguments. For example, a numerical integration subroutine $Int(a, b, f)$ has as arguments two numbers, a and b, and an entire function f (a table for f) and approximates the value of $\int_a^b f(x)dx$.

In this chapter we present increasingly more inclusive formalisms including that of Kleene (1959) and Moschovakis (1969), both using a Kleene-like (Kleene 1959) index-based foundation for partial computable functionals. The function inputs in both versions *are allowed to be nontotal*.

The Kleene version of computing with *total function inputs* is also detailed, leading to the concept of *computing relative to a set*, *Turing degrees* and the *priority method* of Muchnik (1956), Friedberg (1957) in the modern version devised by Sacks (1963). The two versions of the "first recursion theorem" due to Kleene and Moschovakis are also proved in this chapter.

Finally, we also classify here all the *arithmetical relations* (first encountered in Chap. 8) according to their *complexity*. We also give a different proof of Tarski's theorem on the non definability of truth (the first proof was given in Sect. 12.8.1).

In Chaps. 14 and 15 we present some complexity results of all computable functions in the former and of all primitive recursive functions and some interesting subclasses in the latter —for example, the class,[27] of feasibly computable functions of Cobham, that is, those computable in deterministic polynomial time (measured as a function of input length). This class has a neat number-theoretic, program-independent characterisation.

---

[25] Such "oracles" are Hilbert and Bernays (1968), Tourlakis (2003a). Gödel never published a proof of his Second theorem.

[26] Imagine that the function f is passed as a "graph", *not* as a "program", $f = \{(a, b), (a', b'), (a'', b''), \ldots\}$. This graph is in general an infinite set.

[27] Recursion theorists use the terms "class" and "set" synonymously. Thus, "class" in this volume does *not* have the same meaning that set theorists assign to the term as a possibly *non-set collection*.

Some interesting results of Chap. 15 include that,

1. in general, FORTRAN-like programs —more accurately, loop programs of Chap. 3— that allow nesting of the loop instruction equal to just *three* have highly *impractical* run times, certainly as high as

$$
\left. 2^{2^{\cdot^{\cdot^{2}}}} \right\} x \; 2\text{'s}
$$

and

2. even if we restrict loop (program) nesting level to just *two*, the *equivalence problem* of programs, that is, the *program correctness* problem, is unsolvable for such loop programs! In fact, it is not even semi-recursive.

We have included in this volume a good amount of *complexity theory* that will likely be mostly skipped in all but the most advanced courses on computability, as such courses proceed at a much faster pace. There are a few "high-level complexity" results already in Chap. 14 that use diagonalisation for their proof. Later, quite a bit is developed in Chap. 15, including an account of Cobham's class of *feasibly computable functions*; and a thorough look at the hierarchy theory of the primitive recursive functions culminating in the rather startling fact that we cannot algorithmically solve the *correctness problem* of FORTRAN-like programs even if we restrict the *nesting of loops* to just *two levels*. FORTRAN-like languages have as an abstract counterpart the *loop programs* of Meyer and Ritchie (1967) that we study in Chap. 3 and again in Chap. 15.

If I used this book in 3rd or 4th undergraduate course on computability I would skim very quickly over the mathematical "review" chapter, and then cover Chaps. 1, 2, 3, 5, parts of Chap. 6 —say, the first three sections— Chap. 8 (but instructor-simplified further). I would also want to cover the section on the recursion theorem, and also as much complexity theory, as time permits, from Chap. 15.

In a graduate level class I would add to the above coverage Chaps. 11, 12 and 13, reducing as needed the amount of material that I cover in Chap. 15.

The reader will forgive I hope the many footnotes, which the style police may assess as "bad style"! However, there is always a story within a story, the "... *and another thing* ..." of Douglas Adams (or lieutenant Columbo), that is best delegated to footnotes not to disrupt the flow ... Incidentally the book by Wilder (1963) on the foundations of mathematics would lose most of its effectiveness if it were robbed of its superbly informative footnotes!

The style of exposition that I prefer is informal and conversational and is expected to serve well not only the readers who have the guidance of an instructor, but also those readers who wish to learn computability on their own. I use several devices to promote understanding, such as frequent "pauses" that anticipate questions and encourage the reader to rethink an issue that might be misunderstood if *read* but *not* studied and reflected upon. All pauses start with "**Pause.**" and end with "◄"

Apropos quotes and punctuation we follow the "logical approach" (as Gries and Schneider 1994 call it) where punctuation is put inside the quotation marks if and *only if* it is a logical part of the quoted text. So we would *never* write

> The relation "is a member of" is fundamental in set theory. It is denoted by "∈."

No. "." is *not* part of the symbol! We should write instead

> The relation "is a member of" is fundamental in set theory. It is denoted by "∈".

Another feature of the above reference that I have adopted is the logical use of the em-dash "—" as a parenthesis. It should have an *left version* and a *right version* to avoid ambiguities. The left version is contiguous with the following word but not with the preceding word. The right version is the reverse of this. For example, "computability is *fun* —as long as one has the prerequisites— *and* is definitely useful towards understanding the computing processes and their limitations".

I have included numerous remarks, examples and embedded exercises (the latter in addition to the end-of-chapter exercises) that reflect on a preceding definition or theorem.

Influenced by my teaching —where I love emphasising things— but also by Bourbaki, I use in my books the stylised "winding road ahead" warning, ⬨, that I first saw in Bourbaki (1966).

It delimits a passage that is too *important* to skim over. I am also using ⬨⬨ to delimit passages that I could not resist including.

Frankly, the latter *can* be skipped with no injury to continuity (unless you are curious, or *need* the information right away). Thus, the entire Chap. 0 ought to be enclosed between ⬨⬨ signs but I forgot to do so!

There are 256 end-of-chapter exercises and 49 embedded ones in the text. Many have hints and thus I refrained from (subjectively) flagging them for "level of difficulty". After all, as one of my mentors, Allan Borodin, used to say to us (when I was a graduate student at the University of Toronto), "Attempt *all* exercises. The ones you can do, don't do; do the ones you *cannot do*".

Acknowledgments I wish to thank all those who taught me, including my parents, Andreas Katsaros and Yiannis Ioannidis, who taught me geometry; more recently, Allan Borodin, Steve Cook and Dennis Tsichritzis, who taught me computability; and John Lipson, Derek Corneil and John Mylopoulos, who encouraged me to employ piecewise linear topology in computer science.

Toronto, ON, Canada   George Tourlakis
June 2021

Contents

Chapter 0
Mathematical Background: A Review

Overview

As noted in the Preface, I will be assuming of the reader at least a solid course in discrete math, including simple proof techniques (e.g., by induction, by contradiction) and some elementary logic. This chapter will repeat much of these topics, but I will ask forgiveness if there are no strict linear dependencies between the sections in Chap. 0.

For example I talk about *sets* in the section below, which is about *induction* (but I *deal* with sets in *later* sections), I talk about *proof by contradiction* (to be redone formally in the section on *logic*, later).

There are two licensing considerations for this: One, this section is a *review* of (assumed) *known* to the reader topics (except, probably, for the topic on *topology*). It is not a careful construction of a body of mathematics based on previous knowledge. I am only making sure that we start "on the same page" as it were.

Two, the concept of natural number is fundamentally grounded in our intuition and it is safe, with some care, to start this review with the topics of induction and the *least principle* in the set of natural numbers prior to a review of other topics such as sets and logic.

The most important topics in Chap. 0 are the existence and *use* of *nontotal* functions —and the realisation that we deal in this volume with *both* total(ly defined) functions and non total functions, the two types collectively put under the category of *partial functions*. These play a major role in computability.

Other important topics in our review here are *definitions by induction* in a partially ordered set, and Cantor's *diagonal method* or *diagonalisation*, which is an important tool in computability. We will also do a tiny bit of *point set topology*, only used (sparsely) in Chap. 13.

© Springer Nature Switzerland AG 2022
G. Tourlakis, *Computability*, https://doi.org/10.1007/978-3-030-83202-5_0

0.1   Induction over $\mathbb{N}$

This section presents the simplest forms of induction in the set of the natural numbers $\mathbb{N} = \{0, 1, 2, \ldots\}$, the mathematical habit being to use the ellipsis "$\ldots$" to say, more or less, "you know how to obtain more elements". In this case the "implied" (or, as a mathematician would likely say, "understood") "rule" to get more elements is simply to add 1 to the previous element, and *voilá*, we got a new element!

The induction technique helps in proving statements about (or properties of) — denoted by $P(n)$ for "property of n"— the natural numbers handing us out some extra help, an extra *assumption* called "induction hypothesis". In outline induction, essentially, goes like this:

> *Suppose I want to prove $P(n)$, for all n, and to do so from a set of assumptions (axioms, other theorems) $\mathscr{A}$.*
>
> *Then I can so prove suffice it that I*
>
> 1. (I.H.) *fix n and* add the assumption $P(n)$ to what I already know, that is, to $\mathscr{A}$.
> 2. (I.S.) *prove $P(n + 1)$ —using $\mathscr{A}$ and the I.H.— where "n" is still the n I fixed in step 1.*
> 3. (Basis.) *must be sure to prove (often, but not always, this is trivial) $P(0)$*
>
> That is it! *If I do 1.–3. then I have proved $P(n)$ from hypotheses $\mathscr{A}$, for all n.*
> *We simply say that* "we have proved $P(n)$ by induction on n, over $\mathbb{N}$". *Unless there might be confusion as to what set we are doing induction over (or in) we omit the "over $\mathbb{N}$" part.*
> However, we always specify the variable we do induction on (here it is n).

 Next to the numbering above, in brackets, we gave the acronyms for the three steps: *I.H.* stands for *induction hypothesis*, *I.S.* stands for *induction step*, and the *Basis* step (no acronym) is the easy (usually) proof for the smallest possible value of n —normally 0.

One makes sure that the "fixed n" for which we assume $P(n)$ remains *unspecified*, so the logical steps from $P(n)$ to $P(n + 1)$ (from 1. to 2.) are "general", that is, valid for *any* n whatsoever.

The induction process above translates to

0.1.1 Remark (Induction Principle) If for a property of natural numbers, $Q(m)$ we have

1. $Q(0)$ is true
2. On the assumption that $Q(n)$ is true (for fixed unspecified n) I can prove the truth of $Q(n + 1)$ —or we can say, equivalently, for all n, $Q(n)$ implies $Q(n + 1)$,

then *all* natural numbers have the property $Q(n)$.   □

0.1.2 Example. (The Usual "First Example" in Discrete Math Texts)
    We prove here that

$$\sum_{i=0}^{n} i = \frac{n(n + 1)}{2} \tag{1}$$

Well, (1) states our "$P(n)$" so, following the process, we list

1. Fix n and take (1) as the I.H.
2. Thus (I.S.),

$$\sum_{i=0}^{n+1} i \overset{meaning\ of\ \Sigma}{=} n+1+\sum_{i=0}^{n} i$$
$$\overset{I.H.}{=} n+1+\frac{n(n+1)}{2}$$
$$\overset{arithmetic}{=} \frac{(n+1)(n+2)}{2}$$

3. The Basis says $\sum_{i=0}^{0} i = \frac{0(0+1)}{2}$ or $0=0$ which is true.

Done.   □

0.1.3 Remark.

1. We can also do induction over the set $\{k, k+1, k+2, k+3, \ldots\}$. Steps I.H. and
   I.S. have the same form as above, but we will be sure to be conscious of the fact
   that in the I.H. we choose and fix an $n: n \geq k$ (not $n \geq 0$).
   The Basis, of course, will be verified for $n = k$: $P(k)$.
   Why would we want to do that? Because some properties of numbers start being
   true from some $k > 0$ onwards (see next example).
2. We can also do induction over a finite set of natural numbers, $\{0, 1, 2, \ldots m\}$ or
   even $\{r, r+1, r+2, \ldots m\}$. In this case we will be sure to note that the n we fix
   for the I.H. is $< m$ so that when we look into the $n+1$ case we are still in (over)
   the finite set. In the second case the n we fix for the I.H. satisfies $r \leq n < m$ □

0.1.4 Example. We prove that

$$n+2 < 2^n, \text{ for } n \geq 3 \tag{2}$$

Here the first inequality in (2) is our "$P(n)$", so following the process we list

1. Fix n and take (2) as the I.H.
2. Thus (I.S.),

$$n+1+2 = n+2+1$$
$$< \overbrace{2^n}^{by\ I.H.} +1$$
$$\overset{arithmetic}{<} 2^n + 2^n$$
$$\overset{arithmetic}{=} 2^{n+1}$$

3. The Basis is for $n = 3$: Note that $3 + 2 < 2^3 = 8$ is true. Note also that, for $n < 3, n + 2 < 2^n$ is false, so $n = 3$ is the correct start point for the property and therefore for the induction.

Done.   □

A *convenient* but equivalent variant of induction is this:

0.1.5 Definition. (Course-of-Values or "Strong" Induction, or CVI)

The *course-of-values induction,* also called *strong induction*[1] is the following process *employed to prove a property $P(n)$, for all n in* $\mathbb{N}$:

1. Verify the *Basis, $P(0)$*.
2. (I.H.) Fix an n and assume "$P(k)$, for all $k < n$".
3. (I.S.) From the I.H. (and any other assumptions we may have) prove $P(n)$.

                                                                                                   □

2. and 3. above are often stated as

1′. (I.H.) Fix an n and assume $P(k)$, for all $k \le n$.
2′. (I.S.) From the I.H. (and any other assumptions we may have) prove $P(n + 1)$.

0.1.6 Example. Let us prove by induction that every natural number $n > 1$ has a *prime divisor*. Recall that a *prime* is a number > 1 that has just two divisors: 1 and itself. Examples are 2, 3, 7, 11.

Note that our Basis need be $n = 2$ since the property about prime divisors is stated for $n > 1$.

We will use CVI. We will see why this is the right induction to use as we are progressing in this proof.

- (Basis.) For $n = 2$ the property is true: 2 divides 2, and 2 is prime.
- (I.H.) Fix n and assume the property for all k, such that $2 \le k \le n$.
- (I.S.) How about $n + 1$? Well, if $n + 1$ is a prime then we are done, since $n + 1$ has a prime factor (divisor): itself.
  If $n + 1$ is not prime then $n + 1 = a \cdot^2 b$, where neither of a or b equal 1.
  We would like to apply the induction step, but simple induction would need some acrobatics since we do not expect either of a or b to equal n. So, as we are using CVI, the I.H. assumes the property for all numbers $\le n$ (where $n > 1$). Now each of a and b are $< n + 1$ hence $\le n$.[3] By I.H. each has a prime factor. Concentrate, say, on a. But its prime factor is a factor of $n + 1$ as well. Done.

                                                                                                   □

---

[1] This is a misnomer: The two types of induction, "simple" and "strong" are actually equivalent, as we will soon see.

[2] "·" denotes multiplication here.

[3] As $a \ne 1 \ne b$, we have $a > 1$ and $b > 1$, so each a and b is ≥ 2 as needed —see I.H.

The least number principle may be more widely known and/or more widely believable as obvious:

0.1.7 Definition. (Least Principle) If S is a set of natural numbers that is not empty (i.e., it contains at least one number), then it contains a *least* (smallest) *number s* in the sense that for each x in S we have $s \leq x$.   □

0.1.8 Theorem. *The* (simple) *induction principle (Remark 0.1.1) is equivalent to the least principle (Definition 0.1.7).*

Proof So, from any one principle we can prove the other. Let us prove this statement.

- (*Suppose we have the induction principle.*) Let then S be a nonempty set of natural numbers. We will show it contains a least number.
  We will *argue by contradiction*: So let S have *no least element*. Let us write $\mathscr{S}(n)$ for the property that says

$$\text{``}none \ of \text{ the } k \text{ such that } k \leq n \text{ are in } S\text{''.} \tag{1}$$

  Now following the steps of Remark 0.1.1 we note

  1. $\mathscr{S}(0)$, which means "0 is *not* in S", is true (for if 0 were in S it would clearly be least, contradicting hypothesis).
  2. We take as I.H. $\mathscr{S}(n)$ is true for some fixed unspecified n, that is, (1) is true.
  3. How about $\mathscr{S}(n + 1)$? First off, this claims that

$$\text{``}none \ of \text{ the } k \text{ such that } k \leq n + 1 \text{ are in } S\text{''.}$$

  If this is false, $n + 1$ *is* in S since by the I.H. all the smaller numbers are not. But then $n + 1$ is *least* in S which contradicts our assumption. Thus we must revert to "$n + 1$ is not in S" thus proving $\mathscr{S}(n + 1)$.

  The induction having concluded, we have proved that no n is in S. This contradicts the given fact that S is nonempty.
- (*Suppose we have the least principle.*) Let then $Q(n)$ be a property for which we verified 1.–2. of Remark 0.1.1. We pretend that we do not know what this process achieves.
  We will prove using the least principle (Definition 0.1.7) that we have proved that *the set of all n making $Q(n)$ true is* $\mathbb{N}$, thus establishing the validity of the simple induction principle. For convenience let us use the symbol $\{n \in \mathbb{N} : Q(n)\}$ for the described set,[4] and let S be all natural numbers *outside this set.*
  Suppose $\{n \in \mathbb{N} : Q(n)\} \neq \mathbb{N}$. Then S is nonempty. By the least principle,

---

[4] This notation is learnt in a course on discrete mathematics and we will revisit it in the section on sets.

$$\text{let } s \text{ be smallest in } S \qquad (\dagger)$$

Now $s \neq 0$ since we have verified $Q(0)$ during the process 1. and 2. (Remark 0.1.1). So, $s - 1$ is a natural number *not* in S hence $Q(s - 1)$ is true. But the induction we carried out diligently establishes the *truth of $Q(s)$* (*from that of $Q(s - 1)$*). This contradicts that s is *not* in $\{n \in \mathbb{N} : Q(n)\}$ (by ($\dagger$)).

We have no other exit but to admit we were wrong to assume "the set of numbers, S, outside $\{n \in \mathbb{N} : Q(n)\}$ is not *empty*". Thus *no $n \in \mathbb{N}$ exists outside* $\{n \in \mathbb{N} : Q(n)\}$, that is, all n are in: In short, $Q(n)$ is true, for all n. The induction *did* work!

$\square$

0.1.9 Theorem. *The CVI principle (Remark 0.1.5) is equivalent to the least principle (Definition 0.1.7).*

 So, once this is proved, we have that from any one principle we can prove the other.

Proof Let us prove this theorem.

- (*Suppose the CVI principle works (per Definition 0.1.5).*) *Let then S be a nonempty set of natural numbers.*
  We will show, *arguing by contradiction*, that S contains a least number.

  $$\text{So let instead "} S \text{ have } no \text{ } least \text{ } element". \qquad (1)$$

  We will show that as a result of (1) S is empty after all, contradicting the italicised hypothesis at the outset of this bulleted paragraph.
  We write $\mathscr{S}(n)$ for the property "n is not in S". If we show that $\mathscr{S}(n)$ is true for all n we have shown that $S = \emptyset$ (the symbol for the empty set), thus obtaining our contradiction.
  Now following the steps of Definition 0.1.5 we note

  1. $\mathscr{S}(0)$, which means "0 is *not* in S", is true (for if 0 were in S it would clearly be least, contradicting (1)).
  2. We fix n and take as I.H. that $\mathscr{S}(k)$ is true all $k < n$, that is, $0, 1, 2, \ldots, n - 1$ are not in S.
  3. How about $\mathscr{S}(n)$? True or false? If false then $n \in S$ is least, contradicting (1).
     But then $\mathscr{S}(n)$ is true, proving by CVI that *all n are outside S. So S is empty. Contradiction!

- (*Suppose we have the least principle.*) Let then $Q(n)$ be a property for which we verified 1.–2. of Definition 0.1.5. We show that this process proves the truth of $Q(n)$, for all n. Equivalently, we will use the least principle (Definition 0.1.7) to prove that the set $S = \{n \in \mathbb{N} : Q(n)\}$ is all of $\mathbb{N}$.
  Suppose this is not so. Then there are n in $\mathbb{N}$ but not in S. Let then, by the least principle, s be smallest such number.

Now $s \neq 0$ since we have verified $Q(0)$. So —s being the smallest *not* in S— all of $0, 1, 2, \ldots, s - 1$ *are in* S. That is, all of $Q(0), Q(1), \ldots, Q(s - 1)$ are true. But the CVI, for any n, proves the passage from 1. to 2. in Definition 0.1.5. Therefore, here we have $Q(s)$ is true. That is, $s \in S$; contradiction. Thus the assumption that $S \neq \mathbb{N}$ is untenable: we have $S = \mathbb{N}$.

□

Therefore, trivially (transitivity of equivalence) we have by Theorems 0.1.8 and 0.1.9:

0.1.10 Corollary. *All three principles of proof, simple induction, CVI, and least principle are equivalent.*

 In axiomatic arithmetic due to Peano (*Peano Arithmetic* or *PA*) that we will deal with later on in this volume the simple induction is taken as an axiom,[5] the reason being that it cannot be proved[6] from more elementary statements that can be taken as axioms (like the axiom $n + 1 \neq 0$ of PA).

Perhaps the least principle is most readily intuitively acceptable, since one may think of constructing the least element of a nonempty set S by starting from any element a in S and going "down"

$$\ldots < a''' < a'' < a' < a \tag{1}$$

all the members of the sequence being in S. Such a sequence must *terminate*, intuitively, at what is the *least element*. You see, no subset of $\mathbb{N}$ can be "bottomless" since there is an ultimate bottom (the number 0) in $\mathbb{N}$.

Why is this not a proof? *For one*, we do *not* have *sets* in PA that would permit us to argue along the lines of the above, and *secondly* we only used above our intuition of natural numbers and some sleight of hand. For example, the positive rational numbers $\mathbb{Q}$ have 0 as ultimate bottom, but one can start at a rational a and have a non-ending descending sequence of positive rationals like (1) since for any rational $0 < b$, we have

$$0 < \ldots < b/2^3 < b/2^2 < b/2 < b$$

---

[5] For *any* formula A (a formal substitute for "property") the following formula is an *induction axiom*: $A(0) \wedge (\forall x)(A(x) \rightarrow A(x + 1)) \rightarrow A$. Thus we have infinitely many axioms of the *same form*, one for each formula A. That is, the induction axiom is expressed by a *formula form* or *formula schema*.

[6] *Provably*, for example by using models, the induction schema cannot be proved from the other axioms of PA.

0.2   A Crash Course on *Formal* Logic

The exposition of *computability* in this volume is *informal*, that is, it is *neither* axiomatically founded *nor* reasoned within *formal logic*. As such, the reader will be sufficiently well equipped, for the type of logic we will *mostly* need here for page-after-page reasoning, by having been exposed just to the concepts and use of *informal* logic and proof techniques as these are taught in a solid first year course on discrete mathematics.[7]

Nevertheless, computability can be used —and we will do so— to give an exposition of one of (if not *the*) most important metamathematical results of formal logic, namely, Gödel's (first) Incompleteness theorem. Thus, we need preparation toward *reasoning about* formal logic, and this section is a crash course on such *formal* logic: we need to know *exactly what formal logic looks like* and exactly *how it is used* in order to be able *to reason about it*. It has to be a mathematical object with properties that are analysable mathematically.

We use the term "logic" in this volume to mean "first-order logic", that is, the logic we normally use in mathematics to argue about mathematical objects and their properties and prove theorems about said properties. The qualifier "first-order" will be soon defined. It concerns the informal expressions (that are formalised —i.e., accurately codified— in logic) "for some x" and "for all x" and the restriction we put on the nature of the variable x in such expressions, that is, what kind of objects it may denote. We can "do" most of mathematics using first-order logic; after all we can do so for set theory, which most people will accept as the language *and* foundation of all mathematics.

We understand the term *formal logic* to mean a *symbolic mechanism* that we may *use* to *verify*, and discover, mathematical truths by *syntactic manipulation of symbols alone*. This is what Hilbert had in mind when he conjectured that we can formalise *all* of mathematics: that we can reason within mathematics syntactically (relying only on the *form* of statements). Now, if we set this up correctly —I mean, the logic and the *basic statements* (axioms[8]) of the mathematics we want to do— then it turns out that *all the statements we prove syntactically are true in a certain precise sense*. Unfortunately, as Gödel proved with his incompleteness theorems, we can never prove *all* mathematical truths with syntactic means, a blow to Hilbert's belief. Yet, *in practise*, this syntactic approach to proofs works surprisingly well!

 So, here is our *motivation* to learn about formal logic in this section: To prove Gödel's (first) incompleteness theorem eventually (appears first in Chap. 8 and is

---

[7] After all, Kunen (1978) in his chapter on "Combinatorics" in the Handbook of Mathematical Logic states (about his topic) that "We assume familiarity with naïve set theory ... (Halmos (1960)) ..." and continues "A knowledge of logic is neither necessary, nor even desirable". Presumably, he meant "formal logic". But the latter *is needed* (a stronger qualifier than "*desirable*" if you need to prove results such as Gödel's *incompleteness* theorems. Hence the need for the section you are just reading.)

[8] To be defined carefully soon.

thoroughly retold in Chap. 12). That is, we will study the *what* and *how* of formal logical reasoning, in order to prove, following Gödel, that formal reasoning for theories like Peano arithmetic *always tells the truth*, but will *never* manage to tell *all the truth*.

In order to do logic syntactically, we need an *alphabet* of symbols to use in order to build formulas (and terms).

 Incidentally, *computer programming* is also a formal endeavour. You solve "real world" problems via *programming*, but the latter *requires* a formal approach. At the present state of the art, programming is a precise formal process. It *has* a syntax and you cannot write a program in *any syntax you may please*.

Since the end-use of logic is to be a tool for reasoning about mathematics, one will use *special* symbols *according to intended (mathematical) use*.

0.2.1 Example. Peano arithmetic (PA) —the first-order theory of natural numbers— contains these special symbols:

- S that formalises (that is, formally codifies) the function $\lambda x.x + 1$

 We have used here —and will throughout this volume— the so-called Church's "lambda notation (λ notation)". The symbols "λ" and the following "$\cdot$" are to be viewed as an opening and (its matching) closing bracket pair, more precisely, as "begin" / "end" keywords pair, as in programming. They enclose the names of *all* variables that we intend to change as we provide to them *external* values. In short, they enclose *the list of input variables*.

The expression that immediately follows the period —the end of the input list— is the "rule" that describes how we obtain the output. In this case it is the expression $x + 1$.

- $+$ that formalises the function $\lambda xy.x + y$
- $\times$ that formalises the function $\lambda xy.x \times y$
- $<$ that formalises the relation $\lambda xy.x < y$
- 0 that formalises the number zero.[9]

---

[9] At first sight, using "0" both for the syntactic "code" or "symbol" and the "real" zero may be confusing. However, as a rule, "formalists" —the practitioners of formal methods— do not think it is profitable or even desirable to invent obscure codes just for the sake of making them (the symbols!) "look different". In the end, the context will make clear whether a symbol is "formal" or "the real thing"—the latter of which, come to think of it, is also *denoted* by a symbol: Non-formalist mathematicians do not write "zero"; they write "0", even if they are not formalists. This comment applies to the formal symbols $+$, $\times$, $<$ as well.

Let us not forget that the ancient Greeks were extremely good in geometry but did not shine as much in algebra. You see, for the former branch of mathematics they had a *symbolic language* —the figures— but they had not invented such a language for the latter.

ZFC [10] *Axiomatic Set theory* —the first-order theory of set theory à la ZF (with C)— contains just one special symbol:

- $\in$ that formalises the relation $\lambda xy.x \in y$, read, "x is a member of y".

□

 A *general* study of logic must be *independent* of the theory whose theorems it is called upon to prove. To keep the discussion *general*, we do not disclose, in the definition below, which *exactly* are the *function*, *relation* and *constant* symbols, using generic names such as f, p and c respectively, with natural number subscripts.

0.2.2 Tentative Definition. (The Alphabet of First-Order Logic)

Logical symbols

- The symbols in the set $\{\neg, \vee, (,), =, \forall\}$ are the *logical symbols*, that is, those that are essential and sufficient to "do just logic" —or *pure logic*, as we properly say.
- (variables for *objects*, or *object variables*): The *infinite* sequence $v_0, v_1, \ldots$

Math symbols

These are the symbols needed to *do mathematics*. They are also called *nonlogical* symbols, to distinguish them from logical.

- Zero or more *function symbols*: $f_0, f_1, \ldots$
- Zero or more *relation symbols*: $p_0, p_1, \ldots$
- Zero or more *constant symbols*: $c_0, c_1, \ldots$

□

0.2.3 Final Definition. (The Alphabet of First-Order Logic; Final)

Logical symbols

- $\{\#, v, \neg, \vee, (,), =, \forall\}$
- The *actual ontology* of the *object variables* is that they are the following *strings*, generated by the symbols "v", "$($", "$)$", and "$\#$":

$$(v\#), \ (v\#\#), \ (v\#\#\#), \ldots$$

in short,

$$\text{the strings } (v\#^{n+1}), \text{ for } n = 0, 1, 2, \ldots \tag{2}$$

where $\#^{n+1}$ denotes

---

[10] The initials stand for the proposers of this set theory *version*, Z: Zermelo, F: Fraenkel, while C: stands for "choice", as in "*with* the axiom of choice".

$$\underbrace{\#\#\ldots\#}_{n+1 \text{ times}}$$

Thus, the meta-*name* (*metavariable*) v_i of 0.2.2 stands, for each $i \geq 0$, for the string $(v\#^{i+1})$.

It will be convenient —as we do in algebra, arithmetic, set theory, etc.— to employ *additional meta*variables to stand for actual variables in our discussions *about* logic and *when using* logic in doing mathematics: namely, x, y, z, u, w, with or without subscripts or accents *stand for* or *name* (object) variables.

Math symbols

- We add the set of symbols $\{f, p, c\}$ to the alphabet.

  These, along with # and brackets, will generate all the mathematical symbols a (mathematical) theory needs as we explain below.
- The f_i in Tentative Definition 0.2.2 are meta-*names* for functions symbols. The actual ontology of function symbols is that of *strings* of the *actual form* "$(\#^{n+1} f \#^{m+1})$", for $n \geq 0$ and $m \geq 0$.

  If we arrange *all possible*[11] such symbols in an infinite matrix of entries (n, m) then we obtain the matrix below.

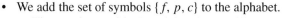

$$
\begin{array}{llll}
(0,0) & (0,1) & (0,2) & (0,3) \ldots \\
 & \nearrow & \nearrow & \nearrow \\
(1,0) & (1,1) & (1,2) \\
 & \nearrow & \nearrow \\
(2,0) & (2,1) \\
 & \nearrow \\
(3,0) \\
\vdots
\end{array}
$$

The arrows indicate one way that we may traverse the matrix, if we want to rearrange the symbols in a *one-dimensional* (potentially) infinite array.

If we call the members of this array by the meta-*names* f_k of bullet two (under the heading Math symbols) in Tentative Definition 0.2.2, then each f_k denotes the entry $(\#^{n+1} f \#^{m+1})$ *iff*[12] the latter is situated at location k of the "linearised" matrix. For example, f_0 denotes $(\# f \#)$.

The $n + 1$ denotes the required (by the function symbol) *number of arguments* —its *arity* as mathematicians call it[13]— while the $m + 1$ indicates

---

[11] This is a mind game for the most extreme case, hence we say "all possible" and "infinite". Recall that, for example, the theory ZFC needs *no function symbols a priori*, so it has none in its alphabet.

[12] if and only if.

[13] This term is not proper English. It was introduced as a "property" of *n-ary* functions: the property to *require n* inputs; to have "arity" n.

(along with $n + 1$) the position of the symbol *in the matrix*: $(\#^{n+1} f \#^{m+1})$ is the entry at "coordinates" (n, m).

"iff" —that is an acronym for "if and only if"— is the informal "is equivalent to" relation between statements and we usually employ it *conjunctionally*, that is, a chain like

$$A \text{ iff } B \text{ iff } C \text{ iff } D$$

means

$$A \text{ iff } B \text{ and } B \text{ iff } C \text{ and } C \text{ iff } D$$

If the chain makes a true statement, then *A* iff *D*.

- Analogously with the previous bullet, the meta-symbol $p_k, k \geq 0$, denotes the entry $(\#^{n+1} p \#^{m+1})$ iff the latter is situated at location k of the "linearised" matrix of all the $(\#^{n+1} p \#^{m+1})$ entries (each such entry occupying location (n, m) in the matrix. As before, $n + 1$ is the arity of the relation symbol (named) p_k.
- The ontology of constant symbols is similar to that of object variables, since arity plays no role. Thus, c_k, for $k \geq 0$, denotes the string "$(c \#^{k+1})$".

□

0.2.4 Remark (On Metanotation) We use *metanotation* to shorten unwieldy mathematical texts so that humans can argue about them —and understand them— without getting attacked by symbols! Thus,

(a) In *specific theories* we will use metanotation borrowed from mathematical practise. For example, in ZFC we will write "∈" rather than $(\#\# p \#)$. In PA we will use $<, S, +, \times$[14] as usual, but note that $<$ in this theory abbreviates $(\#\# p \#)$, while this *actual* predicate symbol not only has a different meaning in set theory —"belongs to a set", rather than "less than"— but even a different *abbreviation* (metasymbol), ∈. Can't we use $<$ in set theory? Of course we *can*! But as it is not a *primitive* mathematical symbol in this theory, it is introduced by a definition as the *actual* symbol, for example, $(\#\# p \#^{m+1})$, for some specific m (this m keeps *distinct* the several symbols that we introduce by definitions in a theory).

   Incidentally the arity of the *metasymbols* $<, S, +, \times$ will be stated, rather than codified *in the* metasymbol. Of course, these arities are, in order, 2, 1, 2, 2.

(b) If ZFC has no constants, nor functions, then does it not use the well known constant "∅" or the function "∩" (by *any* other name if not by these two)?

---

[14] Where $\times$ is simplified to · much of the time —or is even being "implied (i.e., omitted)" when reasoning about multiplication.

Well, what we put in the "*Ur*alphabet" (the quoted term is used here analogously to "Urtext" —an original or the earliest version of a text) of any theory is the bare minimum toward bootstrapping the theory. As the theory progresses (being developed by adding theorems) new constants, functions and relations are introduced by *definitions*!

All these new entities are named at the point of their definition. So, the usual name for the "empty set" is ∅, and *both the entity and its symbol* are introduced by a definition. The same holds for symbols like ∩ and ∪.

It is a good thing that both Tentative Definition 0.2.2 and its final "implementation" Final Definition 0.2.3 have a mechanism to name as many defined symbols as we care to define! In this connection see, for example, Tourlakis (2003a), the section on defined symbols.

(c) How about when we discuss non specific theories, or pure logic, that is, when we are simply developing logic's toolbox and just do logic with no emphasis on any specific mathematics? Are we restricted to rigid notations like v_0, v_k, or worse $(v\#)$, $(\underbrace{v\#\ldots\#})$?
$$ k+1$$
No, we typically use the metanotation

- x, y, z, u, v, w with or without primes or subscripts for variables,
- a, b, c, d with or without primes or subscripts for constants,
- f, g, h with or without primes or subscripts for functions,
- p, q, r with or without primes or subscripts for relations

It is worth emphasising that the above are metasymbols thus, e.g., x stands for *any v_i*, a stands for *any c_i*, etc.   □

Incidentally, going forward we will drop the word "symbol" from expressions like "constant *symbol*", "function *symbol*" and also drop the word "object" from "*object* variable".

The procedure of introducing, say, a new[15] constant in a theory is a bit involved. *For example*, here is in outline what needs to be done to introduce a new constant like the empty set in ZFC.

- First we *prove within ZFC* that a set with the required fundamental "defining property" of the empty set —to have no elements— *exists*, that is, we prove that

$$(\exists y)(\forall x)x \notin y \tag{1}$$

In words, *a set y for which every query "$x \in y$" fails exists.*
- Then we prove that there is *only one* y that makes the formula in (1) above a theorem. At the intuitive level, we note that if two sets, y and z satisfy, for all

---

[15] "New" means "not already present in the *initial* alphabet", what we called "Uralphabet" earlier on.

x, that $x \notin y$ and $x \notin z$, then it is the case that $x \notin y \equiv x \notin z$ and hence $x \in y \equiv x \in z$. But then the axiom (*extensionality*) that states that if two sets have exactly the same elements are equal, yields $y = z$.

- Then we name this unique y with *any name we please* but it is always advisable to use a good meta-name that is widely accepted in the literature for this purpose. In this case, "$\emptyset$" is preferable to "($c\#$)" even if the latter is the "hidden" symbol that ZFC uses —the first available formal name for constants.

   The "definition" involves the addition of a *new formal symbol*, say $\emptyset$, to the alphabet along with a *new axiom* that states the behaviour of the symbol (Tourlakis 2003b).

   In the present case the axiom is

$$y = \emptyset \equiv (\forall x)x \notin y$$

Bourbaki (1966) prefers to view the introduction of new symbols by definitions as *metatheoretical devises* that introduce *abbreviations* of longer mathematical texts; but these new symbols *exist in the metatheory*; outside the logic.

The post-Bourbaki tendency has been to actually augment the languages (formally) every time new symbols are added, and thus view the "definitions" as *axioms* about these formal symbols, not as statements of *an agreement to abbreviate*. Nevertheless, either approach —different as they may be— is to serve convenience only. Adding new symbols formally via axioms or via metatheoretical abbreviations in the end adds *no more power to the theory* —no more theorems (cf. Tourlakis (2003b), pp. 69–75, regarding this observation).

0.2.1   Terms and Formulas

Terms are intuitively the names of objects. Thus in their simplest form they are variables or constants and in general are function *invocation* of functions —or function *calls*, in computer science jargon— that is the name consists of a function (symbol) followed by the requisite[16] number of (object) arguments. This is captured by the following recursive definition.

0.2.5 Definition. (Terms) A *term* is one of:

1. A variable x or a constant c.[17]
2. A string of the form $f t_1 t_2 \cdots t_k$ where f is a function of arity k and the t_i are terms.

                                                                                                □

---

[16] As the function's arity dictates.

[17] Refer to Remark 0.2.4, item (c).

0.2.6 Example. Thus if f and g have arities 2 and 3 respectively, then

$$fgabyfxx$$

is an example of a term.   □

0.2.7 Remark.

(1) We reserve the symbols t and s, with or without primes or subscripts or subscripts, as *metanotation* (*argot*) that denotes arbitrary terms.
(2) Such metanotation is also called *syntactic variables* since they are metavariables used toward discussing syntax.
(3) It is easier to "parse" a term if we use brackets and commas (the latter is not even a formal symbol of our alphabet) to denote complex terms in the "normal" way. That is, *metanotation* or *argot* of the term in the preceding example is the more readable $f(g(a, b, y), f(x, x))$.   □

The reader familiar with the old ALGOL 60 language will recognise that the process of defining formulas here tracks faithfully that for this programming language, namely, function calls (or "function designators", as ALGOL calls them) first, then the simplest possible formulas —the so-called *atomic* formulas— and then the full blown formulas (ALGOL does not define quantification though).

0.2.8 Definition. (Atomic Formulas) An *atomic formula*, in short "*af*" is one of:

1. $t = s$
2. A string of the form $pt_1t_2 \cdots t_k$ where p is a relation of arity k and the t_i are terms.

As in Remark 0.2.7 we usually use the *metanotation* $p(t_1, t_2, \ldots, t_k)$ in preference over $pt_1t_2 \cdots t_k$ as the former is more readable.   □

The syntax for the general case of formulas is defined recursively below.

0.2.9 Definition. (Well-Formed Formulas) A *well-formed formula*, in short "*wff*" is one of:

1. An af.
2. $(\neg A)$, provided A is a wff.
3. $(A \vee B)$, provided *both* A and B are wff.
4. $((\forall x)A)$, provided A is a wff.
   We emphasise that

   - x stands for *any* variable v_i (cf. Remark 0.2.4).
   - Omission of any explicit restriction in the choice of x is *intentional*: It is *not*, in particular, required that x is a substring of A in order to be allowed to form the string "$((\forall x)A)$" *nor* are we disallowed to add "$(\forall x)$" if A already contains "$(\forall x)$" as a substring.

We will denote the set of *all* wff by $\mathbf{WFF}_L$, where L is the alphabet in question. L will be normally omitted, unless we handle several different alphabets in a given discussion.   □

0.2.10 Remark. (Wff Names, and Names of Sets *of Wff)*

1. Again, in the above definition we used A and B as *syntactic variables* that *stand for* wff. However we said things that mathematicians normally say in *argot*: "Provided A is a wff" and "It is *not*, in particular, required that x is a substring of A" as if A is a string of the language. It is not! It is a one-symbol string of the *metatheory* that *denotes* or *stands for* a string of the language. Being clearly aware of this nuance we will continue saying in argot "let A be a wff".

   Consider in analogy, "let p *be* a prime number". Such an expression may occur in working with numbers, or working in computability. Clearly, p is a letter, it is *not* a *number*, let alone a *prime* number! Invariably, "be" is short for "stand for" or "name".

2. In what follows, capital letters from the Latin alphabet stand (are syntactic variables) for arbitrary wff.

3. We will denote *sets* of wff (i.e., subsets of $\mathbf{WFF}_L$) by capital Greek letters that are not the same as any Latin such letters. In practise, we employ $\Gamma, \Delta, \Lambda, \Sigma, \Phi, \Psi, \Omega$ with or without primes or subscripts.   □

0.2.11 Definition. (Immediate Predecessors) The concept of *immediate predecessors* or *i.p.* is defined by tracing Definition 0.2.9.

1. If A is an atomic formula, then it has no i.p.
2. If A is $(\neg B)$, then it has B as an i.p.
3. If A is $((\forall x)B)$, then it has B as an i.p.
4. If A is $(B \vee C)$, then its i.p. are B and C.

                                                                                        □

0.2.12 Remark (Unique Readability of Terms and Wff)   *Unique readability* is terminology describing a phenomenon that applies to terms t and formulas (wff) A, meaning that the last step of putting t or A together can be uniquely described from the syntax of t or A.

   In the case of a wff A unique readability means that its i.p. are uniquely determined by A.

   Similarly for terms: A term t cannot be put together in two different ways, $ft_1 \cdots t_n$ and $gs_1 \cdots s_m$. The reader is referred to any of Bourbaki (1966), Shoenfield (1967), Tourlakis (2008, 2003a) for proofs of these metatheoretical facts.   □

0.2.13 Definition. (Free and Bound *Occurrences* of Variables)   We define the concepts of *free* and *bound occurrences* of variables of A by recursion on the formation of A (Definition 0.2.9).

We emphasise: The concept is not about a *variable*, say, v_i —which may occur as a *substring* in several locations in A— but it rather characterises each specific *occurrence* of a variable separately.

First off, all *occurrences* of variables in a term *by definition* are *called* free.

 We will rather say *are* free (omitting "called").

1. If A is atomic, then *every* variable occurrence in it is (called) by definition free (each of these free occurrences are occurrences *in the terms* used to form A as $t = s$ or $pt_1 \cdots t_m$).
2. If A is $(\neg B)$, then an occurrence of a variable in A is free, iff the same occurrence in B is free.
3. If A is $(B \vee C)$, then an occurrence of a variable in A is free, iff the same occurrence happens to be free in B or in C.
4. If A is $((\forall x)B)$, then the free occurrences of a variable y —which is *distinct* from x— in A are precisely the free occurrences of y in B.
    We say that *all* occurrences of x in A —that is, in $((\forall x)B)$— are *bound*; this is *opposite* terminology to *free*.

$\square$

 0.2.14 Remark. (Notation) The *meta*notations $A[x]$ and $t[x]$ indicate our *interest in* x that *possibility* occurs free in A and, correspondingly t. Which one of the possibly many free occurrences of x are we interested in?

All of them.

Note that our *interest* cannot guarantee the *existence* of *any* occurrences of free x, nor it precludes that other variables may also occur free. In particular then, writing $A[x]$ still allows the possibility that no occurrence of x is free in A.

On the other hand writing $A(x, y, \ldots, z)$ or $t(x, y, \ldots, z)$ states that in fact each of the listed variables inside the round brackets —$x, y, \ldots, z$— occurs free at least once in A or t respectively. $\square$

 0.2.15 Remark. (Belonging to a Quantifier) Consider

$$\left(x = y \vee \left[(\forall x)\left(x = z \vee \{(\forall x)x = c\} \right) \right] \right)$$

 In the interest of visibility, we may use —in metanotation— different shapes and sizes of brackets $(, (, (,], \}$.

There are 5 *occurrences* of x as substrings of this formula. The leftmost *occurrence* is free by cases 1 and 3 above as it is so in the left i.p. of the above formula.

All the other occurrences of x (four of them) are bound due to the right i.p. of the formula above and case 4 of the definition.

There is some nuancing possible: If we imagine the sequence of steps that formed the 2nd i.p., we have a construction-sequence (unwrapping the recursive definition of formulas, Definition 0.2.9)

$$x = c, \quad ((\forall x)x = c), \quad x = z, \quad \left(x = z \vee \{(\forall x)x = c)\} \right),$$

$$\left[(\forall x)\left(x = z \vee \{(\forall x)x = c)\} \right) \right]$$

Thus, the "freedom" of the rightmost x-occurrence was taken away by the rightmost $(\forall x)$. We say that the *bound* occurrence of the 5th x *belongs* to the 2nd quantifier (and so does the 4th occurrence of x, trivially). Similarly, the 3rd bound occurrence of x (and the second, trivially) belong to the first quantifier.

In sum, a specific *bound occurrence of x belongs* to the "$(\forall x)$" of *smallest scope* that includes said occurrence of x.   □

0.2.16 Definition. (Subformulas) The *occurrences* of the subformulas of A are defined by recursion:

1. If A is atomic, then it is the only subformula that occurs in A, or is the only subformula "of A".
2. If A is $(C \vee D)$, then the subformulas of A are precisely: A itself *and* all the subformulas of C and all the subformulas of D.
3. If A is $(\neg C)$, then the subformulas of A are precisely: A itself *and* all the subformulas of C.
4. If A is $((\forall x)C)$, then the subformulas of A are precisely: A itself *and* all the subformulas of C.

If B is a subformula of A but the two are distinct strings, then B is a *proper* subformula of A.   □

0.2.17 Example. In $(x = y \vee x = y)$ the subformula $x = y$ occurs twice, once in the left i.p. and once in the right i.p. Each is a proper subformula of $(x = y \vee x = y)$.   □

0.2.18 Definition. (Defined Connectives) We introduce *in the metatheory* additional connectives, that is, abbreviated *metatheoretical* texts for longer *formal* texts.

- $(A \to B)$ is short for $((\neg A) \vee B)$
- $(A \wedge B)$ is short for $(\neg((\neg A) \vee (\neg B)))$
- $(A \equiv B)$ is short for $((A \to B) \wedge (B \to A))$
- $((\exists x)A)$ is short for $(\neg(\forall x)(\neg A))$

                                      □

0.2.19 Remark (Nomenclature)

- A formula of the type $(\neg A)$ is called an *negation*. It is read "not A".

- A formula of the type $((\forall x)A)$ is called a *universal quantification*. It is read "for all (values of) x, A holds".[18]

  The string that extends from the leftmost bracket to the rightmost is called the *scope* of $(\forall x)$.[19]
- A formula of the type $((\exists x)A)$ is called an *existential quantification*. It is read "for some (value of) x, A holds".

  The string that extends from the leftmost bracket to the rightmost is called the *scope* of $(\exists x)$.
- A formula of the type $(A \lor B)$ is called a *disjunction*. It is read "A or B".
- A formula of the type $(A \land B)$ is called a *conjunction*. It is read "A and B".
- A formula of the type $(A \to B)$ is called an *implication*. It is read "A implies B" or also "if A, then B".
- A formula of the type $(A \equiv B)$ is called an *equivalence*. It is read "A is equivalent to B".[20]

  □

0.2.20 Remark (Reducing the Number of Brackets) As is usual, proofs written by humans are written with as little symbolic burden as is feasible. One simplification of the strict syntactic notation is an agreement on omitting certain brackets in a "controlled way", that is, *anyone* (who is privy to the omitting/reinserting-brackets *process*) can reinsert *all* the omitted brackets correctly (*without having noted which ones were removed*) after said simplification.

For example, a high school student who is asked to insert the omitted brackets properly in something like "$2 + 3 \times 4$" will know that the answer —$(2 + (3 \times 4))$— hinges on an agreement that $\times$ is *stronger* than $+$, hence has the opportunity to use (the shared) 3 in the given expression *before* $+$ does.

Similarly, we are given the strength, or *priority*, of connectives as

$$\neg \ (\forall x) \ (\exists x)$$
$$\land$$
$$\lor$$
$$\to$$
$$\equiv$$

Thus the "unary" connectives have equal priority, the highest, and the priority decreases as row number increases in the above matrix.

As in the arithmetical example above, the stronger (higher priority) connective *binds first*, so, e.g., $\neg A \lor B$ means $((\neg A) \lor B)$.

---

[18] This semantic nomenclature, "values, holds", is mathematical *habit* —even the formalist Bourbaki (1966) uses it— and in no way prejudices against a *formal approach to proofs*!

[19] As in ALGOL or Pascal: The string (program segment) between a **begin** and matching **end** instructions is the scope of the "declarations" that appear immediately after the "**begin**".

[20] Some authors say "equivales", but we will not do so.

We never remove the brackets from $(\forall x)$ and $(\exists x)$. We view these two as *compound symbols*.[21]

What if connectives have equal priorities? Well in the contest between two connectives of equal priority, the one to the right wins. For example:

- $A \to B \to C \to D$ means $(A \to (B \to (C \to D)))$.
- $(\forall x)(\exists y)\neg A$ means $((\forall x)((\exists y)(\neg A)))$. □

0.2.2 Substitution

Much of the work in mathematics and elsewhere where logic (formal or informal) is used, the act of substituting terms for variables is ubiquitous, as in "given a polynomial $2x^3 - 3x - 40$. Find its value if x is 3 —that is, if we substitute x by 3 in this expression."

Seems innocuous enough, but certain precautions are warranted when we do general substitutions (not just substituting variables by constants) in logic.

For example, the formula

$$(\exists y)y \neq x \tag{1}$$

where we omitted a few brackets and used the *argot* "$y \neq x$" for "$\neg y = x$" —says, intuitively, that

"no matter what the value of x may be, there is a y (value) different from x".

This is recognised as being "true", intuitively, for example, *if* we have arbitrary *integer* number values in mind. In particular, it should not matter whether we express (1) with a *different* variable, z, instead of x, like

"no matter what the value of z may be, there is a y different from z".

or, symbolically, $(\exists y)y \neq z$; we are saying the same thing.

But note what happens if we substitute y for the variable x in (1). We obtain $(\exists y)y \neq y$ which says something altogether different than (1)! It says that *there is a value of y that is different from itself*! Absurd!

---

[21] This convention —that is certainly not carved in stone— likely stems from the pre-LaTeX era when manuscripts *were* literally written by hand. Back then authors of books like this used "(z)" for "$(\forall z)$" and "(Ez)" for "$(\exists z)$".

Some authors simply write "$\forall z$" or "$\forall z.$" (note the period, for example, "$\forall z.A$" that implies that $\forall z$ applies —if brackets do not determine otherwise— to all that follows.

What happened is that the y we substituted into x entered "dangerous territory"
—*into the scope of* $(\exists y)$ *and, being the same as the quantified variable* y, it was
captured, that is, it became *bound*.

Such *capture*, in general, changes the original meaning after substitution!

Thus, the *operation* of substitution *of terms for variables* defined below will not be
allowed to go ahead *if capture occurs*.

That is, *the substitution will be aborted* (will be undefined) in that case.

Substitution for terms is plain find-and-replace, as in a word processor, since
terms contain no quantifiers. The "general case" in the definition below describes
precisely what we do in mathematical practise when we substitute, say, 3 for x in
$f(g(x, y), h(y, w, x))$. We get $f(g(3, y), h(y, w, 3))$.

0.2.21 Definition. (Substitution in Terms) The metasymbol $t[x := s]$ denotes the
string that results if we replace *all* occurrences of x in t by s. In particular, the
meta-expression "$[x := s]$" says "replace x *everywhere* by s" and *has a priority
higher than any connective* (in the metalanguage, of course). The find-and-replace
operation is captured by the recursive definition

$$t[x := s] \overset{Def}{\text{ is }} \begin{cases} t & \text{if } t \text{ is } c \\ t & \text{if } t \text{ is } y, \text{ distinct from } x \\ s & \text{if } t \text{ is } x \\ f t_1[x := s] \cdots t_n[x := s] & \text{if } t \text{ is } f t_1 \cdots t_n \end{cases}$$

□

In the definition above we used "is" for "is the same string as", rather than using
"=". This is so because we do not want to confuse the formal "=" with the *informal*
"=".

The formal and informal "=" tell different stories in a context like "$t = s$":

1. The first says that the terms t and s are *provably* (within formal logic) equal.
   This can happen even though they may be very different *as strings*. For example,
   cf. the well known true (for all values of x) algebraic expression "$(x + 1)^2 =
   x^2 + 2x + 1$".
2. The latter meaning is, in the *metatheory*, that t and s are identical as *strings*.
   When we just speak about logic (as we are doing in this definition) we are in the
   metatheory; we will then say "is" for the *informal* "=" between strings.

0.2.22 Exercise. Prove by induction on (the formation of) terms, cf. Definition 0.2.5, that for any terms t and s, *the result of the substitution* $t[x := s]$ *is a
term.*

We say simply, $t[x := s]$ *is a term*. □

0.2.23 Definition. (Substitution in Formulas) The metasymbol $A[x := s]$ denotes the string that results if we replace *all free* occurrences of x in A by s. In particular, the meta-expression "$[x := s]$" says "replace *all free occurrences* of x by s" and has a *priority higher than any connective* (in the metalanguage, of course). The find-and-replace operation is captured by the recursive definition

$$
A[x := s] \overset{\text{Def}}{\text{ is }}
\begin{cases}
t[x := s] = t'[x := s] & \text{if } A \text{ is } t = t' \\
pt_1[x := s] \cdots t_n[x := s] & \text{if } A \text{ is } pt_1 \cdots t_n \\
\neg B[x := s] & \text{if } A \text{ is } \neg B \\
B[x := s] \vee C[x := s] & \text{if } A \text{ is } B \vee C \\
(\forall y)B[x := s] & \text{if } A \text{ is } (\forall y)B \text{ and } : x \text{ does} \\
& \quad not \text{ occur free in } B \\
& \quad or \ y \text{ does } not \text{ occur in } s \\
\text{undefined} & \text{if } A \text{ is } (\forall y)B \text{ and } : x \text{ occurs free in } B \\
& \quad and \ y \text{ occurs in } s
\end{cases}
$$

□

0.2.24 Remark. (Notation) If we have already written $A[x]$ or $t[x]$, then we can simplify the expressions $A[x := s]$ and $t[x := s]$ as $A[s]$ and $t[s]$ respectively.

Similarly, having written $A(\dots, x, \dots)$ or $t(\dots, x, \dots)$, we can simplify the expressions $A[x := s]$ and $t[x := s]$ as $A(\dots, s, \dots)$ and $t(\dots, s, \dots)$ respectively.

□

0.2.25 Remark

(1) Once more, regarding the use of "is": On the left hand side it must be read above as "is the string produced" after the indicated substitution by the relevant recursively invoked case on the right hand side of the large "$\Big\{$".

Similarly, occurrences of "is" on the right hand side denote equality of strings in the metatheory.

(2) In the above definition we took advantage of the high priority of "$[x := s]$" to reduce the number of brackets in the results of substitution that we displayed by cases. For example "$\neg B[x := s]$" says exactly what we intended: We first do the substitution "$B[x := s]$" and *then* apply "$\neg$".

(3) The import of "and x occurs free in B" is that substitution, *if allowed*, will go through. If it is also the case that the quantified variable occurs in the term we are attempting to substitute, then we must abort.

□

0.2.26 Exercise. Prove by induction on the formation of formulas, cf. Definition 0.2.9, that for any formula A and term s, *the result of the substitution $A[x := s]$ is a wff.*

We say simply, $A[x := s]$ *is a wff.*   □

0.2.27 Exercise. Prove by induction on formulas, cf. Definition 0.2.9, that if x has no free occurrences in A, then $A[x := t]$ is just A.   □

0.2.3   Axioms, Rules and Proofs

Intuitively speaking, a "proof" is a *formal* (syntactic) process that generates a *finite* sequence of formulas. The process is so devised that it *preserves* "truth" so that if one is prudent enough to have as *startup assumptions* formulas that are "true" (i.e., well-chosen axioms) then proofs have a good chance[22] to be good syntactic ways to discover all truths!

What's with the annoying quoting of the word "truth"? Well, I hope the quoting indicates that I am referring to the intuitive concept of truth, not to *technical* truth in the context of logic. A mathematically precise definition of truth will not be given in this volume. We will approximate such a definition in Chap. 8, but the full blown "Tarski semantics", i.e., the mathematical definition of truth for first-order logic, we will not get into (as we do not need it). There are several accessible sources to read about this: For example, Shoenfield (1967), Tourlakis (2003a, 2008).

0.2.28 Definition. (Axiom Schemata of *Pure* 1st-Order Logic)
There are two types of (*pure logic*) axioms. Those that speak only about the connectives $\neg$, $\vee$ (Boolean) and those that speak about the connective $\forall$, but also about variables and terms, in general (first-order).

Boolean axioms

For all formulas A, B, C the following are axioms:

1. $A \to A \vee B$
2. $A \vee A \to A$
3. $A \vee B \to B \vee A$
4. $(A \to B) \to (C \vee A) \to (C \vee B)$

---

[22] Well, it is easy to make proofs to *always* tell the truth, but it is impossible to have them tell the *whole* truth in theories like set theory or even number theory (Peano arithmetic). We start looking into this phenomenon and its connection to computability in Chap. 8 and continue in Chap. 12.

1st-Order axioms

For all formulas A, variables x, and terms t, s the following are axioms:

5. $(\forall x)A[x] \rightarrow A[t]$ *(substitution axiom)*
6. $x = x$ *(identity axiom)*
7. $t = s \rightarrow (A[x := t] \equiv A[x := s])$ *(equality axiom)*

□

0.2.29 Definition. (*Primary* Rules of Inference of 1st-Order Logic)

Building proofs is —if we forget for a moment the creative and nondeterministic part of choosing the next step— a straightforward endeavour: Write a sequence of formulas that are either axioms, or hypotheses, or, if not, said formulas are results of rules applied to preceding (in the proof, that is) formulas.

Rules are normally given as fractions, and we have two:

1.

$$\frac{A, A \rightarrow B}{B}$$

called *modus ponens* or *MP*, in short.

The rule (if applied at a proof step) means: If you have already written A and $A \rightarrow B$ —in any order— then you *may* write B in the current step.

2. If x has no free occurrences in A, then

$$\frac{A \rightarrow B}{A \rightarrow (\forall x)B}$$

called $\forall$-*introduction*, in short $\forall$-*intro*. The rule means: If you have already written $A \rightarrow B$ *and the stated restriction on x holds*, then you *may* write $A \rightarrow (\forall x)B$ in the current step.

The restriction can also be stated as "*x does not occur free in the conclusion*".

□

 $\forall$-intro is a special case of the "right $\forall$" (or R$\forall$) rule of Gentzen style logic.

0.2.30 Remark.

(1) Saying "For all formulas A, B and C, variables x, y, and terms t, s" in Definition 0.2.28 —using the metavariables A, B, C, x, y, t, s and a finite number of *shapes* that denote formulas, in 1.–8. above— we manage to describe infinitely many formulas, by the devise of substituting specific formulas in the formula *metavariables*, and similarly specific (object) variables and terms for the variable *metavariables* and term *metavariables*.

These "shapes" we call *schemata* (singular, *schema*).[23] The infinitely many *specific* formulas —in the current case these formulas are *axioms*— that these schemata depict are called *instances* of the (respective) schema.

Thus our first-order axioms are given as *axiom schemata*, and each instance of them is an axiom.

(2) If, for some choice of specific wff A and specific variable x and term t, it so happens that $A[t]$ is undefined, then this is fine: Our schema 5. is *only* supposed to *produce* a specific wff *provided* the A, x, t are such that $A[t]$ goes through. Similar comment for axiom schema 7.

(3) Why *primary* rules? Are there "secondary" rules? Yes, but we call them *derived rules*. These are rules that are "macros" that use the primary rules. Primary rules bootstrap the logic or the mathematical theory. Derived rules make proving theorems user-friendly.

(4) What is the meaning of the qualifier "pure" in Definition 0.2.28? *Pure logic* does not *do* mathematics; it does just logic, that is, proves *logical* or *absolute* theorems that do not *necessarily* enlighten us about *specific* mathematical theories.[24] Correspondingly, the logic that we do just for the sake of logic (using the above axioms and rules of inference) is called pure logic.

(5) A *first-order theory*, or *mathematical theory*, or *applied logic*, is pure logic *augmented* with *mathematical axioms*, synonymously, *nonlogical axioms* or even *special axioms*, so that we can reason about a specific area of mathematics. For example, set theory has among its axioms the axiom

$$(\forall x)(x \in y \equiv x \in z) \rightarrow y = z$$

called *extensionality axiom*. It states, symbolically, that two sets y and z are equal if they contain exactly the same elements (or members).

Peano arithmetic (PA) has $\neg Sx = 0$ as an axiom, among others. It says that, for any number x, its successor, $x + 1$ cannot equal 0. Why? Because PA is about the "arithmetic" of the natural numbers $\mathbb{N} = \{0, 1, 2, 3, \ldots\}$.

(6) In contradistinction to the term "nonlogical" the axioms of pure logic are as a result called the *logical axioms* of first-order logic.   □

---

[23] *Schema* or σχῆμα means shape, figure, or form in Greek.

[24] Every now and then pure logic will offer an insight regarding a specific mathematical theory. For example, pure logic *can* prove for an arbitrary binary (2-ary) predicate p the formula $\neg(\exists y)(pxy \equiv \neg pxx)$ (Levy 1979; Tourlakis 2003a, 2008). If we take p to be the $\in$ of set theory, then the formula becomes $\neg(\exists y)(x \in y \equiv x \notin x)$. In words, there is no set y equal to $\{x : x \notin x\}$. This is what Russell offered as criticism (cf. 0.3.2 below) to Cantor's "naïve" set theory: that some "collections" are *not* sets. This criticism is the content of "Russell's Paradox".

0.2.31 Definition. (Proofs) Within pure first-order logic or a first-order theory, a proof *from* a set of *hypotheses* Γ —also called a Γ-proof— is a finite sequence of formulas

$$A_1, A_2, \ldots, A_n$$

such that, for each $i = 1, 2, \ldots, n$, A_i is one of

1. Axiom (logical or mathematical, as the case may be).
2. Member of Γ.
3. For some k and m, both less than i, A_m is the formula $A_k \to A_i$.[25] In words, A_i *is the result of MP applied to two formulas (A_k and A_m) to the* left *of* A_i.
4. A_i is the formula $X \to (\forall x)Y$ —where x is not free in $X \to (\forall x)Y$— and $X \to Y$ appears to the left of A_i, that is, $X \to Y$ is A_k with $k < i$.

     In words, A_i *is the result of* $\forall$-introduction *applied to a formula that appeared to the* left *of* A_i. □

The axioms and rules are called, Hilbert-style, and the proof definition above is also for a *Hilbert-style proof*.

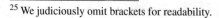

0.2.32 Definition. (Theorems) A *theorem* Γ or Γ-theorem is *any formula that appears* in a Γ-proof.

The metatheoretical notation $\Gamma \vdash A$ or $\vdash_\Gamma A$ is short hand for "A has a proof from Γ".

If Γ is finite, say, $\{Q, P, R\}$ then we prefer to omit the braces and write $Q, P, R \vdash A$ instead. If $\Gamma = \emptyset$ then we write $\vdash A$.

If the theory we are working in is denoted by $\mathcal{T}$, then we write $\vdash_\mathcal{T} A$ —or $\Gamma \vdash_\mathcal{T} A$, or $\vdash_{\mathcal{T}, \Gamma} A$, if we have some additional hypotheses (cf. also Remark 0.2.33 below)— to denote that A is a theorem of $\mathcal{T}$. This is useful in an argument where we are dealing with more than one theory. □

Jargon. In practise we say things like "I will prove $\vdash A$". This means "I will present a formal proof for A —just A— *within the theory* or *within pure logic*, as the case may be, following Definition 0.2.31".

In other words, "I will prove *within the theory* that A has a formal proof (by *exhibiting* the latter)", that is, I will prove the validity of the *metalogical statement* "$\vdash A$".

The first "prove" thus is a promise for action *in the metatheory*, where I will fetch my formal system's *toolbox*, then put it into action and thus *use* it to *generate a formal proof* of A. Such an action, if it succeeds, will verify the metalogical claim "$\vdash A$" —"A is a theorem".

So, *in practise* and in this context we will not be any more careful linguistically than anyone else is in the literature, and will just say "I will prove $\vdash A$" —or imperatively, in an exercise, "prove $\vdash A$"— and everyone will know what I mean!

---

[25] We judiciously omit brackets for readability.

0.2.33 Remark (The Nature of Γ)   We should not confuse the occurrence of the set Γ to the left of $\vdash$ with the mathematical or special axioms of the theory we are working in. It does *not* indicate second guessing the set of mathematical axioms of the theory or an attempt to extending it.

Rather it arises in circumstances such as, for example, applications of the *deduction (meta)theorem*[26] In its simplest form this *meta*theorem states "if I can prove that $A \vdash B$, then I can also prove that $\vdash A \to B$", or, "you want to prove $A \to B$? Then *take A as a hypothesis* —i.e., formally, treat it as an axiom— and prove just B instead".[27] In effect, one adds A to the axioms but only "*temporarily*". One does not change the theory. This is a proof-trick, reminiscent of taking an *Induction Hypothesis*, no more.

Another instance where having added hypotheses (nonempty Γ) arises is when arguing via *proof by contradiction*: "to prove A, just *add $\neg A$ as a new hypothesis* and prove some contradiction (i.e., a formula such as $Q \wedge \neg Q$)".

Several instances where having $\Gamma \neq \emptyset$ will occur start with Proposition 0.2.53 below.

The above discussion illustrates the advisability of defining the concept of "proofs from an *arbitrary* set Γ".   □

0.2.34 Exercise.   (Meta)prove that if

$$A_1, A_2, \ldots, A_i, \ldots, A_n$$

is a proof from some Σ, then so is

$$A_1, A_2, \ldots, A_i$$

□

0.2.35 Exercise.   In view of the preceding exercise (meta)prove that a Δ-theorem is a formula that appears at the *end* of some Δ-proof.[28]   □

0.2.36 Exercise. (Left Weakening)   Prove that if $\Gamma \subseteq \Delta$ and $\Gamma \vdash A$, then $\Delta \vdash A$ as well.   □

---

[26] Even though it is a metatheorem about the behaviour of proofs, it is common in the literature to call it "theorem".

[27] The "A" that one takes as a "temporary" hypothesis is not, in general, what one would consider "axiom material"; it is not in general a provable formula because an implication, $A \to B$, if verified, does not necessarily entail that A can be established as well!

[28] This is good to know, as it stops us from continuing a proof once we have reached the target theorem!

0.2.4   The Boolean Fragment of First-Order Logic and Its Completeness

We do logic, and mathematics, with a view of discovering logical and mathematical "truths". It is therefore important to be able to remove the quotes from the term *truth* and get some sense of what we mean by it.

As already mentioned, with the exception of a brief simplified definition of truth-values of a first-order formula of Peano arithmetic (Chap. 8), we will not get into the details of first-order *semantics* in this volume. However it will be profitable in the interest of *user*-friendliness toward the *practitioner* of proofs to at least define *Boolean semantics*, that is, how truth-values *propagate* with the *Boolean connectives*, ¬, ∨.

For example, we know from first year discrete mathematics courses that if I know the truth-value of each of A and B, then I can *compute* the truth-value —**t** for True *or* **f** for False— of $A \vee B$. Indeed,

$$A \vee B \text{ is } \mathbf{f} \text{ iff both of } A \text{ and } B \text{ are } \mathbf{f} \tag{1}$$

(1) tells us how **t** and **f** propagate from the constituent formulas A and B to the resulting formula $A \vee B$ —the "constituent" formulas are, of course, the i.p. of $A \vee B$ (Definition 0.2.11). *It tells us how* ∨ *acts on truth-values.* In effect, (1) is part of a *recursive definition* of truth-value: It tells us that if I have *already* calculated the (truth-)values of A and B, then here is how to compute the truth-value of $A \vee B$. Similarly, the recursive definition for the other connective is

$$\neg A \text{ is } \mathbf{t} \text{ iff } A \text{ is } \mathbf{f} \tag{2}$$

Both (1) and (2) are incomplete as recursive definitions go. How do you bootstrap the process? The process in each case, for example, obtaining the truth-value of A, will make a *recursive call* with arguments the i.p. of A. *And so on.* Where does the "so on" *end*? That is, what is the *basis* of these recursions?

Both recursive definitions make clear that we want to go backwards undoing the *Boolean structure* of the given formula. That is, when we say "i.p." we look for ∨ or ¬ as the immediately previous connective.

> Thus, the *basis* is when we reach an i.p. that is *atomic* or of the form $(\forall x)A$.

So how do we compute the truth-values of those?

 We don't! Boolean semantics is strictly about how truth-values *propagate* with ¬ and ∨. Our computation of the Boolean truth-value of a C is modulo an *arbitrary assignment* —a *what if*— of values (**t** or **f**) to the subformulas of C that are *atomic* or *of the form* $(\forall x)A$.

From discrete mathematics we recall that if one does *only* Boolean logic then these recursion stoppers are the so-called *Boolean variables*. The alphabet of Boolean logic does *not have notation for terms, predicates and generalisation.*

It is apt to identify and name these recursion stoppers in the first-order case:

0.2.37 Definition. (Prime Formulas) A formula that is either *atomic* or of the form $((\forall x)A)$ is called *prime*.   □

The following proposition is as expected:

0.2.38 Proposition. *Let us call* **Prop**[29] *the set of* all *first-order formulas A defined recursively as*

(a) A is a prime formula of **WFF** *(Definition 0.2.9).*
(b) A is $(\neg B)$, for some B in **Prop***.*
(c) A is $(B \vee C)$, for some B and C in **Prop***.*

Then **WFF=Prop***.*

Proof   There are two directions in showing equality of sets:[30]

• Let A be in **Prop**. Show it is in **WFF**.
  We do induction along the cases of the definition of **Prop** above. This is nothing else but a CVI application on *the number of Boolean connectives in the formula A* below, where for prime formulas this number is 0, since a "$(\forall x)B$" —the recursion stopper— is a *black box*. We cannot see any Boolean connectives inside it.

  (a) *Basis*. A is a prime formula of **WFF**, so it is one of $(\forall x)B$ with $B \in$ **WFF**, or an atomic formula of **WFF**. Either way, it is in **WFF** by Definition 0.2.9.
  (b) A is $(\neg B)$ with B in **Prop**. By the I.H. (one fewer connectives than A) $B \in$ **WFF** hence, by Definition 0.2.9, so is A.
  (c) A is $(B \vee C)$ with B and C in **Prop**. By the I.H. both B and C are in **WFF** hence, by Definition 0.2.9, so is A.

• Let A be in **WFF**. Show it is in **Prop**.
  We do induction along the cases of Definition 0.2.9.

  1. *Basis 1*. A is atomic. Then by (a) in Proposition 0.2.38, A is in **Prop**.
  2. A is $(\neg B)$ with B in **WFF**. By the I.H. B is in **Prop** hence case (b) in Proposition 0.2.38 yields A is in **Prop**.
  3. A is $(B \vee C)$ with B and C in **WFF**. By the I.H. both B and C are in **Prop** hence case (c) in Proposition 0.2.38 yields A is in **Prop**.

---

[29] **Prop** for *propositional* (logic). Boolean logic is also called propositional logic.

[30] This is another instance where sets and logic interdependence creates *apparent* circularities. We know —from our discrete mathematics— that two sets are *equal* iff they contain the same elements. But then to establish such equality just argue that every element of the *first* set is also a member of the second; and conversely. This is exactly what we are doing here.

4. *Basis 2. A* is $((\forall x)B)$ with *B* in **WFF**. By (a) in Proposition 0.2.38, *A* is in **Prop**.

□

0.2.39 Remark. From the foregoing we can now finalise the recursive computation of truth-values of an *A* in **Prop**, no matter how its prime subformulas were *initialised* (arbitrarily, by us) with truth values. Let us indicate (any) such an initialisation by a mapping *V* that maps *all* prime formulas of **Prop** to the set {**f**, **t**}.

We will call such a mapping a *valuation*.

There are infinitely may such mappings! It is *totally up to us* which one we choose in any one investigation since the point is just a "what if". We just want to know the effect of ¬ and ∨ on the overall (computed) truth-value of Prop formulas, relative to an arbitrary choice of *V*. That is,

What if I picked this V ? What will the truth-value of A be?

We denote by $\overline{V}$ the *extension* of *V* —that is, $\overline{V}(A) = V(A)$ if *A* is prime— that computes values for arbitrary formulas of **Prop**.

Thus,

$$\overline{V}(A) = \begin{cases} V(A) & \text{if } A \text{ is } prime \\ \mathbf{t} & \text{if } A \text{ is } \neg B \text{ and } \overline{V}(B) = \mathbf{f} \\ \mathbf{f} & \text{if } A \text{ is } \neg B \text{ and } \overline{V}(B) = \mathbf{t} \\ \mathbf{t} & \text{if } A \text{ is } B \vee C \text{ and } \overline{V}(B) = \mathbf{t} \\ \overline{V}(C) & \text{if } A \text{ is } B \vee C \text{ and } \overline{V}(B) = \mathbf{f} \end{cases}$$

Note that some *prime subformulas* of *A* do not contribute to the computation as they are *unreachable* being invisible to the process of *unpacking the* Boolean *syntax of A, along Boolean connectives only.*

For example, if *A* has a *maximal* prime subformula like $\left((\forall x)(\dots(\forall y)B\dots)\right)$, then the recursion will stop at this (as basis) and will not reach/see the $(\forall y)B$ prime subformula.

By a *maximal prime subformula of A* we mean:

1. A prime subformula *P* of *A* such that
2. There is no prime subformula *Q* of *A* that has *P* as a *proper subformula* (Definition 0.2.16).

□

0.2.40 Definition. (Tautologies) A formula *A* is a *tautology* iff, *for every V*, we have $\overline{V}(A) = \mathbf{t}$. We say that "*A* is a *tautology*", in symbols, $\models_{taut} A$.

We say that *A* is *satisfiable* iff, *for some V*, $\overline{V}(A) = \mathbf{t}$. We say that *V* satisfies *A*. A *set* of formulas *Γ* is satisfiable iff some *V* satisfies *all* $A \in \Gamma$. We say, *V* satisfies *Γ*.

It is profitable to have "relative tautologies", that is, relative to a set of hypotheses: We say that Γ *tautologically implies* A —in symbols $\Gamma \models_{taut} A$— iff, *every* V that satisfies Γ also satisfies A. It is a common convention to write $A, B, C \models_{taut} D$ rather than $\{A, B, C\} \models_{taut} D$. □

0.2.41 Example. (1) Thus for *verification* of tautological implication we do *no work at all* if the set Γ is *not* satisfiable.

For example, $\neg A, A \models_{taut} B$, for *any* B, since *no* V satisfies the left of $\models_{taut}$.

(2) We have trivially $(\forall x)B \models_{taut} (\forall x)B$, since *any* V that satisfies (the left copy of) $(\forall x)B$ also satisfies (the right copy of) $(\forall x)B$.

It is easy to verify that $\models_{taut} x = y \rightarrow x = y \vee z = w$ but also that $\models_{taut} x = x$ is not true, since the prime formula $x = x$ can be *assigned* the value **f** in our "what if"; a **t** is *not* guaranteed. We write $\not\models_{taut} x = x$. □

0.2.42 Remark. It is clear as we chose *not* to restrict the size of Γ in Definition 0.2.40 —Γ can be infinite— that the default, defining a V on *all* prime formulas of **Prop**, is apt.

The next exercise shows that in order to settle just a statement of the type $\models_{taut} A$ we *only* need to know the *finitely many* values of V on the *finitely many* prime subformulas of A. □

0.2.43 Exercise. Given A and two valuations V and V' that *agree on all* the prime subformulas of A, that is

$$V(B) = V'(B)$$

for all such subformulas.

Prove by induction on the number of Boolean connectives of A (working with **Prop**) (Proposition 0.2.38) that $\overline{V}(A) = \overline{V}'(A)$. □

The above justifies the practise of discarding values of a valuation V that do not refer to the prime subformulas of some A when we want to compute $\overline{V}(A)$.

0.2.44 Exercise. Prove (in the metatheory, of course) by induction on the number of hypotheses n that the statements

$$A_1, A_2, \ldots, A_n \models_{taut} B$$

and

$$\models_{taut} A_1 \rightarrow A_2 \rightarrow \ldots \rightarrow A_n \rightarrow B$$

are equivalent (both are true, or both are false).

Hint. Prove and use in your induction this lemma: the statements

$$A_1, A_2, \ldots, A_n \models_{taut} B$$

and

$$A_1, A_2, \ldots, A_{n-1} \models_{taut} A_n \to B$$

are equivalent. Note that the lemma does *not* need induction.   □

0.2.45 Definition. (The Boolean Fragment of 1st-Order Logic) The term *Boolean fragment of 1st-order logic* means 1st-order logic of Definitions 0.2.28 and 0.2.29, but with the rule ∀-*introduction* and the 1st-order axiom schemata (5–7 in Definition 0.2.28) *removed*. We can also refer to this fragment as the *propositional segment* of first order logic.

The term *Propositional fragment of 1st-order logic* is equally applicable.   □

The concepts of *proof* and *theorem*, as well as the notation regarding ⊢ remain the same, modulo the adjustment above to rules and axioms, however we will write ⊢$_{PS}$ to indicate theoremhood verified via proofs that *do not employ* either the rule ∀-introduction or 1st-order axioms.

0.2.46 Exercise. Prove that if $\Gamma \vdash_{PS} A$, then $\Gamma \vdash A$ as well.   □

0.2.5   Three Obvious Metatheorems

The following metatheorems apply to 1st-order proofs and, *a fortiori*, to proofs within the Boolean fragment; *and* are totally expected at the intuitive level.

0.2.47 Metatheorem. The concatenation of any two Γ-proofs, in any order, is a Γ-proof.

Proof Exercise! Just verify that each formula in the concatenation result satisfies the four conditions for Γ-proofs (Definition 0.2.31).

It follows at once,

0.2.48 Corollary. *The concatenation of any finite number of Γ-proofs, in any order, is a Γ-proof.*

0.2.49 Metatheorem. Suppose that $\Gamma \vdash B_i$, for $i = 1, 2, \ldots, n$, and also that $B_1, \ldots, B_n \vdash A$. Then $\Gamma \vdash A$.

Proof We have proofs from Γ for each B_i:

$$\boxed{\ldots B_i}$$

Thus,

$$\boxed{\ldots B_1} \ldots \boxed{\ldots B_i} \ldots \boxed{\ldots B_n}$$

is also a Γ-proof by the corollary to Corollary 0.2.48. But then so is the sequence

$$\boxed{\ldots B_1}\ldots\boxed{\ldots B_i}\ldots\boxed{\ldots B_n}\;\boxed{\ldots B_i \ldots A}$$

where the last box is for $B_1, \ldots, B_n \vdash A$.

Why is this a Γ-proof and not a $\Gamma \cup \{B_1, \ldots, B_n\}$-proof? Because we change the justification of each B_i in the last box, from "hypothesis" to "same reason that we gave for B_i in the i-th box". □

 0.2.50 Remark. The import of the above is that we *may* use derived rules!

If, for example, we have established a *theorem schema* of the form $B_1, \ldots, B_n \vdash A$, where the B_i relate in a fixed syntactic way with A, then we *can* use this schema exactly as we use any *primary* rule in order to extend/continue a proof as demonstrated above! So we call it a *derived rule* and is used just as straightforwardly as MP is!

Pause. Is $A, A \to B \vdash B$ a theorem schema? Why?◄

Careful, though! MP is a primary, not a derived rule!

Sometimes we nickname Metatheorem 0.2.49 the "transitivity of $\vdash$ metatheorem". This is appropriate, especially if we have one B_i: If $\Gamma \vdash B$ and $B \vdash A$, then $\Gamma \vdash A$ the theorem states. □

Finally, as we expected and hoped, we can use *previously proved theorems* in a new proof, *without having to insert their proof*! Below we quote A as a hypothesis (but give no proof!) in establishing $\Gamma, A \vdash B$.

The metatheorem concludes that we are as good as having proved B from Γ alone. Why? Because Γ can generate the "hypothesis" A for us. We do not have to pluck it out of the blue.

0.2.51 Metatheorem. Suppose that $\Gamma \vdash A$, and $\Gamma, A \vdash B$. Then $\Gamma \vdash B$.

 The notation Γ, A is the preferred one for abbreviating $\Gamma \cup \{A\}$. Another "shorthand" notation is $\Gamma + A$.

Proof We have proofs

$$\boxed{\ldots A}$$

from Γ and

$$\boxed{\ldots A \ldots B}$$

from Γ, A. So

$$\boxed{\ldots A}\;\boxed{\ldots A \ldots B}$$

is a proof from Γ —not Γ, A— because we can justify the presence of A in the second box *exactly the same way* as we did in the first box; *not* as a hypothesis!

0.2.6   Soundness and Completeness of the Boolean Fragment of 1st-Order Logic

In what follows all our proofs (in this subsection) within pure (or applied, when the case calls for) logic will be fully annotated Hilbert-style proofs. To facilitate annotation we will arrange them vertically on the page, one formula per line.

The following proposition is useful to record and applies to all of 1st-order logic, not just to its Boolean fragment.

0.2.52 Proposition. *If $\vdash A \to B$, then $A \vdash B$.*

Proof   The quick proof is to just say: "by MP".
   Here is a deliberate proof.

$$(1) \quad A \qquad\qquad\qquad \langle\text{hyp}\rangle$$
$$(2) \quad A \to B \qquad\quad \langle\text{theorem}\rangle$$
$$(3) \quad B \qquad\qquad \langle(1, 2) + \text{MP}\rangle$$

$\square$

0.2.53 Proposition. (Transitivity of $\to$) *For any wff A, B and C, we have $A \to B, B \to C \vdash_{PS} A \to C$.*

Proof   In our annotation below (in $\langle \ldots \rangle$ brackets) "hyp" means hypothesis. The stuff we assume to the left of $\vdash$.

$(1) \quad A \to B$ $\qquad\qquad\qquad\qquad\qquad\qquad\qquad\qquad\qquad\qquad\qquad \langle\text{hyp}\rangle$

$(2) \quad B \to C$ $\qquad\qquad\qquad\qquad\qquad\qquad\qquad\qquad\qquad\qquad\qquad \langle\text{hyp}\rangle$

$(3) \quad (B \to C) \to (\neg A \vee B \to \neg A \vee C) \qquad \langle\text{axiom 4. of Definition 0.2.28}\rangle$

$(4) \quad (A \to B) \to (A \to C) \qquad\qquad \langle(2, 3) + \text{MP; abbrev. } X \to Y \text{ used}\rangle$

$(5) \quad A \to C \qquad\qquad\qquad\qquad\qquad\qquad\qquad\qquad \langle(1, 4) + \text{MP}\rangle$

$\square$

We have just learnt a useful derived rule! We will apply it in the proof of the next proposition.

0.2.54 Proposition. *For any wff A, we have $\vdash_{PS} A \to A$.*

Proof

$$(1) \quad A \vee A \to A \qquad\qquad\qquad \langle\text{axiom}\rangle$$
$$(2) \quad A \to A \vee A \quad \langle\text{axiom where } B \text{ was taken to be } A\rangle$$
$$(3) \quad A \to A \qquad\qquad \langle(1, 2) + \text{trans. of } \to\rangle$$

$\square$

0.2.55 Proposition. *For any wff A and B, we have* $\vdash_{PS} A \to B \vee A$.

Proof

$$
\begin{array}{lll}
(1) & A \to A \vee B & \langle\text{axiom}\rangle \\
(2) & A \vee B \to B \vee A & \langle\text{axiom}\rangle \\
(3) & A \to B \vee A & \langle(1, 2) + \text{trans of} \to\rangle
\end{array}
$$

□

0.2.56 Corollary. *For any wff A and B, we have* $\vdash_{PS} A \to B \to A$.

Proof In the above use $\neg B$ instead of B.   □

0.2.57 Corollary. *For any wff A and B, we have* $A \vdash_{PS} B \to A$.

Proof The above Corollary and Proposition 0.2.52.   □

0.2.58 Proposition. (Symmetry of $\vee$ **rule)** *For any wff A and B, we have* $A \vee B \vdash_{PS} B \vee A$.

Proof Definition 0.2.28 number 3. and Proposition 0.2.52.   □

0.2.59 Proposition. *For any wff A, we have* $\vdash_{PS} A \to \neg\neg A$.

Proof

$$
\begin{array}{lll}
(1) & \neg A \to \neg A & \langle\text{Proposition 0.2.54}\rangle \\
(1') & \neg\neg A \vee \neg A & \langle\text{what (1) means}\rangle \\
(2) & \neg A \vee \neg\neg A & \langle(1') + \text{Proposition 0.2.58}\rangle \\
(3) & A \to \neg\neg A & \langle\text{abbrev. (2)}\rangle
\end{array}
$$

□

0.2.60 Lemma. *For all wff A, B and C, we have* $A \to B \vdash_{PS} C \vee A \to C \vee B$.

Proof Definition 0.2.28 number 4. and Proposition 0.2.52.

0.2.61 Corollary. *For all wff A, B and C, we have* $A \to B \vdash_{PS} A \vee C \to B \vee C$.

Proof Suffices to show that $C \vee A \to C \vee B \vdash_{PS} A \vee C \to B \vee C$.

$$
\begin{array}{lll}
(1) & A \vee C \to C \vee A & \langle\text{axiom}\rangle \\
(2) & C \vee A \to C \vee B & \langle\text{hyp}\rangle \\
(3) & A \vee C \to C \vee B & \langle(1, 2) + \text{Proposition 0.2.53}\rangle \\
(4) & C \vee B \to B \vee C & \langle\text{axiom}\rangle \\
(5) & A \vee C \to B \vee C & \langle(3, 4) + \text{Proposition 0.2.53}\rangle
\end{array}
$$

□

0.2.62 Proposition. *For any wff A, we have* $\vdash_{PS} \neg\neg A \to A$.

Proof

$$
\begin{array}{lll}
(1) & \neg A \to \neg\neg\neg A & \langle \text{Proposition } 0.2.59 \rangle \\
(2) & A \vee \neg A \to A \vee \neg\neg\neg A & \langle (1) + \text{Lemma } 0.2.60 \rangle \\
(3) & \neg A \vee A & \langle \text{Proposition } 0.2.54 \rangle \\
(4) & A \vee \neg A & \langle (3) + \text{Proposition } 0.2.58 \rangle \\
(5) & A \vee \neg\neg\neg A & \langle (2, 4) + \text{MP} \rangle \\
(6) & \neg\neg\neg A \vee A & \langle (5) + \text{Proposition } 0.2.58 \rangle
\end{array}
$$

The last line is $\neg\neg A \to A$.   □

0.2.63 Proposition. (The Contrapositive) *For all wff A and B, we have* $\vdash_{PS}$ $(A \to B) \to (\neg B \to \neg A)$. *We say that* $\neg B \to \neg A$ *is the* contrapositive *of* $A \to B$.

Proof

$$
\begin{array}{lll}
(1) & B \to \neg\neg B & \langle \text{Proposition } 0.2.59 \rangle \\
(2) & \neg A \vee B \to \neg A \vee \neg\neg B & \langle (1) + \text{Lemma } 0.2.60 \rangle \\
(3) & \neg A \vee \neg\neg B \to \neg\neg B \vee \neg A & \langle \text{axiom } 3. \rangle \\
(4) & \neg A \vee B \to \neg\neg B \vee \neg A & \langle (2, 3) + \text{Proposition } 0.2.53 \rangle
\end{array}
$$

The last line when abbreviated says $(A \to B) \to (\neg B \to \neg A)$.   □

0.2.64 Proposition. (Proof by Cases I) *For any wff A, B, C and D, we have* $A \to B, C \to D \vdash_{PS} A \vee C \to B \vee D$.

Proof

$$
\begin{array}{lll}
(1) & A \to B & \langle \text{hyp} \rangle \\
(2) & C \to D & \langle \text{hyp} \rangle \\
(3) & A \vee C \to B \vee C & \langle (1) + \text{Corollary } 0.2.61 \rangle \\
(4) & B \vee C \to B \vee D & \langle (2) + \text{Lemma } 0.2.60 \rangle \\
(5) & A \vee C \to B \vee D & \langle (3, 4) + \text{Proposition } 0.2.53 \rangle
\end{array}
$$
□

0.2.65 Corollary. (Proof by Cases II) *For any wff A, B and C, we have* $A \to C, B \to C \vdash_{PS} A \vee B \to C$.

Proof Proposition 0.2.64 gives us $A \to C, B \to C \vdash_{PS} A \vee B \to C \vee C$. Thus, axiom 2. (0.2.28) and transitivity of $\to$ rest the case.   □

0.2.66 Proposition. *For any wff A, B and C, we have* $\vdash_{PS}$ $((A \vee B) \vee C) \rightarrow$ $(A \vee (B \vee C))$.

Proof   We will invoke Proposition 0.2.64 a few times and also Corollary 0.2.65.
 From

$$\vdash_{PS} A \rightarrow A$$

by Proposition 0.2.54 and

$$\vdash_{PS} B \rightarrow B \vee C$$

by axiom 1. of Definition 0.2.28 we obtain (Proposition 0.2.64)

$$\vdash_{PS} A \vee B \rightarrow A \vee (B \vee C) \tag{1}$$

By Proposition 0.2.55 we have

$$\vdash_{PS} C \rightarrow B \vee C \tag{2}$$

and again

$$\vdash_{PS} B \vee C \rightarrow A \vee (B \vee C) \tag{3}$$

(2) and (3) and Proposition 0.2.53 yield

$$\vdash_{PS} C \rightarrow A \vee (B \vee C) \tag{4}$$

(1) and (4) and Corollary 0.2.65 yield

$$(A \vee B) \vee C \rightarrow A \vee (B \vee C) \qquad \square$$

We can easily prove the converse imitating the above proof (Exercise!)

0.2.67 Corollary. *For any wff A, B and C, we have* $\vdash_{PS} (A \vee (B \vee C)) \rightarrow ((A \vee B) \vee C)$.

0.2.68 Proposition. *For all A, B and C we have* $\vdash_{PS} (A \rightarrow (B \rightarrow C)) \rightarrow (B \rightarrow (A \rightarrow C))$.

Proof   Translating the short form $\rightarrow$, we are asked to prove

$$\vdash_{PS} (\neg A \vee (\neg B \vee C)) \rightarrow (\neg B \vee (\neg A \vee C)) \tag{1}$$

Here is a proof:

(1) $\neg A \vee (\neg B \vee C) \rightarrow (\neg A \vee \neg B) \vee C$ ⟨Corollary 0.2.67⟩

(2) $\neg A \vee \neg B \rightarrow \neg B \vee \neg A$ ⟨axiom⟩

(3) $C \rightarrow C$ ⟨Proposition 0.2.54⟩

(4) $(\neg A \vee \neg B) \vee C \rightarrow (\neg B \vee \neg A) \vee C$ ⟨(2, 3) + Proposition 0.2.64⟩

(5) $\neg A \vee (\neg B \vee C) \rightarrow (\neg B \vee \neg A) \vee C$ ⟨(1, 4) + Proposition 0.2.53⟩

(6) $(\neg B \vee \neg A) \vee C \rightarrow \neg B \vee (\neg A \vee C)$ ⟨Proposition 0.2.66⟩

(7) $\neg A \vee (\neg B \vee C) \rightarrow \neg B \vee (\neg A \vee C)$ ⟨(5,6) + Proposition 0.2.53⟩

The last line is (1).   □

0.2.69 Proposition. *For all A and B we have*

$$\vdash_{PS} \left(A \rightarrow (A \rightarrow B) \right) \rightarrow (A \rightarrow B)$$

Proof Removing the abbreviation "$\rightarrow$", we want to prove,

$$\vdash_{PS} \left(\neg A \vee (\neg A \vee B) \right) \rightarrow (\neg A \vee B) \qquad (1)$$

(1) $\neg A \vee (\neg A \vee B) \rightarrow (\neg A \vee \neg A) \vee B$ ⟨Corollary 0.2.67⟩

(2) $\neg A \vee \neg A \rightarrow \neg A$ ⟨2. from Definition 0.2.28⟩

(3) $B \rightarrow B$ ⟨Proposition 0.2.54⟩

(4) $(\neg A \vee \neg A) \vee B \rightarrow \neg A \vee B$ ⟨(2, 3) + Proposition 0.2.64⟩

(5) $\neg A \vee (\neg A \vee B) \rightarrow \neg A \vee B$ ⟨(1, 4) + Proposition 0.2.53⟩   □

0.2.70 Metatheorem. (Soundness of the Boolean Fragment) If $\Gamma \vdash_{PS} A$, then also $\Gamma \models_{taut} A$.

Proof Induction on proof length n.

Basis. $n = 1$. Then the proof is the one-member sequence A. Only cases 1. and 2. of Definition 0.2.31 apply:

1. A is a Boolean axiom (Definition 0.2.28, 1.–4.) Then any valuation V that satisfies Γ will satisfy A as well since all V satisfy the latter.
2. $A \in \Gamma$. Then any valuation V that satisfies Γ will in particular satisfy A.

   I.H. We assume the metatheorem for proof lengths $\leq n$.
   I.S. Here we have a proof

$$\ldots, A \qquad (1)$$

of length $n + 1$. If A is in Γ or is a Boolean axiom, then we are done by the basis. Otherwise, we look at the case where MP was applied to B and $B \rightarrow A$ *to the left* of A to obtain A. Pick now a valuation V that satisfies Γ.

By the I.H. $\overline{V}(B) = \mathbf{t} = \overline{V}(B \rightarrow A)$, but then $\overline{V}(A) = \mathbf{t}$ since otherwise we would have $\overline{V}(B \rightarrow A) = \mathbf{f}$.

Note that the rule generalisation is not applicable.   □

Soundness helps to verify that a formula is unprovable in the Boolean fragment. For example, $\neg A \wedge A$, for any A, is unprovable, since if we had $\vdash_{PS} \neg A \wedge A$, then we would also have $\models_{taut} \neg A \wedge A$; not so.

Before we state and prove the completeness (Post's theorem) of the Propositional fragment of 1st-order logic, we will need to prove a Boolean version of the *Deduction theorem*, namely:

0.2.71 Metatheorem. (Propositional Deduction Theorem) If $\Gamma, A \vdash_{PS} B$, then also $\Gamma \vdash_{PS} A \rightarrow B$.

Proof Induction on the length n of proof of B from Γ, A.

Basis. $n = 1$. Then B is *one of*

- In Γ. Then $\Gamma \vdash_{PS} B$. We also have $B \vdash_{PS} A \rightarrow B$ (Corollary 0.2.57), hence $\Gamma \vdash_{PS} A \rightarrow B$ by Metatheorem 0.2.49.
- An axiom. This is argued exactly as above.
- A. Then $\vdash_{PS} A \rightarrow A$, where the second A is B. By Exercise 0.2.36, $\Gamma \vdash_{PS} A \rightarrow A$.

I.H. Assume the claim for proofs of length $\leq n$.

Now perform the *I.S.*

Here is a Γ, A-proof of B of length $n + 1$:

$$\ldots, B \tag{1}$$

We focus on the case that B is the result of MP on formulas to its left, because we dealt with the other cases in the basis above.

So let X and $X \rightarrow B$ appear to the left of B in (1). By the I.H. we have

$$\Gamma \vdash_{PS} A \rightarrow X \tag{2}$$

and

$$\Gamma \vdash_{PS} A \rightarrow (X \rightarrow B) \tag{3}$$

that is, we have a proof that we continue —(3)–(5)— as follows:

$\vdots$

(1) $A \to X$ ⟨I.H.: proved from Γ⟩

(2) $A \to (X \to B)$ ⟨I.H.: proved from Γ⟩

(3) $X \to (A \to B)$ ⟨(2) + Proposition 0.2.68 via Proposition 0.2.52⟩

(4) $A \to (A \to B)$ ⟨(1, 3) + Proposition 0.2.53⟩

(5) $A \to B$ ⟨(4) + Proposition 0.2.69 via Proposition 0.2.52⟩   □

 The deduction theorem is used in practice without notice. It simplifies the task of proving $A \to B$ (from some hypotheses Γ) by moving A to the hypotheses side, with the words such as "assume A and prove B". This is in principle easier than the original task, as B is less complex than $A \to B$, *and* we have an enriched set of hypotheses; we have added A.

0.2.72 Metatheorem. (Completeness of the Boolean Fragment)
If $\Gamma \models_{taut} A$, then also $\Gamma \vdash_{PS} A$.

 This is also known as Post's (meta)theorem for Boolean logic. It is a powerful proof-tool in 1st-order logic.

Proof

> We prove the contrapositive: If $\Gamma \nvdash_{PS} A$, then $\Gamma \nvDash_{taut} A$.

The proof is via a construction for which we prove several observations that will lead to the definition of a V that satisfies Γ but not A.

$$\text{So, assume } \Gamma \nvdash_{PS} A \tag{1}$$

We start with a construction (inductive definition of sets Δ_n for each $n \geq 0$) that makes use of a *fixed enumeration of all wff* of **WFF**.

Here is how we can effect the enumeration we want.

Fix first an alphabetic order of the symbols of the alphabet given in Tentative Definition 0.2.2; say, fix the order in which we enumerated the symbols in the aforementioned definition:

$$\{\#, v, \neg, \vee, (,), =, \forall, f, p, c\}$$

We now build *two* lists, List$_1$ and List$_2$ simultaneously as described below.

The generation of List$_2$ is subsidiary to that of List$_1$.

In List$_1$ we list *all* strings over the first-order pure logic alphabet above by *string length*, and in each length-group, *lexicographically* (alphabetically).

This can clearly be done algorithmically, or *effectively* as we alternatively say, by an enumerating computer program that either runs forever —intuitively it is best to so view the program's behaviour— or runs long enough to produce the n-th string, for any n, and then stop. We say we have an *effective* listing of all strings, placed in List$_1$.

Every string that we place in List$_1$ *we test for membership in* **WFF**.

Pause. Can we do that? Yes! See if you can devise a recursive procedure to do so in your favourite programming language, using the recursive Definition 0.2.9.◄

If the test returns "yes", then *copy the string* in List$_2$ as well.

The enumeration of *all* of **WFF** (or **Prop**) that we obtain in List$_2$ we will denote by

$$H_0, H_1, H_2, \ldots \tag{2}$$

We next define by recursion on n[31] a sequence of sets Δ_n, all of which include Γ.

$$\Delta_0 \overset{Def}{=} \Gamma \tag{3}$$

and, for all $n \geq 0$

$$\Delta_{n+1} \overset{Def}{=} \begin{cases} \Delta_n \cup \{H_n\} & \text{if } \Delta_n \cup \{H_n\} \nvdash_{PS} A \\ \Delta_n \text{ otherwise} \end{cases} \tag{4}$$

We also define Δ as the union of all Δ_n, that is, Δ contains all the formulas found in all the various n. This is expressed by the usual set-theoretic notation of *union*, $\bigcup$.

$$\Delta \overset{Def}{=} \bigcup_{n \geq 0} \Delta_n \tag{5}$$

We next state and prove some simple claims asserting the properties of Δ and the various Δ_n.

Claim 1. $\Delta_n \subseteq \Delta_{n+1}$, for all n. This is immediate since either case in (4) contains Δ_n as a subset.

---

[31] We will study such recursions more deeply in this chapter, but suffice it to say at this point that the reader has seen such recursions in a course on discrete mathematics, for example. One such example is defining non negative integer exponents of a positive number a by $a^0 = 1$ and $a^{n+1} = a \cdot a^n$. This can also be thought of as the definition of the sequence $a^0, a^1, a^2, \ldots, a^n, \ldots$ Another example of defining a sequence is more involved: The Fibonacci sequence $F_0, F_1, F_2, \ldots, F_n, \ldots$ given by the recurrence equations $F_0 = 0$, $F_1 = 1$ and, for $n \geq 1$, $F_{n+1} = F_n + F_{n-1}$.

Claim 2. $\Gamma \subseteq \Delta$. Immediate from **Claim 1** and $\Gamma = \Delta_0 \subseteq \Delta_1 \subseteq \ldots \subseteq \Delta_n \subseteq \ldots \subseteq \Delta$.

Claim 3. For all n, $\Delta_n \nvdash_{PS} A$. Induction on n. The *basis* for $n = 0$ is by (3) and (1). *I.H.* Assume that $\Delta_n \nvdash_{PS} A$, for a fixed n. Then we have $\Delta_{n+1} \nvdash_{PS} A$, by the top case in (4) directly, if applicable, otherwise by the bottom case and the I.H.

Claim 4. $\Delta \nvdash_{PS} A$. Otherwise, $\Delta_n \vdash_{PS} A$ for some n, contradicting **Claim** 3.

Claim 5. For *every* B in **Prop** (same as **WFF**!), one of B or $\neg B$ are in Δ. Let B be H_n in the enumeration (2), and $\neg B$ be H_m. Suppose that the construction (4) missed both. Then it must be that $\Delta_n \cup \{B\} \vdash_{PS} A$ and $\Delta_m \cup \{\neg B\} \vdash_{PS} A$. By Metatheorem 0.2.71, the first of the preceding two statements gives $\Delta_n \vdash_{PS} B \to A$ and the second gives $\Delta_m \vdash_{PS} \neg B \to A$. By left weakening (Exercise 0.2.36) each of the above we get

$$\Delta \vdash_{PS} B \to A \qquad\qquad (6)$$

and

$$\Delta \vdash_{PS} \neg B \to A \qquad\qquad (7)$$

hence by Corollary 0.2.65, $\Delta \vdash_{PS} \neg B \vee B \to A$. Since $\vdash_{PS} \neg B \vee B$ (Proposition 0.2.54), MP gives $\Delta \vdash_{PS} A$. This contradicts **Claim** 4.

Claim 6. In **Claim** 5 it must be one or the other, B or $\neg B$, but *not both*. Indeed, if not, then $\Delta \vdash_{PS} B$ and $\Delta \vdash_{PS} \neg B$, hence $\Delta \vdash_{PS} B \vee A$ and $\Delta \vdash_{PS} \neg B \vee A$, applying in each case axiom 1 (Definition 0.2.28) as a rule,[32] via Proposition 0.2.52. Now employing the $\to$-abbreviation we have (6) and (7) above, which lead to a contradiction.

Claim 7. Δ is *deductively closed*, meaning,

$$\text{if } \Delta \vdash_{PS} B, \text{ then } B \in \Delta$$

Well, if B is not in Δ, then $\neg B$ *is* in Δ, by **Claim** 5. Thus, $\Delta \vdash_{PS} \neg B$ as well and we are led to (6) and (7) —and therefore to a contradiction— as we showed in the previous claim or indeed in **Claim** 5.

We are ready to define a *valuation V* on the *prime formulas* of **Prop**, that will satisfy Γ but *not A*, thus concluding our proof.

0.2.73 Definition. (The Valuation V)

$$V(P), \text{ for prime } P, \overset{Def}{=} \begin{cases} \mathbf{t} & \text{if } P \in \Delta \\ \mathbf{f} & \text{otherwise} \end{cases}$$

$\square$

---

[32] This rule, $X \vdash X \vee Y$, is called *expansion*.

The Main Lemma toward the end of the Post Metatheorem proof is:

0.2.74 Main Lemma. Under the valuation V, for all B in **Prop**, we have $\overline{V}(B) = \mathbf{t}$ iff $B \in \Delta$.

Proof [of Main Lemma] We do induction on the formation of Prop formulas (cf. Proposition 0.2.38).

Basis. For B prime the statement is the definition of V in Definition 0.2.73. Nothing else to do here.

I.H. Assume the claim for the immediate predecessors of B:

I.S. Prove the claim for B. Two cases:

- B is $\neg C$.

  So let $\overline{V}(B) = \mathbf{t}$. Then $\overline{V}(C) = \mathbf{f}$. By the I.H. $C \notin \Delta$. By **Claim 5**, $B \in \Delta$.

  Conversely, let $B \in \Delta$, that is, $\neg C$ is. But then $C \notin \Delta$ by **Claim 6**. By the I.H. $\overline{V}(C) = \mathbf{f}$, hence $\overline{V}(B) = \mathbf{t}$.

- B is $C \vee D$.

  So let $\overline{V}(B) = \mathbf{t}$. Then $\overline{V}(C) = \mathbf{t}$ or $\overline{V}(D) = \mathbf{t}$. We will argue just the first case (for C) as the other is entirely similar. By the I.H. $C \in \Delta$, hence $\Delta \vdash_{PS} C$. By axiom 1 of Definition 0.2.28, via Proposition 0.2.52, we have $\Delta \vdash_{PS} C \vee D$. By **Claim 7**, $C \vee D \in \Delta$, that is, $B \in \Delta$.

  Conversely, let $B \in \Delta$. We argue that C or D (at least one) will be in Δ. Well, if neither is, then, by **Claim 5**, $\neg C$ *and* $\neg D$ are. Then $\Delta \vdash_{PS} \neg C$ and $\Delta \vdash_{PS} \neg D$ by **Claim 7**, hence also $\Delta \vdash_{PS} \neg C \vee A$ and $\Delta \vdash_{PS} \neg D \vee A$ as before (axiom 1 of Definition 0.2.28, via Proposition 0.2.52). These conclusions translate to $\Delta \vdash_{PS} C \rightarrow A$ and $\Delta \vdash_{PS} D \rightarrow A$, hence $\Delta \vdash_{PS} C \vee D \rightarrow A$, by Corollary 0.2.65, or $\Delta \vdash_{PS} B \rightarrow A$. It follows (MP) that $\Delta \vdash A$. A contradiction.

  So one of C or D *are* in Δ after all. Say, C is. By the I.H. $\overline{V}(C) = \mathbf{t}$ hence $\overline{V}(B) = \mathbf{t}$. **End of the proof of the Main Lemma** □

So the main lemma certifies that V satisfies Δ, and hence also Γ, since the latter is a subset of the former (**Claim 2**).

But V does not satisfy A (i.e. $\overline{V}(A) = \mathbf{f}$), since $\Delta \nvdash_{PS} A$, hence $A \notin \Delta$.

These two concluding sentences establish $\Gamma \nvDash_{taut} A$, which along with (1) proves our Metatheorem. **End of the proof of Metatheorem 0.2.72** □

0.2.75 Corollary. If $\vDash_{taut} A$, then $\vdash_{PS} A$, hence also $\vdash A$ using no quantifier rules and no first-order axioms.

0.2.76 Remark. (Tautological Implication and Provability) Post's theorem is applied to generate convenient derived rules of inference:

> Any schema of the form $A_1, \ldots, A_n \vDash_{taut} B$ leads to the derived rule $A_1, \ldots, A_n \vdash B$, where the implied by the symbol "$\vdash$" proof utilises no quantifier rules and no first-order axioms.

Here is why:

- First, $A_1, \ldots, A_n \models_{taut} B$ iff $A_1, \ldots, A_{n-1} \models_{taut} A_n \to B$. Indeed, say the left hand side ("lhs") of iff is true and prove the rhs: Let all the A_i, for $i = 1, \ldots, n-1$, be true. Then $A_n \to B$ is true trivially if A_n is false, but is true also if A_n is true due to the lhs of "iff".
- Thus

$$A_1, \ldots, A_n \models_{taut} B \text{ iff } \models_{taut} A_1 \to A_2 \to \ldots \to A_n \to B$$

The above is proved by induction on $n \geq 1$ using the result of the previous bullet (Exercise!)
- Thus, the assumption $A_1, \ldots, A_n \models_{taut} B$ leads to $\models_{taut} A_1 \to A_2 \to \ldots \to A_n \to B$, hence —by Corollary 0.2.75— to

$$\vdash A_1 \to A_2 \to \ldots \to A_n \to B \tag{1}$$

- Assume now $A_1, \ldots, A_n$. Using (1) and n applications of modus ponens we eliminate all occurrences of "$A_i \to$" from left to right (one at a time) and we are left with B. That is, we *proved B*. □

0.2.7 *Useful Theorems and Metatheorems of* (**Pure**) *First-Order Logic*

This subsection contains a set of ad hoc tools (some are derived rules) usable toward facilitating proofs.

0.2.77 Exercise. Prove that if there is no free occurrence of x in A, then $\vdash A \to (\forall x)A$.

Hint. This has a two-line Hilbert proof. □

0.2.78 Theorem. (Generalisation) *The theorem schema*

$$A \vdash (\forall x)A$$

is a very handy derived *rule we will call* generalisation, *in short* Gen.

Proof

(1)	A		$\langle$hyp$\rangle$
(2)	$z = z \to A$		$\langle(1) + \text{Post}\rangle$
(3)	$z = z \to (\forall x)A$	$\langle(2) + \forall\text{-intro; OK: no free } x \text{ in this line}\rangle$	
(4)	$z = z$		$\langle\text{axiom 6 (Definition 0.2.28)}\rangle$
(5)	$(\forall x)A$		$\langle(3, 4) + \text{MP}\rangle$

0.2.79 Proposition. *We have the derived* specialisation *rule —in short,* Spec—

$$(\forall x)A[x] \vdash A[t]$$

as long as the operation $A[x := t]$ *is defined.*

Proof   By the substitution axiom and Proposition 0.2.52.   □

0.2.80 Corollary. *We also have the special case of* Spec, $(\forall x)A \vdash A$

Proof   Take x for t in Proposition 0.2.79.
  Pause. Why is x substitutable into x in A without capture?◄   □

0.2.81 Theorem. *For any term* $t[z]$ *we have* $\vdash x = y \to t[x] = t[y]$.

Proof   By the axiom of equality,

$$\vdash x = y \to (t[x] = t[y] \equiv t[y] = t[y]) \tag{1}$$

Consider next

I.	$z = z$	$\langle$axiom$\rangle$
II.	$(\forall z)z = z$	$\langle$I. + Gen$\rangle$
III.	$t[y] = t[y]$	$\langle$II. + Spec$\rangle$
IV.	$x = y \to (t[x] = t[y] \equiv t[y] = t[y])$	$\langle$(1)$\rangle$
IIV.	$x = y \to t[x] = t[y]$	$\langle$III. + IV. + Post$\rangle$

□

0.2.82 Remark. (Using the Post Metatheorem)   The above proof is the first that employs Post's theorem of the previous subsection. The words we can use are "by tautological implication" (cf. the remark enclosed within markings at the end of Post's Metatheorem above).

  We can also just say "by Post" or simply say "Post" whenever it is not ungrammatical (for example in some contexts we may say "...+ Post". Here the justification "III. + IV. + Post" is apt).

  Here we noted and applied, without offering details, the following

$$A \to (B \equiv C), C \models_{taut} A \to B$$

□

0.2.83 Definition. (Universal Closure) The *universal closure* of a formula A —let us denote it by A^u— is obtained by appending the *canonical* prefix "$(\forall x_1)(\forall x_2)\ldots(\forall x_n)$" in front of A, where the listed x_i are all the free variables of A.

  "Canonical" means that the x_i in the prefix "$(\forall x_1)(\forall x_2)\ldots(\forall x_n)$" are arranged in ascending lexicographic order.   □

0.2.84 Proposition. *If* A^u *is the universal closure of* A *then* $\vdash A$ *iff* $\vdash A^u$.

Proof   By Theorem 0.2.78 and Corollary 0.2.80.   □

0.2.85 Proposition. *If* $\vdash_\Gamma A[x]$, *then* $\vdash_\Gamma A[t]$, *as long as the operation* $A[x := t]$ *is defined.*

Proof   Here is a short Hilbert-style proof:

(1) $A[x]$ ⟨proved⟩
(2) $(\forall x)A[x]$ ⟨(1) + Gen⟩
(3) $A[t]$ ⟨(2) + Spec⟩   □

0.2.86 Metatheorem. (First-Order Deduction Theorem) If $\Gamma, A \vdash B$ can be verified by a proof —*as* originally *defined in Definition 0.2.31*— *where no variables occurring free in A were used in an application of* $\forall$*-intro*, then $\Gamma \vdash A \to B$.

 The hedging in the statement of the metatheorem, namely, the restriction that we have a Γ, A-proof of B that is written faithfully according to the Definition 0.2.31 is technically motivated.

> *What we do* not *want to take place in a* Γ, *A-proof is that* $\forall$-intro *or any of* the rules *derived from it* —*e.g., like* Gen *or the substitution rule Proposition 0.2.85*— *involved some free variable that occurs in A.*

By stressing that the Γ, A-proof we are interested in did not employ $\forall$-intro on a variable x that occurs free in A, *and that the proof is written according to the original definition*, means that *we do not have to enumerate* what else we cannot do in said proof: Obviously it will not contain any *derived* rules like *Gen* or Proposition 0.2.85.

One can state the restriction more simply "...a Γ, A-proof where all free occurrences of variables in A *were treated as constants* throughout the proof ..."

Or, more extremely, one can take A to be *closed* —that is, a formula that has *no free variables*. Such a formula is also called a *sentence*. See corollary below.

Proof   Much of the proof was done in that of Metatheorem 0.2.71, where we did induction on the length of a proof that verifies $\Gamma, A \vdash B$, and we covered all cases already, except the one where

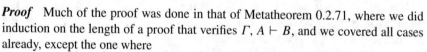

$$\ldots, B$$

is a proof of length $n + 1$ *written according to the restriction stated in the metatheorem*, where B is $X \to (\forall x)Y$, obtained by $\forall$-intro from a previous formula, $X \to Y$, in the proof.

We know two things about the used x:

1. Seeing that x was the bound variable generated by $\forall$-intro, we know that x does not occur free in X.
2. Having followed the restriction of the metatheorem statement, x does not occur free in A either.

Now, the I.H. yields $\Gamma \vdash A \to X \to Y$ (omitting redundant bracketing). We thus reason as follows via a Hilbert-style proof:

(1) $A \to X \to Y$ ⟨from Γ, by I.H.⟩

(2) $A \land X \to Y$ ⟨(1) + Post⟩

(3) $A \land X \to (\forall x)Y$ ⟨(2) + ∀-intro; OK, x is not free in $A \land X$⟩

(4) $A \to X \to (\forall x)Y$ ⟨(3) + Post⟩

(4′) $A \to B$ ⟨alias of line (4); Done.⟩   □

0.2.87 Corollary. *If A is closed and $\Gamma, A \vdash B$, then also $\Gamma \vdash A \to B$.*

0.2.88 Remark. It is clear that we have no right to conclude $\vdash A \to (\forall x)A$ from $A \vdash (\forall x)A$ and say "by the deduction theorem" while doing so. Why "no right"? Because we pay no attention to the *restriction* in the statement of the metatheorem if we do so. Still the question remains since a proof *other than via Metatheorem 0.2.86* might *exist*! We need a definitive counterexample:

Can I prove $\vdash A \to (\forall x)A$ by any means whatsoever other *than Metatheorem 0.2.86?*

A *counterexample is obtained semantically here at the very intuitive* informal level (the reader recalls that our incursion into first-order semantics in this volume is very brief, in outline, and occurs in Chap. 8).

The counterexample relies on *first-order soundness*, that "if $\vdash A$, then A is *true in* all *its possible interpretations into a concrete mathematical formula*". The preceding italicised qualification has a short term: A is *valid*.

Now, A is abstract so to assign truth to it we do so only after we *interpret it in a specific way* —normally, within informal mathematics.

So,

here is why $\vdash A \to (\forall x)A$ is a false claim

if there *are* free occurrences of x in A (compare with Exercise 0.2.77).

Let us *interpret* A to be $x = 0$ where x is a *natural number variable*. By first-order soundness, if we do have $\vdash x = 0 \to (\forall x)x = 0$, then this entails that $x = 0 \to (\forall x)x = 0$ is true *for all natural number values of* x. But not so for x equal to 0,[33] since the formula $0 = 0 \to (\forall x)x = 0$ is clearly false.   □

0.2.89 Definition. (Consistency) A logic, whether *pure* or *applied* (cf. Remark 0.2.30) is called *consistent* iff it does *not* prove *all* the wff over its language. Otherwise it is called *inconsistent*.   □

---

[33] We can substitute 0 only in the free occurrence of x.

0.2.90 Remark. Since formal logic is called upon to *certify* (but also *discover*) mathematical and logical truths *syntactically* via proofs, it is clear that an *inconsistent* logic or theory is a *useless tool* towards effecting such certification, as it certifies *everything* as "true"!

Sometimes we will say that the *axioms* of a theory or a logic *are* consistent or not, since along with the rules of inference they are responsible for deriving all theorems.

We say that a theory is given *intentionally* via axioms and rules —that is, as a *theorem-generator*— or *extensionally*, as the *set of all its theorems*, with no hint of how they are all generated (an example of a theory given extensionally is the CA[34] of Definition 12.8.3).   □

0.2.91 Remark. (*Soundness vs. Correctness*) Soundness, as was the case for the Boolean fragment of our logic, is a simple phenomenon that arises from the fact that the *first-order rules of inference* propagate truth, *and the first-order* logical axioms *are valid* (true no matter how interpreted). We will sample some first-order axioms and verify their validity in Chap. 8 and also verify that the rule ∀-intro propagates truth just as MP does, the latter was verified in this section without much fanfare (cf. proof of Metatheorem 0.2.70).

The important aspect of soundness is that truth propagates with the two primary rules, and hence, just like the logical axioms are, all first-order pure (or absolute) *theorems* are valid.

But the concept bifurcates when we take the path towards exploring *mathematical* theories. Here, normally, there is a particular intended *informal* mathematical theory (e.g. the theory of the natural numbers) that our *formal* theory is called upon to rigorously codify. A well-built formal theory will ensure that its *mathematical axioms* are *true in the intended interpretation*,[35] and thus all its theorems will be true in the intended interpretation as well, due to truth propagating via the rules. We cannot speak of soundness of a mathematical theory, since, for example, Peano arithmetic's axiom $\neg Sx = 0$ is true in the standard (or, *intended*) interpretation but is not *valid*.[36]

Instead we speak of *correctness*: A theory is *correct*[37] means that its *mathematical axioms* are true *under the intended interpretation*.

> Due to the fact that *truth propagates with the inference rules*, all the *theorems* of a correct theory are also true under the intended interpretation.

It can be verified that PA is correct.

---

[34] In fact, Gödel's first incompleteness theorem sates in one of its formulations that there is no way to generate all of CA as theorems proved from a "recognisable" set of axioms. By "recognisable" I mean, intuitively, that you can tell using a finite process if a formula is an axiom or not.

[35] I used "true" rather than "valid" as the latter refers to truth in *all* interpretations.

[36] A counterexample to "$x + 1 \neq 0$" in the set $\mathbb{Z}$ —the set of all integers, negative, positive or zero— is taking $x = -1$ in which case we have $x + 1 = 0$.

[37] Boolos et al. (2003), Smullyan (1992), or Tourlakis (2003a).

Trivially, a correct theory is consistent for there are false sentences, which, of course, it cannot prove. The converse is not true; consistency does not guarantee correctness. In Chap. 12 we will see an example of an *incorrect* theory —yet, consistent— that can be derived from PA by adding a single axiom. □

0.2.92 Theorem. (Consistency Test) *Let a logic or theory $\mathcal{T}$ be given via a set of axioms Γ. Then $\mathcal{T}$ is inconsistent iff, for some A, Γ proves both A and $\neg A$.*

Proof *If*-part. Say $\Gamma \vdash A$ and $\Gamma \vdash \neg A$. I will show that *every* wff B is a theorem.
  Well, $A, \neg A \models_{taut} B$. Thus, by the assumption and Remark 0.2.82, $\Gamma \vdash B$. So, Γ (or $\mathcal{T}$) is inconsistent.
  Only if-part. Say $\mathcal{T}$ is inconsistent, that is, every B is a Γ-theorem. Then for any A that we may pick, both A and $\neg A$ are theorems of $\mathcal{T}$. □

0.2.93 Theorem. (Proof by Contradiction) *We have $\Gamma \vdash A$ —where A is closed— iff $\Gamma + \neg A$ is inconsistent.*

Proof

- *If*-part. Let $\Gamma + \neg A$ prove all formulas. Then it can prove $B \wedge \neg B$, for any closed B that we pick. By the deduction theorem, $\Gamma \vdash \neg A \rightarrow B \wedge \neg B$ (Corollary 0.2.87). Trivially, $\neg A \rightarrow B \wedge \neg B$ tautologically implies A, hence $\Gamma \vdash A$.
- *Only if*-part. Let $\Gamma \vdash A$. Then also $\Gamma, \neg A \vdash A$ by Exercise 0.2.36 and $\Gamma, \neg A \vdash \neg A$ by the definition of proof. We are done by Theorem 0.2.92. □

0.2.94 Theorem. (One-Point-Rule) *If x does not appear in t, then we have the two equivalent absolute theorems*

$$\vdash (\forall x)(x = t \rightarrow A) \equiv A[x := t] \tag{1}$$

and

$$\vdash (\exists x)(x = t \wedge A) \equiv A[x := t]$$

 A *"ping-pong"* argument or a *"push-pull"* argument is based on the absolute theorem

$$\vdash (X \equiv Y) \equiv (X \rightarrow Y) \wedge (Y \rightarrow X) \tag{1}$$

We have (1) by $\models_{taut} (X \equiv Y) \equiv (X \rightarrow Y) \wedge (Y \rightarrow X)$ and Metatheorem 0.2.72. By (1), we have proved $X \equiv Y$ iff we have proved $(X \rightarrow Y) \wedge (Y \rightarrow X)$, the latter iff we have proved $X \rightarrow Y$ *and* $Y \rightarrow X$ —where the latter iff holds by $\models_{taut}$ in each direction.

Proof of (1) (The $\exists$ case is left as an exercise based on (1) and Definition 0.2.18). We do a ping-pong argument to prove the equivalence.

$\rightarrow$

(1) $(\forall x)\left(x = t \rightarrow A\right)$
$\rightarrow \left(t = t \rightarrow A[x := t]\right)$ ⟨substitution axiom⟩

(2) $t = t$ ⟨identity axiom + Proposition 0.2.85⟩

(3) $(\forall x)\left(x = t \rightarrow A\right) \rightarrow A[x := t]$ ⟨(1, 2) + Post⟩

$\leftarrow$

(1) $x = t \rightarrow (A \equiv A[x := t])$ ⟨equality axiom⟩

(2) $A[x := t] \rightarrow x = t \rightarrow A$ ⟨(1) + Post⟩

(3) $A[x := t] \rightarrow (\forall x)\left(x = t \rightarrow A\right)$ ⟨(2) + ∀-intro.

                                     OK, no free x in $A[x := t]$⟩

 The "one-point-rule" is ever-present in the practise of (informal) mathematics in this volume and throughout the literature, so much so that normally it is not quoted by this name or by any name at all. In this volume —Chap. 12— it will also be used formally, as it is stated in Theorem 0.2.94.

0.2.95 Proposition. (∀ **over** →) *For any A and B, we have*

$$\vdash (\forall x)(A \rightarrow B) \rightarrow (\forall x)A \rightarrow (\forall x)B \qquad (\ddagger)$$

Proof By Metatheorem 0.2.86 twice, it suffices to prove instead

$$(\forall x)(A \rightarrow B), (\forall x)A \vdash (\forall x)B \qquad (*)$$

(1) $(\forall x)(A \rightarrow B)$ ⟨hyp⟩

(2) $(\forall x)A$ ⟨hyp⟩

(3) $A \rightarrow B$ ⟨(1) + Spec⟩

(4) A ⟨(2) + Spec⟩

(5) B ⟨(3, 4) + MP⟩

(6) $(\forall x)B$ ⟨(5) + Gen⟩

Gen applied in step (6) is faithful to the requirements of Metatheorem 0.2.86 since the only generalisation in the proof (at step (6)) refers to x occurrences that are not free in either one of the formulas moved to the left of $\vdash$ (cf. ($\ddagger$) and ($*$)). Thus the application of the deduction theorem, that ($*$) guarantees (because it implies) ($\ddagger$), is correct.

In future applications of Metatheorem 0.2.86 we will simply note that applications of ∀-intro or its derivative, generalisation, during a proof from "*Γ, A*" were faithful to the metatheorem's restrictions.

0.2.96 Remark   Here is a Metatheorem 0.2.86-free alternative proof of Proposition 0.2.95.

(1)	$(\forall x)(A \to B) \to (A \to B)$	⟨axiom of substitution⟩
(2)	$(\forall x)A \to A$	⟨axiom of substitution⟩
(3)	$\Big((\forall x)(A \to B)\Big) \wedge (\forall x)A \to (A \to B) \wedge A$	⟨(1, 2) + Post⟩
(4)	$(A \to B) \wedge A \to B$	⟨tautology + Post⟩
(5)	$\Big((\forall x)(A \to B)\Big) \wedge (\forall x)A \to B$	⟨(3, 4) + Post⟩
(6)	$\Big((\forall x)(A \to B)\Big) \wedge (\forall x)A \to (\forall x)B$	⟨(5) + ∀-intro⟩
(7)	$(\forall x)(A \to B) \to (\forall x)A \to (\forall x)B$	⟨(6) + Post⟩

□

0.2.97 Proposition. (∀-Monotonicity) *For any A and B, we have* $A \to B \vdash (\forall x)A \to (\forall x)B$.

Proof

(1)	$A \to B$	⟨hyp⟩
(2)	$(\forall x)(A \to B)$	⟨(1) + Gen⟩
(3)	$(\forall x)(A \to B) \to (\forall x)A \to (\forall x)B$	⟨Proposition 0.2.95⟩
(4)	$(\forall x)A \to (\forall x)B$	⟨(2, 3) + MP⟩

□

0.2.98 Lemma. (∀ over ≡) *For any A and B, we have* $A \equiv B \vdash (\forall x)A \equiv (\forall x)B$.

Proof

(1)	$A \equiv B$	⟨hyp⟩
(2)	$A \to B$	⟨(1) + Post⟩
(3)	$A \leftarrow B$	⟨(1) + Post⟩
(4)	$(\forall x)A \to (\forall x)B$	⟨(2) + ∀-Mon⟩
(5)	$(\forall x)A \leftarrow (\forall x)B$	⟨(3) + ∀-Mon⟩
(6)	$(\forall x)A \equiv (\forall x)B$	⟨(4, 5) + Post⟩

□

0.2.99 Exercise. Prove ∃-monotonicity (cf. Definition 0.2.18), that is, $A \to B \vdash (\exists x)A \to (\exists x)B$.   □

0.2.100 Exercise. Prove that ∃ *distributes* over ≡ just as ∀ does, that is, $A \equiv B \vdash (\exists x)A \equiv (\exists x)B$.   □

0.2.101 Theorem. (Variant Theorem) *Let z be a new variable* (fresh, *as we say*) *for* $(\forall x)A$ *—that is, it occurs neither free nor bound in said formula. Then* $\vdash (\forall x)A[x] \equiv (\forall z)A[z]$, *where we wrote* $A[x]$ *for A and* $A[z]$ *for* $A[x := z]$.

We call $(\forall x)A[x]$ and $(\forall z)A[z]$ *variants* of each other.

Proof   This is going to be another ping-pong argument as this is the normal way to prove equivalences with Hilbert-style proofs, which are "directional".

- $\rightarrow$ direction

    (1) $(\forall x)A[x] \rightarrow A[z]$ ⟨substitution axiom⟩
    (2) $(\forall x)A[x] \rightarrow (\forall z)A[z]$ ⟨(1) + ∀-intro; OK: no free z in $(\forall x)A\lfloor x \rfloor$⟩

- $\leftarrow$ direction The proof will start with the substitution axiom invocation

$$(\forall z)A[z] \rightarrow A[x := z][z := x] \tag{*}$$

Note however that $A[x := z][z := x]$ is just A as it can be proved —on the assumption that z is fresh for A— by induction on (the formation of) A:

1. A atomic.

    Case 1. A is $t = s$. Then $[x := z]$ replaces all x by the *new* z. There will be z (free, of course) occurrences *precisely where* $t = s$ had x occurrences before the substitution $[x := z]$. Thus, $[z := x]$ turns those z occurrences back to x restoring the original $t = s$.
    Case 2. A is $\phi t_1 t_2 \dots t_n$. Precisely as above.

2. A is $(B \vee C)$. Then z is fresh for each of B and C and the (I.H.) yields that $B[x := z][z := x]$ is B and $C[x := z][z := x]$ is C. We are done since $(B \vee C)[x := z][z := x]$ is $B[x := z][z := x] \vee C[x := z][z := x]$.

3. A is $((\forall y)B)$. Then z is fresh for B, and is not y.

    Without loss of generality we assume that y is not x either, else $((\forall y)B)[x := z]$ is just $((\forall y)B)$ and thus $((\forall y)B)[x := z][z := x]$ is also $((\forall y)B)$.

    Freshness guarantees that there is no subformula $((\forall z)C)$ in B, thus doing $B[x := z]$ encounters no capture. Similarly, no free z is in the scope of a $(\forall x)$ in $B[x := z]$ (why?) thus $B[x := z][z := x]$ is permitted and results in B by the I.H. But then $((\forall y)B)[x := z][z := x]$ is $((\forall y)B)$ by the way substitution works.

4. A is $(\neg B)$. Similar to the above, but significantly simpler.

In view of the above digression, our first line below is the simplified (∗) from above.

    (1)$(\forall z)A[z] \rightarrow A[x]$ ⟨substitution axiom⟩

    (2)$(\forall z)A[z] \rightarrow (\forall x)A[x]$ ⟨(1) + ∀-intro; OK: no free x in $(\forall z)A\langle z\rangle$⟩

We are done.

                                                                                              □

0.2.102 Exercise. (Variant Theorem for ∃) Let z be a fresh variable for $(\exists x)A$ —that is, it occurs neither free nor bound in said formula. Then $\vdash (\exists x)A[x] \equiv (\exists z)A[z]$, where we wrote $A[x]$ for A and $A[z]$ for $A[x := z]$.

0.2.103 Theorem. (Equivalence Theorem) *Let the formula A be a subformula of formula C. Let B be another formula. Then we have $A \equiv B \vdash C \equiv C'$, where C' is obtained from C by replacing* any *occurrences of A as a subformula of C with B.*

Proof Induction on the formula C following the cases of Definition 0.2.16.

1. If C is atomic, then it is the only subformula of C, hence it is A. The theorem then reads "$A \equiv B \vdash A \equiv B$", which is correct.
2. If C is $(D \vee E)$ and A is a subformula of C, then A is a subformula of D or of E or both. By the I.H. we have $A \equiv B \vdash D \equiv D'$ and $A \equiv B \vdash E \equiv E'$.[38] Thus by Post we have $A \equiv B \vdash C \equiv C'$.
3. If C is $(\neg D)$ and A is a subformula of C, then A is a subformula of D. By the I.H. we have $A \equiv B \vdash D \equiv D'$. Thus by Post we have $A \equiv B \vdash C \equiv C'$.
4. If C is $((\forall x)D)$ and A is a subformula of C, then A is a subformula of D. By the I.H. we have $A \equiv B \vdash D \equiv D'$. Thus by 0.2.98 we have $A \equiv B \vdash C \equiv C'$. □

0.2.104 Remark. (Leibniz Rule) In some of the literature this *metatheorem* of logic is called the *Leibniz rule* and is taken as a *primitive* rule of inference of the logic (Dijkstra 1968; Gries and Schneider 1994; Tourlakis 2008). □

0.2.105 Theorem. *We have $\vdash x = y \to y = z \to x = z$.*

Proof I have by Axiom 7 $\vdash x = y \to (x = z \equiv y = z)$. By tautological implication from this I get $\vdash x = y \to (y = z \to x = z)$,[39] which, apart from redundant bracketing here, is the same as the statement of the theorem. □

0.2.106 Exercise. Prove also that $\vdash x = y \to y = x$. □

There is a lot to be said in this section, but after all, it is only a review! Still, the following will be useful in the future

0.2.107 Theorem. (Dual of Axiom Definition 0.2.28(5)) $\vdash A[x := t] \to (\exists x)A$.

Proof We have $\vdash (\forall x)\neg A \to \neg A[x := t]$. By tautological implication we have $\vdash A[x := t] \to \neg(\forall x)\neg A$. But this states what we wanted to prove. □

0.2.108 Corollary. (Theorem 0.2.107 As a Rule) $A[x := t] \vdash (\exists x)A$.

Proof By Theorem 0.2.107 and MP. □

0.2.109 Theorem. (∃-Intro (Derived) Rule) *For any formulas A and B we have that $A \to B \vdash (\exists x)A \to B$ as long as x is not free in the right hand side of "$\vdash$".*

---

[38] If, say, A is *not* a subformula of E, then E' is E, thus still $A \equiv B \vdash E \equiv E'$ holds.

[39] $A \to (B \equiv C) \models_{taut} A \to (C \to B)$.

Proof

(1) $A \rightarrow B$ $\langle \text{hyp} \rangle$

(2) $\neg B \rightarrow \neg A$ $\langle 1 + \text{Post} \rangle$

(3) $\neg B \rightarrow (\forall x)\neg A$ $\langle 2 + \forall\text{-intro; using restriction on } x \rangle$

(4) $\neg(\forall x)\neg A \rightarrow B$ $\langle 3 + \text{Post} \rangle$

(4′) $(\exists x)A \rightarrow B$ $\langle 4 + \text{Definition 0.2.18; using the abbreviation "}\exists\text{"} \rangle$ □

0.3   A Bit of Set Theory

This section is about *sets*, not *set theory*. Set theory is a very broad and deep topic and nowadays is practised axiomatically (e.g., Levy (1979); Tourlakis (2003b)). Excellent texts that cover much of the informal (also called "naïve") set theory of Georg Cantor are Kamke (1950); Halmos (1960)).

We will introduce here frequently used in the literature useful notation and some very simple facts. The parts on diagonalisation and inductive (recursive) definition of functions will be the only nontrivial topics.

The concept of set is left undefined (even in axiomatic set theory) just as "point" and "line" are in axiomatic Euclidean (plane) geometry. We learn what points and lines *do* —and similarly what sets *do*— and their behaviour tells us what they are.

Still Cantor had been quite descriptive using dictionary synonyms in an attempt to *describe* sets. So, *collection, aggregate* (or even *class*[40]) were proposed as explanatory terms for the term *set*. Cantor's set theory contains two types of objects: sets, and *atoms* also called *Urelements*. The latter are objects that are not of set *type* such as the natural numbers of our intuition[41] (for modern examples of texts on set theory *with* atoms see Barwise (1975); Tourlakis (2003b)).

The fundamental relation (predicate) of the language of set theory is "$\in$" pronounced "is a member of". Thus $x \in y$ states that the object x (possibly an atom or a set) is a member of the set y. $x \notin y$ states the *opposite*: x is not a member of y —that is, $x \notin y \equiv \neg x \in y$.[42]

Equally fundamental is the axiom of equality of sets: "*Two sets are equal if they contain precisely the same elements*". That is,

$$(\forall x)(x \in y \equiv x \in z) \rightarrow y = z \tag{1}$$

---

[40] In modern set theory some classes are *not* sets.

[41] It turns out that atom-less set theory is so powerful that it can construct sets that *behave* as natural numbers (finite ordinals). This does not change the fact that the non set theorist believes that there are many objects in mathematics that are not sets.

[42] We do not bracket atomic formulas.

(1) is called the axiom of *Extensionality* because it bases equality of two sets on the actual elements *they contain* —the *extension* of the sets— without worrying about *how* the sets were obtained (that is, the *intention* is ignored).

(1) has similar mathematical flavour as Euclid's axiom on parallels: "From a point A not on a line l we can draw exactly one line l' —on the plane defined by A and l— that is parallel to l". Just as there are "realities" where Euclid's axiom is false (e.g., in Riemann's geometry) in the same way one my think of a different reality for sets where, for example, there is an inner "structure" in a set that "relates" its various members, and that this relation or connectedness matters when comparing this set with another.

Our sets have no internal "connectors" between their elements.

So the set given as the collection of just 1 and 2, denoted by $\{1, 2\}$ is *equal* to the set given as "all numbers x satisfying the equation $x^2 - 3x + 2 = 0$".

We denote sets by listing if they are small, e.g., $\{1, 1, 2, 1\}$, $\{2, 11, \#, \$\}$, or by a *defining property* if they are very large, or indeed *infinite* (a technical term that we will define). Thus, the example with the equation can be denoted by

$$\{x : x^2 - 3x + 2 = 0\} \tag{2}$$

in effect collecting together to form a set all the x that make the property "$x^2 - 3x + 2 = 0$" true. The general form of (2) is

$$\{x : P(x)\} \tag{3}$$

or also

$$\{x \,|\, P(x)\} \tag{3'}$$

In (3) or (3') we collect precisely *all* x that make $P(x)$ true.

 Thus $P(x)$ is the *entrance test* for any x.

Thus,

$$x \in \{x : P(x)\} \equiv P(x) \tag{4}$$

Extensionality implies that $\{1, 1, 2, 1\} = \{1, 2\} = \{1, 2, 1, 1\}$. Neither order nor multiplicity of elements matter to equality.

Sets are not "flat". They may contain other sets. E.g., $\{\{\{1\}\}\}$.

Extensionality is deliberately given with an "if", not an "iff". Logic (with no benefit of any set theory) can prove the converse of (1):

$$y = z \rightarrow (\forall x)(x \in y \equiv x \in z)$$

See Exercise 0.8.6.

In set theory we use every letter of the alphabet, upper or lower case, as a set (or atom) variable, with or without primes or subscripts. We write "let $A = \{x \mid P(x)\}$" *naming* a long expression denoting a set —$\{x \mid P(x)\}$— by a short one —A.

0.3.1 Definition. (Subsets) For sets A and B we write $A \subseteq B$ as an abbreviation of $(\forall x)(x \in A \rightarrow x \in B)$.

If moreover we have $A \neq B$, then we write $A \subset B$ or *more clearly*[43] $A \subsetneqq B$.   □

By the complementary actions of generalisation and specialisation, we simplify the definition of $A \subseteq B$ using the *unquantified* $x \in A \rightarrow x \in B$ ("for all x" is understood, when we make a claim that involves a free variable).

Note that $A = B$ iff $A \subseteq B \wedge B \subseteq A$. Translating the right hand side of iff in the first-order language we readily prove this equivalence using extensionality. Exercise 0.8.7.

0.3.2 Example. [Russell's Paradox] When we name a set $\{x : P(x)\}$ —$A = \{x : P(x)\}$— given by some property $P(x)$, we have

$$x \in A \equiv P(x) \tag{†}$$

This is so by (4) above and extensionality.

Cantor allowed any defining property $P(x)$ as a set builder. Russell found a problem with this: Take $x \notin x$ for $P(x)$ and consider the "set" R below. Is this *really* a set?

$$R = \{x : x \notin x\} \tag{‡}$$

Well, (‡) implies (by (†) above) that

$$x \in R \equiv x \notin x$$

Now if R *is* a set, then it is *legitimate* to substitute it into the set/atom variable x. We thus obtain

$$R \in R \equiv R \notin R$$

A contradiction. So R is *not* a set!   □

Paradoxes[44] of set theory, like this one, led to the genesis of axiomatic set theory. However not everybody practises their set theory axiomatically, and in our case we do not need much more than a few definitions and results. Thus the practise of naïve set theory is still relevant, albeit one does so by exercising care about the size of sets

---

[43] In some of the literature one writes $A \subset B$ meaning what we called $A \subseteq B$.
[44] Statements that defy opinion or belief: παρά (against) δοκώ (to believe).

one builds/defines. Enormous sets may lead to trouble: Indeed the Russell "set" R (that we saw in not a set at all) *is* enormous: It contains *everything* (*every* object out there) since it turns out, under reasonable assumptions, that $x \notin x$ is true for all x. That is, the entrance test "$x \notin x$" *does not bar any object* from entering R.

Careful naïve set theory uses relatively small sets as *sets of reference*. If we stay *within* those sets when practising set theory, then we are safe!

In particular, *working inside* such a *reference set*, U, every collection A we use is a sub collection of U, so not larger than U. Therefore it is a set.

Moreover, in the context of work inside a reference set U, when we write $(\forall x)$ we mean "$(\forall x)$ in U", in symbols, $(\forall x \in U)$ or $(\forall x)_{\in U}$. Similarly for $(\exists x)$.

$\mathbb{N}$, or $\mathbb{N}^n$ (symbol to be introduced soon) for various n are relevant *sets of reference* for most of this volume.

0.3.3 Definition. (Union, Intersection, Difference, Complement)   For any sets A and B, we define

$$A \cup B \stackrel{Def}{=} \{x : x \in A \vee x \in B\}$$

$A \cup B$ is called the *union* of A and B.

We also define the *intersection* of A and B by

$$A \cap B \stackrel{Def}{=} \{x : x \in A \wedge x \in B\}$$

Next, define *set difference*:

$$A - B \stackrel{Def}{=} \{x : x \in A \wedge x \notin B\}$$

A special case, when $A = U$ is called the *set complement* of B: $U - B$ denoted by $\overline{B}$. □

Pause. Should we worry that $A \cup B$, which in principle is bigger than both A and B, might be big enough to cause paradoxes?◄

No, since we are working in a reference set U (also called <u>a</u> universe as opposed to the troublesome <u>the</u> universe $R = \{x : x \notin x\} = \{x : x = x\}$) thus we have $A \subseteq U$ and $B \subseteq U$. But then (Exercise!) $A \cup B \subseteq U$; so it is a set.

Incidentally, trivially, $A \cap B \subseteq A$ (and $\subseteq B$) so this result is "smaller" than the two operators which produce it. Definitely it is set if any one of the collections A and B is a subset of U.

Note that $A - B \subseteq A$, so $A - B$ is a set if A is. B will normally be in the reference set, but regardless, the preceding sentence is correct.

We extend the operations of union and intersection to many sets.

0.3.4 Definition. (Families of Sets and Their Unions and Intersections)   Suppose A is a set such that *all* its members are also sets. Then the *union* of A is defined (and denoted) below:

$$\bigcup A \stackrel{Def}{=} \{x : (\exists B \in A)x \in B\} \tag{1}$$

Then the *intersection* of A is defined (and noted in symbols) below:

$$\bigcap A \stackrel{Def}{=} \{x : (\forall B \in A)x \in B\} \tag{2}$$

$\square$

If we are working in a reference set, then the definitions (1) and (2) are modified to

$$\bigcup A \stackrel{Def}{=} \{x \in U : (\exists B \in A)x \in B\} \tag{1'}$$

and

$$\bigcap A \stackrel{Def}{=} \{x \in U : (\forall B \in A)x \in B\} \tag{2'}$$

Thus, for example, $(2')$ implies

$$x \in \bigcap A \equiv x \in U \wedge (\forall B \in A)x \in B \tag{2''}$$

0.3.5 Definition. (More Notation) Sometimes we can assign natural numbers, n, or other objects, a, as indices to every member of a family A. In such a case we may elaborate with the notation

$$A = \{B_a : a \in I\}$$

where I is the index set where the indices come from. Correspondingly, we can use the alternative notation for unions and intersections

$$\bigcup_{a \in I} B_a \stackrel{Def}{=} \bigcup A$$

and

$$\bigcap_{a \in I} B_a \stackrel{Def}{=} \bigcap A$$

If the index set I is $\mathbb{N}$ then we have two more available notations to us for each of $\bigcup$ and $\bigcap$:

$$\bigcup_{n\geq 0} B_n = \bigcup_{n=0}^{\infty} B_n$$

and

$$\bigcap_{n\geq 0} B_n = \bigcap_{n=0}^{\infty} B_n$$

□

0.3.6 Definition. (An Empty Set) The set $\{x : x \neq x\}$ contains no elements since nothing passes the entrance test. It is *an empty set*, we say.

Let $P(x)$ be some other property of x that is false for all x —that is, no x has the property. Then $\{x : P(x)\}$ is also an empty set. Since $P(x)$ is always false, just like $x \neq x$, we have $x \neq x \equiv P(x)$ and hence $x \in \{x : x \neq x\} \equiv x \in \{x : P(x)\}$. By extensionality, $\{x : x \neq x\} = \{x : P(x)\}$. Thus *all empty sets* $\{x : P(x)\}$ are equal to the "canonical" one, $\{x : x \neq x\}$.

 In short, *there is only one empty set*. We denote it by $\emptyset$.

□

In preparation for the subsection on functions and relations we shall need the concept of an *ordered pair*. It has two elements (maybe the same) but their position, first or second distinguishes them regardless.

Is this a *new* object of a different *type*? No, as Kuratowski has showed, we can implement an ordered pair as a set.

 0.3.7 Example. (Kuratowski's Ordered Pair) We prove that

$$\{\{a\}, \{a, b\}\} = \{\{x\}, \{x, y\}\} \tag{1}$$

implies $a = x$ and $b = y$. By (1) and Theorem 0.2.81,

$$\{a\} = \bigcap\{\{a\}, \{a, b\}\} = \bigcap\{\{x\}, \{x, y\}\} = \{x\} \tag{2}$$

(2) trivially implies

$$a = x \tag{3}$$

Thus we can write (1) as

$$\{\{a\}, \{a, b\}\} = \{\{a\}, \{a, y\}\} \tag{1'}$$

hence (using again Theorem 0.2.81, this time with (1'))

$$\{a, b\} = \bigcup\{\{a\}, \{a, b\}\} = \bigcup\{\{a\}, \{a, y\}\} = \{a, y\} \tag{4}$$

We have cases:

1. $a = b$. Then $y = a = b$ by extensionality (actually, by $\{a\} \supseteq \{a, y\}$). We are done in this case due to the by-product $y = b$.
2. $a \neq b$. Then $a \neq y$ as well else $\{a, b\} \subseteq \{a\}$ gives $a = b$ contradicting hypothesis. Thus

$$\{b\} = \{a, b\} - \{a\} = \{a, y\} - \{a\} = \{y\}$$

which yields $b = y$.   □

0.3.8 Definition. (The Ordered Pair) We let

$$(a, b) \overset{Def}{=} \{\{a\}, \{a, b\}\}$$

and call the set that (a, b) *names* the *ordered pair*. a is the first *component* of the pair and b is the second component.   □

For the record, we repeat the findings of the preceding example.

0.3.9 Proposition. *If $(a, b) = (x, y)$, then $a = x$ and $b = y$.*

We can now define ordered triple, quadruple, and in general, n-tuple. For example, if I set

$$(a, b, c) \overset{Def}{=} ((a, b), c)$$

we can verify that

$$(a, b, c) = (x, y, z) \rightarrow a = x \wedge b = y \wedge c = z$$

(Exercises 0.8.9)

0.3.10 Definition. (n-Tuples) The *ordered n-tuple*, or simply *n-tuple*, is defined by stages. At stage 1, we define the 1-tuple, (a), by $(a) = a$. At stage $n + 1$ we use the n-tuple $(x_1, \ldots, x_n)$ and the ordered pair and define

$$(x_1, \ldots, x_n, x_{n+1}) \overset{Def}{=} \left((x_1, \ldots, x_n), x_{n+1} \right)$$

This is a simple inductive definition that actually defines a process of how to build an n-tuple by building the i-tuples of the previous stages as building blocks.   □

It is easy to prove by induction on n that $(x_1, \ldots, x_n) = (y_1, \ldots, y_n)$ implies $x_i = y_i$, for $1 \leq i \leq n$.

We often use *vector notation*, $\vec{x}_n$ for $x_1, x_2, \ldots, x_n$ and $(\vec{x}_n)$ for the n-tuple.

This is useful to note: For $n > 2$, an n-tuple is just a *pair*: $((a_1, \ldots, a_{n-1}), a_n)$.

We can now introduce cartesian products.

0.3.11 Definition. (Cartesian Products) Given two sets A and B,

$$A \times B \overset{Def}{=} \{(x, y) : x \in A \wedge y \in B\}$$

In general

$$A_1 \times \ldots \times A_n \overset{Def}{=} \{(x_1, \ldots, x_n) : x_i \in A_i, \text{for } 1 \le i \le n\} \tag{1}$$

If $A = A_1 = A_2 = \ldots = A_n$, then we write A^n for $A_1 \times \ldots \times A_n$.   □

0.3.12 Example. $\{0\} \times \{1\} = \{(0, 1)\}$ and $\{1\} \times \{0\} = \{(1, 0)\}$. Thus $\{0\} \times \{1\} \neq \{1\} \times \{0\}$. We can state then that "in general, $\times$ on sets is not commutative".   □

Finally, before we go to the next section, we define the *power set*.

0.3.13 Definition. The set of all subsets of a set A is called the power set of A and is denoted by $\mathscr{P}(A)$ or 2^A.   □

Thus, $2^A = \{x : x \subseteq A\}$.

0.3.14 Example. $2^\emptyset = \{\emptyset\}$. $2^{\{0,1\}} = \{\emptyset, \{0, 1\}, \{0\}, \{1\}\}$.   □

0.4   Relations and Functions

0.4.1 Definition. (Functions) A function f —*intentionally*— is an input / output "black box" *with* a rule. This *rule* —represents the *intention*; we *do not* need to think of it as a program, which it could be— will produce zero or more outputs for every given input that we take from *a set of legal inputs* for the given function. The latter set is called the *left field* of the function. The outputs go to a chosen set of possible outputs, the *right field*. The picture is

$$\text{left field} \ni a \longrightarrow \boxed{\text{A function } f} \longrightarrow \left\{ \begin{array}{c} \text{no output} \\ \text{or} \\ \text{one or more outputs} \end{array} \right\} \in \text{right field}$$

We have a number of remarks:

1. f might produce an output b (among others) on input a. We write $f(a) \ni b$ but more often we write $f(a) \simeq b$. In this case we say "f *is defined on input* a" and write $f(a) \downarrow$. That is, $f(a) \downarrow \equiv (\exists y) f(a) \simeq y$.
2. We write $f(a) \uparrow$ for $\neg(\exists y) f(a) \simeq y$. We say "$f$ is undefined on input a".

3. We define the domain of f, dom(f) to stand for $\{x : f(x) \downarrow\}$.
4. We define the range of f, ran(f) to stand for $\{y : (\exists x) f(x) \simeq y\}$.
5. We write $f : A \to B$ to indicate that A is the left field and B is the right field. Trivially, dom(f) $\subseteq A$. We call f *total* iff dom(f) $= A$. Otherwise we call it *nontotal*.

   Clearly a function is total or not. If we do not know or do not care, then we say that the function is *partial*. Clearly, *all* functions are *partial*.
6. Trivially, ran(f) $\subseteq B$. We call f *onto* iff ran(f) $= B$. Otherwise we call it *non onto*.
7. The special type of function that we defined here (that may have multiple outputs *on a given input*) we call *partial multi-valued function*, or *p.m.v.* function. We will work with such functions in Chap. 13, but also see the definition of relations below.
8. If f gives *at most one* output for *any* input, then we call it a *single-valued function*. We soon will drop the qualifier single-valued.

   All definitions 2.–6. apply to single-valued functions as well.   □

0.4.2 Example. We call *number-theoretic all* functions (p.m.v. or single-valued) that have as left field $\mathbb{N}^n$ (for various n) and right field $\mathbb{N}$.

The (single-valued) function

$$\mathbb{N}^2 \ni (x, y) \longrightarrow \boxed{\text{the quotient } \lfloor x/y \rfloor \text{ function}} \longrightarrow \left\{\begin{array}{c} \text{no output} \\ \text{or} \\ \lfloor x/y \rfloor \end{array}\right\} \in \mathbb{N}$$

that we will soon learn to denote simply by

$$\text{``}\lambda xy. \lfloor x/y \rfloor\text{''}$$

is *nontotal* (mathematically speaking) as division by $y = 0$ is *undefined*. This function has infinitely many points of non definition.[45]

The (single-valued) function $\lambda x.x + 1$ is total.   □

0.4.3 Definition. (Functions Extensionally; Relations) Given a function f (p.m.v. or single-valued) the set $\{(x, y) : f(x) \simeq y\}$ is called its *graph*.[46]

We can forget about the *intensional* origin of a function (such as $\lambda xy. \lfloor x/y \rfloor$, for example) and just view it as a *tabulation* of its graph (tabulating an entry (x, y) for *each* pair that makes $f(x) \simeq y$ true).

Then a table for the graph of a p.m.v. function has *no restrictions*, but a table for the graph of a single-valued function *will* have a restriction:

---

[45] For this function we will learn a way to get around its points of non definition; cf. Example 2.3.10.

[46] For obvious reasons. Just plot a function : $\mathbb{N} \to \mathbb{N}$ on the Cartesian x-y plane.

Whenever (x, y) and (x, y') appear in the table, it must be $y = y'$ (this is the single-valued-ness condition).

In discrete mathematics, algebra, and even set theory literature, one defines a *relation* extensionally, as a *table —possibly infinite— of* pairs, with *no* restrictions (that is, the table represents the graph of a p.m.v. function).

Similarly, one defines a *function* extensionally —*without* a qualifier such as "single-valued", this taken for granted— to be a *table of pairs* with the single-valued-ness assured by the restriction above.

Thus, for the record:

1. A p.m.v. function is a *relation* —an unrestricted set of pairs. If the left and right field of the relation are the same, say, A, then we say that *the relation is on A*.
2. A single-valued function is (what we simply call) a *function*. It is a *restricted* (by single-valued-ness) set of pairs.
   For a function f we usually write $f(x) = y$ for $f(x) \simeq y$, but see also Definition 0.4.5 below. □

0.4.4 Remark. The 1. and 2. in the definition above are *reversible*, since an unrestricted table *is* the graph of some p.m.v. function. The function is given by the "rule" (mathematical rule! No expectation or claim that we have a *computational* rule): Given x, look for all pairs that start with x: For every (x, y) found in the table, output y.

Similarly, a restricted table is the graph of a single-valued function.

From now on, single-valued functions are called just *functions*; p.m.v. functions are called *relations* (until Chap. 13).

We normally use upper case letters P, Q, R, S, T with or without primes or subscripts as names for relations, and similarly, normally use f, g, h, k with or without primes or subscripts as names for functions. □

By the -remark below Definition 0.3.10, there is no loss of generality to restrict attention to the study of 2-ary —actually called *binary*— relations.

0.4.5 Definition. (Kleene's Equality) Let f and g be two functions. Then $f(x) \simeq g(y)$ means

$$f(x) \uparrow \wedge g(y) \uparrow \vee (\exists u)(f(x) = u \wedge g(y) = u)$$

That is, two *function calls* have the *same result* if both fail to return, or both return the same value. □

0.4.6 Definition. (1-1 Functions and 1-1 Correspondence) Let $f : A \to B$ be a function. It is *1-1* iff, for all a, b, c, $f(a) = c = f(b)$ implies $a = b$.

It is a *1-1 correspondence* between A and B iff it is 1-1, onto and total.

If $f : A \to B$ is a 1-1 correspondence we write $A \sim B$. □

0.4.7 Example. $f : \{0, 1, 2, 3\} \to \{6, 7, 8, 9\}$ given by $f = \{(0, 9), (3, 6)\}$ is 1-1 but neither total, nor onto.

g with the same left/right fields given by $g = \{(0, 7), (1, 8), (2, 9), (3, 6)\}$ is a 1-1 correspondence.   □

0.4.8 Remark. (Notation) Recall that a relation *extensionally* is, essentially, a set of *pairs* since $(x_1, \dots, x_n) = ((x_1, \dots x_{n-1}), x_n)$, if $n > 2$. We will accept also *unary* relations, when its members are not all pairs. Example, $\{1, (1, 2), 4\}$.

A relation R that *is* a set of pairs is called *binary*. We call it *n*-ary only if we care to note that its members are *n*-tuples, $n > 2$.

It is *customary to use the notation a R b* for $(a, b) \in R$ in imitation of $a < b$, $a = b, a \in b$.   □

0.4.9 Definition. (Converses and Images) We define a few useful terms for p.m.v functions (relations) and single-valued functions (simply put, *functions*).

1. The converse of a function or relation, f^{-1} or R^{-1}, are given extensionally as

$$R^{-1} \stackrel{Def}{=} \{(x, y) : y R x\}$$

and

$$f^{-1} \stackrel{Def}{=} \{(x, y) : y f x\}$$

2. If f is $f : A \to B$ and $X \subseteq A$ while $Y \subseteq B$ we define the *image* of X and the *inverse image* of Y under f by

$$f[X] = \{f(x) : x \in X\} = \{y : (\exists x \in X) f(x) = y\}$$

and

$$f^{-1}[Y] = \{x : (\exists y \in Y) f(x) = y\}$$

□

0.4.10 Proposition. *If $f : A \to B$ is a 1-1 correspondence, then $f^{-1} : B \to A$ is 1-1 correspondence.*

The above can be expressed as "if $A \sim B$, then $B \sim A$".

Proof Exercise 0.8.15.   □

0.4.11 Example. Let $f : \{1, \{1, 2\}, 2\} \to \{a, b, c\}$ given by the table

1	a
2	c
{1,2}	b

Then $f(\{1, 2\}) = b$ but $f[\{1, 2\}] = \{a, c\}$. □

We can compose relations and functions:

0.4.12 Definition. (Composition of Relations and Functions) Let $R : A \to B$ and $S : B \to C$ be two relations or functions. We can define the relation $R \circ S$ by

$$x R \circ S y \text{ iff } (\exists z)(x R z \wedge z R y)$$

□

0.4.13 Proposition. *If both R and S above are functions (single-valued), then so is $R \circ S$ (single-valued).*

Proof Exercise 0.8.16. □

0.4.14 Proposition. *Composition of functions and relations is associative:* $(R \circ S) \circ T = R \circ (S \circ T)$.

Proof Exercise 0.8.17. □

0.4.15 Example. Composition of functions or relations is not commutative: That is, in general, $R \circ S \neq S \circ R$. Here is a *counterexample* for the claim

$$\text{false claim: } R \circ S = S \circ R$$

Take $R = \{(1, 2)\}$ and $S = \{(2, 3)\}$. Then $R \circ S = \{(1, 3)\}$ and $S \circ R = \emptyset$. We say that "$S \circ R$ is the *empty relation* (or function)" since it is a relation table (or function table) that contains nothing. □

0.4.16 Remark (Notation) Consistent with the "$f(x)$" notation, we note for two functions

$$A \xrightarrow{f} B \xrightarrow{g} C$$

that

$$(f \circ g)(a) = b \text{ iff } a\, f \circ g\, b \text{ iff } (\exists z)(a f z \wedge z g b)$$
$$\text{iff } (\exists z)(f(a) = z \wedge g(z) = b) \text{ iff } g(f(a)) = b$$

In short,

$$(f \circ g)(a) = g(f(a)) \downarrow \text{ if } f(a) \downarrow ^{47} and\ g(f(a)) \downarrow ^{48} (f \circ g)(a) \uparrow \text{ otherwise}$$

---

[47] This is the part "$(\exists z)a f z$" above.
[48] This is the part "$(\exists z)z g b$" above.

Clearly, if $f(a) \uparrow$ then $g(f(a)) \uparrow$ as well, i.e., $(f \circ g)(a) \uparrow$.

Note the order reversal: $f \circ g$ *versus* $g(f(x))$. *There is no mystery here. In each case, f takes the input first. In each of the two notations we put f where the input is: $xf \circ gy$ vs. $g(f(x)) = y$.*

For peace of mind we introduce yet a third *notation: "gf stands for $f \circ g$".*

Thus, $(gf)(x) = (f \circ g)(x) = g(f(x))$.   □

0.4.17 Definition. (Inverses) Suppose

$$A \xrightarrow{f} B \xrightarrow{g} A$$

and thus also

$$B \xrightarrow{g} A \xrightarrow{f} B$$

If $f \circ g = \lambda x.g(f(x))$, $x \in A$,[49] then we write

$$f \circ g = gf = 1_A, \text{ where } 1_A : A \to A \text{ is } \lambda x.x$$

and say "f is a *right inverse* of g" and "g is a *left inverse* of f".

 Thus *left* or *right* (inverse) is defined *with respect to the "gf" notation.*

Similar nomenclature applies if we have $g \circ f = 1_B$.   □

0.4.18 Example. (Non Uniqueness of Left and Right Inverses) Take $A = \{1, 2, 3, 4\}$ and $B = \{a, b\}$. Define the following functions (check that they *are* functions): $f_1 = \{(1, a), (2, a), (3, b)\}$, $f_2 = \{(1, b), (2, a), (3, b)\}$, $g_1 = \{(a, 2), (b, 3)\}$, and $g_2 = \{(a, 1), (b, 3)\}$.

Note that $f_1 g_1 = f_1 g_2 = f_2 g_1 = 1_B$.

This example shows cases where neither the left nor the right inverses are unique.
                                                                                       □

What can we say about left and right inverses?

0.4.19 Proposition. *Suppose*

$$A \xrightarrow{f} B \xrightarrow{g} A$$

and $gf = 1_A$. Then

(1) *f is total and 1-1.*
(2) *g is onto (A).*

---

[49] The easy way to determine where the x comes from is to consider where the inputs of the "inner" function, f, are coming from.

Proof

(1) f: 1_A is total, that is, $1_A(x) \downarrow$ for all $x \in A$. By Remark 0.4.16, $f(x) \downarrow$, for all $x \in A$, so f is total. We next check 1-1 ness: Since f is total, the test simplifies to $f(a) = f(b) \rightarrow a = b$ (see Definition 0.4.6). Thus let $f(a) = f(b)$. Then, applying g,

$$a = g(f(a)) = g(f(b)) = b$$

(2) g: To prove that g is onto we must show that for any $a \in A$ we can solve $g(x) = a$ for x. Indeed, $f(a) \downarrow$. So take $x = f(a)$. We calculate: $g(x) = g(f(a)) = a$.

□

0.4.20 Remark Not every function $f : A \rightarrow B$ has a left or right inverse. (Why?) Functions that have neither left nor right inverses exist. Exhibit one (Exercise 0.8.18).

□

0.4.21 Lemma. *Let $f : A \rightarrow B$ be a function. Then $f 1_A = f$ and $1_B f = f$.*

Proof   Exercise 0.8.19.

□

0.4.22 Theorem. *If $f : A \rightarrow B$ has both a left and right inverses, $g : B \rightarrow A$ and $h : B \rightarrow A$, then they are equal and unique.*

Proof   Say $gf = 1_A$ —hence f is total and 1-1— and $fh = 1_B$ —hence f is onto. Thus f is a 1-1 correspondence. By Proposition 0.4.10 f^{-1} is also a 1-1 correspondence.

Now,

$$h = 1_A h = (gf)h = g(fh) = g 1_B = g \tag{1}$$

For uniqueness we show $g = h = f^{-1}$. We use (1). We will show that

$$f^{-1} f = 1_A \tag{2}$$

that is, "f^{-1}" plays the role of "g". Then we will note that by (1), $h = f^{-1}$. But $h = g$, so $g = f^{-1}$ as well.

It remains to actually obtain (2):

$$a f^{-1} f b \text{ iff } (\exists z)(a f^{-1} z f b)^{50} \text{ iff } (\exists z)(z f a \wedge z f b) \text{ iff (single-valued-ness) } a = b$$

□

---

[50] Written conjunctionally.

0.4.1   Equivalence Relations and Partial Order

Two types of binary relations, *equivalence relations* and *partial order* relations play an important role in mathematics. These are equipped with postulated properties.

0.4.23 Definition.   A binary relation R on a set A is

1. *Reflexive* iff aRa holds, for *all* $a \in A$.
2. *Irreflexive* iff there is no a satisfying aRa.
3. *Symmetric* iff, for all a, b, $aRb \rightarrow bRa$.
4. *Antisymmetric* iff, for all a, b, $aRb \wedge bRa \rightarrow a = b$.
5. *Transitive* iff, for all a, b, c, $aRb \wedge bRc \rightarrow aRc$.   □

0.4.24 Example.

1. For any set $A \neq \emptyset$, 1_A is reflexive. How about $A = \emptyset$?
2. The relation $=$ on $\mathbb{N}$ has all the properties listed in Definition 0.4.23 except irreflexivity.
3. $\leq$ on $\mathbb{N}$ has reflexivity, antisymmetry and transitivity.
4. $<$ on $\mathbb{N}$ has irreflexivity and transitivity.
5. For any integer $m > 1$, the relation of pairs (x, y) on $\mathbb{Z}$ (the set of all integers) defined by

$$m \text{ divides } x - y$$

   is reflexive, symmetric and transitive. It is called a "*congruence* modulo m" and it is usually denoted by $x \equiv y \ (m)$ or $x \equiv y \mod m$.   □

0.4.25 Definition. (Equivalence Relations)   A relation on a set A is an equivalence relation iff it is reflexive, symmetric and transitive.

   For each $x \in A$ and equivalence relation $R : A \rightarrow A$ we define the set

$$[x]_R \overset{Def}{=} \{y \in A : yRx\}$$

and call it the *equivalence class* determined by, or with representative, x.

   If R is understood we normally simply write $[x]$.   □

   Thus $x \equiv y \mod m$ is an equivalence relation.

   We have an easy but important proposition:

0.4.26 Proposition.   *The equivalence classes of an equivalence relation $R : A \rightarrow A$ have the following properties:*

(1) $[x] = [y]$ *iff* xRy.
(2) $[x] \cap [y] \neq \emptyset \rightarrow [x] = [y]$.
(3) *For all* x, $[x] \neq \emptyset$.
(4) $A = \bigcup_{x \in A}[x]$.

Proof Exercise 0.8.22.   □

The other important relation in mathematics is *partial order*. It can be defined as a *strict* order, like "<" on $\mathbb{N}$, or as *non-strict* order such as "≤" on $\mathbb{N}$. The two versions are closely connected thus we will give the lead to one of the two.

 We will trade a minor risk for confusion in our notation with notational user-friendliness and write "<" (resp. ≤) for an *arbitrary strict order* (resp. for an *arbitrary non-strict order*) —that is, the symbol will *not* only refer to the concrete "less than" (resp. "less than or equal") on the natural numbers. It will refer to *any order*, as postulated below.

0.4.27 Definition. (Partial order and POsets) A relation < on an *arbitrary* set A is a (*strict*) *partial order* iff it is *irreflexive* and *transitive*.

We informally denote the pair of "ingredients" A and < by $(A, <)$ and call the pair a *partially ordered set*, acronym *POset*.

By abuse of language we may say that the *set* A is a POset, if the "<" we have in mind is clear.   □

 0.4.28 Example. (Why "Partial"?) Of course, $(\mathbb{N}, <)$ is a partially ordered set, *this time* "<" meaning "less than" on $\mathbb{N}$.

Another example is for, say the set $A = \{1, 2\}$ to define a strict order $\subsetneq$ on its power set $2^A = \{\emptyset, \{1\}, \{2\}, \{1, 2\}\}$. It is trivial to verify that $\subsetneq$ is irreflexive and transitive, hence $(2^A, \subsetneq)$ is a POset.

Note that for $\{1\}$ and $\{2\}$ all of the statements $\{1\} = \{2\}, \{1\} \subsetneq \{2\}, \{1\} \supsetneq \{2\}$ are *false*. That is, $\{1\}$ and $\{2\}$ are *not comparable*.

The existence of *non comparable* elements in *specific* POsets justifies calling the *general* order *partial* —for partially defined.

If all the members of a POset are comparable, then the order is called a *total order*.   □

0.4.2   Big-O, Small-o, and the "Other" ∼

This notation is due to the mathematician E. Landau and is in wide use in number theory, but also in computer science in the context of measuring (bounding from above) the computational complexity of algorithms, for all "very large inputs".

0.4.29 Definition. Let f and g be two total functions of one variable, where $g(x) > 0$, for all x. Then

1. $f = O(g)$ —also written as $f(x) = O(g(x))$— read "f is big-oh g", means that there are positive constants C and K in $\mathbb{N}$ such that

$$x > K \text{ implies } |f(x)| \le Cg(x) \tag{1}$$

 We also say that

$$|f(x)| \le Cg(x) \text{ holds (is true) } \textit{almost} \text{ everywhere''} \qquad (2)$$

in the sense that (1) may be false *at most* for $x = 0, 1, \dots, K$ —a *finite set*. We abbreviate "*almost everywhere*" by "*a.e.*". This concept, a.e., will be reintroduced in Definition 4.1.10.

We can replace the "a.e" in (2) or the "$x > K$" in (1) by "for all $x \ge 0$" easily by adding an additive constant $L = \max\left(\{f(0), \dots, f(K)\}\right) + 1$ to (1). We get

$$x \ge 0 \text{ implies } |f(x)| \le Cg(x) + L \qquad (1')$$

2. $f = o(g)$ —also written as $f(x) = o(g(x))$— read "f is small-oh g", means that

$$\lim_{x \to \infty} \frac{f(x)}{g(x)} = 0$$

3. $f \sim g$ —also written as $f(x) \sim g(x)$— read "f is of the same order as g", means that

$$\lim_{x \to \infty} \frac{f(x)}{g(x)} = 1$$

□

"$\sim$" between two sets A and B, as in $A \sim B$, means that there is a 1-1 correspondence $f : A \to B$. Obviously, the context will protect us from confusing this $\sim$ with the one introduced just now, in Definition 0.4.29.

Both definitions 2. and 3. require some elementary understanding of differential calculus. Case 2. says, intuitively, that as x gets extremely large, then the fraction $f(x)/g(x)$ gets extremely small, infinitesimally close to 0. Case 3. says, intuitively, that as x gets extremely large, then the fraction $f(x)/g(x)$ gets infinitesimally close to 1; that is, the function outputs are infinitesimally close to each other.

0.4.30 Example.

1. $x = O(x)$ since $x \le 1 \cdot x$ for $x \ge 0$.
2. $x \sim x$, since $x/x = 1$, and stays 1 as x gets very large.
3. $x = o(x^2)$ since $x/x^2 = 1/x$ which trivially goes to 0 as x goes to infinity.
4. $2x^2 + 1000^{1000}x + 10^{350000} = O(x^2)$. Indeed

$$\frac{2x^2 + 1000^{1000}x + 10^{350000}}{3x^2} = 2/3 + 1000^{1000}/x + 10^{350000}/x^2 < 1$$

for $x > K$ for some well chosen K. Note that $1000^{1000}/x$ and $10^{350000}/x^2$ will each be $< 1/6$ for all sufficiently large x-values: we will have $2/3 +$

$1000^{1000}/x + 10^{350000}/x^2 < 2/3 + 1/6 + 1/6 = 1$ for all such x-values. Thus $2x^2 + 1000^{1000}x + 10^{350000} < 3x^2$ for $x > K$ as claimed.

In many words, in a polynomial, the order of magnitude is determined by the highest power term.   □

The last example motivates

0.4.31 Proposition. *Suppose that $f(x) \geq 0$ for all $x > L$, hence $|f(x)| = f(x)$ for all $x > L$. Now, if $f(x) \sim g(x)$, then $f(x) = O(g(x))$.*

Proof   The assumption says that

$$\lim_{x \to \infty} \frac{f(x)}{g(x)} = 1$$

From "calculus 101" (1st year differential calculus) we learn that this implies that for some K, $x > K$ entails

$$\left| \frac{f(x)}{g(x)} - 1 \right| < 1$$

hence

$$-1 < \frac{f(x)}{g(x)} - 1 < 1$$

therefore, $x > \max(K, L)$ implies $f(x) < 2g(x)$.   □

0.4.32 Proposition. *Suppose that $f(x) \geq 0$ for all $x > L$, hence $|f(x)| = f(x)$ for all $x > L$. Now, if $f(x) = o(g(x))$, then $f(x) = O(g(x))$.*

Proof   The assumption says that

$$\lim_{x \to \infty} \frac{f(x)}{g(x)} = 0$$

From "calculus 101" we learn that this implies that for some K, $x > K$ entails

$$\left| \frac{f(x)}{g(x)} \right| < 1$$

hence

$$-1 < \frac{f(x)}{g(x)} < 1$$

therefore, $x > \max(K, L)$ implies $f(x) < g(x)$.   □

These two propositions add to our toolbox:

0.4.33 Example.

1. $\ln x = o(x^r)$ for any positive real r. Here "ln" stands for $\log_e$ where e is the Euler constant

$$2.7182818284590452353602874713526624977572470937\ldots$$

Seeing that both numerator and denominator

$$\lim_{x\to\infty} \frac{\ln x}{x^r}$$

go to ∞, we have here (if we do not do anything to mitigate) an impasse: We have a "limit" that is indeterminate:

$$\frac{\infty}{\infty}$$

So, we will use "l'Hôpital's rule" (*the limit of the fraction is equal to the limit of the fraction of the derivatives*):

$$\lim_{x\to\infty} \frac{\ln x}{x^r} = \lim_{x\to\infty} \frac{1/x}{rx^{r-1}} = \lim_{x\to\infty} \frac{1}{rx^r} = 0$$

2. $\ln x = O(\log_{10}(x))$. In fact, you can go from one log-base to the other:

$$\log_e(x) = \frac{\log_{10}(x)}{\log_{10}(e)}$$

The claim follows from Proposition 0.4.31 since trivially $\ln x \sim \log_{10}(x)/\log_{10}(e)$. For that reason —and since multiplicative constants are hidden in big-O notation— complexity- and algorithms-practitioners omit the base of the logarithm and write things like $O(\log n)$ and $O(n \log n)$.   □

 0.4.34 Remark. We often use notation such that $O(f(x)) + O(g(y))$. As this refers to two bounds of the form $Cf(x) + K$ and $C'g(x) + K'$ that each bounds some (unnamed) functions $f'(x)$ and $g'(y)$, we have that

$$f'(x) + g'(y) \le Cf(x) + K + C'g(x) + K'$$
$$\le (C + C')(f(x) + g(x)) + K + K', \text{ for all } x, y, \text{ that is,}$$
$$= O(f(x) + g(x))$$

So we write

$$f'(x) + g'(x) = O(f(x) + g(x))$$

Similar comment for an expression $O(f(x)) \times O(g(y))$ (cf. Exercise 0.8.3).   □

0.4.3   Induction Revisited; Inductive Definitions

0.4.35 Definition. (Minimal and Minimum Elements) Let a POset $(A, <)$ be given.

An $a \in A$ is *minimal* iff $\neg (\exists x \in A) x < a$.

We introduce the abbreviation $x \leq y \overset{Def}{\equiv} x < y \vee x = y$ and define:

An element b in A is *minimum* iff $(\forall x \in A) b \leq x$.   □

0.4.36 Example.   For example, in Example 0.4.28 ∅ is *minimal*. It is also *minimum*.

In the POset $(A, \subsetneq)$ where $A = \{\{1\}, \{2\}, \{1, 2\}\}$ the $\{1\}$ and $\{2\}$ are both *minimal*. None of them is *minimum*, as this would require, for example, $\{1\} \subsetneq \{2\}$.   □

Let us pursue the relation $\leq$ we defined from the general $<: A \rightarrow A$ in Definition 0.4.35.

0.4.37 Lemma.   *Given* $<: A \rightarrow A$. *Define* $\leq$ *as in Definition 0.4.35. Then* $\leq$ *is reflexive, antisymmetric and transitive: A non-strict order.*

Proof

- *Reflexivity*: Trivial since $x \leq y$ is $x < y \vee x = y$ and thus $a \leq a$ is $a < a \vee a = a$.
- *Antisymmetry*. Say we have $a \leq b \leq a$. We have *Cases*:

  1. $a = b$. We are done!
  2. $a \neq b$. Then the parts of the hypothesis $a \leq b$ and $b \leq a$ cannot be satisfied by equality, so we have $a < b < a$. By transitivity of $<$ we get $a < a$ and we have just contradicted irreflexivity of $<$. Thus this case does not apply and we are done.
                    □

- *Transitivity*. Say $a \leq b \leq c$. If $a = b = c$ we are done. If $a = b < c$ or $a < b = c$ again we conclude $a < c$ hence $a \leq c$ by Definition 0.4.35. It remains to consider $a < b < c$. But then $a < c$ by transitivity of $<$ and we are done as in the previous subcase.

Thus a strict order trivially leads to a non-strict order by adding the missing pairs (a, a) for all $a \in A$. Correspondingly, if we started with a non-strict order (reflexive, antisymmetric and transitive) then removing the pairs (a, a) should give as a strict order! Exercise 0.8.23.

0.4.38 Definition. (MC and IC)   Let $(A, <)$ be a POset. We say that

1. $(A, <)$ *satisfies*, or *has*, the *minimal condition*, acronym *MC*, iff every nonempty subset B of A has at least one minimal element.
2. $(A, <)$ *satisfies*, or *has*, the *inductiveness condition*, acronym *IC*, iff, for every property $P(x)$, we have that $P(x)$ is true, for all $x \in A$, if

   a. $P(a)$ is true for all *minimal elements of* A, and
   b. For all b, $P(b)$ is true if $P(x)$ is true for all $x < b$.
                    □

0.4.39 Example. Our familiar POset $(\mathbb{N}, <)$ has both MC (this is the least principle) and IC (this is CVI). □

0.4.40 Theorem. *MC and IC are equivalent for any POset $(A, <)$. If it has one, it has the other.*

Proof Let us formulate MC in terms of a *defining property* $Q(x)$ for the various subsets $B = \{x : Q(x)\}$ of A. Thus, MC states

$$\overbrace{(\exists x)Q(x)}^{B \neq \emptyset} \rightarrow \underbrace{(\exists x)\Big(Q(x) \wedge \neg(\exists z)(z < x \wedge Q(z))\Big)}_{some \ x \in B \ and \ there \ is \ no \ smaller \ z \in B} \qquad (1)$$

We show that (1) is equivalent to IC, stated in terms of a property $P(x)$ below. We do so by translating $\exists$ into $\forall$ using the well known Definition 0.2.18 and taking Q as a name for $\neg P$.

$$(\exists x)Q(x) \rightarrow (\exists x)\Big(Q(x) \wedge \neg(\exists z)(z < x \wedge Q(z))\Big) \equiv$$

$$\neg(\forall x)P(x) \rightarrow \neg(\forall x)\neg\Big(\neg P(x) \wedge (\forall z)\neg(z < x \wedge \neg P(z))\Big) \overset{contrapositive}{\equiv}$$

$$(\forall x)\neg\Big(\neg P(x) \wedge (\forall z)\neg(z < x \wedge \neg P(z))\Big) \rightarrow (\forall x)P(x) \overset{deMorgan}{\equiv}$$

$$(\forall x)\Big((\forall z)\neg(z < x \wedge \neg P(z)) \rightarrow P(x)\Big) \rightarrow (\forall x)P(x) \overset{deMorgan}{\equiv}$$

$$\overbrace{(\forall x)\Big(\underbrace{(\forall z)(z < x \rightarrow P(z))}_{CVI \ I.H.} \rightarrow \underbrace{P(x)}_{CVI \ I.S.} \Big) \rightarrow (\forall x)P(x)}^{CVI \ Schema}$$

One thing before we leave this equivalence proof: Where is the CVI *Basis*? Well, when x is *minimal*, then the part $(\forall z)(z < x \rightarrow P(z))$ is vacuously true ($z < x$ is false). But then to establish the implication *I.H.*→*I.S.* we need to verify that $P(x)$ *is* true. So, the schema will make us *do* the *Basis* to take care of the "boundary" case where x is minimal. □

 0.4.41 Example. The POset $(\mathbb{N}^{n+1}, <)$, where $(x, \vec{z}_n) < (y, \vec{w}_n)$ iff $x < y$ and $\vec{z}_n = \vec{w}_n$, has MC and hence also IC: For MC simply note that for any nonempty subset B of $\mathbb{N}^{n+1}$ an element is minimal iff no element in B has a smaller first component.

This observation is useful for our primitive recursions later in this volume. With the help of the next theorem it shows that every primitive recursion *defines* a *unique* function. □

0.4.42 Remark (More Notation)

1. Let f and g be two functions, viewed extensionally. If $f \subseteq g$ we say that g *extends* f, or f is a *restriction* of g. If $f : A \to B$ and $S \subseteq A$, the symbol $f \restriction S$ denotes all those pairs in f with first component in S:

$$f \restriction S \stackrel{Def}{=} \{(x, f(x)) : x \in S\}$$

2. The symbol B^A denotes all *total* functions $f : A \to B$.
3. The symbol $\mathcal{P}(A : B)$ denotes all *partial* functions $f : A \to B$.
4. Let $(A, <)$ be a POset and $a \in A$. The set $\{x \in A : x < a\}$ we denote by $\mathring{S}_a$. The set $\{x \in A : x \leq a\}$ we denote by S_a. "S" for segment. $\mathring{S}_a$ is the *open segment* determined by a while S_a is the *closed segment* so determined.   □

0.4.43 Theorem. *Let $(A, <)$ be a POset with MC and let X be any set. Let the functions $h : A \to X$ and $g : A \times \mathcal{P}(A : X) \to X$ be given. Then there* exists *a unique $f : A \to X$ which satisfies the inductive definition:*

(I) $f(b) \simeq h(a)$, *for all minimal $a \in A$.*
(II) $f(b) \simeq g(b, f \restriction \mathring{S}_b)$, *for all* non *minimal $b \in A$, where "$\simeq$" is as in Definition 0.4.5.*

Proof

(1) *Uniqueness.* We will contradict that there are *two* distinct functions f and f' satisfying (I) and (II) of the theorem. Well, if they do exist, then, by MC, the set $\{x : f(x) \not\simeq f'(x)\}$ has a minimal element b. This is not a minimal element of the entire A, since on those we have $f(b) \simeq h(b) \simeq f'(b)$. But then equation (II) applies and have our contradiction:

$$f(b) \simeq g(b, f \restriction \mathring{S}_b)^{51} \simeq g(b, f' \restriction \mathring{S}_b) \simeq f'(b)$$

(2) For the *existence* we build f by approximations $f_a : S_a \to X$, for all $a \in A$. These approximations fulfil (I) and (II). The existence *of all these approximations* is proved by IC. For any minimal element a of A, $S_a = \{a\}$. Thus we construct

$$f_a = \begin{cases} \{(a, h(a))\} \text{ if } h(a) \downarrow \\ \emptyset \qquad\qquad\qquad \text{othw} \end{cases}$$

The only relevant condition to check for f_a is (I), and that trivially holds.
   Next we pick a *non minimal* element b and *prove* that $f_b : S_b \to X$ — satisfying (I) and (II)— exists.
   We take the I.H. that $f_a : S_a \to X$ meeting (I) and (II) exists, for all $a < b$.

---

[51] By minimality of b.

Since f_a and f_b satisfy the same inductive definition —(I) and (II)— on the set S_a, the uniqueness we already proved (for A, but the same proof applies to any S_x) implies

$$f_a(x) \simeq f_b(x), \text{ for } x \le a \text{ or } f_a = f_b \upharpoonright S_a \tag{†}$$

We can now construct $f_b : S_b \to X$. Its pairs are all those in the f_a, for $a < b$, plus the pair equation (II) gives to $f_b(b)$. That is, we define (construct)

$$f_b = \left\{ \left(b, g\left(b, f_b \upharpoonright \mathring{S}_b \right) \right) \right\} \cup \bigcup_{a<b} f_a$$

where, by the foregoing, $f_b \upharpoonright \mathring{S}_b = \bigcup_{a<b} f_a$.

Now that we have existence of all $f_a : S_a \to X$ we construct f by $f = \bigcup_{a \in A} f_a$. The reader can easily prove that the so defined f fulfils (I) and (II). Exercise 0.8.24.

□

0.5   On the Size of Sets

How can we tell if a set is *infinite*? And are *all* infinite sets *"the same size"*? There is a direct charactersisation of infinite sets due to Dedekind that says "a set is infinite iff it is in 1-1 correspondence with one of its proper subsets". To make sense of this requires the *Axiom of Choice*, which in this elementary review we will avoid. Thus we give instead a direct definition of a *finite* set and then define an infinite set as one that is *not* finite!

0.5.1   Finite Sets

0.5.1 Definition. (Finite and Infinite Sets) A set A is *finite* iff it is *either* empty, *or* is in 1-1 correspondence with $\{x \in \mathbb{N} : x \le n\}$. This "normalized" small set of natural numbers we usually denote by $\{0, 1, 2, \ldots, n\}$.

If a set is *not* finite, then it is, *by definition, infinite*.   □

0.5.2 Example. For any n, $\{0, \ldots, n\}$ is finite since, trivially, $\{0, \ldots, n\} \sim \{0, \ldots, n\}$ using the identity $(\lambda x. x)$ function on the set $\{0, \ldots, n\}$.   □

 0.5.3 Remark. One must be careful when one attempts to explain finiteness via counting by a human.

For example, Achilles[52] could count *infinitely many objects* by constantly accelerating his counting process as follows:

He procrastinated for a *full second*, and then counted the first element. Then, he counted the second object *exactly after* 1/2 a second from the first. Then he got to the third element $1/2^2$ seconds after the previous, ..., he counted the n th item at exactly $1/2^{n-1}$ seconds after the previous, and so on *forever*.

Hmm! It was *not* "forever", was it? After a total of 2 seconds he was done!

You see (as you can easily verify from your calculus knowledge (limits)),[53]

$$1 + \frac{1}{2} + \frac{1}{2^2} + \ldots + \frac{1}{2^{n-1}} + \ldots = \frac{1}{1 - 1/2} = 2$$

So "time" is not a good determinant of finiteness! $\square$

0.5.4 Theorem. *If* $X \subsetneq \{0, \ldots, n\}$, *then there is no* onto *function* $f : X \to \{0, \ldots, n\}$.

 No such f; whether total or not; total-ness is immaterial.

Proof First off, the claim holds if $X = \emptyset$, since then any such f equals $\emptyset$ and its range is empty.

Let us otherwise proceed by way of contradiction, and assume that the theorem is *wrong*: That is, *assume that* it *is* possible to have such onto functions, *for some n and well chosen X*.

Since we assumed that there are such $n > 0$, suppose then that the *smallest n* that allows this to happen is, say, n_0, and let X_0 be a *corresponding* set "X" that works with said n_0, that is,

$$\text{Assume that we have an onto } f : X_0 \to \{0, \ldots, n_0\} \tag{1}$$

Thus $X_0 \neq \emptyset$, by the preceding remark, and therefore $n_0 > 0$, since otherwise $X_0 = \emptyset$.

Let us call H be the set of all x such that $f(x) = n_0$, in short, $H = f^{-1}[\{n_0\}]$. $\emptyset \neq H \subseteq X_0$; the $\neq$ by ontoness.

Case 1. $n_0 \in H$. Then removing all pairs (a, n_0) from f —all these have $a \in H$— we get a new function $f' : X_0 - H \to \{0, 1, \ldots, n_0 - 1\}$, which *is still onto* as we only removed inputs that cause output n_0.

   This contradicts minimality of n_0 since $n_0 - 1$ works too!

Case 2. $n_0 \notin H$.

   If $n_0 \notin X_0$, then we argue exactly as in Case 1 and we just remove the base "H" of the cone (in the picture) from X_0.

   Otherwise, we have two subcases:

---

[52] OK, he was a demigod; but only "demi".

[53] $1 + \frac{1}{2} + \frac{1}{2^2} + \ldots + \frac{1}{2^{n-1}} = \frac{1 - 1/2^n}{1 - 1/2}$. Now let n go to infinity at the limit.

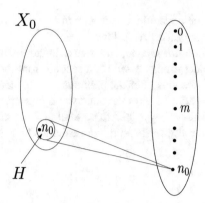

- $f(n_0) \uparrow$. Then (almost) we act as in Case 1: The new "X_0" is $(X_0 - H) - \{n_0\}$, since if we leave n_0 in, then the new "X_0" will not be a subset of $\{0, 1, \ldots, n_0 - 1\}$. We get a contradiction as in Case 1.
- The picture below —that is, $f(n_0) = m$ for some m.

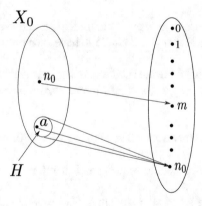

We simply transform the picture to the one below, "correcting" f to have $f(a) = m$ and $f(n_0) = n_0$, that is, *defining a new "f"* that we will call f' by

$$f' = \Big(f - \{(n_0, m), (a, n_0)\} \Big) \cup \{(n_0, n_0), (a, m)\}$$

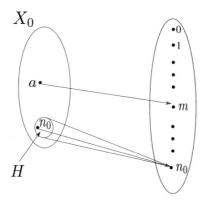

We get a contradiction per Case 1. □

0.5.5 Corollary. (Pigeon-Hole Principle) *If* $m < n$, *then* $\{0, \ldots, m\} \not\sim \{0, \ldots, n\}$.

Proof If the conclusion fails then we have an onto $f : \{0, \ldots, m\} \to \{0, \ldots, n\}$, contradicting Theorem 0.5.4. □

0.5.6 Theorem. *If A is finite due to* $A \sim \{0, 1, 2, \ldots n\}$ *then the "witness" n of this fact is unique: We cannot also* $A \sim \{0, 1, 2, \ldots m\}$ *with* $n \neq m$.

Proof If $\{0, 1, 2, \ldots n\} \sim A \sim \{0, 1, 2, \ldots m\}$, then $\{0, 1, 2, \ldots n\} \sim \{0, 1, 2, \ldots m\}$ by Exercise 0.8.25, hence $n = m$, otherwise we contradict Corollary 0.5.5. □

0.5.7 Definition. Let $A \sim \{0, \ldots, n\}$. Since n is uniquely determined by A we say that A has $n + 1$ elements and write $|A| = n + 1$. □

0.5.8 Corollary. *For any* $n > 0$, *there is no onto function from* $\{0, \ldots, n\}$ *to* $\mathbb{N}$.

Proof Fix an n. By way of contradiction, let $g : \{0, \ldots, n\} \to \mathbb{N}$ be onto. Let

$$Y \overset{Def}{=} \{x \leq n : g(x) > n + 1\}$$

Now let

$$X \overset{Def}{=} \{0, \ldots, n\} - Y$$

and

$$g' \overset{Def}{=} g - Y \times \mathbb{N}$$

The "$g - Y \times \mathbb{N}$" above is an easy way to say "remove all pairs from g that have their first component in Y".

Thus, $g' : X \to \{0, \ldots, n, n+1\}$ is onto, contradicting Theorem 0.5.4 because $X \subseteq \{0, \ldots, n\} \subset \{0, \ldots, n, n+1\}$.   □

0.5.9 Corollary. $\mathbb{N}$ *is infinite.*

Proof By Definition 0.5.1 the opposite case requires that there is an n and a function $f : \{0, 1, 2, \ldots, n\} \to \mathbb{N}$ that is a 1-1 correspondence. Impossible, since any such f cannot be *onto*.

 Our mathematical definitions have led to what we hoped they would: That $\mathbb{N}$ *is* infinite as we intuitively understand, notwithstanding Achilles's accelerated counting!

0.5.2 Some Shades of Infinity

0.5.10 Definition. A set S is called *countable* or *enumerable* iff there is a *total* and onto function function $f : \mathbb{N} \to S$. In other words, iff $\mathrm{ran}(f) = S$.   □

(1) So if a set S has been established to be countable via a total function f, then its members (possibly with repetitions —see the first bullet in the example below) can be arranged as an infinite array of entries indexed by the natural numbers (in ascending order of indices), with no natural number missing as an index.

$$f(0), f(1), f(2), \ldots$$

Here the indices appear in functional notation, as arguments, but we could have written instead

$$f_0, f_1, f_2, \ldots$$

After all, we normally write the Fibonacci function entries as F_n, F_3, etc., rather than $F(n)$, $F(3)$, etc.

(2) Some of the literature uses "enumerable" for "countable *and* infinite". We will use countable and enumerable as synonyms, and if we have an infinite set in mind, we may emphasise this, depending on the context.

0.5.11 Example.

- $\{3\}$ is countable. Indeed, take for f the constant function: for all x, $f(x) = 3$.
- $\mathbb{N}$ is countable. Indeed, take for f the identity function: $\lambda x.x$.
- The set of even natural numbers is countable. Indeed, take for f the function: $\lambda x.2x$.

□

0.5.12 Remark.

1. Suppose A and B are countable. How about $A \cup B$? Well, by assumption we have enumerations for A and B,

$$a_0, \quad a_1, \quad\quad a_2, \quad a_3, \quad\quad a_4, \ldots$$
$$b_0, b_1, \quad\quad\quad b_2, b_3, \quad\quad b_4, \ldots$$

Then

$$a_0, b_0, a_1, b_1, a_2, b_2, \ldots, a_n, b_n, \ldots$$

enumerates $A \cup B$.

2. Suppose now we have a countable sets of sets $\{A_0, A_1, \ldots\}$ each of the A_i being countable. How about $\bigcup_{i=0}^{\infty} A_i$?

Well, we are given an infinite matrix, the rows corresponding to each A_i and the columns running along each enumeration of the A_i, that is,

$$a_{(0,i)}, a_{(1,i)}, a_{(2,i)}, a_{(3,i)}, \ldots$$

Let us consider the matrix of all the A_i enumerations:

$$
\begin{array}{llllll}
A_0: & a_{(0,0)} & a_{(1,0)} & a_{(2,0)} & a_{(3,0)} & \cdots \\
 & & \nearrow & \nearrow & \nearrow & \nearrow \\
A_1: & a_{(0,1)} & a_{(1,1)} & a_{(2,1)} & a_{(3,1)} & \cdots \\
 & & \nearrow & \nearrow & \nearrow & \nearrow \\
A_2: & a_{(0,2)} & a_{(1,2)} & a_{(2,2)} & a_{(3,2)} & \cdots \\
 & & \nearrow & \nearrow & \nearrow & \nearrow \\
A_3: & a_{(0,3)} & a_{(1,3)} & a_{(2,3)} & a_{(3,3)} & \cdots \\
\vdots & \cdots & & & & \\
 & & \nearrow & \nearrow & \nearrow & \nearrow \\
A_i: & a_{(0,i)} & a_{(1,i)} & a_{(2,i)} & a_{(3,i)} & \cdots \\
 & & \nearrow & \nearrow & \nearrow & \nearrow \\
\vdots & \cdots & & & &
\end{array}
$$

Then traversing the matrix as indicated by the arrows we put all its members on a straight line (on a linear ordering), their position number in this line being their position in the enumeration of the union. Worth noting: Traversing as indicated, on each "north-east" *arrow sequence* we encounter entries $a_{(k,m)}$ for which the sum of "coordinates", $k + m$, is constant. □

0.5.13 Definition. A set S is called *uncountable* iff it is not countable. □

0.5.3   Diagonalisation

We start with an example.

0.5.14 Example. Suppose we have a 3×3 matrix

$$
\begin{array}{ccc}
1 & 1 & 0 \\
1 & 0 & 1 \\
0 & 1 & 1
\end{array}
$$

and we are asked: Find a *sequence* of three numbers, *from* $\{0, 1\}$, that does not *match any* row of the above matrix —i.e., is *different from all rows*.

   Sure, you reply. Take 0   0   0.

   That is correct. But what if the matrix were big, say, 100×100, or $10^{10^{10^{10}}} \times 10^{10^{10^{10}}}$, or even *infinite*?

   Is there a *finitely describable technique* that can produce an "unfit" row for any square matrix, no matter how big; even for an infinite one?

   Yes. *It is Cantor's "diagonal method" or "diagonalisation"* which he introduced in his Set Theory.

   He noticed that any row that fits in the matrix as the, say, i-th row, intersects the main diagonal at *the same spot that the i-th column does*.

> That is, at entry (i, i) of the *matrix*, which is the i-th entry of the *row*   (†)

   Thus if we take the main diagonal —a sequence of entries that has the same length as any row— and *change every one of its entries*, then it will not fit anywhere as a row! Because *no row can have an entry that is different than the entry at the location where it intersects the main diagonal!*

   The implementation of this idea gives the answer 0   1   0 *to our original question.*

   While the array 1000   11   3 also follows the principle of *making every entry on the diagonal different than the original*, and works, we were constrained in this example to "using only 0 or 1" entries, otherwise one could also "cheat" and provide "42 42 42" as an example that does not fit.

   More seriously, in a case of a very large or infinite matrix it is best to have a *simple technique* for "changing the diagonal entries" that *works* even if we do not know much about the elements of the matrix. Read on!   □

0.5.15 Example. We have an infinite matrix of 0-1 entries. Can we produce an *infinite sequence* of 0-1 entries that does not match any row in the matrix?

Pause. What is an *infinite sequence*? Our intuitive understanding of the term is captured *mathematically* by the concept of a *total* function f with left field (and hence, being total, domain) $\mathbb{N}$. The n-th member of the sequence is $f(n)$. Intuitively, the sequence is visualised as $f(0), f(1), f(2), \ldots, f(n), \ldots$ ◀

> Yes, *take the main diagonal and flip* every *entry* : $0\ to\ 1\ and\ 1\ to\ 0$ (‡)

Note that the diagonal entries have *matrix coordinates* (i, i) for $i = 0, 1, 2, \ldots$

So if we attempt to place the modified diagonal as row i of the matrix, then we will *fail*:

1. The i-th entry of the placed row will be overlaid on the (i, i) entry of the matrix, by comment (†) in Example 0.5.14 above.
2. But the *original* matrix entry at (i, i) is *different* than the i-th entry of the *modified* diagonal. The placement will not fit!

What we just said is true for *any* i, so the modified diagonal fits *nowhere* in the matrix as a row.

You can see clearly now why the Cantor technique demonstrated here and in the previous example is called "*the diagonal method*" or "*diagonalisation*". It uses the diagonal of a matrix in a clever and simple way.   □

 0.5.16 Remark. What Example 0.5.15 showed is that, using diagonalisation to an infinite 0-1 matrix, *we can define an infinite 0-1 array that fits nowhere as a row of the matrix.*   □

0.5.17 Example. We have an infinite matrix of entries from $\mathbb{N}$ (many may be > 1). Can we produce an infinite sequence of $\mathbb{N}$-entries that does not match any row in the matrix? Yes, *take the main diagonal and* change *every entry from a to $a + 1$*.[54]

If the original diagonal has an a in row i, the constructed row has an $a + 1$ in column i, so it will not fit as row i since the row position i coincides with the matrix diagonal at entry (i, i), but the latter holds a while the row we attempt to fit contains $a + 1$ in that spot.

So the modified diagonal fits nowhere, i being arbitrary.

Seeing that an infinite numerical array is the (sorted by input) sequence of outputs of a *total* function f with left and right fields equal to $\mathbb{N}$, this example can be viewed also like this:

We have a *sequence* of such one-argument functions, say

---

[54] From a to 2^a works too, or from a to $a + 42$; as do infinitely many other "flipping schemes".

$$f_0, f_1, f_2, f_3, \ldots$$

then these define the infinite matrix

$$f_0(0) \; f_0(1) \; f_0(2) \; \ldots \; f_0(i) \; \ldots$$
$$f_1(0) \; f_1(1) \; f_1(2) \; \ldots \; f_1(i) \; \ldots$$
$$\vdots$$
$$f_i(0) \; f_i(1) \; f_i(2) \; \ldots \; f_i(i) \; \ldots$$
$$\vdots$$

The flipping procedure (change a to $a + 1$) above constructs a function d —a (doomed not to fit) candidate *row*— given, for all x, by

$$d(x) = 1 + f_x(x) \tag{1}$$

The function d as an array is

$$d(0), d(1), \ldots$$

and —as we saw— does not fit in the matrix anywhere as a row.
    That is,

> *we have constructed a function d that cannot be any of the f_i*

Let us see this construction from a more direct point of view, in terms of functions, rather than sequences and matrices. This obscures a bit the diagonalisation at play, but it is simpler to write down.
    We show that the constructed function d in (1) above *cannot* be an f_i.

By contradiction, say instead that

$$d = f_i, \text{ for } some \; i \tag{2}$$

Then

$$f_i(i) \overset{(2)}{=} d(i) \overset{(1)}{=} 1 + f_i(i) \tag{3}$$

a contradiction, since the f_i are total hence $f_i(i)$ is a natural number a. Thus, (3), which says $a = 1 + a$, implies the contradiction $0 = 1$.   □

 0.5.18 Remark. The above example says, in particular, that we *cannot* have an enumeration

$$f_0, f_1, f_2, f_3, \ldots \tag{4}$$

of *all* total functions $f : \mathbb{N} \to \mathbb{N}$. Why? If we could, then we just learnt that the defined d is not one of them! But this is a contradiction to (4) containing *all* total unary functions $f : \mathbb{N} \to \mathbb{N}$, it fails to include *one such*: d.

In short, we have

0.5.19 Theorem. (Uncountability of $\mathbb{N}^{\mathbb{N}}$) $\mathbb{N}^{\mathbb{N}}$ *is uncountable.*

□

0.5.20 Example. (Cantor's Original Theorem, Somewhat Amended)
Let S denote the set of *all* infinite sequences of 0s and 1s.
Can we arrange *all* of S in an infinite matrix —one element per row?

No, since Example 0.5.15 shows how to *construct* an infinite 0-1 sequence that is not possibly a row of the matrix, contradicting the qualifier "all" above.

So we cannot do that. That is, this S is uncountable or *not* enumerable.

By the way, the amendment we made to Cantor's theorem here is twofold.

*   He did not care about 0-1 sequences; he wanted to show that the set of reals in the unit interval, $[0, 1] \overset{Def}{=} \{x \in \mathbb{R} : 0 \le x \le 1\}$, is *uncountable*. Well, any real in this set *is* essentially an infinite "binary sequence" that starts with a dot "." ("is essentially" means "is represented by")
*   Cantor actually used base-10, not base-2, representation of the reals in the interval $[0, 1]$. □

Remark 0.5.18 and this example show that *uncountable sets exist.* Here is a more interesting one, in the context of computability theory.

0.5.21 Example. (0.5.15 Retold) Consider the set $2^{\mathbb{N}}$ of *all total* functions from $\mathbb{N}$ to $\{0, 1\}$. Is this countable?
Well, if there is an enumeration of these one-variable functions

$$f_0, f_1, f_2, f_3, \ldots \tag{1}$$

consider the function $D : \mathbb{N} \to \{0, 1\}$ given, for all x, by

$$D(x) = 1 - f_x(x) \tag{2}$$

Clearly, this D *must* appear in the listing (1) since it has the correct left and right fields, *and* is total. Say $D = f_i$ for some i. Then

$$f_i(i) \overset{D=f_i}{=} D(i) \overset{(2)}{=} 1 - f_i(i)$$

A contradiction.

As in Example 0.5.17, for the contradiction to occur it is crucial that the f_i are total! For if, say, $f_i(i) \uparrow$ then we get no contradiction since then both sides of $f_i(i) = 1 - f_i(i)$ respond in the same *way:* Undefined; therefore are equal.   □

0.5.22 Remark. **Worth Emphasising.** Here is how we constructed the *d* functions, directly above and in Example 0.5.17: We have a list of *in principle available indices* for *d*. We want to make sure that *none applies*. A convenient method to do that is to inspect each available index, *i*, and using the diagonal method do this: Ensure that *d* differs from *each* f_i at input *i*, setting $d(i) = 1 - f_i(i)$ (or $1 + f_i(i)$ in Example 0.5.17).

This ensures that $d \neq f_i$; period. We say that *we cancelled the index i* as a possible "*f*-index" of *d*.

Since the process is applied *for each i, we have cancelled* all *possible indices for d* in these two examples: For no *i* can we have $d = f_i$.

We come back to *index cancellation* in Chap. 14.   □

Here is an example of an *uncountable set of sets*, presented in two stages:

 0.5.23 Example. (Part I (Cantor))   Let

$$S_0, S_1, S_2, \ldots \tag{†}$$

be *any* countable set of subsets of $\mathbb{N}$, that is, *members* of the *power set* of $\mathbb{N}$, $\mathscr{P}(\mathbb{N})$. According to our Definitions 0.5.10 we have a total function $f : \mathbb{N} \to \mathscr{P}(\mathbb{N})$ such that $f(i) = S_i$, for all $i \in \mathbb{N}$.

Now, Cantor showed that

we can always construct a subset $D \subseteq \mathbb{N}$ that is *not* one of the S_i

(‡)

Here is how: Let

$$D \overset{Def}{=} \{x \in \mathbb{N} : x \notin S_x\} \tag{1}$$

First off, $D \subseteq \mathbb{N}$ by definition (1). Next, *assume* by way of contradiction, that

$$D = S_i, \text{ for some } i \tag{2}$$

Is $i \in D$ true, or false?

Well, $i \in D \equiv i \in S_i$, by (2). On the other hand, by (1), $i \in D \equiv i \notin S_i$. The last two equivalences yield that $i \in S_i \equiv i \notin S_i$, a contradiction. We must assume that (2) does not hold.   □

0.5.24 Remark The above must be a hidden diagonalisation argument! Witness that we are stating something negative ("flipping") in connection with identifying two indices —$x \notin S_x$— forming a "diagonalisation" of the formula of two variables $y \in S_x$.

This is only a "clue" that diagonalisation is at play. To see that indeed it was, involve the *characteristic function* —denoted by c_S— of the subset S (any subset) of $\mathbb{N}$. This is defined, for all $x \in \mathbb{N}$, by

$$c_S(x) = \begin{cases} 0 & \text{if } x \in S \\ 1 & \text{if } x \notin S \end{cases}$$

Thus each $S \subseteq \mathbb{N}$ is identifiable with a unique c_S, which in turn is associated with a unique 0-1 sequence:

$$i \in S \text{ iff } 0$$
$$\downarrow$$
$$\ldots, \quad 0 \text{ or } 1 \quad , \ldots$$
$$\uparrow$$
$$i\text{-th location}$$

So the enumeration (†) can be represented by the *infinite matrix* of the 0-1 representations (infinite 0-1 strings) obtained via c_{S_i} for each S_i, where the representation of S_i constitutes row i.

What is the characteristic function of D as a 0-1 string? Well, by (1), for all x,

$$x \in D \equiv x \notin S_x$$

This translates, for all x, to

$$c_D(x) = 1 - c_{S_x}(x)$$

In other words, the x-th (0-1) entry of D is what we obtain by flipping the diagonal entry (x, x) from 0 to 1 *and conversely. Diagonalisation!* □

0.5.25 Example. (Part II (Cantor)) The power set $\mathscr{P}(\mathbb{N})$ is uncountable. Well, if we had an enumeration of *all* of $\mathscr{P}(\mathbb{N})$,

$$S_0, S_1, S_2, \ldots \qquad (*)$$

then there would be a member of $\mathscr{P}(\mathbb{N})$, D, that *cannot* appear in the enumeration (Example 0.5.23). □

0.6 Inductively and Iteratively Defined Sets

Much of what we do in computability involves defining objects inductively. The process is to start with a set of *initial objects* and a set of *operations*, where we define a k-ary *operation on a set* S to be a function f with inputs from S and outputs in S, specifically, $f : S^k \to S$, or more generally —relaxing the requirement of obtaining unique outputs for all inputs— a k-ary operation on a set is a relation $R \subseteq S^{k+1}$ where, when we write $R(\vec{x}_k, y)$ meaning $(\vec{x}_k, y) \in R$, we view the $\vec{x}_k$ as the input variables, and y as the output variable. Thus R is in fact a p.m.v. function in the sense of Definition 0.4.1.

0.6.1 Definition. We say that a set S is *closed under* a k-ary operation f or R just in case we are speaking of a function $f : S^k \to S$ or a relation $R \subseteq S^{k+1}$ with the understanding that the $(k+1)$st is the output variable.

An operation as defined here, acting on a finite number of arguments k when "called",[55] is called *finitary*.   □

0.6.2 Definition. Given a set of *initial objects* $\mathcal{I}$ and a *set* of *finitary operations* $\mathcal{O} = \{O_a : a \in A\}$, the object $Cl(\mathcal{I}, \mathcal{O})$ is called the *closure of $\mathcal{I}$ under $\mathcal{O}$* —or the set *inductively defined by the pair* $(\mathcal{I}, \mathcal{O})$— and *denotes* the $\subseteq$-smallest *set* S that satisfies

1. $\mathcal{I} \subseteq S$.
2. S is *closed under all operations in* $\mathcal{O}$, or simply said, *closed under $\mathcal{O}$* or even $\mathcal{O}$-*closed*.
3. The "smallest" qualifier means: Any set T that satisfies 1. and 2. also satisfies $S \subseteq T$.

The set $\mathcal{O}$ may be infinite. In this volume we will only be interested in finitary operations O_i in the sense of Definition 0.6.1, for various $k \geq 1$. We will then call the set $\mathcal{O}$ itself *finitary*.   □

Does $Cl(\mathcal{I}, \mathcal{O})$ *exist* for any given $\mathcal{I}$ and finitary $\mathcal{O}$? Yes. But first,

0.6.3 Theorem. *For any choice of $\mathcal{I}$ and $\mathcal{O}$, if $Cl(\mathcal{I}, \mathcal{O})$ exists, then it is unique.*

Proof Say $S = Cl(\mathcal{I}, \mathcal{O}) = T$. Then, letting S pose as a closure, we get $S \subseteq T$ from Definition 0.6.2. Then, letting T pose as a closure, we get $T \subseteq S$, again from Definition 0.6.2. Thus $S = T$.

0.6.4 Theorem. *For any choice of $\mathcal{I}$ and $\mathcal{O}$ with the restrictions of Definition 0.6.2 the set $Cl(\mathcal{I}, \mathcal{O})$ exists.*

Proof We note a few things.

1. The set

$$T = \mathcal{I} \cup \bigcup_{O_a \in \mathcal{O}} ran(O_a)$$

---

[55] In programming jargon.

satisfies $\mathcal{I} \subseteq T$ and T is $\mathcal{O}$-closed since $\bigcup_{O_a \in \mathcal{O}} \text{ran}(O_a)$ contains all outputs of all O_a.

2. Therefore, the family of sets $G = \{S : \mathcal{I} \subseteq S \wedge S \text{ is } \mathcal{O}\text{-closed}\}$ contains the set T as a member. If we set

$$C \stackrel{Def}{=} \left(\bigcap G\right) \tag{1}$$

then C is a subset of T. By (1), since all $S \in G$ contain $\mathcal{I}$ and are $\mathcal{O}$-closed, C does too. But (1) also implies $C \subseteq S$ for all $S \in G$. So C is $\subseteq$-smallest.

Thus, $C = \text{Cl}(\mathcal{I}, \mathcal{O})$. This proves existence.

The reader may be wondering what is the reason for building T. Well, since $\bigcap \emptyset =$ the class of *all* sets, which is *not* a set by Russell's paradox, we want to be sure that $G \neq \emptyset$.

Moreover, in a *family of sets*, as the name suggests, one must collect sets rather than non-set collections like Russell's. So, if one wants to cross all the "t"s and dot all the "ι"s in the proof above, one has to argue that our T is a set. Briefly, within an axiomatic setting of set theory we would argue like this: Since each O_a is a $(k+1)$-ary relation on some set S ($R \subseteq S^{k+1}$) it is a set by the *subset axiom*. Moreover, $\text{ran}(O_a) \subseteq S$ hence is a set by the subset axiom. Thus the collection of all $\text{ran}(O_a)$ is a set as well by the *collection axiom* of set theory via the map $O_a \mapsto \text{ran}(O_a)$ and the fact that $\mathcal{O}$ is given to be a set. Thus T is a set, since an elementary result of set theory is that for any set B, $\bigcup B$ is a set as well.

0.6.1 Induction on Closures

Definition 0.6.2 immediately provides a method to prove properties of members of $\text{Cl}(\mathcal{I}, \mathcal{O})$ by *induction over the closure*, as we say.

0.6.5 Theorem. *To prove that all $z \in \text{Cl}(\mathcal{I}, \mathcal{O})$ satisfy a relation $P(z)$ it suffices to prove*

1. *$P(z)$ is true if $z \in \mathcal{I}$*
   and
2. *$P(z)$ propagates with each operation in $\mathcal{O}$, that is, whenever $O_i(\vec{x}_k, z)$ is true and $P(x_i)$ is true, for all x_i (true for the inputs), then $P(z)$ is true (true for the output).*

Proof Assume 1. and 2. Let $\mathbb{T} = \{z : P(z)\}$. Now, by 1. and 2. above, $\mathcal{I} \subseteq \mathbb{T}$ and $\mathbb{T}$ is $\mathcal{O}$-closed.

We would like now to conclude —by Definition 0.6.2— that $\text{Cl}(\mathcal{I}, \mathcal{O}) \subseteq \mathbb{T}$, which translates into $P(z)$ (holds) for all $z \in \text{Cl}(\mathcal{I}, \mathcal{O})$.

We need a small extra step since said definition speaks of "a *set* T", while $\mathbb{T}$ might fail to be a set, for example, if $P(z)$ is "$z \notin z$".

We can use the T of the proof of Theorem 0.6.4 to get around this: $S = T \cap \mathbb{T}$ is a set since $S \subseteq T$. Moreover, we have that T contains $\mathcal{I}$ and is $\mathcal{O}$-closed. Thus so is S and thus $\mathrm{Cl}(\mathcal{I}, \mathcal{O}) \subseteq S \subseteq \mathbb{T}$.   □

0.6.2   Induction vs. Iteration

We now view $\mathrm{Cl}(\mathcal{I}, \mathcal{O})$ as obtained iteratively.

0.6.6 Definition. Given a "package" consisting of a set of *initial objects* $\mathcal{I}$ and a set of operations $\mathcal{O}$ as in Definition 0.6.2, we define a $(\mathcal{I}, \mathcal{O})$-*derivation* —or simply *derivation*, if the $(\mathcal{I}, \mathcal{O})$ is understood— to be a finite sequence of objects

$$d_1, d_2, \ldots, d_i, \ldots, d_{n-1}, d_n \tag{1}$$

such that, for all $i = 1, \ldots, n$, d_i is

1. a member of $\mathcal{I}$

   or

2. $O_a(d_{j_1}, d_{j_2}, \ldots, d_{j_k}, d_i)$ holds, that is, the result of an operation from $\mathcal{O}$ acting on inputs $d_{j_1}, d_{j_2}, \ldots, d_{j_k}$ that appeared at the left of d_i in (1), that is, $j_r < i$, for $r = 1, \ldots, k$.

As such, the concept of derivation is an abstraction of that of proof (cf. Definition 0.2.31) from "axioms" $\mathcal{I}$ via "rules" $O_a \in \mathcal{O}$.

The set of all $(\mathcal{I}, \mathcal{O})$-derived objects is defined to be

$$S \overset{Def}{=} \{x : x \text{ } appears \text{ in some } (\mathcal{I}, \mathcal{O})\text{-derivation}\} \qquad\qquad □$$

As was the case with Γ-proofs, the following two propositions are trivial and their proof will be an easy exercise.

0.6.7 Proposition. *If $d_1, d_2, \ldots, d_n, e_1, e_2, \ldots, e_m$ is an $(\mathcal{I}, \mathcal{O})$-derivation, then so is $d_1, d_2, \ldots, d_n$.*

0.6.8 Proposition. *An object D is $(\mathcal{I}, \mathcal{O})$-derived, iff it appears as the* last *member of some $(\mathcal{I}, \mathcal{O})$-derivation.*

0.6.9 Proposition. *If $d_1, d_2, \ldots, d_n$ and $e_1, e_2, \ldots, e_m$ are $(\mathcal{I}, \mathcal{O})$-derivations, then so is their* concatenation $d_1, d_2, \ldots, d_n, e_1, e_2, \ldots, e_m$.

We prove that defining a set S as a $(\mathcal{I}, \mathcal{O})$-closure is equivalent to defining S as the set of all $(\mathcal{I}, \mathcal{O})$-*derived* objects.

0.6.10 Theorem. *For any* initial set of objects *and* finitary operations on objects ($\mathcal{I}$ and $\mathcal{O}$) we have that $\mathrm{Cl}(\mathcal{I}, \mathcal{O}) = \{x : x$ is $(\mathcal{I}, \mathcal{O})$-derived$\}$.

Proof Let us set $S = \{x : x$ is $(\mathcal{I}, \mathcal{O})$-derived$\}$ and show that $\mathrm{Cl}(\mathcal{I}, \mathcal{O}) = S$.

$Cl(\mathcal{I}, \mathcal{O}) \subseteq S$:

We exploit the smallest-ness of $Cl(\mathcal{I}, \mathcal{O})$, that is, we do induction on $Cl(\mathcal{I}, \mathcal{O})$. To this end, all we need to do is show that

- $\mathcal{I} \subseteq S$: Immediate since if $x \in \mathcal{I}$, then

$$x$$

  is a one member derivation (sequence) for x.
- S is closed under each $R(\vec{x}_n, y) \in \mathcal{O}$: Indeed, let each of the x_i have a derivation,[56] $\boxed{\ldots x_i}$, where we implicitly invoked Proposition 0.6.8, and let us concatenate all these derivations as follows

$$\boxed{\ldots x_1} \ldots \boxed{\ldots x_i} \ldots \boxed{\ldots x_n} \tag{1}$$

By Proposition 0.6.9, (1) is a derivation as well. But then so is

$$\boxed{\ldots x_1} \ldots \boxed{\ldots x_i} \ldots \boxed{\ldots x_n} y \tag{2}$$

since y is obtained by an application of $R(\vec{x}_n, y)$ on the already derived x_i. That is, y is derived, hence $y \in S$. So S is R-closed, but R was arbitrary! This rests the case.

$Cl(\mathcal{I}, \mathcal{O}) \supseteq S$:

Let $x \in S$. In this direction we do *induction on the* length n *of a derivation* of x to show that $x \in Cl(\mathcal{I}, \mathcal{O})$.

Basis. $n = 1$. In this case $x \in \mathcal{I}$, but then $x \in Cl(\mathcal{I}, \mathcal{O})$ by the definition of closure.

I.H. Assume the claim for x derived with length $\leq n$.

I.S. Prove that the claim holds when x has a derivation of length $n + 1$.

So we have a derivation (relying on Proposition 0.6.8)

$$\ldots, a_1, \ldots a_i, \ldots, a_k, \ldots, x$$

If x that sits on the $(n + 1)$st position of the sequence is initial, then we are done by the basis. Else, say x is the result of an operation (relation) $Q \in \mathcal{O}$, applied on the indicated a_i, that is, say that $Q(a_1, \ldots, a_i, \ldots, a_k, x)$ is true.

By the I.H., which is applicable since the a_i are derived with lengths $\leq n$, we have that all the a_i are in $Cl(\mathcal{I}, \mathcal{O})$. But this set is closed under all relations contained in $\mathcal{O}$. Hence $x \in Cl(\mathcal{I}, \mathcal{O})$. $\square$

---

[56] The prefix "$(\mathcal{I}, \mathcal{O})$-" will be omitted from the term "$(\mathcal{I}, \mathcal{O})$-derivation" during the rest of the proof.

0.7   A Bit of Topology

This section will be of use in Chap. 13. Topology is a generalisation of, say, Euclidean geometry.

Where in the latter we are interested in the exact shape of figures (triangle, square, circle) and the relation of *congruence* (equality) between figures, in topology shape does not matter but properties that *matter* are, for example, *connectedness* — meaning intuitively the ability to go from any point A to a point B in the same figure via a path that has all its parts *in the figure*— presence or absence of *holes* in (planar) figures, and presence or absence of *cavities* in (solid) figures. The important relation between topological figures is the *homeomorphism*, that is, a 1-1 onto mapping f between two figures that is *continuous* in both directions, f and f^{-1}. If two figures have such a mapping that connects them we say they are *homeomorphic*, intuitively having the same properties that mater to topology. Good general introductions to topology are Hu (1964), Kelley (1955), but there will be no need for the reader to consult further as we will only use elementary facts from topology in Chap. 13, which we will cover in this section.

At the centre is the concept of a "topological space", that is, "a set equipped with a topology":

0.7.1 Definition. (Topological Space) A *topological space* is a "package" of two objects, denoted as a pair $(X, \mathcal{O})$.[57] X —the *space*— is a set and $\mathcal{O}$ is a subset of the power set 2^X that we *want*[58] to contain precisely the *open sets* of X. We say that X is a set *equipped* with the *topology* $\mathcal{O}$.

The *open sets* need to satisfy just four conditions (axioms) in order to be "well-chosen":

(1) $\emptyset$ is open.
(2) X is open.
(3) $\mathcal{O}$ is closed under arbitrary unions, that is, if $\mathcal{A} \subseteq \mathcal{O}$, then $\bigcup \mathcal{A} \in \mathcal{O}$.
(4) $\mathcal{O}$ is closed under nonempty *finite* intersections, that is, if $\mathcal{A} \subseteq \mathcal{O}$ is finite, then $\bigcap \mathcal{A} \in \mathcal{O}$.

It is customary to call the members of the space X "*points*".

0.7.2 Example. Thus,

(a) For any set X we *may* assign it (equip it with) the topology $\mathcal{O} = \{\emptyset, X\}$. This $\mathcal{O}$ is called the *trivial* or *indiscrete* topology.

   The reader can verify the axioms 1)–4) for this topology.

(b) At the other extreme, if X is a set we may choose *every* subset of X to be open: $\mathcal{O} = 2^X$. This topology is called the *discrete topology* and X equipped with this $\mathcal{O}$ —or *topologised* with this $\mathcal{O}$, as we say— is a discrete space. The

---

[57] We have encountered such "packages" before, e.g., $\text{Cl}(\mathcal{I}, \mathcal{O})$ in Definition 0.6.6.

[58] We *choose* the topology, as we say.

terminology stems from the fact that every point a of X leads to the open set $\{a\}$.

We may say "every 1-point set is open".

(c) More familiar to the reader (especially those who have had a solid course in calculus) is the usual topology on the real line $\mathbb{R}$.

The concept of an *open interval* is well known: For any two reals a and b such that $a < b$ the symbol (a, b) denotes the set $\{x \in \mathbb{R} : a < x < b\}$. Then the topology normally assigned to $\mathbb{R}$ has as open sets those that are arbitrary unions of intervals, along with $\emptyset$ and $\mathbb{R}$ too, of course. It is clear that the union of an arbitrary set of sets of the form $\bigcup\{\dots, (a, b), \dots\}$ is a set of the same form, thus axiom 3 for a topology is satisfied. On the other hand, by distributivity of $\cap$ over $\cup$, in an intersection like

$$\left(\bigcup\{\dots, (a, b), \dots\}\right) \cap \left(\bigcup\{\dots, (u, v), \dots\}\right)$$

after applying said distributivity we will end up with a result like

$$\bigcup\{\dots, (a, b) \cap (u, v), \dots\}$$

for all possible a, b, u, v. Now each $(a, b) \cap (u, v)$ is empty or an open interval. Thus we have axiom 4 above.

We have just witnessed the concept of a *basis* for a topology.   □

0.7.3 Definition. (Topological Basis) Given a topological space $(X, \mathcal{O})$. A *basis* for the topology $\mathcal{O}$ is a subset $\mathcal{B}$ of $\mathcal{O}$ such that $U \in \mathcal{O}$ iff it is a(n arbitrary) union of members of $\mathcal{B}$. In symbols,

$$x \in U \in \mathcal{O} \equiv (\exists V \in \mathcal{B})(x \in V \subseteq U)$$

Members of $\mathcal{B}$ are called *basic open sets*.   □

0.7.4 Definition. (Continuous Functions or Maps) Given two topological spaces $(X, \mathcal{O}_X)$ and $(Y, \mathcal{O}_Y)$. A total function $f : X \to Y$ is called *continuous* iff, for any open set U of the space Y, the inverse image $f^{-1}[U]$ is open in the space X.

One often calls the continuous functions "*maps*".   □

0.7.5 Proposition. *A total function $f : X \to Y$, where X is topologised by $\mathcal{O}$ and Y is topologised by a basis $\mathcal{B}$ is continuous iff for every basic open set V we have that $f^{-1}[V]$ is open in X.*

Proof

- (*Only if* part). Let f be continuous. Since any *basic open* set V is also *just open*, we get that $f^{-1}[V]$ is open in X.
- (*If* part). Suppose U is open in Y. Then $U = \bigcup_{a \in I} A_a$ where all A_a are basic open sets. Now

$$f^{-1}[U] = f^{-1}[\bigcup_{a \in I} A_a] = \bigcup_{a \in I} f^{-1}[A_a]$$

and we are done.

□

0.7.6 Definition. (Closed Sets) In a topological space $(X, \mathcal{O})$ a *closed* set A is by definition one whose complement $\overline{A}$ ($= X - A$) is open.   □

 0.7.7 Example. In a discrete space every 1-point set is open, since every set is. Thus every 1-point set is also closed. In fact, every set is also closed (since the complement is an open set).

Sets that are both open and closed are called *closed-open* and sometimes *clopen*.

□

In this volume we will sometimes look at the phenomenon of computation of *functionals* (Chap. 13) from a topological point of view. Such functionals are (partial) functions that may have as inputs numbers but also (partial) functions $\alpha : \mathbb{N} \to \mathbb{N}$. A typical call to such a functional $\mathscr{F}$ might look in its simplest form like $\mathscr{F}(x, \alpha)$, where x is a number.

Thus we will want to topologise the *set of all partial functions* α from $\mathbb{N}$ to $\mathbb{N}$. To allow for non total functions and still keep the semblance of totalness for mathematical convenience, topologically speaking we will depict the absence of output —$\alpha(x) \uparrow$— by adding the *new symbol* "$\Uparrow$" to the copy of $\mathbb{N}$ *on the output side* (the *right field*) and use it as output whenever we have no output.[59]

Thus, a partial function $\alpha : \mathbb{N} \to \mathbb{N}$ becomes a *total* function

$$\alpha : \mathbb{N} \to \mathbb{N} \cup \{\Uparrow\}$$

where, if we have $\alpha(x) \uparrow$ (undefined) originally —i.e., when the right field is $\mathbb{N}$— for some x, then we simply *define* the output to be the *new symbol* $\Uparrow$ and may write $\alpha(x) = \Uparrow$.

 This is only a mathematical devise to facilitate topological considerations — to which end we pad our α to be total— and *we nowhere imply* that we can (algorithmically) tell, for the arbitrary $\alpha : \mathbb{N} \to \mathbb{N}$, whether for some input x we have (or not) $\alpha(x) = \Uparrow$.[60]

Now we can topologise the space $X = \{\alpha : \alpha : \mathbb{N} \to \mathbb{N} \cup \{\Uparrow\}\}$, denoted usually $\left(\mathbb{N} \cup \{\Uparrow\}\right)^{\mathbb{N}}$.[61]

---

[59] We prefer not to conflate the notation for a symbol that says "there is no output" with that of a symbol that *denotes* the "undefined value".

[60] Nor is it the case that we can pad an arbitrary computable α to make it total while remaining computable, as we will learn in this volume. But we are not talking about computable α here.

[61] Opting for the somewhat awkward "$\mathcal{P}(\mathbb{N} : \mathbb{N})$" in Chap. 13 as we prefer not to use "$\Uparrow$" within computability beyond this section.

To $\mathbb{N} \cup \{\Uparrow\}$ we assign *almost* the discrete topology: Every subset of $\mathbb{N}$ is open, and so is $\mathbb{N} \cup \{\Uparrow\}$. Thus no nontrivial (that is, not the whole space or $\emptyset$) open set contains $\Uparrow$.

This topology in turn induces the *product topology*, as we say, to $\left(\mathbb{N} \cup \{\Uparrow\}\right)^{\mathbb{N}}$ the latter set viewed as

$$\left(\mathbb{N} \cup \{\Uparrow\}\right) \times \left(\mathbb{N} \cup \{\Uparrow\}\right) \times \left(\mathbb{N} \cup \{\Uparrow\}\right) \times \ldots \tag{1}$$

with one copy of $\left(\mathbb{N} \cup \{\Uparrow\}\right)$ in each position $0, 1, 2, 3, \ldots$ in the product above.

Of course, the "product" is a user-friendly visualisation of the set of all infinite length "vectors" (meaning sequences or, equivalently, functions α with domain $\mathbb{N}$)

$$\alpha = \alpha(0), \alpha(1), \alpha(2), \ldots, \alpha(i), \ldots$$

each $\alpha(i)$ being one of: a *number* or $\Uparrow$ and *is situated in the i-th copy* of $\mathbb{N} \cup \{\Uparrow\}$ in (1).

The particular product topology we choose is given by a *basis*: The basis consists of *all the sets* of the form depicted below

$$A_0 \times A_1 \times A_2 \times A_3 \times \ldots \times A_i \times \ldots \tag{2}$$

where *only finitely many A_i are subsets* of $\mathbb{N}$ —i.e., $\Uparrow \notin A_i$— *all the rest* being copies of $\left(\mathbb{N} \cup \{\Uparrow\}\right)$.

We say that the $A_i \neq \mathbb{N} \cup \{\Uparrow\}$ above *determine* the basic open set in (2).

Here and also later in this volume we will denote *finite* functions $\alpha : \mathbb{N} \to \mathbb{N}\{\Uparrow\}$ —that is, functions of finite domain— by the dedicated letters $\sigma, \tau, \xi, \theta$.

0.7.8 Definition. (Intervals) The *interval determined by θ*, or interval θ, is the set $\{\alpha : \theta \subseteq \alpha\}$, indicated by using a "hat": $\widehat{\theta} = \{\alpha : \theta \subseteq \alpha\}$. $\square$

If we have a basic open set of the space $\left(\mathbb{N} \cup \{\Uparrow\}\right)^{\mathbb{N}}$ as in (2), *determined by*

$$A_{i_k} : k = 0, 1, \ldots, r$$

then let θ be a finite function $\theta = \{\theta(i_k) \in A_{i_k} : k = 0, 1, \ldots, r\}$.

For any such θ we have $\widehat{\theta} \subseteq A_0 \times A_1 \times A_2 \times A_3 \times \ldots A_i \times \ldots$, and clearly the union of all these $\widehat{\theta}$ equals the set in (2) (where the A_{i_k} determine the basic open set that (2) depicts).

Therefore, *the set of sets $\widehat{\theta}$ determined by finite functions θ,*

$$\widehat{\theta} = \{\alpha : \theta \subseteq \alpha\}$$

is also a basis for the open sets of $\left(\mathbb{N} \cup \{\Uparrow\}\right)^{\mathbb{N}}$.

 This is the basis we shall use in this volume for the topology of the space $\left(\mathbb{N} \cup \{\Uparrow\}\right)^{\mathbb{N}}$.

Note that $\theta \in \widehat{\theta}$ *since* $\theta \subseteq \theta$.

The following is a useful elementary result:

0.7.9 Proposition. (Characterisation of Open Sets)

A set $U \subseteq \left(\mathbb{N} \cup \{\Uparrow\}\right)^{\mathbb{N}}$ *is open iff*

1. *If* $\alpha \in U$ *then* $(\exists\theta)(\theta \in U \wedge \theta \subseteq \alpha)$
   and
2. *If* $\alpha \in U \wedge \alpha \subseteq \beta$, *then* $\beta \in U$.

Proof

- (*Only if* case).

  1. Say U is open and $\alpha \in U$. Then α is in some basic open set that is a *subset* of U, in symbols, $\alpha \in \widehat{\theta} \subseteq U$. This yields two things: One, $\theta \subseteq \alpha$. Two, $\theta \in U$ (see paragraph above). We got 1.

  2. Say $\alpha \in U$ and $\alpha \subseteq \beta$. The first conjunct implies, via 1. (with a θ selected as above), $\alpha \in \widehat{\theta} \subseteq U$, hence (1.) $\alpha \supseteq \theta \in U$.
     If now $\alpha \subseteq \beta$, then $\theta \subseteq \alpha \subseteq \beta$, hence $\beta \in \widehat{\theta} \subseteq U$, and we are done.

- (*If* case). We start with 1. and 2. So let $\alpha \in U$. The *goal* is to find a $\widehat{\theta} \subseteq U$ such that $\alpha \in \widehat{\theta}$, that is, $\theta \subseteq \alpha$.
  By 1. we have a θ such that $\theta \in U \wedge \theta \subseteq \alpha$. OK, let us see if also $\widehat{\theta} \subseteq U$: So let $\beta \in \widehat{\theta}$. Thus, $\theta \subseteq \beta$. We are looking at $\theta \in U \wedge \theta \subseteq \beta$. By 2. we have $\beta \in U$. $\square$

With the above tools we can discuss the continuity of functions

$$\mathscr{G} : \left(\mathbb{N} \cup \{\Uparrow\}\right)^{\mathbb{N}} \to \left(\mathbb{N} \cup \{\Uparrow\}\right)^{\mathbb{N}}$$

In particular, such a $\mathscr{G}$ is continuous iff, for any θ, $\mathscr{G}^{-1}[\widehat{\theta}]$ is open (Proposition 0.7.5).

It is useful to also include functions

$$\mathscr{F} : \mathbb{N} \times \left(\mathbb{N} \cup \{\Uparrow\}\right)^{\mathbb{N}} \to \mathbb{N} \tag{1}$$

the so called *functionals* of Chap. 13 —that have both number and function $\alpha : \mathbb{N} \to \mathbb{N}$ inputs— in such topological considerations. Here we have restricted attention to the case where there is only one of each, number and function inputs. We say the functional has *rank* $(1, 1)$.

Since to $\mathbb{N}$ we assign the discrete topology and to $\mathbb{N} \times \left(\mathbb{N} \cup \{\Uparrow\}\right)^{\mathbb{N}}$ we assign the product topology with basic open sets $A \times B$ —where $A \subseteq \mathbb{N}$ and $B \subseteq \left(\mathbb{N} \cup \{\Uparrow\}\right)^{\mathbb{N}}$ is open— we define

0.7.10 Definition. A partial functional $\mathscr{F}$ is called *partial continuous* iff for all $n \in \mathbb{N}$ the set $\mathscr{F}^{-1}[\{n\}]$ is open. □

0.7.11 Example. Let $S \subseteq \mathbb{N} \times \left(\mathbb{N} \cup \{\Uparrow\}\right)^{\mathbb{N}}$. Then S is open iff it is the domain of a partial continuous functional. Indeed,

- (*If* case.) Say $S = \operatorname{dom}(\mathscr{F})$, where $\mathscr{F}$ is partial continuous. But $\operatorname{dom}(\mathscr{F}) = \bigcup_{n \geq 0} \mathscr{F}^{-1}[\{n\}]$, an open set.
- (*Only if* case.) Say S is open. Define $\mathscr{F}$ by

$$\mathscr{F}(x, \alpha) = \begin{cases} 0 & \text{if } (x, \alpha) \in S \\ \uparrow & \text{othw} \end{cases} \tag{1}$$

Thus $S = \operatorname{dom}(\mathscr{F})$, an open set. On the other hand, the only $n \in \mathbb{N}$ where $F^{-1}[\{n\}]$ is nonempty is $n = 0$. Now, by (1), $S = F^{-1}[\{0\}]$ is an open set, and all other inverse images (for $n \neq 0$) are $\emptyset$, hence open as well. Therefore $\mathscr{F}$ is partial continuous. □

0.8 Exercises

1. Prove that our logic proves $\vdash (\forall x)(\forall x)A \equiv (\forall x)A$ and also $\vdash (\exists x)(\exists x)A \equiv (\forall x)A$.
2. Let us use *proof by contradiction* to establish $\vdash x = y$. So assume $\neg x = y$. Then by Proposition 0.2.85 we obtain $\vdash \neg x = x$. Along with the axiom $x = x$ and Post we get $\vdash \neg x = x \wedge x = x$; our contradiction.
   Do you believe this? If not, why not?
   Criticise the proof: What *exactly* went wrong there, if anything?
3. Fill in the details of Remark 0.4.34.
4. Prove the claim in Exercise 0.2.77.
5. Prove Exercise 0.2.102.
6. Prove the converse of extensionality.
7. Prove that $A = B$ iff $A \subseteq B$ and $B \subseteq A$.
8. What say A is the empty family of sets, that is, $A = \emptyset$. What are $\bigcup A$ and $\bigcap A$ if we are not working in a reference set? What are the answers if we do work inside a reference set U?

9. Prove

$$(a, b, c) = (x, y, z) \rightarrow a = x \wedge b = y \wedge c = z$$

10. Prove that $(x_1, \ldots, x_n) = (y_1, \ldots, y_n)$ implies $x_i = y_i$, for $1 \leq i \leq n$.

11. Prove that for any sets $A, B, \emptyset = A \times B$ iff $A = \emptyset$ or $B = \emptyset$.

12. Prove that if $f : A \rightarrow B$ and $g : B \rightarrow C$ are (single-valued) functions then so is $f \circ g : A \rightarrow C$.

13. Prove that for any set A we have $\emptyset \subseteq A$.

14. Prove that for any function $f : X \rightarrow Y$ and $A \subseteq Y$ ans $B \subseteq Y$, then

(a) $f^{-1}[A \cup B] = f^{-1}[A] \cup f^{-1}[B]$
(b) $f^{-1}[A \cap B] = f^{-1}[A] \cap f^{-1}[B]$
(c) If $A \subseteq B$, then $f^{-1}[B - A] = f^{-1}[B] - f^{-1}[A]$. Does this work if $A \not\subseteq B$?

15. Prove Proposition 0.4.10.

16. Prove Proposition 0.4.13.

17. Prove Proposition 0.4.14.

18. Exhibit a function that has neither a left nor a right inverse.

19. Prove Lemma 0.4.21.

20. Prove that the symmetry rule for relations R is stronger than it seems: It implies (as stated) that $aRb \equiv bRa$, for all a, b.

21. Prove that a relation R is transitive iff $R \circ R = R$.

22. Prove Proposition 0.4.26.

23. Let $\leq : A \rightarrow A$ be reflexive, antisymmetric and transitive. Prove that $<$ defined by $a < b \overset{Def}{\equiv} a \leq b \wedge a \neq b$ is a strict order.

24. Prove that the f constructed in the proof of Theorem 0.4.43 satisfies the equations (I) and (II) of the theorem.

25. Prove the transitivity of 1-1 correspondences: If $A \sim B \sim C$, then $A \sim C$.

26. Prove that if A has a non-ending enumeration $a_0, a_1, a_3, \ldots$, then it also has one where every a_i is repeated infinitely many times.

27. Prove that a subset of a countable set A is countable.
   Hint. Define precisely an enumeration of the subset in terms of a given enumeration of A.

28. Prove that if A or B are countable, then so is $A \cap B$.

29. Prove that if A is countable, then so is $A - B$.

30. Show that for any X, $X \not\sim 2^X$. By contradiction, assume that X can index all of 2^X in a 1-1 and onto way, that is, assume a total, 1-1 and onto $f : X \rightarrow 2^X$.
   Hint. What can you say about the set $\mathcal{D} = \{x \in X : x \notin X_{f(x)}\}$?

31. Prove that $\lambda xy.2^x 3^y$ has infinite range.

32. Prove that $A \sim B$ implies $2^A \sim 2^B$.

Chapter 1
A Theory of Computability

Overview

Computability is the part of logic and theoretical computer science that gives
a mathematically precise formulation to the concepts of *algorithm*, *mechanical
process*, and *calculable* or *computable function* (or relation).

Its development was strongly motivated, in the 1930s, by Hilbert's *program*
to found mathematics on a (provably[1]) consistent (i.e., free from contradiction)
axiomatic basis, in particular by his belief that the *Entscheidungsproblem*, or
decision problem, for axiomatic theories —that is, the problem "is this formula
a *theorem* of that theory?"— was solvable by a mechanical procedure. Naturally,
towards investigating this belief an important ingredient was to establish what *was*
a mechanical procedure.

Now, since antiquity, mathematicians have invented "mechanical procedures",
e.g., Euclid's algorithm for the "greatest common divisor",[2] and had no problem
recognising such procedures when they encountered them. But how do you mathematically *prove* the *nonexistence* of such a mechanical procedure for a particular
problem, such as the Entscheidungsproblem, for example? You cannot just try *all*
such procedures on the problem to see if one solves it or no one does. There are
infinitely many mechanical procedures!

We encounter the same difficulty towards establishing any result of the type
"there is *no* algorithm that solves problem *X*". Problem"*X*" could be, e.g., to check
if an arbitrary program, say written in C, will ever print the number 42.

You need a *mathematical formulation* of what *is* a "mechanical procedure" or
"algorithm" in order to hope that you can address the Entscheidungsproblem, or
problem *X* above. By the way, Church proved (Church 1936a,b) that Hilbert's fundamental *Entscheidungsproblem* admits no solution by functions that are calculable
within any of the known equivalent mathematical frameworks of computability, and

---

[1] Metamathematically.

[2] That is, the largest positive integer that is a common divisor of two given integers.

© Springer Nature Switzerland AG 2022

G. Tourlakis, *Computability*, https://doi.org/10.1007/978-3-030-83202-5_1

in this volume we will see that problem X has no algorithmic solution either, if we use the term "algorithm" in the technical, mathematical sense of this volume.

Intensive activity by many researchers, notably Post (1936, 1944), Kleene (1936), Church (1936b), Turing (1936, 1937), and some 30 years later (Markov 1960), led in the first instance in the 1930s to several alternative formulations, each purporting to mathematically characterise the concepts *algorithm, mechanical procedure*, and *calculable function*. All these formulations were soon proved to be equivalent; that is, the calculable functions admitted by any one of them were the same as those that were admitted by any other. This led Alonzo Church to formulate his *conjecture*, known as "Church's Thesis", that any *intuitively* computable function is also calculable within any of these equivalent mathematical frameworks of calculability or computability.

The eventual introduction of computers provided further motivation for research on the various mathematical frameworks of computation, "models of computation" as we often say, and computability is nowadays a deep and very broad research field. The model of computation that I will present here, due to Shepherdson and Sturgis (1963), is a later model that has been informed by developments in computer science, in particular, by the advent of so-called *high level*[3] programming languages.

1.1   A Programming Framework for the Computable Functions

So, what *is* a *computable function*, mathematically speaking?

There are two main ways to approach this question. One is to *define a programming formalism* —that is, *a programming language* that is technology-independent and, in particular, capable of handling integers of *any* length— and say: "*a function is computable precisely if* it can be 'programmed' in *this* programming language".

Examples of such programming languages are the *Turing Machines* (or TMs) of Turing and the *unbounded register machines* (or URMs) of Shepherdson and Sturgis. Note that the term *machine* in each case is a misnomer, as both the TM and the URM formulations are really programming languages, the first being very much like assembly language of "real" computers, the latter reminding us more of (subsets of) Algol (or Pascal or C).

The other main way is to define a set of computable functions *directly* —without using a programming language as the agent of definition— by a *construction* that resembles *a mathematical proof*, called a *derivation*.

---

[3] The level is "higher" the more the programming language is distanced from machine-dependent details.

In this approach we say a *"function is computable precisely if* it is derivable".

 Either way, a computable function is generated by a *finite devise* (program, or derivation).

In the by-derivation approach we start by accepting some set of *initial functions* $\mathcal{I}$ that are immediately recognisable as being "intuitively computable" —for example, no one would dispute that the function, which on any input x always returns 0, is computable— and choose a set $\mathcal{O}$ of function-building operations that preserve the "computable" property. This approach was originally due to Dedekind (1888) for what we nowadays call *primitive recursive functions*. It evolved later Kleene (1936) into what we nowadays call *partial recursive functions*.

The introduction of computable functions by derivations is very elegant mathematically (cf. Tourlakis (1984)), but is less intuitively immediate, whereas the programming approach has the attraction of appearing to be "natural" to the reader who has done some computer programming.[4]

We now embark on defining the high level programming language *URM*.

The **alphabet** of the language is

$$\leftarrow, +, \dot{-}, :, X, 0, 1, 2, 3, 4, 5, 6, 7, 8, 9, \textbf{if}, \textbf{else}, \textbf{goto}, \textbf{stop}, \P \qquad (1)$$

Just like any other high level programming language, URM manipulates the contents of *variables*.[5]

(1) These variables are restricted to be of *natural number type*.
(2) Since this programming language is *for theoretical analysis only* —rather than for practical implementation— every variable is allowed to hold *any natural number* whatsoever, without limitations to its size, hence the "UR" in the language name ("unbounded register").
(3) The *syntax* of the variables is simple: A variable (name) is a string that starts with X and continues with zero or more 1:

$$\text{URM variable set:} \quad X, X1, X11, X111, X1111, \ldots, X1^{n}, \ldots \qquad (2)$$

where

$$1^{n} \stackrel{Def}{=} \overbrace{1 \ldots 1}^{n \ 1s}, \text{ where } 1^{0} \stackrel{Def}{=} \lambda, \text{ the empty string}$$

---

[4] We will show in Sect. 5.6 that the two approaches are equivalent.

[5] Shepherdson and Sturgis (1963) called the variables "registers".

(4) Nevertheless, as is customary for the sake of convenience, we will utilise the bold face lower case letters **x**, **y**, **z**, **u**, **v**, **w**, with or without subscripts or primes as *metavariables* in most of our discussions of the URM, and in examples of specific programs (where yet more, convenient metanotations for variables may be employed).

1.1.1 Definition. (URM Programs) A *URM program* —or just URM— is a finite (ordered) sequence of instructions of the following five types, separated by the symbol ¶:

$L : \mathbf{x} \leftarrow a$

$L : \mathbf{x} \leftarrow \mathbf{x} + 1$

$L : \mathbf{x} \leftarrow \mathbf{x} \dot{-} 1$ $\hspace{8cm}$ (3)

$L : \mathbf{stop}$

$L : \mathbf{if}\ \mathbf{x} = 0\ \mathbf{goto}\ M\ \mathbf{else}\ \mathbf{goto}\ R$

where L, M, R, a, written in decimal notation, are in $\mathbb{N}$, and **x** is some variable, and ¶ is the *carriage* (hard) *return* symbol, acting as an *instruction separator*. The effect of this separator is consistent with our habit of writing URM programs vertically, one line at a time —naturally making ¶ invisible.

We call instructions of the last type *if-statements*.

Each instruction in a URM program *must be numbered* by its *position number*, L, in the program, where ":" separates the position number from the instruction. We call these numbers *labels*. Thus, the label of the first instruction is always "1". The instruction **stop** must *occur only once* in a program, as the *last instruction*. An *if-statement* may not refer to a label that does not exist in the program.

The semantics of each instruction is given below. $\hspace{6cm}$ □

1.1.2 Definition. (URM Instruction and Computation Semantics)

A URM **computation** is a **sequence of actions** caused by the execution of the instructions of the URM as detailed below.

Every computation **begins** with the instruction labeled "1" as the *current* instruction. The semantic action of instructions of each type is defined if and only if they are current, and is as follows:

(i) $L : \mathbf{x} \leftarrow a$. Action: The value of **x** becomes the (natural) number a. Instruction $L + 1$ will be the next current instruction.

(ii) $L : \mathbf{x} \leftarrow \mathbf{x} + 1$. Action: This causes the value of **x** to increase by 1. The instruction labeled $L + 1$ will be the next current instruction.

(iii) $L : \mathbf{x} \leftarrow \mathbf{x} \dot{-} 1$. Action: This causes the value of **x** to decrease by 1, *if* it was originally non zero. Otherwise it remains 0. The instruction labeled $L + 1$ will be the next current instruction. We revisit the operation "$\dot{-}$" in Example 1.1.10 below.

(iv) L : **stop**. Action: No variable (referenced in the program) changes value. The next current instruction is still the one labeled L.

(v) L : **if x** $= 0$ **goto** M **else goto** R. Action: No variable (referenced in the program) changes value. The next current instruction is numbered M if **x** $= 0$; otherwise it is numbered R.

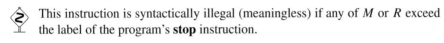

This instruction is syntactically illegal (meaningless) if any of M or R exceed the label of the program's **stop** instruction.

□

1.1.3 Definition. (Halting or Convergent Computation) We say that a computation *terminates*, or *halts*, iff it ever makes the instruction **stop** *current* (as we say, it "reaches" **stop**).

Note that the semantics of "L : **stop**" require the computation to continue *for ever*, but it does so in a trivial manner where *no variable changes value, and the current instruction remains the same*: *The contents of all the URM's variables have* converged[6] to stable values *and no instruction other than **stop** can be executed once **stop** is reached.*

Thus a halting computation is also said to have *converged*; it is a *convergent computation* we say.   □

One usually gives names to URM programs, or as we just say, "to URMs", such as M, N, P, Q, R, F, H, G.

1.1.4 Definition. (URM as an Input/Output Agent) A URM computation is an input/output "black box", and *computes a function* that we denote by

$$M_{\mathbf{y}}^{\vec{\mathbf{x}}_n}$$

in *this precise sense*:

We choose and designate as *input variables* of M the following: $\mathbf{x_1}, \ldots, \mathbf{x_n}$. Then we choose and designate *one* variable **y** of M as *the output variable* (**y** may be one of the $\mathbf{x}_i$).

The function $M_{\mathbf{y}}^{\vec{\mathbf{x}}_n}$, for every $\vec{a}_n$ from $\mathbb{N}^n$ that is "read" into $\vec{\mathbf{x}}_n$, will produce *at most one* value b in **y** according to the computation initialised as follows, *provided* the computation *converges*.

(1) We *initialise the computation*, by doing two things:

    (a) We *initialise* the input variables $\mathbf{x_1}, \ldots, \mathbf{x_n}$ with the input values $a_1, \ldots, a_n$. We *initialise* all other variables of M to be 0.

    (b) We next make the instruction labeled "1" current, and thus start the computation.

---

[6] They stay the same from this event —reaching **stop**— onwards.

(2) If the computation terminates, that is, if at some point the instruction **stop** becomes current, then the value of **y** at that point (and hence at any future point, by (iv) above), is the value of the function $M_{\mathbf{y}}^{\vec{\mathbf{x}}_n}$ for input $\vec{a}_n$.   □

1.1.5 Definition. (Computable Functions) A function $f : \mathbb{N}^n \to \mathbb{N}$ of n variables $x_1, \ldots, x_n$ is called *partial computable* iff for some URM, M, we have $f = M_{\mathbf{y}}^{\mathbf{x_1}, \ldots, \mathbf{x_n}}$. The set of all partial computable functions is denoted by $\mathcal{P}$. The set of all the *total* functions in $\mathcal{P}$ —that is, those that are defined on *all inputs* from $\mathbb{N}$— is the set of *computable* functions and is denoted by $\mathcal{R}$. The term *recursive* is used in the literature synonymously with the term *computable*.   □

1.1.6 Remark. Note that since a URM is a *theoretical*, rather than practical, model of computation we do *not* include *human-computer interface* considerations in the computation. Thus, the "input" phases just happen during initialisation —they are *not* part of the computation.[7] The output phase is implicit. Mathematically, the output value is in **y**. We do not need a print statement![8]

That is why we have dispensed with both **read** and **write** instructions and speak instead of *initialisation* in (1) of Definition 1.1.4.   □

1.1.7 Example. Let M be the program

$$1 : \mathbf{x} \leftarrow \mathbf{x} + 1$$

$$2 : \mathbf{stop}$$

Then $M_{\mathbf{x}}^{\mathbf{x}}$ is the function f given, for all $x \in \mathbb{N}$, by $f(x) = x + 1$, the *successor* function that we will denote by S in this book.   □

1.1.8 Remark (λ Notation) To avoid saying verbose things such as "$M_{\mathbf{x}}^{\mathbf{x}}$ is the function f given, for all $x \in \mathbb{N}$, by $f(x) = x + 1$", we will often use Church's λ-notation and write instead "$M_{\mathbf{x}}^{\mathbf{x}} = \lambda x.x + 1$". A first encounter of this notation was in Example 0.2.1.

In general, the notation "$\lambda \cdots .$" marks the beginning of a sequence of input variables "$\cdots$" by the symbol "λ", and the end of the sequence by the symbol "." What comes after the period "." is the "expression" that indicates what the output is, for the listed input. The template for λ-notation thus is

$$\lambda \text{"input"}.\text{"output-expression"}$$

---

[7] This convention is analogous with that of the older programming formalism, the Turing Machine. There, we input a string in a sequence of consecutive squares —each square holds one symbol— of the "2-way infinite" tape, and make all the other (infinitely many squares) "blank".

[8] Again, this is akin to how we code output for Turing Machines (e.g., Davis (1958a); Tourlakis (1984)). If and when a TM computation halts, the (numerical) output is the number of all the (not necessarily consecutive) occurrences of the symbol "1" on the tape.

Relating to the above example, we note that $f = \lambda x.x + 1 = \lambda y.y + 1 = \lambda z.f(z)$ is correct —although the rule "$y + 1$" is more informative than "$f(z)$".

To the left and right of each "=" we have (a symbol for) the *table*, which *denotes* a function, and we are saying that *all these tables are the same*.

Note that any x, y, z that appear in $\lambda \ldots xyz \ldots .output$ are "apparent" variables and are not free for substitutions —that is, they are *not available* to substitute expressions into them (they are *bound*). Moreover, $f = f(x)$ is incorrect as we have distinct *data types* to the left and right of "=", namely, a table to the left and a number to the right (albeit an unspecified number).

In programming circles, the distinction between *function definition* or *declaration*, $\lambda \vec{x}.f(\vec{x})$, and *function invocation* (or *call*, or *application*, or "use") —what we call a *term*, $f(\vec{x})$, in logic parlance— is well established. The definition part, in programming, uses various notations depending on the programming language and corresponds to writing a program that implements the function, just as we did with M here. This program finitely represents the (potentially infinite) function (table).

However, there is a double standard in notation when it comes to relations. Extensionally, a relation R is a table (i.e., set) of n-tuples. Its counterpart in formal logic is a formula. But where in formal logic we rather infrequently write a formula A as $A[x]$ —doing so only if we want to draw attention to our interest in its (free) variable x— in the metatheory we most frequently write a relation R as $R(\vec{x}_n)$, *without* employing λ notation, to draw attention to its "input slots", which here are $x_1, \ldots, x_n$ (i.e., its "free variables").

Since stating "$R(\vec{a}_n)$", by convention, is short for stating "$\vec{a}_n \in R$", we have two notations for a relation: *logical* or *relational*, on one hand, i.e., $R(\vec{x}_n)$, and, on the other hand, *set-theoretic*, i.e., $\vec{x}_n \in R$. Both notations have no benefit of λ notation.

There are exceptions to this practice, for example, when we define one relation from another one via the process of "freezing" some of the original relation's inputs, or adding don't care inputs. For example, writing $x < y$ (the standard "less than" on $\mathbb{N}$) means that *both* x and y are meant to be the inputs; we have a table of ordered pairs. However, we may write $\lambda x.x < y$ to convey that y is fixed and that the input is just x. Clearly, a different relation arises for each y; we have an infinite family of tables: For $y = 0$ we have the empty table; for $y = 1$ one that contains just 0; for $y = 2$ one that contains just 0, 1; etc.

Similarly, $\lambda xz.x < y$ creates a relation of two inputs x and z, for each y. For each y, it is the set of all pairs (x, z) such that $x < y$. These pairs are infinitely many for each $0 \leq x < y$. □

1.1.9 Definition. A function $f : \mathbb{N}^n \to \mathbb{N}$, that is, one that takes its inputs from $\mathbb{N}$ and its outputs lie in $\mathbb{N}$, is called a *number-theoretic function*. □

1.1.10 Example. Let M be the program

$$1 : \mathbf{x} \leftarrow \mathbf{x} \doteq 1$$

$$2 : \mathbf{stop}$$

Then M_x^x is the function $\lambda x.x \dotdiv 1$, the *predecessor* function. The operation $\dotdiv$ is called "proper subtraction" and is in general defined by

$$x \dotdiv y = \begin{cases} x - y & \text{if } x \geq y \\ 0 & \text{otherwise} \end{cases}$$

The definition of $\dotdiv$ ensures that $\lambda xy.x \dotdiv y$ is a *number-theoretic* function. $\square$

Pause. Why are we restricting computability theory to number-theoretic functions? Surely, in practice we can compute with negative numbers, rational numbers, and with nonnumerical entities, such as graphs, trees, etc. Theory ought to reflect, and explain, our practices, no?◄

It does. Negative numbers and rational numbers can be coded by natural number pairs. Computability of number-theoretic functions can handle such pairing (and unpairing or decoding). Moreover, finite objects such as graphs, trees, and the like that we manipulate via computers can be also coded (and decoded) by natural numbers. After all, the internal representation of all data in computers is, at the hardware level, via natural numbers represented in binary notation.

1.1.11 Example. Let M be the program

$$1 : \mathbf{x} \leftarrow 0$$

$$2 : \mathbf{stop}$$

Then M_x^x is the function $\lambda x.0$, the *zero function* that we will denote by Z in this volume. $\square$

In Definition 1.1.5 we spoke of partial computable and total computable functions. We retain the qualifiers *partial* and *total* for all number-theoretic functions, even for those that might *not* be computable. Indeed, *total* vs. *nontotal* (no hyphen) has been defined with respect to some assumed *left field* for any function. Recall that for a function $f : A \to B$, A is the left and B is the right fields (Definition 0.4.1). Our number-theoretic functions all have right field $\mathbb{N}$. Their left fields on a case-by-case manner, are $\mathbb{N}^n$ for some $n \geq 1$.

The set union of all total *and* nontotal number-theoretic functions is the set of all *partial (number-theoretic) functions.* Thus *partial* is *not* synonymous with *nontotal*.

1.1.12 Example. The *unconditional* **goto** instruction, namely, "$L :$ **goto** L'", can be simulated by $L :$ **if x** $= 0$ **goto** L' **else goto** L'. $\square$

1.1.13 Example. Let M be the program

$$1 : \mathbf{x} \leftarrow 0$$

$$2 : \mathbf{goto} \ 1$$

$$3 : \mathbf{stop}$$

Then $M_{\mathbf{x}}^{\mathbf{x}}$ is the empty function $\emptyset$, sometimes written as $\lambda x.\uparrow$, which is a slight abuse of notation as $\uparrow$ is not an object or number, rather is an *attribute*: *undefined*.[9]

Thus the empty function is partial computable but not total. We have just established $\emptyset \in \mathcal{P} - \mathcal{R}$, *hence* $\mathcal{R} \subsetneq \mathcal{P}$. $\square$

1.1.14 Example.   Let M be the program segment

$$k - 1 : \mathbf{x} \leftarrow 0$$

$$k \quad : \textbf{if } \mathbf{z} = 0 \textbf{ goto } k + 4 \textbf{ else goto } k + 1$$

$$k + 1 : \mathbf{z} \leftarrow \mathbf{z} \,\dot{-}\, 1$$

$$k + 2 : \mathbf{x} \leftarrow \mathbf{x} + 1$$

$$k + 3 : \textbf{goto } k$$

$$k + 4 : \ldots$$

What it does, by the time the computation reaches instruction $k + 4$, is to have set the value of $\mathbf{z}$ to 0, and to make the value of $\mathbf{x}$ equal to the value that $\mathbf{z}$ had when instruction $k - 1$ was current. In short, the above sequence of instructions simulates the following sequence

$$L : \mathbf{x} \leftarrow \mathbf{z}$$

$$L + 1 : \mathbf{z} \leftarrow 0$$

$$L + 2 : \ldots$$

where the semantics of $L : \mathbf{x} \leftarrow \mathbf{z}$ is the standard one from everyday programming: It requires that upon execution of the instruction the value of $\mathbf{z}$ is copied into $\mathbf{x}$, but the value of $\mathbf{z}$ remains unchanged. $\square$

1.1.15 Exercise.   Write a program segment that simulates faithfully $L : \mathbf{x} \leftarrow \mathbf{z}$; i.e., copy the value of $\mathbf{z}$ into $\mathbf{x}$ *without* causing $\mathbf{z}$ to change as a side-effect. $\square$

Because of the above, without loss of generality, one may assume that any input variable, $\mathbf{x}$, of a program M is *read-only*. This means that its value remains invariant throughout any computation of the program. Indeed, if $\mathbf{x}$ is not so, a new input variable, $\mathbf{x}'$, can be introduced as follows, to relieve $\mathbf{x}$ from its input role: Add at the very beginning of M the (macro) instruction $1 : \mathbf{x} \leftarrow \mathbf{x}'$ of Exercise 1.1.15, where $\mathbf{x}'$ is a variable that does not occur in M. Adjust all the following labels consistently, including, of course, the ones referenced by if-statements —a tedious

---

[9] In Definition 0.4.1 we introduced the notations "$f(\vec{x}) \downarrow$" as synonymous to $\vec{x} \in \text{dom}(f)$ and "$f(\vec{x}) \uparrow$" as synonymous to $\vec{x} \notin \text{dom}(f)$, where $\text{dom}(f)$ is the *domain* of f.

but straightforward task. Call M' the so-obtained URM. Clearly, $M'^{\mathbf{x}',\mathbf{y}_1,\dots,\mathbf{y}_n}_{\mathbf{z}} = M^{\mathbf{x},\mathbf{y}_1,\dots,\mathbf{y}_n}_{\mathbf{z}}$, and M' does not change $\mathbf{x}'$.[10]

1.1.16 Example. (Composing Computable Functions)

Suppose that $\lambda x\vec{y}.f(x,\vec{y})$ and $\lambda \vec{z}.g(\vec{z})$ are partial computable, and say $f = F_{\mathbf{u}}^{\mathbf{x},\vec{\mathbf{y}}}$ while $g = G_{\mathbf{x}}^{\vec{\mathbf{z}}}$.

Since we can rewrite any program, renaming its variables at will, we assume *without loss of generality* that $\mathbf{x}$ *is the only variable common* to F and G.

We program as follows:

1. Concatenate the programs G and F in that order —using ¶ as "glue".
2. Remove the last instruction of G (it is k : **stop**, for some k) —call the program segment that results from this deletion G'.
3. Renumber the instructions of F as $k, k+1, \dots$ (and, as a side-effect, modify the references that if-statements of F make —adding $k-1$ to each such reference) in order to give to

$$G'$$
$$F$$

the correct program structure and the intended semantics.

Then,

$$\lambda \vec{y}\vec{z}.f(g(\vec{z}),\vec{y}) = \left(\frac{G'}{F}\right)_{\mathbf{u}}^{\vec{\mathbf{y}},\vec{\mathbf{z}}}$$

Note that *all non-input variables of F will still hold 0* as soon as the execution of $G'\P F$ makes the first instruction of F current *for the first time*. This is because none of these variables can be changed by G' under our assumption, thus ensuring that F *works as designed.*   □

Thus, we have, by repeating the above a finite number of times:

1.1.17 Proposition. *If* $\lambda\vec{y}_n.f(\vec{y}_n)$ *and* $\lambda\vec{z}.g_i(\vec{z})$, *for* $i = 1,\dots,n$, *are partial computable, then so is* $\lambda\vec{z}.f(g_1(\vec{z}),\dots,g_n(\vec{z}))$.

For the record, we will define *composition* to mean the somewhat rigidly defined operation used in Proposition 1.1.17, that is:

---

[10] Within the macro, $\mathbf{x}'$ does change, cf. Exercise 1.1.15. However *after* exiting the macro, $\mathbf{x}'$ is nowhere referenced *and* holds its original value. This is what is meant by "without loss of generality, one may assume that any input variable, $\mathbf{x}'$, of a program M is *read-only*. The value of $\mathbf{x}'$ remains invariant throughout any computation of the program".

1.1.18 Definition. Given any partial functions (*computable or not*) $\lambda \vec{y}_n . f(\vec{y}_n)$ and $\lambda \vec{z} . g_i(\vec{z})$, for $i = 1, \ldots, n$, we say that $\lambda \vec{z} . f(g_1(\vec{z}), \ldots, g_n(\vec{z}))$ is the result of their *composition*. □

We characterised the definition as "rigid". Indeed, note that it requires

- *All* the arguments y_j of f *must* be substituted by some $g_j(\ldots)$ —unlike Example 1.1.16, where we *substituted* a function invocation (cf. terminology in Definition 1.1.8) into x *only* in f there, and substituted nothing into any of the variables $\vec{y}$.
- For *all calls* $g_i(\ldots)$ substituted into the y_i of f, their argument list, "...", *must be the same*, say $\vec{z}$.

As we will show in examples later, this rigidity is only apparent.

We can rephrase Proposition 1.1.17, saying simply that

1.1.19 Theorem. $\mathcal{P}$ is closed under *composition*.

1.1.20 Corollary. $\mathcal{R}$ is closed under composition.

Proof Let f, g_i be in $\mathcal{R}$.

Then they are in $\mathcal{P}$, hence so is $h = \lambda \vec{y} . f\left(g_1(\vec{y}), \ldots, g_m(\vec{y})\right)$ by Theorem 1.1.19.

By assumption, the f, g_i are total. So, for any $\vec{y}$, we have $g_i(\vec{y}) \downarrow$ —a number. Hence also $f\left(g_1(\vec{y}), \ldots, g_m(\vec{y})\right) \downarrow$. That is, h is total, hence, being in $\mathcal{P}$, it is also in $\mathcal{R}$. □

Composing a number of times that *depends on the value of an input variable* —or as we may say, a variable number of times— is *iteration*. The general case of iteration is called *primitive recursion*.

1.1.21 Definition. (Primitive Recursion) A number-theoretic function f is defined by *primitive recursion* from given functions $\lambda \vec{y} . h(\vec{y})$ and $\lambda x \vec{y} z . g(x, \vec{y}, z)$ provided, *for all x and $\vec{y}$*, its values are given by the two equations below:

$$f(0, \vec{y}) = h(\vec{y})$$
$$f(x + 1, \vec{y}) = g(x, \vec{y}, f(x, \vec{y}))$$

h is the *basis function*, while g is the *iterator*. A *unique* f that satisfies the above schema *exists*, because of Theorem 0.4.43 and Example 0.4.41. Moreover, if both h and g are total, then so is f as it can easily be shown by induction on x.

It will be useful to use the notation $f = prim(h, g)$ to indicate in shorthand that f is defined as above from h and g (note the order). □

Note that $f(1, \vec{y}) = g(0, \vec{y}, h(\vec{y}))$, $f(2, \vec{y}) = g(1, \vec{y}, g(0, \vec{y}, h(\vec{y})))$, $f(3, \vec{y}) = g(2, \vec{y}, g(1, \vec{y}, g(0, \vec{y}, h(\vec{y}))))$, etc. Thus the "$x$-value", 0, 1, 2, 3, etc., equals the number of times we compose g with itself (i.e., the number of times we iterate g).

1.1.22 Example. (Iterating Computable Functions)

Suppose that $\lambda x \vec{y} z.g(x, \vec{y}, z)$ and $\lambda \vec{y}.h(\vec{y})$ are partial computable, and, say, $g = G_{\mathbf{z}}^{\mathbf{i}, \vec{\mathbf{y}}, \mathbf{z}}$ while $h = H_{\mathbf{z}}^{\vec{\mathbf{y}}}$.

By earlier remarks we may assume:

(i)   The only variables that H and G have in common are $\mathbf{z}, \vec{\mathbf{y}}$.
(ii) The variables $\vec{\mathbf{y}}$ are read-only in both H and G.
(iii) $\mathbf{i}$ is read-only in G.
(iv) $\mathbf{x}$ does not occur in any of H or G.

We can now argue that the following program, let us call it F, computes f defined as in Definition 1.1.21 from h and g, where $\boxed{H'}_{\mathbf{z}}^{\vec{\mathbf{y}}}$, with input/output variables shown, is program H with the **stop** instruction removed, and $\boxed{G'}_{\mathbf{z}}^{\mathbf{i}, \vec{\mathbf{y}}, \mathbf{z}}$ is program G that has the **stop** instruction removed, and instructions are renumbered (and if-statements are adjusted) as needed:

$$\boxed{H'}_{\mathbf{z}}^{\vec{\mathbf{y}}} \qquad\qquad \text{/* This is } \mathbf{z} \leftarrow h(\vec{\mathbf{y}}) \text{ */}$$

$r:$ $\mathbf{i} \leftarrow 0$

$r+1:$ **if** $\mathbf{x} = 0$ **goto** $k+m+2$ **else goto** $r+2$

$r+2:$ $\mathbf{x} \leftarrow \mathbf{x} \mathbin{\dot-} 1$

$$\boxed{G'}_{\mathbf{z}}^{\mathbf{i}, \vec{\mathbf{y}}, \mathbf{z}} \qquad\qquad \text{/* This is } \mathbf{z} \leftarrow g(\mathbf{i}, \vec{\mathbf{y}}, \mathbf{z}) \text{ */}$$

$k:$ $\mathbf{i} \leftarrow \mathbf{i} + 1$

$k+1:$ $\mathbf{w}_1 \leftarrow 0$

     $\vdots$ $\vdots$

$k+m:$ $\mathbf{w}_m \leftarrow 0$

$k+m+1:$ **goto** $r+1$

$k+m+2:$ **stop**

The instructions $\mathbf{w}_i \leftarrow 0$ set explicitly to zero all the variables of G' other than $\mathbf{i}, \mathbf{z}, \vec{\mathbf{y}}$ to ensure correct behaviour of G'. Note that the $\mathbf{w}_i$ are *implicitly* initialised to zero *only* the first time G' is executed. Clearly, $f = F_{\mathbf{z}}^{\mathbf{x}, \vec{\mathbf{y}}}$.   □

We have at once:

1.1.23 Proposition. *If f, g, h relate as in Definition 1.1.21 and h and g are in $\mathcal{P}$, then so is f. We say that $\mathcal{P}$ is* closed under primitive recursion.

1.1.24 Example. (Unbounded Search) Suppose that $\lambda x \vec{y}.g(x, \vec{y})$ is partial computable, and, say, $g = G_{\mathbf{z}}^{\mathbf{x}, \vec{\mathbf{y}}}$. By earlier remarks we may assume that $\vec{\mathbf{y}}$ and $\mathbf{x}$ are read-only in G and that $\mathbf{z}$ is *not* one of them.

Consider the following program F, where $\boxed{G'}_{\mathbf{z}}^{\mathbf{x}, \vec{\mathbf{y}}}$ is program G with the **stop** instruction removed, and instructions have been renumbered (and if-statements adjusted) as needed so that its first instruction has label 2. Input output variables are indicated.

$$
\begin{array}{ll}
1: & \mathbf{x} \leftarrow 0 \\
& \boxed{G'}_{\,\mathbf{z}}^{\mathbf{x},\vec{\mathbf{y}}} \quad \text{/* This is } \mathbf{z} \leftarrow g(\mathbf{x},\vec{\mathbf{y}}) \text{ */} \\
k: & \text{if } \mathbf{z} = 0 \text{ goto } k+l+3 \text{ else goto } k+1 \\
k+1: & \mathbf{w}_1 \leftarrow 0 \text{ /* Setting all non-input variables to 0; cf. Example 1.1.22 */} \\
\vdots & \vdots \\
k+l: & \mathbf{w}_l \leftarrow 0 \text{ /* Setting all non-input variables to 0; cf. Example 1.1.22 */} \\
k+l+1: & \mathbf{x} \leftarrow \mathbf{x}+1 \\
k+l+2: & \text{goto } 2 \\
k+l+3: & \text{stop}
\end{array}
$$

Let us set $f = F_{\mathbf{x}}^{\vec{\mathbf{y}}}$. Note that, for any $\vec{a}$, $f(\vec{a}) \downarrow$ precisely if the URM F, initialised with $\vec{a}$ as the input values in $\vec{\mathbf{y}}$, ever reaches **stop**. This condition becomes true as long as *both* conditions, (1) and (2) below, are fulfilled:

(1) Instruction k *just found* that $\mathbf{z}$ holds 0. This value of $\mathbf{z}$ is the result of an execution of G (i.e., G' with the **stop** instruction added) with input values $\vec{a}$ in $\vec{\mathbf{y}}$ and, say, b in $\mathbf{x}$ —that is, $g(b,\vec{a}) = 0$ *for the first time*.

(2) In none of the previous iterations (with $\mathbf{x}$-value $< b$) G' (essentially, G) got into a non-ending computation (*infinite loop*) —else we would never reach **stop**. Thus, for all $x < b$, it is $g(x,\vec{a}) > 0$, which is equivalent to "for all $x < b$, it is $g(x,\vec{a}) \downarrow$" by (1).

Correspondingly, the computation of F will never halt for an input $\vec{a}$ iff

- Either G' loops for ever in some iteration,
  or
- It halts in every iteration of $\mathbf{x}$-value b, *but* no b makes $g(b,\vec{a}) = 0$ for the chosen $\vec{a}$.

Thus, for all $\vec{a}$,

$$
f(\vec{a}) = \begin{cases} \min\left\{ x : g(x,\vec{a}) = 0 \wedge (\forall y)\big(y < x \rightarrow g(y,\vec{a}) \downarrow\big) \right\} \\ \uparrow \quad \text{if min does not exist} \end{cases} \tag{μ}
$$

1.1.25 Definition. The operation on partial functions g given for all $\vec{a}$ by (μ) above is called *unbounded search* (along the variable x) and is denoted by the symbol $(\mu x)g(x,\vec{a})$. The function $f = \lambda \vec{y}.(\mu x)g(x,\vec{y})$ is defined precisely when the minimum exists.

Sometimes in the literature one encounters the terminology "f is defined from g by μ-*recursion*", meaning $f = \lambda \vec{x}.(\mu y)g(y,\vec{x})$.   □

The result of Example 1.1.24 yields at once:

1.1.26 Proposition. $\mathcal{P}$ *is closed under unbounded search; that is, if* $\lambda x \vec{y}.g(x,\vec{y})$ *is in* $\mathcal{P}$, *then so is* $\lambda \vec{y}.(\mu x)g(x,\vec{y})$.

Why "unbounded" search? Because we do not know *a priori* how many times we have to go around the loop. This depends on the behaviour of g.

1.1.27 Example. Is the function $\lambda \vec{x}_n.x_i$, where $1 \leq i \leq n$, in $\mathcal{P}$? Yes, and here is a program, M, for it:

$$
\begin{array}{lll}
1: & \mathbf{w}_1 \leftarrow 0 & \\
\vdots & \vdots & \\
i: & \mathbf{z} \leftarrow \mathbf{w}_i \ \{\textbf{Comment}.\ \text{Cf. Exercise 1.1.15}\} \\
\vdots & \vdots & \\
n: & \mathbf{w}_n \leftarrow 0 & \\
n+1: & \textbf{stop} &
\end{array}
$$

$\lambda \vec{x}_n.x_i = M_{\mathbf{z}}^{\vec{\mathbf{w}}_n}$. To ensure that M indeed *has* the $\mathbf{w}_i$ as variables we reference them in instructions at least once, in any manner whatsoever. We standardly denote $\lambda \vec{x}_n.x_i$ by U_i^n. We call the U_i^n functions the generalised identities. $\square$

Before we get more immersed into *partial functions* recall Definition 0.4.5.

The referred to definition is unnecessary if we work with total functions. Note also that when we write $f = g$ we state *set equality* and thus "$=$" is the correct symbol regardless of whether f, g are total or not.

1.1.28 Lemma. *If $f = prim(h, g)$ and h and g are* total, *then so is f.*

Proof Let f be given by:

$$
f(0, \vec{y}) = h(\vec{y})
$$
$$
f(x+1, \vec{y}) = g(x, \vec{y}, f(x, \vec{y}))
$$

We do induction on x to prove

$$\text{"For all } x, \vec{y}, \text{ it is } f(x, \vec{y}) \downarrow \text{"} \tag{*}$$

Basis. $x = 0$:   Well, $f(0, \vec{y}) = h(\vec{y})$, but $h(\vec{y}) \downarrow$ for all $\vec{y}$, so

$$f(0, \vec{y}) \downarrow \text{ for all } \vec{y} \tag{**}$$

As Induction *Hypothesis* (I.H.) take that

$$f(x, \vec{y}) \downarrow \text{ for all } \vec{y} \text{ and fixed } x \tag{†}$$

Do the Induction *Step* (I.S.) to show

$$f(x+1, \vec{y}) \downarrow \text{ for all } \vec{y} \text{ and the fixed } x \text{ of (†)} \tag{‡}$$

Well, by (†) and the assumption on g,

$$g\big(x, \vec{y}, f(x, \vec{y})\big) \downarrow, \text{ for all } \vec{y} \text{ and the fixed } x \text{ of (†)}$$

which says the same thing as (‡). Having proved the latter and the Basis, (∗∗), we have proved (∗) by induction on x.   □

1.1.29 Corollary. $\mathcal{R}$ *is closed under primitive recursion.*

Proof Let h and g be in $\mathcal{R}$. Then they are in $\mathcal{P}$. But then $prim(h, g) \in \mathcal{P}$ as we showed in Proposition 1.1.23. By Lemma 1.1.28, $prim(h, g)$ is total. By definition of $\mathcal{R}$, as the subset of $\mathcal{P}$ that contains all total functions of $\mathcal{P}$, we have $prim(h, g) \in \mathcal{R}$.   □

Why all this dance to get $prim(h, g)$ in $\mathcal{R}$? Because to prove $f \in \mathcal{R}$ you need *Two* things: That

1. $f \in \mathcal{P}$ *And*
2. f is total

    But aren't all the total functions in $\mathcal{R}$ anyway?

    No! They need to be computable too! I.e., of the form $M_{\vec{y}}^{\vec{x}_n}$ for some URM M.

    Aren't *all* total functions computable?

    No! See next section, and heed the last sentence in the last *-remark there!*

1.2   A Digression Regarding $\mathcal{R}$

It is an easy to believe (and verify) fact[11] that

> *We can computationally test if a string over the URM alphabet is a syntactically correct URM or not.*

But then we can enumerate, and even do so computationally, all URMs as follows:

First, fix an alphabetical order in the finite URM alphabet. Then run the following process *with no end*. The process lists the members of *two* lists in parallel: *List1* contains all strings over the URM alphabet, while *List2* contains all strings that are URMs.

---

[11] Via a pseudo program, or indeed one written in C or JAVA.

1. Enumerate into *List1* all strings over the URM alphabet, by increasing length, and in each length group lexicographically. *Incidentally, this enumeration can be done via a program.*
2. *For each string generated* above **do**: *Test* it whether it is a URM or not. If *not*, **Goto** 1.
   If *yes*, then copy the string into *List2* and then **Goto** 1.

You can now enumerate all partial *recursive functions of one variable!*
Simply, build a *List3* along with *List2*:

For each URM M added to *List2*, enumerate all $M_{\mathbf{y}}^{\mathbf{x}}$ for *all* pairs of variables $(\mathbf{x}, \mathbf{y})$ found in M —this is a finite number of $M_{\mathbf{y}}^{\mathbf{x}}$— and place in *List3*. Do so lexicographically with respect to the pairs $(\mathbf{x}, \mathbf{y})$ for each M, recalling that $\mathbf{x}$ is really a string of the form $X1^n$, for $n \geq 0$. Examples of $(\mathbf{x}, \mathbf{y})$ pairs: "$(X, X11)$", "$(X11, X11), (X11111111, X111)$". *This subsidiary enumeration, for each M, also can be done algorithmically!*

 So, the set of all $\mathcal{P}$ functions of one variable is countable.

The above sentence says less than what we proved in outline: We actually proved that the enumeration can be effected by, say, a C program since all steps can be done computationally, intuitively speaking.

Now, as $\mathcal{R} \subsetneq \mathcal{P}$, we conclude that, *in Cantor's sense*,

$$\text{the set of all } \mathcal{R} \text{ functions of one variable is countable.} \qquad (\dagger)$$

Indeed, in the enumeration of the $\mathcal{P}$ functions of one variable just *omit* the non total ones! Cf. Exercise 0.8.27.

 We saw that the set of *all* total functions of one variable, let's call it $\mathcal{T}^{(1)}$, is *uncountable* (Theorem 0.5.19).

Thus, $\mathcal{R} \subsetneq \mathcal{T}^{(1)}$ since a set cannot be both countable and uncountable.

That is, there are total functions f that are not URM-computable; these f are in $\mathcal{T}^{(1)} - \mathcal{R}$.

This is why we always warn: When one wants to prove that $f \in \mathcal{R}$ one must do *two* things! One is that f is computable (is in $\mathcal{P}$). *Never take this for granted!*

 We proved a "weak" unsolvability result here, that total uncomputable functions exist. Why weak? Because this argument does not illuminate what any of these functions may look like. We will do much better in the following sections and identify meaningful, specific functions that are uncomputable.

I conclude this remark by showing that "most" —in Cantor's sense— total functions are uncomputable. That is, if we let $A = \mathcal{T}^{(1)} - \mathcal{R}$, then A is uncountable. Why? Because $\mathcal{T}^{(1)} = A \cup \mathcal{R}$.

Can A be countable? If so, then so is $A \cup \mathcal{R}$ (Remark 0.5.12), contradicting that $\mathcal{T}^{(1)}$ is uncountable.

1.3 Exercises

1. Show that the URM can simulate the instruction

$$\begin{array}{|l}
\textbf{repeat } Y \textbf{ times} \\
\quad \vdots \\
\textbf{end}
\end{array}$$

2. Show that the URM can simulate the instruction below, however equipped with the nonstandard semantics: The number of iterations depends *only* on the value Y had immediately before entering the loop. That is, changing the Y-value *inside* the loop —for example, doing $Y \leftarrow Y + 1$— does *not* affect the number of iterations.

$$\begin{array}{|l}
\textbf{repeat } Y \textbf{ times} \\
\quad \vdots \\
\textbf{end}
\end{array}$$

3. Show that the URM can simulate the instruction

$$\begin{array}{|l}
\textbf{repeat } Y \textbf{ times while } (w \neq 0) \\
\quad \vdots \\
\textbf{end}
\end{array}$$

Chapter 2
Primitive Recursive Functions

Overview

We saw that the successor, zero, and the *generalised identity* functions —which we name S, Z and U_i^n respectively— are in $\mathcal{P}$; thus, not only are they "intuitively computable", but they are so *in a precise mathematical sense*: each is computable by a URM.

We have also shown that "computability" of functions is *preserved* by the operations of *composition*, *primitive recursion*, and *unbounded search*.

The *primitive recursive functions* are an important tool in computability theory, providing coding tools, and insights in every-day programming (with FORTRAN-like loops). These we introduce in this chapter and we will come back to them very often. They are the tools Gödel used in proving his Incompleteness theorems.

Most people introduce the primitive recursive functions —as Dedekind did— via *derivations*, just as one introduces the theorems of logic via proofs, as we will detail in the definition below.

2.1 Definitions and Basic Results

2.1.1 Definition. ($\mathcal{PR}$-**Derivations**; $\mathcal{PR}$-**Functions**) The set

$$\mathcal{I} = \left\{ S, Z, \left(U_i^n \right)_{n \geq i > 0} \right\}$$

is the set of *Initial $\mathcal{PR}$* functions.

A $\mathcal{PR}$-*derivation* is a finite (ordered!) sequence of number-theoretic functions[1]

---

[1] That is, *left field* is $\mathbb{N}^n$ for some $n > 0$, and *right field* is $\mathbb{N}$.

© Springer Nature Switzerland AG 2022

G. Tourlakis, *Computability*, https://doi.org/10.1007/978-3-030-83202-5_2

$$f_1, f_2, f_3, \ldots, f_i, \ldots, f_n \qquad\qquad (1)$$

such that, for *each i*, one of the following holds

1. $f_i \in \mathcal{I}$.
2. $f_i = prim(f_j, f_k)$ and $j < i$ and $k < i$ —that is, f_j, f_k appear *to the left of* f_i.
3. $f_i = \lambda \vec{y}.g\big(r_1(\vec{y}), r_2(\vec{y}), \ldots, r_m(\vec{y})\big)$, and *all* of the $\lambda \vec{y}.r_q(\vec{y})$ and $\lambda \vec{x}_m.g(\vec{x}_m)$ appear *to the left of* f_i in the sequence.

   Any f_i in a derivation is called a $\mathcal{PR}$-derived function.

 In this chapter and until we introduce the $\mathcal{P}$-derivations in Sect. 5.6 we will omit the "$\mathcal{PR}$-" prefix attached to the terms "derived" and "derivation".

   The set of primitive recursive functions, $\mathcal{PR}$, is *all those that are derived.* That is,

$$\mathcal{PR} \overset{Def}{=} \{f : f \text{ is derived}\} \qquad\qquad \square$$

   The above definition defines essentially what Dedekind called "recursive" functions. Subsequently they were renamed *primitive recursive* allowing the unqualified term *recursive* to be synonymous with (total) *computable* and apply to the functions of $\mathcal{R}$.

 We know from Sect. 0.6.1 that the definition by derivations is equivalent to the *inductive definition*: "$\mathcal{PR}$ is defined to be the closure of $\mathcal{I}$ under the two operations, primitive recursion and composition." It is interesting that Dedekind (1888) did not use the inductive definition, nor did the very complete more recent book on primitive recursion by Péter (1967).

2.1.2 Lemma. *The concatenation of two derivations is a derivation.*

Proof Proposition 0.6.9. $\qquad\qquad\qquad\qquad\qquad\qquad\qquad\qquad\qquad\qquad\qquad\qquad\qquad$ $\square$

2.1.3 Corollary. *The concatenation of any finite number of derivations is a derivation.*

2.1.4 Lemma. *If*

$$f_1, f_2, f_3, \ldots, f_k, f_{k+1}, \ldots, f_n$$

is a derivation, then so is $f_1, f_2, f_3, \ldots, f_k$.

Proof In $f_1, f_2, f_3, \ldots, f_k$, every f_m, for $1 \le m \le k$, satisfies 1.–3. of Definition 2.1.1 since all conditions are in terms of what f_m *is*, or what lies *to the left of* f_m. Chopping the "tail" $f_{k+1}, \ldots, f_n$ in no way affects what lies to the left of f_m, for $1 \le m \le k$. $\qquad\qquad\qquad\qquad\qquad\qquad\qquad\qquad$ $\square$

2.1.5 Corollary. $f \in \mathcal{PR}$ *iff f appears at the end of some derivation.*

Proof

(a) The *If.* Say $g_1, \ldots, g_n, \boxed{f}$ is a derivation. Since f occurs in it, $f \in \mathcal{PR}$ by
   Definition 2.1.1.
(b) The *Only If.* Say $f \in \mathcal{PR}$. Then, by Definition 2.1.1,

$$g_1, \ldots, g_m, \boxed{f}, g_{m+2}, \ldots, g_r \tag{1}$$

for some derivation like the (1) above.
   By Lemma 2.1.4, $g_1, \ldots, g_m, \boxed{f}$ is also a derivation.   □

2.1.6 Theorem. $\mathcal{PR}$ *is closed under* composition and primitive recursion.

Proof

• Closure under *primitive recursion*. So let $\lambda \vec{y}.h(\vec{y})$ and $\lambda x \vec{y} z.g(x, \vec{y}, z)$ be in $\mathcal{PR}$.
  Thus we have derivations

$$h_1, h_2, h_3, \ldots, h_n, \boxed{h} \tag{1}$$

and

$$g_1, g_2, g_3, \ldots, g_m, \boxed{g} \tag{2}$$

Then the following is a derivation by Proposition 0.6.9.

$$h_1, h_2, h_3, \ldots, h_n, \boxed{h}, g_1, g_2, g_3, \ldots, g_m, \boxed{g}$$

Therefore so is

$$h_1, h_2, h_3, \ldots, h_n, \boxed{h}, g_1, g_2, g_3, \ldots, g_m, \boxed{g}, prim(h, g)$$

by applying step 2 of Definition 2.1.1.
   This implies $prim(h, g) \in \mathcal{PR}$ by Definition 2.1.1.
• Closure under *composition*. So let $\lambda \vec{y}.h(\vec{x}_n)$ and $\lambda \vec{y}.g_i(\vec{y})$, for $1 \leq i \leq n$, be in
  $\mathcal{PR}$. By Definition 2.1.1 we have derivations

$$\boxed{\ldots, \boxed{h}} \tag{3}$$

and

$$\boxed{\ldots, \boxed{g_i}}, \text{ for } 1 \leq i \leq n \tag{4}$$

By Corollary 2.1.3,

$$\ldots,\boxed{h}\Big|,\Big|\ldots,\boxed{g_1}\Big|,\ldots,\Big|\ldots,\boxed{g_n}$$

is a derivation, and by Definition 2.1.1, case 3, so is

$$\Big|\ldots,\boxed{h}\Big|,\Big|\ldots,\boxed{g_1}\Big|,\ldots,\Big|\ldots,\boxed{g_n}\Big|,\lambda\vec{y}.h(g_1(\vec{y}),\ldots,g_n(\vec{y}))$$

This implies $\lambda\vec{y}.h(g_1(\vec{y}),\ldots,g_n(\vec{y})) \in \mathcal{PR}$ by Definition 2.1.1.   □

2.1.7 Remark. How do you prove that some $f \in \mathcal{PR}$?

Answer By building a derivation

$$g_1,\ldots,g_m,\boxed{f}$$

After a while it becomes *easier* because you might *know* an h and g in $\mathcal{PR}$ such that $f = prim(h, g)$, or you might know some $g, h_1, \ldots, h_m$ in $\mathcal{PR}$, such that $f = \lambda\vec{y}.g(h_1(\vec{y}),\ldots,h_m(\vec{y}))$. *If so, just apply Theorem 2.1.6.*

How do you prove that *all* $f \in \mathcal{PR}$ have a property Q —that is, for all f, $Q(f)$ is true?

Answer By doing *induction on the derivation length* of f.   □

Here are two examples of the above questions and their answers.

2.1.8 Example.

(1) To demonstrate the first Answer above (Remark 2.1.7), show (prove) that $\lambda xy.x + y \in \mathcal{PR}$. Well, observe that

$$0 + y = y$$
$$(x + 1) + y = (x + y) + 1$$

Does the above look like a primitive recursion? Well, almost. However, the first equation should have a function call "$H(y)$" on the right hand side but instead has just y —the input! Also the second equation should have a right hand side like "$G(x, y, x + y)$". We can do that! Take $H = \lambda y.U_1^1(y)$ and $G = \lambda xyz.S\big(U_3^3(x, y, z)\big)$. Be sure to agree that

- H and G recast the two equations above in the correct *form*:

$$0 + y = U_1^1(y)$$
$$(x + 1) + y = SU_3^3\big(x, y, (x + y)\big)$$

- The functions U_1^1 (initial) and SU_3^3 (composition) are in $\mathcal{PR}$.

By Theorem 2.1.6 so is $\lambda xy.x + y$.

In terms of derivations, we have produced the derivation:

$$U_1^1, S, U_3^3, SU_3^3, \underbrace{prim\left(U_1^1, SU_3^3\right)}_{\lambda xy.x+y}$$

 Note the "SU_3^3" with no brackets around U_3^3. It is normal practise to omit brackets, as in $SSZx$, etc., to enhance readability. Of course, $SSZx$ means $S\Big(S\big(Z(x)\big)\Big)$.

(2) To demonstrate the second Answer above (Remark 2.1.7), show (prove) that every $f \in \mathcal{PR}$ is *total*. Induction on derivation length, n, where f occurs.

Basis. $n = 1$. Then f is the only function in the derivation. Thus it must be one of S, Z, or U_i^m. But all these are total.

I.H. (Induction Hypothesis) Assume that the claim is true for all f that occur in derivations of lengths $n \leq l$. That is, *we assume that all such f are total.*

I.S. (Induction Step) Prove that the claim is true for all f that occur in derivations of lengths $n = l + 1$.

$$g_1, \ldots, g_i, \boxed{f}, g_{i+2}, \ldots, g_{l+1} \tag{1}$$

- Case where f is not the last function in the derivation (1). Then dropping the tail $g_{i+2}, \ldots, g_{l+1}$ we have f appear in a derivation of length $\leq l$ and thus it is total by the I.H.

  The interesting case is when f is the last function of a derivation of length $l + 1$ as in (2) below:

$$g_1, \ldots, g_l, \boxed{f} \tag{2}$$

  We have three subcases:

  - $f \in \mathcal{I}$. But we argued this under *Basis*.
  - $f = prim(h, g)$, where h and g are among the $g_1, \ldots, g_l$. By the I.H. h and g are total. But then so is f by Lemma 1.1.28.
  - $f = \lambda \vec{y}.h\Big(q_1(\vec{y}), \ldots, q_t(\vec{y})\Big)$, where the functions h and $q_1, \ldots, q_t$ are among the $g_1, \ldots, g_l$. By the I.H., h and $q_1, \ldots, q_t$ are total. But then so is f by Corollary 1.1.20, when we proved that $\mathcal{R}$ is closed under composition.

$\square$

2.1.9 Example. If $\lambda xyw.f(x, y, w)$ and $\lambda z.g(z)$ are in $\mathcal{PR}$, how about $\lambda xzw.f(x, g(z), w)$? It is in $\mathcal{PR}$ since

$$\lambda xzw.f(x, g(z), w) = \lambda xzw.f(U_1^3(x, z, w), g(U_2^3(x, z, w)), U_3^3(x, z, w))$$

and the U_i^n are primitive recursive. The reader will see at once that to the right of "=" we have correctly formed compositions as expected by the "rigid" definition of composition given in Definition 1.1.18. In particular, $g(U_2^3(x, z, w))$ that is substituted into y in f is a legitimate composition, so by $\{g, U_2^3\} \subseteq \mathcal{PR}$, we have $gU_2^3 \in \mathcal{PR}$.[2] We have the following similar results, for the same function f above:

(1) $\lambda yw.f(2, y, w)$ is in $\mathcal{PR}$. Indeed, this function can be obtained by composition, since

$$\lambda yw.f(2, y, w) = \lambda yw.f\left(SSZ(U_1^2(y, w)), y, w\right)$$

Clearly, using $SSZ\left(U_2^2(y, w)\right)$ above works as well.

(2) $\lambda xyw.f(y, x, w)$ is in $\mathcal{PR}$. Indeed, this function can be obtained by composition, since

$$\lambda xyw.f(y, x, w) = \lambda xyw.f\left(U_2^3(x, y, w), U_1^3(x, y, w), U_3^3(x, y, w)\right)$$

In this connection, note that while $\lambda xy.g(x, y) = \lambda yx.g(y, x)$ —since the two expressions around "=" are the same, but just use different variable names throughout— yet $\lambda xy.g(x, y) \neq \lambda xy.g(y, x)$ in general. For example, $\lambda xy.x \dot{-} y$ asks that we subtract the second input (y) from the first (x), but $\lambda xy.y \dot{-} x$ asks that we subtract the first input (x) from the second (y).

(3) $\lambda xy.f(x, y, x)$ is in $\mathcal{PR}$. Indeed, this function can be obtained by composition, since

$$\lambda xy.f(x, y, x) = \lambda xy.f\left(U_1^2(x, y), U_2^2(x, y), U_1^2(x, y)\right)$$

(4) $\lambda xyzwu.f(x, y, w)$ is in $\mathcal{PR}$. Indeed, this function can be obtained by composition, since

$$\lambda xyzwu.f(x, y, w) =$$

$$\lambda xyzwu.f(U_1^5(x, y, z, w, u), U_2^5(x, y, z, w, u), U_4^5(x, y, z, w, u))$$

$\square$

The above examples are summarised, named, and generalised in the following straightforward exercise:

2.1.10 Exercise. (Grzegorczyk Substitution; Grzegorczyk (1953))
    $\mathcal{PR}$ is *closed under* the following operations:

---

[2] The reduced brackets notation "gU_2^3" was first encountered in Example 2.1.8 in a ⊘-delimited remark.

"closed under" means that if the "input function(s)" is in $\mathcal{PR}$, then so is the "output function".

(i) *Substitution of a function call for a variable*:
    From $\lambda \vec{x}\,y\vec{z}.f(\vec{x}, y, \vec{z})$ and $\lambda \vec{w}.g(\vec{w})$ obtain $\lambda \vec{x}\vec{w}\vec{z}.f(\vec{x}, g(\vec{w}), \vec{z})$.
(ii) *Substitution of a constant for a variable*:
    From $\lambda \vec{x}\,y\vec{z}.f(\vec{x}, y, \vec{z})$ obtain $\lambda \vec{x}\vec{z}.f(\vec{x}, k, \vec{z})$.
(iii) *Interchange of two variables*:
    From $\lambda \vec{x}\,y\vec{z}\,w\vec{u}.f(\vec{x}, y, \vec{z}, w, \vec{u})$ obtain $\lambda \vec{x}\,y\vec{z}\,w\vec{u}.f(\vec{x}, w, \vec{z}, y, \vec{u})$.
(iv) *Identification of two variables*:
    From $\lambda \vec{x}\,y\vec{z}\,w\vec{u}.f(\vec{x}, y, \vec{z}, w, \vec{u})$ obtain $\lambda \vec{x}\,y\vec{z}\vec{u}.f(\vec{x}, y, \vec{z}, y, \vec{u})$.
(v) *Introduction of "don't care" variables*:
    From $\lambda \vec{x}.f(\vec{x})$ obtain $\lambda \vec{x}\vec{z}.f(\vec{x})$.   □

The Grzegorczyk *substitutions* are also called Grzegorczyk *operations*.

By Exercise 2.1.10, composition can simulate the Grzegorczyk operations if the initial functions $\mathcal{I}$ are present. Of course, (i) alone can in turn simulate composition. With these comments out of the way, we see that the "rigidity" of the definition of composition is gone.

2.1.11 Example. The definition of primitive recursion is also rigid. However this also is an illusion.

Take $p(0) = 0$ and $p(x + 1) = x$ —this one defining $p = \lambda x.x \,\dot{-}\, 1$ does not fit the schema.

The schema requires the defined function to have one more variable than the basis, *so no one-variable function can be directly defined via primitive recursion*.

We can get around this.

Define first $\widetilde{p} = \lambda xy.x \,\dot{-}\, 1$ as follows: $\widetilde{p}(0, y) = 0$ and $\widetilde{p}(x + 1, y) = x$. Now this can be dressed up to fit the syntax of the primitive recursive schema,

$$\widetilde{p}(0, y) = Z(y)$$

$$\widetilde{p}(x + 1, y) = U_1^3(x, y, \widetilde{p}(x, y))$$

that is, $\widetilde{p} = prim(Z, U_1^3)$. Then we can get p by (Grzegorczyk) substitution: $p = \lambda x.\widetilde{p}(x, 0)$. Incidentally, this shows that both p and $\widetilde{p}$ are in $\mathcal{PR}$:

- $\widetilde{p} = prim(Z, U_1^3)$ is in $\mathcal{PR}$ since Z and U_1^3 are, then invoking Theorem 2.1.6.
- $p = \lambda x.\widetilde{p}(x, 0)$ is in $\mathcal{PR}$ since $\widetilde{p}$ is, then invoking Exercise 2.1.10.   □

2.1.12 Example. Another rigidity in the definition of primitive recursion is that, *apparently*, one can use only the first variable *as the iterating variable*.

Not so. This too is an illusion.

Consider, for example, $sub = \lambda xy.x \,\dot{-}\, y$. Clearly, $sub(x, 0) = x$ and $sub(x, y + 1) = p(sub(x, y))$ is correct semantically, but the *format* is wrong:

We are not supposed to iterate along the second variable! Well, define instead $\widetilde{sub} = \lambda xy.y \,\dot{-}\, x$, as follows:

$$\widetilde{sub}(0, y) = U_1^1(y)$$

$$\widetilde{sub}(x + 1, y) = p\big(U_3^3(x, y, \widetilde{sub}(x, y))\big)$$

Then, using variable swapping (Grzegorczyk operation (iii)), we can get sub: $sub = \lambda xy.\widetilde{sub}(y, x)$. Clearly, both $\widetilde{sub}$ and sub are in $\mathcal{PR}$, the first by Theorem 2.1.6, the second by Exercise 2.1.10.   □

2.1.13 Exercise. Prove that $\lambda xy.x \times y$ is primitive recursive. Of course, we will usually write multiplication $x \times y$ in "implied notation", xy.   □

2.1.14 Example. The very important "*switch*" (or "if-then-else") function $sw = \lambda xyz.$if $x = 0$ then y else z is primitive recursive. It is directly obtained by primitive recursion on initial functions: $sw(0, y, z) = y$ and $sw(x + 1, y, z) = z$.   □

2.1.15 Proposition. $\mathcal{PR} \subseteq \mathcal{R}$.

Proof Start with proving $\mathcal{PR} \subseteq \mathcal{P}$ and then use Example 2.1.8, part (2). Or, prove directly by induction on derivation length that $\mathcal{PR} \subseteq \mathcal{R}$.   □

 Indeed, the above inclusion is proper, as we will see later.

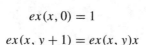 *2.1.16 Example.* Consider the function ex given by

$$ex(x, 0) = 1$$

$$ex(x, y + 1) = ex(x, y)x$$

Thus, if $x = 0$, then $ex(x, 0) = 1$, but $ex(x, y) = 0$ for all $y > 0$. On the other hand, if $x > 0$, then $ex(x, y) = x^y$ for all y.

Note that x^y is "mathematically" undefined when $x = y = 0$.[3] Thus, by Example 2.1.8, part (2), the exponential cannot be a primitive recursive function!

This is rather silly, since the computational process for the exponential is extremely straightforward; thus it is a shame to declare the function non-$\mathcal{PR}$. After all, we know *exactly where and how it is undefined* and we can remove this undefinability by *redefining "x^y" to mean $ex(x, y)$ for all inputs.*

Clearly $ex \in \mathcal{PR}$. In computability we do this kind of redefinition a lot in order to remove easily recognisable points of "non definition" of calculable functions. We will see further examples, such as the remainder, quotient, and logarithm functions.

Caution! We cannot always remove points of non definition of a calculable function and still obtain a computable function. That is, there are functions $f \in \mathcal{P}$ that *have no recursive extensions.* This we will show later.   □

---

[3] In first-year university calculus we learn that "0^0" is an "indeterminate form".

2.2 Primitive Recursive and Recursive Relations

Developing a theory of recursive and primitive recursive *relations* gives us a lot of added flexibility as relations can be acted upon with additional operations $(\neg, \vee, \rightarrow, (\exists y)_{<z})$ that enhance the power of our expressing and reasoning about computational processes.

2.2.1 Definition. A relation $R(\vec{x})$ is (*primitive*) *recursive* iff its *characteristic function*,

$$c_R = \lambda\vec{x}.\begin{cases} 0 & \text{if } R(\vec{x}) \\ 1 & \text{if } \neg R(\vec{x}) \end{cases}$$

is (primitive) recursive ("c" in the function name for "characteristic"). The set of all primitive recursive (respectively, recursive) *relations* is denoted by $\mathcal{PR}_*$ (respectively, $\mathcal{R}_*$). □

Computability theory practitioners often call relations *predicates*.

It is clear that one can go from relation to characteristic function and back in a unique way, since $R(\vec{x}) \equiv c_R(\vec{x}) = 0$. In fact, the concept of characteristic function just recodes the "normal" outputs of a relation $R(\vec{x})$, from **true** or **false** to 0 or 1 respectively, turning relations into number theoretic (single-valued) functions (cf. also Definition 0.4.1).

The concept of relation simplifies the further development of the theory of primitive recursive functions.

The following is useful:

2.2.2 Proposition. $R(\vec{x}) \in \mathcal{PR}_*$ iff some $f \in \mathcal{PR}$ exists such that, for all $\vec{x}$, $R(\vec{x}) \equiv f(\vec{x}) = 0$.

Proof For the *if*-part, I want $c_R \in \mathcal{PR}$. This is so since $c_R = \lambda\vec{x}.1 \dotminus (1 \dotminus f(\vec{x}))$ (using Grzegorczyk substitution and $\lambda xy.x \dotminus y \in \mathcal{PR}$; cf. Example 2.1.12).

For the *only if*-part, $f = c_R$ will do. □

2.2.3 Corollary. $R(\vec{x}) \in \mathcal{R}_*$ iff some $f \in \mathcal{R}$ exists such that, for all $\vec{x}$, $R(\vec{x}) \equiv f(\vec{x}) = 0$.

Proof By the above proof, Proposition 2.1.15, and the fact that $\mathcal{R}$ is closed under composition and hence under Grzegorczyk operations. □

2.2.4 Corollary. $\mathcal{PR}_* \subseteq \mathcal{R}_*$.

Proof By the above corollary and Proposition 2.1.15. □

2.2.5 Theorem. $\mathcal{PR}_*$ *is closed under the Boolean operations.*

Proof It suffices to look at the cases of $\neg$ and $\vee$, since $R \rightarrow Q \equiv \neg R \vee Q$, $R \wedge Q \equiv \neg(\neg R \vee \neg Q)$ and $R \equiv Q$ is short for $(R \rightarrow Q) \wedge (Q \rightarrow P)$.

($\neg$) Say, $R(\vec{x}) \in \mathcal{PR}_*$. Thus (Definition 2.2.1), $c_R \in \mathcal{PR}$. But then $c_{\neg R} \in \mathcal{PR}$, since $c_{\neg R} = \lambda \vec{x}.1 \dot{-} c_R(\vec{x})$, by Grzegorczyk substitution and $\lambda xy.x \dot{-} y \in \mathcal{PR}$.

($\vee$) Let $R(\vec{x}) \in \mathcal{PR}_*$ and $Q(\vec{y}) \in \mathcal{PR}_*$. Then $\lambda \vec{x}\vec{y}.c_{R \vee Q}(\vec{x}, \vec{y})$ is given by

$$c_{R \vee Q}(\vec{x}, \vec{y}) = \text{if } c_R(\vec{x}) = 0 \text{ then } 0 \text{ else } c_Q(\vec{y})$$

and therefore is in $\mathcal{PR}$.   □

2.2.6 Remark Alternatively, for the $\vee$ case above, note that $c_{R \vee Q}(\vec{x}, \vec{y}) = c_R(\vec{x}) \times c_Q(\vec{y})$ and invoke Exercise 2.1.13.   □

2.2.7 Corollary. $\mathcal{R}_*$ *is closed under the Boolean operations.*

Proof As above, mindful of Proposition 2.1.15, and the fact that $\mathcal{R}$ is closed under Grzegorczyk operations.   □

2.2.8 Example. The relations $x \leq y, x < y, x = y$ are in $\mathcal{PR}_*$.

Note that $x \leq y \equiv x \dot{-} y = 0$ and invoke Proposition 2.2.2. Finally invoke Boolean closure and note that $x < y \equiv \neg y \leq x$ while $x = y$ is equivalent to $x \leq y \wedge y \leq x$.   □

It is common practice to use $R(\vec{x})$ and $c_R(\vec{x})$ (almost) interchangeably.

For example, we can write "$f = \lambda \vec{x}yz.\text{if } R(\vec{x}) \text{ then } y \text{ else } z$", which is the same as "$f = \lambda \vec{x}yz.\text{if } c_R(\vec{x}) = 0 \text{ then } y \text{ else } z$" since $R(\vec{x}) \equiv c_R(\vec{x}) = 0$.

Thus, if $R(\vec{x}) \in \mathcal{PR}_*$ (or $\in \mathcal{R}_*$), then $c_R \in \mathcal{PR}$ (or $\in \mathcal{R}$), and thus f is in $\mathcal{PR}$ (or $\in \mathcal{R}$), by Grzegorczyk operations as depicted below:

$$c_R(\vec{x})$$
$$\downarrow$$
$$f = \lambda \vec{x}yz.\text{if } \quad x \quad = 0 \text{ then } y \text{ else } z$$

2.2.9 Proposition. *If $R(\vec{x}, y, \vec{z}) \in \mathcal{PR}_*$ and $\lambda \vec{w}.f(\vec{w}) \in \mathcal{PR}$, then $R(\vec{x}, f(\vec{w}), \vec{z})$ is in $\mathcal{PR}_*$.*

Proof By Proposition 2.2.2, let $g \in \mathcal{PR}$ such that

$$R(\vec{x}, y, \vec{z}) \equiv g(\vec{x}, y, \vec{z}) = 0, \text{ for all } \vec{x}, y, \vec{z}$$

Then

$$R(\vec{x}, f(\vec{w}), \vec{z}) \equiv g(\vec{x}, f(\vec{w}), \vec{z}) = 0, \text{ for all } \vec{x}, \vec{w}, \vec{z}$$

By Proposition 2.2.2, and since $\lambda \vec{x}\vec{w}\vec{z}.g(\vec{x}.f(\vec{w}),\vec{z}) \in \mathcal{PR}$ by Grzegorczyk operations, we have that $R(\vec{x}, f(\vec{w}), \vec{z}) \in \mathcal{PR}_*$. □

2.2.10 Proposition. *If $R(\vec{x}, y, \vec{z}) \in \mathcal{R}_*$ and $\lambda\vec{w}.f(\vec{w}) \in \mathcal{R}$, then $R(\vec{x}, f(\vec{w}), \vec{z})$ is in $\mathcal{R}_*$.*

Proof Similar to that of Proposition 2.2.9. □

2.2.11 Corollary. *If $f \in \mathcal{PR}$ (respectively, in $\mathcal{R}$), then its graph, $z = f(\vec{x})$ is in $\mathcal{PR}_*$ (respectively, in $\mathcal{R}_*$).*

Proof Using the relation $z = y$ and Proposition 2.2.9. □

2.2.12 Exercise. Using unbounded search, prove that if $z = f(\vec{x})$ is in $\mathcal{R}_*$ and f is total, then $f \in \mathcal{R}$. □

2.2.13 Definition. (Bounded Quantifiers)
The abbreviations $(\forall y)_{<z} R(z, \vec{x})$ and $(\exists y)_{<z} R(z, \vec{x})$ stand for $(\forall y)\big(y < z \to R(z, \vec{x})\big)$ and $(\exists y)\big(y < z \wedge R(z, \vec{x})\big)$, respectively, and similarly for the non-strict inequality "$\leq$". □

2.2.14 Theorem. *$\mathcal{PR}_*$ is closed under bounded quantification.*

Proof It suffices to look at the case of $(\exists y)_{<z}$ since $(\forall y)_{<z} R(y, \vec{x}) \equiv \neg(\exists y)_{<z} \neg R(y, \vec{x})$.

Let then $R(y, \vec{x}) \in \mathcal{PR}_*$ and let us give the *name* $Q(z, \vec{x})$ to $(\exists y)_{<z} R(y, \vec{x})$ for convenience. We note that $Q(0, \vec{x})$ is false (Why?) and (logic) $Q(z + 1, \vec{x}) \equiv Q(z, \vec{x}) \vee R(z, \vec{x})$.

Thus, as the following primitive recursion shows, $c_Q \in \mathcal{PR}$.

$$c_Q(0, \vec{x}) = 1$$

$$c_Q(z + 1, \vec{x}) = c_Q(z, \vec{x})c_R(z, \vec{x})$$

□

2.2.15 Corollary. *$\mathcal{R}_*$ is closed under bounded quantification.*

2.3 Bounded Search

We introduce here a very powerful tool (operator on functions) in the reasoning about primitive recursive and recursive functions that makes many constructions of functions significantly easier than they would be if were restricted to the original tools only, primitive recursion and composition. We also introduce *definition by cases*, that also helps in the construction of functions and the certification of their (primitive, partial, or total) recursiveness.

2.3.1 Definition. (Bounded Search) Let f be a *total* number-theoretic function of $n + 1$ variables. The symbol $(\mu y)_{<z} f(y, \vec{x})$, for all $z, \vec{x}$, stands for

$$\begin{cases} \min\{y : y < z \wedge f(y, \vec{x}) = 0\} & \text{if } (\exists y)_{<z} f(y, \vec{x}) = 0 \\ z & \text{otherwise} \end{cases}$$

So, *unsuccessful search returns the first number to the right of the search-range.*

We define "$(\mu y)_{\leq z}$" to mean "$(\mu y)_{<z+1}$".   □

2.3.2 Theorem. $\mathcal{PR}$ *is closed under the bounded search operation* $(\mu y)_{<z}$. *That is, if* $\lambda y\vec{x}.f(y, \vec{x}) \in \mathcal{PR}$, *then* $\lambda z\vec{x}.(\mu y)_{<z} f(y, \vec{x}) \in \mathcal{PR}$.

Proof Set $g = \lambda z\vec{x}.(\mu y)_{<z} f(y, \vec{x})$ for convenience. Then the following primitive recursion settles it:

$g(0, \vec{x}) = 0$

Why 0 above?

$g(z + 1, \vec{x}) = $ if $(\exists y)_{<z} f(y, \vec{x}) = 0$ then $g(z, \vec{x})$

else if $f(z, \vec{x}) = 0$ then z else $z + 1$ □

2.3.3 Corollary. $\mathcal{PR}$ *is closed under the bounded search operation* $(\mu y)_{\leq z}$.

2.3.4 Exercise. Prove the corollary.   □

2.3.5 Corollary. $\mathcal{R}$ *is closed under the bounded search operations* $(\mu y)_{<z}$ *and* $(\mu y)_{\leq z}$.

Consider now a set of *mutually exclusive* relations $R_i(\vec{x})$, $i = 1, \ldots, n$, that is, $R_i(\vec{x}) \wedge R_j(\vec{x})$ is *false* for each $\vec{x}$ as long as $i \neq j$.

Then we can define a function f *by cases* R_i from given functions f_j by the requirement (for all $\vec{x}$) given below:

$$f(\vec{x}) = \begin{cases} f_1(\vec{x}) & \text{if } R_1(\vec{x}) \\ f_2(\vec{x}) & \text{if } R_2(\vec{x}) \\ \ldots & \ldots \\ f_n(\vec{x}) & \text{if } R_n(\vec{x}) \\ f_{n+1}(\vec{x}) & \text{otherwise} \end{cases}$$

where, as is usual in mathematics, "if $R_j(\vec{x})$" is short for "if $R_j(\vec{x})$ is true" and "otherwise" is the condition $\neg(R_1(\vec{x}) \vee \cdots \vee R_n(\vec{x}))$. We have the following result:

2.3.6 Theorem. (Definition by Cases) *If the functions* f_i, $i = 1, \ldots, n+1$ *and the relations* $R_i(\vec{x})$, $i = 1, \ldots, n$ *are in* $\mathcal{PR}$ *and* $\mathcal{PR}_*$, *respectively, then so is* f *above.*

Proof By repeated use of if-then-else via Grzegorczyk operations. That is, note that

$$f(\vec{x}) = \text{if} \qquad R_1(\vec{x}) \text{ then } f_1(\vec{x})$$

$$\text{else if} \qquad R_2(\vec{x}) \text{ then } f_2(\vec{x})$$

$$\cdots$$

$$\text{else if} \qquad R_n(\vec{x}) \text{ then } f_n(\vec{x})$$

$$\text{else} \qquad f_{n+1}(\vec{x}) \qquad\qquad\qquad \square$$

2.3.7 Corollary. *Same statement as above, replacing $\mathcal{PR}$ and $\mathcal{PR}_*$ by $\mathcal{R}$ and $\mathcal{R}_*$, respectively.*

2.3.8 Corollary. *Same statement as above, replacing $\mathcal{PR}$ and $\mathcal{PR}_*$ by $\mathcal{P}$ and $\mathcal{R}_*$, respectively.*

 It is worth acknowledging in words the content of Corollary 2.3.8: $\mathcal{P}$ is closed under definition by *recursive cases*.

The tools we now have at our disposal allow easy certification of the primitive recursiveness of some very useful functions and relations. But first a definition:

2.3.9 Definition. $(\mu y)_{<z} R(y, \vec{x})$ means $(\mu y)_{<z} c_R(y, \vec{x})$.
Similarly, for *recursive predicates* $Q(y, \vec{x})$, unbounded search $(\mu y) Q(y, \vec{x})$ means $(\mu y) c_Q(y, \vec{x})$. $\qquad\qquad\square$

 Thus, if $R(y, \vec{x}) \in \mathcal{PR}_*$ (resp. $\in \mathcal{R}_*$), then $\lambda z\vec{x}.(\mu y)_{<z} R(y, \vec{x}) \in \mathcal{PR}$ (resp. $\in \mathcal{R}$), since $c_R \in \mathcal{PR}$ (resp. $\in \mathcal{R}$).

2.3.10 Example. The following are in $\mathcal{PR}$ or $\mathcal{PR}_*$ as appropriate:

(1) $\lambda xy.\lfloor x/y \rfloor^4$ (the quotient of the division x/y). This is another instance of a nontotal function with an "obvious" way to remove the points where it is undefined (recall also the case of $\lambda xy.x^y$). Thus the symbol is *extended* to *mean*

$$(\mu z)_{\leq x}\big((z + 1)y > x\big) \qquad\qquad (*)$$

for all x, y.
Pause. *Why* is the above expression correct?
Because setting $z = \lfloor x/y \rfloor$ we have

$$z \leq \frac{x}{y} < z + 1$$

by the definition of "$\lfloor \ldots \rfloor$". Thus, z is *smallest* such that $x/y < z + 1$, or such that $x < y(z + 1)$. ◀

---

[4] For any real number x, the symbol "$\lfloor x \rfloor$" is called the *floor* of x. It succeeds in the literature (with the same definition) the so-called "greatest integer function, $[x]$", i.e., the *integer part* of the real number x. Thus, *by definition*, $\lfloor x \rfloor \leq x < \lfloor x \rfloor + 1$.

It follows that, for $y > 0$, the search in $(*)$ yields the "normal math" value for $\lfloor x/y \rfloor$, while it re-defines $\lfloor x/0 \rfloor$ as $= x + 1$.

(2) $\lambda xy.rem(x, y)$ (the remainder of the division x/y). $rem(x, y) = x \dotminus y\lfloor x/y \rfloor$.

(3) $\lambda xy.x \mid y$ (x divides y). $x \mid y \equiv rem(y, x) = 0$; now apply Proposition 2.2.2. Note that if $y > 0$, we cannot have $0 \mid y$ —a good thing!— since $rem(y, 0) = y$. Our redefinition of $\lfloor x/y \rfloor$ yields $0 \mid 0$ but we can live with this in practice.

(4) $Pr(x)$ (x is a prime). $Pr(x) \equiv x > 1 \wedge (\forall y)_{<x}(y \mid x \rightarrow y = 1 \vee y = x)$.

(5) $\pi(x)$ (the number of primes $\leq x$).[5] The following primitive recursion certifies the claim: $\pi(0) = 0$, and $\pi(x + 1) = $ if $Pr(x + 1)$ then $\pi(x) + 1$ else $\pi(x)$.

(6) $\lambda n.p_n$ (the nth prime). First note that the graph $y = p_n$ is primitive recursive: $y = p_n \equiv Pr(y) \wedge \pi(y) = n + 1$. Next note that, for all n, $p_n \leq 2^{2^n}$ (see Exercise 2.4.1 below), thus $p_n = (\mu y)_{\leq 2^{2^n}}(y = p_n)$, which settles the claim.

(7) $\lambda nx.\exp(n, x)$ (*the* exponent of p_n in the prime factorization of x). $\exp(n, x) = (\mu y)_{<x}\neg(p_n^{y+1} \mid x)$.

  Pause. Is x a good bound? Yes! $x = \ldots p_n^y \ldots \geq p_n^y \geq 2^y > y$. ◀

(8) $Seq(x)$ (says that x's prime number factorization contains 2, and has no gaps; no prime, between 2 and the largest in the factorisation, is missing.)
$$Seq(x) \equiv x > 1 \wedge (\forall y)_{\leq x}(\forall z)_{\leq x}(Pr(y) \wedge Pr(z) \wedge y < z \wedge z \mid x \rightarrow y \mid x). \qquad \square$$

 2.3.11 Remark. What makes $\exp(n, x)$ "*the* exponent of p_n in the prime factorization of x", rather than *an* exponent, is Euclid's prime number factorization theorem: *Every number $x > 1$ has a unique factorization —within permutation of factors— as a product of primes.* $\qquad \square$

2.4 Coding Sequences; Special Recursions

2.4.1 Exercise. Prove by induction on n, that for all n we have $p_n \leq 2^{2^n}$.

Hint Consider, as Euclid did,[6] $p_0 p_1 \cdots p_n + 1$. If this number is prime, then it is greater than or equal to p_{n+1} (Why?). If it is composite, then none of the primes up to p_n divide it. So any prime factor of it is greater than or equal to p_{n+1} (Why?). $\qquad \square$

2.4.2 Definition. (Coding Sequences) A sequence of numbers, $a_0, \ldots, a_n, n \geq 0$, is *coded* by the number denoted by the symbol $\langle a_0, \ldots, a_n \rangle$ and defined as

$$\prod_{i \leq n} p_i^{a_i+1} \qquad \square$$

For *coding* to be useful, we need a simple *decoding* scheme. By Remark 2.3.11 there is no way to have $z = \langle a_0, \ldots, a_n \rangle = \langle b_0, \ldots, b_m \rangle$, *unless*

---

[5] The π-function plays a central role in number theory, figuring in the so-called *prime number theorem*. See, for example, LeVeque (1956).

[6] In his proof that there are infinitely many primes.

(i) $n = m$

   and

(ii) for $i = 0, \ldots, n, a_i = b_i$.

Thus, it makes sense to correspondingly define the *decoding expressions*:

(*i*) $lh(z)$ (pronounced "length of z") as shorthand for $(\mu y)_{<z} \neg (p_y | z)$.

   Note that if p_y is the first prime *not* in the decomposition of z, and $Seq(z)$ holds, then since numbering of primes starts at 0, the length of the coded sequence z is indeed y.

   Pause. Is the bound z sufficient? *Yes!*

$$z = 2^{a+1} 3^{b+1} \cdots p_{y \dot- 1}^{exp(y \dot- 1, z)} \geq \underbrace{2 \cdot 2 \cdots 2}_{y \ times} = 2^y > y$$

◀

(*ii*) $(z)_i$ is *shorthand* for $exp(i, z) \dot- 1$

Note that

(a) $\lambda i z.(z)_i$ and $\lambda z.lh(z)$ are in $\mathcal{PR}$.
(b) If $Seq(z)$, then $z = \langle a_0, \ldots, a_n \rangle$ for some $a_0, \ldots, a_n$. In this case, $lh(z)$ equals the number of distinct primes in the decomposition of z, that is, the length $n + 1$ of the coded sequence. Then $(z)_i$, for $i < lh(z)$, equals a_i. For larger i, $(z)_i = 0$. Note that if $\neg Seq(z)$ then $lh(z)$ need not equal the number of distinct primes in the decomposition of z. For example, 10 has 2 primes, but $lh(10) = 1$.

 The tools lh, $Seq(z)$, and $\lambda i z.(z)_i$ are sufficient to perform *decoding, primitive recursively*, once the truth of $Seq(z)$ is established. This coding/decoding scheme is essentially that of Gödel (1931), and will be the one we use throughout these notes.

2.4.1 Concatenating (Coded) Sequences; Stacks

Consider the two sequences $a_0, \ldots, a_m$ and $b_0, \ldots, b_n$. Is there an easy (e.g., primitive recursive) computational way to obtain $\langle a_0, \ldots, a_m, b_0, \ldots, b_n \rangle$ from $\langle a_0, \ldots, a_m \rangle$ and $\langle b_0, \ldots, b_n \rangle$? Yes, there is: Note that by Definition 2.4.2 we have

$$\langle a_0, \ldots, a_m \rangle = \prod_{i \leq m} p_i^{a_i + 1} \tag{1}$$

and

$$\langle b_0, \ldots, b_n \rangle = \prod_{i \leq n} p_i^{b_i + 1}$$

Thus, let us write $z = \langle a_0, \ldots, a_m \rangle$ and $w = \langle b_0, \ldots, b_n \rangle$. Then we have

$$\langle a_0, \ldots, a_m, b_0, \ldots, b_n \rangle = \left(\prod_{i \leq m} p_i^{a_i+1} \right) \prod_{i \leq n} p_{i+m+1}^{b_i+1}$$

$$= \langle a_0, \ldots, a_m \rangle \prod_{i \leq n} p_{i+m+1}^{b_i+1} \qquad \text{using (1)}$$

$$= z \prod_{i < lh(w)} p_{i+lh(z)}^{\exp(i,w)} \qquad \text{notation from 2.3.10}$$

The last right hand side expression above is meaningful regardless of whether z and w code sequences —i.e., even if one or both of $Seq(z)$ and $Seq(w)$ are false.

Thus we define

2.4.3 Definition. The function $\lambda zw.z * w$ given, for *all* z, w in $\mathbb{N}$, by

$$z * w \stackrel{Def}{=} z \prod_{i < lh(w)} p_{i+lh(z)}^{\exp(i,w)}$$

is called the *concatenation of natural numbers* function. $\square$

Our discussion above yields at once

2.4.4 Proposition. *If* $x = \langle a_0, \ldots, a_m \rangle$ *and* $y = \langle b_0, \ldots, b_n \rangle$, *then* $x * y = \langle a_0, \ldots, a_m, b_0, \ldots, b_n \rangle$.

But is $\lambda xy.x * y$ easily computable? Well, yes, it is if the qualifier "easily" means primitive recursively.

This will need a simple lemma:

2.4.5 Lemma. *If* $\lambda x \vec{y}.f(x, \vec{y}) \in \mathcal{PR}$ *then so is* $\lambda z\vec{y}. \prod_{x<z} f(x, \vec{y})$.

Proof Set $g = \lambda z\vec{y}. \prod_{x<z} f(x, \vec{y})$ for convenience.

Then the following primitive recursion settles it.

$$g(0, \vec{y}) = 1$$
$$g(z+1, \vec{y}) = g(z, \vec{y})f(z, \vec{y})$$

$\square$

2.4.6 Corollary. *If* $\lambda x \vec{y}.f(x, \vec{y}) \in \mathcal{R}$ *then so is* $\lambda z\vec{y}. \prod_{x<z} f(x, \vec{y})$.

2.4.7 Corollary. $\lambda xy.x * y \in \mathcal{PR}$.

Proof Simply use Lemma 2.4.5, the discussion and notation following Definition 2.4.2, and the results of Examples 2.1.16 and 2.3.10 to see that $\lambda w. \prod_{i<lh(w)} p_{i+lh(z)}^{\exp(i,w)} \in \mathcal{PR}$. $\square$

2.4.8 Example. $lh(1) = 0$. Thus even though $Seq(1)$ is false, 1 is a good candidate to code the *empty sequence*. More evidence that this is so is the fact that $1 * z = z * 1 = z$ for all z (Exercise, using Definition 2.4.3).   □

The reader most likely is familiar with the *stack* data structure from elementary programming.

2.4.9 Definition. (Stacks) A stack is an *ordered list* of *natural numbers*, where we may *add to* (or *delete from*) its data only at *one* end of the list, called the *top of the stack*. The length of the stack at a given moment is the number of elements it contains.

In this volume our stacks will allow us to inspect (read) an element anywhere along their length.

The terminology for adding to the stack is to *push* on the stack; that of deleting is to *pop* from the stack.

Adding creates a new top (the old top is "below" the new one after the addition). Deletion, if the stack is not *empty*, will remove the *current* top item. The new top is the item that was below the item we deleted, before the deletion.   □

2.4.10 Remark. In a course in programming a stack may contain elements of any data type that the programming language in hand allows. But we only use natural numbers in this volume.

In our computability theory it is easy to implement a stack using prime power coding. If the elements of the stack s are in the sequence $a_0, a_1, \ldots, a_n$, with a_n being at the top, then

$$s = \langle a_0, a_1, \ldots, a_n \rangle$$

The top, expressed as a function of s, is $(s)_{lh(s) \dot{-} 1}$ if the stack is not empty ($\neq 1$), else it is undefined.

The three operations —in C-like pseudo-code— are implemented as follows:

- *read the top and assign to* **x**: if $s > 1$ (s non-empty), then $\mathbf{x} \leftarrow (s)_{lh(s) \dot{-} 1}$
- *push* $b \in \mathbb{N}$ *to the stack*: $s \leftarrow s * \langle b \rangle$
- *pop the stack*: If $s > 1$, then $s \leftarrow \left\lfloor s / p_{lh(s) \dot{-} 1}^{\exp(lh(s) \dot{-} 1, s)} \right\rfloor$.

   If we want to save into the variable **x** what we popped, then we do:
   If $s > 1$, then

   $\{$

   1. $\mathbf{x} \leftarrow (s)_{lh(s) \dot{-} 1}$
   2. pop s

   $\}$

                                                              □

2.4.2 Course-of-Values Recursion

Primitive recursion defines a function "at $n+1$" in terms of its value "at n". However we also have examples of "recursions" (or "recurrences"), one of the best known perhaps being the *Fibonacci sequence*, $0, 1, 1, 2, 3, 5, 8, \ldots$, that is given by $F_0 = 0$, $F_1 = 1$ and (for $n \geq 1$) $F_{n+1} = F_n + F_{n-1}$, where the value at $n + 1$ depends on both the values at n and $n - 1$.

We may also have recursions where the value at $n + 1$ depends, in general, on the entire *history*, or *course-of-values*, of the function values at inputs $n, n - 1, n - 2, \ldots, 1, 0$. The easiest way to represent the entire history of values of a *total* number-theoretic function f "at (input) x", namely, the set $\{f(0, \vec{y}), f(1, \vec{y}), \ldots, f(x, \vec{y})\}$, is to code it by a single number!

2.4.11 Definition. (Course-of-Values Recursion) We say that f, of $n + 1$ arguments, is defined from two total functions —namely, the *basis* function $\lambda \vec{y}_n.b(\vec{y}_n)$ and the *iterator* $\lambda x \vec{y}_n z.g(x, \vec{y}_n, z)$— by *course-of-values recursion* if, for all x, $\vec{y}_n$, the following equations hold:

$$\begin{cases} f(0, \vec{y}_n) & = b(\vec{y}_n) \\ f(x + 1, \vec{y}_n) & = g(x, \vec{y}_n, H(x, \vec{y}_n)) \end{cases} \tag{1}$$

where $\lambda x \vec{y}_n.H(x, \vec{y}_n)$ is the *history function*, which is given "at x" (for all $\vec{y}_n$) by

$$\langle f(0, \vec{y}), f(1, \vec{y}), \ldots, f(x, \vec{y})\rangle \qquad \square$$

2.4.12 Exercise. Prove that f given by (1) is total.

Hint Use strong induction on x. $\qquad \square$

The major result in this subsection is that both $\mathcal{PR}$ and $\mathcal{R}$ are closed under course-of-values recursion.

2.4.13 Theorem. $\mathcal{PR}$ *is closed under course-of-values recursion.*

Proof So, let b and g be in $\mathcal{PR}$. We will show that $f \in \mathcal{PR}$. It suffices to prove that the history function H is primitive recursive, for then $f = \lambda x \vec{y}_n.\left(H(x, \vec{y}_n)\right)_x$ and we are done by Grzegorczyk substitution. To this end, the following equations —true for all x, $\vec{y}_n$— settle the case:

$$H(0, \vec{y}_n) = \langle b(\vec{y}_n)\rangle$$

$$H(x + 1, \vec{y}_n) = H(x, \vec{y}_n) p_{x+1}^{g(x, \vec{y}_n, H(x, \vec{y}_n))+1} \qquad \square$$

The same proof with trivial adjustments yields:

2.4.14 Corollary. $\mathcal{R}$ *is closed under course-of-values recursion.*

2.4.15 Example. The Fibonacci sequence, $(F_n)_{n \geq 0}$, is given by

$$F_0 = 0$$

$$F_1 = 1$$

otherwise, for $n > 0$,

$$F_{n+1} = F_n + F_{n-1}$$

The sequence can be viewed as the function $\lambda n . F_n$. As such it is in $\mathcal{PR}$.

Indeed, letting H_n be the history of the sequence at n —that is, $\langle F_0, \ldots, F_n \rangle$ — we have the following course-of-values recursion for $\lambda n . F_n$ in terms of functions known to be in $\mathcal{PR}$.

$$F_0 = 0$$

$$F_{n+1} = \text{if } n = 0 \text{ then } 1$$

$$\text{else } (H_n)_n + (H_n)_{n \dot- 1} \qquad \square$$

2.4.3 Simultaneous Primitive Recursion

This recursion is instrumental towards the study of URM program and *loop program* computations. Loop programs are introduced in a later chapter.

2.4.16 Definition. Given total functions h_i, g_i, for $i = 0, 1, \ldots, k$. We say that the following equations-schema defines —for all $x, \vec{y}$— the new functions f_i from the given functions by a *simultaneous primitive recursion*.

$$\begin{cases} f_0(0, \vec{y}) & = h_1(\vec{y}) \\ \vdots \\ f_k(0, \vec{y}) & = h_k(\vec{y}) \\ f_0(x+1, \vec{y}) & = g_0(x, \vec{y}, f_0(x, \vec{y}), \ldots, f_k(x, \vec{y})) \\ \vdots \\ f_k(x+1, \vec{y}) & = g_k(x, \vec{y}, f_0(x, \vec{y}), \ldots, f_k(x, \vec{y})) \end{cases} \qquad (2)$$

$\square$

Hilbert and Bernays (1968) proved the following:

2.4.17 Theorem. *If all the h_i and g_i are in $\mathcal{PR}$ (resp. $\mathcal{R}$), then so are all the f_i obtained by the schema (2) of simultaneous recursion.*

Proof Define, for all x and $\vec{y}$,

$$F(x, \vec{y}) \stackrel{\text{Def}}{=} \langle f_0(x, \vec{y}), \ldots, f_k(x, \vec{y}) \rangle$$

$$H(\vec{y}_n) \stackrel{\text{Def}}{=} \langle h_0(\vec{y}), \ldots, h_k(\vec{y}) \rangle$$

$$G(x, \vec{y}, z) \stackrel{\text{Def}}{=} \langle g_0(x, \vec{y}, (z)_0, \ldots, (z)_k), \ldots, g_k(x, \vec{y}, (z)_0, \ldots, (z)_k) \rangle$$

We readily have that $H \in \mathcal{PR}$ (resp. $\in \mathcal{R}$) and $G \in \mathcal{PR}$ (resp. $\in \mathcal{R}$) depending on where we assumed the h_i and g_i to be. We can now rewrite schema (2) as

$$\begin{cases} F(0, \vec{y}) & = H(\vec{y}) \\ F(x+1, \vec{y}) & = G\left(x, \vec{y}, F\left(x, \vec{y}\right)\right) \end{cases} \tag{3}$$

▶ The second line of (3) is condenced from

$$\begin{aligned} F(x+1, \vec{y}) \quad & = \langle f_0(x+1, \vec{y}), \ldots, f_k(x+1, \vec{y}) \rangle \\ & = \left\langle g_0\left(x, \vec{y}, f_0(x, \vec{y}), \ldots, f_k(x, \vec{y})\right), \ldots, g_k\left(\ldots\right) \right\rangle \\ & = \left\langle g_0\left(x, \vec{y}, (F(x, \vec{y}))_0, \ldots, (F(x, \vec{y}))_k\right), \ldots, g_k\left(\ldots\right) \right\rangle \end{aligned}$$

By the above remarks, $F \in \mathcal{PR}$ (resp. $\in \mathcal{R}$) depending on where we assumed the h_i and g_i to be. In particular, this holds for each f_i since, for all $x, \vec{y}$, $f_i(x, \vec{y}) = (F(x, \vec{y}))_i$ (Grzegorczyk operations). □

2.4.18 Example. We saw one way to justify that $\lambda x.rem(x, 2) \in \mathcal{PR}$ in Example 2.3.10. A direct way is the following. Setting $f(x) = rem(x, 2)$, for all x, we notice that the sequence of outputs (for $x = 0, 1, 2, \ldots$) of f is

$$0, 1, 0, 1, 0, 1 \ldots$$

Thus, ignoring the result from Example 2.3.10, the following primitive recursion shows that $f \in \mathcal{PR}$:

$$\begin{cases} f(0) & = 0 \\ f(x+1) & = 1 \dot{-} f(x) \end{cases}$$

Here is a way, via simultaneous recursion, to obtain a proof that $f \in \mathcal{PR}$, *without using any arithmetic*! Notice the infinite "matrix"

$$0\ 1\ 0\ 1\ 0\ 1 \ldots$$
$$1\ 0\ 1\ 0\ 1\ 0 \ldots$$

Let us call g the function that has as its sequence of outputs the entries of the second row —obtained by shifting the first row by one position to the left. The first row still represents our f. Now

$$
\begin{cases}
f(0) & = 0 \\
g(0) & = 1 \\
f(x+1) & = g(x) \\
g(x+1) & = f(x)
\end{cases}
\tag{1}
$$

□

2.4.19 Example. We saw one way to justify that $\lambda x. \lfloor x/2 \rfloor \in \mathcal{PR}$ in Example 2.3.10. A direct way is the following.

$$
\begin{cases}
\left\lfloor \dfrac{0}{2} \right\rfloor & = 0 \\[2mm]
\left\lfloor \dfrac{x+1}{2} \right\rfloor & = \left\lfloor \dfrac{x}{2} \right\rfloor + rem(x, 2)
\end{cases}
$$

where rem is in $\mathcal{PR}$ by Example 2.4.18.

Alternatively, here is a way that can do it —via simultaneous recursion— and with only the knowledge of how to add 1. Consider the matrix

$$
\begin{array}{l}
0\ 0\ 1\ 1\ 2\ 2\ 3\ 3 \ldots \\
0\ 1\ 1\ 2\ 2\ 3\ 3\ 4 \ldots
\end{array}
$$

The top row represents $\lambda x. \lfloor x/2 \rfloor$, let us call it "$f$". The bottom row we call g and is, again, the result of shifting row one to the left by one position. Thus, we have a simultaneous recursion

$$
\begin{cases}
f(0) & = 0 \\
g(0) & = 0 \\
f(x+1) & = g(x) \\
g(x+1) & = f(x) + 1
\end{cases}
\tag{2}
$$

□

2.4.4 Simulating a URM Computation

We have defined in Definition 1.1.2 what a URM computation is. This subsection will deal with mathematically *simulating* a computation.

We can do so by pencil and paper if we are sufficiently patient to engage in a potentially infinite task.

What we need to do is keep track of the values of *all the variables* of the given URM as the computation process *reaches each* instruction. As we saw in Definition 1.1.2, *at most one* variable changes at each instruction execution (instructions of types 1–3), while the **if-statement** and **stop** change no variable. Moreover, the **stop** instruction *practically ends* the computation, even if the latter, in theory, continues forever: *All variable contents have converged to* final *values once the execution reaches* **stop**.

In this subsection we will see good justification for allowing all computations to go forever, even after they reached the instruction **stop**.

The simulation simply *requires us to record* what *each* variable's value is when we *reach* an instruction "$L : \ldots$"

Thus we record an array of values for said instruction,

$$L; a_1, a_2, \ldots$$

where L is the label or *instruction counter* and the a_i are the values of *all* the (finitely many!) x_i, the latter variables of the given URM being *ordered* in an *arbitrary* manner, but *fixed* for the length of the simulation.

This array is a snapshot of the computation, or an *instantaneous description*, in short *ID*.

One way to order the variables in the head row of Table 2.1 is lexicographically, seeing that each is a string $X1^n$, for $n \geq 0$.

The semantics in Definition 1.1.2 allow us to build the next snapshot, since we know which instruction of M is labelled L.

A sequence of *discrete events* —such as *reaching* and recording an instruction in the course of a computation of a URM M— when arranged in a sequence are implicitly associated with their position-numbers in the sequence, $0, 1, 2, \ldots$

The event in position i we say occurred in *step* or *time i*. The very first event occurs in time 0, it is the event at which instruction 1 is reached.

In Table 2.1, for each instruction *reached, we record the full ID*.

Our main tool for the simulation is a growing matrix like this:

Table 2.1 Simulation table

y	IC	...	$\mathbf{x}$	...	Comment
0	1	...	a	...	$a = 0$ if $\mathbf{x}$ noninput
...	...	...	...	...	
i	L	...	b	...	Use Definition 1.1.2 to *write down* the $(i+1)$st row (L', c, etc.)
$i+1$	L'	...	c	...	
...	...	...	...	...	

Deliberately, the bottom bounding line is missing to convey that this matrix is growing —one row at a time— downwards *forever*.

The very first entry is the ID corresponding to reaching instruction 1 *for the very first time*. It is the *initial ID*:

$$1; a_1, a_2, \ldots$$

where each a_i corresponding to an *input* variable $\mathbf{x}_i$ is some input number, while all the other a_j equal 0.

So, at each *time*, or *step*, $y = 0, 1, 2, \ldots$, row y contains (a pointer "L" to) the instruction we need to do *next*, and what values *each* variable of M holds, *before* we execute said (next) instruction.

Let us fix a URM M. Say, $\mathbf{x}_i$, for $i = 1, \ldots m$, are *all* its variables, of which, without loss of generality, the first n are input ($n \le m$).

We define $\lambda y \vec{a}_n . X_i(y, \vec{a}_n)$, for $i = 1, \ldots m$, by

$$X_i(y, \vec{a}_n) = \text{the } \mathbf{x}_i \text{ contents at } time\ y, \text{ if } \vec{a}_n \text{ was the } input. \tag{1}$$

Similarly, we define $\lambda y \vec{a}_n . IC(y, \vec{a}_n)$ by

$$IC(y, \vec{a}_n) = \text{the current label at time } y, \text{ if } \vec{a}_n \text{ was the } input. \tag{2}$$

It turns out that these functions are definitionally very simple; enough so to be primitive recursive. To ensure that we can *prove* this, first off we must ensure they are *total*. This explains our choice to trivially continue a computation forever, even though it has reached **stop**. This convention causes all computations to have infinite length, but the *convergent* ones —that is, those that do reach **stop**— have all contents of variables converge to final values.

Thus $\lambda y \vec{a}_n . X_i(y, \vec{a}_n)$ and $\lambda y \vec{a}_n . IC(y, \vec{a}_n)$ are total.

2.4.20 Simulation Theorem.

Let M be a URM with variables $\mathbf{x}_1, \mathbf{x}_2, \ldots \mathbf{x}_{n+1}, \mathbf{x}_{n+2}, \ldots, \mathbf{x}_m$, of which $\mathbf{x}_i$, for $i = 1, \ldots, n$, are input variables while $\mathbf{x}_1$ is *also* the output variable.

With reference to Definition 1.1.2 and the above discussion, the *simulating functions* $\lambda y \vec{a}_n . X_i(y, \vec{a}_n)$, for $i = 1, \ldots m$, and $\lambda y \vec{a}_n . IC(y, \vec{a}_n)$ are in $\mathcal{PR}$.

Proof We have the following simultaneous recursion that defines the simulating functions:

$$X_i(0, \vec{a}_n) = a_i, \text{ for } i = 1, \ldots, n$$
$$X_i(0, \vec{a}_n) = 0, \text{ for } i = n + 1, \ldots, m$$
$$IC(0, \vec{a}_n) = 1$$

For $y \geq 0$ and $i = 1, \ldots, m$,

$$
X_i(y+1, \vec{a}_n) = \begin{cases}
c & \text{if } IC(y, \vec{a}_n) = L \wedge \text{``}L : \mathbf{x}_i \leftarrow c\text{''} \in M \\
X_i(y, \vec{a}_n) + 1 & \text{if } IC(y, \vec{a}_n) = L \wedge \text{``}L : \mathbf{x}_i \leftarrow \mathbf{x}_i + 1\text{''} \in M \\
X_i(y, \vec{a}_n) \dotminus 1 & \text{if } IC(y, \vec{a}_n) = L \wedge \text{``}L : \mathbf{x}_i \leftarrow \mathbf{x}_i \dotminus 1\text{''} \in M \\
X_i(y, \vec{a}_n) & \text{othw}
\end{cases}
$$

The above if-parts are of necessity wordy and indirect, as we can only give descriptive conditions because *we have not "looked inside" M to be able to discuss a specific instruction-set.* We will do so in two examples below.

In general, suffice it to say that for each $\mathbf{x}_i$ we scan the program M from top to bottom and write a case for each instruction of types 1–3 that involves $\mathbf{x}_i$ (Definition 1.1.1):

Thus, e.g., *if* the $k_j : \mathbf{x}_i = \mathbf{x}_i + 1$ (for $j = 1, \ldots, r$) are the *only* instructions in M that involve the "+1" instruction with $\mathbf{x}_i$, and *if* there are *no* " $\dotminus$ 1" instructions for this variable, *but* we have exactly two of these:

$$R : \mathbf{x}_i \leftarrow 2$$

$$Q : \mathbf{x}_i \leftarrow 5$$

then we will add the cases below (two for the above two instructions and r for the "+1" instructions):

$$
X_i(y+1, \vec{a}_n) = \begin{cases}
2 & \text{if } IC(y, \vec{a}_n) = R \\
5 & \text{if } IC(y, \vec{a}_n) = Q \\
\vdots & \vdots \\
X_i(y, \vec{a}_n) + 1 & \text{if } IC(y, \vec{a}_n) = k_j \\
\vdots & \vdots \\
X_i(y, \vec{a}_n) & \text{othw}
\end{cases}
$$

The descriptive text

$$\text{``and `}L : \mathbf{x}_i \leftarrow \mathbf{x}_i + 1\text{' is in } M\text{''}$$

has been replaced by the precise r cases "if $IC(y, \vec{a}_n) = k_j$", one for each $j = 1, \ldots, r$.

The recurrence for IC is

$$
IC(y+1, \vec{a}_n) = \begin{cases}
L' & \text{if } IC(y, \vec{a}_n) = L \wedge \text{``} L : \textbf{if } x_i = 0 \textbf{ goto } L' \\
& \textbf{else goto } L''' \text{''} \in M \wedge X_i(y, \vec{a}_n) = 0 \\
L'' & \text{if } IC(y, \vec{a}_n) = L \wedge \text{``} L : \textbf{if } x_i = 0 \textbf{ goto } L' \\
& \textbf{else goto } L''' \text{''} \in M \wedge X_i(y, \vec{a}_n) > 0 \\
k & \text{if } IC(y, \vec{a}_n) = k \text{ and ``} k : \textbf{stop} \text{'' is in } M \\
IC(y, \vec{a}_n) + 1 & \text{othw}
\end{cases}
$$

Again to remove the descriptive inexact nature of the conditions above, in general we will write two conditions for each if-statement in the program. Say we have only two if-statements in our M as shown below. Moreover, let us assume that **stop** has label k.

$$
\begin{array}{cccc}
\vdots & \vdots & & \\
L : & \text{if } x_5 = 0 & \textbf{goto } L' & \textbf{else goto } L'' \\
\vdots & \vdots & & \\
R : & \text{if } x_{45} = 0 & \textbf{goto } R' & \textbf{else goto } R'' \\
\vdots & \vdots & &
\end{array}
$$

Then we can make the definition of IC precise:

$$
IC(y+1, \vec{a}_n) = \begin{cases}
L' & \text{if } IC(y, \vec{a}_n) = L \wedge X_5(y, \vec{a}_n) = 0 \\
L'' & \text{if } IC(y, \vec{a}_n) = L \wedge X_5(y, \vec{a}_n) > 0 \\
R' & \text{if } IC(y, \vec{a}_n) = R \wedge X_{45}(y, \vec{a}_n) = 0 \\
R'' & \text{if } IC(y, \vec{a}_n) = R \wedge X_{45}(y, \vec{a}_n) > 0 \\
k & \text{if } IC(y, \vec{a}_n) = k \\
IC(y, \vec{a}_n) + 1 & \text{othw}
\end{cases}
$$

Since the iterator functions only utilise the functions $\lambda z.a$, $\lambda z.z + 1$, $\lambda z.z \doteq 1$, $\lambda z.z$, and predicates $\lambda z.z = 0$, and $\lambda z.z > 0$ —all in $\mathcal{PR}$ and $\mathcal{PR}_*$— it follows that all the simulating functions are in $\mathcal{PR}$, as claimed. □

2.4.21 Example. Let M be the program below

$$1 : \textbf{if } \mathbf{x}_2 = 0 \textbf{ goto } 5 \textbf{ else goto } 2$$

$$2 : \mathbf{x}_1 \leftarrow \mathbf{x}_1 + 1$$

$$3 : \mathbf{x}_2 \leftarrow \mathbf{x}_2 \div 1$$

$$4 : \textbf{if } \mathbf{x}_1 = 0 \textbf{ goto } 1 \textbf{ else goto } 1$$

$$5 : \textbf{stop}$$

Let us assume that $\mathbf{x}_2$ is the input variable. The simulating equations take the concrete form below, where a denotes the input value:

$$X_1(0, a) = 0$$

$$X_2(0, a) = a$$

$$IC(0, a) = 1$$

For $y \geq 0$ we have

$$X_1(y + 1, a) = \begin{cases} X_1(y, a) + 1 & \text{if } IC(y, a) = 2 \\ X_1(y, a) & \text{otherwise} \end{cases}$$

$$X_2(y + 1, a) = \begin{cases} X_2(y, a) \div 1 & \text{if } I(y, a) = 3 \\ X_2(y, a) & \text{otherwise} \end{cases}$$

$$IC(y + 1, a) = \begin{cases} 5 & \text{if } IC(y, a) = 1 \wedge X_2(y, a) = 0 \\ 2 & \text{if } IC(y, a) = 1 \wedge X_2(y, a) > 0 \\ 1 & \text{if } IC(y, a) = 4 \wedge X_1(y, a) = 0 \\ 1 & \text{if } IC(y, a) = 4 \wedge X_1(y, a) > 0 \\ 5 & \text{if } I(y, a) = 5 \\ IC(y, a) + 1 & \text{otherwise} \end{cases}$$

$\square$

2.4.22 Example. Let M be the program below

$$1 : \mathbf{x}_1 \leftarrow \mathbf{x}_1 + 1$$

$$2 : \mathbf{x}_1 \leftarrow \mathbf{x}_1 \div 1$$

$$3 : \mathbf{x}_1 \leftarrow a$$

$$4 : \textbf{if } \mathbf{x}_1 = 0 \textbf{ goto } 1 \textbf{ else goto } 1$$

$$5 : \textbf{stop}$$

where x_1 is the input variable. As of necessity, x_1 is the output variable as well. The simulation equations are

$$X_1(0, a) = 0$$

$$IC(0, a) = 1$$

For $y \geq 0$ we have

$$X_1(y + 1, a) = \begin{cases} X_1(y, a) + 1 & \text{if } IC(y, a) = 1 \\ X_1(y, a) \div 1 & \text{if } IC(y, a) = 2 \\ a & \text{if } IC(y, a) = 3 \\ X_1(y, a) & \text{otherwise} \end{cases}$$

$$IC(y + 1, a) = \begin{cases} 1 & \text{if } IC(y, a) = 4 \wedge X_1(y, a) = 0 \\ 1 & \text{if } IC(y, a) = 4 \wedge X_1(y, a) > 0 \\ 5 & \text{if } I(y, a) = 5 \\ IC(y, a) + 1 & \text{otherwise} \end{cases}$$

□

Here is a ubiquitous application:

2.4.23 Theorem. *For every URM $M_{x_1}^{\vec{x}_n}$ —where the input/output choice has been indicated— there is a primitive recursive predicate $T_M(\vec{a}_n, y)$ —depending on M— and a primitive recursive "output function" $\lambda y \vec{a}_n.out_M(y, \vec{a}_n)$ —also depending on M— that behave as follows*

$$T_M(\vec{a}_n, y) \equiv \text{the URM } M_{x_1}^{\vec{x}_n} \text{ has reached \textbf{stop} in } y \text{ steps}$$

and

$$M_{x_1}^{\vec{x}_n} = \lambda \vec{a}_n.out_M\left((\mu y)T_M(\vec{a}_n, y), \vec{a}_n\right) \tag{1}$$

Proof Let X_i be the simulating functions for the variables of M and IC as above. Let **stop** be the k-th instruction of M. Then

$$T_M(\vec{a}_n, y) \equiv IC(y, \vec{a}_n) = k$$

and

$$out_M(y, \vec{a}_n) = X_1(y, \vec{a}_n), \text{ for all } y, \vec{a}_n$$

Now, if $(\exists y)T_M(\vec{a}_n, y)$, then $(\mu y)T_M(\vec{a}_n, y)$ finds the smallest such y.

Of course, $\lambda \vec{a}_n.(\mu y)T_M(\vec{a}_n, y) \in \mathcal{P}$. The validity of (1) is now obvious if we take $out_M = X_1$. $\qquad\square$

2.4.24 Remark. The predicates below are all in $\mathcal{PR}_*$ by closure operations (Theorems 2.2.5 and 2.2.14)

- $\neg T_M(\vec{a}_n, y)$; says "URM $M_{\mathbf{x}_1}^{\vec{\mathbf{x}}_n}$ with input $\vec{a}_n$ *did not yet reach* **stop** in y steps". Could it not be that $T_M(z, \vec{a}_n)$, for $z < y$? No, because of the **stop** semantics (Definition 1.1.2).

-

$$T_M(\vec{a}_n, y) \wedge (\forall z)_{<y} \neg T_M(\vec{a}_n, z)$$

says; "URM $M_{\mathbf{x}_1}^{\vec{\mathbf{x}}_n}$ with input $\vec{a}_n$ *first* reached **stop** in y steps".

- $(\exists z)_{<y} T_M(\vec{a}_n, z)$; says "URM $M_{\mathbf{x}_1}^{\vec{\mathbf{x}}_n}$ with input $\vec{a}_n$ reached **stop** *before* y steps".

In short, the three bullets above can be rephrased in plainer English (in order):

- The URM $M_{\mathbf{x}_1}^{\vec{\mathbf{x}}_n}$ with input $\vec{a}_n$ requires $> y$ steps to converge (if it does converge).
- The URM $M_{\mathbf{x}_1}^{\vec{\mathbf{x}}_n}$ with input $\vec{a}_n$ converges in a minimum of y steps.
- The URM $M_{\mathbf{x}_1}^{\vec{\mathbf{x}}_n}$ with input $\vec{a}_n$ converges in $< y$ steps. $\qquad\square$

2.4.25 Corollary. *Let $M_{\mathbf{x}_1}^{\vec{\mathbf{x}}_n}$ be as above and let $\lambda \vec{a}_n.f(\vec{a}_n) = M_{\mathbf{x}_1}^{\vec{\mathbf{x}}_n}$. Then with T_M as in Theorem 2.4.23 we have, for all $\vec{a}_n$,*

$$f(\vec{a}_n) \downarrow \equiv (\exists y)T_M(\vec{a}_n, y)$$

and

$$f(\vec{a}_n) = out_M\left((\mu y)T_M(\vec{a}_n, y), \vec{a}_n\right)$$

2.4.5 Pairing Functions

Coding of sequences $a_0, a_1, \ldots, a_n$, for $n \geq 1$, has a special case; pairing functions, that is, the case of $n = 1$. This special case is important in various theoretical considerations, especially when one wants to distance oneself from prime-power coding due to the latter's inefficiency, as compared to coding pairs with polynomial functions (rather than exponential).

2.4.26 Definition. A *total, 1-1* function $J : \mathbb{N} \times \mathbb{N} \to \mathbb{N}$ is called a *pairing function*. $\qquad\square$

2.4.27 Remark. A decoder is a pair of **total** functions K, L from $\mathbb{N} \to \mathbb{N}$ such that, for any z that is equal to $J(x, y)$ for some x and y, they "compute" said x and y:

$$K(z) = x$$

and

$$L(z) = y$$

That is, K and L decode appropriate z values.

On a non-code z, K and L give some *nonsense answer* —but answer they will; they are total!— just as the decoder $\lambda zi.(z)_i$ does when $Seq(z)$ is false.

One usually encounters the (capital) letters K, L in the literature as (generic) names for projection functions of some (generic) pairing function. In turn, the generic symbol for the latter is J rather than "f". We will conform to this notational convention in what follows.   □

The set of "tools" consisting of a pairing function J and its two projections K and L is a coding/decoding scheme for sequences of length two. We want to have computable such schemes and indeed there is an abundance of *primitive recursive* pairing functions that also have primitive recursive projections.

Some of those we will indicate in the examples below and others we will let the reader to discover in the exercises section.

2.4.28 Example. The function $J = \lambda xy.\langle x, y \rangle$ is pairing. Good decoders/projections are $K = \lambda z.(z)_0$ and $L = \lambda z.(z)_1$. All three are already known to us as members of $\mathcal{PR}$.

This J is not onto. For example, $5 \notin \operatorname{ran}(J)$. Nevertheless, K and L are total —because $\lambda iz.(z)_i$ is; indeed is in $\mathcal{PR}$.   □

2.4.29 Example. The function $J = \lambda xy.2^x 3^y$ is pairing. As its projections we normally take $K = \lambda z.\exp(0, z)$ and $L = \lambda z.\exp(1, z)$ (cf. Example 2.3.10). All three are already known to us as members of $\mathcal{PR}$.

This J is not onto either. Again, $5 \notin \operatorname{ran}(J)$. Nevertheless, K and L are total —because $\lambda iz.\exp(i, z)$ is; indeed is in $\mathcal{PR}$.   □

2.4.30 Example. The function $J = \lambda xy.2^x(2y + 1)$ due to Grzegorczyk is pairing. Indeed, since 2 is the only even prime, "$2^x(2y + 1)$" is a "forgetful" depiction of a number's prime-number decomposition, where all powers of odd primes are lumped together in "$2y + 1$". Clearly it is in $\mathcal{PR}$ since we have addition, multiplication, successor and $\lambda x.2^x$ in $\mathcal{PR}$.

$K = \lambda z.\exp(0, z)$ works: If $z = 2^x(2y + 1)$, then $Kz = x$.[7] What is L in closed form? (*Hint.* You may use functions from Example 2.3.10.)   □

 2.4.31 Example. In Examples 2.4.29 and 2.4.30 we note that $J(x, y) \geq x$ and $J(x, y) \geq y$, for all x, y. Thus an alternative way to prove that the related K and L are in $\mathcal{PR}$ is to compute as follows:

---

[7] In Example 2.1.8 we agreed to omit brackets whenever we can get away with it, as in SSx and here as in Kx.

$$Kz = (\mu x)_{\leq z}(\exists y)_{\leq z}(J(x, y) = z) \tag{1}$$

and

$$Lz = (\mu y)_{\leq z}(\exists x)_{\leq z}(J(x, y) = z) \tag{2}$$

Equipped with Theorems 2.2.14 and 2.3.2, and Definition 2.3.9, we see that (1) and (2) establish that K and L are in $\mathcal{PR}$.

Caution In some cases, e.g., in the Axt classes K_2, K_3 of Chap. 15 (*before* one shows that they are equal to Grzegorczyk's $\mathcal{E}^3, \mathcal{E}^4$) we do not have bounded quantification or search, thus solving for K, L *explicitly* is our only option.   □

2.4.32 Example. Here is a pairing function that does not require exponentiation. Let $J(x, y) = (x + y)^2 + x$. Clearly, $J \in \mathcal{PR}$.

So let us set $z = (x + y)^2 + x$ and solve this "equation" for x and y (uniquely, hopefully). Well, $(x + y)^2 \leq z < (x + y + 1)^2$. Thus $x + y \leq \sqrt{z} < x + y + 1$, hence

$$x + y = \lfloor \sqrt{z} \rfloor \tag{1}$$

Then, $z = \lfloor \sqrt{z} \rfloor^2 + x$ and therefore $Kz = z \dotminus \lfloor \sqrt{z} \rfloor^2$. By (1), $Lz = \lfloor \sqrt{z} \rfloor \dotminus Kz$.

As in Example 2.4.31, the J here satisfies $J(x, y) \geq x$ and $J(x, y) \geq y$.

The primitive recursiveness of K, L also follows from the calculations $Kz = (\mu x)_{\leq z}(\exists y)_{\leq z}(J(x, y) = z)$ and $Lz = (\mu y)_{\leq z}(\exists x)_{\leq z}(J(x, y) = z)$.   □

Why bother about pairing functions when we have the coding of sequences scheme of the previous subsection? Because prime-power coding is computationally "expensive", while quadratic schemes such as that of the previous example allow us to significantly reduce the "computational complexity (cost)" of coding/decoding.

 Apart from the issue of efficiency, coding via pairing functions is also applicable to computability on abstract domains (other than $\mathbb{N}$) where one goes to some extreme length to find a pairing scheme (not based on algebraic operations) and then uses it to do *coding of sequences* (cf. for example, Fenstad (1980)).

But back to our context, can we code *arbitrary* length sequences efficiently? Yes, because any J, K, L scheme can lead to a coding/decoding scheme for sequences $a_1, \ldots, a_n$, $n \geq 2$, for both the cases of a fixed or variable n (Definition 2.4.35 and Example 2.4.37 below).

2.4.33 Example. The pairing functions given by

$$J'(x, y) = \left\lfloor \frac{(x + y)(x + y + 1)}{2} \right\rfloor + x \tag{1}$$

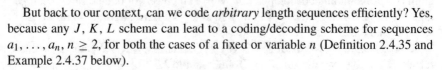

and

$$J''(x, y) = \left\lfloor \frac{(x + y)(x + y + 1)}{2} \right\rfloor + y \tag{2}$$

are onto. Let us show so for (2) in an easy but roundabout way. What (2) does is to give the *position* of the pair (x, y) in a *traversal* or *linearisation* (enumeration!) of the infinite matrix below walking along the *finite north-east diagonals*

$$
\begin{array}{cccccc}
(0, 0) & (0, 1) & (0, 2) & (0, 3) & (0, 4) & (0, 5) & \ldots \\
(1, 0) & (1, 1) & (1, 2) & (1, 3) & (1, 4) & (1, 5) & \ldots \\
(2, 0) & (2, 1) & (2, 2) & (2, 3) & (2, 4) & (2, 5) & \ldots \\
(3, 0) & (3, 1) & (3, 2) & (3, 3) & (3, 4) & (3, 5) & \ldots \\
\vdots & \vdots & \vdots & \vdots & \vdots & \vdots
\end{array}
$$

Thus the traversal produces

$$(0, 0), (1, 0), (0, 1), (2, 0), (1, 1), (0, 2), \ldots$$

the enumeration of *all* of $\mathbb{N} \times \mathbb{N}$.

So it is onto. But what is a "closed form" of it? We argue that it is (2) above.

Indeed, the n-th diagonal starts with the south-west entry $(n, 0)$ and ends with the north-east entry $(0, n)$.

(a) The components of *each* pair on the so determined diagonal add to n.
(b) The diagonal contains $n + 1$ entries.
(c) Thus we have $1 + 2 + 3 + \cdots + n = n(n + 1)/2$ entries *before* the diagonal that starts at $(n, 0)$.
(d) Therefore the j-th entry ($j = 0, 1, 2, \ldots, n$) on the diagonal "$(n, 0) \to (0, n)$" is in the position

$$n(n + 1)/2 + j \tag{3}$$

of the enumeration, counting $(0, 0)$ as the 0-th element, of course.

Let us express (3) differently: We are actually given (x, y) and we want its *position*, $J''(x, y)$ in the enumeration.

We relate this to points (a)–(d) above.

• The diagonal where (x, y) belongs starts at $(x + y, 0)$ and y is the y-th element on it, counting $(x + y, 0)$ as the 0-th.
• Thus, connecting with (a)–(d), $x + y = n$ and $y = j$.
• So, (3) morphs into (2), inserting $\lfloor \ldots \rfloor$ for proper form, even though

$$(x + y)(x + y + 1)/2$$

is an integer. $\qquad\qquad\qquad\qquad\qquad\qquad\qquad\qquad\qquad\qquad\qquad\qquad\quad\square$

2.4.34 Example. *(Minsky (1967) and Schwichtenberg (1969))* The last example is another non onto pairing function that reappears in Chap. 15 in Schwichtenberg's sharpening of the comparison between the $\mathcal{K}$ hierarchy and the Grzegorczyk hierarchy. We set

$$J = \lambda xy.2^{x+y+2} + 2^{y+1}$$

Intuitively the 1-1-ness is almost immediate: If $z = 2^{x+y+2} + 2^{y+1}$, then to solve for x and y do:

1. Convert z to binary notation. If the notation does not contain *exactly* two "1 bits", where the rightmost one is *not* at location zero, then z is *not* in the range of J; *Exit.*
2. Note the locations of the two ones l (left) and r (right).
3. Solve $l = x + y + 2$ and $r = y + 1$ *uniquely* for x and y.

Mathematically, note that

$$2^{x+y+2} \leq z < 2^{x+y+3}$$

thus

- $x + y + 2 = \lfloor \log_2 z \rfloor$
- $y + 1 = \left\lfloor \log_2 \left(z \dot{-} 2^{\lfloor \log_2 z \rfloor} \right) \right\rfloor$
- Thus, $Lz = \left\lfloor \log_2 \left(z \dot{-} 2^{\lfloor \log_2 z \rfloor} \right) \right\rfloor \dot{-} 1$ and $Kz = \lfloor \log_2 z \rfloor \dot{-} (Lz + 2)$.

The reader can verify that $\lambda z. \lfloor \log_2 z \rfloor \in \mathcal{PR}$. □

2.4.35 Definition. Given a primitive recursive pairing scheme J, K, L.

For any fixed $n \geq 2$ we define by induction on n the symbol $[\![a_1, \ldots, a_n]\!]^{(n)}$:
$[\![x, y]\!]^{(2)} = J(x, y)$; and $[\![x, y_1, \ldots, y_n]\!]^{(n+1)} = J\left(x, [\![y_1, \ldots, y_n]\!]^{(n)} \right)$. □

2.4.36 Exercise. Prove by induction on $n \geq 2$ that,

- The function $\lambda \vec{x}_n . [\![a_1, \ldots, a_n]\!]^{(n)}$ is total
- The function $\lambda \vec{x}_n . [\![a_1, \ldots, a_n]\!]^{(n)}$ is 1-1
- The function $\lambda \vec{x}_n . [\![a_1, \ldots, a_n]\!]^{(n)}$ is primitive recursive.

Therefore it codes n-tuples into numbers. □

2.4.37 Example. By the second and third bullets above, we can define Π_i^n, for $i = 1, \ldots, n$ —the projections— to satisfy

$$\text{If } z = [\![a_1, \ldots, a_n]\!]^{(n)}, \text{ then } \Pi_i^n(z) = a_i, \text{ for } i = 1, \ldots, n \qquad (1)$$

We may use an inductive definition with n as the induction variable:

$$\Pi_1^2 = K$$
$$\Pi_2^2 = L$$

and, for $n \geq 2$

$$\Pi_1^{n+1} = K$$
$$\Pi_{i+1}^{n+1} = \Pi_i^n L \quad , \text{for } i = 1, \ldots, n$$

Whether n is a fixed number or a variable it will be nice to have an explicit definition of the Π_i^n in terms solely of K and L (cf. Exercise 2.5.9) $\quad\square$

2.4.38 Exercise. By induction on n, show that

- $\Pi_i^n \left([\![a_1, \ldots, a_n]\!]^{(n)} \right) = a_i$, for $i = 1, \ldots, n$.
- The Π_i^n are total.
- The Π_i^n are onto.
- The Π_i^n are in $\mathcal{PR}$. $\quad\square$

2.5 Exercises

1. Let q be the quotient and r the remainder of the division a/b, that is,

$$a = bq + r, \text{ and } 0 \leq r < b \tag{1}$$

Recall that the *greatest common divisor*, gcd, of two positive natural numbers a and b is the maximum, d, among the *common* divisors of a and b. We write $d = \gcd(a, b)$.

- Prove that if a, b, r relate as in (1) then $\gcd(a, b) = \gcd(b, r)$.
  Base on this observation an algorithm that finds $\gcd(a, b)$.
- How many steps does this algorithm take as a function of a and b?
- Prove

  (a) $\gcd(a, a) = a$
  (b) If $a > b$, then $\gcd(a, b) = \gcd(a - b, b)$
  (c) $\gcd(a, b) = \gcd(b, a)$.

  Derive an algorithm on (a)–(c), which uses subtraction only (and if-tests). How many steps as a function of a and b does this algorithm take to completion?

2. Conclude Example 2.4.30.
3. Show that Grzegorczyk's pairing function is *onto* $\mathbb{N} - \{0\}$. What small modification will turn it into an onto $\mathbb{N}$ pairing function?
4. Prove that $\lambda n.n!$ —where $n! = 1 \times 2 \times 3 \times \cdots \times n$, the *factorial*— is in $\mathcal{PR}$.

5. Prove that the "switch" or "if-then-else" function $sw = \lambda xyz.if\ x = 0\ then\ y\ else\ z$ is in $\mathcal{PR}$.

6. Do Exercise 2.4.38.

7. Find K and L in closed form (it will involve the square root as was the case in Example 2.4.32) for each of the pairing functions J' and J'' of Example 2.4.33.

8. Do Exercise 2.4.36.

9. Find closed forms for the Π_i^n of Example 2.4.37 in terms of K and L only.

10. Prove that $\lambda z. \lfloor \log_2 z \rfloor \in \mathcal{PR}$. As $\log_2 0$ makes no sense mathematically (undefined), arrange (in the manner of Example 2.3.10) that we "redefine" $\lfloor \log_2 0 \rfloor \overset{Def}{=} 0$.

11. Prove that a course-of-values recursion defines one and only one number-theoretic function.

12. Without using definition by cases, show that both $\lambda xy.\max(x, y)$ and $\lambda xy.\min(x, y)$ are in $\mathcal{PR}$.

13. Define a new kind of bounded search, $(\overset{\circ}{\mu}y)_z$ by

$$(\overset{\circ}{\mu}y)_{<z} f(y, \vec{x}) \overset{Def}{=} \begin{cases} \min\{y : y < z \wedge f(y, \vec{x}) & \text{if } (\exists y)_{<z} f(y, \vec{x}) = 0 \\ 0 & \text{othw} \end{cases}$$

Prove that $\mathcal{PR}$ and $\mathcal{R}$ are closed under $(\overset{\circ}{\mu}y)_{<x}$.

14. Assume the class of functions $\mathcal{C}$ is closed under $(\overset{\circ}{\mu}y)_{<x}$ and substitution. Then its class of predicates, $\mathcal{C}_* \overset{Def}{=} \{f(\vec{x}) = 0 : f \in \mathcal{C}\}$, is closed under $(\exists y)_{<z}$.

15. Are the assumptions in 14 enough to also work with $(\mu y)_{<z}$?

16. Given h, g and j in $\mathcal{PR}$ such that $j(y, \vec{x}) \leq y$, for all $y, \vec{x}$. Then the function f defined by the schema

$$f(0, \vec{x}) = h(\vec{x})$$
$$f(y + 1, \vec{x}) = g(y, \vec{x}, f(j(y, \vec{x}), \vec{x}))$$

is in $\mathcal{PR}$.

Hint. This is a special course-of-values recursion.

17. Suppose the *graph* of f, $\lambda y\vec{x}.y = f(\vec{x})$, is primitive recursive. Let also be known that

- f is total
- $f(\vec{x}) \leq g(\vec{x})$ and $g \in \mathcal{PR}$.

Then show that $f \in \mathcal{PR}$. Are both stated conditions in the two bullets required? Why?

18. Prove that the function

$$
\left. \lambda x.2^{2^{\cdot^{\cdot^{\cdot^{2}}}}} \right\} x2\text{'s}
$$

is in $\mathcal{PR}$.

19. Prove that the sets $\emptyset$ and $\mathbb{N}$ are in $\mathcal{PR}_*$.

20. Prove that every finite set $\{a_1, \ldots, a_n\}$ is primitive recursive.

21. Prove that $\lambda x. \left\lfloor 10^x \sqrt{2} \right\rfloor \in \mathcal{PR}$.

22. Prove that $\lambda x. \Big(\text{the } (x+1)\text{-th digit in the decimal expansion of } \sqrt{2} \Big)$ is in $\mathcal{PR}$.

23. Prove that $\lambda x. \left\lfloor \sqrt{x} \right\rfloor \in \mathcal{PR}$ without using bounded search.

Chapter 3
Loop Programs

Overview
The *loop programs* were introduced by Meyer and Ritchie (1967) as a program theoretic counterpart to the number theoretic introduction of the set of primitive recursive functions $\mathcal{PR}$. This programming formalism among other things connects the definitional (or structural) complexity of primitive recursive functions with their (run time) computational complexity, but this latter phenomenon will be addressed in Chap. 15. In the present chapter we will offer an easy informal proof that $\mathcal{PR}$ is an *incomplete formalism* for the concept of "computable function".

3.1 Syntax and Semantics of Loop Programs

Loop programs are very similar to programs written in FORTRAN, but have a number of simplifications, notably they lack an unrestricted do-while instruction (equivalently, they lack a goto instruction). What they do have is

(1) Each program references (uses) a finite number of variables that we denote metamathematically by single letter names (upper or lower case is all right) with or without subscripts or primes.[1]
(2) Instructions are of the following types (X, Y could be any variables below, including the case of two identical variables):

   (i) $X \leftarrow 0$
  (ii) $X \leftarrow Y$
 (iii) $X \leftarrow X + 1$

---

[1] The precise syntax of variables will be given in the last section of this chapter, but even after this fact we will continue using signs such as X, A, Z', Y''_{34} for variables —i.e., we will continue using metanotation.

© Springer Nature Switzerland AG 2022
G. Tourlakis, *Computability*, https://doi.org/10.1007/978-3-030-83202-5_3

(iv) **Loop** $X\ldots$ **end**, where "$\ldots$" represents a *sequence of syntactically valid instructions* (which, taken together, will be called in Definition 3.1.1 below, a "loop program"). The **Loop** part is matched or balanced by the **end** part as it will become evident by the inductive definition below (Definition 3.1.1).

Informally, the structure of loop programs can be defined by induction:

3.1.1 Definition. Every instruction of *type* (i)–(iii) *standing by itself* is a (*one-line*) *loop program*. If we already have two loop programs P and Q, then so are the strings in (a) and (b) below (these strings have also user-friendlier depictions as indicated):

(a) P;Q, built by *superposition* (concatenation) and normally denoted vertically, without the separator ";", like this:

$$P$$

$$Q$$

and, for any variable X (that *may or may not* occur in P),

(b) **Loop** X; P; **end**, called *loop closure* (of P), and normally written vertically without separators ";" like this:

Loop X

$$P$$

end

□

3.1.2 Definition. (Loop Program Semantics) *The set of all loop programs will be denoted by L.*

The informal semantics of loop programs is given by induction on the definition (above) of loop programs, as follows:

The computation starts with the top instruction of the program (just as is the case of the URM) and continues as described below. We imagine a *computation pointer* (or *instruction counter*) walking along the instructions —from the top toward the bottom of the program except as directed by a *loop instruction*; see below)— always pointing at the instruction the program will do *next* (so-called "current instruction") just as our finger would point if we were tracing a loop program "by hand".

1. *If* the *computation pointer* of a program overshoots the last instruction (physically) and points to the *first spot* immediately below the program —we say it points at the *empty instruction* there— *then the program stops.*
2. If the program R is just *one* of

- $X \leftarrow 0$. Then X gets the value 0 stored in it, the computation pointer now points to the empty instruction following and the program stops (exits, halts).
- $X \leftarrow Y$. Then X gets the value of Y stored in it. *The value of Y does not change.* The computation pointer now points to the empty instruction following and the program stops.
- $X \leftarrow X + 1$. Then X gets the value $k + 1$ if k was stored in it *before the execution of the instruction.* The computation pointer now points to the empty instruction following and the program stops.

3. If the program R is $P; Q$ —usually written without a ";", vertically,

$$R : \left\{ \begin{matrix} P \\ Q \end{matrix} \right.$$

we assume (I.H.) that we know the semantics of P and Q. So, we point to the top (first) instruction of P and start the computation. *If* and when the computation pointer reaches the first instruction of Q —that is, if P as *standalone* stops computing eventually— then the computation continues with the instructions of Q, and stops if and when it overshoots the last instruction of Q.

4. Let the program R be **Loop** $X;$ $P;$ **end** —usually written without any ";", vertically,

$$R : \left\{ \begin{matrix} \textbf{Loop } X \\ P \\ \textbf{end} \end{matrix} \right.$$

The semantics of R is in terms of the semantics of program superposition and on the I.H. that we understand how P computes: If k is the *original* value of X *before* the computation pointer *enters* the loop, then

- If $k = 0$, then the loop is *skipped*, so the computation pointer points to the empty instruction (below the end). The computation stops (halts).
- If $k > 0$, then we compute as if we had k copies of P on top of each other,

$$R : k \left\{ \begin{matrix} P \\ P \\ \vdots \\ P \end{matrix} \right.$$

This description makes clear that it is irrelevant whether X occurs in P. It is the value k —the *pre-entry value of X*— that *determines* how many times the computation goes around the loop.

$\square$

Before we get to the next definition we are reminded of the notation $P_Y^{\vec{X}_n}$ from the URM case: Here the symbol denotes (similarly) the *function* computed *by loop program P* if we use $\vec{X}_n = X_1, X_2, \ldots, X_n$ as the input variables and Y as the output variable. When we *initialise —externally to the program—* the input variables above with values $a_1, a_2, \ldots, a_n$ (the action may be *nicknamed* "we read the values a_j into the respective X_j") we also initialise *all* non input variables *that occur in the program* to zero, just as we do in the case of the URM.

Let us settle an important but easy result.

3.1.3 Theorem. *Every loop program terminates regardless of its specific initialisation.*

Proof Refer to the semantics (Definition 3.1.2) and do induction on the formation of programs R: Let R be

- $X \leftarrow X + 1$. This program stops.
- $X \leftarrow Y$. This program stops.
- $X \leftarrow 0$. This program stops.

- Assume that P *and* Q (regardless of initialisations) stop.
  Then trivially so does

$$R : \begin{cases} P \\ Q \end{cases}$$

because P will eventually exit (by I.H.) and thus the current instruction will be the first instruction of Q. But this also eventually halts (by I.H.)
- Assume that P (regardless of initialisations) stops.
  Then trivially so does

$$R : \begin{cases} \textbf{Loop } X \\ P \\ \textbf{end} \end{cases}$$

Indeed

1. X holds 0 initially, then the loop is skipped and the program halts.
2. So let X have $k > 0$ as initial value. Then by Definition 3.1.2 the stack of k programs P will eventually exit because each P does (by I.H.):
   Thus after placing the computation pointer on the *first* instruction of the *first* copy of P, it will eventually (I.H.) point to the first instruction of *the second copy*. Continuing like this, the instruction pointer eventually will reach and point to the first instruction of *the last copy*. But by the I.H. the last copy of P also halts.

We glossed over an induction on k here. The reader is encouraged to fill in the missing details (Exercise 3.5.1).

$\square$

We have at once:

3.1.4 Corollary. *Every* $P_Y^{\vec{X}} \in \mathcal{L}$ *is total.*

Here are some examples of loop programs:

3.1.5 Example.

1.

$$R : \left\{ X \leftarrow X + 1 \right.$$

$R_X^X = \lambda x.x + 1$ (refer to the same notation in Example 1.1.7) Of course, $\lambda x.x + 1 = \lambda y.y + 1 = \lambda X.X + 1$. In the λ-notations here x, y, X are apparent variables not open for substitution.

2.

$$R : \left\{ X \leftarrow Y \right.$$

Then $R_Y^X = \lambda x.0$ and similarly, $R_X^Y = \lambda y.y = \lambda x.x$ and $R_X^{XY} = \lambda xy.y$.

3. If we call the following program P, what is P_Y^Z?

$$X \leftarrow 0$$

$$Y \leftarrow 0$$

Loop Z

$$Y \leftarrow X$$

$$X \leftarrow X + 1$$

end

Note that if Z is k initially then X will hold k at the very end, since X starts as 0 and we increment it by 1, k times. Y always "remembers" the value of X *before* the increment. Thus, $P_Y^Z = \lambda z.z \doteq 1$. We managed to subtract by adding!

4. If R is

$$R : \left\{ \begin{array}{l} \textbf{Loop } Y \\ \quad \textbf{Loop } Z \\ \quad\quad X \leftarrow X + 1 \\ \quad \textbf{end} \\ \textbf{end} \end{array} \right.$$

Then $R_X^{YZ} = \lambda xy.x \times y$.   □

We define the set of loop programmable functions, $\mathcal{L}$:

3.1.6 Definition. The symbol $\mathcal{L}$ stands for $\{P_Y^{\vec{X}_n} : P \in L\}$.   □

3.2 $\mathcal{PR} \subseteq \mathcal{L}$

3.2.1 Theorem. $\mathcal{PR} \subseteq \mathcal{L}$.

Proof By induction over $\mathcal{PR}$ and brute-force (loop) programming:

Basis $\lambda x.x + 1$ is P_X^X where P is $X \leftarrow X + 1$. Similarly, $\lambda \vec{x}_n.x_i$ is $P_{X_i}^{\vec{X}_n}$ where P is

$$X_1 \leftarrow X_1; X_2 \leftarrow X_2; \ldots; X_n \leftarrow X_n$$

The case of $\lambda x.0$ is as easy.

How does one compute $\lambda \vec{x}\vec{y}.f(g(\vec{x}), \vec{y})$ if g is $G_Z^{\vec{X}}$ and f is $F_W^{Z\vec{Y}}$? One uses

$$\begin{pmatrix} G \\ F \end{pmatrix}_W^{\vec{X}\vec{Y}}$$

We ensure that Z is the only variable common between F and G.

Thus, in particular, none of the non-input variables of F are in G —nor are the $\vec{Y}$— so, immediately after G passes control to F, all non-input variables of F are still 0 and F computes correctly where its input Z gets the computed value from G while $\vec{Y}$ holds the original (unaltered by G) input.

The general case $\lambda \vec{x}_m.f(g_1(\vec{x}_m), \ldots, g_n(\vec{x}_m))$ is a corollary by repeating the process above.

Finally, we indicate in *pseudo-code* how to compute $f(x, \vec{y}_n)$ where

$$f(0, \vec{y}_n) = h(\vec{y}_n)$$
$$f(x + 1, \vec{y}_n) = g(x, \vec{y}_n, f(x, \vec{y}_n))$$

assuming we have loop programs H and G for h and g respectively. The pseudo-code is

$$z \leftarrow h(\vec{y}_n) \qquad \textbf{Computed as } H_Z^{\vec{Y}_n}$$

$$i \leftarrow 0$$

Loop x

$$z \leftarrow g(i, \vec{y}_n, z) \qquad \textbf{Computed as } G_Z^{I, \vec{Y}_n, Z}$$

$$i \leftarrow i + 1$$

end

Once again one has to eliminate side-effects. For starters, H and G must have no common variables, except $\vec{Y}_n$. Further, neither H nor G are allowed to change $\vec{y}_n$. G must not change i either —that is, the variables I, $\vec{Y}_n$ are *read-only. Every* non input variable W of G must be explicitly set to 0 ($W \leftarrow 0$) at the end of G —by a *modified* G if the original was not doing this— so that G correctly computes "according to its spec" every time we enter this sub program while we are looping around the loop x times.

Note that, correctly, the last value of i used by G is $x - 1$ (case $x > 0$). $\qquad \square$

3.3 $\mathcal{L} \subseteq \mathcal{PR}$

To handle the converse of Theorem 3.2.1 we will simulate the computation of a loop program P by an array of primitive recursive functions. The schema of simultaneous recursion will play a central role.

3.3.1 Definition. For any $P \in L$ and any variable Y in P, the symbol P_Y is an abbreviation of $P_Y^{\vec{X}_n}$, where $\vec{X}_n$ are *all* the variables that occur in P. $\qquad \square$

3.3.2 Theorem. $\mathcal{L} \subseteq \mathcal{PR}$.

Proof The plan is to use *induction over the definition of L* (Definition 3.1.1) —thus using L as a proxy for $\mathcal{L}$ because the latter was not inductively defined.

We will prove that for every $P \in L$ and any Y in P, $P_Y \in \mathcal{PR}$.

Why *is the above plan sufficient, for what we want*, which is to show statement (1) below?

$$\text{For all } P \in L, \text{ and all } \vec{X}_n \text{ and } Y \text{ in } P, \text{ we have } P_Y^{\vec{X}_n} \in \mathcal{PR} \qquad (1)$$

Because, say we picked a $P \in L$ and a Y in P as output variable. Say $\vec{X}_n, \vec{W}_m$ is the set of *all* variables in P. But then, if our plan succeeds we have that

$$P_Y = P_Y^{\vec{X}_n, \vec{W}_m} \in \mathcal{PR}$$

If we now set $P_Y = \lambda \vec{X}_n \vec{W}_m . f(\vec{X}_n, \vec{W}_m)$, we have $f \in \mathcal{PR}$, and —by Grzegorczyk substitution— we also have

$$\lambda \vec{X}_n . f(\vec{X}_n, \underbrace{0, \ldots, 0}_{m \text{ zeros}}) \in \mathcal{PR}$$

But

$$\lambda \vec{X}_n . f(\vec{X}_n, \underbrace{0, \ldots, 0}_{m \text{ zeros}}) = P_Y^{\vec{X}_n}$$

On to our plan then!

For the *basis*, we have cases:

- P is $X \leftarrow 0$. Then $P_X = \lambda x.0 \in \mathcal{PR}$.
- P is $X \leftarrow Y$. Then $P_X = \lambda xy.y \in \mathcal{PR}$, while $P_Y = \lambda xy.y \in \mathcal{PR}$.
- P is $X \leftarrow X + 1$. Then $P_X = \lambda x.x + 1 \in \mathcal{PR}$

Let next do the induction step:

(A) P is $Q; R$.

    (i) Case where *no variables are common* between Q and R. Let the Q variables be $\vec{z}_k$ and the R variables be $\vec{u}_m$.

        - What can we say about $\left(Q; R\right)_{z_i}$?

            If we let $\lambda \vec{z}_k . f(\vec{z}_k) = Q_{z_i}$, then $f \in \mathcal{PR}$ by the I.H. But then, so is $\lambda \vec{z}_k \vec{u}_m . f(\vec{z}_k)$ by Grzegorczyk operations. But this is $\left(Q; R\right)_{z_i}$.

        - Similarly we argue for $\left(Q; R\right)_{u_j}$.

    (ii) Case where $\vec{y}_n$ are common between Q and R. Let $\vec{z}$ and $\vec{u}$ be the *non-common* variables as above.

        ▶ Thus the set of variables of $\left(Q; R\right)$ is $\vec{y}_n \vec{z}_k \vec{u}_m$

Now, pick an output variable w_i.

- If w_i is among the z_j, then we are back to the first bullet of case (i).
- So let the w_i be a component of the vector $\vec{y}_n \vec{u}_m$ instead. This case is fully captured by the figure below.

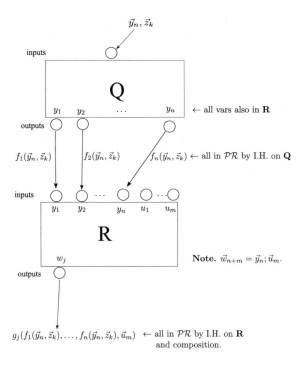

(B) P is **Loop** x; Q; **end**.

There are two subcases: x in Q, or not.

(a) x is not in Q:

So, let $\vec{y}_n$ be all the variables of Q; x is not being one of them. Let

$$\lambda x \vec{y}_n . f_0(x, \vec{y}_n) \text{ denote } P_x \qquad (2)$$

and, for $i = 1, \ldots, n$,

$$\lambda x \vec{y}_n . f_i(x, \vec{y}_n) \text{ denote } P_{y_i} \qquad (3)$$

Moreover, let

$$\lambda \vec{y}_n . g_i(\vec{y}_n) \text{ denote } Q_{y_i} \qquad (4)$$

By the I.H., the g_i are in $\mathcal{PR}$ for $i = 1, 2, \ldots, n$.

We want to prove that the functions in (2) and (3) are also in $\mathcal{PR}$. Since $f_0 = \lambda x \vec{y}_n . x$ (Why?), we only deal with the f_i for $i > 0$.

The plan is to set up a simultaneous recursion that produces the f_i from the g_i.

Now imagine the computation of P with input $x, y_1, \ldots, y_n$. We have two sub-subcases:

- $x = 0$.   In this sub-subcase, the loop is skipped and no variables are changed by the program. In terms of (2) and (3), what I just said translates into

$$f_0(0, \vec{y}_n) = 0 \tag{5}$$

and

$$f_i(0, \vec{y}_n) = y_i, \text{ for } i = 1, \dots, n \tag{6}$$

- $x = k + 1$, i.e., positive.   The effect of P is

$$k \text{ copies} \left\{ \begin{array}{l} Q \\ Q \\ Q \\ \vdots \\ Q \end{array} \right. \\ Q \tag{7}$$

What is $f(k + 1, \vec{y}_n)$? Well, consult the picture below:

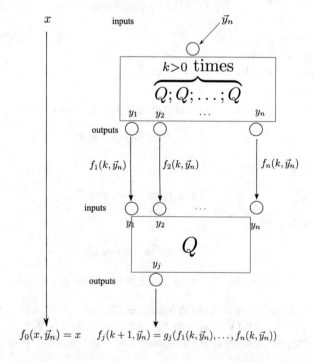

$$f_0(x, \vec{y}_n) = x \qquad f_j(k+1, \vec{y}_n) = g_j(f_1(k, \vec{y}_n), \dots, f_n(k, \vec{y}_n))$$

We now have a simultaneous primitive recursion that yields the f_i from the g_i. The latter being in $\mathcal{PR}$ by the I.H. on Q, so are the former.

(b) x is in Q:

So, let x, $\vec{y}_n$ be all the variables of Q, where we use x as an alias for y_0. Let

$$\lambda x \vec{y}_n . f_0(x, \vec{y}_n) \text{ denote } P_x \qquad (8)$$

and, for $i = 1, \ldots, n$,

$$\lambda x \vec{y}_n . f_i(x, \vec{y}_n) \text{ denote } P_{y_i} \qquad (9)$$

Moreover, let

$$\lambda x \vec{y}_n . g_0(x, \vec{y}_n) \text{ denote } Q_x \qquad (10)$$

$$\lambda x \vec{y}_n . g_i(x, \vec{y}_n) \text{ denote } Q_{y_i} \qquad (11)$$

By the I.H., the g_i are in $\mathcal{PR}$ for $i = 1, 2, \ldots, n$.

We want to prove that the functions in (8) and (9) are also in $\mathcal{PR}$ by employing an appropriate simultaneous recursion. The basis equations are the same as (5) and (6).

For $x = k+1$ we simply consult the figure below, to yield the recurrence equations

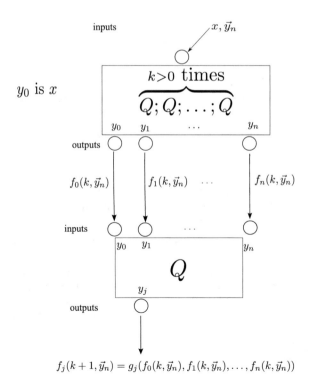

$$f_j(k+1, \vec{y}_n) = g_j(f_0(k, \vec{y}_n), f_1(k, \vec{y}_n), \ldots, f_n(k, \vec{y}_n))$$

$$f_j(k + 1, \vec{y}_n) = g_j(f_0(k, \vec{y}_n), f_1(k, \vec{y}_n), \ldots, f_n(k, \vec{y}_n)), j = 0, \ldots, n$$

As the g_j are in $\mathcal{PR}$, so are the f_j.

$\square$

All in all, we have that

$$\mathcal{PR} = \mathcal{L}$$

3.4 Incompleteness of $\mathcal{PR}$

For this section we will start by presenting the alphabet used to write loop programs.

$$\Sigma = \{X, 0, 1, \leftarrow, +, \textbf{Loop}, \textbf{end}, \P, ; \}$$

Bold face above is used to denote that words like **end** are to be looked upon as *single symbols*.

The variables are formally (syntactically) the strings of the form

$$X \underbrace{11\ldots1}_{n \text{ 1s}} \text{ —that is, } X1^n\text{— for } n \geq 1$$

but we will continue using metanotation for variables, as before: X, Y, Z, X'', y_{23}, z, x.

As with the URMs, we view the carriage control symbol $\P$ as invisible inter-instruction "glue" that has the indirect visual effect of forcing the next instruction to be written on a the next line.

We can now argue that $\mathcal{PR}$ cannot possibly contain *all* the *intuitively computable total* functions. We see this as follows:

(A) The *language* —that is, the set of loop programs L, and their syntax— are described sufficiently simply in Sect. 3.1 for the reader to accept that we can mechanically test whether a string over the alphabet Σ is a correctly formed program or not. In fact, see Exercise 3.5.2.

(B) We algorithmically build the list, $List_1$, of *all* strings over Σ: List by length; in each length group lexicographically.[2]

(C) Simultaneously to building $List_1$ we build $List_2$ as follows: For every string α that was put in $List_1$, copy it into $List_2$ iff $\alpha \in L$ (which we can test by (A)).

(D) Simultaneously to building $List_2$ build $List_3$: For every P (program) copied in $List_2$ copy all the finitely many strings P_Y^X (for all choices of X and Y in

---

[2] Fix the ordering of Σ on p.164 as listed above.

P) alphabetically. To this end, think of P_Y^X as the string "$P; X; Y$". Thus, the ordering is decided by the part "$X; Y$", which really is $X1^n; X1^m$ for some $n \geq 1$ and $m \geq 1$.

At the end of all this we have an algorithmic list of *all* the functions $\lambda x. f(x)$ of $\mathcal{PR}$, listed by their aliases, the P_Y^X. Let us call this list

$$f_0, f_1, f_2, \ldots, f_x, \ldots$$

By Cantor's diagonal method we define a new function d for all x as follows:

$$d(x) = f_x(x) + 1 \tag{1}$$

Two observations:

1. d is total (obvious, since each f_x is) and *intuitively* computable. Indeed, to compute $d(a)$ generate the lists long enough until you have the a-th item (counting as in $0, 1, 2, \ldots, a$) in $List_3$. This item has the format P_Y^X. I.e., we have found a *loop program* with its *designated input (one) and output variables*.
   Start this program with input the value a (in X). On termination add 1 to what Y holds and return. This is $d(a)$.
2. d is not in the list! For otherwise, $d = f_i$ for some $i \geq 0$. We get a contradiction:

$$f_i(i) \overset{\text{by } d=f_i}{=} d(i) \overset{\text{by (1) above}}{=} f_i(i) + 1$$

3.5 Exercises

1. Perform the omitted induction (on k) in the proof outline given in Theorem 3.1.3.
2. Write a program in C, or in your preferred programming language, to parse an input string over the alphabet Σ on p.164 and return "yes" (or "true") if the string is in L, but returns "no" (or "false") if it is not.
3. Using the incompleteness of $\mathcal{PR}$ methodology, build a relation $R(x)$ that is not in $\mathcal{PR}_*$.
4. Using the incompleteness of $\mathcal{PR}$ methodology, propose an intuitively computable function $\lambda xy. H(x, y)$ that *strictly* majorises every unary function in $\mathcal{PR}$ in this sense: Let $\lambda x. g(x) \in \mathcal{PR}$. Then for some i, $g(x) < H(i, x)$, for all x.
5. Can *any* function that so majorises (as the H of 4 does) the unary functions of $\mathcal{PR}$ be primitive recursive? Why?
6. Show constructively that we can program $if\ x = 0\ then\ y\ else\ z$ using a program in L.

7. Given the semantics of the **Loop-end** instruction pair, write a detailed URM program that simulates

Loop X

P

end

where P is a loop-program.

State your implementation strategy as a preamble to your solution; do not just offer unexplained program code.

8. Add line numbers to all lines of the loop program syntax. Can you now write a loop program "macro" (a set of instructions) that simulates **goto** L, where L is a line number? If yes, please show exactly how. If not, please prove why not.

9. Modify the loop program syntax as follows:

   (a) Add line numbers to all lines of a loop program.
   (b) Add the instruction **goto** L to the loop program instruction-set, where L is a line number.

   Via a two-way simulation show that the new loop-program formalism is *exactly* as powerful as that of the URM. That is, show that the two programming formalisms compute exactly the same functions, namely, those in $\mathcal{P}$.

10. Write a loop program that computes $\lambda x. \lfloor x/k \rfloor$.

11. Write a loop program that computes $\lambda x. rem(x, k)$.

12. Write a loop program that computes $\lambda x. \lfloor \sqrt{x} \rfloor$.

13. Suppose that all of $\lambda x.h(x)$, $\lambda x.i(x)$, $\lambda y.j(y)$ and $\lambda xyz.g(x, y, z)$ are in $\mathcal{PR}$, and $j(y) < y$, for all $y > 0$.

   Then the function f given by the following schema

$$f(x, 0) = h(x)$$

$$f(x, y) = g\Big(x, y, f\big(i(x), j(y)\big)\Big)$$

   is in $\mathcal{PR}$.

   Hint. Use a loop program to compute f. It is profitable to tabulate a few values of f for a small y. This is a course-of-values recursion where we modify the "parameter" (x) (Péter 1967).

14. Show that if h and g are in $\mathcal{PR}$, then so is f given by

$$f(0, y) = h(y)$$

$$f(x + 1, y) = g\Big(x, f\big(x, y + 1\big)\Big)$$

15. Consider the alphabet

$$A = \{1 \,,; \,, U \,, Z \,, S \,, C \,, c \,, R \,, r\}$$

We define the set of *names* $\mathscr{F}$ and their *ranks* over the alphabet A by the inductive definition below, that is, $\mathscr{F}$ is the *closure* of the *initial objects* set given in (I)–(III), under the *string operations* given in (IV) and (V).

16. Initial *names* and their *ranks*

(I) Z of *rank* 1.

(II) S of *rank* 1.

(III) $U1^n$; 1^m has rank n, $1 \le m \le n$, where 1^n means

$$\overbrace{11\dots 1}^{n}$$

Operations on *names* and their ranks.

(IV) From names F and G_i, $1 \le i \le n$, where F has rank n and the all the G_i have the same rank, say, m, the name formed as

$$CFG_1G_2\cdots G_nc$$

has rank m. This corresponds to *composition*, if the F and the G_i name $\mathcal{PR}$-derived functions.

(V) From names H and G, of ranks n and $n+2$ respectively, we build the name

$$RHGr$$

of rank $n+1$. This corresponds to *primitive recursion*, if the H (basis) and the G (iterator) name $\mathcal{PR}$-derived functions.

Intuitively, a name codes all the information contained in a primitive recursive derivation of the last member of the derivation, and the *rank* of the name is the *arity* of the derived function.

Prove

- $\mathscr{F}$ can be listed algorithmically.
- There is a 1-1 *algorithmic* (in the *informal* sense) correspondence between $\mathscr{F}$ and the set of all primitive recursive derivations.
- $\mathcal{PR}$ can be (informally speaking) algorithmically listed.
- The set of all unary $\mathcal{PR}$ functions can by listed by a total intuitively computable $\lambda i x.\xi(i, x)$.
- Prove that $\xi \notin \mathcal{PR}$.

Chapter 4
The Ackermann Function

Overview

The "Ackermann function" was proposed, of course, by Ackermann. The version here is a simplification by Robert Ritchie.

It provides us with an example of a *recursive* function that is *not* in $\mathcal{PR}$. Unlike the example in Chap. 3, which provided an alternative such function by diagonalisation, the proof that the Ackermann function is not primitive recursive is by a *majorisation argument*. Simply, it is too big to be primitive recursive!

But this function is more than just *intuitively* computable! It *is* computable —no hedging— as we will show by showing it to be a member of $\mathcal{R}$.

Another thing it does is that it provides us with an example of a function $\lambda \vec{x}. f(\vec{x})$ that is "hard to compute" (in the sense $f \notin \mathcal{PR}$) but whose *graph* —$\lambda y \vec{x}. y = f(\vec{x})$— is nevertheless "easy to compute" ($\in \mathcal{PR}_*$).[1]

4.1   Growth Properties

4.1.1 Definition. The Ackermann function, $\lambda nx. A_n(x)$, is given, for all $n \geq 0$, $x \geq 0$ by the equations

$$A_0(x) = \quad x + 2$$

$$A_{n+1}(x) = \quad A_n^x(2)$$

---

[1] Here the colloquialisms "easy to compute" and "hard to compute" are aliases for "primitive recursive" and "not primitive recursive", respectively. This is a hopelessly coarse rendering of *easy/hard* and a much better gauge for the *runtime complexity* of a problem is on which side of $O(2^n)$ it lies. However, our gauge will have to do for now: All I want to leave you with is that for some functions it is *easier* to compute the graph —to the quantifiable extent that it is in $\mathcal{PR}_*$— than the function itself, meaning that the latter fails to be primitive recursive.

© Springer Nature Switzerland AG 2022
G. Tourlakis, *Computability*, https://doi.org/10.1007/978-3-030-83202-5_4

where h^x is function iteration.   □

For any $\lambda y.h(y)$, the function $\lambda xy.h^x(y)$ is given by the primitive recursion

$$h^0(y) = y$$

$$h^{x+1}(y) = h\left(h^x(y)\right)$$

It is obvious then that if $h \in \mathcal{PR}$ then so is $\lambda xy.h^x(y)$.

 The λ-notation makes it clear that both n and x are *arguments* of the Ackermann function. While we could have written $A(n, x)$ instead, it is notationally less challenging to use the chosen notation. We refer to the n as the *subscript argument*, and to x as the *inner argument*.

4.1.2 Remark An alternative way to define the Ackermann function, extracted directly from Definition 4.1.1, is as follows:

$$A_0(x) = \qquad x + 2$$

$$A_{n+1}(0) = \qquad\qquad 2$$

$$A_{n+1}(x + 1) = \quad A_n(A_{n+1}(x))$$
□

4.1.3 Lemma.   *For each $n \geq 0$, $\lambda x.A_n(x) \in \mathcal{PR}$.*

Proof   Induction on n: For the basis, clearly $A_0 = \lambda x.x+2 \in \mathcal{PR}$. Assume now the case for (arbitrary, fixed) n —i.e., $A_n \in \mathcal{PR}$— and go to that for $n + 1$. Immediate from Remark 4.1.2, last two equations.   □

It turns out that the function blows up in size far too fast with respect to the argument n. We now quantify this remark.

The following unassuming lemma is the key to proving the growth properties of the Ackermann function. It is also the least straightforward to prove, as it requires a *double induction* —at once on n and x— as dictated by the fact that the "recursion" of Remark 4.1.2 does not leave any argument fixed.

 The above shows in particular that, for all n and all x, $A_n(x) \downarrow$. That is, $\lambda nx.A_n(x)$ is total.

4.1.4 Lemma.   *For each $n \geq 0$ and $x \geq 0$, $A_n(x) > x + 1$.*

Proof   We start an induction on n:
n-Basis. $n = 0$: $A_0(x) = x + 2 > x + 1$; true.
n-I.H.[2] For all x and a fixed (but unspecified) n, *assume $A_n(x) > x + 1$.*

---

[2] To be precise, what we are proving is "$(\forall n)(\forall x)A_n(x) > x+1$". Thus, as we start on an induction on n, its I.H. is "$(\forall x)A_n(x) > x + 1$" for a fixed unspecified n.

n-I.S.[3] For all x and the above fixed (but unspecified) n, we must *prove* $A_{n+1}(x) > x + 1$.

We do the n-I.S. by induction on x:

   x-Basis. $x = 0$: $A_{n+1}(0) = 2 > 1$; true.

   x-I.H. For the above fixed n, we now fix an x (but leave it unspecified) for which we *assume* $A_{n+1}(x) > x + 1$.

   x-I.S. For the above fixed (but unspecified) n and x, *prove* $A_{n+1}(x + 1) > x + 2$:

$$A_{n+1}(x + 1) = A_n(A_{n+1}(x)) \quad \text{by Remark 4.1.2}$$
$$> A_{n+1}(x) + 1 \quad \text{by } n\text{-I.H.}$$
$$> x + 2 \quad \text{by } x\text{-I.H.} \qquad \square$$

4.1.5 Lemma. $\lambda x.A_n(x) \nearrow$.

"$\lambda x.f(x) \nearrow$" means that the (total) function f is *strictly increasing*, that is, $x < y$ implies $f(x) < f(y)$, for any x and y. Clearly, to establish the property one just needs to check for the arbitrary x that $f(x) < f(x + 1)$.

Proof   We handle two cases separately.

   A_0: $\lambda x.x + 2 \nearrow$; immediate.

   A_{n+1}: $A_{n+1}(x + 1) = A_n(A_{n+1}(x)) > A_{n+1}(x) + 1$ —the ">" by Lemma 4.1.4. $\qquad \square$

4.1.6 Lemma. $\lambda n.A_n(x + 1) \nearrow$.

Proof $A_{n+1}(x + 1) = A_n(A_{n+1}(x)) > A_n(x + 1)$ —the ">" by Lemmata 4.1.4 (left argument > right argument) and 4.1.5. $\qquad \square$

The "$x + 1$" in Lemma 4.1.6 is important since $A_n(0) = 2$ for all n. Thus $\lambda n.A_n(0)$ is increasing but *not* strictly (constant).

4.1.7 Lemma. $\lambda y.A_n^y(x) \nearrow$.

Proof $A_n^{y+1}(x) = A_n(A_n^y(x)) > A_n^y(x)$ —the ">" by Lemma 4.1.4. $\qquad \square$

4.1.8 Lemma. $\lambda x.A_n^y(x) \nearrow$.

Proof   Induction on y: For $y = 0$ we want that $\lambda x.A_n^0(x) \nearrow$, that is, $\lambda x.x \nearrow$, which is true. We next take as I.H. that

$$A_n^y(x + 1) > A_n^y(x) \tag{1}$$

We want

$$A_n^{y+1}(x + 1) > A_n^{y+1}(x) \tag{2}$$

---

[3] To be precise, the step is to prove —from the basis and I.H.— "$(\forall x)A_{n+1}(x) > x + 1$" for the n that we fixed in the I.H. It turns out that this is best handled by induction on x.

But (2) follows from (1) and Lemma 4.1.5, by applying A_n to both sides of ">".   □

4.1.9 Lemma.   *For all* n, x, y, $A_{n+1}^y(x) \geq A_n^y(x)$.

Proof   Induction on y: For $y = 0$ we want that $A_{n+1}^0(x) \geq A_n^0(x)$, that is, $x \geq x$, which is true. We now take as I.H. that

$$A_{n+1}^y(x) \geq A_n^y(x)$$

We want

$$A_{n+1}^{y+1}(x) \geq A_n^{y+1}(x)$$

This is true because

$$A_{n+1}^{y+1}(x) = A_{n+1}\Big(A_{n+1}^y(x)\Big)$$

by Lemma 4.1.6

$$\geq A_n\big(A_{n+1}^y(x)\big)$$

Lemma 4.1.5 and I.H.

$$\geq A_n^{y+1}(x)$$ □

4.1.10 Definition.   Given a predicate $P(\vec{x})$, we say that $P(\vec{x})$ *is true almost everywhere* —in symbols "$P(\vec{x})$ a.e."— iff the set of (vector) inputs that make the predicate *false* is *finite*. That is, the set $\{\vec{x} : \neg P(\vec{x})\}$ is finite.

A statement such as "$\lambda x y . Q(x, y, z, w)$ a.e." can also be stated, less formally, as "$Q(x, y, z, w)$ a.e. *with respect to x and y*".   □

4.1.11 Lemma. $A_{n+1}(x) > x + l$ a.e. with respect to x.

Thus, in particular, $A_1(x) > x + 10^{350000}$ a.e.

Proof   In view of Lemma 4.1.6 and the note following it, it suffices to prove

$$A_1(x) > x + l \text{ a.e. with respect to } x$$

Well, since

$$A_1(x) = A_0^x(2) = \overbrace{(\cdots(((y+2)+2)+2)+\cdots+2)}^{x\ 2\text{'s}} \|_{\text{evaluated at } y = 2} = 2 + 2x$$

we ask: Is $2 + 2x > x + l$ a.e. with respect to x? It is so for all $x > l - 2$ (only $x = 0, 1, \ldots, l - 2$ fail).   □

4.1.12 Lemma. $A_{n+1}(x) > A_n^l(x)$ *a.e. with respect to x.*

Proof   If one (or both) of l and n is 0, then the result is trivial. For example,

$$A_0^l(x) = \overbrace{(\cdots(((x+2)+2)+2)+\cdots+2)}^{l\ 2\text{'s}} = x + 2l$$

We are done by Lemma 4.1.11.

Let us then assume that $l \geq 1$ and $n \geq 1$. We note that (straightforwardly, via Definition 4.1.1)

$$A_n^l(x) = A_n(A_n^{l-1}(x))$$

$$= A_{n-1}^{A_n^{l-1}(x)}(2) = A_{n-1}^{A_{n-1}^{A_n^{l-2}(x)}(2)}(2) = A_{n-1}^{A_{n-1}^{A_{n-1}^{A_n^{l-3}(x)}(2)}(2)}(2)$$

The straightforward observation that *we have a "ladder" of k A_{n-1}'s precisely when the topmost exponent is $l - k$* can be ratified by induction on k (left to the reader). Thus we state

$$A_n^l(x) = \left.\begin{array}{l} {}^{A_{n-1}^{\cdot^{\cdot^{A_{n-1}^{A_n^{l-k}(x)}(2)}}}} \\ A_{n-1}^{\cdot^{\cdot^{\cdot (2)}}} \end{array}\right\} k\ A_{n-1}$$

In particular, taking $k = l$,

$$A_n^l(x) = \left.\begin{array}{l} {}^{A_{n-1}^{\cdot^{\cdot^{A_{n-1}^{A_n^{l-l}(x)}(2)}}}} \\ A_{n-1}^{\cdot^{\cdot^{\cdot (2)}}} \end{array}\right\} l\ A_{n-1} = \left.\begin{array}{l} {}^{A_{n-1}^{\cdot^{\cdot^{A_{n-1}^{A_{n-1}^x(2)}}}}} \\ A_{n-1}^{\cdot^{\cdot^{\cdot (2)}}} \end{array}\right\} l\ A_{n-1} \qquad (*)$$

Let us now take $x > l$.

Thus, by $(*)$,

$$A_{n+1}(x) = A_n^x(2) = \left.\begin{array}{l} {}^{A_{n-1}^{\cdot^{\cdot^{A_{n-1}^2(2)}}}} \\ A_{n-1}^{\cdot^{\cdot^{\cdot (2)}}} \end{array}\right\} x\ A_{n-1} \qquad (**)$$

By comparing $(*)$ and $(**)$ we see that the first "ladder" is topped (after l A_{n-1} "steps") by x and the second is topped by

$$\left.\begin{array}{l} {}^{A_{n-1}^{\cdot^{\cdot^{A_{n-1}^2(2)}}}} \\ A_{n-1}^{\cdot^{\cdot^{\cdot (2)}}} \end{array}\right\} x-l\ A_{n-1}$$

Thus —in view of the fact that $A_n^y(x)$ increases with respect to each of the arguments n, x, y— we conclude by asking ...

"Is $\quad {}^{x-l}A_{n-1}\left\{ \begin{matrix} .^{.^{A_{n-1}^2(2)}} \\ A_{n-1}^{.^{.^{.}}} \end{matrix} \right. {}^{.^{.}}(2) > x$ a.e. with respect to x?"

... and answering, "Yes", because by $(**)$ this is the same question as "is $A_{n+1}(x - l) > x$ a.e. with respect to x?", which we answered affirmatively in Lemma 4.1.11. □

4.1.13 Lemma. *For all n, x, y, $A_{n+1}(x + y) > A_n^x(y)$.*

Proof

$$A_{n+1}(x + y) = A_n^{x+y}(2)$$
$$= A_n^x\left(A_n^y(2)\right)$$
$$= A_n^x\left(A_{n+1}(y)\right)$$
$$> A_n^x(y) \quad \text{by Lemmata 4.1.4 and 4.1.8}$$
□

4.2   Majorisation of $\mathcal{PR}$ Functions

We say that a function f *majorizes* another function, g, iff $g(\vec{x}) \leq f(\vec{x})$ for all $\vec{x}$. The following theorem states precisely in what sense "*the Ackermann function majorizes all the functions of $\mathcal{PR}$*".

4.2.1 Theorem. *For every function $\lambda\vec{x}.f(\vec{x}) \in \mathcal{PR}$ there are numbers n and k, such that for all $\vec{x}$ we have $f(\vec{x}) \leq A_n^k(\max(\vec{x}))$.*

Proof   The proof is by induction with respect to $\mathcal{PR}$. Throughout I use the abbreviation $|\vec{x}|$ for $\max(\vec{x})$ as this is notationally friendlier.

For the basis, f is one of:

- *Basis.*

  Basis 1. $\lambda x.0$. Then $A_0(x)$ works ($n = 0, k = 1$).
  Basis 2. $\lambda x.x + 1$. Again $A_0(x)$ works ($n = 0, k = 1$).
  Basis 3. $\lambda\vec{x}.x_i$. Once more $A_0(x)$ works ($n = 0, k = 1$): $x_i \leq |\vec{x}| < A_0(|\vec{x}|)$.

- *Propagation with composition.* Assume as I.H. that

$$f(\vec{x}_m) \leq A_n^k(|\vec{x}_m|) \tag{1}$$

and

$$\text{for } i = 1, \ldots, m, \; g_i(\vec{y}) \leq A_{n_i}^{k_i}(|\vec{y}|) \tag{2}$$

Then

$$f(g_1(\vec{y}), \ldots, g_m(\vec{y})) \le A_n^k(|g_1(\vec{y}), \ldots, g_m(\vec{y})|), \text{ by (1)}$$
$$\le A_n^k(|A_{n_1}^{k_1}(|\vec{y}|), \ldots, A_{n_m}^{k_m}(|\vec{y}|)|), \text{ by Lemma 4.1.8 and (2)}$$
$$\le A_n^k\left(|A_{\max n_i}^{\max k_i}(|\vec{y}|)|\right), \text{ by Lemmas 4.1.8 and 4.1.9}$$
$$\le A_{\max(n,n_i)}^{k+\max k_i}(|\vec{y}|), \text{ by Lemma 4.1.9}$$

- *Propagation with primitive recursion.* Assume as I.H. that

$$h(\vec{y}) \le A_n^k(|\vec{y}|) \tag{3}$$

and

$$g(x, \vec{y}, z) \le A_m^r(|x, \vec{y}, z|) \tag{4}$$

Let f be such that

$$f(0, \vec{y}) = h(\vec{y})$$
$$f(x+1, \vec{y}) = g(x, \vec{y}, f(x, \vec{y}))$$

I claim that

$$f(x, \vec{y}) \le A_m^{rx}\left(A_n^k(|x, \vec{y}|)\right) \tag{5}$$

I prove (5) by induction on x:

For $x = 0$, I want $f(0, \vec{y}) = h(\vec{y}) \le A_n^k(|0, \vec{y}|)$. This is true by (3) since $|0, \vec{y}| = |\vec{y}|$.

As an I.H. assume (5) for fixed x.

The case for $x + 1$:

$$f(x+1, \vec{y}) = g(x, \vec{y}, f(x, \vec{y}))$$
$$\le A_m^r(|x, \vec{y}, f(x, \vec{y})|), \text{ by (4)}$$
$$\le A_m^r\left(\left|x, \vec{y}, A_m^{rx}\left(A_n^k(|x, \vec{y}|)\right)\right|\right), \text{ by the I.H. (5), and Lemma 4.1.8}$$
$$= A_m^r\left(A_m^{rx}\left(A_n^k(|x, \vec{y}|)\right)\right), \text{ by Lemma 4.1.8 and } A_m^{rx}\left(A_n^k(|x, \vec{y}|)\right) \ge |x, \vec{y}|$$
$$= A_m^{r(x+1)}\left(A_n^k(|x, \vec{y}|)\right)$$
$$\le A_m^{r(x+1)}\left(A_n^k(|x+1, \vec{y}|)\right), \text{ by Lemma 4.1.8}$$

With (5) proved, let me set $l = \max(m, n)$. By Lemma 4.1.9 I now get

$$f(x, \vec{y}) \leq A_l^{rx+k}(|x, \vec{y}|) \underset{\text{Lemma 4.1.13}}{<} A_{l+1}(|x, \vec{y}| + rx + k) \qquad (6)$$

Now, $|x, \vec{y}| + rx + k \leq (r + 1)|x, \vec{y}| + k$ thus, (6) and Lemma 4.1.5 yield

$$f(x, \vec{y}) < A_{l+1}((r + 1)|x, \vec{y}| + k) \qquad (7)$$

To simplify (7) note that there is a number q such that

$$(r + 1)x + k \leq A_1^q(x) \qquad (8)$$

for all x. Indeed, this is so since (easy induction on y) $A_1^y(x) = 2^y x + 2^y + 2^{y-1} + \cdots + 2$. Thus, to satisfy (8), just take $y = q$ large enough to satisfy $r + 1 \leq 2^q$ and $k \leq 2^q + 2^{q-1} + \cdots + 2$.

By (8), the inequality (7) yields, via Lemma 4.1.5,

$$f(x, \vec{y}) < A_{l+1}(A_1^q(|x, \vec{y}|)) \leq A_{l+1}^{1+q}(|x, \vec{y}|)$$

(by Lemma 4.1.9) which is all we want.

$\square$

 4.2.2 Remark. Reading the proof carefully we note that the *subscript argument* of the *majorant*[4] is precisely the maximum depth of nesting of primitive recursion that occurs in a derivation of f.

Pause In which derivation? There are infinitely many. ◀

Indeed, the initial functions have a majorant with subscript 0; composition has a majorant with subscript no more than the maximum subscript of the component parts —*no increase*; primitive recursion has a majorant with a subscript that is bigger than the *maximum* subscript of the h- and g-majorants by precisely 1. $\square$

4.2.3 Corollary. $\lambda nx.A_n(x) \notin \mathcal{PR}$.

Proof By contradiction: If $\lambda nx.A_n(x) \in \mathcal{PR}$ then also $\lambda x.A_x(x) \in \mathcal{PR}$ (identification of variables —the so-called *diagonalisation of $A_n(x)$*). By the theorem above, for some n, k, $A_x(x) \leq A_n^k(x)$, for all x, hence, by Lemma 4.1.12

$$A_x(x) < A_{n+1}(x), \text{ a.e. with respect to } x \qquad (1)$$

On the other hand, $A_{n+1}(x) < A_x(x)$ a.e. with respect to x —indeed for all $x > n + 1$ by 4.1.6— which contradicts (1). $\square$

---

[4] The function that does the majorising.

4.3 The Graph of the Ackermann Function Is in $\mathcal{PR}_*$

How does one compute a yes/no answer to the question

$$\text{``}A_n(x) = z?\text{''} \tag{1}$$

Thinking "recursively" (in the programming sense of the word), we will look at the question by considering three cases, according to the definition in the Remark 4.1.2:

(a) If $n = 0$, then we will directly check (1) as "is $x + 2 = z?$".
(b) If $x = 0$, then we will directly check (1) as "is $2 = z?$".
(c) In all other cases, i.e., $n > 0$ and $x > 0$, we shall naturally ask *two* questions (both *must* be answerable "yes" for (1) to be true):[5] "Is there a w such that $A_{n-1}(w) = z$ **and** also $A_n(x - 1) = w?$"

 Steps (a)–(c) are entirely analogous to steps in a proof. Just as in a proof we verify the truth of a statement via syntactic means, here we are verifying the truth of $A_n(x) = z$ by such means.

Steps (a) and (b) correspond to writing down axioms. Step (c) corresponds to attempting to prove B by applying MP (modus ponens) where we are looking for an A such that we have a proof of both A and $A \rightarrow B$. In fact, closer to the situation in (c) above is a proof step where we want to prove $X \rightarrow Y$ and are looking for a Z such that both $X \rightarrow Z$ and $Z \rightarrow Y$ are known to us theorems. Z plays a role entirely analogous to that of w above.

Assuming that we want to pursue the process (a)–(c) by pencil and paper or some other equivalent means, it is clear that the pertinent data that we are working with are *ordered triples* of numbers such as n, x, z, or $n - 1, w, z$, etc. That is, the *letter* "A", the *brackets*, the *equals sign*, and the *position of the arguments* (subscript vs. inside brackets) are just *ornamentation*, and the string "$A_i(j) = k$", *in this section's context*, does not contain any more information than the *ordered* triple "(i, j, k)".

Thus, to "compute" an answer to (1) we need to write down enough triples, in stages (or steps), as needed to justify (1): At each stage we may write a triple (i, j, k) down just in case *one* of (i)–(iii) holds:

(i) $i = 0$ and $k = j + 2$
(ii) $j = 0$ and $k = 2$
(iii) $i > 0$ and $j > 0$, and for some w, we have already written down the two triples $(i - 1, w, k)$ and $(i, j - 1, w)$.

 4.3.1 Remark. Since "(i, j, k)" abbreviates "$A_i(j) = k$", Lemma 4.1.4 implies that $j < k$. □

---

[5] Note that $A_n(x) = A_{n-1}(A_n(x - 1))$.

Our theory is more competent with numbers (than with pairs, triples, etc.) preferring to *code* tuples into single numbers. Thus if we were to carry out the pencil and paper algorithm *within our theory*, then we should code all these triples, which we write down step by step, by single numbers: We will use our usual prime-power coding, $\langle i, j, k \rangle$, to do so.

The verification process for $A_n(x) = z$, described in (a)–(c), is a sequence of steps of types (a), (b) or (c) that *ends* with the (coded) triple $\langle n, x, z \rangle$.

We will code such a sequence. We note that our computation is "tree-like", since a "complicated" triple such as that of case (iii) above *requires* two similar others to be already written down, each of which in turn will require two *earlier* similar others, etc., until we reach "leaves" [cases (i) or (ii)] that can be dealt with ouright without passing the buck.

This "tree", just like the tree of a mathematical proof, can be "linearised" and thus be arranged in a *sequence* of coded triples $\langle i, j, k \rangle$ so that the presence of a "$\langle i, j, k \rangle$" implies that all its dependencies appear *earlier* (to its left).

We will code the entire proof sequence by a single number, u, using prime-power coding.

The major result in this subsection is the theorem below, that given any number u, we can primitively recursively check whether or not it is a code of an Ackermann function computation:

4.3.2 Theorem. *The predicate*

$$Comp(u) \stackrel{Def}{=} u \ codes \ an \ Ackermann \ function \ computation$$

is in $\mathcal{PR}_*$.

Proof The auxiliary predicates $\lambda vu.v \in u$ and $\lambda vwu.v <_u w$ mean

$$u = \langle \ldots, v, \ldots \rangle \qquad (v \ \text{is member of the coded sequence})$$

and

$$u = \langle \ldots, v, \ldots, w, \ldots \rangle \qquad (v \ \text{appears before} \ w \ \text{in the code} \ u)$$

respectively. Both are in $\mathcal{PR}_*$ since

$$v \in u \equiv Seq(u) \wedge (\exists i)_{<lh(u)}(u)_i = v$$

and

$$v <_u w \equiv Seq(u) \wedge (\exists i, j)_{<lh(u)}\big((u)_i = v \wedge (u)_j = w \wedge i < j\big)$$

The right hand side of "$\equiv$" below rests the case of the proof.

$$Comp(u) \equiv Seq(u) \wedge (\forall v)_{\leq u}\left(v \in u \rightarrow Seq(v) \wedge lh(v) = 3 \wedge\right.$$

{**Comment** : Case (i), p. 177} $\left\{(v)_0 = 0 \wedge (v)_2 = (v)_1 + 2 \vee\right.$

     {**Comment** : Case (ii)} $(v)_1 = 0 \wedge (v)_2 = 2 \vee$

     {**Comment** : Case (iii)} $\left[(v)_0 > 0 \wedge (v)_1 > 0 \wedge\right.$

$$(\exists w)_{<(v)_2}\left(\langle (v)_0 \doteq 1, w, (v)_2\rangle <_u v \wedge \langle (v)_0, (v)_1 \doteq 1, w\rangle <_u v\right)\Big]\Big\}\Big)$$

Remark 4.3.1 justifies the bound on $(\exists w)$ above. $\square$

Thus $A_n(x) = z$ iff $\langle n, x, z\rangle \in u$ for some u that satisfies $Comp$. In short

$$A_n(x) = z \equiv (\exists u)(Comp(u) \wedge \langle n, x, z\rangle \in u) \tag{1}$$

If we succeed in finding a bound for u that is a primitive recursive function of n, x, z then we will have succeeded showing:

4.3.3 Theorem. $\lambda nxz.A_n(x) = z \in \mathcal{PR}_*$.

Proof We assume a computation u that as soon as it verifies $A_n(x) = z$ quits, that is, it only includes $\langle n, x, z\rangle$ (at the very end) and all the needed *predecessor* coded triples, but nothing else. How big can u be?

Note that

$$u = \cdots p_r^{\langle i,j,k\rangle+1} \cdots p_l^{\langle n,x,z\rangle+1} \tag{2}$$

for appropriate l ($=lh(u) - 1$). For example, if all we want is to verify $A_0(10) = 12$, then $u = p_0^{\langle 0,10,12\rangle+1}$.

Similarly, if all we want is to verify $A_1(1) = 4$, then —since the "recursive calls" here are to $A_0(2) = 4$ and $A_1(0) = 2$— two possible u-values work: $u = p_0^{\langle 0,2,4\rangle+1} p_1^{\langle 1,0,2\rangle+1} p_2^{\langle 1,1,4\rangle+1}$ or $u = p_0^{\langle 1,0,2\rangle+1} p_1^{\langle 0,2,4\rangle+1} p_2^{\langle 1,1,4\rangle+1}$.

How big need l be? No bigger than needed to provide distinct *positions* ($l + 1$ such) in the computation, for all the "needed" triples $\langle i, j, k\rangle$. Since z is the largest possible output (and larger than any input) that is computed, there are no more than $(z + 1)^3$ triples possible, so $l + 1 \leq (z + 1)^3$. Therefore, (2) yields

$$u \leq \cdots p_r^{\langle z,z,z\rangle+1} \cdots p_l^{\langle z,z,z\rangle+1}$$

$$= \left(\Pi_{i \leq l} p_i\right)^{\langle z,z,z\rangle+1}$$

$$\leq p_l^{(l+1)(\langle z,z,z\rangle+1)}$$

$$\leq p_{(z+1)^3}^{(z+1)^3(\langle z,z,z\rangle+1)}$$

Setting $g = \lambda z.p_{(z+1)^3}^{(z+1)^3((z,z,z)+1)}$ we have $g \in \mathcal{PR}$ and we are done by (1):

$$A_n(x) = z \equiv (\exists u)_{\leq g(z)}(Comp(u) \wedge \langle n, x, z \rangle \in u) \qquad \qquad \square$$

Worth noting: If f is total and $y = f(\vec{x})$ is in $\mathcal{PR}_*$, then it does *not* necessarily follow that $f \in \mathcal{PR}$, as 4.3.3 exemplifies. On the other hand, if f is total and $y = f(\vec{x})$ is in $\mathcal{R}_*$, then, trivially, $f \in \mathcal{R}$ since $f = \lambda \vec{x}.(\mu y)(y = f(\vec{x}))$.

What is missing from the preceding expression is a *primitive recursive bound* on the search (μy), and this absence does not allow us to conclude that f is primitive recursive even when its graph is. For example, such a bound is impossible in the Ackermann case as we know from its growth properties.

4.3.4 Theorem. $\lambda nx.A_n(x) \in \mathcal{R}$.

Proof $\lambda nx.A_n(x) = \lambda nx.(\mu z)(z = A_n(x))$. But $\lambda nxz.z = A_n(x)$ is in $\mathcal{PR}_*$, thus $\lambda nx.A_n(x) \in \mathcal{P}$. But this function is total! $\qquad \qquad \square$

4.4   Exercises

1. Explore applying the technique of this chapter to proving the claim of Exercise 3.5.13. In particular, do an arithmetisation and show that the graph of f is in $\mathcal{PR}_*$. Moreover see if you can majorise f by A_n for a fixed n.
2. In a similar way explore Exercise 3.5.14.
3. Prove that, for $n \geq 0$, $A_n(x) < A_x(2)$ a.e.
   As corollaries also prove

   (a) If $\lambda x.f(x) \in \mathcal{PR}$, then $f(x) < A_x(2)$ a.e.
   (b) $\lambda x.A_x(2) \notin \mathcal{PR}$.

4. Show that it is impossible to obtain $\mathcal{PR}$ as the closure of a finite set of initial functions *under substitution only*.
   Hint. Prove that the functions of such a closure would be majorised by A_n for some fixed n.
5. Péter (1967) Let g_1, g_2 and h be in $\mathcal{PR}$. Prove that the "double unnested" recursion below produces an $f \in \mathcal{PR}$.

$$f(0, y) = g_1(y)$$
$$f(x + 1, 0) = g_2(x)$$
$$f(x + 1, y + 1) = h(x, y, f(x + 1y), f(x, y + 1))$$

Hint. Use the methodology of this chapter to show

- $f(x, y) \leq A_n^k(x + y)$, for some n, k and all x, y.

- Imitate the proof that $z = A_n(x) \in \mathcal{PR}_*$ to show the graph of f is in $\mathcal{PR}_*$
- Combine the observations in the preceding bullets to conclude that $f \in \mathcal{PR}$.

6. Prove 5 following a different *Hint*: Note that the value $f(x, y)$ is computed from values $f(i, j)$ such that $i + j + 1 = x + y$. Use then the "history function" $H(u) = \prod_{i \le u} p_i^{f(i,u-i)}$ and show that $H \in \mathcal{PR}$.

7. Refer to the informal universal function $\lambda i x. f_i(x)$ of Sect. 3.4.

- Prove that it is informally computable, but not in $\mathcal{PR}$.
- Prove that for any fixed n, there is an m, such that $A_n(x) < f_m(x)$, for all x.
- Prove that for any fixed m, there is an n, such that $f_m(x) < A_n(x)$ a.e.

Chapter 5
(Un)Computability via Church's Thesis

Overview

The aim of Computability is to *mathematically capture* (for example, via URMs) the *informal* notions of "algorithm" and "computable function" (or "computable relation").

Several mathematical models of computation, that were very different in their syntactic details and semantics, have been proposed in the 1930s by several researchers (Post, Church, Kleene, Turing), and, more recently, by Markov (1960) and Shepherdson and Sturgis (1963). They were promptly *proved to be equivalent*, that is, they all computed exactly the same number-theoretic functions, those in $\mathcal{P}$ that we introduced earlier.

This provided "empirical" evidence that prompted Church to state his *belief*, known as "*Church's Thesis*", that

> Every *informal* algorithm (pseudo-program) that we propose for the computation of a function can be implemented in each of the known mathematical models of computation. In particular, *it can be "programmed" as a URM.*

 We note that at the present state of our understanding of the concept of "algorithm" or "algorithmic process", *there is no known way* to define —via a pseudo-program of sorts— an "intuitively computable" function on the natural numbers, *which is outside of $\mathcal{P}$*. Thus, $\mathcal{P}$ appears to *formalise* the *largest* —i.e., most inclusive— set of "intuitively computable" functions (on the natural numbers) known.

Church's Thesis is not a theorem. It cannot be, as it "connects" precise mathematical objects (URM, $\mathcal{P}$) with imprecise *informal* ones ("algorithm", "computable function").

However, in use it provides the operational convenience and pedagogical advantage of concentrating on the *high level* of detail (essentially, no much detail at all) of why a construction does what we say it does —or why a mathematical definition produces a computable function— without overdoing the mathematics in the process.

© Springer Nature Switzerland AG 2022

G. Tourlakis, *Computability*, https://doi.org/10.1007/978-3-030-83202-5_5

In the *early chapters* of this book we will use Church's Thesis —to which we will refer, in short, as "CT"— to justify that various constructions we jot down yield computable functions. This will make arguments less detailed and more connected with experience, thus also easier; and shorter.

In the literature, Rogers (1967) relies on CT almost exclusively, despite the advanced level of his book. Cutland (1980) also delegates to Church's thesis the more involved proofs.

On the other hand, Davis (1958a), Tourlakis (1984, 2012) never use CT, and give all the necessary mathematical constructions (and proofs) in their full gory details. We will follow the fully mathematical approach in the later more advanced chapters, but even when we do use Church's thesis we will, invariably, also provide a fully mathematical proof.

5.1   The "Big Bang" for Logic and Computability

We have noted that computability is the part of logic that gives a mathematically precise formulation to the concepts *algorithm, mechanical procedure, computation*, and *calculable* or *computable* function.

Obtaining such a mathematical formulation, on one hand, directly responds to *Hilbert's Program* that included the determination of freedom from contradiction (consistency) of an axiomatic theory by "mechanical means".

On the other hand, mathematical logic was centrally involved in Hilbert's work as he strongly advocated *formalism* as a means to avoid paradoxes that sloppy semantic reasoning entails. Interestingly,[1] he also believed that one should be able to determine by *mechanical means* whether a formula in a theory is a theorem or not. This was his *decision problem* or *Entscheidungsproblem*.

Thus, on one hand Hilbert's Program acted catalytically to advances in logic: Gödel published his Incompleteness theorems (Gödel 1931) —each of which dealt a serious blow to Hilbert's Program: Formalised mathematics cannot certify (prove) all truths. For example, in formal arithmetic known as Peano arithmetic (PA) there are *infinitely many true sentences* (true in the standard interpretation over $\mathbb{N}$) that are *not* PA theorems. This was the content of the first incompleteness theorem. The second one says that PA cannot prove its own consistency, a fact which intuitively translates to *consistency —contrary to Hilbert's belief— cannot be proved by mechanical finitary means*.

As if this was not damaging enough to Hilbert's Program, Church (1936a) proved that the *Entscheidungsproblem* is not solvable by mechanical means —more precisely, the decision problem of PA is not so solvable.

---

[1] "Interestingly", because why would a leading mathematician want to mechanise theorem proving, or if not mechanise "proving", at least mechanise *detecting* provability back in the 1930s?

The terminology "mechanical procedure" or "means" was central in the work above. Intuitively the terminology was understood to mean a finitely described procedure similar to today's computer programs. But there were no computers or programs in the 1930s, however mathematicians could *recognise* an "*algorithm*" if they saw one. Arguably, an algorithm is a synonym for "mechanical procedure" and, for example, there was Euclid's algorithm for finding the "greatest (meaning *largest*) common divisor" of two natural numbers.

But if one wants to argue that a *mechanical process* or *algorithm* does *not exist* for a certain problem, one needs much more than the ability to recognise algorithms. Mathematicians need the ability to determine which tasks or problems admit such processes or algorithms and which do not —and to understand the "why"— and, on the other hand, to classify functions into two groups, those that are calculable (by a mechanical process!) and those that are not.

Powerful tools have been developed in such a theory of "mechanical procedures", *computability* —for some of which the reader is expected to become a competent user— that are used to prove that many tasks, indeed *uncountably*[2] infinitely many, do not admit mechanical procedures. We will see that one such problem is that of "program correctness": To *determine* whether an *arbitrary program* —say, written in C— is *faithful to its design specifications* for *all inputs*;[3] in short, if it is "correct".

Thus, to show that a problem admits a mechanical procedure solution the idea is straightforward: just *find* one.[4] This is a "programming" problem, and can be handled with some patience and ingenuity —in principle.

But how can I be *sure* that a mechanical procedure for a particular problem does *not exist*? Surely we cannot propose to try each one from the set of infinitely many mechanical procedures on our problem until we verify that none of them works?

Pause Can you convince yourself that there are infinitely many syntactically correct, say, C programs? ◄

To prove the negation of an existential statement such as this you need a *mathematical formulation* of what a "mechanical procedure" *precisely is* and develop and exploit mathematical tools to prove such "negative results".

Such tools will be mostly what we will be manufacturing in this volume. In fact there is more *Un*computability in in our theory than there is *computability*. As we remarked in the preface, this work is about the *limitations of the computing processes*, mostly.

Intensive activity by many pioneers of computability Post (1936, 1944); Kleene (1943); Church (1936b); Turing (1936, 1937); Markov (1960) led in the 1930s to several alternative formulations of the concepts of *computable* function and relation,

---

[2] In Cantor's sense of uncountable sets such as the set of the reals vs. countable sets such as the set of the natural numbers. Cantor explained the precise reason why the former set is "more infinite" than the latter. See how this applies to the set of functions that are *not* calculable: Sect. 1.2.

[3] Clearly, *not* by running the program on *all* inputs! We will not live long enough to see the answer!

[4] ... and prove that it works to specification!

each purporting to mathematically *capture* the concepts *algorithm, mechanical procedure*, and *computable function*.

All these formulations were proved to be pairwise equivalent; that is, the calculable functions admitted by any one of these formulations were the same as those that were admitted by any other. This led Alonzo Church to formulate his conjecture, widely known as "Church's Thesis", that *we did it*:

We *did* capture the *intuitive* concept of calculable function, by, say the mathematical concept of the *set of functions* $\mathcal{P}$.[5]

The eventual introduction of computers further fuelled the study of and research on the various mathematical frameworks of computation, "models of computation" as we often say, and *computability* is nowadays a vibrant and very extensive field.

5.2   A Leap of Faith: Church's Thesis

In this chapter we will derive results using Church's thesis when we need to prove that a certain computable function exists. Think of it as writing "pseudo code" only towards solving a programming problem.

Church's thesis, or CT as we will call it in this chapter, states

> Every *informal* algorithm (pseudo-program) that we propose for the computation of a function can be implemented (*made mathematically precise*, in other words) in each of the known mathematical models of computation. In particular, *it can be "programmed" as a URM*.

We noted that as early as the late 1950s there were dissenters towards Church's thesis, e.g., Kalmár (1957). A more recent critique was noted in the context of the so-called relativised computability[6] (with *partial* oracles) —where infinite-size *inputs* such as functions on the natural numbers are allowed. It was observed in

---

[5] I should be clear that even though this "thesis" has the flavour of a "completeness *theorem*" in the realm of computability, it is not.

In logic a *mathematical* definition for the intuitive (experiential) concept of *validity* or *universal truth* is given: a formula is *universally true* (in a technical mathematical sense) iff it is true in *all interpretations* ("interpretation" is also a technical, mathematical term). Gödel's *Completeness theorem* then states that every valid formula of logic has a syntactic (also called *formal*; depending only on form) *proof* —which is a *finite syntactic* object— without using any *mathematical* axioms.

But we have no *mathematical* definition for the intuitive (experiential) concept of "computable" function *a priori* —we are searching for one! Thus, the best we can do here is to *speculate* about the above mentioned *equivalent mathematical* formulations of "computable" function that each (fully) *captures* the intuitive notion of computable function.

In other words, Church's Thesis is an empirically formed belief rather than a provable result. It is not surprising that some researchers in this field, for example, Péter (1967) and Kalmár (1957) pointed out that it is conceivable that the *intuitive* concept of calculability may in the future be extended to exceed the power of the various mathematical models of computation that we currently know.

[6] Relativised computability is covered in Chap. 13.

Tourlakis (1986) via a specific example that CT fails: The example constructed an *intuitively computable functional*, as we call functions that have function inputs, that is not (*mathematically*) computable in the extant theories of computability with partial oracles. It is the function that *compares the lengths of two computations*. The reason of its uncomputability (mathematically) is that said function is *non monotone* with respect to the oracle argument, while the "standard" theories of computability with partial oracles, e.g., Davis (1958a), Kleene (1959), and Moschovakis (1969), compute only monotone functionals. Tourlakis (1986) introduced a mathematical model of *non monotone* computability where the aforementioned counterexample to Church's Thesis does not apply.

All told, as far as we know, no lasting critique has been offered to date towards the use of CT *to obtain informal shortcuts in obtaining results about* $\mathcal{P}$. In effect, at the present state of "evidence" the latter set appears to *formalise* the *largest* —i.e., most inclusive— set of "intuitively computable" functions (on the natural numbers) known.

While we use CT arguments in this chapter, in the remaining chapters we will offer mathematically exact proofs, possibly prefixed by an informal argument.

 Here is the template of **how** to use CT:

- We **completely** present —that is, no essential detail is missing— an algorithm in *pseudo-code*.
  Pause. "pseudo-code" does not mean "sloppy-code"! ◄
- We then say: By CT, there is a URM that implements our algorithm. Hence the function that our pseudo code computes is in $\mathcal{P}$.

5.3   The Effective List of All URMs

 For ease of reference we repeat some introductory material from Sect. 1.1. The new material in this section starts with Remark 5.3.2 below.

We recall from the definition of URM programs —introduced in Sect. 1 and in particular in 1.1.1— that these programs are *strings* over a finite alphabet A:

$$A = \{\leftarrow, +, \dot{-}, :, X, 0, 1, 2, 3, 4, 5, 6, 7, 8, 9, \mathbf{if}, \mathbf{else}, \mathbf{goto}, \mathbf{stop}, \P\}$$

Just like any other high level programming language, URM manipulates the contents of *variables. All variables are of natural number type.*

X and 1 finitely generate the variables of the URM programs as

$$X, X1, X11, \ldots X1^{n}, \ldots \tag{2}$$

where

$$1^n \overset{Def}{=} \overbrace{1 \ldots 1}^{n \; 1s}, \text{ and } 1^0 \overset{Def}{=} \lambda, \text{ the empty string}$$

while the symbols $0, 1, 2, \ldots, 9$ finitely generate natural *number constants* (in decimal notation) that we generically denote by a, b, c, with or without subscripts and primes, and *instruction labels* that we generically denote by L, R, P, with or without subscripts and primes.

As is customary for the sake of convenience, we will also utilise the bold face lower case letters $\mathbf{x}, \mathbf{y}, \mathbf{z}, \mathbf{u}, \mathbf{v}, \mathbf{w}$, with or without subscripts or primes as *meta* names that stand for unspecified strings of the type $X1^n$ in most of our discussions of the URM, and in examples of specific example programs (where yet more, convenient metanotations for variables may be employed such as X, Y, Z, with or without subscripts or primes).

We have defined that

5.3.1 Definition. (URM Programs) A *URM program* is a *finite (ordered) sequence* of instructions (or commands) of the following five types:

$$
\begin{aligned}
&L : \mathbf{x} \leftarrow a \\
&L : \mathbf{x} \leftarrow \mathbf{x} + 1 \\
&L : \mathbf{x} \leftarrow \mathbf{x} \dot{-} 1 \\
&L : \mathbf{stop} \\
&L : \mathbf{if}\ \mathbf{x} = 0\ \mathbf{goto}\ M\ \mathbf{else\ goto}\ R
\end{aligned}
\tag{3}
$$

where L, M, R, a, written in decimal notation, are in $\mathbb{N}$, and $\mathbf{x}$ is some variable. We call instructions of the last type *if-statements*.

Any two consecutive instructions in a syntactically correct URM program are separated ("glued") by the ¶ symbol that serves as an instruction separator.

 We chose ¶, the "hard return" symbol, for the role of instruction separator in order to be consistent with the expositional practise of *writing programs vertically*, one instruction per line. Thus, as in ordinary text, ¶ is invisible in the programs that we will write in these notes, but causes us to write the next instruction on the next line.

Each instruction in a URM program *must be numbered* by its *position number*, L, in the program, where ":" separates the position number from the instruction. We call these numbers *labels*. Thus, the label of the first instruction must always be "1". The instruction **stop** must *occur only once* in a program, as the *last instruction*. It is syntactically illegal for the *if-statement* $L :$ **if** $\mathbf{x} = 0$ **goto** M **else goto** R to refer to labels M and R that are not actually labels that occur in the program where the *if-statement* appears.   □

5.3.2 Remark. It is obvious from 1.1.1 that we can algorithmically check the syntactic correctness of a URM program. Further, if we assign a number to each alphabet symbol as in the matrix below then we can view each URM as a string of

$\leftarrow$	$+$	$\dot{-}$	$:$	X	0	1	2	3	4	5	6	7	8	9	**if**	**else**	**goto**	**stop**	¶
1	2	3	4	5	6	7	8	9	10	11	12	13	14	15	16	17	18	19	20

symbols from the (or, "over the", as we say) set

$$\{1, 2, 3, 4, 5, 6, 7, 8, 9, 10, 11, 12, 13, 14, 15, 16, 17, 18, 19, 20\}$$

which we interpret as a number base-21. Conversely, any number, which when viewed base-21 has no zero digits, represents a string over A.[7]

Therefore, a question like "is the predicate $URM(z)$ —that states 'z is a *string* over A that parses correctly as a URM'— decidable?" can be dealt with by our computability theory despite the fact that our theory deals only with number theoretic predicates and functions.

In fact, by CT and the opening sentence in this remark we can answer, "yes", viewing the string z as a number, written base-21, and the predicate $URM(z)$ as a number-theoretic predicate.[8]   □

Now we can show that we can algorithmically, or as we also say, *effectively* enumerate all URMs. This process is not different than the process we followed to enumerate all loop programs in 3.4.

5.3.3 Theorem. *The set of all URMs can be effectively enumerated, in the sense that there is a total computable function E of one variable such that*

- *For each z, we have $URM(E(z))$*
- *If $URM(w)$ is true, then for some z, $E(z) = w$*

Proof Consider the pseudo program below whose computation results in a *non-ending* enumeration of all URMs in a "standard" listing (sequence) $List_2$:

(A) We can algorithmically build the list $List_1$ of *all* strings over A: *List by increasing length* and in each length-group enumerate *in lexicographic order*.
(B) Simultaneously to building $List_1$, build $List_2$ as follows: For every string w placed in $List_1$, copy it into $List_2$ iff $URM(w)$ is true (cf. Remark 5.3.2).

A modification of the above pseudo-program can ensure that $E(z)$ is the z-th URM in the enumeration for all $z \in \mathbb{N}$:

---

[7] If we allow the digit zero then we lose the 1-1 correspondence between number "codes" and strings. For example, if we assigned code 0 to $\leftarrow$ then the strings $\leftarrow$, $\leftarrow\leftarrow$, $\leftarrow\leftarrow\leftarrow$, etc., *all* have numerical code 0. So 0 does not decode uniquely to a string under these circumstances.

[8] I.e., a subset of $\mathbb{N}$ in the one-variable case.

proc $E(z)$

(A') **Comment**. Given $z \geq 0$ as argument. The procedure $E(z)$ will output the z-th URM from $List_2$.

(B') $w \leftarrow 0$; **Comment**. Keeps track of how many strings u we placed in $List_2$.

(C') Algorithmically build the list $List_1$ as described above;

(D') Simultaneously to building $List_1$ build $List_2$ as follows:

For every string u placed in $List_1$

if $URM(u)$ is **true**, then **do**

$\Big\{$

- copy u into $List_2$;
- **if** $w = z + 1$, **then Return**(u) **else** $w \leftarrow w + 1$;

$\Big\}$

By CT, the above procedure defines a (total) computable function $\lambda z.E(z)$ such that $E(z)$ is the z-th URM. Why total? Because there are infinitely many URMs, thus, for every z there *will* be a z-th URM to be listed.   □

5.3.4 Corollary. *The set of all partial computable functions* of one variable *can be effectively enumerated using their URMs as proxies, that is, we enumerate the functions as* $N_{x'}^x$, *where N runs over the list of all URMs and the* **x** *and* **x'** *run over all choices of pairs of of input-output variables from among the variables of N.*

Every computable function f is some $N_{x'}^x$ and thus occupies *at least one* position i in the listing. Why not exactly one? Because for every N we can add at its end, but *before* the **stop** instruction, one or more instructions $\mathbf{z} \leftarrow 1$ where $\mathbf{z}$ is *fresh* (a new variable). Any one of the modified N, call it N', satisfies $N_{x'}^x = N'^x_{x'}$. Thus every function $f \in \mathcal{P}$ has infinitely many programs that compute it. This entails that the enumeration of Corollary 5.3.4 is with infinitely many repetitions, for each unary $f \in \mathcal{P}$.

Proof This has the status of a corollary since the proof is an easy modification of the theorem's proof. The obvious idea is to enumerate the $N_{x'}^x$ by using $List_2$ as source, and for *each N generated*, to list *all* strings $N00\mathbf{x}00\mathbf{x'}$ —which stand for $N_{x'}^x$, for all **x** and **x'** in N— *lexicographically* with respect to the "tail" $00\mathbf{x}00\mathbf{x'}$. Incidentally, for each N we have finitely many such strings.

Thus, if the enumerating function is called F, we have the pseudo-program below that computes $F(z)$ for $z \geq 0$. $F(z)$ is the z-th $N_{x'}^x$ in the listing of all computable partial functions of one argument.

proc $F(z)$

(A') **Comment**. Given $z \geq 0$ as argument. The procedure $F(z)$ will output the z-th unary partial computable function $N_{x'}^x$ —as $N00\mathbf{x}00\mathbf{x'}$— placed in $List_3$.

(B') $w \leftarrow 0$; **Comment**. Keeps track of how many strings u we placed in $List_3$.

(C') Algorithmically build the list $List_1$ as described earlier;

(D') Simultaneously to building $List_1$ also build $List_3$ as follows:

For every string u placed in $List_1$
if $URM(u)$ is **true**, then **do**
{

    for each pair of variables $\mathbf{x}$ and $\mathbf{x}'$ in the URM u, **do**
    {

        i. add the string $u00\mathbf{x}00\mathbf{x}'$ in $List_3$, arranging each of these additions in lexicographic order of the strings $00\mathbf{x}00\mathbf{x}'$.
        ii. **if** $w = z + 1$, **then Return**$(u00\mathbf{x}00\mathbf{x}')$ **else** $w \leftarrow w + 1$;

    }

}

By CT, the above procedure defines a (total) computable function $\lambda z.F(z)$ such that $F(z)$ is the z-th unary computable partial function. □

5.3.5 Corollary. *The set of all partial computable functions* of n variables *can be* effectively *enumerated using their URMs as proxies, that is, we enumerate them as* $N_{\mathbf{x}'}^{\vec{\mathbf{x}}_n}$, *where N runs over the list of all URMs and the $\vec{\mathbf{x}}_n$ and $\mathbf{x}'$ represent all pairs of choice of input (n-vector)-output variables from among the variables of N.*

Proof Trivial modification of the proof of Corollary 5.3.4. Here we enumerate, for each URM N that we find (in our URM enumeration), all functions $N_{\mathbf{x}'}^{\vec{\mathbf{x}}_n}$, for all choices of $\vec{\mathbf{x}}_n$ and $\mathbf{x}'$ in N. The symbol $N_{\mathbf{x}'}^{\vec{\mathbf{x}}_n}$ is rendered in one dimension as the string $N00\mathbf{x}_10\mathbf{x}_20\ldots0\mathbf{x}_n00\mathbf{x}'$. □

5.4 The Universal Function Theorem

The following is an extremely useful tool in the development of computability theory. It is Kleene's "*universal function theorem*".

5.4.1 Theorem. (Universal Function Theorem) *There is a* partial computable *two-variable function h with this property: For any one-variable function $f \in \mathcal{P}$, there is a number $i \in \mathbb{N}$ (dependent on f) such that $h(i, x) \simeq f(x)$ for all x. Equivalently, $\lambda x.h(i, x) = f$.*

Recall (0.4.5) that "$\simeq$" for partial function *calls*, $f(\vec{x})$ and $g(\vec{y})$, means the usual —equality of numbers— if both side are *defined*. $f(\vec{x}) \simeq g(\vec{y})$ is also true if both sides are *undefined*. In symbols,

$$f(\vec{x}) \simeq g(\vec{y}) \text{ iff } f(\vec{x}) \uparrow \wedge g(\vec{y}) \uparrow \vee (\exists z)\Big(f(\vec{x}) = z \wedge g(\vec{y}) = z\Big)$$

The "universality" of h lies in the fact that it (or the URM that computes it) acts like a "stored program" (i.e., general purpose or universal) "computer": To compute a function f we present both a *"program"* for f —*coded* as the number i— and the input *data* (the x) to h and then we let it crank along.

Proof   Each $\lambda x.f(x) \in \mathcal{P}$ is a $M_{\mathbf{y}}^{\mathbf{x}}$, by definition.

In Corollary 5.3.4 we proved that we can algorithmically enumerate *all* $\lambda x.f(x) \in \mathcal{P}$, with repetitions, by algorithmically enumerating all strings of the form $N_{\mathbf{x'}}^{\mathbf{x}}$ using the computable enumerator $\lambda i.F(i)$ that maps $i \in \mathbb{N}$ to the $N_{\mathbf{x'}}^{\mathbf{x}}$ —where N runs over all URMs.

Now we have three things to do:

1. *Define* $\lambda i x.h(i, x)$. Well, by the last sentence in the statement of the corollary, for each $i \in \mathbb{N}$, define $\lambda x.h(i, x)$ *to be* $F(i)$ from the proof of Corollary 5.3.4; that is, the $N_{\mathbf{x'}}^{\mathbf{x}}$ at *location* i in the enumeration of all $N_{\mathbf{x'}}^{\mathbf{y}}$.
2. Show that $h \in \mathcal{P}$. *Here is how the universal h is computed*

   - Given input i and x.
   - Call $F(i)$. This returns a unary computable function $N_{\mathbf{x'}}^{\mathbf{x}}$, for some N and variables $\mathbf{x}$ and $\mathbf{x'}$ in N.
   - Now run program N with x inputed into the input program-variable $\mathbf{x}$. If and when N stops, then we return the value held in the program-variable $\mathbf{x'}$ of N.

   By CT, $h \in \mathcal{P}$.
3. *Universality*: Given $\lambda x.f(x) \in \mathcal{P}$. Thus, $f = N_{\mathbf{x'}}^{\mathbf{x}}$, for some N and variables $\mathbf{x}$ and $\mathbf{x'}$ in N. By Corollary 5.3.4, there is a z such that $F(z) = N_{\mathbf{x'}}^{\mathbf{x}}$. By 1. above, h fulfils $\lambda x.h(z, x) = f$.
   $\qquad\qquad\qquad\qquad\qquad\qquad\qquad\qquad\qquad\qquad\qquad\qquad\qquad\qquad$ □

We will next introduce a standard notation due to Rogers (1967):

5.4.2 Definition.   In all that follows, ϕ_i will denote the i-ith unary function in *the algorithmic list of* all $M_{\mathbf{y}}^{\mathbf{x}}$.
   We rephrase: $\phi_i = F(i)$.
   $\qquad\qquad\qquad\qquad\qquad\qquad\qquad\qquad\qquad\qquad\qquad\qquad\qquad\qquad\qquad$ □

5.4.3 Remark.

(1) Equipped with the above definition we can rephrase the Universal Function Theorem 5.4.1 as

$$h(i, x) \simeq \phi_i(x), \text{ for all } i \text{ and } x$$

or even (better)

$$\lambda x.f(x) \in \mathcal{P} \text{ iff, for some } i \in \mathbb{N}, \text{ we have } f = \phi_i$$

It is worth "parsing" this "iff" above:
$\rightarrow$ direction: The hypothesis means $f = N_{\mathbf{v}}^{\mathbf{u}}$ for some N. If $N_{\mathbf{v}}^{\mathbf{u}}$ occupies location i in the list, then, by 5.4.2, $f = \phi_i$.
$\leftarrow$ direction: The hypothesis $f = \phi_i$ means that $f = N_{\mathbf{v}}^{\mathbf{u}}$, where $N_{\mathbf{v}}^{\mathbf{u}}$ occupies location i in the list. But, $f = N_{\mathbf{v}}^{\mathbf{u}}$ says that f is indeed computable; in $\mathcal{P}$.

(2) $\lambda i x.\phi_i(x) \in \mathcal{P}$ because $\lambda i x.h(i, x) \in \mathcal{P}$.

(3) Intuitively, Theorem 5.4.1 says that our theory is powerful enough to allow us to program a "compiler" for one-argument functions of $\mathcal{P}$: Indeed, a URM M with I/O convention such that $h = M_{\mathbf{z}}^{\mathbf{uv}}$ is such a compiler. In order to compute $\phi_x(y)$ we input the "program" x in $\mathbf{u}$ and the "data" y in $\mathbf{v}$ and, if and when the computation ends, $\mathbf{z}$ will hold the value $\phi_x(y)$.

(4) Calling x the "program" for $\lambda y.\phi_x(y)$ is not exact, but *is eminently apt*: x is just a number, not a set of URM instructions; but this number is the *address* (location) of *a URM program for $\lambda y.\phi_x(y)$. Given the address, we* can retrieve *this program from a list via a computational procedure, F of Corollary 5.3.4, in a finite number of steps!*

(5) *In the literature the address x in ϕ_x is called a ϕ-index. So, if $f = \phi_i$ then i is one of the infinitely many addresses where we can find how to program f.* □

5.4.4 Corollary. *For each $n \geq 1$, there is a partial computable $(n + 1)$-variable function $H^{(n+1)}$ with this property: For any n-variable function $f \in \mathcal{P}$, there is a number $i \in \mathbb{N}$ such that $H^{(n+1)}(i, \vec{x}_n) \simeq f(\vec{x}_n)$ for all $\vec{x}_n$. Equivalently, $\lambda \vec{x}_n.H^{(n+1)}(i, \vec{x}_n) = f$.*

Proof As that for h, but using the enumeration of the $N_{\mathbf{x}'}^{\vec{x}_n}$ instead (cf. Corollary 5.3.5). $H^{(2)} = h$. □

Correspondingly we extend Rogers' notation:

5.4.5 Definition. In all that follows, $\phi_i^{(n)}$ —that is, in terms of the notation in the preceding corollary also $\lambda \vec{x}_n.H^{(n+1)}(i, \vec{x}_n)$— will denote the i-th n-ary function in *the algorithmic list of all $M_{\mathbf{y}}^{\vec{x}_n}$*. Thus, $\phi_i^{(1)}$ is ϕ_i by definition (compare with 5.4.2). □

5.5 The Kleene T-Predicate and the Normal Form Theorems

5.5.1 Definition. (The Kleene T Predicate)
For any fixed $n > 0$, we define $T^{(n)}(z, \vec{a}_n, y)$ by

$$T^{(n)}(z, \vec{a}_n, y) \overset{Def}{\equiv} \text{the } z\text{-th URM } M_{\mathbf{x}_1}^{\vec{x}_n} \text{ (Corollary 5.3.5) on input } \vec{a}_n \text{ converges in } y \text{ steps}$$

If $n = 1$, then we write $T(z, a, y)$ for $T^{(1)}(z, a, y)$. □

5.5.2 Lemma. *For each $n > 0$, $T^{(n)}(z, \vec{a}_n, y)$ is in $\mathcal{PR}_*$.*

Proof Refer to Corollaries 2.4.25, 5.4.4 and Definition 5.4.5.

Let $M_{\mathbf{x}_1}^{\vec{x}_{n+1}}$ compute the universal $(n + 1)$-variable function $H^{(n+1)}$ of Corollary 5.4.4. This is universal for all n-argument partial recursive functions $\phi_i^{(n)}$:

$$H^{(n+1)}(i, \vec{a}_n) \simeq \phi_i^{(n)}(\vec{a}_n), \text{ for all } i \text{ and } \vec{a}_n$$

Our $T^{(n)}$ here is the T_M of 2.4.23, for the $(n+1)$-input URM

$$M_{\mathbf{x}_1}^{\vec{\mathbf{x}}_{n+1}} = \lambda z \vec{a}_n . H^{(n+1)}(z, \vec{a}_n)$$

that is,

$$T^{(n)}(z, \vec{a}_n, y) \text{ iff } T_M(z, \vec{a}_n, y) \text{ iff } IC(y, z, \vec{a}_n) = k$$

where k labels **stop** in $M_{\mathbf{x}_1}^{\vec{\mathbf{x}}_{n+1}}$.

Thus $T^{(n)}(z, \vec{a}_n, y)$ is in $\mathcal{PR}_*$.   □

The Kleene *Normal Form theorem* is a fundamental result and tool in computability. It states,

5.5.3 Theorem. (Kleene Normal Form) *For each fixed $n > 0$ we have an out $\in$ $\mathcal{PR}$ and, for all $z, \vec{a}_n$,*

(1) $\phi_z^{(n)}(\vec{a}_n) \downarrow \equiv (\exists y) T^{(n)}(z, \vec{a}_n, y)$ *and*
(2) $\phi_z^{(n)}(\vec{a}_n) \simeq out\Big((\mu y) T^{(n)}(z, \vec{a}_n, y), z, \vec{a}_n\Big)$

Proof Refer to 2.4.25, Corollary 5.4.4, 5.4.5, and Lemma 5.5.2.

Let $M_{\mathbf{x}_1}^{\vec{\mathbf{x}}_{n+1}}$ compute the universal $(n+1)$-variable function $H^{(n+1)}$ of Corollary 5.4.4. By 2.4.25, we have for all $z, \vec{a}_n$,

$$H^{(n+1)}(z, \vec{a}_n) \downarrow \equiv (\exists y) T_M(z, \vec{a}_n, y) \tag{*}$$

and

$$H^{(n+1)}(z, \vec{a}_n) \simeq out_M\Big((\mu y) T_M(z, \vec{a}_n, y), z, \vec{a}_n\Big) \tag{**}$$

where "T_M" and "out_M" are those of 2.4.25.

By the last remark in the proof of Lemma 5.5.2, "$T^{(n)}$ here is the T_M of 2.4.23" (here associated with $M_{\mathbf{x}_1}^{\vec{\mathbf{x}}_{n+1}}$). Replacing $H^{(n+1)}(z, \vec{a}_n)$ by $\phi_z^{(n)}(\vec{a}_n)$ in (*) and (**) and setting

$$out \stackrel{Def}{=} out_M$$

we obtain (1) and (2) of the theorem statement.   □

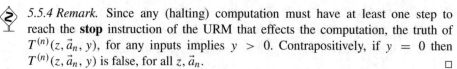

5.5.4 Remark. Since any (halting) computation must have at least one step to reach the **stop** instruction of the URM that effects the computation, the truth of $T^{(n)}(z, \vec{a}_n, y)$, for any inputs implies $y > 0$. Contrapositively, if $y = 0$ then $T^{(n)}(z, \vec{a}_n, y)$ is false, for all $z, \vec{a}_n$.   □

5.6   A Number-Theoretic Definition of $\mathcal{P}$

We know that $\mathcal{P}$ contains Z, S and all the U_i^n, for $n > 0$ and $1 \leq i \leq n$, and is closed under *composition*, (μy) (*μ-recursion*) and *prim*. Let us define then

5.6.1 Definition. ($\mathcal{P}$-Derivations)   The set

$$\mathcal{I} = \left\{ S, Z, \left(U_i^n \right)_{n \geq i > 0} \right\}$$

is the set of *Initial $\mathcal{P}$-functions*.[9]

   A *$\mathcal{P}$-derivation* is a finite (ordered!) sequence of number-theoretic functions,

$$f_1, f_2, \ldots, f_i, \ldots, f_n$$

where, for each i , one of the following holds

1. $f_i \in \mathcal{I}$.
2. $f_i = prim(f_j, f_k)$ and $j < i$ and $k < i$ —that is, f_j, f_k appear *to the left of* f_i.
3. $f_i = \lambda \vec{y}.g(r_1(\vec{y}), r_2(\vec{y}), \ldots, r_m(\vec{y}))$, and *all* of the $\lambda \vec{y}.r_q(\vec{y})$ and $\lambda \vec{x}_m.g(\vec{x}_m)$ appear *to the left of* f_i in the sequence.
4. $f_i = \lambda \vec{x}.(\mu y) f_r(y, \vec{x})$, where $r < i$.

Any f_i in a derivation is called a $\mathcal{P}$-derived function. The symbol $\widetilde{\mathcal{P}}$, stands for the set of $\mathcal{P}$-derived functions. That is,

$$\widetilde{\mathcal{P}} \overset{Def}{=} \{f : f \text{ is } \mathcal{P}\text{-derived}\} \qquad \qquad \square$$

The aim is to show that $\mathcal{P}$ is the set of all $\mathcal{P}$-derived functions as the terminology in Definition 5.6.1 ought to clearly betray. Of course, we could also have said that $\widetilde{\mathcal{P}}$ is the *closure* of $\mathcal{I}$ above, under the operations *composition* and *primitive recursion* and *unbounded search* (cf. Theorem 0.6.10).
   We will achieve our aim by proving $\mathcal{P} = \widetilde{\mathcal{P}}$.

   First a lemma:

5.6.2 Lemma. $\mathcal{PR} \subseteq \widetilde{\mathcal{P}}$.

Proof   Let $f \in \mathcal{PR}$. Then f is $\mathcal{PR}$-derived. But then it is also $\widetilde{\mathcal{P}}$-derived —a $\widetilde{\mathcal{P}}$-derivation need not necessarily use the (μy)-step 4 in Definition 5.6.1. So, $f \in \widetilde{\mathcal{P}}$.
                                                                                                      $\square$

5.6.3 Theorem. $\mathcal{P} = \widetilde{\mathcal{P}}$.

---

[9] Same as the set of initial $\mathcal{PR}$-unctions of 2.1.1.

Proof

Case $\mathcal{P} \supseteq \widetilde{\mathcal{P}}$: This is by an easy induction on the length of derivation of an $f \in \widetilde{\mathcal{P}}$. The basis (length=1) is since $\mathcal{I} \subseteq \mathcal{P}$. The induction steps 2–4 (from Definition 5.6.1) follow from the closure properties of $\mathcal{P}$.

Case $\mathcal{P} \subseteq \widetilde{\mathcal{P}}$: Let $\lambda \vec{x}_n . f(\vec{x}_n) \in \mathcal{P}$. By 5.5.3, for some i,

$$f = \lambda \vec{x}_n . out\Big((\mu y) T^n(i, \vec{x}_n, y), i, \vec{x}_n \Big) \tag{1}$$

By the lemma, the right hand side of (1) is in $\widetilde{\mathcal{P}}$ (recall also Theorem 2.4.23 and Lemma 5.5.2). So is f, then. $\qquad\qquad\qquad\qquad\qquad\qquad\qquad\qquad\qquad\square$

Among other things, Theorem 5.6.3 allows us to prove properties of $\mathcal{P}$ by induction on $\mathcal{P}$-derivation length, and to show that $f \in \mathcal{P}$ via a way other than URM-programming: Place f in a $\mathcal{P}$-derivation.

The number-theoretic characterisation of $\mathcal{P}$ given here was one of the foundations of computability proposed in the 1930s, due to Kleene.

5.7 The S-m-n Theorem

A fundamental theorem in computability is the *Parametrisation* or *Iteration* or also "*S-m-n*" theorem of Kleene. In fact, the *S-m-n*-theorem along with the universal function theorem and a handful of additional initial computable functions are known to be *sufficient* tools towards founding computability axiomatically —but we will not get into this matter in this volume.

5.7.1 Theorem. (Parametrisation Theorem) *For every* $\lambda x y . g(x, y) \in \mathcal{P}$ *there is a function* $\lambda x . f(x) \in \mathcal{R}$ *such that*

$$g(x, y) \simeq \phi_{f(x)}(y), \text{ for all } x, y \tag{1}$$

Preamble. (1) above is based on these observations: Given a program M that computes the function g as $M_{\mathbf{z}}^{\mathbf{uv}}$ with $\mathbf{u}$ receiving the input value x and $\mathbf{v}$ receiving the input value y —each via an "implicit" read statement— *we can, for any fixed value* x, *construct* a new program *dependent on the value* x, which behaves exactly as M does, because it consists of all of M's instructions, plus one more: The new program $N(x)$ —the notation "(x)" conveying the dependency of N on x— inputs x into $\mathbf{u}$ *explicitly* via an assignment statement *added* at the very top of M as $1 : \mathbf{u} \leftarrow x$.

Of course, if $x \neq x'$, *the programs* $N(x)$ *and* $N(x')$ *differ in their first instruction, so they are different.*

Let us denote, for each value x, the *position* of $N(x)_{\mathbf{z}}^{\mathbf{v}}$ in our standard effective enumeration of all the $N_{\mathbf{w}'}^{\mathbf{w}}$ by the expression $f(x)$, to convey the dependency on x.

Clearly the correspondence $x \mapsto f(x)$ is single-valued, and moreover, by the last remark in the preceding paragraph (in italics), it is a 1-1 function.

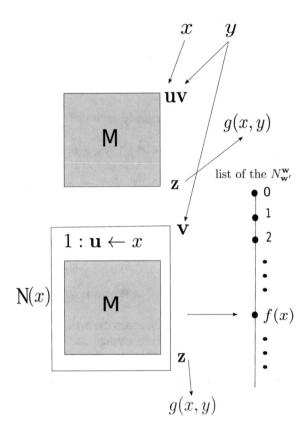

In sum, the new program, $N(x)$, constructed from M and the value x is at location $f(x)$ of the standard listing —in the notation of Corollary 5.3.4, $F(f(x)) = N(x)_{\mathbf{z}}^{\mathbf{y}}$. Thus $N(x)_{\mathbf{z}}^{\mathbf{y}}$ with input y outputs $g(x, y)$ for said x, that is, in the notation introduced in Definition 5.4.2, we have

$$g(x, y) \simeq \phi_{f(x)}(y), \text{ for all } y \text{ and the fixed } x \text{ —that is, for all } x \text{ and } y \qquad (2)$$

Proof Of the S-m-n theorem. The proof is encapsulated by the preceding figure, and much of the argument was already presented in the Preamble located between the two ⬙ signs above (in particular, we have shown (2)).

Below *we just settle the claim that we can compute the address $f(x)$ from x*, that is, $\lambda x. f(x) \in \mathcal{R}$.

So, fix an input x for the variable $\mathbf{u}$ of program M. Next, *construct $N(x)$*. A trivial algorithm exists for the construction:

- Given M and x.
- Modify M into $N(x)$ by adding $1 : \mathbf{u} \leftarrow x$ at the top of M as a *new* "first" instruction. See the above figure.
- Change nothing else in the M-part of $N(x)$, but do *renumber* all the original instructions of M, from "$L : \ldots$" to "$L + 1 : \ldots$".
  Of course, every original M-instruction of the type

$$L : \textbf{if } \mathbf{x} = 0 \textbf{ goto } P \textbf{ else goto } R$$

must also change "in its action part", namely, into

$$L + 1 : \textbf{if } \mathbf{x} = 0 \textbf{ goto } P + 1 \textbf{ else goto } R + 1$$

- Now —to compute $f(x)$— go down the effective list of **all** $N_{\mathbf{w}'}^{\mathbf{w}}$ and keep comparing to $N(x)_{\mathbf{z}}^{\mathbf{y}}$, until you find it in the list and return its address. More explicitly,

$$\textbf{proc } f(x)$$
$$\textbf{for } z = 0, 1, 2, \ldots \qquad\qquad \textbf{do}$$
$$\textbf{if } F(z) = N(x)_{\mathbf{z}}^{\mathbf{y}} \quad \textbf{then return } z$$

- The returned value z is equal to $f(x)$. Note that the if-test in the pseudo code will eventually succeed and terminate the computation, since *all* $N_{\mathbf{x}'}^{\mathbf{x}}$ are in the range of F of Corollary 5.3.4. In particular, this means that f is total.

By Church's thesis the informal algorithm above —described in five bullets— can be realised as a URM. Thus, $f \in \mathcal{R}$. $\qquad\qquad\qquad\qquad\qquad\qquad\qquad\qquad\square$

 Worth Repeating: It must not be lost between the lines what we have already observed: that *the S-m-n function f is 1-1*.

Two important corollaries suggest themselves:

5.7.2 Corollary. *For every $\lambda x \, \vec{y}_n . g(x, \vec{y}_n) \in \mathcal{P}$ there is a function $\lambda x . f(x) \in \mathcal{R}$ such that*

$$g(x, \vec{y}_n) \simeq \phi_{f(x)}(\vec{y}_n), \textit{ for all } x, \vec{y}_n$$

Proof Imitate the proof of Theorem 5.7.1 using the fact that we have an effective enumeration of all n-ary computable partial functions (Corollary 5.3.5). $\qquad\square$

5.7.3 Corollary. *There is a function $S_1^m \in \mathcal{R}$ of 2 variables such that*

$$\phi_i^{(m+1)}(x, \vec{y}_m) \simeq \phi_{S_1^m(i,x)}^{(m)}(\vec{y}_m), \textit{ for all } i, x, \vec{y}_m$$

Proof   The proof is that of Theorem 5.7.1 with a small twist: In the proof of Theorem 5.7.1 we start with a URM M for g. Here instead we have an *address* i of a URM for $\phi_i^{(m+1)}$, the latter being the counterpart of g in the current case.

The program $N(x)$ that we have built in the proof of Theorem 5.7.1 *depends* on the value x that is inputed via an assignment rather a read statement. Said program is a trivial modification of the program M for g, where the first input variable $\mathbf{u}$ loses its "input status" and participates instead in the very first instruction as "$1 : \mathbf{u} \leftarrow x$".

The corresponding program here we will call $N(i, x)$ due to its obvious dependence on i that (indirectly) tells us *which* program "M" for $\phi_i^{(m+1)}$ we start with.

So, the construction of $N(i, x)$ is

1. Fetch the program for $\phi_i^{(m+1)}$ found in location i of the effective listing of all $N_{\mathbf{x}'}^{\vec{\mathbf{x}}_{m+1}}$. Call it $M_{\mathbf{z}}^{\mathbf{u},\vec{\mathbf{v}}_m}$, where we have also indicated its input/output variables.
2. Build $N(i, x)$ by adding $1 : \mathbf{u} \leftarrow x$ before the first instruction of M. Shift all labels of M by 1, so that $N(i, x)$ is syntactically correct (cf. Theorem 5.7.1).
3. The $N(i, x)$ program, with its input/output variables indicated, is $N(i, x)_{\mathbf{z}}^{\vec{\mathbf{v}}_m}$ and can be located in the effective list of all $N(i, x)_{\mathbf{x}'}^{\vec{\mathbf{x}}_m}$ (cf. Corollary 5.3.5).

The argument for the recursiveness of S_1^m has a bit more subtlety than that of $f(x)$ of Theorem 5.7.1 due to the dependency on i. To compute the expression $S_1^m(i, x)$,

- Given i, x.
- Find the program at location i in the effective enumeration of all $N_{\mathbf{x}'}^{\vec{\mathbf{x}}_{m+1}}$. See step 1 in the construction above.
- Build $N(i, x)_{\mathbf{z}}^{\vec{\mathbf{v}}_m}$ as in step 2 above, and locate it in the effective list of all $N(i, x)_{\mathbf{x}'}^{\vec{\mathbf{x}}_m}$ (Corollary cf. 5.3.5).
- **Return** the address you found in the previous step. This is $S_1^m(i, x)$.

By CT and the 1–3 algorithm above, $S_1^m \in \mathcal{R}$. $\square$

5.7.4 Corollary.   *There is a function $S_n^m \in \mathcal{R}$ of $n + 1$ variables such that*

$$\phi_i^{(m+n)}(\vec{x}_n, \vec{y}_m) \simeq \phi_{S_n^m(i,\vec{x}_n)}^{(m)}(\vec{y}_m), \text{ for all } i, \vec{x}_n, \vec{y}_m$$

Proof   This is now easy! In step 1. in the previous proof fetch the program for $\phi_i^{(m+n)}$ —instead of that of $\phi_i^{(m+1)}$— found in location i of the effective listing of all $N_{\mathbf{x}'}^{\vec{\mathbf{x}}_{m+n}}$. Call it $M_{\mathbf{z}}^{\vec{\mathbf{u}}_n,\vec{\mathbf{v}}_m}$, where we have also indicated its input/output variables. The counterpart of step 2. above is now to place the *program segment* below *before* all instructions of M:

$$1 : \mathbf{u}_1 \leftarrow x_1$$
$$2 : \mathbf{u}_2 \leftarrow x_2$$
$$\vdots$$
$$n : \mathbf{u}_n \leftarrow x_n$$

taking all the $\mathbf{u}_i$ off input duty.

The rest is routine and entirely analogous with the preceding proof, thus is left to the reader.   □

The notation of the symbol S_n^m indicates that the first n variables of $\phi_i^{(m+n)}$ are taken off input duty while the last m of the original $m + n$ input variables have still input duty.

5.8   Unsolvable "Problems"; the Halting Problem

Some of the comments below (and Definition 5.8.1) occurred already in earlier sections (Definition 2.2.1). We revisit and introduce some additional terminology (e.g., the term "decidable").

Recall that a number-theoretic *relation* Q is a subset of $\mathbb{N}^n$, where $n \geq 1$. A relation's outputs are $\mathbf{t}$ or $\mathbf{f}$ (or "yes" and "no"). However, a number-theoretic relation *must* have values ("outputs") also in $\mathbb{N}$.

Thus we *re-code* $\mathbf{t}$ and $\mathbf{f}$ as 0 and 1 respectively. This convention is preferred by recursion theorists (as people who do research in computability like to call themselves) and is the opposite of the re-coding that, say, the C language employs (0 for $\mathbf{f}$ and non-zero for $\mathbf{t}$).

5.8.1 Definition. (Decidable Relations)   A relation $Q(\vec{x}_n)$ is *computable*, or *recursive*, or *decidable*, means that the function

$$c_Q = \lambda \vec{x}_n . \begin{cases} 0 & \text{if } Q(\vec{x}_n) \\ 1 & \text{otherwise} \end{cases}$$

is in $\mathcal{R}$.

The set of *all* computable relations we denote by $\mathcal{R}_*$.

By the way, we call the function $\lambda \vec{x}_n . c_Q(\vec{x}_n)$ —which does the re-coding of the outputs of the relation— the *characteristic function* of the relation Q ("c" for "characteristic").   □

Thus, "a relation $Q(\vec{x}_n)$ is computable or decidable" means that some URM computes c_Q. But that means that some URM behaves as follows:

On input $\vec{x}_n$, it halts and outputs 0 iff $\vec{x}_n$ satisfies Q (i.e., iff $Q(\vec{x}_n)$), it halts and outputs 1 iff $\vec{x}_n$ does *not* satisfy Q (i.e., iff $\neg Q(\vec{x}_n)$).

We say that the relation has a *decider*, i.e., the URM that *decides* membership of *any* tuple $\vec{x}_n$ in the relation.

5.8.2 Definition. (Problems)   A "*Problem*" is a formula of the type "$\vec{x}_n \in Q$" or, equivalently, "$Q(\vec{x}_n)$".

Thus, by definition, *a "problem" is a* membership question.   □

5.8.3 Definition. (Unsolvable Problems)   A problem "$\vec{x}_n \in Q$" is called any of
the following:

1. *Undecidable*
2. *Recursively unsolvable*
   or just
3. *Unsolvable*

iff $Q \notin \mathcal{R}_*$ —in words, iff Q is *not* a computable relation.   □

Here is the most famous undecidable problem:

$$\phi_x(x) \downarrow \tag{1}$$

A different formulation of problem (1) is

$$x \in K$$

where

$$K = \{x : \phi_x(x) \downarrow\}^{10} \tag{2}$$

that is, *the set of all numbers x, such that machine M_x on input x has a (halting!)
computation.*
    K we shall call the "*halting set*", and (1) we shall the "*halting problem*".

5.8.4 Theorem.   *The* halting problem *is unsolvable.*

Proof   We show, *by contradiction,* that $K \notin \mathcal{R}_*$.
    Thus we start by assuming the opposite.

$$\text{Let } K \in \mathcal{R}_* \tag{3}$$

that is, we can *decide membership* in K via a URM, or, what is the same, we can
decide truth or falsehood of $\phi_x(x) \downarrow$ for any x:
    Consider then the infinite matrix below, each row of which denotes a function in
$\mathcal{P}$ as an array of outputs, the outputs being a natural number, or the special symbol
"↑" for any undefined entry $\phi_x(y)$.

By Theorem 5.4.1 *each* one argument function of $\mathcal{P}$ sits in some row (as an array of
outputs).

---

[10] All three Rogers (1967), Tourlakis (1984), and Tourlakis (2012) use K for this set, but this
notation is by no means standard. It is unfortunate that this notation clashes with that for the first
projection K of a pairing function J. However the context will manage to fend for itself!

$$\phi_0(0) \ \phi_0(1) \ \phi_0(2) \ \ldots \ \phi_0(i) \ \ldots$$
$$\phi_1(0) \ \phi_1(1) \ \phi_1(2) \ \ldots \ \phi_1(i) \ \ldots$$
$$\phi_2(0) \ \phi_2(1) \ \phi_2(2) \ \ldots \ \phi_2(i) \ \ldots$$
$$\vdots$$
$$\phi_i(0) \ \phi_i(1) \ \phi_i(2) \ \ldots \ \phi_i(i) \ \ldots$$
$$\vdots$$

We will *show* that under the assumption (3) *that we hope to contradict*, the flipped diagonal[11] represents a *partial recursive function* as an array of outputs, and hence *must* fit the matrix along some row i since we have that all ϕ_i (as arrays) are rows of the matrix.

On the other hand, flipping the diagonal is diagonalising, *and thus the diagonal function constructed* cannot *fit. Contradiction!* So, we must blame (3) and thus we have its negation proved: $K \notin \mathcal{R}_*$

In more detail, or as most texts present this, we define the flipped diagonal for all x as

$$d(x) \simeq \begin{cases} \downarrow & \text{if } \phi_x(x) \uparrow \\ \uparrow & \text{if } \phi_x(x) \downarrow \end{cases}$$

Strictly speaking, the above does not *define* d since the "$\downarrow$" in the top case is not a value; it is ambiguous. Easy to fix:

One way to do so is

$$d(x) \simeq \begin{cases} 42 & \text{if } \phi_x(x) \uparrow \\ \uparrow & \text{if } \phi_x(x) \downarrow \end{cases} \tag{4}$$

Here is why the function in (4) is partial computable:
Given x, do:

- Use the decider for K (for $\phi_x(x) \downarrow$, that is) —assumed to exist by (3)— to test which condition obtains in (4); top or bottom.
- If the top condition is true, then we return 42 and stop.
- If the bottom condition holds, then transfer to an infinite loop, for example:

while $1 = 1$ **do**

end

---

[11] Flipping all $\uparrow$ red entries to $\downarrow$ and vice versa. This flipping is a mechanical procedure by *assumption* (3).

By CT, the 3-bullet program has a URM realisation, so d is computable.
  Say now

$$d = \phi_i \tag{5}$$

What can we say about $d(i) \simeq \phi_i(i)$? Well, we have two cases:

Case 1. $\phi_i(i) \downarrow$.   Then we are in the bottom case of (4). Thus $d(i) \uparrow$. But we
  also have $d(i) \simeq \phi_i(i)$ by (5), thus we have just contradicted the case
  hypothesis, $\phi_i(i) \downarrow$.

Case 2. $\phi_i(i) \uparrow$.   We have $d(i) = 42$ in this case, thus, $d(i) \downarrow$. By (5) $d(i) \simeq$
  $\phi_i(i)$, thus again we have contradicted the case hypothesis, $\phi_i(i) \uparrow$.
  So we reject (3).

$\square$

  In terms of *theoretical significance*, the above is perhaps the most significant
unsolvable problem that enables the process of discovering more! How?
  As a first example we illustrate the "program correctness problem" (see below).
But how does "$x \in K$" help? Through the following technique of *reduction*:

Let P be a new *problem* for which we want to see whether $\vec{y} \in P$ can be solved by
a URM. We build a *reduction* that goes like this:
  (1) Suppose that we have a URM M that decides $\vec{y} \in P$, for all $\vec{y}$.
  (2) Then we show how to use M as a subroutine *to also decide $x \in K$, for all x.*
  (3) Since the latter problem is unsolvable, *no such URM M exists!* In short, $P(\vec{y})$
is unsolvable too.

  The *equivalence problem* is

Given two programs M and N can we test to see whether they compute the same function?

Of course, "testing" for such a question *cannot be done by experiment*: We *cannot
just run M and N for all inputs* to see if they get the same output, because, for one
thing, "all inputs" are infinitely many, and, for another, there may be inputs that
cause one or the other program to run forever (infinite loop).

  By the way, the equivalence problem is the general case of the *"program
correctness"* problem which asks

Given a program P and a *program specification* S, does the program *fit* the specification *for
all inputs*?

since we can view a specification as just another formalism to express a function
computation. By CT, all such formalisms, programs or specifications, boil down
to URMs, and hence the above asks whether two given URMs compute the same
function —program equivalence.
  Let us show now that the program equivalence problem cannot be solved by any
URM.

5.8.5 Theorem. (Equivalence Problem) *The equivalence problem of URMs is the problem "given i and j; is $\phi_i = \phi_j$?"*[12]
   This problem is undecidable.

Proof   The proof is by a reduction (see above), hence by contradiction. We will show that if we have a URM that solves it, "yes"/"no", then we have a URM that solves the halting problem too!

$$\text{So let the URM } E \text{ solve the } equivalence problem. \tag{*}$$

Let us use it to answer the question "$a \in K$" —that is, "$\phi_a(a) \downarrow$", for any a.

$$\text{So, fix an } a \text{ that we want to test.} \tag{2}$$

Consider the following two computable functions given by:
For all x:

$$Z(x) = 0$$

and

$$\widetilde{Z}(x) \simeq \begin{cases} 0 & \text{if } x = 0 \wedge \phi_a(a) \downarrow \\ 0 & \text{if } x \neq 0 \end{cases}$$

Both functions are intuitively computable: For Z we already have shown a URM M that computes it. For $\widetilde{Z}$ and input x compute as follows:

- Print 0 and stop if $x \neq 0$.
- On the other hand, if $x = 0$ then, using the universal function h start computing $h(a, a)$, which is the same as $\phi_a(a)$ (cf. Theorem 5.4.1). If this ever halts just print 0 and halt; otherwise let it loop forever.

By CT, $\widetilde{Z}$ is in $\mathcal{P}$, that is, it has a URM program, say $\widetilde{M}$.
   We can *compute* the locations i and j of M and $\widetilde{M}$ respectively by going down the list of all $N_\mathbf{w}^\mathbf{w}$. Thus $Z = \phi_i$ and $\widetilde{Z} = \phi_j$.
   By assumption ($*$) above, we proceed to feed i and j to E. This machine will halt and answer "yes" (0) precisely when $\phi_i = \phi_j$; will halt and answer "no" (1) otherwise. But note that $\phi_i = \phi_j$ iff $\phi_a(a) \downarrow$. We have thus solved the halting problem since a is arbitrary! *This is a contradiction to the existence of URM E.*   □

---

[12] If we set $P = \{(i, j) : \phi_i = \phi_j\}$, then this problem is the question "$(i, j) \in P$?" or "$P(i, j)$?".

5.9 Exercises

1. Prove that the problem $\phi_x(y) \downarrow$ is unsolvable.
2. Let $\lambda xy.\langle x, y \rangle$ be any primitive recursive pairing function with primitive recursive projections Π_1 and Π_2 —that is, if $z = \langle x, y \rangle$, then $\Pi_1 z = x$ and $\Pi_2 z = y$.
   Prove that the set $\{\langle x, y \rangle : \phi_x(y) \downarrow\}$ is not recursive.
3. Prove that the problem $\lambda xyz.\phi_x(y) \simeq z$ is unsolvable.
   Hint. If instead it is recursive, then so is $\phi_x(x) = 1$, by Grzegorczyk substitution. Now apply diagonalisation.
4. Prove that the problem $\lambda xyz.\phi_x(y) > z$ is unsolvable.
5. Prove that the problem $\lambda xy.\phi_x(y) \simeq z$ is unsolvable.
6. Prove that the problem $\lambda x.(\exists y)\phi_x(y) \simeq 42$ —in words, "will the URM at address x ever print 42?" is unsolvable.

Chapter 6
Semi-Recursiveness

Overview

This chapter introduces the *semi-recursive* relations $Q(\vec{x})$. These play a central role in computability. As the name suggests these are kind of "half" recursive. Indeed, we can program a URM M to *verify* membership in such a Q, but if an input is not in Q, then our verifier will not tell us so; it will loop for ever. That is, M *verifies* membership but does *not decide* it in a yes/no manner, that is, by halting and "printing" the appropriate 0 (yes) or 1 (no).

Can't we tweak M into an M' that is a *decider* of such a Q? No, *not in general!* For example, our halting set K has a verifier but (provably) has no decider! This much we know: having a decider for K means $K \in \mathcal{R}_*$, and we know that this is not the case.

Since the "yes" of a verifier M is signaled by halting but the "no" is signaled by looping forever, the definition below does not require the verifier to print 0 for "yes". In this case "yes" equals "halting".

The chapter introduces a general *reduction* technique to prove that relations are not recursive or not semi-recursive according to the case. We prove the closure properties of the set of all semi-recursive relations and also the important *graph theorem*. We prove the projection theorem and also give a characterisation of semi-recursive sets as images of recursive functions. The startling theorem of Rice is included: *all* sets of the form $A = \{x : \phi_x \in \mathcal{C} \subseteq \mathcal{P}\}$ are not recursive, unless $A = \emptyset$ or $A = \mathbb{N}$.

6.1 Semi-Decidable Relations

6.1.1 Definition. (Semi-Recursive or Semi-Decidable Sets)   A relation $Q(\vec{x}_n)$ is *semi-decidable* or *semi-recursive* iff there is a URM, M, which on input $\vec{x}_n$ has a *(halting!)* computation iff $\vec{x}_n \in Q$. *The output of M is unimportant!*

A more mathematically precise way to say the above is:

© Springer Nature Switzerland AG 2022
G. Tourlakis, *Computability*, https://doi.org/10.1007/978-3-030-83202-5_6

A relation $Q(\vec{x}_n)$ is *semi-decidable* or *semi-recursive* iff there is an $f \in \mathcal{P}$ such that

$$Q(\vec{x}_n) \equiv f(\vec{x}_n) \downarrow \tag{1}$$

Since an $f \in \mathcal{P}$ is some $M_{\mathbf{y}}^{\vec{x}_n}$, M is a verifier for Q.

The set of *all* semi-decidable relations we will denote by $\mathcal{P}_*$.[1]   □

6.1.2 Remark. Yet another way to say (1) is:

A relation $Q(\vec{x}_n)$ is *semi-decidable* or *semi-recursive* iff there is an $e \in \mathbb{N}$ such that

$$Q(\vec{x}_n) \equiv \phi_e^{(n)}(\vec{x}_n) \downarrow \tag{2}$$

We call the e in (2) *a semi-recursive index* or *semi-index* of $Q(\vec{x}_n)$.

Of course, every semi-recursive $Q(\vec{x}_n)$ has infinitely many semi-indices since the f in (1) in Definition 6.1.1 is equal to $\phi_e^{(n)}$ for infinitely many e.

If $n = 1$, i.e., $Q \subseteq \mathbb{N}$, then we have the notation (Rogers 1967)

$$Q = W_e$$

for any semi-index e of Q.

Thus,

$$\boxed{x \in W_e \equiv \phi_e(x) \downarrow} \tag{*}$$

and

$$\boxed{x \notin W_e \equiv \phi_e(x) \uparrow} \tag{**}$$

□

We have at once

6.1.3 Theorem. (Kleene Normal Form for Predicates) *A relation $Q(\vec{x}_n)$ is semi-recursive with semi-index $i \in \mathbb{N}$ iff*

$$Q(\vec{x}_n) \equiv (\exists y)T^{(n)}(i.\vec{x}_n, y)$$

Proof *If*-part. By Theorem 5.5.3, $\phi_i^{(n)}(\vec{x}_n) \downarrow \equiv (\exists y)T^{(n)}(i.\vec{x}_n, y)$. Now, invoke Definition 6.1.1.

---

[1] This is not a standard notation in the literature. Most of the time the set of all semi-recursive relations has *no* symbolic name! We are using this symbol in analogy to $\mathcal{R}_*$ —the latter being fairly "standard".

Only if-part. By Definition 6.1.1, $Q(\vec{x}_n) \equiv \phi_i^{(n)}(\vec{x}_n) \downarrow$. Now, invoke Theorem 5.5.3.   □

The following figure shows the two modes of handling a query, "$\vec{x}_n \in A$", by a URM.

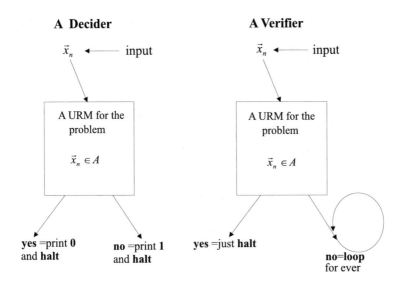

Here is an important semi-decidable set.

 6.1.4 Example. *K is semi-decidable.* We work within Definition 6.1.1. Note that the function $\lambda x.\phi_x(x)$ is in $\mathcal{P}$ via the universal function theorem of 5.4.1, namely, $\lambda x.\phi_x(x) = \lambda x.h(x, x)$ and we know $h \in \mathcal{P}$.

Thus $x \in K \equiv \phi_x(x) \downarrow$ settles it. By Definition 6.1.1 (1) we are done.   □

6.1.5 Example. Any recursive relation A is also semi-recursive.

That is,

$$\mathcal{R}_* \subseteq \mathcal{P}_*$$

Indeed, intuitively, all we need to do to convert a decider for $\vec{x}_n \in A$ into a verifier for $\vec{x}_n \in A$ is to "intercept" the "print 1"-step and convert it into an "infinite loop", **while**$(1)^2$
    {
    }
By CT we can certainly do that via a URM implementation.

A more explicit way (which still invokes CT) is to say, OK: Since $A \in \mathcal{R}_*$, it means that c_A, its characteristic function, is in $\mathcal{R}$.

---

[2] 1 in the C language evaluates as "true". One may also use **while**$(1 = 1)$.

Define a new function f as follows:

$$f(\vec{x}_n) \simeq \begin{cases} 0 & \text{if } c_A(\vec{x}_n) = 0 \\ \uparrow & \text{if } c_A(\vec{x}_n) \simeq 1 \end{cases}$$

This is intuitively computable since c_A is total computable (the "$\uparrow$" is implemented by the same **while** as above).

Hence, by CT, $f \in \mathcal{P}$. But

$$\vec{x}_n \in A \equiv f(\vec{x}_n) \downarrow$$

because of the way f was defined. Definition 6.1.1 rests the case.

One more way to do this: *Totally* mathematical ("formal", as one might say, incorrectly[3]) this time!

$$f(\vec{x}_n) = \text{if } c_A(\vec{x}_n) = 0 \text{ then } 0 \text{ else } \emptyset(\vec{x}_n)$$

That is using the sw function that is in $\mathcal{PR}$ and hence in $\mathcal{P}$, as in

$$\begin{array}{ccccc} & c_A(\vec{x}_n) & & 0 & \emptyset(\vec{x}_n) \\ & \downarrow & & \downarrow & \downarrow \\ f(\vec{x}_n) = \text{if} & z & = 0 \text{ then } u & \text{else} & w \end{array}$$

$\emptyset$ is, of course, the empty function which by Grzegorczyk operations can have any number of arguments we please! For example, we may take

$$\emptyset = \lambda \vec{x}_n.(\mu y)g(y, \vec{x}_n)$$

where $g = \lambda y\vec{x}_n.SZ(y) = \lambda y\vec{x}_n.1$.

In what follows we will often present first the informal way (for example, proofs by Church's Thesis, or just by handwaving) of doing a proof, and then we will present the mathematically rigorous way.   □

An important observation following from the above examples deserves theorem status:

6.1.6 Theorem. $\mathcal{R}_* \subsetneq \mathcal{P}_*$

Proof The $\subseteq$ part of "$\subsetneq$" is Example 6.1.5 above.

---

[3] "Formal" refers to *syntactic* proofs based on axioms. Our "*mathematical*" proofs are mostly *semantic*, they depend on meaning, not just syntax. That is how it is in the majority of mathematical publications. So one should say "mathematical" rather than "formal" in this context.

The $\neq$ part is due to $K \in \mathcal{P}_*$ (Example 6.1.4) and the fact that the halting problem is unsolvable ($K \notin \mathcal{R}_*$).

So, there are sets in $\mathcal{P}_*$ (e.g., K) that are not in $\mathcal{R}_*$. □

What about $\overline{K}$, that is, the *complement*

$$\overline{K} = \mathbb{N} - K = \{x : \phi_x(x) \uparrow\}$$

of K? The following general result helps us handle this question. □

6.1.7 Theorem. *A relation $Q(\vec{x}_n)$ is recursive if both $Q(\vec{x}_n)$ and $\neg Q(\vec{x}_n)$ are semi-recursive.*

Before we proceed with the proof, a remark on notation is in order.

In *set notation* we write the complement of a set, A, of n-tuples as $\overline{A}$. This means, of course, $\mathbb{N}^n - A$, where

$$\mathbb{N}^n = \underbrace{\mathbb{N} \times \cdots \times \mathbb{N}}_{n \text{ copies of } \mathbb{N}}$$

In *relational notation* we write the same thing (complement) as

$$\neg A(\vec{x}_n)$$

Proof We want to prove that some URM, N, *decides*

$$\vec{x}_n \in Q$$

We take two verifiers, M for "$\vec{x}_n \in Q$" and M' for "$\vec{x}_n \in \overline{Q}$,"[4] and run them —on input $\vec{x}_n$— as "co-routines" (i.e., we crank them simultaneously).

If M halts, then we stop everything and print "0" (i.e., "yes").

If M' halts, then we stop everything and print "1" (i.e., "no").

CT tells us that we can put the above —if we want to— into a single URM, N. □

Here is a *mathematical* proof that imitates what we did above. Let $Q(\vec{x}_n) \equiv (\exists y) T^{(n)}(i, \vec{x}_n, y)$ and $\neg Q(\vec{x}_n) \equiv (\exists y) T^{(n)}(j, \vec{x}_n, y)$ for some i and j (by Theorem 6.1.3). Then computing

$$f(\vec{x}_n) \stackrel{Def}{=} (\mu y)\left(T^{(n)}(i, \vec{x}_n, y) \vee T^{(n)}(j, \vec{x}_n, y)\right)$$

---

[4] We can do that, i.e., M and M' exist, since both Q and $\overline{Q}$ are semi-recursive.

runs the verifiers for Q and $\neg Q$ —at input $\vec{x}_n$— found at locations i and j respectively. A smallest y *will* be found as one or the other verifier halts. Thus $f \in \mathcal{R}$. But which verifier halted?

Well, $Q(\vec{x}_n) \equiv T^{(n)}(i, \vec{x}_n, f(\vec{x}_n))$ is true iff the verifier i stops. Incidentally, this equivalence —and the recursiveness of f— show that Q is recursive.   □

6.1.8 Remark The above is really an "iff"-result, because $\mathcal{R}_*$ is *closed under complement* as we showed in Corollary 2.2.7.

Thus, if Q is in $\mathcal{R}_*$, then so is $\overline{Q}$, by closure under $\neg$. By Theorem 6.1.6, both of Q and $\overline{Q}$ are in $\mathcal{P}_*$.   □

6.1.9 Example. $\overline{K} \notin \mathcal{P}_*$.

Why? Well, if instead it were $\overline{K} \in \mathcal{P}_*$, then combining this with Example 6.1.4 and Theorem 6.1.7 we would get $K \in \mathcal{R}_*$, which we know is not true.

Thus, $\overline{K}$ is a extremely unsolvable problem! This problem is so hard it is not even *semi*-decidable!   □

6.2   Some More Diagonalisation

6.2.1 Example. We said in the cautionary remark at the end of Example 2.1.16 "That is, there are functions $f \in \mathcal{P}$ that *have no recursive extensions*. This we will show later."

Now is "later"! We show that $f = \lambda x.\phi_x(x) + 1$ has no total computable extensions. (Thus, in particular, it is not total, but computable it is!) By way of contradiction, assume that

$$f \subseteq g \in \mathcal{R} \tag{1}$$

Then

$$g = \phi_i, \text{ for some } i \tag{2}$$

Let us compute $g(i)$:

$$g(i) = \phi_i(i) \tag{3}$$

by (2). Now, (1) tells a different story:

$$f(i) = \phi_i(i) + 1 \ (\textit{defined, by (3)}), \text{ hence } = g(i) \ (\text{by (1)})$$

We have from the above and (3): $\phi_i(i) + 1 = g(i) = \phi_i(i)$, a contradiction, since all sides are defined.   □

6.2.2 Exercise. Prove that $\lambda x.\phi_x(x)$ has no total computable extensions either. $\square$

We will come back to diagonalisation yet again later, but let us conclude here with two hidden diagonalisations that prove, yet again, the unsolvability of the halting problem.

6.2.3 Example. In Example 6.1.9 we proved $\overline{K} \notin \mathcal{P}_*$ using $K \notin \mathcal{R}_*$. Let us reverse this sequence of events here, and prove $\overline{K} \notin \mathcal{P}_*$ first, deriving from it the unsolvability of the halting problem (if the latter is solvable, then so is $\overline{K}$ by closure of $\mathcal{R}_*$ under $\neg$). Well, if $\overline{K} \notin \mathcal{P}_*$ is false, then

$$x \in \overline{K} \equiv \phi_i(x) \downarrow \tag{1}$$

for some i, hence, by Theorem 6.1.3 and (1),

$$x \in \overline{K} \equiv (\exists y)T(i, x, y) \tag{1'}$$

But $x \in \overline{K}$ says $\phi_x(x) \uparrow$, so (1') becomes

$$\phi_x(x) \uparrow \equiv (\exists y)T(i, x, y) \tag{2}$$

or (by Theorem 5.5.3(1))

$$\neg(\exists y)T(x, x, y) \equiv (\exists y)T(i, x, y) \tag{3}$$

Setting the variable x equal to i throughout in (3) we get a contradiction:

$$\neg(\exists y)T(i, i, y) \equiv (\exists y)T(i, i, y)$$

$\square$

Let us redo Example 6.2.3 using Example 0.5.23 giving Cantor credit for this method!

 6.2.4 Example. Again, we show $\overline{K} \notin \mathcal{P}_*$ directly, not via the halting problem. By (∗∗) in Remark 6.1.2, $x \notin W_x \equiv \phi_x(x) \uparrow$, that is,

$$x \in \overline{K} \equiv x \notin W_x \tag{1}$$

We translate (1) as the equivalent

$$\overline{K} = \{x \in \mathbb{N} : x \notin W_x\} \tag{2}$$

By (2) and Example 0.5.23, $\overline{K}$ is the "D" for the sequence of sets

$$W_0, W_1, W_2, \ldots$$

Thus there is *no i* such as $\overline{K} = W_i$ as we saw in Example 0.5.23. By Remark 6.1.2, this is tantamount to $\overline{K}$ not being semi-recursive.   □

6.3 Unsolvability via Reducibility

We turn our attention now to a *methodology* towards discovering new undecidable problems, and also new non semi-recursive problems, beyond the ones we learnt about so far, which are just, $x \in K$, $\phi_i = \phi_j$ (equivalence problem) and $x \in \overline{K}$. In fact, we will learn shortly that $\phi_i = \phi_j$ is worse than undecidable; just like $\overline{K}$ it is not even semi-decidable.

The tool we will use for such discoveries is the concept of *reducibility* of one set to another:

6.3.1 Definition. (Strong Reducibility) For any two subsets of $\mathbb{N}$, A and B, we write

$$A \leq_m B^5$$

or more simply

$$A \leq B \tag{1}$$

pronounced *A is strongly reducible or m-reducible to B*, meaning that there is a (total) *recursive* function f such that

$$x \in A \equiv f(x) \in B \tag{2}$$

We say that *"the reduction is effected by f"*.

If f is 1-1 we may write $A \leq_1 B$ and say that A is *1-1-reducible* or just *1-reducible* to B.   □

When (1) —equivalently, (2)— holds, then, intuitively, "A is easier than B to either decide or verify" since the question "$x \in A$" is algorithmically transformable (the algorithm being that for computing $f(x)$) to the equivalent question "$f(x) \in B$". If we answer the latter, we have answered the former as well. This observation has a very precise counterpart (Theorem 6.3.3 below). But first,

6.3.2 Lemma. *If $Q(y, \vec{x}) \in \mathcal{P}_*$ and $\lambda\vec{z}.f(\vec{z}) \in \mathcal{R}$, then $Q(f(\vec{z}), \vec{x}) \in \mathcal{P}_*$.*

Proof By Definition 6.1.1 there is a $g \in \mathcal{P}$ such that

---

[5] The subscript m stands for "many one", and refers to f. We do not require it to be 1-1, that is; *many* (inputs) to *one* (output) will be fine.

$$Q(y, \vec{x}) \equiv g(y, \vec{x}) \downarrow \tag{1}$$

Now, for any $\vec{z}$, $f(\vec{z})$ is some number which if we plug into y in (1) throughout we get an equivalence:

$$Q(f(\vec{z}), \vec{x}) \equiv g(f(\vec{z}), \vec{x}) \downarrow \tag{2}$$

But $\lambda \vec{z} \vec{x}.g(f(\vec{z}), \vec{x}) \in \mathcal{P}$ by Grzegorczyk operations. Thus, (2) and Definition 6.1.1 yield $Q(f(\vec{z}), \vec{x}) \in \mathcal{P}_*$.   □

6.3.3 Theorem. *If $A \leq B$ in the sense of Definition 6.3.1, then*

(*i*) *if $B \in \mathcal{R}_*$, then also $A \in \mathcal{R}_*$*
(*ii*) *if $B \in \mathcal{P}_*$, then also $A \in \mathcal{P}_*$*

Proof

Let $f \in \mathcal{R}$ effect the reduction.

(*i*) Let $z \in B$ be in $\mathcal{R}_*$.
  Then for some $g \in \mathcal{R}$ we have

$$z \in B \equiv g(z) = 0$$

and thus

$$f(x) \in B \equiv g(f(x)) = 0 \tag{1}$$

  But $\lambda x.g(f(x)) \in \mathcal{R}$ by composition, so (1) says that "$f(x) \in B$" is in $\mathcal{R}_*$. This is the same as "$x \in A$".
(*ii*) Let $z \in B$ be in $\mathcal{P}_*$. By Lemma 6.3.2, so is $f(x) \in B$. But this says $x \in A$.   □

Taking the "contrapositive", we have at once:

6.3.4 Corollary. *If $A \leq B$ in the sense of Definition 6.3.1, then*

(*i*) *if $A \notin \mathcal{R}_*$, then also $B \notin \mathcal{R}_*$*
(*ii*) *if $A \notin \mathcal{P}_*$, then also $B \notin \mathcal{P}_*$*

We can now use K and $\overline{K}$ as "yardsticks" —or reference "problems"— and discover more undecidable and also *non semi-decidable* problems.

The idea of the corollary is applicable to the so-called "complete index sets".

6.3.5 Definition. (Complete Index Sets) Let $\mathcal{C} \subseteq \mathcal{P}$ and $A = \{x : \phi_x \in \mathcal{C}\}$. A is thus the set of *all* programs (known by their addresses) x that compute any *unary* $f \in \mathcal{C}$: Indeed, let $\lambda x.f(x) \in \mathcal{C}$. Thus $f = \phi_i$ for some i. Then $i \in A$. But this is true of *all* m for which $\phi_m = f$.

We call A a *complete* (all) *index* (programs) set.   □

6.3.6 Example. The set $A = \{x : \mathrm{ran}(\phi_x) = \emptyset\}$ is *not semi-recursive*.

Recall that "range" for $\lambda x.\, f(x)$, denoted by $\mathrm{ran}(f)$, is defined by

$$\{x : (\exists y) f(y) = x\}$$

We will try to show that

$$\overline{K} \leq A \tag{1}$$

If we can do that much, then Corollary 6.3.4, part (ii), will do the rest.
    Well, define

$$\psi(x, y) \simeq \begin{cases} 0 & \text{if } \phi_x(x) \downarrow \\ \uparrow & \text{if } \phi_x(x) \uparrow \end{cases} \tag{2}$$

Here is how to compute ψ:
    Given x, y, ignore y. Call $h(x, x)$ ($\simeq \phi_x(x)$). If the call ever halts, then print "0" and halt everything. If it never halts, then you will never return from the call, which is the correct behaviour specified in (2) for $\psi(x, y)$.
    By CT, ψ is in $\mathcal{P}$, so, by the S-m-n Theorem, there is a recursive h such that

$$\psi(x, y) \simeq \phi_{h(x)}(y), \text{ for all } x, y$$

(a) *Mathematically*, without invoking CT, note that $\psi(x, y) \simeq 0 \times h(x, x)$ thus ψ is in $\mathcal{P}$ and is defined iff $h(x, x) \downarrow$ —that is, iff $\phi_x(x) \downarrow$.
(b) *You may* not *use S-m-n until after you have proved that your* "$\lambda xy.\psi(x, y)$" *is in* $\mathcal{P}$.

We can rewrite this as,

$$\phi_{h(x)}(y) \simeq \begin{cases} 0 & \text{if } \phi_x(x) \downarrow \\ \uparrow & \text{if } \phi_x(x) \uparrow \end{cases} \tag{3}$$

or, rewriting (3) without arguments (as equality of functions, not equality of function calls)

$$\phi_{h(x)} = \begin{cases} \lambda y.0 & \text{if } \phi_x(x) \downarrow \\ \emptyset & \text{if } \phi_x(x) \uparrow \end{cases} \tag{3'}$$

In (3'), $\emptyset$ stands for $\lambda y. \uparrow$, the empty function.

Thus,

$$h(x) \in A \text{ iff ran}(\phi_{h(x)}) = \emptyset \quad \overbrace{\text{iff}}^{\text{bottom case in } 3'} \quad \phi_x(x) \uparrow$$

The above says $x \in \overline{K} \equiv h(x) \in A$, hence $\overline{K} \leq A$, and thus $A \notin \mathcal{P}_*$ by Corollary 6.3.4, part (ii). $\qquad\square$

6.3.7 Example. The set $B = \{x : \phi_x \text{ has finite domain}\}$ is *not semi-recursive*.

This is really easy (once we have done the previous example)! *All we have to do is "talk about" our findings, above, differently!*

We use the same ψ as in the previous example, as well as the same h as above, obtained by S-m-n.

Looking at (3′) above we see that the top case has infinite domain, while the bottom one has finite domain (indeed, empty). Thus,

$$h(x) \in B \text{ iff } \phi_{h(x)} \text{ has finite domain} \quad \overbrace{\text{iff}}^{\text{bottom case in } 3'} \quad \phi_x(x) \uparrow$$

The above says $x \in \overline{K} \equiv h(x) \in B$, hence $B \notin \mathcal{P}_*$ by Corollary 6.3.4, part (ii). $\quad\square$

6.3.8 Example. Let us mine (3′) twice more to obtain two more important undecidability results.

1. Show that $G = \{x : \phi_x \text{ is a constant function}\}$ is undecidable.

   We (re-)use (3′) of Example 6.3.6. Note that in (3′) the top case defines a constant function, but the bottom case defines a non-constant. Thus

   $$h(x) \in G \equiv \phi_{h(x)} = \lambda y.0 \equiv x \in K$$

   Hence $K \leq G$, therefore $G \notin \mathcal{R}_*$.
2. Show that $I = \{x : \phi_x \in \mathcal{R}\}$ is undecidable. Again, we retell what we can read from (3′) in words that are relevant to the set I:

   $$h(x) \in I \overset{\emptyset \notin \mathcal{R}}{\equiv} \phi_{h(x)} = \lambda y.0 \equiv x \in K$$

   Thus $K \leq I$, therefore $I \notin \mathcal{R}_*$. $\qquad\square$

We soon will sharpen the result 2 of the previous example (see Theorem 6.4.7).

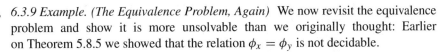

6.3.9 Example. (The Equivalence Problem, Again) We now revisit the equivalence problem and show it is more unsolvable than we originally thought: Earlier on Theorem 5.8.5 we showed that the relation $\phi_x = \phi_y$ is not decidable.

Here we sharpen this to: the relation $\phi_x = \phi_y$ is not semi-decidable.

By Lemma 6.3.2, if the 2-variable predicate above is in $\mathcal{P}_*$ then so is $\lambda x.\phi_x = \phi_y$, i.e., taking a constant for y. Choose then for y a ϕ-index for the *empty function*.

So, if $\lambda xy.\phi_x = \phi_y$ is in $\mathcal{P}_*$ then so is

$$\phi_x = \emptyset$$

which is equivalent to

$$\mathrm{ran}(\phi_x) = \emptyset$$

and thus not in $\mathcal{P}_*$ by Example 6.3.6.   □

 6.3.10 Example. (An Impossibly Hard Problem; Again) Here is a more insightful proof of the non semi-recursivensess of the correctness problem: The functions $\lambda y.1$ and $\lambda y.c_T(x, x, y)$, the latter for $x = 0, 1, 2, \ldots$ are in $\mathcal{PR}$, thus they can be *finitely* given (defined) by loop programs. Here c_T is the characteristic function of $T(x, y, z)$.

Is the problem of determining whether or not

$$\lambda y.1 = \lambda y.c_T(x, x, y) \tag{1}$$

decidable? verifiable? for no matter which x we ask the question?

This asks, in essence, whether we can decide or verify *equivalence of two loop programs*.

Well, no! Why? Because (1) is equivalent to the statement

$$(\forall y)\Big(c_T(x, x, y) = 1\Big) \tag{2}$$

same as

$$(\forall y)\neg T(x, x, y)$$

same as

$$\neg(\exists y)T(x, x, y) \tag{3}$$

But (3) says that $\phi_x(x) \uparrow$ which is *not* verifiable! Nor is (1)!

This argument will recur in Chap. 15 for a tiny subclass of $\mathcal{PR}$.   □

6.3.11 Example. The set $C = \{x : \mathrm{ran}(\phi_x) \text{ is finite}\}$ is *not semi-decidable*.

Here we cannot reuse (3′) above, because *both* cases —in the definition by cases— have functions of *finite range*. We want one case to have a function of finite range, but the other to have *infinite range*.

Aha! This motivates us to choose a different "ψ" (hence a different S-m-n function "k"), and retrace the steps we took above.

Define

$$g(x, y) \simeq \begin{cases} y & \text{if } \phi_x(x) \downarrow \\ \uparrow & \text{if } \phi_x(x) \uparrow \end{cases} \qquad \text{(ii)}$$

Here is an algorithm for g:

Given x, y.

Call the universal function $h(x, x)$ —or, just call $\phi_x(x)$.

If this ever returns, then print "y" and halt everything. If it never returns, then the correct behaviour for $g(x, y)$ is obtained once more: namely, we got $g(x, y) \uparrow$ if $x \in \overline{K}$.

By CT, g is partial recursive.

Before we proceed, here is a mathematically precise reason (no CT!) for the partial recursiveness of g:

$g(x, y) \simeq y + 0 \times \phi_x(x)$, for all x, y. Hence $g \in \mathcal{P}$ by substitution.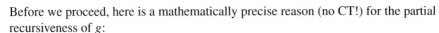

Thus by S-m-n, for some recursive unary k we have

$$g(x, y) \simeq \phi_{k(x)}(y), \quad \text{for all } x, y$$

Hence, by (ii)

$$\phi_{k(x)} = \begin{cases} \lambda y. y & \text{if } x \in K \\ \emptyset & \text{othw} \end{cases} \qquad \text{(iii)}$$

It follows that

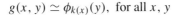

$$k(x) \in C \text{ iff } \phi_{h(x)} \text{ has finite range} \quad \overbrace{\text{iff}}^{\text{bottom case in } iii} \quad x \in \overline{K}$$

That is, $\overline{K} \leq C$ and we are done. □

6.3.12 Exercise. Show that $D = \{x : \text{ran}(\phi_x) \text{ is infinite}\}$ is undecidable. □

6.3.13 Exercise. Show that $F = \{x : \text{dom}(\phi_x) \text{ is infinite}\}$ is undecidable. □

Enough "negativity"! Here is an important "positive result" that helps to prove that certain relations *are* semi-decidable:

6.3.14 Theorem. (Projection Theorem) *A relation $Q(\vec{x}_n)$ is semi-recursive iff there is a recursive (decidable) relation $S(y, \vec{x}_n)$ such that*

$$Q(\vec{x}_n) \equiv (\exists y) S(y, \vec{x}_n) \qquad (1)$$

Q is obtained by "projecting" S along the y-coordinate, hence the name of the theorem.

Proof *If*-part. Trivially,

$$Q(\vec{x}_n) \equiv (\mu y)S(y, \vec{x}_n) \downarrow$$

But $\lambda \vec{x}_n.(\mu y)S(y, \vec{x}_n) \in \mathcal{P}$.

Only if-part. Say $Q(\vec{x}_n)$ has semi-index $i \in \mathbb{N}$. Then, we are done by Theorem 6.1.3 since we have

$$Q(\vec{x}_n) \equiv (\exists y)T^{(n)}(i, \vec{x}_n, y)$$

and $T^{(n)}$ is primitive recursive: Take $\lambda y\vec{x}_n.S(y, \vec{x}_n) \overset{Def}{\equiv} \lambda y\vec{x}_n.T^{(n)}(i, \vec{x}_n, y)$. $\square$

 This theorem sometimes has a qualifier: "*strong* projection theorem", the weak version *not* being an iff result: See 3 under Theorem 6.5.2.

6.3.15 Example. The set $A = \{(x, y, z) : \phi_x(y) = z\}$ is semi-recursive.

Here is a verifier for the above predicate:

Given input x, y, z. **Comment**. Note that $\phi_x(y) = z$ is true iff two things happen: (1) $\phi_x(y) \downarrow$ *and* (2) the computed value is z.

1. Call the universal function h on input x, y.
2. If the call ever returns, then

   - If the returned value $h(x, y)$ equals z then halt everything (the "yes" output).
   - If the returned value is not equal to z, then enter an infinite loop (say "no", by looping).

By CT the above informal verifier can be formalised as a URM M.

But is it correct? Does it verify $\phi_x(y) = z$?

Yes. See **Comment** above. $\square$

 A mathematical proof of the above uses Theorem 5.5.3:

$$\phi_x(y) \simeq z \equiv (\exists w)\Big(T(x, y, w) \land out(w, x, y) = z\Big)$$

The predicate in the big brackets is in $\mathcal{PR}_*$ by Theorem 2.2.5. Now invoke Theorems 6.3.14.

Incidentally, the predicate (in the left hand side of $\equiv$ above) that we have discussed here is not decidable. See Exercise 6.11.1.

6.4   Recursively Enumerable Sets

In this section we explore the rationale behind the alternative name "*recursively enumerable*" or "*computably enumerable*" that is used in the literature for the semi-recursive or semi-computable sets/predicates. Short names for this alternative jargon are "r.e." and "c.e." respectively. To avoid cumbersome codings (of n-tuples, by single numbers) we restrict attention to the one variable case in this section. *I.e., predicates that are subsets of $\mathbb{N}^n$ for the case $n = 1$.*

First we define:

6.4.1 Definition. A set $A \subseteq \mathbb{N}$ is called *computably enumerable* (c.e.) or *recursively enumerable* (r.e.) precisely if *one* of the following cases holds:

- $A = \emptyset$
- $A = \text{ran}(f)$, where $f \in \mathcal{R}$.

□

Thus, the c.e. or r.e. relations are exactly those we can *algorithmically enumerate* as *the set of outputs* of a (*total*) recursive function:

$$A = \{f(0), f(1), f(2), \ldots, f(x), \ldots\}$$

Hence the use of the term "c.e." replaces the non technical term "algorithmically" (in "algorithmically" enumerable) by the technical term "computably".

Note that we had to hedge and ask that $A \neq \emptyset$ for any enumeration to take place, because no recursive function can have an empty range.

Next we prove:

6.4.2 Theorem. ("c.e." or "r.e." vs. Semi-Recursive)
Any non empty semi-recursive relation A $(A \subseteq \mathbb{N})$ is the range of some (emphasis: total) recursive function of one variable.

Conversely, every set A such that $A = \text{ran}(f)$ —where $\lambda x.f(x)$ is recursive— is semi-recursive (and, trivially, nonempty).

In short, the semi-recursive sets are precisely the c.e. or r.e. sets. For $A \neq \emptyset$ this is the content of Theorem 6.4.2 while $\emptyset$ is r.e. by definition, and known to us to be also semi-recursive —indeed in $\mathcal{PR}_* \subseteq \mathcal{R}_* \subseteq \mathcal{P}_*$.

Before we prove the result, here is an example:

6.4.3 Example. The set $\{0\}$ is c.e. Indeed, $f = \lambda x.0$, our familiar function Z, effects the enumeration with repetitions (lots of them!)

$$x = 0 \quad 1 \quad 2 \quad 3 \quad 4 \quad \ldots$$
$$f(x) = 0 \quad 0 \quad 0 \quad 0 \quad 0 \quad \ldots$$

□

Proof
(I) *We prove the first sentence of the theorem.* So, let $A \neq \emptyset$ be semi-recursive.

By the projection theorem 6.3.14 there is a *decidable* (recursive) relation $Q(y, x)$ such that

$$x \in A \equiv (\exists y)Q(y, x), \text{ for all } x \tag{1}$$

Thus,

every $x \in A$ *will have some associated value* y *such that* $Q(y, x)$ *holds.* $\qquad$ (2)

and conversely,

if $Q(y, x)$ *holds for some* y, x *pair, then* $x \in A$. $\qquad$ (2′)

(2) and (2′) jointly rephrase (1), but also suggest the idea of how to *enumerate* all $x \in A$:

We should look for all *pairs* (y, x) *in a* systematic manner, *and for each such pair that is in* Q *we should just output (enumerate) the* x-*component*

But we know how to generate all pairs in a systematic manner, computably.

for $z = 0, 1, 2, 3, \ldots$ **do**

generate the pair $\left((z)_0, (z)_1 \right)$. Here "$y$" is $(z)_0$ and "x" is $(z)_1$

Pause. Will the above generate *all* pairs? Sure: For any x and y, the pair (y, x) is guaranteed to be output when we reach the z-value $\langle y, x \rangle$ $\;(= 2^{y+1}3^{x+1})$. ◀

Thus the enumerating partial recursive function f we want (that has range equal to A) is:

$$f(z) \simeq \begin{cases} (z)_1 & \text{if } Q\left((z)_0, (z)_1 \right) \\ \uparrow & \text{othw} \end{cases} \tag{3}$$

This works for any semi-recursive A, empty or not, and gives a mathematical proof to Corollary 6.4.4 below.

A small tweak takes care of a *nonempty* A with a *total* f: So let $a \in A$ be any. Modify (3) into (4). $\mathrm{ran}(g) = A$ and $g \in \mathcal{R}$.

$$g(z) \simeq \begin{cases} (z)_1 & \text{if } Q\left((z)_0, (z)_1 \right) \\ a & \text{othw} \end{cases} \tag{4}$$

(II) Proof *of the second sentence of the theorem.* So, let $A = \mathrm{ran}(f)$ —where f is recursive. Thus,

$$x \in A \equiv (\exists y)f(y) = x$$

By Grzegorczyk operations, the fact that $z = x$ is in $\mathcal{R}_*$ and the assumption $f \in \mathcal{R}$, the relation $f(y) = x$ is recursive. We are done by the projection theorem. $\qquad\square$

6.4.4 Corollary. *If A is semi-recursive, then $A = \mathrm{ran}(f)$ for some $f \in \mathcal{P}$.*

Do we have a converse? Is the range of any partial recursive function semi-recursive? *Yes.* Theorem 6.6.2.

6.4.5 Corollary. *An $A \subseteq \mathbb{N}$ is semi-recursive iff it is r.e. (c.e.)*

Proof For nonempty A this is Theorem 6.4.2. For empty A we note that this is r.e. by Theorem 6.4.2 but also semi-recursive by $\emptyset \in \mathcal{PR}_* \subseteq \mathcal{R}_* \subseteq \mathcal{P}_*$.   □

Corollary 6.4.5 allows us to prove some non-semi-recursiveness results by Cantor diagonalisation. See below.

6.4.6 Remark. In view of the coincidence of semi-recursive and r.e. sets, one will encounter also the term *r.e.-index* of a semi-recursive or c.e. set A as synonym for *semi-index*. Interestingly the term "r.e.-index" still applies to an x such $A = W_x$ and not to an x such that $A = \mathrm{ran}(\phi_x)$.

We will adhere with the term "semi-index" in this volume.   □

6.4.7 Theorem. *The complete index set $A = \{x : \phi_x \in \mathcal{R}\}$ is not semi-recursive.*

This sharpens the undecidability result for A that we established in Example 6.3.8.

Proof By the equivalence of c.e.-ness and semi-recursiveness we prove that A is not c.e.

If not, note first that $A \neq \emptyset$ since, e.g., $Z \in \mathcal{R}$ and thus at least one ϕ-index is in A (a ϕ-index for Z).

Thus, Theorem 6.4.2 applies and there is an $f \in \mathcal{R}$ such that $A = \mathrm{ran}(f) = \{y : y = f(x),$ for some $x\}$, that is, $y \in A \equiv (\exists x) f(x) = y$.

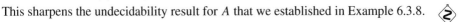

> In words, a ϕ-index y is in A iff it has the form $f(x)$ for some x (1)

Define

$$d = \lambda x.1 + \phi_{f(x)}(x) \tag{2}$$

Since $\lambda x.\phi_{f(x)}(x) \in \mathcal{P}$, we obtain $d \in \mathcal{P}$. But $\phi_{f(x)}$ is total since all the $f(x)$ are ϕ-indices of total functions by (1).

By the same comment,

$$d = \phi_{f(i)}, \text{ for some } i \tag{3}$$

Let us compute $d(i)$: $d(i) \simeq 1 + \phi_{f(i)}(i)$ by (2). Also, $d(i) \simeq \phi_{f(i)}(i)$ by (3), thus

$$1 + \phi_{f(i)}(i) \simeq \phi_{f(i)}(i)$$

which is a contradiction since both sides of "$\simeq$" are defined.   □

One can take as d different functions, for example, either of $d = \lambda x.42 + \phi_{f(x)}(x)$ or $d = \lambda x.1 \dot- \phi_{f(x)}(x)$ works. And infinitely many other choices do.

6.5 Some Closure Properties of Decidable and Semi-Decidable Relations

We already know (Corollaries 2.2.7 and 2.2.15) that

6.5.1 Theorem. $\mathcal{R}_*$ *is closed under all Boolean operations,* $\neg, \wedge, \vee, \Rightarrow, \equiv,$ *as well as under* $(\exists y)_{<z}$ *and* $(\forall y)_{<z}$.

6.5.2 Theorem. $\mathcal{P}_*$ *is closed under* $\wedge$ *and* $\vee$. *It is also closed under* $(\exists y)$, *or, as we say, "under projection". Moreover it is closed under* $(\exists y)_{<z}$ *and* $(\forall y)_{<z}$.
It is not *closed under negation (complement), nor under* $(\forall y)$.

Proof

1. Let $Q(\vec{x}_n)$ be semi-decided by a URM M, and $S(\vec{y}_m)$ be semi-decided by a URM N.

   Here is how to semi-decide $Q(\vec{x}_n) \vee S(\vec{y}_m)$:

   Given input $\vec{x}_n, \vec{y}_m$, we call machine M with input $\vec{x}_n$, and machine N with input $\vec{y}_m$ and let them crank simultaneously (as "co-routines").

   If *either one* halts, then halt everything! This is the case of "yes" (input verified).

   Mathematically, say i and e are semi-indices of Q and S respectively. Then

$$Q(\vec{x}_n) \vee S(\vec{y}_m) \equiv (\exists z)T^{(n)}(i, \vec{x}_n, z) \vee (\exists z)T^{(m)}(e, \vec{y}_m, z)$$

   But (logic), $(\exists z)T^{(n)}(i, \vec{x}_n, z) \vee (\exists z)T^{(m)}(e, \vec{y}_m, z) \equiv (\exists z)\big(T^{(n)}(i, \vec{x}_n, z) \vee T^{(m)}(e, \vec{y}_m, z)\big)$, thus

$$Q(\vec{x}_n) \vee S(\vec{y}_m) \equiv (\exists z)\big(T^{(n)}(i, \vec{x}_n, z) \vee T^{(m)}(e, \vec{y}_m, z)\big)$$

   Done by Theorem 6.3.14.

2. For $\wedge$ it is almost the same, but our halting criterion is different:

   Here is how to semi-decide $Q(\vec{x}_n) \wedge S(\vec{y}_m)$:

   Given input $\vec{x}_n, \vec{y}_m$, we call machine M with input $\vec{x}_n$, and machine N with input $\vec{y}_m$ and let them crank simultaneously ("co-routines").

   If *both* halt, then halt everything!

   Mathematically, let i and e be semi-indices as above. Then

$$Q(\vec{x}_n) \wedge S(\vec{y}_m) \equiv (\exists z)T^{(n)}(i, \vec{x}_n, z) \wedge (\exists z)T^{(m)}(e, \vec{y}_m, z) \qquad (\dagger)$$

Logic will not help here, because $\exists$ does not distribute over $\wedge$, the intuitive reason being that in ($\dagger$) —assuming the left hand side of $\equiv$ to be true— there are in general *distinct* values u and w that make $T^{(n)}(i, \vec{x}_n, u)$ and $T^{(m)}(e, \vec{y}_m, w)$ true.

So, *coding* to the rescue! Let $z = \langle u, w \rangle$. Then $u = (z)_0$ and $w = (z)_1$ and we have

$$Q(\vec{x}_n) \wedge S(\vec{y}_m) \equiv (\exists z)\left(T^{(n)}(i, \vec{x}_n, (z)_0) \wedge T^{(m)}(e, \vec{y}_m, (z)_1)\right)$$

and are done by Theorem 2.2.5 followed by Theorem 6.3.14.

3. *The* $(\exists y) Q(y, \vec{x}_n)$ *Case:*

Let $Q(y, \vec{x}_n)$ be *semi*-decidable. Then, by Theorem 6.3.14, a *recursive* (in fact primitive recursive by Theorem 6.1.3) $P(z, y, \vec{x}_n)$ exists such that

$$Q(y, \vec{x}_n) \equiv (\exists z) P(z, y, \vec{x}_n) \tag{1}$$

It follows that

$$(\exists y) Q(y, \vec{x}_n) \equiv (\exists y)(\exists z) P(z, y, \vec{x}_n) \tag{2}$$

This does not settle the proof, as I cannot readily conclude that

$$(\exists y)(\exists z) P(z, y, \vec{x}_n)$$

is semi-decidable: you see, the projection theorem requires a *single* $(\exists y)$ in front of a decidable predicate!

Coding to the rescue.

Well, instead of saying that there are *two* values y and z that verify (along with $\vec{x}_n$) the predicate $P(z, y, \vec{x}_n)$, I can say the same thing —after letting $w = 2^{z+1}3^{y+1} = \langle z, y \rangle$— if I say $(\exists w) P((w)_0, (w)_1, \vec{x}_n)$. Thus I have

$$(\exists y) Q(y, \vec{x}_n) \equiv (\exists w) P((w)_0, (w)_1, \vec{x}_n) \tag{3}$$

But now $P((w)_0, (w)_1, \vec{x}_n)$ is decidable by the decidability of P and Grzegorczyk operations, and in (3) we quantified the decidable $P((w)_0, (w)_1, \vec{x}_n)$ with just *one* $(\exists w)$. *The projection theorem now applies!*

4. For $(\exists y)_{<z} Q(y, \vec{x})$, where $Q(y, \vec{x})$ is semi-recursive, we first note that

$$(\exists y)_{<z} Q(y, \vec{x}) \equiv (\exists y)\left(y < z \wedge Q(y, \vec{x})\right) \tag{*}$$

By $\mathcal{PR}_* \subseteq \mathcal{R}_* \subseteq \mathcal{P}_*$, $y < z$ is semi-recursive. By closure properties of $\mathcal{P}_*$ established so far in this proof, the right hand side of $\equiv$ in (*) is semi-recursive, thus so is the left hand side.

5. For $(\forall y)_{<z} Q(y, \vec{x})$, where $Q(y, \vec{x})$ is semi-recursive, we first note that (by the projection theorem) a *decidable* P exists such that

$$Q(y, \vec{x}) \equiv (\exists w) P(w, y, \vec{x})$$

By the above equivalence, we need to prove that

$$(\forall y)_{<z}(\exists w)P(w, y, \vec{x}) \text{ is semi-recursive} \tag{**}$$

(**) says that

for *each* $y = 0, 1, 2, \ldots, z - 1$ there is a w-value w_y so that $P(w_y, y, \vec{x})$

Since all those w_y are finitely many (z many!) there is a value u bigger than *all* of them (for example, take $u = \max(w_0, \ldots, w_{z-1}) + 1$). Thus (**) says (i.e., *is equivalent to*)

$$(\exists u)(\forall y)_{<z}(\exists w)_{<u}P(w, y, \vec{x})$$

The part to the right of ($\exists u$) above is *decidable* (by closure properties of $\mathcal{R}_*$, since $P \in \mathcal{R}_*$ —cf. Corollaries 2.2.7 and 2.2.15). We are done by *strong projection*.

6. Why is $\mathcal{P}_*$ not closed under negation (complement)?
   Because we know that $K \in \mathcal{P}_*$, but $\overline{K} \notin \mathcal{P}_*$.
7. Why is $\mathcal{P}_*$ not closed under ($\forall y$)?
   Because

$$x \in K \equiv (\exists y)Q(y, x) \tag{1}$$

for some recursive Q (projection theorem) and by the fact (quoted again above) that $K \in \mathcal{P}_*$.

(1) is equivalent to

$$x \in \overline{K} \equiv \neg(\exists y)Q(y, x)$$

which in turn is equivalent to

$$x \in \overline{K} \equiv (\forall y)\neg Q(y, x) \tag{2}$$

Now, by closure properties of $\mathcal{R}_*$, $\neg Q(y, x)$ is recursive, hence also in $\mathcal{P}_*$ since $\mathcal{R}_* \subseteq \mathcal{P}_*$.

So, if $\mathcal{P}_*$ were closed under ($\forall y$), then the above $(\forall y)\neg Q(y, x)$ would be semi-recursive. But that is $x \in \overline{K}$! $\square$

6.6 Computable Functions and Their Graphs

We prove a fundamental result here, that

6.6.1 Theorem. $\lambda\vec{x}.f(\vec{x}) \in \mathcal{P}$ *iff the graph* $y \simeq f(\vec{x})$ *is in* $\mathcal{P}_*$.

Proof

- ($\rightarrow$, that is, the *Only if*) Let $\lambda \vec{x}. f(\vec{x}) \in \mathcal{P}$. By an easy adaptation of the proof of Example 6.3.15 it follows that $y \simeq f(\vec{x})$ is semi-computable.

  Or, note that since $f = \phi_e^{(n)}$, for some e, we have

$$f(\vec{x}) \simeq y \text{ iff } \phi_e^{(n)}(\vec{x}) \simeq y \text{ iff } (\exists z)\Big(T^{(n)}(e, \vec{x}, z) \wedge y \simeq out(z, e, \vec{x})\Big)$$

- ($\leftarrow$, that is, the *If*) Let $y \simeq f(\vec{x})$ be semi-computable. By the projection theorem

$$y \simeq f(\vec{x}) \equiv (\exists z) Q(z, y, \vec{x}) \tag{1}$$

for some decidable Q. The idea of *how* to find the correct y, *if any*, once we are *given* an $\vec{x}$, is to search (*simultaneously!*) for a z *and* a y that "work" in (1) —meaning that they satisfy $Q(z, y, \vec{x})$ for the given $\vec{x}$.[6] Thus

$$f(\vec{x}) \simeq \Big((\mu w) Q((w)_0, (w)_1, \vec{x})\Big)_1$$

and hence $f \in \mathcal{P}$ by closure properties. Incidentally the process of querying Q systematically[7] for all pairs (z, y) either forever, or until we succeed to make $Q(z, y, \vec{x})$ true, is called *dovetailing*. □

We can now settle

6.6.2 Theorem. *If $A = ran(f)$ and $f \in \mathcal{P}$, then $A \in \mathcal{P}_*$.*

Proof By Theorem 6.6.1 $y \simeq f(x)$ is semi-recursive. By closure properties of $\mathcal{P}_*$, so is $(\exists x) y \simeq f(x)$. But $(\exists x) y \simeq f(x) \equiv y \in ran(f)$, that is, $(\exists x) y \simeq f(x) \equiv y \in A$ since $ran(f) = A$. □

6.7 Some Complex Reductions

This section highlights a more sophisticated reduction scheme that improves our ability to effect reductions of the type $\overline{K} \leq A$.

6.7.1 Example. Prove that $A = \{x : \phi_x \text{ is a constant}\}$ is not semi-recursive. This is not amenable to the technique of saying "OK, if A is semi-recursive, then it is r.e. Let me show that it is not so by diagonalisation". This worked for $B = \{x : \phi_x \text{ is total}\}$ but no obvious diagonalisation comes to mind for A.

Nor can we simplistically say, OK, start by defining

---

[6] We saw this idea in the proof of Theorem 6.4.2.

[7] The "system" here being $w \mapsto ((w)_0, (w)_1)$.

$$g(x, y) \simeq \begin{cases} 0 & \text{if } x \in \overline{K} \\ \uparrow & \text{othw} \end{cases}$$

The problem is that if we plan next to say "*now g is partial recursive* hence by S-m-n, etc.", then the italicised part is wrong. $g \notin \mathcal{P}$, *provably*! For if it is computable, then so is $\lambda x.g(x, x)$ by Grzegorczyk operations. But

$$g(x, x) \downarrow \text{ iff we have the top case, iff } x \in \overline{K}$$

Thus

$$x \in \overline{K} \equiv g(x, x) \downarrow$$

which proves that $\overline{K} \in \mathcal{P}_*$. *Contradiction.*   □

6.7.2 Example. (6.7.1 Continued) Now, "Plan B" is to "*approximate*" the top condition $\phi_x(x) \uparrow$ (same as $x \in \overline{K}$).

The idea is that, "*practically*", if the computation $\phi_x(x)$ after a "huge" number of steps y has still not hit **stop**, this situation approximates —let me say once more— "practically", the situation $\phi_x(x) \uparrow$. This somewhat fuzzy thinking suggests that we try next

$$f(x, y) \simeq \begin{cases} 0 & \text{if the call } \phi_x(x) \text{ has not halted in } y \text{ steps} \\ \uparrow & \text{othw} \end{cases} \tag{1}$$

The "othw" says, of course, that the computation of the call $\phi_x(x)$ —or $h(x, x)$, where h is the universal function— did halt in y steps (possibly fewer).

By Theorem 5.5.3 and Definition 5.5.1, the top condition in (1) has the mathematical formulation

$$(\forall z)_{\leq y} \neg T(x, x, z)$$

or, since $T(x, x, z) \rightarrow T(x, x, z + 1)$ by the semantics of **stop** (Definition 1.1.2), the simpler formulation

$$\neg T(x, x, y)$$

will do. Either way,

the top condition in (1) is primitive recursive (thus so is the bottom)   (‡)

Thus $f \in \mathcal{P}$ via definition by (primitive)recursive cases (cf. Corollary 2.3.8)[8]

We may now apply S-m-n: There is a recursive k such that

$$\phi_{k(x)}(y) \simeq \begin{cases} 0 & \text{if the call } \phi_x(x) \text{ has not returned in } y \text{ steps} \\ \uparrow & \text{othw} \end{cases} \tag{3}$$

Analysis of (3) in terms of the "key" conditions $\phi_x(x) \uparrow$ *and* $\phi_x(x) \downarrow$:

(A) Case where $\phi_x(x) \uparrow$.
   Then, the call $\phi_x(x)$ did *not* return in y steps, *for any y.*
   Thus, by (3), we have $\phi_{k(x)}(y) = 0$, *for all y, that is,*

$$\phi_x(x) \uparrow \implies \phi_{k(x)} = \lambda y.0 \tag{4}$$

(B) Case where $\phi_x(x) \downarrow$. Let $m = smallest$ y such that the call $\phi_x(x)$ returned in m steps. Therefore,

  • for step counts $y = 0, 1, 2, \ldots, m - 1$ the computation of $\phi_x(x)$ has not yet hit **stop**, so the *top* case of definition (3) holds for all these y-values. We get

$$\textbf{input } y \quad = 0, \quad 1, \quad \ldots, \quad m - 1$$

$$\textbf{output } \phi_{k(x)}(y) = 0, \quad 0, \quad \ldots, \quad 0$$

  • for step counts $y = m, m + 1, m + 2, \ldots$ the computation of $\phi_x(x)$ has already halted (it hit **stop**, first at step m),[9] so the *bottom* case of definition (3) holds. We get

$$\textbf{input } y \quad = m, \quad m + 1, \quad m + 2, \quad \ldots$$

$$\textbf{output } \phi_{k(x)}(y) = \uparrow, \quad \uparrow, \quad \uparrow, \quad \ldots$$

In short:

$$\phi_x(x) \downarrow \implies \phi_{k(x)} = \overbrace{(0, 0, \ldots, 0)}^{\text{length } m} \tag{5}$$

In

$$\phi_{k(x)} = \overbrace{(0, 0, \ldots, 0)}^{\text{length } m}$$

---

[8] From the functions $\lambda y.0$ and $\emptyset$.

[9] Cf. semantics of **stop** in Definition 1.1.2.

we depict the function $\phi_{k(x)}$ as an array of m output values.

 Two things: *One*, in English, when $\phi_x(x) \downarrow$, the function $\phi_{k(x)}$ *is not a constant! Not even total!*

Two, m depends on x, of course, when said x brings us to case (**B**) of the analysis. Regardless, the non-constant / non total nature of $\phi_{k(x)}$ is still a fact;

$$\phi_{k(x)} \text{ just the length } m \text{ of the finite array } \overbrace{(0, 0, \ldots, 0)}^{\text{length } m} \text{ changes.}$$

Our analysis yielded:

$$\phi_{k(x)} = \begin{cases} \lambda y.0 & \text{if } \phi_x(x) \uparrow \\ \text{not a constant function} & \text{if } \phi_x(x) \downarrow \end{cases} \qquad (6)$$

We conclude now as follows for $A = \{x : \phi_x \text{ is a constant}\}$:

$$k(x) \in A \text{ iff } \phi_{k(x)} \text{ is a constant iff the top case of (6) applies iff } \phi_x(x) \uparrow$$

That is, $x \in \overline{K} \equiv k(x) \in A$, hence $\overline{K} \leq A$.   □

6.7.3 Example. Prove (again) that $B = \{x : \phi_x \in \mathcal{R}\} = \{x : \phi_x \text{ is total}\}$ is not semi-recursive.

We piggy back on the previous example and the same f through which we found a $k \in \mathcal{R}$ such that

$$\phi_{k(x)} = \begin{cases} \lambda y.0 & \text{if } \phi_x(x) \uparrow \\ \overbrace{(0, 0, \ldots, 0)}^{\text{length } m} & \text{if } \phi_x(x) \downarrow \end{cases} \qquad (7)$$

The above is (6) of the previous example.

We will use different *words* now to describe the bottom case, which we displayed explicitly in (7). Note that $\overbrace{(0, 0, \ldots, 0)}^{\text{length } m}$ is a non-recursive (nontotal) function listed as a *finite* array of outputs. Thus we have

$$\phi_{k(x)} = \begin{cases} \lambda y.0 & \text{if } \phi_x(x) \uparrow \\ \text{nontotal function} & \text{if } \phi_x(x) \downarrow \end{cases} \qquad (8)$$

and therefore

$$k(x) \in B \text{ iff } \phi_{k(x)} \text{ is total iff the top case of (8) applies iff } \phi_x(x) \uparrow$$

That is, $x \in \overline{K} \equiv k(x) \in B$, hence $\overline{K} \leq B$.   □

6.7.4 Example. In Exercise 6.3.12 I asked you to prove that $D = \{x : \text{ran}(\phi_x)$ is infinite} is *not recursive*. This task is directly based on Example 6.3.11.

Here I show that D actually is not even *semi*-recursive, *a fact that furnishes an example of a set that neither it, nor its complement are semi-recursive!*

We (heavily) piggy back on Example 6.7.2 above. We want to find $j \in \mathcal{R}$ such that

$$\phi_{j(x)} = \begin{cases} \text{inf. range} & \text{if } \phi_x(x) \uparrow \\ \text{finite range} & \text{if } \phi_x(x) \downarrow \end{cases} \tag{*}$$

Define ψ (almost) like f of Example 6.7.2 by

$$\psi(x, y) \simeq \begin{cases} y & \neg T(x, x, y) \\ \uparrow & \text{othw} \end{cases}$$

a definition by primitive recursive cases, using partial recursive part-functions, $\lambda y.y$ (top) and $\emptyset$ (bottom). Hence $\psi \in \mathcal{P}$.

The important difference here is that we force infinite range in the top case by outputting the input y.

The question $\psi \in \mathcal{P}$ settled, by S-m-n there is a $j \in \mathcal{R}$ such that

$$\phi_{j(x)}(y) \simeq \begin{cases} y & \text{if the call } \phi_x(x) \text{ has not returned in } y \text{ steps} \\ \uparrow & \text{othw} \end{cases} \tag{†}$$

Analysis of (†) *in terms of the "key" conditions* $\phi_x(x) \uparrow$ *and* $\phi_x(x) \downarrow$:

(I) Case where $\phi_x(x) \uparrow$.

Then, for *all* input values y, $\phi_x(x)$ is still not at **stop** after y steps. Thus by (†), we have $\phi_{j(x)}(y) = y$, for all y, that is,

$$\phi_x(x) \uparrow \implies \phi_{j(x)} = \lambda y.y \tag{1}$$

(II) Case where $\phi_x(x) \downarrow$. Let $m = smallest$ y such that the call $\phi_x(x)$ returned in m steps. Therefore, as before we find that for $y = 0, 1, \ldots, m - 1$ we have $\phi_{j(x)}(y) = y$, that is,

$$\begin{array}{llll} \textbf{input} \quad y & = 0, & 1, \quad \ldots, & m - 1 \\ \textbf{output} \quad \phi_{j(x)}(y) & = 0, & 1, \quad \ldots, & m - 1 \end{array}$$

and as before,

$$\begin{array}{lllll} \textbf{input} \quad y & = m, & m + 1, & m + 2, & \ldots \\ \textbf{output} \quad \phi_{j(x)}(y) & = \uparrow, & \uparrow, & \uparrow, & \ldots \end{array}$$

that is,

$$\phi_x(x) \downarrow \implies \phi_{j(x)} = (0, 1, \ldots, m - 1)\text{—finite range} \qquad (2)$$

(1) and (2) say that we got $(*)$ —p. 231— above. Thus

$j(x) \in D$ iff $\text{ran}(\phi_{j(x)})$ is infinite, iff we have the top case, iff $\phi_x(x) \uparrow$

Thus $\overline{K} \leq D$ via j.   □

6.8 An Application of the Graph Theorem

A definition like

$$f(x, y) \simeq \begin{cases} 0 & \text{if } x \in K \\ \uparrow & \text{othw} \end{cases} \qquad (1)$$

is a special case of the so-called *"definition by positive cases"*. That is,

- The cases listed explicitly (here $x \in K$) are semi-recursive, but the "othw" is *not* semi-recursive. Therefore, as the latter cannot be *verified*, we let the function output be undefined in this case.

 In *any* definition by cases

$$g(\vec{x}) \simeq \begin{cases} \vdots & \vdots \\ g_i(\vec{x}) & \text{if } R_i(\vec{x}) \\ \vdots & \vdots \end{cases}$$

we have

If $R_i(\vec{x})$ then $g_i(\vec{x})$

that is, we only need *verify* $R_i(\vec{x})$ —even if it is (primitive)recursive— to select the answer $g_i(\vec{x})$. However, in the *(primitive) recursive case* the "othw" is the negation of $R_1(\vec{x}) \vee R_2(\vec{x}) \vee \ldots \vee R_m(\vec{x})$, where $R_m(\vec{x})$ is the last explicit condition/case. By closure properties of $(\mathcal{PR}_*)$ $\mathcal{R}_*$, the "othw" case is (primitive) recursive as well.

- In a *definition by positive cases* the g_i are partial recursive.

The general form of such a definition is

$$g(\vec{x}) \simeq \begin{cases} \vdots & \vdots \\ g_i(\vec{x}) & \text{if } R_i(\vec{x}) \\ \vdots & \vdots \\ g_m(\vec{x}) & \text{if } R_m(\vec{x}) \\ \uparrow & \text{othw} \end{cases} \tag{2}$$

where the g_i are in $\mathcal{P}$ and the R_i are in $\mathcal{P}_*$.

 Note that $\mathcal{P}_*$ is *not* closed under negation, thus the "othw" in (2) is not in general semi-recursive. We observe this in the special case of (1) where the "othw" is $x \in \overline{K} \notin \mathcal{P}_*$.

Does a definition like (2) yield a partial recursive g?
Yes:

6.8.1 Theorem. (Definition by Positive Cases) *g in (2), under the stated conditions, is partial recursive.*

Proof We use the graph theorem, so it suffices to prove that

$$y \simeq g(\vec{x}) \text{ is semi-recursive} \tag{3}$$

Now, (3) is true precisely when $g(\vec{x}) \downarrow$ *thus the output is a* number, *say, y as above. For this to happen, some explicit condition $R_i(\vec{x})$ was true and $y \simeq g_i(\vec{x})$ was also true. In short, $y \simeq g_i(\vec{x}) \wedge R_i(\vec{x})$ was true.* Thus we prove (3) by noting

$$y \simeq g(\vec{x}) \equiv y \simeq g_1(\vec{x}) \wedge R_1(\vec{x}) \vee y \simeq g_2(\vec{x}) \wedge R_2(\vec{x}) \vee \ldots \vee y \simeq g_m(\vec{x}) \wedge R_m(\vec{x})$$

The right hand side of $\equiv$ is semi-recursive since each $R_i(\vec{x})$ is (given) and each $y \simeq g_i(\vec{x})$ is ($g_i \in \mathcal{P}$ and Theorem 6.6.1) at which point we invoke closure properties of $\mathcal{P}_*$ (Theorem 6.5.2). □

6.8.2 Exercise. Use Theorem 6.8.1 to give a mathematical proof that ψ defined below is in $\mathcal{P}$.

$$\psi(x, y) \simeq \begin{cases} 0 & \text{if } y = 0 \wedge \phi_x(x) \downarrow \ \vee \ y = 1 \\ \uparrow & \text{othw} \end{cases}$$

□

6.9 Converting Between the c.e. and Semi-Recursive Views Algorithmically

Suppose that a semi-recursive set S (subset of $\mathbb{N}$) is given as a W_i, that is, via a verifier i. Can we algorithmically —given i— construct a program for an *enumerator* of S?

How about the converse?

6.9.1 Theorem. *There is an $h \in \mathcal{R}$ such that, for all $i \in \mathbb{N}$, we have $W_i = ran(\phi_{h(i)})$.*

Proof

1. *First proof.* Now that we have Theorem 6.8.1, we know that ψ below is partial recursive.

$$\psi(i, x) \simeq \begin{cases} x & \text{if } \phi_i(x) \downarrow \text{ Same as: } x \in W_i \\ \uparrow & \text{othw} \end{cases}$$

By S-m-n we have an $h \in \mathcal{R}$ such that $\psi(i, x) = \phi_{h(i)}(x)$, for all i, x. Thus, for any i,

$$x \in ran(\phi_{h(i)}) \text{ iff } \phi_i(x) \downarrow \text{ iff } x \in W_i$$

That is, $W_i = ran(\phi_{h(i)})$.

2. *Second proof.* This is more "elementary" Define

$$\tau(i, z) \simeq \begin{cases} (z)_0 & \text{if } T(i, (z)_0, (z)_1) \text{ Same as: } (z)_0 \in W_i \\ \uparrow & \text{othw} \end{cases}$$

τ is defined by primitive recursive cases, hence is in $\mathcal{P}$. By S-m-n we have an $f \in \mathcal{R}^{10}$ such that $\tau(i, z) = \phi_{f(i)}(z)$, for all i, z. Thus, for any i,

$$(z)_0 \in ran(\phi_{f(i)}) \text{ iff } T(i, (z)_0, (z)_1) \downarrow \text{ iff } (z)_0 \in W_i$$

As $(z)_0$ ranges over the entire $\mathbb{N}$ —$\lambda z.(z)_0$ is onto— we are done: $ran(\phi_{f(i)}) = W_i$.

$\square$

---

[10] We should not use the same letter for two functions obtained in the same proof, for similar roles, unless we are prepared —or care— to prove that they are the same. Hence, f it is!

6.9.2 Theorem. *There is an $f \in \mathcal{R}$ such that, for all $i \in \mathbb{N}$, we have $\mathrm{ran}(\phi_i) = W_{f(i)}$. $x \in \mathrm{ran}(\phi_i) \equiv (\exists z)\Big(T(i,(z)_0,(z)_1) \wedge out((z)_1, i, (z)_0) = x\Big)$*

Proof
Letting $\psi = \lambda i x.(\mu z)\Big(T(i,(z)_0,(z)_1) \wedge out((z)_1, i, (z)_0) = x\Big)$ we have $\psi \in \mathcal{P}$
and

$$x \in \mathrm{ran}(\phi_i) \equiv \psi(i,x) \downarrow \tag{1}$$

By S-m-n there is an $f \in \mathcal{R}$ such that, for all i, x, $\phi_{f(i)}(x) = \psi(i,x)$, thus, from (1),

$$\phi_{f(i)}(x) \downarrow \equiv x \in \mathrm{ran}(\phi_i)$$

or $W_{f(i)} = \mathrm{ran}(\phi_i)$, for all i. $\qquad\square$

6.9.3 Example. (Intersection of Two W_i) Prove that there is an $f \in \mathcal{R}$ such that, for all i and j, $W_i \cap W_j = W_{f(i,j)}$. We have

$$x \in W_i \cap W_j \text{ iff } (\exists y)T(i,x,y) \wedge (\exists y)T(j,x,y)$$

$$\text{iff } (\exists z)\Big(T(i,x,(z)_0) \wedge T(j,x,(z)_1)\Big) \tag{1}$$

$$\text{iff } (\mu z)\Big(T(i,x,(z)_0) \wedge T(j,x,(z)_1)\Big) \downarrow$$

Thus, defining $\psi = \lambda i j x.(\mu z)\Big(T(i,x,(z)_0) \wedge T(j,x,(z)_1)\Big)$ we have

$$x \in W_i \cap W_j \equiv \psi(i,j,x) \downarrow \tag{2}$$

Trivially, $\psi \in \mathcal{P}$, thus, by S-m-n, there is an $f \in \mathcal{R}$ such that $\psi(i,j,x) = \phi_{f(i,j)}(x)$, for all i, j, x. Therefore (2) becomes $x \in W_i \cap W_j \equiv \phi_{f(i,j)}(x) \downarrow$, that is, $W_i \cap W_j = W_{f(i,j)}$. $\qquad\square$

6.9.4 Exercise. (Union of two W_i) There is an $h \in \mathcal{R}$ such that, for all i and j, $W_i \cup W_j = W_{h(i,j)}$. This, unlike 6.9.3, does not need any coding. $\qquad\square$

6.9.5 Proposition. *There is a $\tau \in \mathcal{R}$ such that, for all $x \in \mathbb{N}$, we have $W_{\tau(x)} = \{x\}$.*

Proof Let

$$\psi(x,y) \simeq \begin{cases} 0 & \text{if } y = x \\ \uparrow & \text{othw} \end{cases}$$

The above is a definition by positive cases, hence $\psi \in \mathcal{P}$ and by S-m-n a τ in $\mathcal{R}$ exists such that

$$\phi_{\tau(x)}(y) \simeq \begin{cases} 0 & \text{if } y = x \\ \uparrow & \text{othw} \end{cases}$$

Thus, $W_{\tau(x)} = \mathrm{dom}(\phi_{\tau(x)}) = \{x\}$. $\square$

6.9.6 Proposition. (Inverse images of W_i) *Let $f \in \mathcal{R}$. Then there is a $\sigma \in \mathcal{R}$ such that, for all i, we have $W_{\sigma(i)} = f^{-1}[W_i]$.*

Proof

$$x \in f^{-1}[W_i] \text{ iff } f(x) \in W_i \text{ iff } (\exists y)T(i, f(x), y) \text{ iff } (\mu y)T(i, f(x), y) \downarrow \qquad (1)$$

Let $\psi \stackrel{Def}{=} \lambda ix.(\mu y)T(i, f(x), y)$. Thus, $\psi \in \mathcal{P}$. By S-m-n, there is a $\sigma \in \mathcal{R}$ such that, for all i and x, $\psi(i, x) = \phi_{\sigma(i)}(x)$. From (1) we now obtain $x \in f^{-1}[W_i]$ iff $\phi_{\sigma(i)}(x) \downarrow$, that is, $f^{-1}[W_i] = W_{\sigma(i)}$. $\square$

The above is warmup: It generalises to an $f \in \mathcal{P}$, and if the latter is given by a program, as ϕ_j, then equipped with i and j as inputs we can compute the position in the standard listing, hence also the instructions, for a program for $\phi_j^{-1}[W_i]$.

6.9.7 Corollary. *There is a $\sigma \in \mathcal{R}$ such that, for all i and j, we have $W_{\sigma(i,j)} = \phi_j^{-1}[W_i]$.*

Proof

$$x \in \phi_j^{-1}[W_i] \text{ iff } \phi_j(x) \in W_i$$

$$\text{iff } (\exists y)T(i, \phi_j(x), y)$$

$$\text{iff } (\exists y)(\exists z)\Big(T(i, z, y) \wedge z = \phi_j(x)\Big)$$

$$\text{iff } (\exists y)(\exists z)\Big(T(i, z, y) \wedge (\exists w)\big(T(j, x, w) \wedge out(w, j, x) = z\big)\Big)$$

$$\text{iff } (\exists u)\Big(T(i, (u)_1, (u)_0) \wedge \big(T(j, x, (u)_2) \wedge out((u)_2, j, x) = (u)_1\big)\Big)$$

$$\text{iff } (\mu u)\Big(T(i, (u)_1, (u)_0) \wedge \big(T(j, x, (u)_2) \wedge out((u)_2, j, x) = (u)_1\big)\Big) \downarrow$$

Letting $\psi = \lambda ijx.(\mu u)\Big(T(i, (u)_1, (u)_0) \wedge \big(T(j, x, (u)_2) \wedge out((u)_2, j, x) = (u)_1\big)\Big)$ we next invoke S-m-n and conclude as above with a two argument —i and j— S-m-n function σ. $\square$

6.9.8 Proposition. *There is an $h \in \mathcal{R}$ such that, for all i and j, we have $W_{h(i,j)} = \phi_j[W_i]$.*

Proof

$$x \in \phi_j[W_i] \text{ iff } (\exists y)(x = \phi_j(y) \wedge y \in W_i)$$

$$\text{iff } (\exists y)\Big((\exists z)\big(T(j, y, z) \wedge out(z, j, y) = x\big) \wedge (\exists w)T(i, y, w)\Big)$$

$$\text{iff } (\exists u)\Big(\big(T(j, (u)_0, (u)_1) \wedge out((u)_1, j, (u)_0) = x\big) \wedge T(i, (u)_0, (u)_2)\Big)$$

$$\text{iff } (\mu u)\Big(\big(T(j, (u)_0, (u)_1) \wedge out((u)_1, j, (u)_0) = x\big) \wedge T(i, (u)_0, (u)_2)\Big) \downarrow$$

Letting $\psi = \lambda ijx.(\mu u)\Big(\big(T(j, (u)_0, (u)_1) \wedge out((u)_1, j, (u)_0) = x\big) \wedge T(i, (u)_0, (u)_2)\Big)$ we next invoke S-m-n and conclude as above with a two argument —i and j— S-m-n function h.   □

We can obtain some more elaborate corollaries from Theorem 6.9.1. In particular, we recall that if a semi-recursive set A is nonempty, then we can enumerate it by a recursive function.

Namely, if the semi-recursive set A is nonempty, then it can be enumerated by a recursive function given, essentially,[11] by:

$$f(z) \simeq \begin{cases} (z)_0 & \text{if } T(i, (z)_0, (z)_1) \\ a & \text{othw} \end{cases}$$

where, by assumption, we let $A = W_i$ for some i and let a be an unspecified $a \in A$. Now "a" depends on i.

If we can computably *choose* an "a" in *each* nonempty W_i then we can sharpen Theorem 6.9.1. This choice is easy and we can obtain it by an easy variant of the *selection theorem* which provides a computable *axiom of choice*.[12]

6.9.9 Theorem. (Selection Theorem —Simple Form) *There is a function $sel \in \mathcal{P}$ such that $(\exists y)y \in W_i$ iff $sel(i) \in W_i$.*

 Since "$sel(i) \in W_i$" means "$(\exists z)(z = sel(i) \wedge z \in W_i)$" the truth of the first quoted sentence (for any fixed i) necessitates that $sel(i) \downarrow$ as the second form of the sentence makes clear.

Proof We *dovetail* a computation, that is, we compute by stages such that the $n+1$-st stage of the process computes $\phi_i(y)$ for exactly $n + 1$ computation steps of the URM i, *for each of the inputs* $y = 0, 1, 2, \ldots, n$.

---

[11] In loc. cit. we used the projection theorem instead of the T-predicate.

[12] An easy version of this axiom of set theory states that if we have an indexed family of nonempty sets $(S_a)_{a \in I}$, then there is a total function f with $\mathrm{dom}(f) = I$ such that $f(a) \in S_a$, for all $a \in I$.

As soon as at some stage $n + 1$ we have that one of the $y = 0, \ldots, n$ satisfies $\phi_i(y) \downarrow$, we halt everything and return the (say, the first one) value of y —among the $0, 1, 2, \ldots, n$— values that we detected causing $\phi_i(y) \downarrow$. This y is our $sel(i)$.

Mathematically,

$$sel(i) \overset{Def}{\simeq} \left((\mu z) T\big(i, (z)_0, (z)_1\big) \right)_0 \tag{1}$$

Clearly, we get the *if* direction of the theorem by pure logic Theorem 0.2.107.

For the *only if*, assume $(\exists y) y \in W_i$ is true. That is, $(\exists y)(\exists z) T(i, y, z)$ is true or, equivalently, $(\exists z) T(i, (z)_0, (z)_1)$ is. Better still, the last one is expressed by "$(\mu z) T(i, (z)_0, (z)_1) \downarrow$ is true" or, even better, by

$$\left((\mu z) T(i, (z)_0, (z)_1) \right)_0 \downarrow \quad \text{is true} \tag{2}$$

Now fix attention on this smallest z that we computed in (2). For this z, $(z)_0 = sel(i) \downarrow$ by (1). Hence $T(i, sel(i), (z)_1)$ is true, that is, $sel(i) \in W_i$.   □

 6.9.10 Remark. We can computably select *a* member of $W_i \neq \emptyset$, but not *the* smallest member. Cf. Exercise 7.0.4   □

We now have an improvement on Theorem 6.9.1.

6.9.11 Corollary. *There is an $f \in \mathcal{R}$ such that, for all $i \in \mathbb{N}$, we have $W_i = ran(\phi_{f(i)})$, and moreover if $W_i \neq \emptyset$, then $\phi_{f(i)} \in \mathcal{R}$.*

Proof We reuse a modified τ from the proof of Theorem 6.9.1. Define

$$\tau(i, z) \simeq \begin{cases} (z)_0 & \text{if } T(i, (z)_0, (z)_1) \\ sel(i) & \text{othw} \end{cases}$$

τ is defined by primitive recursive cases, hence is in $\mathcal{P}$. By S-m-n we have an $f \in \mathcal{R}$ such that $\tau(i, z) = \phi_{f(i)}(z)$, for all i, z.

We conclude that $ran(\phi_{f(i)}) = W_i$ exactly as in the proof of Theorem 6.9.1.

And now we add: Assume $W_i \neq \emptyset$. By Theorem 6.9.9 $sel(i) \in W_i$ (defined!) Hence for this i, $\phi_{f(i)} \in \mathcal{R}$.   □

Another way to do the above corollary is to force a strictly increasing enumeration:

6.9.12 Corollary. *There is an $f \in \mathcal{R}$ such that, for all $i \in \mathbb{N}$, we have $W_i = ran(\phi_{f(i)})$, and moreover*

- *$\phi_{f(i)}$ is 1-1*
- *If $\phi_{f(i)}$ is finite, then $dom(\phi_{f(i)}) = \{0, 1, 2, \ldots, m\}$, for some m.*

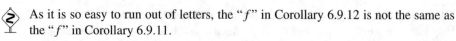 As it is so easy to run out of letters, the "f" in Corollary 6.9.12 is not the same as the "f" in Corollary 6.9.11.

Proof Define $\lambda i z. \psi(i, z)$ by

$$\psi(i, 0) = sel(i)$$

$$\psi(i, z + 1) = \Big((\mu w) \big[T(i, (w)_0, (w)_1) \wedge (w)_0 \notin H(i, z) \big] \Big)_0 \tag{1}$$

Where we wrote $H(i, z)$ for the history $\langle \psi(i, 0), \psi(i, 1), \ldots, \psi(i, z) \rangle$. Since (1) is a course-of-values recursion, $\psi \in \mathcal{P}$ and by S-m-n there is an $f \in \mathcal{R}$ such that

$$\psi(i, z) = \phi_{f(i)}(z), \text{ for all } i \text{ and } z$$

Now the first bullet of the corollary is obvious as we force —in (1)— each new $\phi_{f(i)}(z + 1)$ to be different from all $\phi_{f(i)}(u)$, for $u \leq z$.

For the second bullet, if z_0 is smallest such that $\phi_{f(i)}(z_0) \uparrow$, then (1) implies —for all $u \geq z_0$— that $\phi_{f(i)}(u) \uparrow$, since $H(i, u) \uparrow$, for $u = z_0$, and thus for all $u > z_0$ as well.

Finally, we have $W_i = \text{ran}(\phi_{f(i)})$:

1. If $W_i = \emptyset$, then $\psi(i, 0) \uparrow$ hence $\text{ran}(\phi_{f(i)}) = \emptyset$.
   Otherwise,
2. W_i is finite. Can I have some $b \in W_i - \text{ran}(\phi_{f(i)})$?
   If so, note that, by bullet two, $\text{dom}(\phi_{f(i)}) = \{0, 1, \ldots, m\}$ for some $m \geq 0$. Thus $b \notin H(i, m)$. Let $t = (\mu w)\big(T(i, (w)_0, (w)_1) \wedge b = (w)_0\big)$ be smallest value for any $b \in W_i - \text{ran}(\phi_{f(i)})$. Then the second equation of (1) calculates $\phi_{f(i)}(m + 1) = (t)_0 = b$. This contradicts $b \notin \text{ran}(\phi_{f(i)})$.
3. W_i is infinite. Can I have some $b \in W_i - \text{ran}(\phi_{f(i)})$? If yes, then $b \notin H(i, z)$ is true for all z. So, consider again the t of case 2 above. If t has never picked b in Eq. (1) above to form $\phi_{f(i)}(z + 1) = (t)_0 = b$, then it must be that we have infinitely many "$w < t$" that are picked instead to set $\phi_{f(i)}(z + 1) = (w)_0 \neq b$. But we cannot have infinitely many natural numbers below t. □

In view of Corollary 6.9.12 and Theorem 6.9.2 we obtain at once

6.9.13 Corollary. *There is an $h \in \mathcal{R}$ such that, for all $i \in \mathbb{N}$, we have $\text{ran}(\phi_i) = \text{ran}(\phi_{h(i)})$, and moreover*

- $\phi_{h(i)}$ *is 1-1*
- *If $\phi_{h(i)}$ is finite, then $\text{dom}(\phi_{h(i)}) = \{0, 1, 2, \ldots, m\}$, for some m.*

6.9.14 Corollary. *There is an $h \in \mathcal{R}$ such that, for all $i \in \mathbb{N}$, we have $\text{ran}(\phi_i) = \text{ran}(\phi_{h(i)})$, and moreover if $\text{ran}(\phi_i) \neq \emptyset$, then $\phi_{h(i)} \in \mathcal{R}$.*

That is, not only a c.e. set has a partial recursive enumeration via some ϕ_i but, moreover, if it is nonempty, then we can algorithmically construct a procedure for a *recursive* enumeration, $\phi_{h(i)}$.

Proof By Theorem 6.9.2 there is a $f \in \mathcal{R}$ such that $\text{ran}(\phi_i) = W_{f(i)}$. □

6.10   Some General Rice-Like Tools

This section contains some unifications of tools we presented already in our example reductions, into powerful lemmata (and one theorem) due to Rice and others (Myhill, McNaughton, Shapiro).

6.10.1 Theorem. *Let $A = \{x : \phi_x \in C\}$, where $C \subseteq \mathcal{P}$. If some $f \in C$ has no finite subfunction $\xi: C \ni \xi \subseteq f$, then $\overline{K} \leq A$.*

Proof   This argument generalises the one in Example 6.7.2.[13] Pick an $f \in C$ that has *no* finite subfunction in C, and define ψ by

$$\psi(x, y) \simeq \begin{cases} f(y) & \text{if the call } \phi_x(x) \text{ has not returned in } y \text{ steps} \\ \uparrow & \text{othw} \end{cases}$$

The above is a definition by $\mathcal{PR}_*$ cases, hence $\psi \in \mathcal{P}$. By S-m-n we have an $h \in \mathcal{R}$ such that, for all x, y,

$$\phi_{h(x)}(y) \simeq \begin{cases} f(y) & \text{if the call } \phi_x(x) \text{ has not returned in } y \text{ steps} \\ \uparrow & \text{othw} \end{cases} \tag{1}$$

Analysis:

1. Let $\phi_x(x) \uparrow$. Then the top case in (1) is true for all y, thus

$$\phi_x(x) \uparrow \text{ implies } \phi_{h(x)} = f \tag{2}$$

2. Let $\phi_x(x) \downarrow$. Let $m = $ smallest y such that $\phi_x(x)$ returns in m steps. So,

   - for $y = 0, \ldots, m - 1$ we have (top case of (1))

$$\phi_{h(x)} = \overbrace{(f(0), \ldots, f(m-1))}^{\xi}$$

   - for $y = m, m + 1, \ldots$ we have (bottom case of (1))

$$\phi_{h(x)} = \overbrace{(\uparrow, \uparrow, \ldots)}^{\text{forever}}$$

   In short,

---

[13] $\lambda y.0$ has no finite subfunction that is a constant.

$$\phi_{h(x)} = \overbrace{(f(0), \ldots, f(m-1))}^{\xi} \tag{3}$$

Thus

$$h(x) \in A \text{ iff } \phi_{h(x)} \in A \overset{\text{cannot be } \xi}{\text{ iff }} \overset{(2)}{\phi_{h(x)} = f} \text{ iff } \phi_x(x) \uparrow$$

Thus $\overline{K} \leq A$.   □

6.10.2 Corollary. *Under the conditions of Theorem 6.10.1, $A \notin \mathcal{P}_*$.*

6.10.3 Theorem. *Let $A = \{x : \phi_x \in \mathcal{C}\}$, where $\mathcal{C} \subseteq \mathcal{P}$. If for some $f \in \mathcal{C}$ some computable extension g of f is not also in $\mathcal{C}$, then $\overline{K} \leq A$.*

Proof So, let us fix $f = \phi_m$ and $g = \phi_n$ as stated in the theorem. We want to define g by

$$g(x, y) \simeq \begin{cases} \phi_m(y) & \text{if } \phi_x(x) \uparrow \\ \phi_n(y) & \text{othw} \end{cases}$$

This is not a properly formatted definition (e.g., it is not formatted by positive cases) to help us argue that g is computable. So we elaborate as follows:

$$g(x, y) \simeq \begin{cases} \phi_m(y) & \text{if } \phi_x(x) \text{ does } not \text{ return before } \phi_m(y) \text{ does} \\ \phi_n(y) & \text{if } \phi_x(x) \text{ does return before } \phi_m(y) \text{ does} \end{cases} \tag{1}$$

The top case in (1) happens when either (i) $\phi_m(y)$ returned, and $\phi_x(x)$ did too at the same time, or is still running; or (ii) we have $\phi_m(y) \uparrow \wedge \phi_x(x) \uparrow$. In both subcases we must define $g(x, y)$ as $\phi_m(y)$, albeit the first subcase yields a number, while the second yields $\uparrow$.

Armed with this observation, we rewrite (1) as

$$g(x, y) \simeq \begin{cases} \phi_m(y) & \text{if } (\exists z)\Big(T(m, y, z) \wedge (\forall w)_{<z} \neg T(x, x, w))\Big) \\ \phi_n(y) & \text{if } (\exists z)\Big(T(x, x, z) \wedge (\forall w)_{\leq z} \neg T(m, y, w))\Big) \\ \uparrow & \text{othw} \end{cases} \tag{2}$$

Per our preambular comments, the top case of (1) has been implemented in (2) by splitting it into the top and bottom cases. (2) is a definition by positive cases (Theorem 6.8.1), the conditions top and second being semi-recursive by Theorem 6.3.14. Thus, $g \in \mathcal{P}$ and, by the S-m-n theorem, we have an $h \in \mathcal{R}$ such that, for all x, y,

$$\phi_{h(x)}(y) \simeq \begin{cases} \phi_m(y) & \text{if } (\exists z)\Big(T(m, y, z) \wedge (\forall w)_{<z} \neg T(x, x, w))\Big) \\ \phi_n(y) & \text{if } (\exists z)\Big(T(x, x, z) \wedge (\forall w)_{\leq z} \neg T(m, y, w))\Big) \\ \uparrow & \text{othw} \end{cases} \tag{3}$$

Analysis:

1. Let $\phi_x(x) \uparrow$. First note that the middle case in definition (3) is not applicable here, since said case includes "$(\exists z)\big(T(x, x, z) \wedge \ldots\big)$", which implies $\phi_x(x) \downarrow$. Now, we have two subcases, as in the preamble:

   • Fix some y and let $\phi_m(y) \downarrow$. Then, for some z, $T(m, y, z)$ is true. By our assumption, i.e., $\phi_x(x) \uparrow$, we have $(\forall w)_{<z} \neg T(x, x, w))$ is true for all z, *and thus also for the one that makes $T(m, y, z)$ work.*

     We thus have the top case in (3). This subcase proved:

$$\phi_x(x) \uparrow \Longrightarrow \phi_{h(x)}(y) \simeq \phi_m(y)$$

   • Say, $\phi_m(y) \uparrow$, for the given y. We already noted that the middle case in (3) is not applicable; nor is the top case since $(\exists z)T(m, y, z)$ is now false. Thus we have the bottom case, where we have both $\phi_m(y) \uparrow$ *and* $\phi_x(x) \uparrow$. (3) responds "$\uparrow$" in the bottom case, which is the same response as: "$\phi_m(y)$". Thus, once again, in this subcase of $\phi_x(x) \uparrow$, we have

$$\phi_x(x) \uparrow \Longrightarrow \phi_{h(x)}(y) \simeq \phi_m(y)$$

   We summarise both subcases in a y-independent manner as

$$\boxed{\phi_x(x) \uparrow \Longrightarrow \phi_{h(x)} \simeq \phi_m} \tag{4}$$

2. Let $\phi_x(x) \downarrow$. Here, for some z, $T(x, x, z)$ of the middle condition is true.

   Fix a y. Now, the associated conjunct of $T(x, x, z)$ in the middle case of (3), namely, $(\forall w)_{\leq z} \neg T(m, y, w)$, could be true; or could be false —that is, $(\exists w)_{\leq z} T(m, y, w)$ could be true.

   We deal with these two subcases separately:

   • Here assume $(\forall w)_{\leq z} \neg T(m, y, w)$ holds, for the smallest z that makes $T(x, x, z)$ true. Thus the middle case *is* true, and we have

$$\phi_x(x) \downarrow \Longrightarrow \phi_{h(x)}(y) \simeq \phi_n(y)$$

   • Now assume $(\exists w)_{\leq z} T(m, y, w)$ holds, for the smallest z that makes $T(x, x, z)$ true.

     Hmm. This sends us to the top case, that is,

$$\phi_x(x) \downarrow \Longrightarrow \phi_{h(x)}(y) \simeq \phi_m(y)$$

No worries: $\phi_m(y) \downarrow$ here, so, by the assumption $\phi_m \subseteq \phi_n$ the above can be rewritten as

$$\phi_x(x) \downarrow \Longrightarrow \phi_{h(x)}(y) \simeq \phi_n(y)$$

All in all, summarising in a y-independent manner,

$$\boxed{\phi_x(x) \downarrow \Longrightarrow \phi_{h(x)} \simeq \phi_n} \tag{5}$$

The end game is now easy:

$$h(x) \in A \text{ iff } \phi_{h(x)} \in C \overset{\phi_n \notin C}{\text{ iff }} \phi_{h(x)} = \phi_m \overset{\text{by (4) and (5): top case in (3)}}{\text{ iff }} \phi_x(x) \uparrow$$

Thus, $\overline{K} \leq A$. $\square$

6.10.4 Corollary. *Under the assumptions of Theorem 6.10.3, $A \notin \mathcal{P}_*$.*

6.10.5 Example. An immediate application of Theorem 6.10.3 and its corollary is that $E = \{x : \phi_x = \emptyset\}$ —an instance of the equivalence problem— is not semi-recursive. The reason is that $C = \{\emptyset\}$ here, and every unary computable function extends $\emptyset$. E.g., $\emptyset \subseteq U_1^1$, but $U_1^1 \notin \{\emptyset\}$. $\square$

6.10.6 Corollary. (Rice, Shapiro) *Let $A = \{x : \phi_x \in \mathcal{C}\}$ be semi-recursive. Then $f \in \mathcal{C} \equiv (\exists \text{ finite } \theta)(\theta \subseteq f \wedge \theta \in \mathcal{C}\}$.*

Proof   By the semi-recursiveness of A:

- ($\rightarrow$ direction.) Say $f \in \mathcal{C}$. By Theorem 6.10.1, there is a $\theta \in \mathcal{C}$ (finite) such that $\theta \in \mathcal{C}$.
- ($\leftarrow$ direction.) Say we have $\theta \subseteq f \wedge \theta \in \mathcal{C}$ for a finite θ. By Theorem 6.10.3, it must be that $f \in \mathcal{C}$. $\square$

6.10.7 Theorem. (Rice's Theorem) *The complete index set $A = \{x : \phi \in C\}$ is recursive iff either $A = \emptyset$ or $A = \mathbb{N}$.*

Proof   The *if*-part is trivial, since $\emptyset$ and $\mathbb{N}$ are actually in $\mathcal{PR}_*$.
For the *only if* we assume $A \in \mathcal{R}_*$ and consider two cases:

1. Let $\emptyset$ (the function) be in C. As A is also semi-recursive, by the corollary to Theorem 6.10.3, we must have that every unary $f \in \mathcal{P}$ is in C, since $\emptyset \subseteq f$. Thus $C = \mathcal{P}$, hence $A = \mathbb{N}$.
2. Let $\emptyset \notin C$. Thus, $\emptyset \in \mathcal{P} - C$. Then $\mathcal{P} - C = \mathcal{P}$ and as $\overline{A} = \{x : \phi_x \in \mathcal{P} - C\}$ is semi-recursive, we get $\overline{A} = \mathbb{N}$, that is, $A = \emptyset$. $\square$

Rice called a complete index set A that is either $\mathbb{N}$ or $\emptyset$ "trivial". Thus his theorem says that *of all the infinitely many complete index sets, only the two trivial ones are recursive.*

This theorem is about *program sets; complete index sets.* There are infinitely many nontrivial sets that are recursive or indeed primitive recursive. For example, $Pr(x)$ (the set of primes), $y = A_n(x)$ (the graph of the Ackermann function), which are fairly complex, *are* primitive recursive.

6.11 Exercises

1. Prove that $\phi_x(y) \simeq z$ is unsolvable.
   Hint. If it were recursive then so would be $\phi_{h(x)}(0) = 0$. Refer to Example 6.3.8
   1. now and check whether the immediately preceding predicate is equivalent to $x \in K$.
2. Prove again that $\phi_x(y) \simeq z$ is unsolvable, this time ignoring reducibility and using a direct diagonalisation.
3. Prove that $f = \lambda x.\phi_x(x)$ has no total computable extensions.
4. (Mathematically) prove Theorem 6.5.2, 2 case directly from the definition of semi-recursiveness without the help of either the normal form theorem or the projection theorem.
5. Prove that $P(\vec{x})$ is semi-recursive *iff* for some $f \in \mathcal{P}$ we have $P(\vec{x}) \equiv f(\vec{x}) \simeq 0$.
6. Is the "proof" I quote below correct? If not, where *exactly* does it go wrong?
   "Let $y \simeq f(\vec{x}_n)$ be semi-recursive. Then $y \simeq f(\vec{x}_n) \equiv \psi(y, \vec{x}_n) \simeq 0$ for some $\psi \in \mathcal{P}$.
   Thus $g = \lambda \vec{x}_n.(\mu y)\psi(y, \vec{x}_n)$ is in $\mathcal{P}$. But $g = f$, since the unbounded search finds the y that makes $y \simeq f(\vec{x}_n)$ true, if $f(\vec{x}_n) \downarrow$. Therefore, $f \in \mathcal{P}$."
7. We already know (proved in this chapter) that $\phi_x(y) \simeq z$ is semi-recursive. It is also the case that $\phi_x(y) \not\simeq z$ is semi-recursive, since it is equivalent to

$$(\exists w)(\phi_x(y) \simeq w \wedge z \neq w)$$

   We conclude from Theorem 6.1.7 that $\phi_x(y) \simeq z$ is recursive.
   Wait a minute! We have already proved that $\phi_x(y) \simeq z$ is *not* recursive.
   Either that proof, or this proof are wrong. *Which one is it and exactly in which step?*
8. Prove that $\{x : 0 \in W_x\}$ is semi-recursive. Is it recursive? Why?
9. Prove that $\{x : 0 \in \mathrm{ran}(\phi_x)\}$ is semi-recursive. Is it recursive? Why?
10. Is $\lambda x.x \in W_y$ recursive?
    Hint. Consider cases according to y.
11. Do not use Rice's theorem or the Rice lemmata (Theorems 6.10.1, 6.10.3).
    Is $\{x : \phi_x \in \mathcal{PR}\}$ recursive? r.e.?

12. Do not use Rice's theorem or the Rice lemmata (Theorems 6.10.1, 6.10.3).
    Is $\{x : \phi_x \notin \mathcal{PR}\}$ recursive? r.e.?

13. Do not use Rice's theorem or the Rice lemmata (Theorems 6.10.1, 6.10.3).
    Is $\{x : \phi_x$ not 1-1$\}$ r.e.? How about its complement?

14. Do not use Rice's theorem or the Rice lemmata (Theorems 6.10.1, 6.10.3).
    Is $\{x : \phi_x$ onto$\}$ r.e.? How about its complement?

15. Do not use Rice's theorem or the Rice lemmata (Theorems 6.10.1, 6.10.3).
    Prove that $\{x : W_x = \{0\}\}$ is not r.e.

16. Do not use Rice's theorem or the Rice lemmata (Theorems 6.10.1, 6.10.3).
    Prove that $\{x : W_x \in \mathcal{R}_*\}$ is not r.e. How about its complement?

17. Do not use Rice's theorem or the Rice lemmata (Theorems 6.10.1, 6.10.3).
    Prove that the set $A = \{x : W_x = \mathbb{N}\}$ is not semi-recursive.

18. Do not use Rice's theorem or the Rice lemmata (Theorems 6.10.1, 6.10.3).
    Prove that the set $A = \{x : W_x =$ the set of all even numbers$\}$ is not r.e.

19. Do not use Rice's theorem or the Rice lemmata (Theorems 6.10.1, 6.10.3).
    Prove that the set $A = \{x : \text{ran}(\phi_x)$ has exactly five distinct elements$\}$ is not semi-recursive.

20. Do not use Rice's theorem or the Rice lemmata (Theorems 6.10.1, 6.10.3).
    Prove that the set $D = \{x : \phi_x$ is the characteristic function of some set$\}$ is not semi-recursive.

21. Do not use Rice's theorem or the Rice lemmata (Theorems 6.10.1, 6.10.3).
    Prove that the set $O = \{x : \text{ran}(\phi_x)$ contains only odd numbers$\}$ is not semi-recursive.

22. Do not use Rice's theorem or the Rice lemmata (Theorems 6.10.1, 6.10.3).
    Prove that the set $Pr = \{x : \text{ran}(\phi_x)$ contains only prime numbers$\}$ is not semi-recursive.

23. Do not use Rice's theorem or the Rice lemmata (Theorems 6.10.1, 6.10.3).
    Prove that the set $Prd = \{x : \text{dom}(\phi_x)$ contains only prime numbers$\}$ is not semi-recursive.

24. Prove, *via Rice's lemmata*, that the set of ϕ-indices of each computable function f is not c.e.
    Hint. You may want to consider the total and nontotal cases separately.
    In the following Exercises you may use Rice's theorem or lemmata to given an answer/proof with detailed reasons.

25. Prove that $\{x : W_x = \{0, 1, 2\}\}$ is not r.e. How about its complement?

26. Is $\{x : W_x$ is infinite$\}$ r.e.? How about its complement?

27. Is $\{x : W_x = \emptyset\}$ r.e.?

28. Is the theorem of Rice applicable to the set $K = \{x : \phi_x(x) \downarrow\}$?

29. Given a recursive unary f. Is the theorem of Rice applicable to the set $\{x : \phi_{f(x)}(x) \downarrow\}$? How about $\{x : \phi_x(f(x)x) \downarrow\}$?

30. Is the theorem of Rice applicable to the set $\overline{K} = \{x : \phi_x(x) \uparrow\}$?

31. We know that $\{x : \phi_x(x) = c\}$, c a constant, is r.e. but not recursive. Can the non recursiveness of this set be proved by Rice's theorem?

Chapter 7
Yet Another Number-Theoretic Characterisation of $\mathcal{P}$

Overview

In Davis (1958b), Kleene (1952), Tourlakis (1984) one finds an alternative inductive characterisation of $\mathcal{P}$ based on an augmented set of initial functions, but primitive recursion as a *primary* operation (definitionally) is absent (but, of course, is a derived operation). Moreover unbounded search is *primitive*, but is a different version than the (μy) of Definition 1.1.25, and is *defined on total functions only*. We introduce the tools for coding from Smullyan (1961), Bennett (1962), based on *2-adic* (pronounced "*dyadic*" as opposed to *binary*) concatenation. The inductive characterisation of $\mathcal{R}$ derived in this chapter will be used in Chap. 12.

7.0.1 Definition. (A Different Unbounded Search) We define a new unbounded search $(\widetilde{\mu} y)$ on any *total* number theoretic function $g : \mathbb{N}^{n+1} \to \mathbb{N}$ that produces a number theoretic function $f : \mathbb{N}^n \to \mathbb{N}$ as follows, for all $\vec{x}_n$:

$$f(\vec{x}_n) \simeq \begin{cases} \min\{y : g(y, \vec{x}_n) = 0\} & \text{if } (\exists y)g(y, \vec{x}_n) = 0 \\ \uparrow & \text{othw} \end{cases} \tag{1}$$

We write $f = \lambda \vec{x}_n.(\widetilde{\mu} y)g(y, \vec{x}_n)$.

For a predicate $P(y, \vec{x})$,

$$(\widetilde{\mu} y)P(y, \vec{x}) \stackrel{Def}{=} (\widetilde{\mu} y)c_P(y, \vec{x}), \text{ as usual.} \qquad \square$$

7.0.2 Remark. ((μy) vs ($\widetilde{\mu} y$)) For a *total* g, the part $(\forall z)_{<y}g(y, \vec{x}_n) \downarrow$ in Definition 1.1.25 is true. Thus, trivially, if $g \in \mathcal{R}$, then

$$(\widetilde{\mu} y)g(y, \vec{x}_n) = (\mu y)g(y, \vec{x}_n)$$

G. Tourlakis, *Computability*, https://doi.org/10.1007/978-3-030-83202-5_7

Now, we know that $\lambda \vec{x}_n.(\mu y)g(y, \vec{x}_n) \in \mathcal{P}$ (Proposition 1.1.26), hence also, $\lambda \vec{x}_n.(\widetilde{\mu} y)g(y, \vec{x}_n) \in \mathcal{P}$.

Note that $\widetilde{\mu}$ may well produce nontotal functions. For example, if $g = \lambda y x.1$, then $\lambda x.(\widetilde{\mu} y)g(y, x) = \emptyset$.

How do we program $(\widetilde{\mu} y)g(y, \vec{x}_n)$?

Trivially (but informally), as:

$y \leftarrow 0$

while $g(y, \vec{x}_n) \neq 0$ **do**

$y \leftarrow y + 1$

end

Pause. But there is a catch!◄

In the second instruction above we *must* produce a *program* for g because we do not know an infinite object —that a function in general is— but only do so via a *finite description*: a program. However (cf. Example 6.3.8)

$$\boxed{\textit{We cannot tell (algorithmically) if a program computes a total function!}}$$

Can we work around this? Why not apply the definition of $(\widetilde{\mu} y)$ to *any* function of $\mathcal{P}$, total or not, and write a program to compute $(\widetilde{\mu} y)g(y, \vec{x})$ for any function in $\mathcal{P}$? This program will then, surely(?) work correctly on total functions, in particular, even if we *do not know* a total function when we see (a program of) one?

That will not work either! See the following Proposition 7.0.3.   □ ⚡

7.0.3 Proposition. $\mathcal{P}$ is not *closed under* $(\widetilde{\mu} y)$.

Proof We exhibit a $\mathcal{P}$-function on which application of $(\widetilde{\mu} y)$ produces an *uncomputable* function.

Define ψ by

$$\psi(x, y) \simeq \begin{cases} 0 & \text{if } y = 0 \wedge \phi_x(x) \downarrow \ \vee \ y = 1 \\ \uparrow & \text{othw} \end{cases}$$

The top condition is semi-recursive by the semi-recursiveness of K and closure properties of $\mathcal{P}_*$.

Thus ψ is defined by positive cases and thus it is indeed in $\mathcal{P}$.

Now, *mathematically speaking* —making no claim that we can *compute* it— for any $a \in \mathbb{N}$ there *is* a smallest y that makes $\psi(a, y) = 0$. Specifically,

(1) If $\phi_a(a) \downarrow$, then $y = 0$ is the smallest that makes $\psi(a, y) = 0$.
(2) If $\phi_a(a) \uparrow$, then $y = 1$ is the smallest that makes $\psi(a, y) = 0$.

Summarising (1) and (2), and using the $\widetilde{\mu}$-notation, we have, for all $a \in \mathbb{N}$,

$$(\widetilde{\mu} y)\psi(a, y) \simeq \begin{cases} 0 & \text{if } a \in K \\ 1 & \text{if } a \notin K \end{cases}$$

So, $\lambda x.(\widetilde{\mu}y)\psi(x, y)$ is the characteristic function of K, hence it cannot be recursive! Applying $\widetilde{\mu}$ to ψ led us outside of $\mathcal{P}$. In short, there is no way to program $\lambda x.(\widetilde{\mu}y)\psi(x, y)$ in any of the known (equivalent) mathematical programming frameworks, or, as Church would have said: It is not possible to program this function, *period*! □

7.0.4 Exercise. There is no computable —as a function of i— selector for the minimum element of $W_i \neq \emptyset$. That is, prove that $\lambda i. \min(W_i)$ is not in $\mathcal{P}$. □

7.0.5 Definition. We define $\mathcal{P}_{alt}$ as the closure of

$$\mathcal{I}_{alt} = \left\{ Z, S, \{U_i^n\}_{\substack{n \geq 1 \\ 1 \leq i \leq n}} , \lambda xy.x + y, \lambda xy.x \doteq y, \lambda xy.xy \right\}$$

under *composition*[1] and $\widetilde{\mu}$ applied to total functions. □

The above is a purely mathematical definition, since total computable functions may well be given by their programs, but we are unable —as noted enough times— to recognise such programs for what they are.

We aim to show that $\mathcal{P}_{alt} = \mathcal{P}$. The easy lemma is

7.0.6 Lemma. $\mathcal{P}_{alt} \subseteq \mathcal{P}$.

Proof An implied induction —i.e., details are left to the reader— on the set $\mathcal{P}_{alt}$. Well, all its initial functions are in $\mathcal{P}$, and the latter is closed under composition and $\widetilde{\mu}$ as we saw in Remark 7.0.2. □

7.0.7 Definition. $\mathcal{R}_{alt}$ is defined to be the set of *all total* functions in $\mathcal{P}_{alt}$. □

7.0.8 Remark. (Regular Functions and Predicates) Noting that only $\widetilde{\mu}$ can produce non total functions in $\mathcal{P}_{alt}$, it is clear that if we apply $\widetilde{\mu}$ on functions g in $\mathcal{R}_{alt}$ that satisfy

$$(\forall \vec{x})(\exists y)g(y, \vec{x}) = 0 \tag{1}$$

then $\lambda \vec{x}.(\widetilde{\mu}y)g(y, \vec{x}) \in \mathcal{R}_{alt}$. $\mathcal{R}_{alt}$-functions that satisfy (1) are called *y-regular*.

Analogously, a predicate Q such that $(\forall \vec{x})(\exists y)Q(\vec{x}, y)$ is also called *y-regular*.

If we do not wish to draw attention to the y we say just *regular*, rather than *y-regular*. □

As is usual, we define the predicates of $\mathcal{R}_{alt}$:

7.0.9 Definition. We define

$$\mathcal{R}_{alt,*} \stackrel{Def}{=} \{P(\vec{x}) : c_P \in \mathcal{R}_{alt}\}$$ □

---

[1] In Tourlakis (1984) we omit Z, S and the U_i^n from $\mathcal{I}_{alt}$ and close under Grzegorczyk operations instead.

Corollary 2.2.3 applies to $\mathcal{R}_{alt,*}$, since $\lambda xy.x \doteq y$ is in $\mathcal{R}_{alt}$ and Grzegorczyk operations are applicable. For the record,

7.0.10 Proposition. $P(\vec{x}) \in \mathcal{R}_{alt,*}$ *iff some* $f \in \mathcal{R}_{alt}$ *exists such that, for all* $\vec{x}$, $P(\vec{x}) \equiv f(\vec{x}) = 0$.

The above immediately yields substitution of function calls for predicate arguments:

7.0.11 Proposition. *If* $P(\vec{x}, y, \vec{z}) \in \mathcal{R}_{alt,*}$ *and* $\lambda\vec{w}.f(\vec{w}) \in \mathcal{R}_{alt}$, *then*

$$P(\vec{x}, f(\vec{w}), \vec{z}) \in \mathcal{R}_{alt,*}$$

7.0.12 Exercise. The predicates $x \leq y$, $x < y$, and $x = y$ are in $\mathcal{R}_{alt,*}$. □

Another well-trodden result that is useful in the current context is

7.0.13 Proposition. $\mathcal{R}_{alt,*}$ *is closed under Boolean operations.*

Proof Protagonists in earlier similar proofs were $\lambda xy.xy$ for closure under $\vee$ and $\lambda xy.x \doteq y$ for closure under $\neg$.

□

7.0.14 Theorem. (Bounded Search) $\mathcal{R}_{alt}$ *is closed under bounded search*

$$(\mu y)_{<z} g(y, \vec{x}) \stackrel{Def}{=} \begin{cases} \min\{y : y < z \wedge g(y, \vec{x}) = 0\} \\ z \ if \ min \ does \ not \ exist \end{cases}$$

Proof For $g \in \mathcal{R}_{alt}$, we have $(\mu y)_{<z} g(y, \vec{x}) = (\widetilde{\mu}y)|y - z| \cdot g(y, \vec{x})$, which is a functional form of $(\mu y)_{<z} g(y, \vec{x}) = (\widetilde{\mu}y)\left(y = z \vee g(y, \vec{x}) = 0\right)$. The part "$|y-z|\cdot$" —equivalently "$y = z \vee$"— ensures that if $\neg(\exists y)_{<z} g(y, \vec{x}) = 0$, then the search answers z (it "stops" an unsuccessful search).

We can say this a bit better: If $g \in \mathcal{R}_{alt}$, then so is $\lambda y\vec{x}.|y - z| \times g(y, \vec{x})$ and the latter is y-regular.

Therefore, $\lambda\vec{x}.(\mu y)_{<z} g(y, \vec{x}) \in \mathcal{R}_{alt}$. □

 There is no point in using $\widetilde{\mu}$ for bounded search, as this search applies to total functions anyway.

 Whether we used $(\widetilde{\mu}y)$ or (μy) would make no difference in either the statement of the theorem or its proof. Incidentally, $|x - y| = x \doteq y + y \doteq x$, for all x, y.

As earlier defined, $(\mu y)_{<z} P(y, \vec{x})$ means $(\mu y)_{<z} c_P(y, \vec{x})$. In fact, $(\widetilde{\mu}y)P(y, \vec{x})$ means $(\widetilde{\mu}y)c_P(y, \vec{x})$.

7.0.15 Corollary. (Bounded Quantification) $\mathcal{R}_{alt,*}$ *is closed under bounded quantification.*

Proof Suffices to show that if $P(y, \vec{x}) \in \mathcal{R}_{alt,*}$, then so is $(\exists y)_{<z} P(y, \vec{x})$.
Well,

$$(\exists y)_{<z} P(y, \vec{x}) \equiv (\mu y)_{<z} P(y, \vec{x}) < z \qquad \square$$

7.0.16 Example. Our familiar $x \mid y$ is in $\mathcal{R}_{alt,*}$:

$$x \mid y \equiv x > 0 \wedge (\exists z)_{\leq y} y = xz$$

Incidentally, as before, $(\exists z)_{\leq y}$ and $(\mu z)_{\leq y}$ mean $(\exists z)_{<y+1}$ and $(\mu z)_{<y+1}$ respectively. $\qquad \square$

7.0.17 Proposition. $\mathcal{P}_{alt}$ *is closed under definition by $\mathcal{R}_{alt,*}$ cases:*

$$f(\vec{x}) \simeq \begin{cases} f_1(\vec{x}) & \text{if } P_1(\vec{x}) \\ f_2(\vec{x}) & \text{if } P_2(\vec{x}) \\ \vdots & \vdots \\ f_k(\vec{x}) & \text{if } P_k(\vec{x}) \\ f_{k+1}(\vec{x}) & \text{othw} \end{cases}$$

Proof Let us define $Q(\vec{x})$ as $\neg(P_1(\vec{x}) \vee \ldots P_k(\vec{x}))$. Thus $Q(\vec{x}) \in \mathcal{R}_{alt,*}$ by Proposition 7.0.13. Then

$$f(\vec{x}) = f_1(\vec{x})(1 \dot{-} c_{P_1}(\vec{x})) + \cdots + f_k(\vec{x})(1 \dot{-} c_{P_k}(\vec{x})) + f_{k+1}(\vec{x})(1 \dot{-} c_Q(\vec{x}))$$

$\qquad \square$

We next address the converse of Lemma 7.0.6:

7.0.18 Lemma. $\mathcal{P} \subseteq \mathcal{P}_{alt}$.

Proof [Beginning]

> It suffices to prove that $\mathcal{P}_{alt}$ is closed under primitive recursion

because then it will contain $\mathcal{PR}$ since the latter is so closed, as well as under composition, and is the smallest such set. But then we can invoke the normal form Theorem 5.5.3 to show that $f \in \mathcal{P}$ implies $f \in \mathcal{P}_{alt}$.

Now, we can express $z = f(x, \vec{y})$, where f is given by $f(0, \vec{y}) = h(\vec{y})$ and $f(x + 1, \vec{y}) = g(x, \vec{y}, f(x, \vec{y}))$ as

$$(\exists m_0, m_1, \ldots, m_x)\Big(m_0 = h(\vec{y}) \wedge m_x = z \wedge (\forall i)_{<x} m_{i+1} = g(i, \vec{y}, m_i)\Big) \qquad (1)$$

as a glimpse at the iterative computation of such an f readily suggests. Cf. Example 1.1.22 and Theorem 3.2.1. The latter, especially, clearly indicates that the

$f(i, \vec{y})$ —or the "m_i" values as we call them here— are computed as successive values of the z variable via the statement $z \leftarrow g(i, \vec{y}, z)$ that occurs on p. 158.

The problem is that (1) is not a formula, since the prefix $(\exists m_0, m_1, \ldots, m_x)$ has variable length: it depends on x.

The workaround is to code $m_0, m_1, \ldots, m_x$ by a single number m. However not having our prime power coding scheme (Definition 2.4.2) *a priori* available in $\mathcal{P}_{alt}$ (Why not?) we need to define an alternative coding in $\mathcal{P}_{alt}$ (in fact in $\mathcal{R}_{alt}$, as codings are total), avoiding the functions $\lambda xy.x^y$ and $\lambda n.p_n$ whose definition depends on primitive recursion being in hand. We will develop this alternative coding below, and then return to conclude the proof.   □

Pause. Can we also prove the above (once we have shown that $\mathcal{P}_{alt}$ is closed under primitive recursion) as follows?

Given that $\mathcal{P}_{alt}$ includes the initial functions of $\mathcal{P}$ and both sets are closed under composition and $\widetilde{\mu}$ on total functions, all we have to do is show that $\mathcal{P}_{alt}$ is also closed under primitive recursion. As $\mathcal{P}$ is so already, and being the $\subseteq$-smallest set that contains the initial functions $\mathcal{I}$ and is closed under the three named operations, the set inclusion follows. ◄

If not, *why* not?

7.0.19 Remark. Recall Euclid's theorem, that for any natural numbers $n \geq 0$ and $m > 1$, there *exist unique* q and r, where $0 \leq r < m$, such that

$$n = mq + r \tag{2}$$

Equipped with (2) we can also prove that for any natural numbers $n > 0$ and $m > 1$, there *exist unique* Q and R, where $0 < R \leq m$ such that

$$n = mQ + R \tag{3}$$

Indeed, write $n = a + 1$. Then by (2) we have $a = mq + r$ where $0 \leq r < m$.

Thus we have (3) where $Q = q$ and $R = r + 1$, and the inequality is now $0 < r \leq m$. This settles *existence* of Q and R as stated. For *uniqueness* assume we also have

$$n = m\overline{Q} + \overline{R} \tag{3'}$$

where $0 < \overline{R} \leq m$. (3) and (3') imply

$$m|Q - \overline{Q}| = |R - \overline{R}| \tag{5}$$

Now the maximum distance between R and $\overline{R}$ is $m - 1$, thus, if $Q \neq \overline{Q}$ —and therefore $|Q - \overline{Q}| \geq 1$— we would have the left hand side in (5) be greater than the right hand side. We must conclude that $Q = \overline{Q}$ and thus $R = \overline{R}$.

Now (2) is responsible for the unique representation of a natural number n in base-$m > 0$ notation. The said well known result states that such an n can be written uniquely as

$$n = d_r m^r + d_{r-1} m^{r-1} + \cdots + d_1 m^1 + d_0, \text{ where, for all } i, 0 \le d_i < m \qquad (*)$$

We leave its proof as an exercise, but existence and uniqueness hinge on using (2) to write

$$n = mq + d_0, \text{ for } 0 \le d_0 < m$$

and apply the obvious I.H. to q.

We call the notation $(*)$ "m-ary" (as in "binary", when $m = 2$; but we say "decimal-notation", when $m = 10$). The unique d_i are the m-ary digits of n. We may write $n = \langle d_r d_{r-1} \ldots d_0 \rangle_m$, instead of writing the expression in $(*)$ although for our familiar case of decimal notation we will write just 10 for "ten" rather than $\langle 10 \rangle_{10}$. For the same number in binary notation we will write 1010, rather than $\langle 1010 \rangle_2$, as long as the context removes any ambiguities as to which base we have in mind.   □

7.0.20 Exercise. Prove the unproved claims in Remark 7.0.19.   □

7.0.21 Exercise. Prove by induction, using (3) from Remark 7.0.19, that, given $m > 0$, *any* natural number $n > 0$ can be written *uniquely* as

$$n = d_r m^r + d_{r-1} m^{r-1} + \cdots + d_1 m^1 + d_0, \text{ where, for all } i, 0 < d_i \le m \qquad (**)$$

□

7.0.22 Definition. (*m-adic Notation*) Smullyan (1961), Bennett (1962) retold in Tourlakis (1984). For any $n > 0, m > 0$ in $\mathbb{N}$, we say that $(**)$ of Exercise 7.0.21 is *the m-adic notation* of n. The d_i are the *m-adic digits*. For $m = 2$ tradition has it to say "*dyadic*" notation, digits, etc.

Note that the inequalities here are subtly different than those of $(*)$ in Remark 7.0.19.

The d_i are called the *m-adic digits* of n. We may write $n = \overline{\langle d_r d_{r-1} \ldots d_0 \rangle}_m$ to express the expansion $(**)$.

In particular, the notation $\overline{\langle d_r d_{r-1} \ldots d_0 \rangle}_m$ names a natural number.

The length of the notation is $r + 1$, that is, the number of m-adic digits in it.

We write $|n|$ for the length of n, which is here $|n| = r + 1$.

The bar over the array of digits distinguishes the m-adic from the m-ary notations (no bar for the latter).

The m-adic notation of 0 is *by definition* the *empty string*, thus this notation has length 0 as there are no m-adic digits at all for 0.   □

7.0.23 Example. Note that we can write (∗∗) as

$$n = d_r m^r + d_{r-1} m^{r-1} + \cdots + d_i m^i + d_{i-1} m^{i-1} + \cdots + d_1 m^1 + d_0$$

$$= \left(d_r m^{r-i} + \cdots + d_{i+1} m + d_i \right) m^i + d_{i-1} m^{i-1} + \cdots + d_1 m^1 + d_0$$

$$= \overline{\langle d_r \cdots d_{i+1} d_i \rangle}_m m^i + \overline{\langle d_{i-1} \cdots d_1 d_0 \rangle}_m$$

Let us write x and y for the numbers (denoted by) $\overline{\langle d_r \cdots d_{i+1} d_i \rangle}_m$ and $\overline{\langle d_{i-1} \cdots d_1 d_0 \rangle}_m$. Thus,

1. $i = |y|$
   and
2. $n = x \cdot m^{|y|} + y$

But also, trivially, the m-adic notation of n is the concatenation of the m-adic notations of x and y (in that order).

These two remarks —1. and 2.— immediately suggest a definition and a proposition: □

7.0.24 Definition. (*m-adic Concatenation*) For a given $m > 0$ and any two numbers x and y from $\mathbb{N}$, $x *_m y$ denotes the concatenation of the m-adic notations of x and y, *their m-adic concatenation*, as we say.

That is, if $x = \overline{\langle aa'a'' \cdots \rangle}_m$ and $y = \overline{\langle bb'b'' \cdots \rangle}_m$, then

$$x *_m y = \overline{\langle aa'a'' \cdots bb'b'' \cdots \rangle}_m \qquad (†)$$

When $m = 2$ we will simply write $x * y$ instead of $x *_2 y$.   □

We view "$*_m$" as a *number-theoretic* function of two arguments, given by definition (†) in Definition 7.0.24.

7.0.25 Proposition. $x *_m y = xm^{|y|} + y$, *for all x and y*.

Proof Combine the preceding definition with the example preceding it.   □

7.0.26 Exercise. In what follows $*_m$ denotes m-adic concatenation as defined above.

- Prove that for any fixed $m > 0$, $\lambda xy.x *_m y \in \mathcal{PR}$.
- Prove that $0 *_m u = u = u *_m 0$, for all u.

                                                                               □

7.0.27 Remark. In the temporarily interrupted proof of Lemma 7.0.18 we took leave in order to develop some coding for sequences within $\mathcal{R}_{alt}$.

Following Smullyan (1961), Bennett (1962), retold in Tourlakis (1984), we will define a coding in $\mathcal{R}_{alt}$ that is based on *dyadic concatenation*. The first result in the exercise above does not help, since we will not know whether $\mathcal{PR} \subseteq \mathcal{R}_{alt}$ *before* we get some useful coding going.

So let us continue, trying first of all to show *directly* that $z = x * y$ and *thus also* $\lambda xy.x * y$ are in $\mathcal{R}_{alt}$.

Pause. Why "thus also"? Because if f is total, and $z = f(\vec{x}) \in \mathcal{R}_{alt,*}$, then $\lambda \vec{x}.(\tilde{\mu}z)z = f(\vec{x}) \in \mathcal{R}_{alt}$ ◀

In what follows in this section, our basis of m-adic notation is fixed to be 2.

□

7.0.28 Lemma. *For all x, $2^{|x|} \leq x + 1 < 2^{|x|+1}$, where $|x|$ denotes 2-adic length.*

Proof

- (Case $x > 0$:) Here we have the obvious (we omit the subscript 2 of $\langle \ldots \rangle$, in view of the boxed statement immediately above)

$$\underbrace{\langle 11 \ldots 1 \rangle}_{|x|} \leq x \leq \underbrace{\langle 22 \ldots 2 \rangle}_{|x|}$$

that is

$$2^{|x|} - 1 \leq x \leq 2^{|x|+1} - 2 < 2^{|x|+1} - 1$$

hence $2^{|x|} \leq x + 1 < 2^{|x|+1}$

- (Case $x = 0$:) Here $|x| = 0$ by definition since 0 is represented by the empty sequence λ. Thus, we have $x + 1 = 1 = 2^0 = 2^{|x|} < 2^{|x|+1}$. Another way of writing this is $2^{|x|} \leq x + 1 < 2^{|x|+1}$, verifying the lemma in the $x = 0$ case as well.

□

7.0.29 Lemma. *The predicates below are in $\mathcal{R}_{alt,*}$*

- $Pow2(x) \equiv x$ *is a power of 2*
- $y = m\,Pow2(x)$, *which states that y is the* minimum *power of 2 that is $> x + 1$* —*by Lemma 7.0.28 this power is $y = 2^{|x|+1}$.*
- $z = 2^{|y|}$
- $z = x * y$

Proof

- So, every nontrivial factor of x is also of the form "2^y". We say this by saying, *equivalently*, that *every* factor of x is *divisible by 2*, for if q is a factor and is of the form $2k$, then k is also a factor of the form $2r$, etc.

$$Pow2(x) \equiv x \geq 1 \wedge (\forall y)_{\leq x}(y \mid x \wedge y > 1 \rightarrow 2 \mid y)$$

-

$$y = mPow2(x) \equiv Pow2(y) \wedge x + 1 < y \wedge (\forall z)_{<y}(Pow2(z) \rightarrow z \leq x + 1)$$

- Using Lemma 7.0.28,

$$z = 2^{|x|} \equiv (\exists w)_{\leq 2z}(w = mPow2(x) \wedge z = \lfloor w/2 \rfloor)$$

Pause. Is one of $\lambda x. \lfloor x/2 \rfloor$ (or $y = \lfloor x/2 \rfloor$) in $\mathcal{R}_{alt}$ (or $\mathcal{R}_{alt,*}$)? Is it all right to bound $(\exists w)$ by $\leq 2z$? Why and Why?◀

- $z = x * y$. We have established in Proposition 7.0.25 that $x * y = x2^{|y|} + y$. Thus, using the previous bullet and Proposition 7.0.25,

$$z = x * y \equiv (\exists w)_{\leq z}(\exists u)_{\leq z}(u = 2^{|y|} \wedge w = xu \wedge z = w + y)$$

□

1. By Remark 7.0.27, all functions that entered in Lemma 7.0.29 above as graphs are in $\mathcal{R}_{alt}$. In particular, $\lambda xy.x * y \in \mathcal{R}_{alt}$.
2. Before we state and prove the following corollary, we note that

m-adic concatenation is associative

just as concatenation of strings over an arbitrary alphabet is, because m-adic notation does not allow the 0 digit that will make associativity break down:
Indeed, m-ary concatenation which does include the 0 digit is not associative. We define the latter number theoretically by the same formula as in the m-adic case (cf.Definition 7.0.24) albeit we choose a different symbol to denote m-ary concatenation below

$$x {}^\frown{}^m y \overset{Def}{=} xm^{|y|} + y$$

Note that, for example, in 2-ary (binary) notation, we have $0 {}^\frown{}^2 y = 0.2^{|y|} + y = y$, thus

$$x {}^\frown{}^2 \left(0 {}^\frown{}^2 y\right) = x {}^\frown{}^2 y = x2^{|y|} + y$$

while

$$\left(x {}^\frown{}^2 0\right) {}^\frown{}^2 y = \left(2x\right) {}^\frown{}^2 y = x2^{|y|+1} + y$$

By contrast, $*_m$ behaves sanely, as Exercise 7.0.26, second bullet, shows. Associativity of m-adic concatenation allows us to write unambiguously

$$a_1 *_m a_2 *_m \ldots *_m a_n \tag{1}$$

without brackets since, as is well known, associativity makes the meaning of expression (1) independent of how brackets are inserted.

7.0.30 Corollary. $z = x_1 * x_2 * \cdots * x_n$ *is in* $\mathcal{R}_{alt,*}$ *and* $\lambda \vec{x}_n . x_1 * x_2 * \cdots * x_n \in \mathcal{R}_{alt}$.

Proof   Simultaneous induction on $n \geq 2$ for both the graph case and the function case.

Basis. $n = 2$. This is by the last bullet in Lemma 7.0.29 (graph) and Remark 7.0.27. Assuming the (predicate *and* function) case for n (*I.H.*) we deal with the (predicate) case for $n + 1$ (*I.S.*):

$$z = x_1 * x_2 * \cdots * x_n * x_{n+1} \equiv (\exists w)_{\leq x_1 * x_2 * \cdots * x_n}(w = x_1 * x_2 * \cdots * x_n \wedge z = w * x_{n+1})$$

The function case for $n + 1$ follows via Remark 7.0.27. $\square$

Before we present Smullyan's solution to coding based on 2-adic concatenation —continuing to denote it by $*$ without subscript 2— we will present another lemma and a definition.

7.0.31 Lemma.   *The following predicates are in* $\mathcal{R}_{alt,*}$:

  (i) $x B y$, meaning x is a prefix *of y.*
  (ii) $x E y$, meaning x is a suffix *of y.*
 (iii) $x P y$, meaning x is a part *of y.*

Proof

  (i) $x B y \equiv (\exists z)_{\leq y} y = x * z$
  (ii) $x E y \equiv (\exists z)_{\leq y} y = z * x$
 (iii) $x P y \equiv (\exists z)_{\leq y} (\exists w)_{\leq y} y = w * x * z$

                                                                                                    $\square$

Just as notation like $(\exists y)_{\leq z}$ is introduced for convenience, we will introduce similar abbreviations below:

7.0.32 Definition. (Part of Quantification) Let Q stand for any of B, E, P of 7.0.31. For any relation $R(z, \vec{x})$ we define

$$(\exists z)_{Qw} R(z, \vec{x}) \overset{Def}{\equiv} (\exists z)_{\leq w}\left(z Q w \wedge R(z, \vec{x}) \right)$$

and

$$(\forall z)_{Qw} R(z, \vec{x}) \overset{Def}{\equiv} (\forall z)_{\leq w}\left(z Q w \rightarrow R(z, \vec{x}) \right)$$

                                                                                                    $\square$

7.0.33 Exercise.   Prove that

$$(\exists z)_{Qw} R(z, \vec{x}) \equiv \neg(\forall z)_{Qw} \neg R(z, \vec{x})$$

□

7.0.34 Lemma. $\mathcal{R}_{alt,*}$ *is closed under part of quantification.*

Proof By closure operations of $\mathcal{R}_{alt,*}$, 7.0.32 and Lemma 7.0.31.   □

We are ready to introduce the coding of sequences of numbers. In principle, the concept is straightforward. A number sequence

$$a_0, a_1, \ldots, a_n$$

is coded as the number whose 2-adic notation is the string below

$$\# * \overline{a_0} * \# * \overline{a_1} * \# * \ldots * \# * \overline{a_n} * \#$$

where # is a special separator that we will explain shortly.

In the above $\overline{a_i}$ denotes the 2-adic notation of a_i, but *we will agree to a convention of*

> writing n for either the *number* or *its 2-adic notation; no overline*

from now on. The context will fend off any ambiguity. This is akin to standard practice of writing just n, for a number n, without making a fuss about its decimal notation.

Unfortunately we have no way of having a special "digit" #. Our digits are 1 or 2, nothing else! So the separator, or "glue", # will be a string of the symbols 1 and 2 and therefore, in order to *recognise* it as "*the* separator" —in any particular coded sequence— we must make it *dependent on the sequence we are coding*: We determine the *longest* substring of the form

$$\overbrace{111\ldots1}^{m}$$

—which will call a *tally*[2] — that occurs as a part in at least one among *all* the a_i that we wish to code as a single number, and use as #,

> for *this* sequence,

the number that has as 2-adic expansion the string

---

[2] Or 1-tally, if we need to discuss 2-tallies, etc. But we will not have to.

$$\overbrace{2\,111\ldots 1\,}^{m+1}2$$

Clearly, to carry out our coding, and in particular to perform *decoding*, we need —to begin with— a function in $\mathcal{R}_{alt,*}$ that will return the (number whose dyadic notation is the) longest tally in a given number z. Thus we can attempt to decode a number z^3 by trying to find out *what value of # was used.*

7.0.35 Lemma. *The following predicates and functions are in $\mathcal{R}_{alt,*}$ or $\mathcal{R}_{alt}$ depending on their type.*

(1) $tally(x) \equiv^{Def} x$ is a tally of ones.
(2) $y = maxtal(x)$*; means that y is the maximum-length tally of ones that is part of x.*
(3) $\lambda x.maxtal(x)$ —*the function version.*
(4) $dcode(z)$*; means that z is dyadic code ("dcode") for some sequence of numbers $a_0, a_1, \ldots, a_n$.*
(5) $x \in z$*; says that x is a number among the terms of the numerical sequence that z is a (d)code of.*[4]
(6) $consec(x, y, z)$*; says that x and y are consecutive terms (y after x) of the numerical sequence that z is a (d)code of.*
(7) $First(x, z)$ *means that z codes a sequence that has x as its zeroth term.*
(8) $Last(x, z)$ *means that z codes a sequence that has x as its last term.*

Proof

(1) $tally(x) \equiv \neg 2Px$
(2) $y = maxtal(x) \equiv tally(y) \wedge yPx \wedge \neg 1yPx$[5]
(3) $\lambda x.maxtal(x) \in \mathcal{R}_{alt}$ follows from the above and the "pause" in Remark 7.0.27.
(4) z is a code *for some* $a_0, a_1, \ldots, a_n$ iff its dyadic representation has the form

$$\#\overline{a_0}\#\overline{a_1}\#\ldots\#\overline{a_n}\#^6$$

where $\# = 2 * maxtal(z) * 2.$[7] Thus,

---

[3] Which may or may not be a code; e.g., if $1Bz$ is true, the number z is not a code. Why?

[4] This symbol appears in other parts of this volume and means the same thing, but the coding scheme may be different in other parts of the book.

[5] We will often omit the concatenation operator $*$ and write xy for $x * y$ where the context will not allow the confusion of xy with "implied multiplication" $x \times y$. Incidentally, it should be clear that "$\neg 1yPx$" is an abbreviation of "$\neg\big((1y)Px\big)$". Also note that I could have written "$\neg y1Px$" instead of "$\neg 1yPx$".

[6] In this notation we view # as a *string*.

[7] In this notation we view # as a *number*, nevertheless we felt that including "$*$" explicitly is safer.

$$dcode(z) \equiv \Big(2 * maxtal(z) * 2\Big)Bz \wedge \Big(2 * maxtal(z) * 2\Big)Ez \wedge$$

$$\neg z = \Big(2 * maxtal(z) * 2\Big) \wedge \neg\Big(2 * maxtal(z) * 2 * maxtal(z) * 2\Big)Pz \wedge$$

$$(\exists x, u)_{Pz}\Big((2 * maxtal(z) * 2 * x * 2 * maxtal(z) * 2)Pz \wedge$$

$$\neg maxtal(z)Px \wedge uPx \wedge maxtal(z) = u1\Big)$$

that is, a number is a code iff it has *at least two* # —one at the beginning and one at the end— and no two # overlap. Moreover, at least one of the terms a_i of the coded sequence must contain a tally that is exactly by one shorter than $maxtal(z)$. In particular, we note that $\neg maxtal(z)Pa_i$ is true, for $i = 0, \dots, n$. **Pause**. Why?◄

(5) $x \in z \equiv dcode(z) \wedge \Big(2 * maxtal(z) * 2 * x * 2 * maxtal(z) * 2\Big)Pz \wedge$
      $\neg maxtal(z)Px.$

(6) $consec(x, y, z) \equiv x \in z \wedge y \in z \wedge \Big(2*maxtal(z)*2*x*2*maxtal(z)*2*$
      $y * 2 * maxtal(z) * 2\Big)Pz.$

(7) $First(x, z) \equiv x \in z \wedge \Big(2 * maxtal(z) * 2 * x * 2 * maxtal(z) * 2\Big)Bz.$

(8) $Last(x, z) \equiv x \in z \wedge \Big(2 * maxtal(z) * 2 * x * 2 * maxtal(z) * 2\Big)Ez.$ □

We would also need to refer to the i-th term coded into a dcode z. The obvious solution would be an *iteration* but it begs the question: "*Is $\mathcal{R}_{alt}$ closed under primitive recursion?*" that we are trying to resolve in the last few pages. There is a useful bypass:

Pause. Why not code the location of the term i into the code, using a pairing function?◄

Indeed, why not! We will code the terms a_i of our sequence

$$a_0, a_1, \dots, a_n \tag{1}$$

so that location information is *glued* to each a_i (Smullyan 1961). To do so, we use the pairing function of Example 2.4.32

$$J = \lambda xy.(x + y)^2 + x$$

Of course, $J \in \mathcal{R}_{alt}$. Moreover, so are the projections K and L as seen via bounded search (cf. Example 2.4.32). So, we code the sequence (1) as

$$w = \# * \overline{J(0, a_0)} * \# * \overline{J(1, a_1)} * \# * \dots * \# * \overline{J(n, a_n)} * \# \tag{2}$$

Thus,

$$a_i = L\Big((\widetilde{\mu}y)(y \in w \wedge K(y) = i)\Big)$$

We can now conclude the proof of Lemma 7.0.18.

Proof [**Of Lemma 7.0.18; Conclusion**] The formally incorrect (1) of p.251 becomes the predicate P below:

$$P(m, x, \vec{y}, z) \equiv (\exists m)\Big(dcode(m) \wedge (\exists b)_{Pm}\big(First(b, m) \wedge Kb = 0 \wedge Lb = h(\vec{y})\big)\wedge$$

$$(\exists e)_{Pm}\big(Last(e, m) \wedge Ke = x \wedge Le = z\big)\wedge$$

$$(\forall u, v)_{Pm}\big(consec(u, v, m) \rightarrow$$

$$\big(Kv = Ku + 1 \wedge Lv = g(Ku, \vec{y}, Lu)\big)\Big)$$

The part of P quantified by "$(\exists m)$" —let us call this part $Q(m, x, \vec{y}, z)$— is in $\mathcal{R}_{alt,*}$. Thus, to obtain $\lambda x \vec{y}. f(x, \vec{y})$ we need to find a *pair* of m and z — implemented as the single number $u = J(m, z)$— that satisfies P for the given $x, \vec{y}$ and return z, discarding m:

Thus,

$$f = \lambda x \vec{y}. L\big((\widetilde{\mu}u)Q(Ku, x, \vec{y}, Lu)\big)$$

This search technique that gave us f is exactly the same used in the proof of Theorem 6.6.1.

Having proved that $\mathcal{R}_{alt}$ is closed under primitive recursion we have, as indicated in the first part of the proof, $\mathcal{PR} \subseteq \mathcal{R}_{alt}$, and thus also

- $T^{(n)}(z, \vec{y}_n, w) \in \mathcal{R}_{alt,*}$
- $out \in \mathcal{R}_{alt}$ (cf. Sect. 5.5 and also Theorem 2.4.23)

But then, if $g \in \mathcal{P}$, then for some i, n, we have (Theorem 5.5.3)

$$g = \lambda \vec{y}_n. out((\widetilde{\mu}z)T^{(n)}(i, \vec{y}_n, z), \vec{y}_n)$$

Pause. $(\widetilde{\mu}z)$? Not (μz)?◄

Thus $g \in \mathcal{P}_{alt}$ hence $\mathcal{P} \subseteq \mathcal{P}_{alt}$. □

Having established $\mathcal{P} = \mathcal{P}_{alt}$, and hence $\mathcal{R} = \mathcal{R}_{alt}$, we quote the Normal Form theorem again, using the *alt* subscript and $\widetilde{\mu}$ throughout:

$$\boxed{\lambda \vec{x}_n. f(\vec{x}_n) \in \mathcal{P}_{alt} \text{ iff, for some } i, f = \lambda \vec{x}_n. out((\widetilde{\mu}y)T^{(n)}(i, \vec{x}_n, y), i, \vec{x}_n)}$$

and

$$\boxed{\lambda \vec{x}_n. f(\vec{x}_n) \in \mathcal{R}_{alt} \text{ iff, for some } i, f = \lambda \vec{x}_n. out((\widetilde{\mu}y)T^{(n)}(i, \vec{x}_n, y), i, \vec{x}_n)}$$

In the latter case $T^n(i, \vec{x}_n, y)$ is y-regular.

 7.0.36 Remark. (Worth Noting) Every $f \in \mathcal{P}_{alt}$ or $f \in \mathcal{R}_{alt}$ can be demonstrated to be so by a derivation that applies $(\widetilde{\mu} y)$ once. □

$\mathcal{P}_{alt}$ was defined inductively (as a closure of an initial set under certain operations) in Definition 7.0.5. Is there a similar characterisation for $\mathcal{R}_{alt}$? Yes, as the next corollary (of all the work we did in this chapter) proves.

7.0.37 Corollary. $\mathcal{R}_{alt}$ *is the* closure *of the initial functions of Definition 7.0.5 under composition and* $(\widetilde{\mu} y)$ *applied to total regular functions.*

Proof Let us call $\overline{\mathcal{R}_{alt}}$ the closure defined in the corollary statement, and show $\overline{\mathcal{R}_{alt}} = \mathcal{R}_{alt}$.

$\overline{\mathcal{R}_{alt}} \subseteq \mathcal{R}_{alt}$:

We do induction on $\overline{\mathcal{R}_{alt}}$. $\mathcal{R}_{alt}$ contains the same initial functions as $\overline{\mathcal{R}_{alt}}$, and is closed under the same operations $((\widetilde{\mu} y)$ on regular functions, and composition). But $\overline{\mathcal{R}_{alt}}$ is the smallest such set. Done.

$\mathcal{R}_{alt} \subseteq \overline{\mathcal{R}_{alt}}$:

All functions (and hence predicates) that we constructed in this chapter have *derivations* in $\overline{\mathcal{R}_{alt}}$[8] and thus are in $\overline{\mathcal{R}_{alt}}$ by Theorem 0.6.10. Thus $\overline{\mathcal{R}_{alt}}$ is *closed under primitive recursion* for the same reason that $\mathcal{R}_{alt}$ is, and hence $\mathcal{PR} \subseteq \overline{\mathcal{R}_{alt}}$, since the initial functions of $\mathcal{PR}$ are in $\overline{\mathcal{R}_{alt}}$ and the latter set is closed under composition, and (provably) primitive recursion.
Thus, $T^{(n)}(i, \vec{x}_n, y) \in \overline{\mathcal{R}_{alt,*}}$, for all n and i, and also *out* $\in \overline{\mathcal{R}_{alt}}$.
Let now $\lambda \vec{x}_n . f(\vec{x}_n) \in \mathcal{R}_{alt}$. By the preceding one-line paragraph and by the last boxed statement above, $\lambda \vec{x}_n . f(\vec{x}_n) \in \overline{\mathcal{R}_{alt}}$ via closure properties of $\overline{\mathcal{R}_{alt}}$ (composition and $(\widetilde{\mu} y)$ on total regular functions). □

7.1   Exercises

1. Prove that given $\mathbb{N} \ni b > 1$ and $\mathbb{N} \ni n > 0$ prove by CVI on n that there exists a unique way to write n im b-adic notation.
Hint. Use (3) in Remark 7.0.19.

---

[8] Rather than "*derivations* in $\overline{\mathcal{R}_{alt}}$", the precise terminology is that they have $(\mathcal{J}, \mathcal{O})$-derivations, where $\mathcal{J} = \mathcal{I}_{alt}$ and $\mathcal{O} = \{(\widetilde{\mu} y)$ on regular functions, composition$\}$. Incidentally, we know that the process of proving that a function is in some set, using some initial functions of the set and some primary operations under which it is closed, is one of building a(n implicit) derivation (cf. Theorem 0.6.10).

2. Prove that $n > 1$ is a power of 2 iff every divisor $d > 1$ of n is even.

3. Prove that $(\exists z)_{Qw} R(z, \vec{x}) \equiv \neg(\forall z)_{Qw} \neg R(z, \vec{x})$.

4. Complete the proof of Lemma 7.0.29 by answering the questions posed in the "pause".

Chapter 8
Gödel's First Incompleteness Theorem via the Halting Problem

Overview

It is rather surprising that *Un*provability and *Un*computability are intimately connected. Gödel's original proof of his *first incompleteness theorem* did not use methods of computability —indeed computability theory was not yet developed.

Before we turn to an elementary proof of the incompleteness theorem that is based on the unsolvability of the halting problem —hence our opening remarks— we show here, *in outline*, how *he* did it in Gödel (1931).

He constructed a sentence (closed formula) D^1 in the language of *Peano Arithmetic* (PA) that meant —when *interpreted in the standard model*— "I am not a (PA) theorem".

This idea was based on a variation —namely, "I am lying"— of the "liar's paradox" of the ancient Cretan philosopher Epimenides.[2]

Apropos standard model, this consists of $\mathbb{N}$, as a provider of mathematical objects (natural numbers, in this case) to interpret variables and constants of PA, and also provides functions and relations to interpret the abstract functions and predicates of PA. There is a surprisingly small set of such functions and relations needed up in front. We need a concrete number-theoretic interpretation of the PA symbol "S"; as such we take the interpretation $\lambda x.x + 1$. Then we interpret the PA "$+$" as *plus*, the "$\times$" as *times* and the predicate "$<$" as *less than*. That is it!

The essence of the standard interpretation is that all axioms of PA are true in it. It follows, *since the rules of inference of logic preserve truth*, that *if D is a theorem, then it is true in this interpretation*. Well, if it *is* a theorem, then it is also *false*, because it says: "I am not a theorem".

This contradiction means that D is *not* a theorem! But then it is true! It says just that!

---

[1] "D" for diagonal, since diagonalisation was at play in Gödel's proof.

[2] His original version was "All Cretans are liars".

© Springer Nature Switzerland AG 2022

G. Tourlakis, *Computability*, https://doi.org/10.1007/978-3-030-83202-5_8

Se here is a witness of the *incompleteness* of PA: We exhibited (Gödel did) a *true* sentence that is not syntactically (formally) *provable*.

Gödel's original proof was quite complicated, in particular in the subtask of constructing the *sentence D* in the language of arithmetic (see a more detailed outline in Chap. 9). The incompleteness theorem speaks to the inability of formal *mathematical* theories, such as Peano arithmetic and set theory, to *totally capture* the concept of mathematical *truth*. This does not contradict Gödel's own *completeness* theorem that says "if $\models A$, then $\vdash A$" —in words, if the formula A is true in *all* interpretations, then it has a proof in first-order logic, *without using any mathematical axioms*.

You see, completeness talks about *absolute truth* —in all interpretations possible— while incompleteness speaks about *truth in the "standard" or "intended" model only; mathematical truth.*

In this chapter we will briefly look at the standard model of PA and will show the incompleteness theorem as a consequence of the unsolvability of the halting problem.

8.1   Preliminaries

To set the stage more accurately, we will need some definitions and notation. In order to do arithmetic we need, first of all, a first-order (logical) language which we use to write down formulas and proofs. Before one gets to describing the language (that is, a set of strings)[3] one needs an "alphabet".

This alphabet has two parts: One that is standard in all such languages (*logical symbols*), namely variables (or "object variables") *denoted by* x, y, z, u, v, w with or without accents or subscripts. These are of natural number type in our application of interest, *Peano arithmetic* (PA). The logical part also contains the symbols

$$\neg, \forall, \vee, =, (,) \tag{1}$$

but also the symbols v and # to finitely generate the infinite sequence of variables, $v_0, v_1, \ldots$ that we denote usually as we indicated above. Now, we mentioned that we generate the variables, thus, even the v_i are *metasymbols*. As we saw in Final Definition 0.2.3 on p. 10, v_i, for any $i \geq 0$, is a *short name* for

$$(v \overbrace{\# \ldots \#}^{i+1 \ \#})$$

The logical alphabet thus morphed into

---

[3] See Sect. 0.2.

$$\#, v, \neg, \vee, (,), =, \forall \tag{2}$$

exactly as in Final Definition 0.2.3, p. 10. We will keep the above order of symbols fixed for the benefit of the remaining discussion in this chapter.

Additional useful logical symbols can be *introduced by definitions*. Cf. Definition 0.2.18. *These are not part of our alphabet.*

The other part of the alphabet is *specific to doing arithmetic* and we call it the *mathematical* or *nonlogical* part of the alphabet. It contains the "standard" symbols one uses in arithmetic,

$$0, S, +, \times, < \tag{3}$$

These we can interpret any way we please, being "nonlogical".

However, if our applied logic is being built to do arithmetic, *then we* must *interpret the alphabet symbols in the* intended *way, as indicated in the table below*:

One last central remark before we proceed toward the goal of this chapter, we note that the actual ontology of the symbols in (3) is

1. 0 (the constant zero): $(c\#)$ (cf. Final Definition 0.2.3, p. 10)
2. S (the successor function symbol): $(\#f\#)$
3. $+$ (the addition function symbol): $(\#\#f\#)$
4. $\times$ (the multiplication function symbol): $(\#\#f\#\#)$
5. $<$ (the less-than predicate symbol): $(\#\#p\#)$

Standard Interpretation of the language of arithmetic

Abstract (alphabet) symbol	Concrete interpretation
0	0 (zero)
S	$\lambda x.x + 1$
$+$	$\lambda xy.x + y$
$\times$	$\lambda xy.x \times y$
$<$	$\lambda xy.x < y$

8.1.1 Definition. One forms now in the well known manner (see for example Tourlakis 2003a, 2008) the symbols for "1", "2" and "112056734555". An arbitrary $n \in \mathbb{N}$ has the formal counterpart

$$\underbrace{SS \cdots S}_{n \text{ times}} 0$$

which we *denote* by $\widetilde{n}$. We call the expressions $\widetilde{n}$ *numerals*. We let $\widetilde{0}$ stand for 0 (case of zero S-occurrences above). □

The alphabet has now converged to

$$
\overbrace{\#, v, \neg, \vee, (,), =, \forall,}^{\text{logical part}} \overbrace{0, S, +, \times, <<}^{\text{mathematical part}} \tag{2}
$$

where we will continue using in this chapter the more readable metasymbols $0, S, +, \times, <$ for the mathematical part.

 Analogously with Remark 5.3.2, we fix the above order (2) and assign position numbers 1–13 to the symbols listed. Thus, any string over the alphabet (2) can be viewed as a number written base-14 (without 0 digits) —alternatively in 13-adic notation. Therefore we can apply our computability theory to *sets* of formulas, since we can view them as sets of numbers.[4]

Now, a mathematical axiomatic system (or theory) that we use to formally (i.e., syntactically) prove theorems of arithmetic contains:

1. The first-order alphabet (2).
2. A recursive set of strings: The *well-formed formulas* (wff).[5]
   That is, $\mathbf{WFF} \in \mathcal{R}_*$ or equivalently, $\lambda A.(A \in \mathbf{WFF})$ is in $\mathcal{R}_*$. ($\mathbf{WFF}$ is defined in Definition 0.2.9.)
3. A *recursive* subset, MA, of $\mathbf{WFF}$: The *mathematical* (or nonlogical) axioms for arithmetic. The recursiveness requirement is reasonable: Intuitively it asks that the mathematical axioms be *recognisable*, an obvious requirement if we are to be able to *use them in proofs*![6]
   That is, $\lambda A.(A \in MA)$ is in $\mathcal{R}_*$.
4. A recursive set of *logical* axioms, Λ_1. Recursiveness is again an obvious "practical" requirement. That is, $\lambda A.(A \in \Lambda_1)$ is in $\mathcal{R}_*$.
   In our case Λ_1 is given as Definition 0.2.28.
5. The rules of inference, namely just modus ponens (MP) and $\forall$-intro of Sect. 0.2.
   These rules are recursive, intuitively,[7] in the sense that we have two recursive predicates MP and $AIntro$ such that

---

[4] We encountered and used m-adic notation in Chap. 7 in order to code sequences of numbers. A more thorough use of this notation will occur in Chap. 12.

[5] The reader will agree by looking back to Sect. 0.2 that one can easily write a, say, C-program to check whether or not a string A over the alphabet (2) is a wff. That is, in the sense explained in Remark 5.3.2, and by Church's thesis, the set of all wff *is* recursive.

[6] A particularly famous choice of axioms is due to Peano —the resulting theory then being called *Peano arithmetic*, in short PA. It has axioms that define the behaviour of every mathematical symbol, plus includes the induction axiom schema:

$$
P(0) \wedge (\forall x)(P(x) \to P(Sx)) \to P(x)
$$

The *schema* (or *form*, in English) gives one axiom for each choice of wff P.

[7] We can mechanically check if they are applicable, or if they were applied in a step of some proof.0

(a) $MP(A, B; C)$ is true iff

$$A \in \textbf{WFF} \text{ and } C \in \textbf{WFF} \text{ and } B \text{ is } A \rightarrow C$$

and

(b) $AIntro(A; B)$ is true iff

$$A \in \textbf{WFF} \text{ and, for some } x \textit{ not free in } B \in \textbf{WFF},$$

$$A \text{ is } X \rightarrow Y \text{ and } B \text{ is } X \rightarrow (\forall x)Y$$

Thus MP represents the statement we make when we write the rule as a "fraction":

$$MP(A, B; C) \text{ says: } \frac{A, B}{C}, \text{ where } B \text{ abbreviates } A \rightarrow C$$

and $AIntro$ represents the statement we make when we write the $\forall$-intro rule as a "fraction":

$$AIntro(A; B) \text{ says: } \frac{A : X \rightarrow Y}{B : X \rightarrow (\forall x)Y}, \text{ with } X, Y \text{ and } x \text{ as above.}$$

8.1.2 Remark. How do we *interpret* a formula of Peano arithmetic over $\mathbb{N}$?

Here is a simple procedure that does not get into Tarski semantics. The procedure we outline here (*and* also the Tarski semantics approach) appeared in a different form in Tourlakis (2008). The significant difference between that and the procedure here is dictated by the significant difference between generalisation as introduced in this volume (same as in Shoenfield 1967; Mendelson 1987; Tourlakis 2003a) and generalisation in Tourlakis (2008), where it has a restricted form that we call "*weak*", namely, "if $\vdash A$, then $\vdash (\forall x)A$"[8] rather than the *strong* (or *unconstrained*; in this volume), namely, "$A \vdash (\forall x)A$".

The simplest way to interpret an $A \in \textbf{WFF}$ is to replace all symbols of A, from left to right, by symbols from the metatheory of $\mathbb{N}$ (from "real" mathematics, as we say). This process will build a "concrete" formula of the metatheory that one hopes to know how to evaluate (true, false, or "it depends",[9] the latter in the case the translation has free variables). Of course, as in normal mathematics, true/false means true/false for all values of the free variables of the interpreted formula. For example, $\neg Sx = 0$, interpreted normally as $x + 1 \neq 0$ in the standard interpretation of Peano arithmetic is true (for all x, that is) while $x = y$ is "it depends". *The concrete counterpart of A over $\mathbb{N}$ we will denote by $A^{\mathbb{N}}$.*

---

[8] Also (Bourbaki 1966; Enderton 1972).

[9] There is no "it depends" in Tourlakis (2008) as the concrete interpretation is always without free variables.

The following table describes the translation from the abstract to the concrete, as we traverse A from left to right and transform it (eventually) into $A^{\mathbb{N}}$, one symbol at a time: Every *abstract* symbol ξ becomes a *concrete* symbol $\xi^{\mathbb{N}}$. We call what we are describing here the *standard* or *intended interpretation* but also the *standard* or *intended model*, the word "model" conveying that all mathematical axioms of PA are true in this interpretation. We will not verify this here.

Examples

- $\neg Sx = 0$ becomes under interpretation: $\neg x + 1 = 0$.
- $(\forall x)\neg Sx = 0$ becomes: $(\forall x \in \mathbb{N})\neg x + 1 = 0$—usually written $(\forall x \in \mathbb{N})$ $x + 1 \neq 0$.
- $x + Sy = S(x + y)$ becomes $x + (y + 1) = (x + y) + 1$.
- $(\forall x)(\forall y)(x + Sy = S(x + y))$ becomes $(\forall x \in \mathbb{N})(\forall y \in \mathbb{N})(x + (y + 1) = (x + y) + 1)$.

    I have the redundant brackets in red type to remind the reader that they can be omitted: $x + Sy = S(x + y)$ is an atomic formula that has no outermost brackets.

Abstract (language) symbol	Concrete interpretation (translation)
0	$0^{\mathbb{N}}$, the informal 0
$\tilde{n}$ (any $n \in \mathbb{N}$)	n
x (an arbitrary *free* variable)	x (stays unchanged)
x (an arbitrary *bound* variable)	x (stays unchanged)
(	((stays unchanged)
)	) (stays unchanged)
$\vee$	$\vee$
$\neg$	$\neg$
$\wedge$	$\wedge$
$\rightarrow$	$\rightarrow$
$\equiv$	$\equiv$
Sx	$x + 1$
$x + y$	$x + y$
$x \times y$	$x \times y$
x^y	x^y
$x < y$	$x < y$
$(\forall x)$	$(\forall x \in \mathbb{N})$
$(\exists x)$	$(\exists x \in \mathbb{N})$

It is unfortunate above that the same symbol denotes both the formal and informal 0. The context will come to the rescue when needed!   □

8.1.3 Exercise. Prove by induction on n that $\tilde{n}^{\mathbb{N}} = n$, thus showing that line two in the above table is redundant.

Hint. Employ line one and the translation of Sx from the table.   □

8.1.4 Example. The inductive definition of wff given in Definition 0.2.9 requires three *recursive steps* in the definition of formulas over the minimal[10] alphabet (2) on p. 268:

- *If A and B are formulas, then so is $(A \vee B)$.* So, $(A \vee B)^{\mathbb{N}}$ is $(A^{\mathbb{N}} \vee B^{\mathbb{N}})$ —or $A^{\mathbb{N}} \vee B^{\mathbb{N}}$ removing the redundant outermost brackets. Why? Because processing from left to right we translate "(" to "(" (cf. the table above), then A to $A^{\mathbb{N}}$, then translate $\vee$ to itself, next B to $B^{\mathbb{N}}$, and finally ")" to ")".
- *If A is a formula, then so is $(\neg A)$.* So, $(\neg A)^{\mathbb{N}}$ is $(\neg A^{\mathbb{N}})$. The proof is simpler than the above.
- *If A is a formula, then so is $\Big((\forall x) A \Big)$, for any object variable x.* Upon translation, "("[11] translates to itself, $(\forall x)$ translates to $(\forall x \in \mathbb{N})$, A to $A^{\mathbb{N}}$. Finally, ")" translates as itself.   □

The following is also immediate:

8.1.5 Proposition. *Assume that A has only x free, so it can be also written as $A(x)$ (cf. Remark 0.2.14). Then $A(\widetilde{a})^{\mathbb{N}}$ is $A^{\mathbb{N}}[x := a]$, where "$[x := a]$" was defined in Definition 0.2.21. $A^{\mathbb{N}}[x := a]$ can be also written as $A^{\mathbb{N}}(a)$*

Proof Traversing $A(\widetilde{a})$ from left to right and translating we obtain $A(\widetilde{a})^{\mathbb{N}}$. We may think this as a two pass process, knowing that *each $\widetilde{a}$ among those that occupy the spot of a previously free occurrence of x* that we encountered in the traversal will translate to a.

Then why not translate $A(x)$ first to $A^{\mathbb{N}}(x)$ and then substitute $a = \widetilde{a}^{\mathbb{N}}$ in all the free x in the translated expression.   □

8.1.6 Corollary. *Assume that A is $A(x, y, \ldots)$ (cf. Remark 0.2.14). Then $A(\widetilde{a}, \widetilde{b}, \ldots)^{\mathbb{N}}$ is $A^{\mathbb{N}}(a, b, \ldots)$.*

8.1.7 Remark.

(1) More generally one can trivially adapt the proof of Proposition 8.1.5 to show for any term t, $A(t)^{\mathbb{N}}$ is $A^{\mathbb{N}}[x := t^{\mathbb{N}}] = A^{\mathbb{N}}(t^{\mathbb{N}})$.

(1) We promised in Chap. 0 that we will show at some point that $\forall$-intro propagates truth, just as MP does. This is the point!

  To this end it is profitable to look at interpretations beyond the specific one in the table on p. 270. Let us denote any such interpretation as a pair of two *ingredients*, $\mathcal{D} = (\mathbb{D}, \mathcal{M})$.

  Here $\mathbb{D}$ is the nonempty set where all the variables of the interpreted formula vary over, and where our abstract constants take their fixed values from. $\mathcal{M}$ is the mapping that assigns concrete objects and relations to elements of our language. Rather than write things like $\mathcal{M}(S)$ and $\mathcal{M}(<)$ we will write instead,

---

[10] *Minimal*: Defined connectives are *not* included.

[11] For something like "(" or ")" that we may be using to open or close a parenthetical statement, or are just *talking about* it, we sometimes employ quotes to make clear that it is the latter case.

in analogy to what we did on p. 270, $S^{\mathbb{D}}$ and $<^{\mathbb{D}}$ respectively. $(\forall x)$ will translate as $(\forall x \in \mathbb{D})$.

So fix an arbitrary such interpretation $\mathcal{D} = (\mathbb{D}, \mathcal{M})$, assume that A has no free occurrences of x, and to fix ideas without loss of generality (about which variables are free in A and B)

$$\text{assume that } \Big(A(y, z, \ldots) \to B(x, y, z, \ldots)\Big)^{\mathbb{D}} \text{ is true}$$

True, that is, for all values that $x, y, \ldots$ can take from $\mathbb{D}$.

Restate as (says the same thing)

$$\text{if I fix values into } y, z, \ldots, \text{ then } \Big(A(y, z, \ldots) \to B(x, y, z, \ldots)\Big)^{\mathbb{D}} \text{ is true}$$

True, that is, *for all values that x can take from $\mathbb{D}$*.

The above says the same thing as

$$\text{if I fix values into } y, \ldots, \text{ then } \Big(A(y, z, \ldots) \to (\forall x)B(x, y, z, \ldots)\Big)^{\mathbb{D}} \text{ is true}$$

The values in $y, z, \ldots$ being arbitrary I am done (the interpretation of $A(y, z, \ldots) \to (\forall x)B(x, y, z, \ldots)$ in $\mathcal{D}$ is true.

(3) Let us sample one "complex" logical axiom and prove it is *valid*, that is, true in any interpretation. This will give us confidence in our axioms!

Let us probe axiom 2: $(\forall x)A[x] \to A[t]$. Without loss of generality let us say that A is $A(x, y, z, \ldots)$ and t is $t(u, v, \ldots)$.

"Any interpretation" is $\mathcal{D} = (\mathbb{D}, \mathcal{M})$

> *as long as I do not spell out any specifics about $\mathbb{D}$ or $\mathcal{M}$*

So,

$$\Big((\forall x)A[x] \to A[t]\Big)^{\mathbb{D}}$$

is (cf. Example 8.1.4 and Proposition 8.1.5)

$$(\forall x \in \mathbb{D})A^{\mathbb{D}}(x, y, \ldots) \to A^{\mathbb{D}}\Big(t^{\mathbb{D}}(u, v, \ldots), y, \ldots\Big) \qquad (*)$$

To see that $(*)$ is true, for all values of the free variables, namely,

$$y, z, \ldots; u, v, \ldots, \qquad (\dagger)$$

fix a value from $\mathbb{D}$ —*but don't say*! you need this argument to work *for no matter which values*— for *each* of the free variables above and assume that the left hand side of $\to$ in $(*)$ is true (if it is false, then we have nothing to prove!)

This means that for all values of x from $\mathbb{D}$ and the chosen ones for the variables in the list (†), $A^{\mathbb{D}}(x, y, \ldots)$ is true.

But then so is $A^{\mathbb{D}}\left(t^{\mathbb{D}}(u, v, \ldots), y, \ldots\right)$ since $t(u, v, \ldots)$ holds a value from $\mathbb{D}$ as a result of the input values we chose for $u, v, \ldots$ Said value — $t(u, v, \ldots)$— is a particular $\mathbb{D}$-value of x in $A^{\mathbb{D}}$. So the right hand side of $\to$ in $(*)$ is true too. Done. $\square$

8.1.8 Lemma. *The predicate $Proof(y)$ that is true iff y prime-power codes a PA proof is recursive.*

Proof The proof relies on the assumptions 1.–5. on p. 268. See also the proof of Theorem 4.3.2 for the notations $u \in v$ and $u <_w v$.

$$Proof(y) \equiv Seq(y) \wedge lh(y) \geq 1 \wedge (\forall A)_{\leq y}\Big(A \in y \to A \in \mathbf{WFF} \wedge$$

$$\Big\{ A \in \Lambda_1 \vee A \in MA \vee$$

$$(\exists B, C)_{\leq y}\Big(B <_y A \wedge C <_y A \wedge MP(B, C; A) \vee$$

$$B <_y A \wedge AIntro(B; A)\Big)\Big\}\Big)$$

$\square$

8.1.9 Lemma. (Gödel) *The predicate $Bew(y, A)$*[12] *that is true iff y prime-power codes a proof of A is recursive.*

Proof

$$Bew(y, A) \equiv Proof(y) \wedge A \in y$$ $\square$

8.1.10 Theorem. *The set of PA theorems, Θ, is semi-recursive.*

Proof

$$A \in \Theta \equiv (\exists y)Bew(y, A)$$

Now invoke Lemma 8.1.9 and Theorem 6.3.14. $\square$

8.1.11 Corollary. *The set of all theorems that are* closed (they are sentences) *is semi-recursive as well.*

---

[12] "Bew" for Beweis=Proof.

Proof There exists a mechanical procedure to check if a given formula has free variables or not. By CT, $sentence(A)$ —"A is a sentence"— is recursive, hence semi-recursive. Thus, so is

$$A \in \Theta \wedge sentence(A)$$

by closure properties of $\mathcal{P}_*$.   □

8.2   The First Incompleteness Theorem

Here is a "modern" proof of PA incompleteness, via a simple reduction proof within computability:

8.2.1 Theorem. (Gödel's First Incompleteness Theorem) *There is a true (in the standard interpretation over* $\mathbb{N}$*) but not (formally) provable* sentence *of Peano arithmetic.*

Proof We assume the theorem *false* —that is, *all true sentences are formally provable*— and derive a contradiction. The contradiction will be to show that the halting problem is *decidable* under our assumption.

Here is how to decide $\phi_a(a) \downarrow$ for an arbitrary a.

(1) Given $a \in \mathbb{N}$.
(2) Run *simultaneously* a verifier for $\phi_a(a) \downarrow$ and a verifier for $\phi_{\widetilde{a}}(\widetilde{a}) \uparrow \in \Theta$ —the latter verifier exists by Corollary 8.1.11.
(3) *Decision*: If the first verifier halts, we report "$\phi_a(a) \downarrow$ is true" and stop everything. If the second verifier halts, then we report "$\phi_a(a) \downarrow$ is false" and stop everything.

Clearly, *exactly one* of $\phi_a(a) \downarrow$ and $\phi_a(a) \uparrow$ will be true.

By the assumed falseness of the incompleteness theorem, if $\phi_a(a) \uparrow$ is true, then $\phi_{\widetilde{a}}(\widetilde{a}) \uparrow \in \Theta$.

Thus, one and only one verifier halts, and we therefore have a decider for $x \in K$.   □

 But wait! What does $\phi_{\widetilde{a}}(\widetilde{a}) \uparrow$ have to do with PA? PA is about arithmetic, not computations! We will show in the next section that $\phi_{\widetilde{a}}(\widetilde{a}) \uparrow$ is actually (expressible by) a formula of PA, that is true iff $\phi_a(a) \uparrow$ is.

8.3 $\phi_x(x) \uparrow$ Is Expressible in the Language of Arithmetic

The title of this section means that there is a formula of arithmetic, let us call it $A(x)$, such that, for all $n \in \mathbb{N}$, $\phi_n(n) \uparrow$ is true iff $A(\widetilde{n})$ is true— the latter in the sense of Remark 8.1.2.

8.3.1 Definition. A *relation* $R(x)$ over the natural numbers —that is, a *relation* in the metatheory of formal arithmetic— is *expressed* by a *formula in the language of* PA, $B(x)$, iff for all $n \in \mathbb{N}$, we have

$$R(n) \text{ is true iff } B(\widetilde{n}) \text{ is true (in the standard interpretation)}$$

Suppressing reference to $B(x)$ we can also say that $R(x)$ is *expressible*, in the language of PA.

The definition can be extended in the obvious way in the case of relations of many variables.

A *total function* in the metatheory of formal arithmetic —is *expressible* iff its graph is. □

8.3.2 Remark. This "expressibility" is contingent only upon the *formal language* of *formal arithmetic* and *in no way depends on the axioms and rules of inference* employed within the theory. □

Let us next define —*in computability*, not *within PA*— the set of *Arithmetical* relations on the set of natural numbers.

8.3.3 Definition. The set of *Arithmetical relations* is the smallest set of relations —the capitalisation of "Arithmetical" will persist until Chap. 12— over the set of natural numbers that satisfies:

It contains the "initial" relations (of three variables) $z = x + y, z = x \times y$, and $z = x^y$,[13] where the exponentiation is the *total "ex"* of Example 2.1.16 but we have here reverted to the standard notation.
  Moreover,

(1) If $Q(\vec{x})$ and $P(\vec{y})$ are in the set, then so are $\neg Q(\vec{x})$ and $Q(\vec{x}) \vee P(\vec{y})$.
(2) If $R(y, \vec{x})$ is in the set, then so is $(\forall y)R(y, \vec{x})$.
(3) If $Q(\vec{x})$ is in the set, then so are all its *explicit transformations*.
  Explicit transformations (Smullyan 1961; Bennett 1962) are exactly the following: *substitution* of any *constant* into a variable, *expansion* of the variables-list by "don't care" variables (arguments), *permutation* of variables,

---

[13] The Arithmetical relations have a lot of tolerance for variations in the choice of "initial" relations in their definition: Sometimes as much as *all* of $\mathcal{R}_*$ is taken as the initial Arithmetical relations. Sometimes as little as $z = x + y$ and $z = x \times y$. For technical convenience here we have added the graph of exponentiation here rather than choosing the most minimalist approach.

identification of variables —that is, Grzegorczyk operations (ii)–(v) (cf. Exercise 2.1.10), albeit applied to relations.

A *total* function f is Arithmetical iff its graph is.   □

Clearly the set of Arithmetical relations is closed under the remaining Boolean connectives and $(\exists y)$.

8.3.4 Lemma. *Every Arithmetical relation is expressible in the language of arithmetic, over the alphabet* (2) *of p. 268.*

Proof We proceed by induction along the cases of Definition 8.3.3. The basis contains three cases, $z = x + y$ and $z = x \times y$ —we will be using implied multiplication notation: $z = xy$— and $z = x^y$.

We argue that the metamathematical relation $z = x+y$ is expressed by the *formal* $z = x + y$ of arithmetic.

We use the same metavariables in the theory and metatheory, e.g., x, y and z immediately above, recognising that some authors prefer to typeset, say, the formal counterpart of x as **x**. We will trust the context, as usual.

This requires us (cf. Definition 8.3.1) to establish for all m, n, k:

$$m = n + k \text{ holds iff } \widetilde{m} = \widetilde{n} + \widetilde{k} \text{ is true (in the standard iterpretation)} \qquad (*)$$

Indeed,

$$\widetilde{m} = \widetilde{n} + \widetilde{k} \text{ is true iff } (8.1.2) \ \widetilde{m}^{\mathbb{N}} = \widetilde{n}^{\mathbb{N}} + \widetilde{k}^{\mathbb{N}} \text{ is true}$$

$$\text{iff } (8.1.2) \ m = n + k \text{ is true}$$

The verifications for the relations $z = xy$ and $z = x^y$ are omitted being entirely analogous. We leave it to the reader to verify that if $R(\vec{x})$ and $Q(\vec{y})$ are expressed by the formulae $A(\vec{x})$ and $B(\vec{y})$ respectively, then $\neg R(\vec{x})$ and $R(\vec{x}) \vee Q(\vec{y})$ are expressed by $\neg A(\vec{x})$ and $A(\vec{x}) \vee B(\vec{y})$, respectively.

Next, we show that $(\forall y)R(y, \vec{x}_r)$ is expressed by $(\forall y)A(y, \vec{x}_r)$, if $R(y, \vec{x}_r)$ is expressed by $A(y, \vec{x}_r)$.

We are given, for any $c, b_1, \ldots, b_r$ in $\mathbb{N}$, that

$$R(c, b_1, \ldots, b_r) \text{ is true iff } A(\widetilde{c}, \widetilde{b}_1, \ldots, \widetilde{b}_r) \text{ is true} \qquad (**)$$

Now, we *fix* $b_1, \ldots, b_r$ in $\mathbb{N}$.

$(\forall y)R(y, b_1, \ldots, b_r)$ holds iff *for all* $c \in \mathbb{N}$ we have that $R(c, b_1, \ldots, b_r)$ holds. By $(**)$ this is equivalent to saying, after arbitrarily fixing the $\widetilde{b}_1, \ldots, \widetilde{b}_r$,

$$\text{"for all } c \in \mathbb{N}, A(\widetilde{c}, \widetilde{b}_1, \ldots, \widetilde{b}_r) \text{ is true"}$$

that is,

$$\text{``for all } c \in \mathbb{N}, \left(A(\widetilde{c}, \widetilde{b}_1, \ldots, \widetilde{b}_r) \right)^{\mathbb{N}} \text{ is true''}$$

By Proposition 8.1.5 the latter says

$$\text{``for all } c \in \mathbb{N}, \left(A^{\mathbb{N}}(c, b_1, \ldots, b_r) \right) \text{ is true''}$$

which is the same as "$\left((\forall x) A(x, \widetilde{b}_1, \ldots, \widetilde{b}_r) \right)^{\mathbb{N}}$ is true", and therefore says that (Remark 8.1.2) "$(\forall x) A(x, \widetilde{b}_1, \ldots, \widetilde{b}_r)$ is true in the standard model".

We next look into explicit transformations. Let then $Q(y, \vec{x}_r)$ be defined by the formula $A(y, \vec{x}_r)$. Thus, for any fixed $i \in \mathbb{N}$, $Q(i, \vec{x}_r)$ is clearly defined by $A(\widetilde{i}, \vec{x}_r)$, since for all $a, b_1, \ldots, b_r$ we have $Q(a, b_1, \ldots, b_r)$ is true iff $A(\widetilde{a}, \widetilde{b}_1, \ldots, \widetilde{b}_r)$ is true; in particular, $Q(i, b_1, \ldots, b_r)$ is true iff $A(\widetilde{i}, \widetilde{b}_1, \ldots, \widetilde{b}_r)$ is true.

The case of identifying or permuting variables being trivial, we conclude by looking at the case of adding one "don't care" variable (a case that is extensible by a trivial induction to any fixed number). So let $A(\vec{x}_r)$ define $Q(\vec{x}_r)$ and let z be a *new* variable.

I will argue that $A(\vec{x}_r) \wedge z = z$ defines the relation $R = \lambda z \vec{x}_r . Q(\vec{x}_r)$:
We have, on one hand, for all $b_1, \ldots, b_r$:

$$Q(b_1, \ldots, b_r) \text{ is true iff } A(\widetilde{b}_1, \ldots, \widetilde{b}_r) \text{ is true} \qquad (\text{***})$$

On the other hand, for all $c, b_1, \ldots, b_r$,

$$R(c, b_1, \ldots, b_r) \equiv Q(b_1, \ldots, b_r) \qquad (\dagger)$$

and, since $\widetilde{c} = \widetilde{c}$ is true —since its interpretation is $c = c$— we have

$$A(\widetilde{b}_1, \ldots, \widetilde{b}_r) \text{ is true iff so is } A(\widetilde{b}_1, \ldots, \widetilde{b}_r) \wedge (\widetilde{c} = \widetilde{c})$$

Along with (***) and $(\dagger)$ we get

$$R(c, b_1, \ldots, b_r) \text{ is true iff } A(\widetilde{b}_1, \ldots, \widetilde{b}_r) \wedge (\widetilde{c} = \widetilde{c}) \text{ is true} \qquad \square$$

After all this, it is now clear that to show that $\phi_x(x) \uparrow$ is expressible in the language of arithmetic suffices, due to the preceding lemma, to prove that it is Arithmetical. In turn, since $\phi_x(x) \uparrow \equiv \neg(\exists y) T(x, x, y)$, it suffices to prove that the Kleene predicate is Arithmetical.

It will so follow if we can prove that every function $f \in \mathcal{PR}$ has an Arithmetical graph, for then if c_T is the characteristic function of T (cf. Lemma 5.5.2), we will have that $c_T(x, y, z) = w$ —and therefore $c_T(x, y, z) = 0$ by explicit transformation— is Arithmetical.[14]

Items (7)–(10) below are due to Grzegorczyk (1953).

8.3.5 Lemma. *The following relations are Arithmetical.*

(1) $x = 0$ *(and hence $x \neq 0$)*
(2) $x \leq y$ *(and hence $x < y$)*
(3) $x \mid y$
(4) $Pr(x)$
(5) $Seq(z)$
(6) $Next(x, y)$ *(meaning $x < y$ are consecutive primes)*
(7) $pow(z, x, y)$ *(meaning $x > 1$ and x^y is the highest power of x dividing z)*
(8) $\Omega(z)$ *(meaning z has the form $p_0 p_1^2 p_2^3 \cdots p_n^{n+1}$ for some n)*
(9) $y = p_n$
(10) $z = \exp(x, y)$ *(cf. Example 2.3.10)*

We need not worry about *bounding* our quantifications, for it is not our purpose to show these relations in $\mathcal{PR}_*$. Indeed we know from earlier work that they are in this set. This time we simply want to show that they are Arithmetical according to Definition 8.3.3.

Proof

(1) $x = 0$ (and hence $x \neq 0$): $x = 0$ is an explicit transform of $x = y + z$; $x \neq 0$ is obtained by negation.
(2) $x \leq y$ (and hence $x < y$): This is equivalent to $(\exists z)(x + z = y)$.
(3) $x \mid y$: This is equivalent to $(\exists z)y = xz$ (I am using "implied multiplication" throughout: "xy" rather than "$x \times y$" or "$x \cdot y$").
(4) $Pr(x)$: This is equivalent to $x > 1 \wedge (\forall y)(y \mid x \to y = 1 \vee y = x)$.
(5) $Seq(z)$: This is equivalent to $z > 1 \wedge (\forall x)(\forall y)(Pr(x) \wedge Pr(y) \wedge x < y \wedge y \mid z \to x \mid z)$.
(6) $Next(x, y)$: This is equivalent to $Pr(x) \wedge Pr(y) \wedge x < y \wedge \neg(\exists z)(Pr(z) \wedge x < z \wedge z < y)$.
(7) $pow(z, x, y)$: This is equivalent to $x > 1 \wedge x^y \mid z \wedge \neg x^{y+1} \mid z$.[15]
(8) $\Omega(z)$: This is equivalent to $Seq(z) \wedge \neg 4 \mid z \wedge (\forall x)(\forall y)\big(Next(x, y) \wedge y \mid z \to (\exists w)(pow(z, x, w) \wedge pow(z, y, w + 1))\big)$.
(9) $y = p_n$: This is equivalent to $(\exists z)(\Omega(z) \wedge pow(z, y, n + 1))$.
(10) $z = \exp(x, y)$: This is equivalent to $(\exists w)(pow(y, w, z) \wedge w = p_x)$. □

We can now prove the following theorem that concludes the business of this section.

---

[14] Gödel proved all this without the need to have exponentiation as a primitive operation in arithmetic. However, adopting this operation makes things considerably easier and, as mentioned earlier (footnote 13 on p. 275), it does not change the set of Arithmetical relations.

[15] Again recall that x^y is that of Example 2.1.16, and $x^{y+1} \mid z \equiv (\exists u)(u = y + 1 \wedge x^u \mid z)$. On the other hand, $x^u \mid z \equiv (\exists w)(w = x^u \wedge w \mid z)$.

8.3.6 Theorem. *Every $f \in \mathcal{PR}$ is Arithmetical.*

Proof It suffices to prove that for each $f \in \mathcal{PR}$, its graph $y = f(\vec{x})$ is Arithmetical.

We do induction on the length of derivation of f, or, equivalently, induction over $\mathcal{PR}$ (cf. Sect. 0.6.1):

(1) *Basis (Initial functions).*   There are three graphs to work with here: $y = x + 1$, $y = 0$ and $y = x$ (or, fancily, $y = x_i$; or more fancily, $y = U_i^n(\vec{x}_n)$). They all are explicit transforms of $y = x + z$.

(2) *Composition.*   Say, the property is true for the graphs of $f, g_1, \ldots, g_n$. This is the I.H. How about $y = f(g_1(\vec{x}_m), g_2(\vec{x}_m), \ldots, g_n(\vec{x}_m))$? Well, this graph is equivalent to

$$(\exists u_1) \cdots (\exists u_n)\big(y = f(\vec{u}_n) \wedge u_1 = g_1(\vec{x}_m) \wedge \cdots \wedge u_n = g_n(\vec{x}_m)\big)$$

and we are done by the I.H. (Compare with Theorem 0.2.94).

(3) *Primitive recursion.*   This is the part that benefits from the work we put into Lemma 8.3.5. Here's why: Assume (I.H.) that the graphs of h and g are Arithmetical, and let f be given for all $x, \vec{y}$ by

$$f(0, \vec{y}) = h(\vec{y})$$

$$f(x + 1, \vec{y}) = g(x, \vec{y}, f(x, \vec{y}))$$

Now, to state $z = f(x, \vec{y})$ is equivalent to stating

$$(\exists m_0)(\exists m_1) \cdots (\exists m_x)\Big(m_0 = h(\vec{y}) \wedge z = m_x \wedge$$

$$(\forall w)\big(w < x \rightarrow m_{w+1} = g(w, \vec{y}, m_w)\big)\Big)$$

(i)

The trouble with the "relation" (i) above is that it is not a relation at all, because it has a variable-length prefix: $(\exists m_0)(\exists m_1) \cdots (\exists m_x)$. We invoke coding to salvage the argument. Let us use a single number,

$$m = p_0^{m_0} p_1^{m_1} \cdots p_x^{m_x}$$

to represent all the m_i, for $i = 0, \ldots, x$. Clearly,

$$m_i = \exp(i, m), \text{ for } i = 0, \ldots, x$$

We can now rewrite (i) as

$$(\exists m)\Big(\exp(0, m) = h(\vec{y}) \wedge z = \exp(x, m) \wedge$$

(ii)

$$(\forall w)\big(w < x \rightarrow \exp(w + 1, m) = g(w, \vec{y}, \exp(w, m)))\big)\Big)$$

The above is Arithmetical because of the I.H. Some parts of it are more complicated than others. For example, the part

$$\exp(w + 1, m) = g(w, \vec{y}, \exp(w, m))$$

is equivalent to

$$(\exists u)(\exists v)\big(u = \exp(w + 1, m) \wedge v = \exp(w, m) \wedge u = g(w, \vec{y}, v)\big)$$

The above is Arithmetical by the I.H. and the preceding lemma. This completes the proof.   □

8.3.7 Exercise. Redo the proof of Theorem 8.3.6, this time using the coding of Chap. 7.   □

Chapter 9
The Second Recursion Theorem

Overview

The recursion theorem is attributed to Kleene, but it was embedded in a somewhat different format in Gödel's first incompleteness theorem proof (Gödel 1931), where, via his *substitution function* and the *fixed point lemma*, he constructed a sentence of Peano Arithmetic (PA) that said "I am not a theorem".

The analogues in Kleene's proof of his *second recursion theorem*[1] are the S-m-n functions (instead of the Gödel substitution function) and the *second recursion theorem* itself (instead of the *fixpoint lemma* of Gödel 1931; Carnap 1934).

9.1 Historical Note

Here is what Gödel did in outline:

First, he built an arithmetisation of the symbols, terms, formulas and proofs of arithmetic —assigning to all of them what we now call Gödel *numbers*— so that formulas and proofs can speak about themselves using numbers as proxies. Gödel numbers are defined both in the theory as *numerals* (such as $\tilde{n}$, where n is a number of informal mathematics; cf. p. 267), but also in the metatheory outside PA, where they are just natural numbers.

He also introduced the so-called *substitution function*, s, as a formal object (function) of the theory.[2] If we denote by $gn\{E\}$ the informal (metatheoretical) Gödel number of expression E —say, $g = gn\{E\}$— then $\ulcorner E \urcorner$ is the *formal* Gödel number of expression E, that is, we *define*

$$\ulcorner E \urcorner \stackrel{Def}{=} \tilde{g} \text{ iff } g = gn\{E\}$$

---

[1] He has also stated and a proved a "first recursion theorem". See Chap. 13.

[2] Introduced as a, provably within formal arithmetic, primitive recursive function.

© Springer Nature Switzerland AG 2022
G. Tourlakis, *Computability*, https://doi.org/10.1007/978-3-030-83202-5_9

or

$$\ulcorner E \urcorner \overset{Def}{=} \widetilde{gn\{E\}}$$

The behaviour of $s(x, y)$, for any formula A of exactly one free variable v_0, is the following:

- Formally: $s(\ulcorner A(v_0) \urcorner, \widetilde{n}) = \ulcorner A(\widetilde{n}) \urcorner$ is *provable* inside formal arithmetic.
- Metatheoretically: (we will use the same symbol s for the informal counterpart of the function), the formula "$s(gn\{A(v_0)\}, \widetilde{n}) = gn\{A(\widetilde{n})\}$" is *true*.

That is, (essentially and informally) $s(x, y)$ computes the Gödel number of the one-variable formula of Gödel number x —the one variable being, say, v_0— *after* we substitute v_0 everywhere by the term y.

Now we can prove as Gödel did the *fixpoint lemma*.[3]

9.1.1 Lemma. (Fixpoint, or Diagonalisation Lemma) [4]*For any formula A of Peano Arithmetic that has only v_0 free, there is a number e such that we can prove* $\widetilde{e} = \ulcorner A(\widetilde{e}) \urcorner$.

Proof Throughout the proof s is the formal version. For some $d \in \mathbb{N}$, PA proves $\widetilde{d} = \ulcorner A(s(v_0, v_0)) \urcorner$. Now, $\ulcorner A(s(\widetilde{d}, \widetilde{d})) \urcorner$ is provably (in PA) equal to $s(\widetilde{d}, \widetilde{d})$, by the behaviour of s. Thus, if we set $\widetilde{e} = s(\widetilde{d}, \widetilde{d})$ we are done.   □

 We gave above a version of the fixpoint lemma that most closely compares with the proof of the recursion theorem below. The "standard" statement of the lemma is slightly different, namely,

> For any formula A of Peano Arithmetic that has only v_0 free, there is a sentence B such that we can prove $B \equiv A(\ulcorner B \urcorner)$.

The distance from the above proof to a proof of the "standard" statement is small: Take B to be the sentence $A(s(\widetilde{d}, \widetilde{d}))$. So, the (provable) equivalence

$$A(s(\widetilde{d}, \widetilde{d})) \equiv A(s(\widetilde{d}, \widetilde{d})) \tag{1}$$

can be rewritten, by writing B in two different ways:

$$B \equiv A(\ulcorner B \urcorner)$$

where in the right hand side of (1) we replaced equals by equals, noting the second sentence in the proof of the lemma, which says that PA proves $\ulcorner B \urcorner = s(\widetilde{d}, \widetilde{d})$.

---

[3] Actually the fixed point lemma entered only implicitly in Gödel's proof in Gödel (1931), but Carnap (1934) teased it out and explicitly formulated it.

[4] Also known as the *diagonalisation lemma* (or *diagonal lemma* in Boolos 1979).

To finish this preamble to the recursion theorem, let us —arguing in the metatheory— take for granted that Gödel defined (a primitive recursive) predicate "Bew" (cf. Lemma 8.1.9) such that

$$Bew(x, y)$$

said —analogously with the Kleene predicate $T(i, x, y)$[5]— that there is a proof y (Gödel number, y, of a proof) of x (this is the Gödel number of a formula).

The analogy is quite strong despite superficial differences. In the logic side of the analogy, there is one system of proofs considered: that of Peano arithmetic. Of course, a proof is a *verification* that a formula is a theorem. On the computability side, there are *devices* (plural) for verifying membership in certain sets: these devices are the URMs $M_{\mathbf{w}}^{\mathbf{z}}$ in locations $i = 0, 1, 2, 3, \ldots$ of our standard enumeration, and the counterparts of proofs are the verifying-computations of these URMs. Hence the need for the argument i in T.

To end the story, in informal outline (omitting the "~" on top of numbers), of the original incompleteness proof, let us define

$$\Theta(x) \stackrel{Def}{\equiv} (\exists y) Bew(x, y)$$

$\Theta(x)$ says that x is (the Gödel number of) a theorem.

By the fixpoint lemma, for some e we have,

$$e = gn\{\neg\Theta(e)\} \tag{1}$$

Now, $\neg\Theta(e)$ says "e is not a theorem". In other words, by (1), $\neg\Theta(e)$ says "*I am not a theorem*".

So, can we prove (the formal counterpart of) $\neg\Theta(e)$ in PA?

If yes, then by the *correctness* of PA we have that $\neg\Theta(e)$ is true in the standard interpretation of PA (see also Chap. 8), *since in a correct theory all its theorems are true in the intended interpretation*.

But the fact that the sentence has a proof makes it *false* as well (since it *denies* that its is a theorem!) Contradiction.

We conclude that *we cannot prove the sentence in PA*, so it *is* true, *and* unprovable, *exactly* as it says.

So, Gödel exhibited a *witness* —this sentence $\neg\Theta(e)$— to the incompleteness of PA.

Incidentally, nor $\Theta(e)$ is provable (by correctness of PA), because it is false.

This leads to Gödel's syntactic statement of his incompleteness theorem: *There is a sentence D of PA such that neither it nor its negation are provable.*

---

[5] There is a convergent computation of y steps of the i-th URM that *verifies* input x.

Such formulas are called *undecidable* (by PA). This concept applies to *formulas* of PA, *not sets of numbers* thus it must not be confused with *undecidability* (meaning uncomputability).

9.2   Kleene's Version of the Second Recursion Theorem

Let us prove the recursion theorem and compare with the proof of the fixpoint lemma.

9.2.1 Theorem. (The Second Recursion Theorem) *Let* $\lambda y \vec{x}_n . f(y, \vec{x}_n) \in \mathcal{P}$. *Then there is an* $e \in \mathbb{N}$, *such that* $f(e, \vec{x}_n) = \phi_e^{(n)}(\vec{x}_n)$, *for all* $\vec{x}_n$.

Proof   By assumption, $\lambda z \vec{x}_n . f\left(S_1^n(z, z), \vec{x}_n\right) \in \mathcal{P}$. Thus, by Definition 5.4.5,

$$f\left(S_1^n(z, z), \vec{x}_n\right) \simeq \phi_i^{(n+1)}(z, \vec{x}_n), \text{ for some } i \tag{1}$$

hence

$$f\left(S_1^n(i, i), \vec{x}_n\right) \simeq \phi_i^{(n+1)}(i, \vec{x}_n) \stackrel{\text{(Corollary 5.7.3)}}{\simeq} \phi_{S_1^n(i,i)}^{(n)}(\vec{x}_n) \tag{2}$$

Take $e = S_1^n(i, i)$.   □

The strong resemblance between the results Theorem 9.2.1 and Lemma 9.1.1 and their proofs —the same story in two different languages— should be obvious:

Note first that a ϕ-index of a URM is analogous to a Gödel number of a formula of PA. Let us then *temporarily* call the ϕ-indices "Gödel numbers". We re-trace the proof of Theorem 9.2.1 using this temporary jargon (and corresponding notation).

First off, the role of the S-m-n function, in the jargon of this remark, is that $S_1^n(i, a)$ is the Gödel number[6] of the function obtained from $\phi_i^{(n+1)}$ —function with "Gödel number" i— if we substitute a in the first variable, z, of the latter. This tracks accurately the role of Gödel's "s".

Thus, given $\lambda y \vec{x}_n . f(y, \vec{x}_n) \in \mathcal{P}$, let

$$i = \ulcorner \lambda z \vec{x}_n . f\left(S_1^n(z, z), \vec{x}_n\right) \urcorner \tag{1'}$$

This is (1) of Theorem 9.2.1 proof and tracks well the proof of Lemma 9.1.1 where we considered the formula $A(s(v_0, v_0))$.

---

[6] Functions viewed *extensionally* have, as we know, infinitely many ϕ-indices (see also Corollary 9.4.2 below). However, we allow the definite article "the" here thinking rather of the *intentional* object for our $\lambda z \vec{x}_n . f\left(S_1^n(z, z), \vec{x}_n\right)$ —i.e., the URM at location i— and also of *the* URM at location $S_1^n(i, i)$ that computes $\lambda \vec{x}_n . f\left(S_1^n(i, i), \vec{x}_n\right)$. Cf. Theorem 5.7.1 and its corollaries, in particular Corollary 5.7.3.

What is the Gödel number (cf. footnote 6 on this page) of $\lambda \vec{x}_n . f\left(S_1^n(i,i), \vec{x}_n\right)$?
Well, $S_1^n(i,i)$ by (1) and the substitution of i in z. This is (2) in Theorem 9.2.1
proof. Thus, take $e = S_1^n(i,i)$. This corresponds to the last sentence in the proof of
Lemma 9.1.1.

It has been often remarked that the proof of the recursion theorem is too short
and yet obscure or unexpected; "how does one discover it?", other than by trial
and error (cf. Moschovakis 2010). In fact the last reference shares the anecdote that
when Kleene himself was asked by some of his students the very question in quotes,
allegedly Kleene replied, perhaps partly in jest, "by fiddling with the Iteration (S-
m-n) theorem".

Still, as the above comments make rather clear, the question perhaps ought to
be how Gödel and Carnap Carnap (1934) came up with the *fixpoint theorem* in PA
and related theories. Well, their motive, Gödel's at least, was to formalise within
PA the sentence "I am not a theorem". This and the *substitution function* of Gödel
—his "S-m-n function"— almost deterministically lead to the proof (cf. also Wilder
1963). As we saw, the two proofs, of the fixpoint lemma and of Theorem 9.2.1 are
essentially one.

9.2.2 Corollary. (The Second Recursion Theorem with Parameters I)
Let $\lambda z \vec{y}_m \vec{x}_n . f(z, \vec{y}_m, \vec{x}_n) \in \mathcal{P}$. *Then there is an m-variable* $h \in \mathcal{R}$, *such that*
$f(h(\vec{y}_m), \vec{y}_m, \vec{x}_n) \simeq \phi_{h(\vec{y}_m)}^{(n)}(\vec{x}_n)$, *for all* $\vec{y}_m, \vec{x}_n$.
Moreover, h being an S-m-n function is 1-1.

Proof   The proof is almost the same as that of Theorem 9.2.1.
By assumption, $\lambda z \vec{y}_m \vec{x}_n . f\left(S_{1+m}^n(z, z, \vec{y}_m), \vec{y}_m, \vec{x}_n\right) \in \mathcal{P}$. Thus, by Definition 5.4.5,

$$f\left(S_{1+m}^n(z, z, \vec{y}_m), \vec{y}_m, \vec{x}_n\right) \simeq \phi_i^{(n+m+1)}(z, \vec{y}_m, \vec{x}_n), \text{ for some } i \in \mathbb{N} \qquad (1)$$

hence

$$f\left(S_{1+m}^n(i, i, \vec{y}_m), \vec{y}_m, \vec{x}_n\right) \simeq \phi_i^{(n+m+1)}(i, \vec{y}_m, \vec{x}_n) \overset{\text{(Corollary 5.7.4)}}{\simeq} \phi_{S_{1+m}^n(i,i,\vec{y}_m)}^{(n)}(\vec{x}_n) \qquad (2)$$

Take $h = \lambda \vec{y}_m . S_{1+m}^n(i, i, \vec{y}_m)$.   □

9.2.3 Corollary. (The Second Recursion Theorem with Parameters II)
Let $\lambda z \vec{x}_n . f(z, \vec{x}_n) \in \mathcal{P}$. *Then there is an m-variable* $h \in \mathcal{R}$, *such that*
$f(h(\vec{y}_m), \vec{x}_n) \simeq \phi_{h(\vec{y}_m)}^{(n)}(\vec{x}_n)$, *for all* $\vec{y}_m, \vec{x}_n$.
Moreover, h being an S-m-n function is 1-1.

Proof   The proof is essentially the same as that of Corollary 9.2.2.
By assumption, $\lambda z \vec{y}_m \vec{x}_n . f\left(S_{1+m}^n(z, z, \vec{y}_m), \vec{x}_n\right) \in \mathcal{P}$. Thus,

$$f\left(S_{1+m}^n(z, z, \vec{y}_m), \vec{x}_n\right) \simeq \phi_i^{(n+m+1)}(z, \vec{y}_m, \vec{x}_n), \text{ for some } i \in \mathbb{N} \qquad (1)$$

hence

$$f\left(S_{1+m}^n(i, i, \vec{y}_m), \vec{x}_n\right) \simeq \phi_i^{(n+m+1)}(i, \vec{y}_m, \vec{x}_n) \overset{\text{(Corollary 5.7.4)}}{\simeq} \phi_{S_{1+m}^n(i,i,\vec{y}_m)}^{(n)}(\vec{x}_n) \qquad (2)$$

Take $h = \lambda \vec{y}_m.S_{1+m}^n(i, i, \vec{y}_m)$.   □

9.2.4 Corollary. *If the unary f is in $\mathcal{R}$, then for some e, $\phi_e = \phi_{f(e)}$.*

Proof Apply Theorem 9.2.1 to $f = \lambda yx.\phi_{f(y)}(x)$.   □

9.3   The Fixed Point Theorem of Computability

The recursion theorem's version according to Corollary 9.2.4 is also called the *fixed point theorem* or *fixpoint theorem* (of computability).[7] It is implied by the second recursion theorem as above, but it is worth exploring a direct proof:

Given $f \in \mathcal{R}$. Consider the ψ defined below

$$\psi(x, y) \simeq \begin{cases} \phi_{f(\phi_x(x))}(y) & \text{if } \phi_x(x) \downarrow \\ \uparrow & \text{othw} \end{cases} \qquad (1)$$

The above is a definition by positive cases, the top case defining a function in $\mathcal{P}$ by the *universal* (Theorem 5.4.1 and Remark 5.4.3) theorem (or by the Normal Form Theorem 5.5.3).

By the S-m-n theorem we have an $h \in \mathcal{R}$ such that

$$\phi_{h(x)} \simeq \begin{cases} \phi_{f(\phi_x(x))} & \text{if } \phi_x(x) \downarrow \\ \emptyset & \text{othw} \end{cases} \qquad (1')$$

Let $a \in \mathbb{N}$ such that $\phi_a = h$. Then $(1')$ becomes

$$\phi_{\phi_a(x)} \simeq \begin{cases} \phi_{f(\phi_x(x))} & \text{if } \phi_x(x) \downarrow \\ \emptyset & \text{othw} \end{cases} \qquad (1'')$$

$e \simeq \phi_a(a)$ works:

$$\phi_e = \phi_{\phi_a(a)} \overset{\phi_a(a)=h(a)\downarrow}{=} \phi_{f(\phi_a(a))} = \phi_{f(e)}$$

 9.3.1 Remark.

---

[7] As distinguished from the fixed point theorem of PA.

1. The fixpoint theorem for function $f \in \mathcal{R}$ does not prove the existence of a fixed point of f in the sense "$e \simeq f(e)$" but rather provides two, *in general different*, programs —one of the form "$f(e)$" the other being e— that compute the same function.
2. The proof above goes through unchanged if we replace everywhere the argument y by $\vec{y}_n$. The only trivial change is to superscript those ϕ that have $\vec{y}_n$ as argument with (n).   □

9.3.2 Theorem. (Fixpoint Theorem with Parameters) *For any* $\lambda \vec{x}_n u . f(\vec{x}_n, u) \in \mathcal{R}$ *there is a* $\lambda \vec{x}_n . h(\vec{x}_n) \in \mathcal{R}$ *such that* $\phi_{h(\vec{x}_n)} = \phi_{f\left(\vec{x}_n, h(\vec{x}_n)\right)}$.

Proof We imitate the above proof. Consider the ψ defined below

$$\psi(\vec{x}_n, u, y) \simeq \begin{cases} \phi_{f(\vec{x}_n, \phi_u^{(n+1)}(u, \vec{x}_n))}(y) & \text{if } \phi_u^{(n+1)}(u, \vec{x}_n) \downarrow \\ \uparrow & \text{othw} \end{cases} \tag{1}$$

$\psi \in \mathcal{P}$ due to the analogous argument from above, hence by the S-m-n theorem we have a $\sigma \in \mathcal{R}$ such that

$$\phi_{\sigma(u, \vec{x}_n)}(y) \simeq \begin{cases} \phi_{f(\vec{x}_n, \phi_u^{(n+1)}(u, \vec{x}_n))}(y) & \text{if } \phi_u^{(n+1)}(u, \vec{x}_n) \downarrow \\ \uparrow & \text{othw} \end{cases} \tag{1$'$}$$

Let $a \in \mathbb{N}$ such that $\sigma = \phi_a^{(n+1)}$. Thus, (1$'$) becomes

$$\phi_{\phi_a^{(n+1)}(u, \vec{x}_n)}(y) \simeq \begin{cases} \phi_{f(\vec{x}_n, \phi_u^{(n+1)}(u, \vec{x}_n))}(y) & \text{if } \phi_u^{(n+1)}(u, \vec{x}_n) \downarrow \\ \uparrow & \text{othw} \end{cases} \tag{1$''$}$$

Then $h = \lambda \vec{x}_n . \phi_a^{(n+1)}(a, \vec{x}_n)$ works:

$$\phi_{h(\vec{x}_n)} = \phi_{\phi_a^{(n+1)}(a, \vec{x}_n)} \overset{\phi_a^{(n+1)}(a, \vec{x}_n) = h(\vec{x}_n) \downarrow}{=} \phi_{f(\vec{x}_n, \phi_a^{(n+1)}(a, \vec{x}_n))} = \phi_{f(\vec{x}_n, h(\vec{x}_n))} \qquad □$$

 9.3.3 Remark. By Theorem 5.7.1, h above is 1-1.   □

9.3.4 Corollary. (Double Fixpoint Theorem) Muchnik (1958), Smullyan (1961) *For any f and g of two arguments in $\mathcal{R}$ there are a and b in $\mathbb{N}$ such that $\phi_a = \phi_{f(a,b)}$ and $\phi_b = \phi_{g(a,b)}$.*

Proof By Theorem 9.3.2, let $h \in \mathcal{R}$ such that $\phi_{h(x)} = \phi_{g(x,h(x))}$ for all x. Let a, by Corollary 9.2.4 satisfy

$$\phi_a = \phi_{f(a, h(a))}$$

Take $b = h(a)$.   □

9.3.5 Remark At the beginning of this section we remarked that the fixed point theorem follows from the second recursion theorem. The converse is also true, so the two are equivalent but we should not fail to note that the former uses both the universal *and* S-m-n theorems in its direct proof, while the latter uses only the S-m-n theorem.

The proof "fixpoint theorem $\Longrightarrow$ recursion theorem" is as simple as the other direction (Corollary 9.2.4): Let $\psi(x, \vec{y}_n) \in \mathcal{P}$. By S-m-n, there is an $h \in \mathcal{R}$ such that $\phi_{h(x)}^{(n)}(\vec{y}_n) \simeq \psi(x, \vec{y}_n)$.

Now employ Remark 9.3.1 (item 2.) to conclude that for some $e \in \mathbb{N}$, we have

$$\psi(e, \vec{y}_n) \simeq \phi_{h(e)}^{(n)}(\vec{y}_n) \overset{\text{ibid}}{\simeq} \phi_e^{(n)}(\vec{y}_n) \qquad \qquad \square$$

9.4   Some Applications

Our first application is another proof of Rice's theorem (cf. also Theorem 6.10.7):

9.4.1 Theorem. (Rice's Theorem) *The complete index set $A = \{x : \phi \in \mathcal{C}\}$ is recursive iff either $A = \emptyset$ or $A = \mathbb{N}$.*

Proof *If*-part. Both $\emptyset$ and $\mathbb{N}$ are primitive recursive, hence recursive.

   Only if-part. By contradiction. Let

$$A \in \mathcal{R}_*, \text{ yet } \emptyset \neq A \neq \mathbb{N} \tag{1}$$

By (1), let $a \in A$ and $b \notin A$. Now define f by

$$f(x) = \begin{cases} b & \text{if } x \in A \\ a & \text{othw} \end{cases} \tag{2}$$

It is clear that

1. $f \in \mathcal{R}$ since it is defined from constant functions by recursive cases.
2.

$$x \in A \equiv f(x) \notin A \tag{3}$$

   (by (2))

By Corollary 9.2.4, let e be such that

$$\phi_e = \phi_{f(e)} \tag{4}$$

Then

$$e \in A \text{ iff (def. of } A) \phi_e \in C \text{ iff (4) } \phi_{f(e)} \in C \text{ iff (def. of } A) \ f(e) \in A$$

Thus $e \in A \equiv f(e) \in A$, contradicting (3). $\qquad\square$

Rice called a complete index set, which is empty or $\mathbb{N}$, "trivial". Thus, his theorem, in plain English, is that *of all the infinitely many complete index sets only the two trivial ones are decidable.* Rogers (1967) attributes the above proof to G.C. Wolpin.

9.4.2 Corollary. *Every f of one variable in $\mathcal{P}$ has infinitely many ϕ-indices.*

Proof This is an overkill, but elegant. Let $C = \{f\}$. Now, $\emptyset \neq \{f\} \neq \mathcal{P}$, hence the complete index set $A = \{x : \phi \in C\}$ is nontrivial. By Theorem 9.4.1, A is not recursive, hence is infinite, since all finite sets are recursive. $\qquad\square$

9.4.3 Exercise. Is this true? "all finite sets are recursive."

Can you sharpen to "primitive recursive"? $\qquad\square$

9.4.4 Corollary. *The set $K = \{x : \phi_x(x) \downarrow\}$ is not a complete index set, that is, for no $C \subseteq \mathcal{P}$ can we have $K = \{x : \phi_x \in C\}$.*

Proof Define ψ by

$$\psi(x, y) = \begin{cases} 0 & \text{if } x = y \\ \uparrow & \text{othw} \end{cases}$$

ψ is defined from the partial recursive $\lambda y.0$ and $\emptyset$ by primitive recursive cases, so $\psi \in \mathcal{P}$. By Theorem 9.2.1 there is an $e \in \mathbb{N}$ such that

$$\phi_e = \lambda x.\psi(x, e) \tag{1}$$

$$\text{Let } m \text{ be } m \neq e \text{ but } \phi_m = \phi_e \text{ (Corollary 9.4.2)} \tag{2}$$

We can now contradict

$$K = \{x : \phi_x \in C\} \tag{3}$$

no matter what C might be. So, assume (3).

- By (1), $\phi_e(e) \downarrow$ hence $\phi_e \in C$ by (3)
- By (2), $\phi_m \in C$ as well. But $\phi_m(m) = \psi(m, e) \uparrow$. This contradicts (3), since equality to K requires $\phi_m(m) \downarrow$. $\qquad\square$

9.4.5 Corollary. (A Self-Reproducing Machine) *For some $e \in \mathbb{N}$, $\phi_e = \lambda x.e$.*

That is, the URM M_x^x at location e in our standard enumeration (cf. Theorem 5.4.1) outputs itself (well, the *address* of itself) for all inputs.

Proof Apply Theorem 9.2.1 to $U_2^2 = \lambda xy.y$ to get an e such that $\phi_e(x) = U_2^2(x, e)$, for all x.   □

9.5   Some Unusual Recursions

The recursion theorem allows us to show that recursions that are not primitive, still define partial recursive functions.

Let us approach this by an example first. Recall the Ackermann function of Chap. 4 in the version given in Remark 4.1.2:

$$A_0(x) = x + 2$$

$$A_{n+1}(0) = 2 \qquad\qquad (*)$$

$$A_{n+1}(x + 1) = A_n(A_{n+1}(x))$$

Let us rewrite this with both the n and x variables on the same line, inside the same set of brackets, and making a couple of additional trivial changes:

$$A(n, x) = \begin{cases} \text{if} & n = 0 \text{ then } x + 2 \\ \text{else if} & x = 0 \text{ then } 2 \\ \text{else} & A\left(n \mathbin{\dot-} 1, A(n, x \mathbin{\dot-} 1)\right) \end{cases} \qquad (1)$$

We can show that (1) —*if viewed as an "equation" where we are seeking to solve for an "unknown" function A*— has a partial recursive solution $\lambda nx.\phi_e(n, x)$, for some $e \in \mathbb{N}$ that *we will seek to show that exists*.

We do so using Theorem 9.2.1 as follows:
Define a function F by $(1')$ below

$$F(z, n, x) = \begin{cases} \text{if} & n = 0 \text{ then } x + 2 \\ \text{else if} & x = 0 \text{ then } 2 \\ \text{else} & \phi_z^{(2)}\left(n \mathbin{\dot-} 1, \phi_z^{(2)}(n, x \mathbin{\dot-} 1)\right) \end{cases} \qquad (1')$$

$F \in \mathcal{P}$ as it is defined by (primitive) recursive cases from the functions

$$\lambda znx.x + 2, \ \lambda znx.2, \text{ and } \lambda znx.\phi_z^{(2)}\left(n \mathbin{\dot-} 1, \phi_z^{(2)}(n, x \mathbin{\dot-} 1)\right)$$

all in $\mathcal{P}$. By Theorem 9.2.1, there is an e such that

$$F(e, n, x) = \phi_e^{(2)}(n, x) \qquad (2)$$

for all n, x. We modify (1') using (2):

$$\phi_e^{(2)}(n, x) = \begin{cases} \text{if} & n = 0 \text{ then } x + 2 \\ \text{else if} & x = 0 \text{ then } 2 \\ \text{else} & \phi_e^{(2)}\left(n \dotminus 1, \phi_z^{(2)}(n, x \dotminus 1)\right) \end{cases} \quad (1'')$$

Eureka! Comparing, (1) and (1'') it is clear that letting $A = \phi_e^{(2)}$ we have a solution for (1)

 And it is a partial recursive solution too, being a "$\phi_i^{(n)}$".

Pause Wait! How do I know this is *the* Ackermann function of (1)? There may be many solutions of (1); some might not even be partial recursive! ◄

The question above is outside computability and is "elementary" (this does not mean "easy" all the time,[8] but it is this time).

First off, we never have had actually shown that (1) —or the equivalent formulation (∗)— defines a *unique* function $\lambda nx.A_n(x)$.

 By the way we already know that said function —if it exists — is total, having proved, for each n, that the "n-th branch" of it, $\lambda x.A_n(x) \in \mathcal{PR}$, is. Now each branch does exist by a general theorem on recursive definitions (that applies to primitive recursion). Therefore so does the function $\lambda nx.A_n(x)$. For each input (n, x), use the n-th branch to obtain $A_n(x)$.

Here is a proof of uniqueness (by double induction). Say we also have $\lambda nx.B_n(x)$ satisfying the equations (∗), that is,

$$B_0(x) = x + 2$$
$$B_{n+1}(0) = 2 \quad (**)$$
$$B_{n+1}(x + 1) = B_n(B_{n+1}(x))$$

We show

$$(\forall n)(\forall x)A_n(x) = B_n(x)$$

We start this as an induction on n and let the process force us into a subsidiary induction on x.

n-Basis: $n = 0$: We have $A_0(x) = x + 2 = B_0(x)$. Done.

---

[8] The lay person —also, Sherlock Holmes— will say "elementary" as a synonym for "easy". Mathematicians use the term more technically. They mean that the tools used in the proof are "not advanced" —but the task may still be very hard. E.g., the prime number theorem, that says $\lim_{x \to \infty} \frac{\pi(x)}{x/\log x} = 1$ has a proof that uses complex variable calculus and one that does not. Arguably, the latter is harder, but it is called the "elementary" one.

n-I.H.: *Assume* $A_n(x) = B_n(x)$ for *fixed n* and *all x*.

n-I.S.: *Prove* $A_{n+1}(x) = B_{n+1}(x)$ for *fixed n* and *all x*. This sends us to a subsidiary induction on x. The n is still fixed!

  x-Basis: $x = 0$: We have $A_{n+1}(0) = 2 = B_{n+1}(0)$. Done.

  x-I.H.: *Assume* $A_{n+1}(x) = B_{n+1}(x)$ for *the above fixed n* and *fixed x*.

  x-I.S.: *Prove* $A_{n+1}(x+1) = B_{n+1}(x+1)$ for *the above fixed n* and *fixed x*. Well,

$$A_{n+1}(x+1) \overset{(*)}{=} A_n(A_{n+1}(x)) \overset{n\text{-I.H.}}{=} B_n(A_{n+1}(x)) \overset{x\text{-I.H.}}{=} B_n(B_{n+1}(x)) \overset{(**)}{=} B_{n+1}(x+1)$$

OK, the solution to $(*)$ is *unique*. But does one *exist*? Well, yes, we showed more than that: That a partial recursive solution $\phi_e^{(2)}$ exists. Moreover, since the unique solution is total, we now have rediscovered that $\mathcal{R} \ni \phi_e^{(2)} = \lambda nx.A_n(x)$.

We will now formulate a general result that states for *equations* like the following —in one *function* variable that we named α— that they have partial recursive α-solutions

$$\alpha(\vec{x}_n) \simeq \ldots \alpha \ldots \tag{1}$$

that is, there are $\phi_e^{(n)}$ of the same number of (numerical) arguments as α that when substituted for α everywhere in (1) make the equality stated true.

Above "…" indicate any operations on *functions* that are acceptable to act on functions of $\mathcal{P}$. The function symbol α occurs in the right hand side of "$\simeq$" any (finite) number of occurrences as a *call*, with any arguments.

How can this be made precise?

First a slightly more specific illustration: Say the α that we are defining in terms of itself (recursively) in (1) has two arguments. Let f and g be given unary partial recursive functions. Then α might appear on the right hand side giving (1) the form

$$\alpha(x, y) = \ldots \alpha(f(x), f(g(y))) \ldots \alpha\Big(f(\alpha(x+4, y^2)), \alpha\big(f(x), g(x)\big)\Big) \ldots \tag{2}$$

We are interested in just those cases that, as we say technically, the right hand side of (1) —as a function of $\vec{x}_n$— is *partial recursive in* α.

9.5.1 Definition. A function is *partial recursive in* α iff it is $\mathcal{P}$-derived after we modified the initial functions set, $\mathcal{I}$, in Definition 5.6.1 to *include* α as a new initial function.

Let us denote the set of all such functions $\mathcal{P}^\alpha$. $\square$

 In short, a function is partial recursive in α iff it is obtained by a finite number of partial recursive operations using as initial functions α and those in $\mathcal{I}$ of Definition 5.6.1.

We will look at partial recursiveness "in some α" in detail in Chap. 13.

9.5.2 Lemma. *If* $\alpha \in \mathcal{P}$, *then a function that is partial recursive in* α *is just partial recursive.*

In particular, if we replace α throughout the right hand side of (1) by a partial recursive function $\phi_z^{(n)}$ of the same arity as α, then we end up with a partial recursive function of $n + 1$ variables (z is the extra variable).

Proof Let f be partial recursive in α, and moreover $\alpha \in \mathcal{P}$. We show that $f \in \mathcal{P}$. We do induction on the derivation length, n, of f. Without loss of generality (Lemma 2.1.4), f is at the end of the considered derivation.

Basis. Derivation length 1. Then $f \in \mathcal{I} \cup \{\alpha\} \subseteq \mathcal{P}$.

I.H. Assume the claim for derivation lengths $\leq n$.

I.S. Consider the case of length $n + 1$. By Definition 5.6.1 we have four cases for f. Each case 2–4 preserves membership in $\mathcal{P}$, and we are done by the I.H. Case 1 is already done as the *Basis.*   □

Equation (1) above has now become precise. It says that its right hand side is a function in $\mathcal{P}^\alpha$. In turn, Lemma 9.5.2 says that if we replace α (everywhere on the right hand side of "$\simeq$") by $\phi_z^{(n)}$ of the same arity as α, then the expression E on the right hand side in (1) depends on z and $\vec{x}_n$, and that $\lambda z \vec{x}_n . E \in \mathcal{P}$.

α in (1) acts as a function *variable* to solve for. A solution h for α is a *specific function* that makes (1) true for all $\vec{x}_n$ if we replace *all* —to the left and right of "$\simeq$"— occurrences of α by h.

We show that if the right hand side of (1) is partial recursive in α, then *(1) always has a partial recursive solution h for α.* That is,

$$(\exists e)\left(\text{if we replace } \alpha \text{ in (1) everywhere by } \phi_e^{(n)}, \text{ then "}\simeq\text{" true for all } \vec{x}_n\right) \qquad (3)$$

In order to force (3) to happen we use the recursion theorem.

Here are the steps:

(I) Let z be a new variable (i.e., it does not occur in (1)) and replace the left hand side "$\alpha(\vec{x}_n)$" by $G(z, \vec{x}_n)$.

(II) Replace all occurrences (of the letter) α in the right hand side by $\phi_z^{(n)}$. We obtain an expression E on the right hand side (of "$\simeq$") and we already noted that $\lambda z \vec{x}_n . E \in \mathcal{P}$ (Lemma 9.5.2).

(III) (1) has become

$$G(z, \vec{x}_n) \simeq \ldots \phi_z^{(n)} \ldots \qquad (1')$$

(IV) By the recursion theorem there is an e such that

$$G(e, \vec{x}_n) \simeq \phi_e^{(n)}(\vec{x}_n), \text{ for all } \vec{x}_n$$

Thus, (1') yields

$$\phi_e^{(n)}(\vec{x}_n) \simeq \ldots \phi_e^{(n)} \ldots \qquad (1'')$$

That is, substituting everywhere in (1) the function variable α by $\phi_e^{(n)}$ we have solved (1) —i.e., we *forced the equality of both sides of* "$\simeq$" to be true, for all $\vec{x}_n$— and with a $\mathcal{P}$-solution at that!

9.5.3 Remark.

1. The above process (I)–(IV) does *not* settle any questions about the uniqueness, totalness, etc., of the solution $\phi_e^{(n)}$.
2. The technique is *constructive*, since if we have a URM program for G we can *construct* —cf. Example 1.1.16— a URM program a for $\lambda z \vec{x}_n.G(S_1^n(z, z), \vec{x}_n)$. But then the program e we want is $S_1^n(a, a)$ according to the recursion theorem 9.2.1.
3. If we denote the expression "$\dots \alpha \dots$" in (1) by

$$\mathscr{F}(\vec{x}_n, \alpha) \tag{4}$$

then (1) is the problem of finding a fixpoint α of (4). "Finding" succeeds in the strongest possible terms by the remark 2. above.
4. Finding fixpoint solutions is not the exclusive province of recursion theory. Arithmetic and (projective) geometry also employ the fixpoint solutions method, and utilise for it the nickname "*method of the false position*".

   We outline how the *false position technique* applies to geometry (see figure below).

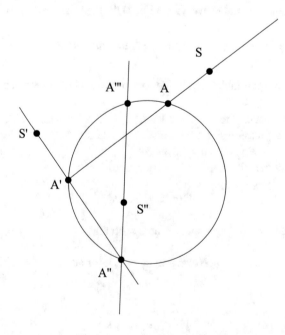

Thus, we are given a circle and three points, S, S' and S'' on its plane. We want to construct (draw) a triangle whose three vertices lie on the given circle and its three sides each contain one of the three given points.

We start the construction by making three tries, that will most likely be in "false position" —that is, unlikely to solve the problem. To avoid cluttering the figure I display just *one attempt*. It has *three* steps, one for each given point.

Step 1 Through S I draw a line (just guessing which way to draw) that cuts the circle at A and A'.

Step 2 I continue by drawing a line through S' *and* A'. It cuts the circle again at A''.

Step 3 I continue by drawing a line through S'' *and* A''. It cuts the circle again at A'''. *Now if A''' were coincident with A, then I should consider myself very lucky*: The three lines I drew (in such a lucky manner) do form a triangle —$AA'A''$— that has its vertices on the circle, and each of its sides contain one of the points S, S' and S''.

Well, we do not despair: If I repeat the 3-step construction choosing a different line through S, then, the second time I would obtain different candidate vertices, say, B, B', B'' and B''', where the latter, if I am still not lucky, would *not* coincide with B.

OK, we try once more, obtaining C, C', C'', C''', the latter in the worst case not coinciding with the starting point C. *This is the last time* that we do the construction for two reasons. The underlying (projective geometry) theory (e.g., Veblen and Young 1916) tells us that:

- The correspondence established by the 3-step construction, that maps point A to A'''; B to B'''; C to C'''; and in general (starting point X) X to X''' is a *projectivity π on the circle*.
- Any projectivity is *uniquely determined* by three pairs of corresponding points, that is, our π is fully determined by

$$\pi(A) = A'''$$
$$\pi(B) = B''' \tag{5}$$
$$\pi(C) = C'''$$

This is why we only need to guess —that is, repeat the construction *Steps* (1)–(3) with starting points A, B and C— *three times only*.

- If some point M on the circle satisfies $\pi(M) = M$, then it is called a *double point* of the projectivity, which is nothing else but a fixpoint of π.
- Given a projectivity (5) we may construct the double points M and N. Those (each) are such that if we start the construction by drawing the first line from S to M (resp. N), then M (resp. N) *will* coincide with M''' (resp. N''') and we will have a triangle satisfying all the required properties.

The *construction* of the double/fixed points is a simple ruler and compass one.

We see now the analogy between how we solved problem (1) with how we (as indicated in outline immediately above) applied the technique of "false position" to solve a geometric problem.

In solving (1) we "forced" α and $\mathscr{F}(\vec{x}_n, \alpha)$ to become equal (as functions) —a fixed point of $\mathscr{F}$— while the "false position" in this case is that, in general, given an arbitrary function α it will be $\alpha \neq \mathscr{F}(\vec{x}_n, \alpha)$.   □

9.6   Exercises

1. Let g be recursive. Show that

$$h(n, \vec{x}_k) \simeq \begin{cases} 0 & \text{if } g(n, \vec{x}_k) = 0 \\ h(n+1, \vec{x}_k) + 1 & \text{otherwise} \end{cases} \tag{1}$$

   has a computable h-solution $\phi_e^{(k+1)}$ that satisfies $\phi_e^{(k+1)}(0, \vec{x}_k) \simeq (\widetilde{\mu}y)g(y, \vec{x}_k)$, for all $\vec{x}_k$.

2. Let g be recursive. Show that

$$H(n, \vec{x}_k) \simeq \begin{cases} n & \text{if } g(n, \vec{x}_k) = 0 \\ H(n+1, \vec{x}_k) & \text{otherwise} \end{cases} \tag{1}$$

   has a computable H-solution $\phi_e^{(k+1)}$ that satisfies $\phi_e^{(k+1)}(0, \vec{x}_k) \simeq (\widetilde{\mu}y)g(y, \vec{x}_k)$, for all $\vec{x}_k$.

3. Prove that for some m, $\phi_m(x) \simeq \lfloor \sqrt[m]{x} \rfloor$, for all x.
4. Prove that for some m, $\phi_m(x) \simeq \lfloor \log_m x \rfloor$, for all x.
5. Prove that for some m, $\phi_m(x) \simeq m^x$, for all x.

Chapter 10
A Universal (non-$\mathcal{PR}$) Function for $\mathcal{PR}$

Overview

This chapter shows how to assign a numerical code to every function f of $\mathcal{PR}$ so that the code encapsulates a specific derivation of f. This is similar to Exercise 3.5.15, but here we give a number-theoretic version.

In turn, this tool allows a different proof than either that of said exercise, or those via Ackermann's function or the informal one in Sect. 3.4 for the result $\mathcal{PR} \subsetneq \mathcal{R}$: We show here that there is a universal function for $\mathcal{PR}$, and diagonalise (alternatively, majorise).

In more detail, unpacking said code provides us with the tools to *compute* f. In effect, we will build a *total* computable function $Eval$ (stands for "evaluate") such that, *essentially*, for every $\lambda \vec{x}_n.f(\vec{x}_n) \in \mathcal{PR}$, we have $Eval(\widehat{f}, \vec{x}_n) = f(\vec{x}_n)$ for all $\vec{x}_n$, where $\widehat{f}$ is some code that encapsulates a derivation for f. Thus $Eval$ is universal for $\mathcal{PR}$. Provably, it will not be in $\mathcal{PR}$.

10.1 $\mathcal{PR}$ Derivations and Their Arithmetisation

$\mathcal{PR}$ has been defined as the set of all derived functions using as initial functions $\mathcal{I} = \{\lambda x.0, \lambda x.x + 1\} \cup \{\lambda \vec{x}_n.x_i : 1 \leq i \leq n \wedge n \in \mathbb{N}\}$ and as operations *composition* and *primitive recursion*. We will assign "derivation codes" to each $\mathcal{PR}$ function, using our usual prime-power coding, namely,

$$\langle x_0, \ldots, x_n \rangle = \prod_{i \leq n} p_i^{x_i+1}$$

Recall also $\lambda xy.x * y$ from Definition 2.4.3 (and the related result Corollary 2.4.7). We next "arithmetise" $\mathcal{PR}$-functions and their *derivations*, assigning a numerical code to each specific derivation, and hence to each derived function. Thus, this code once "unpacked" tells us how the function was put together from simpler functions, and hence how to compute it.

The following table assigns *recursively* code numbers —that we will call Gödel numbers[1]— (middle column) to all functions in $\mathcal{PR}$. In the table, $\widehat{f}$ indicates a code of f.

Function	Code	Comment
$C_c^n = \lambda \vec{x}_n.c$	$\langle 0, n, 0, c \rangle$	$n > 0 \wedge c \geq 0$
$S_i^n = \lambda \vec{x}_n.x_i + 1$	$\langle 0, n, 1, i \rangle$	$1 \leq i \leq n$
$U_i^n = \lambda \vec{x}_n.x_i$	$\langle 0, n, 2, i \rangle$	$1 \leq i \leq n$
Composition: $f(g_1(\vec{y}_m), \ldots, g_n(\vec{y}_m))$	$\langle 1, m, \widehat{f}, \widehat{g}_1, \ldots, \widehat{g}_n \rangle$	f *must* be n-ary *all* g_i *must* be m-ary
$prim(h, g)$ "basis" h and "iterator" g	$\langle 2, n+1, \widehat{h}, \widehat{g} \rangle$	h *must* be n-ary g *must* be $(n+2)$-ary

The recursive definition depicted in the table works on codes, not directly on functions. After all, each function has infinitely many different derivations. Thus *what* we are actually doing is *defining recursively* a subset of $\{z : Seq(z)\}$ —let us call it Pri (primitive recursive indices).

Note that the first three positions in a code a denote respectively

1. $(a)_0$: Basis cases ($=0$), or not ($=1$ or $=2$).
2. $(a)_1$: Number of inputs.
3. $(a)_i$, for $i \geq 2$: Type of action (see left column above, and also Definition 10.1.2 below).

10.1.1 Definition. *Pri* is a *closure* that satisfies:
   Initial codes.

   (i) $\{\langle 0, n, 0, c \rangle : n > 0 \wedge c \geq 0\} \subseteq Pri$
  (ii) $\{\langle 0, n, 1, i \rangle : n > 0 \wedge 1 \leq i \leq n\} \subseteq Pri$
 (iii) $\{\langle 0, n, 2, i \rangle : n > 0 \wedge 1 \leq i \leq n\} \subseteq Pri$
          Operations on codes.
  (iv) *Coding composition:* $\langle 1, m, a, b_1, \ldots, b_n \rangle \in Pri$, *If* a and the b_i are all in Pri and moreover $(a)_1 = n$ and $(b_i)_1 = m$, for $i = 1, \ldots, n$. We say that $a, b_1, \ldots, b_n$ are *immediate predecessors* or *i.p.* of $\langle 1, m, a, b_1, \ldots, b_n \rangle$.
   (v) *Coding primitive recursion:* $\langle 2, n+1, a, b \rangle \in Pri$, *If* a and the b are in Pri and moreover $(a)_1 = n$ and $(b)_1 = n + 2$. We say that a and b are *immediate predecessors* or *i.p.* of $\langle 2, n+1, a, b \rangle$.

$\square$

By the uniqueness of prime number decomposition, the rules defining the set Pri are *unambiguous*, i.e., if, say, $\langle 1, m, a, b_1, \ldots, b_n \rangle = \langle 1, k, a', b'_1, \ldots, b'_k \rangle$, then $m = k$ and $a = a'$, while $b_i = b'_i$, for $i = 1, \ldots, m = k$.[2] In other words,

- A code is uniquely *one* of the initial ones ((i)–(iii)) *or* a result of a rule (iv) *or* (v), all the "or" here being *exclusive or*.
- The input codes for rules (iv) and (ii) are uniquely determined by the outputs of the rules, $\langle 1, m, a, b_1, \ldots, b_n \rangle$ and $\langle 2, n+1, a, b \rangle$. Thus we can say *the* i.p. from now on.

Hence the set of Pri codes is partially ordered by the relation $C \prec C'$ meaning that there is a chain of codes

$$C = c_1 \text{ i.p. of } c_2 \text{ i.p. of } c_3 \text{ i.p. of } \ldots \text{ i.p. of } c_n = C' \tag{1}$$

This partial order has MC (Definition 0.4.38) since minimal elements can be found starting with any code C' and going down (leftwards) a chain like (1) that will terminate.

By the theorem on inductive definitions 0.4.43 we may define an interpretation (or *semantics*) *function* from Pri to $\mathcal{PR}$ by induction on *(the definition of) Pri*.

Thus, we define the semantics of codes as follows:

10.1.2 Definition. (Semantics of Pri) We define a (nontotal: some $a \in \mathbb{N}$ are not codes[3]) function $\lambda a.[a]$ for each $a \in Pri$, that is, a function that maps codes into (derivations, inductively) of functions of $\mathcal{PR}$:

Basis 1	$[\langle 0, n, 0, c \rangle]$	$= C_c^n$
Basis 2	$[\langle 0, n, 1, i \rangle]$	$= S_i^n$
Basis 3	$[\langle 0, n, 2, i \rangle]$	$= U_i^n$
induction step 1	$[\langle 1, m, a, b_1, \ldots, b_n \rangle] = \lambda \vec{y}_m.[a]([b_1](\vec{y}_m), \ldots, [b_n](\vec{y}_m))$	
induction step 2	$[\langle 2, n+1, a, b \rangle]$	$= \lambda x \vec{y}_n.prim([a], [b])$

$\square$

 The notation "$[e]$" for the primitive recursive function of index e is similar to, but distinct from, Kleene's notation $\{i\}$ for the partial recursive function of index i.

We can now make the intentions implied in the above table official:

10.1.3 Theorem. $\mathcal{PR} = \{[a] : a \in Pri\}$.

Proof $\subseteq$-part. Induction on $\mathcal{PR}$ to show that every $f \in \mathcal{PR}$ has a realisation as $[a]$, for some a.

---

[2] Of course, $\langle r, \ldots \rangle \neq \langle r', \ldots \rangle$ if $r \neq r'$.

[3] However, this mapping is total if the left field is restricted to Pri.

Basis: Examine all initial functions f (Definition 2.1.1). For example, we show that $Z = C_0^1 = [a]$ for some a: Indeed by the *Basis 1* case in Definition 10.1.2 we have that $a = \langle 0, 1, 0, 0 \rangle$ works.

The I.S. shows that the claim *propagates* with

1. Forming a primitive recursion: Let $f \in \mathcal{PR}$ by virtue of $f = prim(h, g)$, where $\{h, g\} \subset \mathcal{PR}$.

    By the I.H., there are a and b in *Pri* such that $h = [a]$ and $g = [b]$. Then $f = [\langle 2, n + 1, a, b \rangle]$ by case *Induction step 2* in Definition 10.1.2.
2. Forming a composition: *Exercise*.

$\supseteq$-part. By induction on the recursive definition of *Pri* (Definition 10.1.1) we show that $a \in Pri$ implies $[a] \in \mathcal{PR}$.

Basis: If $a \in Pri$ because $a = \langle 0, n, 1, i \rangle$, then $\lambda x_n.[\langle 0, n, 1, i \rangle] = \lambda x_n.x_i + 1 \in \mathcal{PR}$. In short, $[\langle 0, n, 1, i \rangle] \in \mathcal{PR}$ (Definition 10.1.2). Similarly for the other cases where $(a)_0 = 0$.

Say, a code has the form $\langle 2, n + 1, a, b \rangle \in Pri$, where a and b are in *Pri*. By the I.H. over *Pri* and the fact that a and b are the immediate predecessors of $\langle 2, n + 1, a, b \rangle$, we have $[a] \in \mathcal{PR}$ and $[b] \in \mathcal{PR}$. Thus, by closure properties of $\mathcal{PR}$, $prim([a], [b]) \in \mathcal{PR}$. Since by the semantics in line *Induction step 2* of Definition 10.1.2 state: $[\langle 2, n + 1, a, b \rangle] = \lambda x \vec{y}_n.prim([a], [b])$ we conclude that $[\langle 2, n + 1, a, b \rangle] \in \mathcal{PR}$.   □

 10.1.4 Remark. Thus, $f \in \mathcal{PR}$ iff, for some $a \in Pri$, $f = [a]$.   □

10.1.5 Example. Every function $f \in \mathcal{PR}$ has infinitely many *Pri*-codes. Indeed, let $f = [\widehat{f}]$. Since $f = \lambda \vec{x}_n.u_1^1(f(\vec{x}_n))$, we obtain $f = [\langle 1, n, \langle 0, 1, 1, 2 \rangle, \widehat{f} \rangle]$. Since $\langle 1, n, \langle 0, 1, 1, 2 \rangle, \widehat{f} \rangle > \widehat{f}$, the claim follows.   □

10.1.6 Theorem. *The relation $x \in Pri$ is primitive recursive.*

Proof We show (by a course-of-values recursion) that c_{Pri} —the characteristic function of *Pri*— is in $\mathcal{PR}$.

$$c_{Pri}(0) = 1^4$$

and, for $x \geq 0$,

---
[4] 0 is not a code.

$$c_{Pri}(x+1) = \begin{cases} 0 & \text{if } (\exists n, c)_{\le x}\Big(x+1 = \langle 0, n, 0, c\rangle \wedge 0 < n \wedge 0 \le c\Big) \\ 0 & \text{if } (\exists n, i)_{\le x}\Big(x+1 = \langle 0, n, 1, i\rangle \wedge i \le n \wedge 1 \le i\Big) \\ 0 & \text{if } (\exists n, i)_{\le x}\Big(x+1 = \langle 0, n, 2, i\rangle \wedge i \le n \wedge 1 \le i\Big) \\ 0 & \text{if } (\exists n, a, b)_{\le x}\Big(x+1 = \langle 2, n+1, a, b\rangle \wedge \\ & \quad c_{Pri}(a) = 0 \wedge c_{Pri}(b) = 0 \wedge (a)_1 = n \wedge (b)_1 = n+2\Big) \\ 0 & \text{if } (\exists m, n, a, b)_{\le x}\Big(x+1 = \langle 1, m, a\rangle * b \wedge c_{Pri}(a) = 0 \wedge \\ & \quad (a)_1 = n \wedge Seq(b) \wedge lh(b) = n \wedge \\ & \quad (\forall i)_{<n}\big(c_{Pri}((b)_i) = 0 \wedge ((b)_i)_1 = m\big)\Big) \\ 1 & \text{othw} \end{cases}$$

$\square$

10.2   The S-m-n Theorem for $\mathcal{PR}$

We show that we can parametrise a $\mathcal{PR}$ definition, that is, if we have a $\mathcal{PR}$-derivation for $f(x, y)$ then we can computationally obtain a $\mathcal{PR}$-derivation for $\lambda y.f(a, y)$ as a (computable) function of a. In fact the "computable" function that produces the new derivation is primitive recursive.

This is hardly surprising if we think the matter through via loop programs in analogy to our approach of proving the S-m-n theorem for URM-computable functions.

But let us proceed mathematically here working with the Pri indices.

10.2.1 Theorem. *There there is a $\mathcal{PR}$-function Sb_1 such that the following is true*

$$[Sb_1(a, y)](x_1, x_2, \ldots, x_m) = [a](y, x_1, \ldots, x_m)$$

Proof One can easily verify that

$$Sb_1(a, y) = \langle 1, m, a, \langle 0, m, 0, y\rangle, \langle 0, m, 2, 1\rangle, \langle 0, m, 2, 2\rangle, \ldots, \langle 0, m, 2, m\rangle\rangle \text{ [5]}$$

works.

$\square$

---

[5] Of course, $Sb_1(a, y)$ is in Pri, for all $a \in Pri$ and all $y \in \mathbb{N}$.

10.2.2 Definition. The $n + 1$-ary functions Sb_n are defined for each fixed $n \geq 1$ by induction on n:

1. Sb_1 is that from Theorem 10.2.1.
2. Assuming Sb_n has been defined we let

$$Sb_{n+1}(a, \vec{y}_{n+1}) = Sb_1(Sb_n(a, \vec{y}_n), y_{n+1})$$

$\square$

10.2.3 Corollary. (Primitive Recursive S-m-n Theorem) *For each $n \geq 1$, $Sb_n \in \mathcal{PR}$ and we have that, for all $m \geq 1$, the following holds.*

$$[Sb_n(a, \vec{y}_n)](\vec{x}_m) = [a](\vec{y}_n, \vec{x}_m)$$

10.2.4 Theorem. (Primitive Recursion Theorem) *For any $\lambda z \vec{x}_n . f(z, \vec{x}_n) \in \mathcal{PR}$ there is a number $e \in \mathbb{N}$ such that the following holds:*

$$f(e, \vec{x}_n) = [e](\vec{x}_n)$$

Proof We rephrase and contextualise the proof of Theorem 9.2.1. Set

$$g(z, \vec{x}_n) \overset{Def}{=} f(Sb_1(z, z), \vec{x}_n) \tag{1}$$

$g \in \mathcal{PR}$ by Grzegorczyk substitution. By Theorem 10.1.3 fix $a \in \mathbb{N}$ such that

$$g(z, \vec{x}_n) = [a](z, \vec{x}_n)$$

holds. By Theorem 10.2.1 we note

$$g(z, \vec{x}_n) = [a](z, \vec{x}_n) = [Sb_1(a, z)](\vec{x}_n) \tag{2}$$

(1) and (2) yield

$$f(Sb_1(a, a), \vec{x}_n) = [Sb_1(a, a)](\vec{x}_n)$$

which is what we want if we let $e = Sb_1(a, a)$. $\square$

10.3 The Universal Function *Eval* for $\mathcal{PR}$

We now turn to computing a *universal function*, *Eval*, for *all* primitive recursive functions, which by Theorem 10.1.3 is *precisely* the set of all functions $[a]$ for $a \in Pri$. *Eval* computes and returns $[a](\vec{x})$ for all *compatible choices* of $[a]$ and $\vec{x}$.

Since we do not allow variable length argument lists in any given function, *Eval* will take two arguments, a and b. If $a \notin Pri$ —checked primitive recursively by Theorem 10.1.6— it returns 0. If $a \in Pri$, then *Eval* —before it acts further— will want to verify that $Seq(b)$ *and* $(a)_1 = lh(b)$. That is, b prime-power codes a vector input of the *correct length*, $n = lh(b)$, for the primitive recursive function $[a]$. If not, *Eval* will return 0.

Else, if this test checks out, *Eval* will return $[a]\Big((b)_0, \ldots, (b)_{n \dot- 1}\Big)$.

Pause Why can't we have a *different Eval*, say, $Eval_k$, for each argument list length, k?◄

Because in the recursive definition of $Eval_k$ we will be obliged to make recursive calls to $Eval_n$ with various n.

Just think of the case $a = \langle 1, k, a, b_1, \ldots, b_n \rangle$.

Starting with a call $Eval_k(\langle 1, k, a, b_1, \ldots, b_n \rangle, \vec{x}_k)$ we will end up calling $Eval_n(a, \vec{y}_n)$. However, in general $k \neq n$. Take, for example, the case of $SU_1^3(x, y, z)$. Thus, computing

$$Eval_3\Big(\langle 1, 3, \langle 0, 1, 1, 1 \rangle, \langle 0, 3, 2, 1 \rangle\rangle, x, y, z\Big)$$

will require the calls

$$Eval_3(\langle 0, 3, 2, 1 \rangle, x, y, z) \text{ and } Eval_1(\langle 0, 1, 1, 1 \rangle, u)$$

10.3.1 Theorem. *The function* $\lambda ab.Eval(a, b)$ *defined for all a and b as*

$$Eval(a, b) = \begin{cases} [a]\Big((b)_0, \ldots, (b)_{(a)_1 \dot- 1}\Big) & \text{if } a \in Pri \wedge Seq(b) \wedge (a)_1 = lh(b) \\ 0 & \text{othw} \end{cases}$$

is in $\mathcal{R}$.

Proof First we write "pseudo code", in the form of a *recursive C-like procedure*, to compute $Eval(a, b)$. "/*...*/" encloses the comment "...."

$Eval(a, b)$ /* Returns $[a]\big((b)_0, \ldots, (b)_{(a)_1 \dot{-} 1}\big)$,

or 0; as outlined above. */

if $a \in Pri \wedge Seq(b) \wedge (a)_1 = lh(b)$ **do**

$\{$

    if $(\exists n, c)_{<a}\big(a = \langle 0, n, 0, c \rangle \wedge n \geq 1 \wedge c \geq 0\big)$ **return**(c)

    if $(\exists n, i)_{<a}\big(a = \langle 0, n, 1, i \rangle \wedge n \geq i \wedge i \geq 1\big)$ **return**$((b)_{i \dot{-} 1} + 1)$

    if $(\exists n, i)_{<a}\big(a = \langle 0, n, 2, i \rangle \wedge n \geq i \wedge i \geq 1\big)$ **return**$((b)_{i \dot{-} 1})$

    if$(\exists n, h, g)_{<a}(\exists x, y)_{<b}\Big($

      $a = \langle 2, n+1, h, g \rangle \wedge Seq(y) \wedge b = \langle x \rangle * y\Big)$ **do**

      $\{$

      $z \leftarrow Eval(h, y)$ /* z is local; it receives "$[h](\vec{q}_n)$" where $y = \langle \vec{q}_n \rangle$ */

      for $i = 1$ **to** x **do** /* If $x = 0$ initially, then the loop is skipped */

        $\{$

        $z \leftarrow Eval(g, \langle i \dot{-} 1 \rangle * y * \langle z \rangle)$ /* " $= [g](i-1, \vec{q}_n, z)$" where $y = \langle \vec{q}_n \rangle$ */

        $\}$

        return(z)

      $\}$

    if $(\exists m, n, f, g)_{<a}\big(a = \langle 1, m, f \rangle * g \wedge Seq(g) \wedge lh(g) = (f)_1 \wedge (f)_1 = n\big)$ **do**

      return$(Eval(f, \prod_{i<n} p_i^{Eval((g)_i, b)+1}))$

$\}$

/*Case where the very first if-test fails*/   **return**(0)

By CT the above recursive procedure defines a URM (total) computable function $\lambda ab.Eval(a, b)$ that computes $[a]((b)_0, \ldots, (b)_{(a)_1 \dot{-} 1})$ when $a \in Pri$ and $(a)_1 = lh(b))$, returning 0 otherwise.   □

 10.3.2 Remark. We mathematise the above proof via the recursion theorem 9.2.1: We rewrite the definition of the $Eval(a, b)$ function above as

$Eval(a, b) =$

if $a \in Pri \wedge Seq(b) \wedge (a)_1 = lh(b)$ **then**

$\Bigg\{$

    if $(\exists n, c)_{<a}\big(a = \langle 0, n, 0, c \rangle \wedge n \geq 1 \wedge c \geq 0\big)$ **then** 0

    else if $(\exists n, i)_{<a}\big(a = \langle 0, n, 1, i \rangle \wedge n \geq i \wedge i \geq 1\big)$ **then** $(b)_{i \dotminus 1} + 1$

    else if $(\exists n, i)_{<a}\big(a = \langle 0, n, i, 2 \rangle \wedge n \geq i \wedge i \geq 1\big)$**then**$(b)_{i \dotminus 1}$

    else if $(\exists n, h, g)_{<a}(\exists x, y)_{<b}\Big($

    $a = \langle 2, n + 1, h, g \rangle \wedge Seq(y) \wedge b = \langle x \rangle * y\Big)$ **then**

    $\Big((\mu m)\Big(Seq(m) \wedge (m)_0 = Eval(h, y) \wedge$

        $(\forall i)_{<x}(m)_{i+1} = Eval\big(g, \langle i \rangle * y * \langle (m)_i \rangle\big)\Big)\Big)_x^{\;6}$

    else if $(\exists m, n, f, g)_{<a}\big(a = \langle 1, m, f \rangle * g \wedge Seq(g) \wedge$

        $lh(g) = (f)_1 \wedge (f)_1 = n\big)$

    then $Eval\Big(f, \prod_{i<n} p_i^{Eval((g)_i, b)+1}\Big)$

    else 0 **Comment**. If we are this far, this "else" never applies.

$\Bigg\}$

else   0

By the recursion theorem, $Eval \in \mathcal{R}$ (cf. Sect. 9.5).   □

10.3.3 Exercise. Justify the conclusion in the last sentence above, that is, show that there is[7] a *unique total* solution $Eval$ to the recursive definition given above. A good approach is to show that *any* solution "$E(a, b)$" of the recurrence will satisfy —for all a, b— under some conditions $E(a, b) = [a]((b)_0, \dots, (b)_{lh(b) \dotminus 1})$ and otherwise $E(a, b) = 0$; and E will be total.   □

10.3.4 Theorem. *There is a recursive function not in $\mathcal{PR}$.*

---

[6] This is the procedural form of the existential statement (ii) in the proof of Theorem 8.3.6, although in said proof we used a *variant* of the coding Definition 2.4.2 used here.

[7] The "there is" is furnished by the recursion theorem.

Proof Let $\lambda x.f(x) \in \mathcal{PR}$. Then so is $\lambda x.f(x) + 1$. By Theorem 10.3.1, for some $i \in Pri$ such that $(i)_1 = 1$, we have

$$f(x) + 1 = Eval(i, \langle x \rangle), \text{ for all } x \qquad (1)$$

Thus,

$$f(x) < Eval(i, \langle x \rangle), \text{ for some } i \text{ and all } x \qquad (2)$$

Now, if $Eval \in \mathcal{PR}$ then so would be $\lambda x.Eval(x, \langle x \rangle)$ by Grzegorczyk operations. Then, by (2),

$$Eval(x, \langle x \rangle) < Eval(j, \langle x \rangle), \text{ for some } j \text{ and all } x$$

in particular, $Eval(j, \langle j \rangle) < Eval(j, \langle j \rangle)$. A contradiction.

So $Eval$ is an example of a recursive function not in $\mathcal{PR}$.   □

The above proof was a majorisation argument in the style of Corollary 4.2.3. This chapter is about the technique in obtaining $Eval$, rather than about the result Theorem 10.3.4 itself, since the latter we have obtained already twice (Chaps. 3 and 4).

10.3.5 Exercise. Redo the proof of Theorem 10.3.4 by a pure diagonalisation without a majorisation component.   □

10.4   Exercises

1. Complete the missing details from the proof of Theorem 10.1.3.
2. Elaborate the proof of Theorem 10.2.1.
3. Prove Corollary 10.2.3.
4. Can we use the primitive recursive recursion theorem in Remark 10.3.2 rather than Theorem 9.2.1? Why?

Chapter 11
Enumerations of Recursive and Semi-Recursive Sets

Overview

We saw that the semi-recursive sets are computably enumerable and vice versa (Sect. 6.4). In this chapter we will look more closely at enumerations of semi-recursive —especially the special cases of recursive and finite— sets.

11.1   Enumerations of Semi-Recursive Sets

Suppose $A = \{x : W_x \in \mathcal{C}\}$ where $\mathcal{C}$ is some set of semi-recursive subsets of $\mathbb{N}$. Suppose also that A is semi-recursive, and therefore $A = \mathrm{ran}(f)$, where $f \in \mathcal{P}$. Then this $\mathcal{C}$ is enumerated by A —or, equivalently, by f— since

$$W_x \in \mathcal{C} \text{ iff } x \in A \tag{1}$$

and since $A = \mathrm{ran}(f)$, we can write (1) as

$$W_x \in \mathcal{C} \text{ iff } x \in \mathrm{ran}(f) \tag{2}$$

Since all sets in $\mathcal{C}$ are semi-recursive, each $X \in \mathcal{C}$ is a W_x and that x is in A. Note that every such X is a W_x for infinitely many x (cf. Corollary 9.4.2), each x being in A by the immediately previous sentence.

Not every class of semi-recursive sets can be so enumerated as the two theorems below and the specific examples that follow them indicate.

11.1.1 Theorem. *Let $A = \{x : W_x \in \mathcal{C}\}$ where $\mathcal{C}$ is some set of semi-recursive subsets of $\mathbb{N}$. If some semi-recursive sets C and D are such that $\mathcal{C} \ni C \subseteq D \notin \mathcal{C}$, then $\overline{K} \leq A$ hence A is not semi-recursive.*

© Springer Nature Switzerland AG 2022
G. Tourlakis, *Computability*, https://doi.org/10.1007/978-3-030-83202-5_11

Proof This is actually a direct consequence of Theorem 6.10.3.

Let C' denote $\{\psi \in \mathcal{P} : \text{dom}(\psi) \in \mathcal{C}\}$ —recall that $W_x = \text{dom}(\phi_x)$ by definition. Thus, $A = \{x : \phi_x \in C'\}$. Let next

$$\psi = \lambda x.\begin{cases} 0 & \text{if } x \in C \\ \uparrow & \text{othw} \end{cases}$$

and

$$\tau = \lambda x.\begin{cases} 0 & \text{if } x \in D \\ \uparrow & \text{othw} \end{cases}$$

By Theorem 6.8.1, ψ and τ are in $\mathcal{P}$ and, trivially, $\text{dom}(\psi) = C$ while $\text{dom}(\tau) = D$. By assumption on C and D we have $C' \ni \psi \subseteq \tau \notin C'$. Thus, by Theorem 6.10.3 $\overline{K} \leq A$ and therefore A is not semi-recursive. □

11.1.2 Remark The contrapositive formulation of the above is useful in a proof by contradiction that a set

$$A = \{x : W_x \in \mathcal{C}\} \tag{1}$$

is not c.e.

Given A by (1). If it is semi-recursive, then for any c.e. sets C and D such that $\mathcal{C} \ni C \subseteq D$ we must have $D \in \mathcal{C}$. □

11.1.3 Theorem. *Let* $A = \{x : W_x \in \mathcal{C}\}$ *where* $\mathcal{C}$ *is some set of semi-recursive subsets of* $\mathbb{N}$. *Suppose that, for some* $C \in \mathcal{C}$, *the condition* (1) *below holds.*

$$\text{for all finite sets } D, \text{ if } D \subseteq C, \text{ then } D \notin \mathcal{C} \tag{1}$$

Then $\overline{K} \leq A$ *and hence* A *is not semi-recursive.*

Proof This is actually a direct consequence of Theorem 6.10.1: Indeed let once more $C' \overset{Def}{=} \{\psi \in \mathcal{P} : \text{dom}(\psi) \in \mathcal{C}\}$. Hence

$$A = \{x : \phi_x \in C'\} \tag{2}$$

Now, $C = \text{dom}(f)$ for some $f \in \mathcal{P}$, thus $f \in C'$. Without loss of generality, we take f such that $f(x) \simeq 0 \equiv f(x) \downarrow$ —for if f does not have this property we simply use $\lambda x.0 \cdot f(x)$ instead of f.

Pick now any finite $D \subseteq C$. By (1), $D \notin \mathcal{C}$. Define

$$\xi = \lambda x.\begin{cases} 0 & \text{if } x \in D \\ \uparrow & \text{othw} \end{cases}$$

Since $\mathrm{dom}(\xi) = D$ by definition, we have $\mathrm{dom}(\xi) \notin C$ by (1), hence $\xi \notin C'$ by definition of C'. But $\xi \subseteq f \in C'$. Invoking Theorem 6.10.1 for the set in (2) we have $\overline{K} \leq A$. □

11.1.4 Remark Once again, we note the contrapositive formulation of the above. If a set

$$A = \{x : W_x \in C\} \tag{1}$$

is c.e., then every $C \in \mathcal{C}$ must have some finite subset D that is in $\mathcal{C}$ as well. □

11.1.5 Example.

1. The set $A = \{x : W_x \text{ is infinite}\}$ is not r.e. (equivalently, semi-recursive). If $\mathcal{C}$ is the set of all infinite r.e. sets, then for each $C \in \mathcal{C}$, if $D \subseteq C$ is finite, then it is not in $\mathcal{C}$. Now invoke Theorem 11.1.3.
2. As an immediate consequence of 1, we have that set $A = \{x : W_x = \mathbb{N}\}$ is not r.e.
3. The set $A = \{x : W_x \text{ is finite}\}$ is not r.e. If $\mathcal{C}$ is the set of all finite r.e. sets, then pick any $C \in \mathcal{C}$. Now, $\mathbb{N}$ is r.e. but infinite. So it is not in $\mathcal{C}$. But $C \subseteq \mathbb{N}$. Now invoke Theorem 11.1.1.
4. The set $A = \{x : W_x = \emptyset\}$ is not r.e. This is a special case of item 3.
5. The set $A = \{x : W_x \in \mathcal{R}_*\}$ is not r.e. Note that $\emptyset \in \mathcal{C} = \mathcal{R}_*$. Now, $\emptyset \subseteq K$ (K *is* a "W_e") but $K \notin \mathcal{R}_*$. Now invoke Theorem 11.1.1. □

11.2 Enumerations of Recursive Sets

A recursive set $A \neq \emptyset$ being also semi-recursive has an enumeration given by a recursive function $f : A = \mathrm{ran}(f)$ (cf. 6.4.2). We can do much better than this result that does not use the fact that A is recursive.

11.2.1 Theorem. *A recursive set $A \neq \emptyset$ has an* increasing *enumerating $f \in \mathcal{R}$, that is, $\mathrm{ran}(f) = A$ and, for all x and y, $x \leq y$ implies $f(x) \leq f(y)$.*

If moreover A is infinite, then there is a $g \in \mathcal{R}$ that is strictly increasing —for all x and y, $x < y$ implies $f(x) < f(y)$— and $\mathrm{ran}(g) = A$.

Proof We have two cases:

- A is *finite*. Say $A = \{a_0, a_1, \ldots, a_n\}$, where $a_0 < a_1 < \ldots < a_n$. Define

$$f(x) = \begin{cases} a_0 & \text{if } x = 0 \\ a_1 & \text{if } x = 1 \\ \vdots & \vdots \\ a_n & \text{if } x \geq n \end{cases} \tag{1}$$

f is given by definition by primitive recursive cases hence $f \in \mathcal{PR} \subsetneq \mathcal{R}$. Moreover, by inspection of ran$(f) = A$ and

$$f(0) < f(1) < \ldots < f(n) = f(n+1) = f(n+2) = f(n+3) = \ldots$$

- A is *infinite*. Suppose A is given by (a program for) its characteristic function c_A.[1] Define g by primitive recursion

$$g(0) = (\mu y)c_A(y)$$
$$g(x+1) = (\mu y)(y > g(x) \wedge c_A(y) = 0) \tag{2}$$

What (2) does is —after having chosen the smallest element of A for $g(0)$— that at stage $x+1$ it picks the smallest element that exceeds all previous picks and assigns it as the value of $g(x+1)$.

Since A is infinite, a $y > g(x)$ always exists, thus we have $g(x+1) \downarrow$ for all x. Also $g(0) \downarrow$ as $A \neq \emptyset$. Coupling this with the fact that $g \in \mathcal{P}$ as the latter class is closed under primitive recursion, we have that $g \in \mathcal{R}$. The second equation of (2) also entails that $g(x+1) > g(x)$, for all x, thus establishing the strictly increasing property of g.

We claim that ran$(g) = A$. Say $a \in A - \text{ran}(g)$. As g is strictly increasing and $g(0)$ is the smallest element of A, a will satisfy

$$f(x) < a < g(x+1) \tag{3}$$

for some $x \geq 0$. Let a be the smallest in the range indicated in (3). Then (3) contradicts the second equation of (2) as a was *not* picked as the value for $g(x+1)$.

$\square$

The following is kind of a converse:

11.2.2 Theorem. *If $A \subseteq \mathbb{N}$ has an* increasing *enumerating function $f \in \mathcal{R}$, then $A \in \mathcal{R}_*$.*

Proof As $f \in \mathcal{R}$, $A \neq \emptyset$. We have two cases

- A is finite. Then it is recursive, indeed primitive recursive.
- A is infinite. Then f is unbounded, that is, for any $n \in \mathbb{N}$ there is an x such that $n \leq f(x)$. Thus we can decide the problem $n \in A$:
  Just generate the sequence of inputs (of f) $x = 0, 1, 2, \ldots$ *until the first time* that we detect that $n \leq f(x)$ is true. Once this is known then $n \in A$ if "$\neq$" is resolved as "=", while we have $n \notin A$ if "$\neq$" is resolved as "<".
  In short,

---

[1] A recursive set can be "given" by a decider that computes its characteristic function, or by a verifier. See more on finite descriptions of recursive sets in the next subsection.

$$n \in A \equiv f\Big((\mu x)n \le f(x)\Big) = n$$

As the right hand side of $\equiv$ is in $\mathcal{R}_*$, then so is $n \in A$.

$\square$

By Corollary 6.4.4 every c.e. set is the range of an $f \in \mathcal{P}$ while every nonempty c.e. set can be enumerated by a $g \in \mathcal{R}$ (Theorem 6.4.2). Can we extract (algorithmically) *nontrivial recursive* subsets of c.e. sets given as $\mathrm{ran}(\phi_x)$ or — equivalently (cf. Theorems 6.9.2 and 6.9.1)— as $\mathrm{dom}(\phi_x)$, that is, W_x? The qualifier "nontrivial" aims to avoid unhelpful answers like "Ø is such a recursive subset".

11.2.3 Theorem. *Suppose $A = \mathrm{ran}(\phi_x)$ is infinite. Then there is an $h \in \mathcal{R}$, such that*

1. *$\mathrm{ran}(\phi_{h(x)}) \subseteq A$*
2. *If ϕ_x is recursive, then $\phi_{h(x)}$ is strictly increasing, hence $\mathrm{ran}(\phi_{h(x)})$ is an infinite recursive subset of A.*

Proof The idea is to choose larger and larger output values of ϕ_x as the values for the subsequence $\phi_{h(x)}$. It will help the intuition in understanding the definition below to *think of ϕ_x as a recursive function* —which in general it is not!

$\lambda z.\psi(x, z)$ below will end up —via S-m-n— being the sought $\phi_{h(x)}$.

So, define ψ by primitive recursion

$$\begin{aligned} \psi(x, 0) &= \phi_x(0) \\ \psi(x, z + 1) &= \phi_x\Big((\mu y)(\phi_x(y) > \psi(x, z))\Big) \end{aligned} \tag{1}$$

If ϕ_x is recursive for some x, then the definition (1) builds a $\lambda z.\psi(x, z)$ with infinite range since the set A is unbounded and hence, for any z, we can find a y such that satisfies $\phi_x(y) > \psi(x, z)$.

Of course, ϕ_x is *not* recursive *in general*, nor can we test algorithmically to learn for which values of x it is (cf. Example 6.3.8).

The task is then twofold:

We must settle that:

- Definition (1) makes sense for all values of the arguments (x, z), and also that $\psi \in \mathcal{P}$.

Now, the iterator function in the second equation of (1) is

$$\lambda xzw.\phi_x\Big((\mu y)(\phi_x(y) > w)\Big)$$

The relation of three arguments

$$\phi_x(y) > w \tag{2}$$

is semi-recursive[2] but not recursive. Thus we cannot compute the minimum y that satisfies the relation for any given x, w (cf. 7.0.4).

The idea is to use a different condition than (2) that allows a proof that $\psi \in \mathcal{P}$ easily, and which condition is equivalent to (2) when ϕ_x is total.

Thus we will do the following: Note that if $\phi_x(y) \downarrow$, then

$$\phi_x(y) > w \equiv w + 1 \dotdiv \phi_x(y) \simeq 0 \tag{3}$$

So let us work with "$w + 1 \dotdiv \phi_x(y) \simeq 0$" rather than "$\phi_x(y) > w$" in the definition of ψ.

Now, $g = \lambda xyw.w + 1 \dotdiv \phi_x(y) \in \mathcal{P}$ and so is $\widetilde{g} = \lambda xw.(\mu y)g(x, y, w)$.

 Note that even if, for some x and w, $\min\{y : g(x, y, w) = 0\}$ exists, $\widetilde{g}(x, w)$ might fail to attain it (due to $\widetilde{g}(x, w) \uparrow$) if the condition "$(\forall z)_{<y}g(x, z, w) \downarrow$" of the definition of (μy) does not hold.

Let us now reformulate (1) more carefully as (1′) below:

$$\psi'(x, 0) = \phi_x(0)$$
$$\psi'(x, z + 1) = \phi_x\Big(\widetilde{g}(x, \psi'(x, z))\Big) \tag{1′}$$

By closure of $\mathcal{P}$ under primitive recursion, $\psi' \in \mathcal{P}$ and by S-m-n, there is an $h \in \mathcal{R}$ such that

$$\psi'(x, z) = \phi_{h(x)}(z), \text{ for all } x, z$$

- Firstly, the presence of "$\phi_x($" as a prefix in each equation of (1′) makes it clear that $\operatorname{ran}(\phi_{h(x)}) \subseteq \operatorname{ran}(\phi_x) = A$.
  This inclusion does not need the assumption that A is infinite.

- Secondly, if for some x we have $\phi_x \in \mathcal{R}$, then $\phi_{h(x)}$ has also property 2 stated in the theorem:
  Indeed, whenever $\phi_x \in \mathcal{R}$, it is the case that $\widetilde{g}(x, \psi'(x, z)) = \min y$ such that $\phi_x(y) > \psi'(x, z)$, because of (3).
  Thus —from (1′) and (3)— $\phi_{h(x)}(z + 1) = \phi_x\Big((\mu y)(\phi_x(y) > \phi_{h(x)}(z))\Big)$, that is, $\phi_{h(x)}(z + 1) > \phi_{h(x)}(z)$, for all z, and thus (Theorem 11.2.2) $\operatorname{ran}(\phi_{h(x)}) \in \mathcal{R}_*$ for such x.

  □

11.2.4 Corollary. *There is a $\sigma \in \mathcal{R}$ such that $W_{\sigma(x)} \subseteq W_x$ for all x, and whenever W_x is infinite $W_{\sigma(x)}$ is recursive and infinite.*

---

[2] $\phi_x(y) > w \equiv (\exists u)(\phi_x(y) = u \wedge u > w)$.

Proof By Theorem 6.9.1, there is a τ such that $W_x = \mathrm{ran}(\phi_{\tau(x)})$ and $W_x \neq \emptyset$ implies that $\phi_{\tau(x)} \in \mathcal{R}$. Now rerun the above proof using $\phi_{\tau(x)}$ rather than ϕ_x as the starting point to obtain the counterpart of $\phi_{h(x)}$ of the theorem; here it will be $\phi_{h(\tau(x))}$.

Thus, $\mathrm{ran}\big(\phi_{h(\tau(x))}\big) \subseteq W_x$, and if W_x is infinite, then $\mathrm{ran}\big(\phi_{h(\tau(x))}\big)$ is recursive and infinite.

Invoking Theorem 6.9.2 to obtain an $f \in \mathcal{R}$ such that, for all x, $\mathrm{ran}(\phi_x) = W_{f(x)}$, we set $\sigma = \lambda x. f(h(\tau(x)))$ and rephrase the previous sentence in terms of W_i notation. $\qquad\square$

11.2.5 Corollary. *There is an $h \in \mathcal{R}$ such that, for all x,*

1. $\mathrm{ran}(\phi_{h(x)}) \subseteq \mathrm{ran}(\phi_x)$.
2. $\mathrm{ran}(\phi_{h(x)}) \in \mathcal{R}_*$.
3. *If ϕ_x is defined and strictly increasing on the segment $S = \{0, 1, \ldots, u\}$, then $\phi_{h(x)}(z) = \phi_x(z)$, for all $z \in S$.*
4. *If ϕ_x is total and strictly increasing, then $\phi_{h(x)} = \phi_x$.*

Proof The conclusions 1–3 are extracted from the proof of Theorem 11.2.3. Consider $(1')$ (ibid) from which we obtained $\phi_{h(x)}$.

Immediately, 1 here is the 1 in the cited theorem.

For 2, if $\mathrm{ran}(\phi_{h(x)})$ is *finite*, then we are done.

Else $\mathrm{ran}(\phi_{h(x)})$ is *infinite*.

Firstly, this implies it is total. Indeed, suppose $\phi_{h(x)}(z + 1) \downarrow$ —this is true for infinitely many z. Since

$$\phi_{h(x)}(z + 1) = \phi_x\Big((\mu y)\big(\phi_{h(x)}(z) + 1 \dot{-} \phi_x(y)\big)\Big) \qquad (*)$$

by $(1')$ (second equation) in the proof of Theorem 11.2.3, we see that the case $\phi_{h(x)}(z) \uparrow$ would contradict assumption, hence we also have $\phi_{h(x)}(z) \downarrow$. The z being arbitrary, we conclude as claimed.

Secondly, given that both $\phi_{h(x)}$ and $\phi_{h(x)}(z + 1)$ are defined —and the latter's value is $\phi_x(y)$ where y satisfies $\phi_x(y) > \phi_{h(x)}$— we conclude that $\phi_{h(x)}$ is strictly increasing. Thus $\mathrm{ran}(\phi_{h(x)})$ is recursive.

Regarding 3: Let $\lambda y. \phi_x(y) \downarrow$ for $y = 0, 1, 2, \ldots, u$, and also strictly increasing on this initial segment of its domain.

Take the I.H. that for a fixed $z < u$ we have $\phi_{h(x)}(z) \downarrow$ and $\phi_{h(x)}(z) = \phi_x(z)$. For the I.S. at input $z + 1 \leq u$, we look at equation $(*)$. We have

- (I.H.) $\phi_{h(x)}(z) = \phi_x(z)$.
- $\big(\phi_{h(x)}(z) + 1 \dot{-} \phi_x(v)\big) \downarrow$ for all $v \leq z$, and hence $(\exists y)\big(\phi_{h(x)}(z) + 1 \dot{-} \phi_x(y) \simeq 0\big)$ is true, since $\phi_x(z + 1) > \phi_x(z) = \phi_{h(x)}(z)$. Thus the unbounded search $(\mu y)(\ldots)$ in $(*)$ will pick $y = z + 1$, and $(*)$ will now say: $\phi_{h(x)}(z + 1) = \phi_x(z + 1)$.

Clearly, 4 is a trivial consequence of 3. $\qquad\square$

11.2.1 Finite Descriptions of Recursive Sets

How is a recursive set "*given*"? This means, of course, *given* by a *finite* description. Well, one way is if a decider —equivalently, a ϕ-index of the characteristic function of— the recursive set is given. Or a verifier is given, or perhaps an enumerator (that is, a ϕ_x, is given as "x", that has said recursive set as its range).

The enumerator and verifier descriptions being equivalent (cf. Theorems 6.9.1 and 6.9.2) we essentially have two of the three mentioned ways.

Can we go algorithmically from one type of finite description of a recursive set to the other? Yes and no.

11.2.6 Proposition. *There is an $h \in \mathcal{R}$ such that if ϕ_x is the characteristic function of a recursive set, then $h(x)$ is a semi-index (a verifier) of that set:* $W_{h(x)} = \{y : \phi_x(y) = 0\}$.

Proof The intuition for this is straightforward: If f is recursive and is the characteristic function of a recursive set A, then let us focus one one decider, $\phi_x = f$. We can modify this decider to loop forever for an input y instead of returning 1 ($y \notin A$): The modified program is

$$\text{if } \phi_x(y) = 0 \text{ then return } 0$$

$$\text{else get into an infinite loop}$$

The above readily translates to

$$\psi(x, y) = \begin{cases} 0 & \text{if } \phi_x(y) = 0 \\ \uparrow & \text{othw} \end{cases}$$

Above we have a definition by positive cases, hence $\psi \in \mathcal{P}$. By S-m-n we get an $h \in \mathcal{R}$ such that $\psi(x, y) = \phi_{h(x)}(y)$, for all x, y. Thus, if ϕ_x is a characteristic function of A, then $A = \{y : \phi_x(y) = 0\} = \text{dom}(\phi_{h(x)}) = W_{h(x)}$. $\square$

11.2.7 Definition. A *characteristic index* is a ϕ-index of some recursive characteristic function. If ϕ_x is the characteristic function of a set A, then x is one of the (infinitely many) characteristic indices *of the set* A. $\square$

Thus the proposition above is the positive result: We can computationally go (h does the computation) from a characteristic index of a set to a semi-index of the set.

The negative result is that we cannot go from a semi-index of a recursive set to a characteristic index. Thus, a recursive set can be finitely "given", or can be "named" by a characteristic index or by a semi-index. We cannot computably go from one to the other in both directions.

11.2.8 Proposition. *There is* no $f \in \mathcal{P}$ *such that whenever W_x is recursive, we have $f(x) \downarrow$ and $c_{W_x} = \phi_{f(x)}$.*

Proof In 6.3.6, display (3′), we have an $h \in \mathcal{R}$ such that

$$\phi_{h(x)} = \begin{cases} \lambda y.0 & \text{if } x \in K \\ \emptyset & \text{if } x \in \overline{K} \end{cases}$$

or, in W_i notation,

$$W_{h(x)} = \begin{cases} \mathbb{N} & \text{if } x \in K \\ \emptyset & \text{if } x \in \overline{K} \end{cases} \tag{1}$$

Both alternatives for $W_{h(x)}$ in (1) are recursive, so, if we *had* an f as described here, then $\lambda x. f(h(x)) \in \mathcal{R}$ and

$$1 \in \mathrm{ran}\big(\phi_{f(h(x))}\big) \equiv x \in \overline{K} \tag{2}$$

because in the top case of (1) we have $1 \notin \mathrm{ran}\big(\phi_{f(h(x))}\big)$. Expanding (2) we get a contradiction to the non semi-recursiveness of $\overline{K}$, as the left hand side of $\equiv$ is

$$(\exists y)(\exists z)(T(f(h(x)), y, z) \land out(z, f(h(x)), y) = 1)$$

which is semi-recursive by closure properties of $\mathcal{P}_*$.   □

11.2.2   Finite Descriptions of Finite Sets

Finite sets are special cases of recursive sets, so they can be finitely described by characteristic indices or semi-indices. We can also directly code the members of the finite set by a single number, obtaining a *third kind* of finite representation.

I say "kind of" since there are several ways to code n numbers into a single number. One is to code $D = \{a_1, a_2, \ldots, a_n\}$ by a number z whose prime number factorisation contains the prime p_m (with exponent equal to 1) iff m is in D. To avoid ambiguities, D is given without repetitions.

So D is coded as

$$p_{a_1} p_{a_2} p_{a_3} \cdots p_{a_n}$$

Thus, $\{0, 3\}$ is coded as $p_0 p_3 = 2 \cdot 7 = 14$. Of course, permuting the elements of D does not change the code.

If $D = \emptyset$, then we assign the code 1 in the above coding scheme.

Another way, perhaps the most prominent in the literature, e.g., in Rice (1956), Kreider and Ritchie (1965), Rogers (1967), is to code D (once again, listed without repetitions) by

$$w = 2^{a_1} + 2^{a_2} + 2^{a_3} + \cdots + 2^{a_n}$$

The above is a code that is also insensitive to permutations (of the set, hence of the summands 2^{a_i}). When w is expressed in binary notation it will have a 1 precisely in the locations a_i.

11.2.9 Remark.

(1) If $D = \varnothing$, then we assign the code 0 in the above coding scheme.
(2) Whether we use primes or binary representation we will not say most of the time, and this will only occur a few times in proofs. Assuming the coding is fixed, then the notation D_x is almost standard, and denotes *the finite set of code x.*
(3) Not all integers are prime factorisation codes of finite sets as the lack of exponents makes clear. We can primitive recursively check a number z for being such a code or not.

$$cancode(z) \equiv z = 1 \vee z > 1 \wedge (\forall x)_{\leq z}(Pr(x) \wedge x|z \to \neg x^2|z)$$

□

11.2.10 Definition. (Canonical Index of a Finite Set) Assume that we have chosen a particular naming (coding) scheme for finite sets A. We will call *canonical index* of A any x that so codes the elements of A according to our scheme. We will indicate this fact by writing $A = D_x$. In general, "D_x" denotes a finite set of canonical index x. □

It is easy to see that if we have the canonical index x for a finite set A, then we can compute a *characteristic index*, and hence also (cf. Proposition 11.2.6) a semi-index for the set.

11.2.11 Proposition. *There is a $\tau \in \mathcal{R}$ such that if D_x is given via a canonical index x, then $\phi_{\tau(x)} \downarrow$ and denotes a characteristic index of D_x.*

Proof We assume prime factorisation coding for convenience to avoid invoking Church's Thesis. We define

$$\psi(x, y) = \begin{cases} 0 & \text{if } p_y|x \\ 1 & \text{othw} \end{cases} \tag{1}$$

$\psi \in \mathcal{PR} \subseteq \mathcal{P}$ and S-m-n provides the τ we want: If x is a canonical (prime factorisation) index of some set, then by (1) $\phi_{\tau(x)}$ is the characteristic function of D_x and thus $\tau(x)$ is a characteristic index of the set. □

11.2.12 Corollary. *There is a $\sigma \in \mathcal{R}$ such that if D_x is given by a canonical index x, then $D_x = W_{\sigma(x)}$.*

Proof Apply the above, and then invoke Proposition 11.2.6. $\sigma = \lambda x.h(\tau(x))$. □

11.2.13 Proposition. *There is no* $f \in \mathcal{P}$ *such that if* x *is a characteristic index of a finite set, then* $f(x) \downarrow$ *and* $f(x)$ *is the cardinality of* D_x.

Proof The proof is in the style of proof that we gave in Example 6.7.2.

Our URMs compute forever after they actually "stop" (reach the **stop** instruction). This will not help here, thus we modify our Kleene predicate for the purposes of the present proof, so that $T(x, x, y)$ can be true, given x, for at most *one* value of y. Thus we first define a "new T", "$\widetilde{T}$", by

$$\widetilde{T}(x, w, y) \equiv T(x, w, y) \wedge (\forall z)_{<y} \neg T(x, w, z) \tag{1}$$

Clearly, if $\widetilde{T}(x, w, y)$ is true, then $\widetilde{T}(x, w, u)$ is false if $y < u$, since otherwise we would have $T(x, w, y)$ false by (1), and therefore also $\widetilde{T}(x, w, y)$ would be false. On the other hand, if $\widetilde{T}(x, w, y)$ is true, (1) immediately implies that no u below y satisfies $\widetilde{T}(x, w, u)$ either. Now we define ψ:

$$\psi(x, y) = \begin{cases} 1 & \text{if } \neg\widetilde{T}(x, x, y) \\ 0 & \text{othw} \end{cases} \tag{1}$$

This ψ is in $\mathcal{PR}$ thus, by S-m-n, we have some $h \in \mathcal{R}$ so that (1) becomes

$$\phi_{h(x)}(y) = \begin{cases} 1 & \text{if } \neg\widetilde{T}(x, x, y) \\ 0 & \text{othw} \end{cases} \tag{1'}$$

Now as in Example 6.7.2 we analyse two cases:

(a) $\phi_x(x) \uparrow$. Then the top case of (1') is true no matter what y. Thus $\phi_{h(x)}(y) = 1$ for all y, that is

$$\phi_x(x) \uparrow \Longrightarrow {}^3\phi_{h(x)} = \lambda y.1 \tag{2}$$

(b) $\phi_x(x) \downarrow$. Then for some *unique* value of y, say y_0, $\widetilde{T}(x, x, y_0)$ is true. Thus, for all other values of y, $\neg\widetilde{T}(x, x, y)$ is true. So the bottom case is true for $y = y_0$ and we have that

$$\phi_x(x) \downarrow \Longrightarrow \phi_{h(x)} \text{ outputs: } \overbrace{1, 1, \ldots, 1,}^{\text{all 1}} \overset{y = y_0}{\underset{\downarrow}{0}} , \underbrace{1, 1, \ldots}_{\text{all 1}} \tag{3}$$

---

[3] This is our informal "implies". Used here since it is more easily visible when the writing around it is rather dense.

We have concluded that $\phi_{h(x)}$ is the characteristic function of a finite set at all times (i.e., for all x): Of the set $\emptyset$ if $x \in \overline{K}$, and of a singleton —that is, $\{y_0\}$— if $x \in K$.

Thus, if f existed and worked as hoped, $f(h(x))$ would be 0 if $x \in \overline{K}$ (cardinality of $\emptyset$) and 1 if $x \in K$ (which happens to be the cardinality of $\{y_0\}$). That is, $f(h(x)) = 0 \equiv x \in \overline{K}$. But $fh \in \mathcal{R}$. Contradiction. $\square$

11.2.14 Corollary. *There is no* $g \in \mathcal{P}$ *such that if* x *is a characteristic index of a finite set* $S = \phi_x^{-1}[\{0\}]$, *then* $g(x) \downarrow$ *and* $S = D_{g(x)}$.

Proof Working with the same ψ as in the proof above, we have $g(h(x)) \downarrow$ for all x and thus $\lambda x.g(h(x)) \in \mathcal{R}$. Coding finite sets using prime factorisation —and assuming that the corollary statement is false— we thus have $x \in \overline{K} \equiv g(h(x)) = 1$ (the prime-power code of $\emptyset$). But $g(h(x)) = 1 \in \mathcal{R}_*$. Contradiction. $\square$

 If g —let's call it g' to allow for this change of viewpoint— is supposed to work using coding via binary expansions instead, then we still get the contradiction: $x \in \overline{K} \equiv g'(h(x)) = 0$ (the binary-expansion code of $\emptyset$) since $g'(h(x)) = 0 \in \mathcal{R}_*$.

11.2.15 Corollary. *There is no* $\tau \in \mathcal{P}$ *such that if* x *is a semi-index of a finite set, then* $\tau(x) \downarrow$ *and* $W_x = D_{\tau(x)}$.

Proof If such τ exists, then we can obtain a g that works for Corollary 11.2.14 hence obtain a contradiction.

Indeed, let h be *as in Proposition 11.2.6* (*not* as in the proof above). Thus given a characteristic index x of a finite set S we obtain a semi-index $S = W_{h(x)}$.

Thus if τ worked with semi-indices, contrary to the claim of this corollary, then we would have $\tau h(x) \downarrow$ and $S = D_{\tau h(x)}$, contradicting Corollary 11.2.14. $\square$

11.3   Enumerations of Sets of Semi-Recursive Sets

In set theory or logic, or mathematics in general, once we have shown that a set S is countable then we have shown, specifically, that an onto function $f : \mathbb{N} \to S$ exists. Moreover mathematics (logic, set theory) do not care very much about the nature of the members of S.

In computability we want more. We want our enumerations to be *computable* functions. We also have a constraint: *Our functions must output natural numbers.* Thus, we want to have numerical "names" (finite descriptions) of the objects we are enumerating via computable functions.

We have had quite a bit of practice in this chapter with *names* or *finite descriptions of sets*. We now want to make a few comments, and conclude with the very interesting Rice-Myhill-Shapiro-MacNaughton theorem, and also the unexpected answer on the possibility to enumerate all recursive sets (one answer we have seen in Example 11.1.5, but there is another answer as well! See below.)

Suppose C is a set (or class) of semi-recursive sets. Then this can be enumerated, in principle, using semi-indices as finite descriptions of the sets in C, or use, alternatively, ϕ-indices of *enumerator*(s) of each set in the class.

Recursive sets being semi-recursive can be enumerated, in principle, using the naming mentioned above, but also can be enumerated by enumerating ϕ-indices of their characteristic functions.

Finite sets being recursive, admit the same methodologies of enumeration as recursive sets do, but also can be enumerated using as names their canonical indices.

We have the following definition that aims to capture the types of the various classes C that we care to enumerate their contents.

11.3.1 Definition. (Rice)

1. A *recursively/computably enumerable* —or r.e./c.e.— *class* C of *semi-recursive* sets W_x is one for which there is a c.e. set I of indices, such that

$$A \in C \equiv (\exists x)(x \in I \land W_x = A) \tag{1}$$

2. A *characteristically enumerable class* C of *recursive* sets is one for which there is a c.e. set I of indices, such that

   a. $x \in I$ implies that $\phi_x \in \mathcal{R}$ and $\operatorname{ran}(\phi_x) \subseteq \{0, 1\}$.
   b. $A \in C \equiv (\exists x)(x \in I \land \phi_x^{-1}[\{0\}] = A)$.

3. A *canonically enumerable class* C of *finite* sets is one for which there is a c.e. set I of indices, such that $A \in C \equiv (\exists x)(x \in I \land D_x = A)$.

4. A *completely recursively/computably enumerable* —or *completely* r.e./c.e.— *class* C of *semi-recursive* sets W_x is one for which there is *a c.e. set* I of indices, such that

$$W_x \in C \equiv x \in I \tag{2}$$

□

11.3.2 Remark.

(a) Definition 1 above can be reformulated as follows. There is an $f \in \mathcal{R}$ such that

$$A \in C \equiv (\exists x)(W_{f(x)} = A)$$

(b) Definition 2 above can be reformulated as follows. There is an $f \in \mathcal{R}$ such that

   a. $x \in \mathbb{N}$ implies that $\phi_{f(x)} \in \mathcal{R}$ and $\operatorname{ran}(\phi_{f(x)}) \subseteq \{0, 1\}$.
   b. $A \in C \equiv (\exists x)(\phi_{f(x)}^{-1}[\{0\}] = A)$.

(c) Definition 3 above can be reformulated as follows. There is an $f \in \mathcal{R}$ such that

$$A \in \mathcal{C} \equiv (\exists x)(D_{f(x)} = A)$$

(d) How are Definitions 1 and 4 in Definition 11.3.1 different? The latter is directly related to our familiar complete index set concept (Definition 6.3.5). Indeed (2) can be rephrased as $I = \{x : W_x \in \mathcal{C}\}$. *All* indices of sets in $\mathcal{C}$ are found in I. On the other hand, Definition 1 only requires *some* x such that $W_x \in \mathcal{C}$ to be in I.

(e) We saw in Example 11.1.5 that $\mathcal{R}_*$ is not a completely c.e. class because $\{x : W_x \in \mathcal{R}_*\}$ is not semi-recursive. See however the Theorem 11.3.4 below!

(f) The class of *all* W_x sets is a completely c.e. class since the the index set here is $I = \mathbb{N}$. W_x is r.e. iff $x \in \mathbb{N}$. $\qquad\qquad\square$

11.3.3 Example. An r.e. class that is not characteristically enumerable.

The class of finite sets $\mathcal{C}$ that contains, for each $x \in \mathbb{N}$, the finite set D below

$$D = \begin{cases} \{x\} & \text{if } x \in \overline{K} \\ \{x, x+1\} & \text{if } x \in K \end{cases}$$

is an r.e. class. Indeed, define f by

$$f(0, x) = x$$

$$f(y+1, x) = x + \begin{cases} 1 & \text{if } \phi_x(x) \downarrow \\ \uparrow & \text{othw} \end{cases} \tag{1}$$

The above is, formally, a primitive recursion hence $f \in \mathcal{P}$ (Theorem 6.8.1 is used in the second equation). As there is no recursive call to f in (1), the definition can easily be recast as a definition by cases: one for $y = 0$ and another one for $y > 0$. But we need not do that.

By S-m-n there is an $h \in \mathcal{R}$ such that

$$\lambda y.f(y, x) = \phi_{h(x)}, \text{ for all } x$$

Hence our class of finite sets $\mathcal{D}$ includes all the $\operatorname{ran}(\phi_{h(x)})$ below (that is, for all $x \geq 0$)

$$\operatorname{ran}(\phi_{h(x)}) = \begin{cases} \{x\} & \text{if } x \in \overline{K} \\ \{x, x+1\} & \text{if } x \in K \end{cases}$$

Using Theorem 6.9.2 we have a $\sigma \in \mathcal{R}$ such that

$$W_{\sigma h(x)} = \begin{cases} \{x\} & \text{if } x \in \overline{K} \\ \{x, x+1\} & \text{if } x \in K \end{cases} \tag{2}$$

σh effects the recursive enumeration of the W-versions of our finite sets (Remark cf. 11.3.2). Is $\mathscr{D}$ also characteristically enumerable?

If so, let $g \in \mathcal{R}$ be such that

- $x \in \mathbb{N}$ implies that $\phi_{g(x)} \in \mathcal{R}$ and $\mathrm{ran}(\phi_{f(x)}) \subseteq \{0, 1\}$.
- $D \in \mathscr{D} \equiv (\exists x)(\phi_{g(x)}^{-1}[\{0\}] = D)$.

Thus we have

$$x \in \overline{K} \equiv (\exists y)(\phi_{g(y)}(x) = 0 \wedge \phi_{g(y)}(x + 1) = 1) \tag{3}$$

Since $\phi_{g(y)}(z) = w \equiv (\exists u)(T(g(y), z, u) \wedge out(u, g(y), z) = w)$, (3) implies that $\overline{K}$ is r.e. $\qquad\square$

11.3.4 Theorem. $\mathcal{R}_*$ is an r.e. class of r.e. sets.

Proof We start by using (ϕ-indices of) enumerators as the *naming apparatus* of W_x sets. In what follows, $h \in \mathcal{R}$ is that of Corollary 11.2.5.

- We claim that $\{\mathrm{ran}(\phi_{h(x)}) : x \geq 0\}$ is the set of *all* recursive sets. By 2. in the quoted corollary,

$$\{\,\mathrm{ran}(\phi_{h(x)}) : x \geq 0\} \subseteq \mathcal{R}_*$$

We need the converse inclusion.

-

1. Let A be finite (special case of recursive). Assume $A = \{a_0, a_1, \ldots, a_n\}$ with $a_1 < a_2 < \ldots < a_n$. Then f below enumerates A in strict order on $\{0, 1, \ldots, n\}$:

$$f(x) = \begin{cases} a_0 & \text{if } x = 0 \\ a_1 & \text{if } x = 1 \\ a_2 & \text{if } x = 2 \\ \vdots & \quad\vdots \\ a_n & \text{if } x = n \\ \uparrow & \text{if } x > n \end{cases}$$

   $f = \phi_x$ for some x. By 3 in Corollary 11.2.5 $f = \phi_x = \phi_{h(x)}$. Thus $A \in \{\mathrm{ran}(\phi_{h(x)}) : x \geq 0\}$.
2. Let B be infinite. By Theorem 11.2.1 there is a strictly increasing recursive g such that $\mathrm{ran}(g) = B$. If we pick an x such that $g = \phi_x$, we also have $g = \phi_{h(x)}$ by 4. in Corollary 11.2.5. Thus $B \in \{\mathrm{ran}(\phi_{h(x)}) : x \geq 0\}$.

- Having established

$$\{\mathrm{ran}(\phi_{h(x)}) : x \geq 0\} = \mathcal{R}_* \tag{1}$$

we need only express the result in terms of W_i notation. To this end, Theorem 6.9.2 yields a $\sigma \in \mathcal{R}$ such that, for all $x \in \mathbb{N}$, we have $\mathrm{ran}(\phi_x) = W_{\sigma(x)}$. We can thus rewrite (1) as

$$\{W_{\sigma h(x)} : x \geq 0\} = \mathcal{R}_*$$

and we are done.

$\square$

The following is due to Rice (1956), and also Shapiro, McNaughton and Myhill.

11.3.5 Theorem. (Characterisation of Completely r.e. Classes of Sets) *A class $\mathcal{C}$ of semi-recursive sets W_x is completely r.e. iff there is a canonically enumerable class $\mathcal{D}$ of finite sets, such we have*

$$A \in \mathcal{C} \equiv (\exists D)(D \in \mathcal{D} \wedge D \subseteq A) \tag{1}$$

Rice (above reference) called the class $\mathcal{D}$ a *key array* of $\mathcal{C}$.

Proof By Definition 11.3.1, 4, $\mathcal{C}$ is completely r.e. means that $W_x \in \mathcal{C}$ is semi-recursive, or

$$\{x : W_x \in \mathcal{C}\}$$

is.

- *If* direction. So assume that we have (1). Let us rewrite it a bit more helpfully:

$$W_x \in \mathcal{C} \equiv (\exists y)(D_y \in \mathcal{D} \wedge D_y \subseteq W_x) \tag{1'}$$

Now $D_y \in \mathcal{D} \equiv (\exists x)(y = f(x))$, where $f \in \mathcal{R}$ enumerates the r.e. set I —i.e., $I = \mathrm{ran}(f)$— that indexes the finite sets of $\mathcal{D}$ according to Definition 11.3.1, 3. Thus, $D_y \in \mathcal{D}$ is semi-recursive. The same is true for $D_y \subseteq W_x$: Indeed,

$$D_y \subseteq W_x \equiv (\forall z)_{<y}(z \geq 1 \wedge p_z | y \rightarrow \phi_x(z) \downarrow)$$

or

$$D_y \subseteq W_x \equiv (\forall z)_{<y}(z \geq 1 \wedge p_z | y \rightarrow (\exists w)T(x, z, w)) \tag{2}$$

The right hand side of $\equiv$ in (2) is semi-recursive by closure properties of $\mathcal{P}_*$. These observations —and closure properties of $\mathcal{P}_*$ once more— rest the case via (1') that $W_x \in \mathcal{C}$ is semi-recursive.

- *Only if* direction. Now we are given that

$$\mathcal{P}_{\mathcal{C}} = \{x : W_x \in \mathcal{C}\} \in \mathcal{P}_* \tag{3}$$

We must prove the equivalence in (1).

To this end, let me first find a *canonically enumerable class* —called $\mathcal{D}$— of finite sets such that (1) holds. A good guess is to take as $\mathcal{D}$ the set of *all* finite sets in $\mathcal{C}$.

$$\mathcal{D} \stackrel{Def}{=} \{D : D \text{ finite} \land D \in \mathcal{C}\}$$

To see that $\mathcal{D}$ is canonically enumerable note that Corollary 11.2.12 furnishes a $\sigma \in \mathcal{R}$ such that —for all x that are prime factorisation codes— we have $D_x = W_{\sigma(x)}$. Thus,

$$\sigma(x) \in \mathcal{P}_{\mathcal{C}} \equiv W_{\sigma(x)} \in \mathcal{C}$$

hence the r.e. set $\sigma^{-1}[\mathcal{P}_{\mathcal{C}}]$ (Proposition 6.9.6) contains at least one canonical code for each $D \in \mathcal{C}$, but also some non-codes. Then the index set of canonical codes we want for $\mathcal{D}$ is

$$I = \sigma^{-1}[\mathcal{P}_{\mathcal{C}}] \cap \{z : cancode(z)\} \quad \text{(Remark cf. 11.2.9)}$$

an r.e. set by closure properties of $\mathcal{P}_*$.

Now we prove (1).

(a) $\rightarrow$ *direction*: Let $A \in \mathcal{C}$. By (3) and Theorem 11.1.3 there is a finite $\mathcal{C} \ni D \subseteq A$. Thus $D \in \mathcal{D}$. We showed $(\exists D)(D \in \mathcal{D} \land D \subseteq A)$, i.e., the right hand side of (1).

(b) $\leftarrow$ *direction*: Assume we have $(\exists D)(D \in \mathcal{D} \land D \subseteq A)$ (the right hand side of (1)). By definition of $\mathcal{D}$, $D \in \mathcal{C}$. By Theorem 11.1.1 and (3) it is $A \in \mathcal{C}$. $\square$

We can mimic the definitions of the various enumerations of classes of sets in the domain of sets of functions.

11.3.6 Definition. (Coding Finite Functions) First let us agree on a *coding* of *finite functions* that is an extension of the prime factorisation coding of finite sets.

We code a finite set of input/output pairs (a, b) —such as a finite function ξ is— by a number z iff its prime number factorisation includes p_a^{b+1} as a factor, for all (a, b). If $\xi = \emptyset$ its code is 1.

If u is the (prime power) code of some finite function, then we can name the function ξ_u —but also $[u]$— "ξ" or "θ" generically indicating a finite function. As some contexts may negate the latter convention, the notation $[u]$ is preferable.

The notation ξ_u is analogous to the notation D_u for a finite set of canonical index u.

We will call the code of a finite function its *canonical code*. $\square$

11.3.7 Proposition. *There is a $\sigma \in \mathcal{R}$ such that $\phi_{\sigma(z)} = [z]$.*

Proof We define

$$\psi(z, y) = \begin{cases} (z)_y & \text{if } z \geq 1 \wedge p_y | z \\ \uparrow & \text{othw} \end{cases}$$

$\psi \in \mathcal{P}$ and S-m-n provides the σ we want:

$$\phi_{\sigma(z)}(y) = \begin{cases} (z)_y & \text{if } z \geq 1 \wedge p_y | z \\ \uparrow & \text{othw} \end{cases} \tag{1}$$

If z is a canonical index of a finite function ξ, then by (1) $\phi_{\sigma(z)} = \xi$.   □

11.3.8 Corollary. *Every $z \geq 1$ determines a finite function ξ_z.*

Proof For $z = 1$, $\xi_z = \emptyset$. Otherwise we can retrieve all the input/output pairs of ξ. Indeed, $\xi_z(y) = \psi(z, y)$ from the proof of Proposition 11.3.7.   □

11.3.9 Definition. (Rice)

1. A *recursively/computably enumerable* —or r.e./c.e.— *class $\mathcal{C}$ of partial recursive functions* is one for which there is a c.e. set I of indices, such that

$$f \in \mathcal{C} \equiv (\exists x)(x \in I \wedge \phi_x = f) \tag{1}$$

Equivalently, there is a $g \in \mathcal{R}$ such that

$$f \in \mathcal{C} \equiv (\exists x)\phi_{g(x)} = f \tag{1'}$$

2. A *canonically enumerable class $\mathcal{C}$ of finite* functions is one for which there is a c.e. set I of indices, such that $\xi \in \mathcal{C} \equiv (\exists x)(x \in I \wedge [x] = \xi)$.
   Equivalently, $\mathcal{C}$ is a canonically enumerable class of *finite* functions iff for some $g \in \mathcal{R}$,

$$\xi \in \mathcal{C} \equiv (\exists x)[g(x)] = \xi$$

3. A *completely recursively/computably enumerable* —or *completely* r.e./c.e.— *class $\mathcal{C}$ of partial recursive* functions ϕ_x is one for which there is a c.e. set I of indices, such that

$$\phi_x \in \mathcal{C} \equiv x \in I \tag{2}$$

                    □

11.3.10 Theorem. (Characterising Completely r.e. Sets of Functions) Myhill and Shepherdson (1955), McNaughton, Shapiro. *A class $\mathscr{C}$ of partial computable functions ϕ_x is completely r.e. iff there is a canonically enumerable class $\mathscr{D}$ of finite functions, such we have*

$$f \in \mathscr{C} \equiv (\exists \xi)(\xi \in \mathscr{D} \wedge \xi \subseteq f) \tag{1}$$

Proof By Definition 11.3.9, 3, $\mathscr{C}$ is completely r.e. means that the predicate $\phi_x \in \mathscr{C}$ is semi-recursive, being equivalent to $x \in I$ where

$$I = \{x : \phi_x \in \mathscr{C}\}$$

- *If* direction. So assume that we have (1) and show that $\phi_x \in \mathscr{C}$ is semi-recursive. Now

$$\phi_x \in \mathscr{C} \equiv (\exists y)(\xi_y \in \mathscr{D} \wedge \xi_y \subseteq \phi_x) \tag{1'}$$

Observe that $\xi_y \in \mathscr{D} \equiv (\exists x)(y = f(x))$, where $f \in \mathcal{R}$ enumerates the r.e. set A —i.e., $A = \mathrm{ran}(f)$— that indexes the finite functions of $\mathscr{D}$ according to Definition 11.3.9, 2.

Thus, $\xi_y \in \mathscr{D}$ is semi-recursive. The same is true for $\xi_y \subseteq \phi_x$: Indeed,

$$\xi_y \subseteq \phi_x \equiv y = 1 \vee y > 1 \wedge (\forall z)_{<y}(p_z|y \to \phi_x(z) = \exp(z, y) \dot{-} 1)$$

or

$$\xi_y \subseteq \phi_x \equiv y = 1 \vee y > 1 \wedge (\forall z)_{<y}(p_z|y \to$$
$$(\exists w)\Big(T(x, z, w) \wedge out(w, x, z) = \exp(z, y) \dot{-} 1\Big) \tag{2}$$

The right hand side of $\equiv$ in (2) is semi-recursive by closure properties of $\mathcal{P}_*$. These observations —and closure properties of $\mathcal{P}_*$ once more— rest the case via (1') that $\phi_x \in \mathscr{C}$ is semi-recursive.

- *Only if* direction. This is entirely analogous to the proof of Theorem 11.3.5 and is left as *Exercise*. The key lemmata here are Theorems 6.10.1 and 6.10.3. $\square$

11.4 Exercises

1. Prove that $\phi_x(y) > w$ is not recursive.
   Hint. Grzegorczyk operations followed by direct diagonalisation.
2. We know that $\mathcal{R}$ is not a *completely* r.e. class of functions. Is it an r.e. class of functions?
3. We saw that there is an $f \in \mathcal{R}$ such that $A \in \mathcal{R}_* \equiv (\exists y)W_{f(y)} = A$ (11.3.4). Doesn't the following argument contradict the above result?

Let $S = \{x : x \notin W_{f(x)}\}$. Since $x \in S \equiv x \notin W_{f(x)}$, $S \in \mathcal{R}_*$ (closure of $\mathcal{R}_*$ under complement). Thus, $S = W_{f(m)}$, for some m. But

$$m \in W_{f(m)} \overset{S=W_{f(m)}}{\equiv} m \in S \overset{\text{Def of } S}{\equiv} m \notin W_{f(m)}$$

4. Complete the proof of Theorem 11.3.10.
5. Is every infinite recursive subset of $\mathbb{N}$ the range of a strictly increasing *primitive* recursive function? Why?
6. Is $\mathcal{PR}$ an r.e. class of functions?
   Hint. Consider using a ϕ-indexing via a modified URM model: It is a a a loop program programming language, with a **goto**.
7. Is $\mathcal{P} - \mathcal{R}$ an r.e. class of functions?
   Hint. Recall that $0, 1, 2, 3, \ldots$ are all the indices of the $\mathcal{P}$ functions. List *all* pairs of numbers generate by $z \mapsto ((z)_0, (z)_1)$ and view the first component as a ϕ-index and the second one as an *input*.
   Can you arrange, for any $(z)_0$ listed, to modify the program $(z)_0$ so that it loops forever on input $(z)_1$. If so, does this help to exclude all $\mathcal{R}$ functions, but to exclude no $\mathcal{P}$ functions?
8. Prove that $\mathcal{P} - \mathcal{PR}$ is an r.e. class.
9. Prove that the class of finite functions is an r.e. class.
10. Prove that the class of infinite domain functions is an r.e. class.

Chapter 12
Creative and Productive Sets; Completeness and Incompleteness

Overview

This chapter introduces the very important topic of productive and creative sets, the *simple* sets of Post, and also takes an in depth look at the phenomenon of the incompleteness of "strong" mathematical theories like Peano Arithmetic. It turns out that there is a connection between the logic and the computability topics and we give several mathematically complete proofs of the first incompleteness theorem (Gödel 1931) including the version by Rosser (1936). We also prove Church's theorem of the undecidability of the set of theorems of logic (Church 1936a,b; this provides a negative answer to Hilbert's *Entscheidungsproblem*). Our inevitable arithmetisation of logic towards obtaining the above results is based on concatenation, in the style of Smullyan (1992), Bennett (1962), Quine (1946).

12.1 Definitions

Productive sets were introduced by Dekker (1955). They are kind of *effectively non-c.e.* sets in the sense that for each such set S there is an algorithm that computes a counterexample to *any* claim such as "$S = W_e$", for any $e \in \mathbb{N}$. The algorithm, taking e as input, produces an $a \in S - W_e$. It is more convenient to give the definition in terms of set inclusion rather than equality:

12.1.1 Definition. (Productive Sets) A set $S \subseteq \mathbb{N}$ is called *productive* iff for some $f \in \mathcal{R}$, whenever $W_i \subseteq S$ we have $f(i) \in S - W_i$. We say that f is a productive function for S. □

Thus, $f(i)$ provides a constructive counterexample to the claim $W_i \subseteq S$ and hence also to this one: $W_i = S$.

12.1.2 Example. (Productive Sets Exist) $\overline{K}$ is productive with productive function $\lambda x.x$. Indeed, note

© Springer Nature Switzerland AG 2022
G. Tourlakis, *Computability*, https://doi.org/10.1007/978-3-030-83202-5_12

$$x \in K \equiv x \in W_x \tag{1}$$

hence

$$x \in \overline{K} \equiv x \notin W_x \tag{1'}$$

So let $W_i \subseteq \overline{K}$. By (1), $i \in W_i$ is false (otherwise $i \in K$), that is, by (1'), $i \in \overline{K}$. All told $i \in \overline{K} - W_i$. $\square$

12.1.3 Theorem. *Productive sets are not semi-recursive.*

Proof By the above remark. $\square$

By the above, productive sets are clearly infinite. In fact,

12.1.4 Theorem. *Every productive set contains an infinite c.e. set.*

Proof Let A be productive and $f \in \mathcal{R}$ be a productive function for it. Now $\emptyset \subseteq A$. Let e be a semi-index of $\emptyset$. Then $f(e) \in A - W_e$. Now add $f(e)$ to W_e to obtain $W_{e'}$. It is $f(e') \in A - W_{e'}$. Add $f(e')$ to the latter to obtain $W_{e''}$. And so on. This builds the infinite r.e. set

$$\{f(e), f(e'), f(e''), f(e'''), \ldots\}$$

The mathematical details are only slightly more complex:
The general step is adding $f(e^{(n)})^1$ to $W_{e^{(n)}}$, that is, forming $W_{e^{(n)}} \cup \{f(e^{(n)})\}$, where

$$f(e^{(n)}) \in A - W_{e^{(n)}} \tag{1}$$

By Proposition 6.9.5,

$$\{f(e^{(n)})\} = W_{\tau\left(f(e^{(n)})\right)}$$

Now, the construction in the informal paragraph above defines

$$W_{e^{(n+1)}} \stackrel{Def}{=} W_{e^{(n)}} \cup \{f(e^{(n)})\} \stackrel{6.9.4}{=} W_{h\left(e^{(n)}, \tau\left(f(e^{(n)})\right)\right)} \tag{2}$$

and hence the primitive recursion below shows that $\lambda n.e^{(n)} \in \mathcal{R}$.

$$e^{(0)} = e$$

$$e^{(n+1)} = h\left(e^{(n)}, \tau\left(f(e^{(n)})\right)\right)$$

---

[1] Here I use the superscript (n) as an alias of n primes.

$\text{ran}\big(\lambda n. f(e^{(n)})\big)$ is the c.e. set we want, noting that, by (1) and (2), it is infinite. □

12.1.5 Remark. The above proof also goes through by picking up *any* $W_e \subseteq A$. The proof constructs an infinite r.e. subset of A that is *disjoint* from W_e (by (1) and (2)). □

12.1.6 Corollary. *Every productive set has an infinite recursive subset.*

Proof By Corollary 11.2.4. □
   The following is useful:

12.1.7 Corollary. *Every productive set has an 1-1 productive function.*

Proof Let $f \in \mathcal{R}$ be a productive function for A. We will construct a 1-1 productive function $f11$ for A. The *basic* idea is to form $f11$ by course of values recursion to force 1-1 ness:

$$f11(0) = f(0)$$
$$f11(x+1) = \text{ some appropriate } y \notin \{f11(0), f11(1), \dots, f11(x)\}$$

where "$\{f11(0), f11(1), \dots, f11(x)\}$" is implemented as usual by the history function of $f11$, $\widehat{f11}$, namely

$$\widehat{f11}(x) = \langle f11(0), f11(1), \dots, f11(x)\rangle$$

and "$\in \langle \dots \rangle$" we have encountered before, e.g., in the proof of Theorem 4.3.2.
   Next we tweak the recursive definition to force $f11$ to be a productive function of A. Thus, for the recursion step above, at $x+1$, we want to ensure that the chosen "y" is in $A - W_{x+1}$ if $W_{x+1} \subseteq A$.
   Well, by Theorem 12.1.4 and Remark 12.1.5, if $W_{x+1} \subseteq A$, then the finite sequence of length $x+2$,

$$f(e^{(x+1)}), f(e^{(x+2)}), \dots, f(e^{(2x+2)}) \tag{1}$$

has distinct points all of which are in $A - W_{x+1}$. Since the sequence

$$f11(0), f11(1), \dots, f11(x) \tag{2}$$

has length $x+1$ there is a *smallest member* of (1) not in (2). We choose that as our "y" at the recursion step above.
   Does this work? All that we need verify is that $f11$ is productive for A: At $x=0$ it is so by the basis of the recursion. At $x+1$, *if* $W_{x+1} \subseteq A$, then

- By Theorem 12.1.4 and Remark 12.1.5 all points in (1) are distinct, and
- are in $A - W_{x+1}$

Thus having —at this step— set $f11(x+1)$ equal to an $f(e^{(x+i)})$, for some $i \geq 1$, we fulfill $f11(x+1) \in A - W_{x+1}$ by the second bullet above. □

12.1.8 Exercise. Equipped with the verbal proof above the reader will have no trouble filling the missing details and expressing the recursion mathematically, thus avoiding Church's thesis.   □

Interestingly, the seemingly more general definition of productive sets via partial productive functions below turns out to be equivalent to Definition 12.1.1.

12.1.9 Alternative Definition. A set $S \subseteq \mathbb{N}$ is called *productive* with partial computable productive function ψ iff whenever $W_i \subseteq S$ then we have $\psi(i) \downarrow$ and $\psi(i) \in S - W_i$.   □

12.1.10 Theorem. *Definitions 12.1.1 and 12.1.9 are equivalent.*

Proof

1. If S is according to Definition 12.1.1 then it is so by Definition 12.1.9 as well since $f \in \mathcal{R} \subseteq \mathcal{P}$.
2. This time S is productive as in Alternative Definition 12.1.9 via a partial (possibly non total) productive function ψ, and we need to find a total productive function f.

   Start with a function η, defined below

$$\eta(x, y, z) \simeq \begin{cases} \phi_y(z) & \text{if } \psi(x) \downarrow \\ \uparrow & \text{othw} \end{cases}$$

As the above is a definition by positive cases, $\eta \in \mathcal{P}$ and we have by S-m-n a function $h \in \mathcal{R}$ such that

$$\phi_{h(x,y)}(z) \simeq \begin{cases} \phi_y(z) & \text{if } \psi(x) \downarrow \\ \uparrow & \text{othw} \end{cases}$$

or

$$\phi_{h(x,y)} = \begin{cases} \phi_y & \text{if } \psi(x) \downarrow \\ \varnothing & \text{othw} \end{cases}$$

and by Theorem 9.3.2 there is a $\sigma \in \mathcal{R}$ such that

$$\phi_{\sigma(y)} = \phi_{h(\sigma(y),y)} = \begin{cases} \phi_y & \text{if } \psi(\sigma(y)) \downarrow \\ \varnothing & \text{othw} \end{cases}$$

In terms of domains,

$$W_{\sigma(y)} = \begin{cases} W_y & \text{if } \psi(\sigma(y)) \downarrow \\ \varnothing & \text{othw} \end{cases} \tag{1}$$

We note that $\psi\sigma$ is total: Indeed, if $\psi\sigma(y) \uparrow$, *for some* y, *then* $W_{\sigma(y)} = \emptyset \subseteq S$, hence $\psi\sigma(y) \downarrow$ *by 12.1.9; a contradiction.*

Thus letting $f = \psi\sigma$ we got our total productive function: Let $W_y \subseteq S$. By (1) —since $\psi(\sigma(y)) \downarrow$— we have $W_{\sigma(y)} = W_y \subseteq S$. Since ψ is a (partial) productive function, $\psi(\sigma(y)) \in S - W_{\sigma(y)}$. Translating in terms of f, $f(y) \in S - W_y(= S - W_{\sigma(y)})$.

$\square$

R.e. sets that have a productive complement were called *creative* by Post in recognition of the fact that the set of Peano Arithmetic (PA) theorems is a creative set, and the latter set requires more than a "mechanical" process to recognise its theorems (in fact, even the sub-theory ROB[2] of PA, introduced in Sect. 12.6, has a creative set of theorems). This comment is really a spoiler for Church's theorem on the unsolvability of the Entscheidungsproblem (Theorem 12.11.19), that is, that the set of theorems of any consistent extension of ROB is not recursive.

12.1.11 Definition. (Creative Sets) A set A is called *creative* iff it is semi-recursive, and $\overline{A}$ is productive.

$\square$

We immediately have that a creative set is *not* recursive, since its complement is not r.e.

12.1.12 Example. K is creative, since we know that K is r.e. and that $\overline{K}$ is productive.

$\square$

A creativity concept that is very relevant to the incompletable-ness phenomenon that we study in Sect. 12.11 is that of the *effective inseparability* of r.e. sets.

12.1.13 Definition. A pair of *disjoint* sets A and B are *effectively inseparable* just in case a $\psi \in \mathcal{P}$ (associated with the pair) exists such that whenever

$$A \subseteq W_a \wedge B \subseteq W_b \wedge W_a \cap W_b = \emptyset^3$$

we will have $\psi(a, b) \downarrow$ and $\psi(a, b) \in \overline{W}_a \cap \overline{W}_a$.

We will call ψ a *separation function* of A and B.

$\square$

If we omit "disjoint" from the definition we should note that the third conjunct in the hypothesis is false, trivialising the definition.

Note that ψ effectively provides examples of numbers that are not in either A or B.

12.1.14 Proposition. *If the disjoint r.e. sets A and B are an effectively inseparable pair then each is creative.*

Proof Let ψ be a separation function for A and B. It suffices to show that both $\overline{A}$ and $\overline{B}$ are productive. We check $\overline{A}$ since the case for $\overline{B}$ is entirely analogous:

---

[2] It is Peano Arithmetic with the induction schema removed.

[3] This condition forces $A \cap B = \emptyset$.

Let $W_x \subseteq \overline{A}$ and let $W_e = A$ and $W_{e'} = B$, for some e and e' in $\mathbb{N}$. By Theorem 6.9.4 there is an $h \in \mathcal{R}$ such that

$$W_{h(x,e')} = W_x \cup W_{e'} \tag{1}$$

Thus we have $A \subseteq W_e$ and $B \subseteq W_{h(x,e')}$ and $W_e \cap W_{h(x,e')} = \emptyset$. By assumption on A and B, we have $\psi(e, h(x, e')) \downarrow$ and $\psi(e, h(x, e')) \in \overline{A} \cap \overline{W}_{h(x,e')}$, or

$$\psi(e, h(x, e')) \in \overline{A} - W_{h(x,e')} \overset{(1)}{\subseteq} \overline{A} - W_x$$

Thus $\lambda x . \psi(e, h(x, e'))$ is a partial productive function for $\overline{A}$.   □

12.1.15 Proposition. *Effectively inseparable r.e. pairs exist.*

Proof The standard example is to take $A = \{x : \phi_x(x) = 0\}$ and $B = \{x : \phi_x(x) = 1\}$. Since

$$\phi_x(x) \simeq y \equiv (\exists z)(T(x, x, z) \wedge out(z, x, x) = y)$$

we have that both sets are semi-recursive by the projection theorem. They are also trivially disjoint. Let $A \subseteq W_u$ and $B \subseteq W_v$ where

$$W_u \cap W_v = \emptyset \tag{1}$$

and using the abbreviation "$\Phi_u(x)$" for "$(\mu z)T(u, x, z)$" define

$$\chi(u, v)(x) \simeq \begin{cases} 1 & \text{if } \Phi_u(x) < \Phi_v(x) \\ 0 & \text{if } \Phi_v(x) < \Phi_u(x) \\ \uparrow & \text{othw} \end{cases} \tag{2}$$

Since $\Phi_u(x) < \Phi_v(x) \equiv (\exists z)(T(u, x, z) \wedge \neg T(v, x, z))$, the above is a definition by positive cases hence $\chi \in \mathcal{P}$ and by S-m-n we have an $f \in \mathcal{R}$ such that we may rewrite (2) as

$$\phi_{f(u,v)}(x) \simeq \begin{cases} 1 & \text{if } \Phi_u(x) < \Phi_v(x) \\ 0 & \text{if } \Phi_v(x) < \Phi_u(x) \\ \uparrow & \text{othw} \end{cases} \tag{3}$$

We claim that f is a separation (total) function. Indeed, towards proving $f(u, v) \in \overline{W}_u \cap \overline{W}_v$ we proceed by contradiction:

1. Let $f(u, v) \in W_u$. By (1), $f(u, v) \notin W_v$. Thus the top condition in (3) is true, hence $\phi_{f(u,v)}(f(u, v)) \simeq 1$, which is a contradiction ($f(u, v) \in B$).

2. Let $f(u, v) \in W_v$. By (1), $f(u, v) \notin W_u$. Thus the middle condition in (3) is
true, hence $\phi_{f(u,v)}(f(u, v)) \simeq 0$, which is also a contradiction.
□

Looking ahead to Definition 12.11.43, we easily prove

12.1.16 Proposition. *If A and B are c.e. and effectively inseparable, then they are also recursively inseparable.*

Proof Arguing by contradiction we note that if $S \in \mathcal{R}_*$ separates A and B, then $A \subseteq S$ and $B \subseteq \overline{S}$. Both S and $\overline{S}$ are r.e. and $S \cup \overline{S} = \mathbb{N}$. This does not leave any room for any associated separation function ψ for A and B to provide an element in $\overline{S} \cap \overline{\overline{S}} = \emptyset$.
□

12.1.17 Remark. The above implies that the A and B of Proposition 12.1.15 are recursively inseparable. A direct proof is given in Lemma 12.11.44.
□

12.2   Strong Reducibility Revisited

We recall the definition of *many-one* reducibility —$\leq_m$— from Definition 6.3.1 that was central in our proofs of several unsolvability results. Here we will say a bit more about it.

12.2.1 Definition. (Strong Reducibility) The notation $A \leq_m B$ —read A is *many-one reducible* (or *m-reducible*) to B— means that, for some recursive f,

$$x \in A \equiv f(x) \in B^4 \tag{1}$$

We say that A is *1-1 reducible*, or *1-reducible* to B —in symbols $A \leq_1 B$— iff $A \leq_m B$ via f and f is 1-1.

We write just $A \leq B$ (no subscript) when we are indifferent to the kind of subscript.
□

12.2.2 Remark.

1. Thus if $A \leq_1 B$ then also $A \leq_m B$.
2. All our reducibilities obtained by the S-m-n theorem are 1-reducibilities since the S-m-n functions are 1-1.
□

12.2.3 Proposition. $A \leq_m B$ (resp. $A \leq_1 B$) via f is equivalent to any one of the following:

1) $A = f^{-1}[B]$
2) $f[A] \subseteq B$ and $f[\overline{A}] \subseteq \overline{B}$
3) $c_A = c_B f$

---

[4] Recall that writing "$P(x)$" we claim "$P(x)$ is true for all x". In particular the definition of $\leq_m$ in (1) is for all x.

Proof A straightforward exercise.

For (2), and starting from $A \leq_m B$ or $A \leq_1 B$ via f, note that Definition 12.2.1 requires $x \in A \equiv f(x) \in B$, that is, $x \in A \rightarrow f(x) \in B$ and $x \in \overline{A} \rightarrow f(x) \in \overline{B}$. The former yields the first inclusion stated in (2) while the latter yields the second.

Conversely, the first inclusion yields $x \in A \rightarrow f(x) \in B$ and the second yields $x \in \overline{A} \rightarrow f(x) \in \overline{B}$, thus $x \in A \equiv f(x) \in B$.   □

12.2.4 Exercise. Let A be recursive and $\emptyset \neq A \neq \mathbb{N}$. Prove that $A \leq_m \overline{A}$.

Hint. Cf. proof of Theorem 9.4.1.   □

12.2.5 Proposition. *Either of $A \leq_m B$ or $A \leq_1 B$ and the assumption that A is productive imply that B is also productive.*

Proof Let $A \leq_m B$ or $A \leq_1 B$ via f (1-1-ness is neither necessary nor assumed). Let $g \in \mathcal{R}$ be a productive function for A.

By assumption and Proposition 12.2.3,

$$A = f^{-1}[B] \tag{1}$$

Let $W_x \subseteq B$. Hence, $f^{-1}[W_x] \subseteq f^{-1}[B] = A$. By Proposition 6.9.6 we have an $h \in \mathcal{R}$ such that $W_{h(x)} = f^{-1}[W_x] \subseteq A$. By productiveness, $gh(x) \in A - W_{h(x)} = f^{-1}[B] - f^{-1}[W_x]$. Thus $fgh(x) \in B - W_x$ so that B is productive with productive function fgh.   □

12.2.6 Corollary. *If A is creative, B is r.e. and $A \leq B$, then B is creative.*

Proof The assumption yields $\overline{A} \leq \overline{B}$ thus $\overline{B}$ is productive.   □

12.2.7 Example. In particular, $\overline{K} \leq A$ implies that A is productive. Thus all our reducibilities $\overline{K} \leq A$ in Chap. 5 proved more than the non r.e.-ness of A: They proved it is productive.   □

The following proposition will be handy in a naïve semantic proof of Gödel's *first incompleteness theorem.*

12.2.8 Proposition. *If B is semi-recursive and $A \cap B$ is productive, then A is also productive.*

Proof Let f be a total productive function for $A \cap B$ and let $B = W_e$, for some e. We will construct a productive function for A.

Let $W_x \subseteq A$. Therefore $W_x \cap W_e \subseteq A \cap B$, or, by Example 6.9.3

$$W_{h(x)} \subseteq A \cap B$$

for some $h \in \mathcal{R}$. By the choice of f,

$$f(h(x)) \in A \cap B - W_{h(x)} = A \cap B - W_x \cap B \tag{1}$$

(1) implies that $f(h(x)) \in A$ (and in B), but then it must be $f(h(x)) \notin W_x$. fh is the productive function for A. □

12.2.9 Remark. As remarked in Sect. 6.3, $A \leq B$ says intuitively that B is "at least as hard" as A when determining their membership relation because, in particular, $A \leq B$ means $x \in A \equiv f(x) \in B$ for some recursive f, so *if I can solve membership for B, then I can do so for A*. Well, almost: $\overline{K} \not\leq K$ since K is r.e. yet $\overline{K}$ is not (cf. Theorem 6.3.3), yet intuitively, if I can decide (yes/no) membership in K, then I should be able to use the *same procedure* with minor changes (yes to no, no to yes) to decide $x \in \overline{K}$.

This observation also yields the fact that $\leq$ is not a total order since, for example, K and $\overline{K}$ are incomparable: $K \not\leq \overline{K}$ from the above remarks and the trivial "$A \leq B$ iff $\overline{A} \leq \overline{B}$".

So the reducibility relation does not perfectly match the "B is 'at least as hard' as A" intuition.

This remark's content, less specifically, can be stated as "there are sets S such that $S \not\leq \overline{S}$". □

12.2.10 Exercise. Show that $\emptyset$ and $\mathbb{N}$ are incomparable with respect to each of $\leq_m$ and $\leq_1$. □

12.2.11 Exercise. Show that each of $\leq_m$ and $\leq_1$ is transitive and reflexive. □

By the above exercise, viewing the relations as sets of pairs (of sets) we note that $\leq_m \cap \geq_m$ and $\leq_1 \cap \geq_1$ are equivalence relations.

Intuitively, $A \leq B \wedge B \leq A$ says that A and B are "equally hard".

12.2.12 Definition. We write $\equiv_m$ for $\leq_m \cap \geq_m$ and $\equiv_1$ for $\leq_1 \cap \geq_1$.

We call the the equivalence classes of $\equiv_m$ the *many-one degrees* or *m-degrees* while the equivalence classes of $\equiv_1$ are the *1-1 degrees* or *1-degrees*.

Each equivalence class contains "pairwise equally hard" sets according to the "metric" $\leq_m$ or $\leq_1$.

If a degree contains a recursive (resp. r.e.) set it is called a *recursive* (resp. r.e.) *m-* or *1-degree*.

The degree (equivalence class) where a set A belongs is indicated by $deg_m(A)$ (resp. $deg_1(A)$) if we are talking about an *m*-degree (resp. 1-degree).

An arbitrary degree, with no reference to one of its set members, will be denoted by bold face lowercase Greek letters, with or without subscripts or primes: $\boldsymbol{\alpha}, \boldsymbol{\alpha}_3, \boldsymbol{\beta}''', \boldsymbol{\gamma}, \boldsymbol{\delta}''_{42}$. □

By Theorem 6.3.3 a degree that contains a recursive set contains *only* recursive sets. Same comment for a degree that contains an r.e. set; all its sets will be r.e.

12.2.13 Definition. For two *m*-degrees (resp. 1-degrees) $\boldsymbol{\alpha}$ and $\boldsymbol{\beta}$ we define

$$\boldsymbol{\alpha} \leq_m \boldsymbol{\beta} \text{ (resp. } \boldsymbol{\alpha} \leq_1 \boldsymbol{\beta})$$

to mean

> For some $A \in \alpha$ and $B \in \beta$, it is $A \leq_m B$ (resp. $A \leq_1 B$)

We will denote the POset of all m- (resp. 1-) degrees by $(\mathfrak{D}_m, \leq_m)$ (resp. $(\mathfrak{D}_1, \leq_1)$) or simply $\mathfrak{D}_m$ (resp. $\mathfrak{D}_1$) since the ordering in each case is understood from the subscript of $\mathfrak{D}$. □

12.2.14 Proposition. *Definition 12.2.13 is independent of the choice of representatives in the degrees α and β.*

Proof We look at the case $\leq_m$, the case of $\leq_1$ being entirely analogous.

Let $A \in \alpha \leq_m \beta \ni B$. Let also $A' \in \alpha$ and $B' \in \beta$.

Then $A' \leq_m A \leq_m B \leq_m B'$ hence $A' \leq_m B'$ by transitivity of $\leq_m$. Thus the conclusion $\alpha \leq_m \beta$ is supported by A' and B' as well. □

12.2.15 Proposition. *For any two m-degrees α and β there is an m-degree γ that is smallest such that $\alpha \leq_m \gamma$ and $\beta \leq_m \gamma$, in the sense that if also $\alpha \leq_m \delta$ and $\beta \leq_m \delta$, then $\gamma \leq_m \delta$.*

Proof Let $A \in \alpha$ and $B \in \beta$. We will show that

$$\gamma \stackrel{Def}{=} deg_m(A \oplus B)$$

works, where we wrote $A \oplus B$ for the *disjoint union* of A and B, namely,

$$A \oplus B \stackrel{Def}{=} \{2x : x \in A\} \cup \{2x + 1 : x \in B\}$$

- By the trivial $A \leq_m A \oplus B$ via $\lambda x.2x$ and $B \leq_m A \oplus B$ via $\lambda x.2x + 1$ we have $\alpha \leq_m \gamma$ and $\beta \leq_m \gamma$.
- Let also $\alpha \leq_m \delta$ and $\beta \leq_m \delta$, that is, for some $D \in \delta$ we have $A \leq_m D$ via f and $B \leq_m D$ via g.
  If so we have $A \oplus B \leq_m D$ via h given by

$$h(x) = \begin{cases} f(\lfloor x/2 \rfloor) & \text{if } 2 \mid x \\ g(\lfloor x/2 \rfloor) & \text{if } \neg 2 \mid x \end{cases} \tag{1}$$

 Regarding the definition of h note that $\lfloor x/2 \rfloor$ is a simpler way to say $\lfloor (x - 1)/2 \rfloor$ when x is odd.

To conclude, let $x \in A \oplus B$.

Case 1. Let $x = 2m$ where $m \in A$. Then $f(m) \in D$ hence $f(\lfloor x/2 \rfloor) \in D$. That is, $h(x) \in D$. Conversely, if $h(x) \in D$ —still $x = 2m$— then $h(x) = f(m) \in D$ and thus $m \in A$ from $A \leq_m D$ via f. So $x \in A \oplus B$.

Case 2. Let $x = 2m + 1$ where $m \in B$. Then $g(m) \in D$ hence $g(\lfloor x/2 \rfloor) \in D$, and we continue and conclude analogously to Case 1. □

In algebra jargon, we are saying above that the set of m-degrees, $\mathfrak{D}_m$, forms, under the partial order $\leq_m$, an *upper semi-lattice*, that is, a *POset* such that any two of its members α and β have a *least upper bound* (l.u.b.), that is, a γ above each exists that is smallest in the sense of the proposition above.

12.3 Completeness

A "hardest" among r.e. sets with respect to the $\leq_m$ (resp. $\leq_1$) ordering is called m-complete (resp. 1-complete).

12.3.1 Example. Let K_0 stand for $\{\langle x, y \rangle : \phi_x(y) \downarrow\}$. Since

$$z \in K_0 \equiv Seq(z) \wedge \phi_{(z)_0}((z)_1) \downarrow$$

K_0 is semi-recursive. Let now A be an arbitrary r.e. set. We have $A = W_e$ for some e. Note that

$$x \in A \text{ iff } x \in W_e \text{ iff } \langle e, x \rangle \in K_0$$

Thus $A \leq_1 K_0$ via $\lambda x. \langle e, x \rangle$. In other words, K_0 is 1-complete.   □

12.3.2 Definition. A set A is *m-complete* (resp. 1-complete) iff two conditions hold:

1. A is r.e.
2. For any r.e. set B, we have $B \leq_m A$ (resp. $B \leq_1 A$).   □

Trivially, if A is 1-complete, then it is also m-complete. We will settle the question on whether the converse is true shortly.

12.3.3 Example. It turns out that K is 1-complete as well: First, we know that it is semi-recursive. Let now A be an arbitrary r.e. set. We have $A = W_e$ for some e. Define

$$\psi(x, y) \simeq \begin{cases} y & \text{if } x \in A \\ \uparrow & \text{othw} \end{cases}$$

$\psi \in \mathcal{P}$ as the above is a definition by positive cases. Thus for some $h \in \mathcal{R}$ we have

$$\phi_{h(x)}(y) \simeq \begin{cases} y & \text{if } x \in A \\ \uparrow & \text{othw} \end{cases}$$

Hence $\phi_{h(x)}(h(x)) \downarrow \equiv x \in A$, that is, $A \leq_1 K$ via h which being an S-m-n function is 1-1.   □

12.3.4 Remark.

1. The above two examples yield $K \equiv_1 K_0$ by Definition 12.2.12.
2. Since $A \leq_1 B$ implies $A \leq_m B$ we have that $A \equiv_1 B$ implies $A \equiv_m B$. Therefore, if A is 1-complete, then it is also m-complete.
3. Is the converse true as well? That if A is m-complete, then it is also 1-complete?
4. Also: 2. says that $\leq_1 \subseteq \leq_m$ and $\equiv_1 \subseteq \equiv_m$. Are these inclusions equalities? We will address both questions, 3. and 4. soon.
5. $\mathbf{0}'_m$ and $\mathbf{0}_m$ (resp. $\mathbf{0}'_1$ and $\mathbf{0}_1$) denote the m-degree (resp. 1-degree) of K and of the recursive sets $\emptyset \neq A \neq \mathbb{N}$ respectively. By completeness of K we have $\mathbf{0}_m \leq_m \mathbf{0}'_m$. On the other hand $\mathbf{0}'_m \leq_m \mathbf{0}_m$ does not hold, as it would imply $K \leq_m A$, rendering A non recursive. Thus $\mathbf{0}_m <_m \mathbf{0}'_m$.
6. The recursive sets $\emptyset$ and $\mathbb{N}$ are the sole occupants of the m-degrees $deg_m(\emptyset)$ and $deg_m(\mathbb{N})$ since there is no set A such that $\emptyset \leq_m A$ or $A \leq_m \emptyset$ and, taking complements, no set B such that $\mathbb{N} \leq_m A$ or $A \leq_m \mathbb{N}$.

   Thus these two m-degrees are isolated points in the m-degree ordering; they compare to nothing. $\square$

12.3.5 Proposition. *Any m-complete and any 1-complete set A is creative.*

Proof Thus, A is r.e. and its completeness implies that $K \leq_m A$ or $K \leq_1 A$. We are done by Corollary 12.2.6. $\square$

The converse is also true.

12.3.6 Theorem. (Myhill (1955)) *Any creative set A is 1-complete, and hence also m-complete.*

Proof So let A be creative and let also $B \neq \emptyset$ be an arbitrary c.e. set. We will show that

$$B \leq_1 A \tag{1}$$

via an appropriately constructed 1-1 function $f \in \mathcal{R}$. Let also $g \in \mathcal{R}$ be a 1-1 productive function for $\overline{A}$ (cf. Corollary 12.1.7), that is, we have

$$W_x \subseteq \overline{A} \rightarrow g(x) \in \overline{A} - W_x \tag{2}$$

We define by the recursion theorem (Corollary 9.2.2) a 1-1 function $\lambda x.h(x) \in \mathcal{R}$ such that

$$\phi_{h(x)}(y) \simeq \begin{cases} 0 & \text{if } x \in B \wedge y \simeq g(h(x)) \\ \uparrow & \text{othw} \end{cases} \tag{3}$$

or

$$W_{h(x)} = \begin{cases} \{g(h(x))\} & \text{if } x \in B \\ \emptyset & \text{othw} \end{cases}$$

Thus, $f = gh$ is 1-1 and recursive. Now

1. $x \notin B$ implies $W_{h(x)} = \emptyset \subseteq \overline{A}$ thus, by the choice of g, $f(x) \simeq g(h(x)) \in \overline{A} - \emptyset = \overline{A}$, that is, $f(x) \notin A$.
2. On the other hand, $x \in B$ implies $W_{h(x)} = \{g(h(x))\}$. Can $f(x) \notin A$? If so, $\overline{A} \supseteq \{g(h(x))\} = W_{h(x)}$. By productiveness, $gh(x) \notin W_{h(x)} = \{g(h(x))\}$; a contradiction.

We have shown $x \in B \equiv f(x) \in A$, that is, $B \leq_1 A$.   □

Before application of the recursion theorem the right hand side of (3) above defines the function

$$\psi = \lambda zxy. \begin{cases} 0 & \text{if } x \in B \wedge y \simeq g(z) \\ \uparrow & \text{othw} \end{cases}$$

thus Corollary 9.2.2 yields a 1-1 $h \in \mathcal{R}$ such that $\psi(h(x), x, y) \simeq \phi_{h(x)}(y)$, for all, x, y.

12.3.7 Theorem. *A set $S \subseteq \mathbb{N}$ is creative iff it is m-complete iff it is 1-complete.*

Proof Using the symbol $\Longrightarrow$ *conjunctionally*, we have

$$S \text{ 1-complete} \Longrightarrow S \text{ m-complete} \overset{\text{Proposition 12.3.5}}{\Longrightarrow} S \text{ creative} \overset{\text{Theorem 12.3.6}}{\Longrightarrow}$$

$$S \text{ 1-complete}$$

□

12.4   Recursive Isomorphisms

12.4.1 Definition. A *recursive isomorphism* is a *recursive 1-1* and *onto* function $f : \mathbb{N} \to \mathbb{N}$. We say that a recursive isomorphism f *sends* or *maps* a set A to a set B if we also have $f[A] = B$.

In this case we write $A \sim B$[5] —borrowing the notation from that for "1-1 correspondence"— pronouncing this as "*A is recursively isomorphic to B*".   □

F. Klein suggested that mathematics is the study of invariants under selected groups of transformations (that is, 1-1 and onto functions). For example, (plane) projective geometry is about studying the invariants under projection from a point and (dually) section with a line, topology is interested in the invariants under homeomorphisms,

---

[5] The alternative notation $A \equiv B$ we would rather avoid for fear of confusion with other uses of "$\equiv$". But then we are guilty of using the symbol $\sim$ in three different ways, two of which were introduced in Chap. 0.

and it can be argued that recursion theory is interested in those properties that remain invariant under $\sim$ —for example, recursiveness and semi-recursiveness.

In this very short section we will state and prove just one result, Myhill's theorem below. Significantly, it implies that, *extensionally*, we have *exactly one* formal arithmetic since each such arithmetic is recursively isomorphic to K (cf. Corollary 12.11.34 below).

12.4.2 Definition. A *finite* 1-1 correspondence f between two sets A and B is a finite subset of $\mathbb{N} \times \mathbb{N}$ that is a 1-1 function

$$f = \{(x_1, y_1), (x_2, y_2), (x_3, y_3), \ldots, (x_n, y_n)\}$$

such that $x_i \in A$ iff $y_i = f(x_i) \in B$, for all $1 \leq i \leq n$.

12.4.3 Lemma. *If $A \leq_1 B$, and f is a finite 1-1 correspondence between A and B, then, for any arbitrarily chosen $x \notin dom(f)$, we can find* constructively *an appropriate $y \notin ran(f)$ to extend f to the finite 1-1 correspondence $f \cup \{(x, y)\}$ between A and B.*

Proof Let $A \leq_1 B$ be effected by a 1-1 $g \in \mathcal{R}$.

Now, as the lemma suggests, choose $x \notin dom(f)$. Then proceed to construct a $y \notin ran(f)$ such that $x \in A$ iff $g(x) = y \in B$. We do this iteratively, as follows:

1. *Step 0*:   Obtain $y = g(x)$. If $y \notin ran(f)$, then *we are done*. Note that the specific pair (x, y) satisfies $x \in A$ iff $y \in B$. **Exit!**
2. If *not done*, then we need to perform *further Steps*:
3. *Assume* (I.H.) that we have so far *computationally* (using f and g) *constructed* a sequence, one sequence term per step, of

$$\boxed{k \; distinct \; \text{points } in \text{ ran } (g) \cap \text{ran } (f)} \; y', y'', \ldots, y^{(k)} \tag{1}$$

where the superscript "(k)" denotes k primes while, for any a, $a^{(0)} = a$ *by definition* (no primes).

$$\boxed{\textbf{Note} : \text{The "}y\text{" that we want } must \text{ be in } \text{ran}(g) - \text{ran}(f)} \tag{2}$$

Furthermore, by (1) the points $x^{(r)} = f^{-1}(y^{(r)})$, for $r = 1, 2, \ldots, k$, are *distinct*, since f^{-1} is 1-1. Thus also the points

$$x = x^{(0)}, x', x'', \ldots, x^{(k)} \text{ are distinct} \tag{3}$$

since $x \notin dom(f) = ran(f^{-1})$.

Finally, our I.H. that captures the history of our construction so far posits that we have computed the $y^{(i)}$ in (1) *specifically* by setting

$$y^{(i+1)} = g(x^{(i)}), \text{ for } 0 \leq i < k \tag{4}$$

4. By the comment in the box in (1) we have not yet found the "y", which the lemma is talking about, among the points in (1). Cf. also Note (2).
   Thus we do the *Next Step*:
   Set $y^{(k+1)} = g(x^{(k)})$.

   By 1-1 ness of g and (3), $y^{(k+1)}$ is not in the list (1) $\tag{5}$

   Thus,

   a. if $k = n$, *then we are done*, since there are only n distinct points $y_i \in \mathrm{ran}(f)$, thus we found a point $y = y^{(k+1)}$ *not* in $\mathrm{ran}(f)$, which satisfies $x \in A$ iff $y \in B$,[6] *and* we have thus extended f by adding $(x, y^{(k+1)}) = (x, y)$. **Exit!**
   b. If $k < n$ we may still have $y^{(k+1)} \in \mathrm{ran}(f)$. *If not, then this is the y we want; done.* **Exit!**
   Else, we continue the construction: Set $k \leftarrow k + 1$ and **Goto 2**, noting that (1) remains true by (5), and hence (3) remains true. □

12.4.4 Theorem. (Myhill (1955)) $\sim\, =\, \equiv_1$.

Proof

- $\sim\,\subseteq\,\equiv_1$. Let f be a recursive isomorphism between A and B. Then $A = f^{-1}[B]$ and therefore $A \leq_1 B$ (cf. Proposition 12.2.3). Using f^{-1} instead, we also get $B \leq_1 A$. Thus $A \equiv_1 B$.
- $\sim\,\supseteq\,\equiv_1$. The proof relies straightforwardly on the lemma above. Since the lemma cannot guarantee that the union of all extensions obtained by it is onto $\mathbb{N}$, we need to alternate between choosing an $x \notin \mathrm{dom}(f)$ and choosing a $y \notin \mathrm{ran}(f)$ so that all of $\mathbb{N}$ is considered with a *chosen* element both on the left field and right field sides of the under construction recursive isomorphism. Naturally, when we pick a $y \notin \mathrm{ran}(f)$ we work in a mirror image manner with respect to the proof of Lemma 12.4.3 using the $B \leq_1 A$ part of the hypothesis.
  The proof proceeds by stages. To fix ideas, let $A \leq_1 B$ via the 1-1 $g \in \mathcal{R}$ while $B \leq_1 A$ via the 1-1 $h \in \mathcal{R}$.
  The procedure:

  1. Start with the finite 1-1 correspondence between A and B $\{(0, g(0))\}$ For $n \geq 0$ we add *pairs* by stages to the current finite correspondence, one at a time.

---

[6] We observe, using conjunctional equivalence "$\Longleftrightarrow$",

$$x \in A \overset{g}{\Longleftrightarrow} y' \in B \overset{f^{-1}}{\Longleftrightarrow} x' \in A \overset{g}{\Longleftrightarrow} y'' \in B \overset{f^{-1}}{\Longleftrightarrow} x'' \in A \overset{g}{\Longleftrightarrow} y''' \in B \ldots$$

2. **Stage $2n - 1$.**   If no pair has (yet) n as the *second* component, then choose n as "y" and apply Lemma 12.4.3 with the assumption $B \leq_1 A$ (in lieu of $A \leq_1 B$) to add the pair (x, y) for an appropriate x.
3. **Stage $2n$.**   If no pair has (yet) n as the *first* component, then choose n as "x" and apply Lemma 12.4.3 with the assumption $A \leq_1 B$ (exactly as in the lemma) to add the pair (x, y) for an appropriate y.

By Church's Thesis our construction of pairs can be rendered mathematically precisely to yield a recursively enumerable (infinite) relation

$$j = \{(a_0, b_0), (a_1, b_1), (a_2, b_2), (a_3, b_3), \ldots\}$$

that by construction is a 1-1 correspondence $j : \mathbb{N} \to \mathbb{N}$.

By Theorem 6.6.1, $j \in \mathcal{R}$ (it is trivially total).

Does j carry A onto B? That is, is $x \in A$ iff $j(x) \in B$? Yes, since by the lemma, at every stage, we have $a_i \in A$ iff $b_i = j(a_i) \in B$ and $b_r \in B$ iff $j^{-1}(a_r) \in A$ (i.e., j is onto B). That is, we have $A \sim B$.   □

12.5   A Simple Set

12.5.1 Definition. (Simple Sets (Post)) A set $A \subseteq \mathbb{N}$ is called *simple* if it satisfies the following three conditions

(1) A is semi-recursive.
(2) $\overline{A}$ is infinite
(3) $\overline{A}$ contains no infinite W_e.   □

 By condition (3) a simple set cannot be recursive.

12.5.2 Theorem. (Post (1944)) *Simple sets exist.*

Proof To satisfy (3) we ensure that the construction of A intersects *every* infinte W_e —we don't mind if it does so with finite W_e too.[7] Other than that, the idea is to make A (intuitively) "sparse" so the it has an infinite complement and enumerate its elements algorithmically, so it is semi-recursive.

In short, for every x and e we put x in A if $x \in W_e$ (so $A \cap W_e \neq \emptyset$) *and* also (sparseness) $x > 2e$. This is done by listing all pairs (x, e) and picking what meets our conditions. As the $x \in W_e$ questions are not necessarily decidable we ask the question infinitely many times, once for each number of *steps taken* in a computation by the verifier of W_e (dovetailing; cf. Theorem 6.9.9):

---

[7] Which is good for the *construction* of A since the question "is W_e infinite?" is not recursive.

$$f(e) = \Big((\mu z) T(e, (z)_0, (z)_1) \wedge (z)_0 > 2e \Big)_0 \qquad (i)$$

f is in $\mathcal{P}$. So its *range* $A \overset{Def}{=} \mathrm{ran}(f)$ is semi-recursive. We have met (1) of the definition.

Say W_e is *infinite*. Then (i) *will* yield an $f(e) \in A \cap W_e$. We have met requirement (3).

Moreover, the fact that $f(e) > 2e$ implies that *no member of* $A - f(e)$ — *is in the* second half *of the ordered set*

$$\{0, 1, 2, \ldots, 2e\}$$

Thus, *for all* e, as many as e numbers are in $\overline{A}$. Therefore we met the requirement (2). □

12.5.3 Corollary. *There is a non recursive but semi-recursive set that is* not

(1) *Creative*
(2) *1- or m-complete*

Proof

(1) Say A is simple. If it is creative, then $\overline{A}$ is productive. Impossible since a productive set contains an *infinite* semi-recursive subset (Theorem 12.1.4).
(2) A simple set cannot be 1- or m-complete either, since it then would be creative (Proposition 12.3.5).

□

12.6 Robinson's Arithmetic

Robinson's Arithmetic, in short ROB, (due to Robinson (1950)) is a *formal theory of arithmetic* —that is, a first-order logic where certain *mathematical* (or *nonlogical*) *axioms* have been *added* so that we can *syntactically* (as we say, *formally*) *certify* truths about the set of natural numbers $\mathbb{N}$.

It is a significant fragment of the well known *Peano Arithmetic*, or *PA*, the latter using the same formal language as ROB and all the latter's axioms, but also adding to those a single *axiom schema*, that of mathematical induction.

We will give syntactic proofs of Gödel's and Rosser's Incompleteness theorems in this chapter (Sect. 12.11), but we will also prove Church's theorem on the recursive unsolvability of Hilbert's *Entscheidungsproblem* for ROB and PA, and indeed for *any* consistent extension of ROB.

An inconsistent theory (Definition 0.2.89) has all formulas as theorems, therefore its *Entscheidungsproblem* is trivially recursive, since we can test easily whether a string over an alphabet is a wff or not.

12.6.1 Definition. (ROB and Other Formal Arithmetics) We will call *formal arithmetic*, somewhat non specifically, *any first-order mathematical theory over a language $L_{\mathscr{A}}$ for arithmetic*, which has a *recursive set of axioms* that *contains at least* the ROB mathematical axioms below (Robinson 1950).

By a "*language*, $L_{\mathscr{A}}$, for arithmetic" we mean a first-order language augmented (at a minimum) with the mathematical (nonlogical) symbols

$$0, S, +, \times, <$$

The underlying logic itself, for any formal arithmetic, is understood to have — without loss of generality— just two rules of inference, *modus ponens* (MP)

$$\frac{A, A \to B}{B}$$

and ∀-*introduction* (∀-intro)

$$\frac{A \to B}{A \to (\forall x)B}, \text{ provided there is no free } x \text{ in the conclusion}$$

Both rules are "strong", that is, we do not need to know or care how the "numerators" are *obtained* in order to "compute" the "denominators".[8]

As is usual, the object variables are denoted by convenient *names*; as *metavariables*, so writing "x" we do not have a specific object variable in mind, and x and y stand for any actual variables, or even for the same actual variable. In the few times we will be forced to write down some of the latter, we will take the point of view that these "actual" object variable really are the members of the infinite sequence

$$v_0, v_1, \ldots, v_i, \ldots^9$$

The specific formal arithmetic over $L_{\mathscr{A}}$ whose theorems are derived by precisely *the following mathematical axioms we will call ROB. We all also call by the same name just the set of axioms.*

It is the *formally correct* thing to do to fully parenthesise *arithmetical expressions* (that is, *terms*, cf. Definition 0.2.5) when written in *infix* (operation symbol placed

---

[8] A seemingly more general definition of formal arithmetic does not specify the rules but only requires them to be recursive. Once the concept of recursiveness is *mathematically* extended to non numerical sets and functions (Sect. 12.8.1) —but see also earlier informal discussions on the matter, e.g., in Remark 5.3.2 and Theorem 5.3.3— we will see at once that MP and ∀-intro are indeed recursive.

[9] These "specific" variable names are really *meta*-names themselves! Recall (cf. Final Definition 0.2.3) that each v_n stands for ($v\#\ldots\#$) where the string of # has length $n + 1$.

between the operands) in order to avoid ambiguities. So we write *formally* $(x + y)$ and $(x \times y)$ for terms associated with the functions ("operations") of arity 2, namely, $+$ and $\times$.

However, as it is a habit of the mathematician to do so, one normally *agrees* that $\times$ is *stronger* (has higher priority) than $+$, and therefore, e.g., $x + y \times z$ is *informal* for the *formal* $(x + (y \times z))$. S, as a 1-ary (unary) function, it is agreed to be of the highest priority, thus, one writes Sx normally, with no bracketing. However, if S is to have $(x + y)$ as an argument then we must write $S(x + y)$ not only formally —$(x + y)$ is formal for $x + y$— but also informally.

In displaying the axioms below we use the above agreed upon informal notation and we use brackets only if needed.

ROB-1. (*Axioms for S*)

> **S1.** $\quad \neg Sx = 0$ (for any variable x)
> **S2.** $\quad Sx = Sy \to x = y$ (for any variables x, y)

ROB-2. (*Axioms for $+$*)

> **+1.** $\quad x + 0 = x$ (for any variable x)
> **+2.** $\quad x + Sy = S(x + y)$ (for any variables x, y)

ROB-3. (*Axioms for $\times$*)

> **$\times$1.** $\quad x \times 0 = 0$ (for any variable x)
> **$\times$2.** $\quad x \times Sy = (x \times y) + x$ (for any variables x, y)

ROB-4. (*Axioms for $<$*)

> **$<$1.** $\quad \neg x < 0$ (for any variable x)
> **$<$2.** $\quad x < Sy \equiv x < y \lor x = y$ (for any variables x, y)
> **$<$3.** $\quad$ (Trichotomy) $x < y \lor x = y \lor y < z$ (for any variables x, y).

$\qquad\qquad\qquad\qquad\qquad\qquad\qquad\qquad\qquad\qquad\qquad\qquad\qquad\qquad$ □

 12.6.2 Remark. Two examples of formal arithmetics are ROB and PA.

We said above "over a *language* $L_{\mathscr{A}}$ for arithmetic, ...", referring actually to the *alphabet* of the language, while the language itself is rather the set of formulas and terms over said alphabet. However, this abuse of terminology is common in the literature, and in fact one often even omits the logical part of the alphabet and attaches the name "language" only to the set of non logical (mathematical) symbols: E.g., we say "$\{\in\}$ is the language of set theory ZFC", "$\{0, S, +, \times, <\}$ is the language of ROB", "$\{0, S, +, \times, <\}$ is the language of PA". □

It turns out that ROB is sufficiently "rich" to support (or suffer from!) Gödel's first incompleteness theorem, which, in its *semantic* version, says that there are *closed formulas* (*sentences*) over the language L_{ROB}, which even though they are true in the *standard interpretation* (cf. Chap. 8) they are not *syntactically* (*i.e., formally*) *verifiable* as such (i.e., are not provable) within ROB.

In fact, ROB has infinitely many such sentences, and it is actually *incompletable*. That is, it is *not possible* to render it complete by adding *some* recursive set of mathematical axioms to it. *In this chapter we will see a few syntactic versions of the theorem.*

It is (almost) trivial to exhibit some simple sentences that are not provable in ROB, even though they are standardly true: E.g., $(\forall x)\neg x < x$, $(\forall x)(\forall y)x + y = y + x$.[10] On the other hand, Gödel's theorem is a deep result that rather points to the *incompletable*-ness of ROB.

 12.6.3 Remark. We noted that the name ROB will also apply to the *list of the axioms.* This is because it is *customary* to name a theory by the *same name* regardless of whether it is given *extensionally* —that is, as the *set of all its theorems*— or *intensionally*, that is, by *a recursive (or even finite) description* of *how to* obtain *all its theorems*. The intentional description is most often given as a set of mathematical axioms and rules of inference.

We will also consider in this chapter some consistent theories, $\mathcal{T}$, which are *extensions* of ROB in the sense that $\mathcal{T}$ —viewed extensionally— contains *all* the theorems of ROB, and its language $L_{\mathcal{T}}$ contains L_{ROB}.

But these theories need *not* be *formal arithmetics*: A formal arithmetic *must* have —by Definition 12.6.1— an intensional nature that includes a *recursive* set of axioms. Not *every* consistent extension of ROB can be generated by a recursive set of axioms. For example, we will see that CA^{all} of Definition 12.11.22 cannot be, and hence nor can CA itself (cf. Definition 12.8.3). □

12.7   Arithmetical Relations and Functions, Revisited

12.7.1 Definition. (Arithmetical Relations and Functions, Refined) Following Smullyan (1992) let us call *arithmetical*,[11] lower case "a", the set of relations and functions obtained from Definition 8.3.3, by

- *Dropping* the relation $z = x^y$ from the list of initial relations.
- Making the "A" in the adjective "Arithmetical" lower case throughout in the definition. □

12.7.2 Proposition. *If $P(y, \vec{x})$ and $\lambda\vec{z}.f(\vec{z})$ are arithmetical/Arithmetical, then so is $P(f(\vec{z}), \vec{x})$.*

Proof By the informal Theorem 0.2.94 we have

$$P(f(\vec{z}), \vec{x}) \equiv (\exists y)\Big(y = f(\vec{z}) \wedge P(y, \vec{x})\Big)$$

We are done by Definitions 8.3.3 and 12.7.1. □

---

[10] We can prove these *in* PA however using (formal) induction! Cf. Tourlakis (2003a).

[11] Actually, Smullyan uses the adjective "arithmétic", accent on "me", instead of "arithmetical". But it seems that the majority of the literature uses "arithmetical".

12.7.3 Theorem. *All functions in $\mathcal{R}_{alt}$ are ~~a~~rithmetical,[12] hence so are all relations of $\mathcal{R}_{alt,*}$.*

Proof Induction over $\mathcal{R}_{alt}$ via Corollary 7.0.37:

(1) *Basis (initial functions of $\mathcal{R}_{alt}$).*

- $\lambda xy.x + y$. We want the graph $z = x + y$ to be arithmetical. It is so by Definition 12.7.1.
- $\lambda xy.x \times y$. We want $z = x \times y$ to be arithmetical. It is so by Definition 12.7.1.
- $\lambda xy.x \mathbin{\dot{-}} y$. We want $z = x \mathbin{\dot{-}} y$ to be arithmetical. We have

$$z = x \mathbin{\dot{-}} y \equiv x < y \wedge z = 0 \vee x = z + y$$

Seeing that $z = 0$ is an explicit transform of $z = x + y$ and

$$x < y \equiv \neg y \leq x \equiv \neg(\exists w)x = y + w \equiv (\forall w)\neg x = y + w$$

we are done by Definition 12.7.1 (via (1), (2), (3) of Definition 8.3.3).

(2) *The induction steps on the obvious I.H.*

- *Composition.* Let $f = \lambda \vec{y}.h(g_1(\vec{y}), \ldots, g_n(\vec{y}))$. Thus[13]

$$z = f(\vec{y}) \equiv (\exists u_1) \ldots (\exists u_n)\Big(z = h(\vec{u}_n) \wedge u_1 = g_1(\vec{y}) \wedge \cdots \wedge u_n = g_n(\vec{y})\Big)$$

We are done by the (obvious) I.H. and (1) and (2) of Definition 8.3.3.

- *Case of closure under $(\widetilde{\mu}y)$.* Let $f = \lambda \vec{z}_n.(\widetilde{\mu}y)g(y, \vec{z}_n)$, where g is y-regular. Thus,[14]

$$w = f(\vec{z}_n) \equiv (\forall \vec{x}_n)(\exists y)g(y, \vec{x}_n) = 0 \wedge g(w, \vec{z}_n) = 0 \wedge (\forall u)_{<w}g(u, \vec{z}_n) \neq 0$$

Thus $w = f(\vec{z}_n)$ is arithmetical by (1), (2), (3) of Definition 8.3.3.

Turning to relations, since $R(\vec{x}) \in \mathcal{R}_{alt,*}$ means $R(\vec{x}) \equiv c_R(\vec{x}) = 0$, where $c_R \in \mathcal{R}_{alt}$ the result follows from what we proved above, applying an explicit transformation to the graph of c_R, $c_R(\vec{x}) = y$. □

In Chap. 8 (Sect. 8.3) we proved that every $f \in \mathcal{PR}$ and every $P \in \mathcal{PR}_*$ are *Arithmetical*. This and the inclusion $\mathcal{PR} \subseteq \mathcal{R}_{alt}$ trivially imply that every $f \in \mathcal{PR}$ and every $P \in \mathcal{PR}_*$ are *arithmetical* as well. It also implies that $z = x^y$ is arithmetical. Thus,

12.7.4 Corollary. *The sets of Arithmetical and arithmetical relations coincide.*

---

[12] I will have to stop underlining "a" soon.

[13] Using the informal Theorem 0.2.94 yet again.

[14] $(\forall \vec{x}_n)$ means $(\forall x_1) \ldots (\forall x_n)$.

12.8   Expressibility Revisited

In Chap. 8 we defined what it means to say that a number-theoretic relation $Q(x)$ or *total* function f is *expressible* (Definition 8.3.1) in the language of PA (or ROB), $L_{PA} = L_{ROB} = \{0, S, +, \times, <\}$. We also defined the Arithmetical relations and functions (Definition 8.3.3) —and the corresponding lower case version, *arithmetical* relations and functions in Definition 12.7.1— and showed that the *Arithmetical* and hence also the *arithmetical* relations are all expressible in the language of PA (ROB), cf. Lemma 8.3.4.

We have a converse:

12.8.1 Theorem. *Every number theoretic relation that is expressible in the language $L_{ROB} = \{0, S, +, \times, <\}$ is arithmetical (hence also Arithmetical; Corollary 12.7.4).*

Proof   Induction on the formation of formulas of L_{ROB}.

1. Case of atomic, $t = s$ or $t < s$, where t, s stand for any terms.

   Thus (see discussion on interpretations in Remark 8.1.2) the expressible relations by the various instances of $t = s$ have the form $f(\vec{x}) = g(\vec{y})$ where f and g are in $\mathcal{PR}$, in fact are derived from the functions $\lambda xy.x + y$, and $\lambda xy.x \times y$ *using Grzegorczyk operations only*. Similarly, expressible relations by the various instances of $t < s$ have the form $f(\vec{x}) < g(\vec{y})$ with similarly constructed f and g in $\mathcal{PR}$. We are done in this case since every $\mathcal{PR}$ function is arithmetical by the above remarks, and noting that, e.g.,

$$f(\vec{x}) < g(\vec{y}) \equiv (\exists u, v)(u = f(\vec{x}) \wedge v = g(\vec{y}) \wedge u < v)$$

2. The claim propagates with the formation of formulas, that is, from A to $\neg A$ and $(\forall x)A$ and from A and B to $A \vee B$.

   This step is straightforward. For example, let $R(y, \vec{x}_n)$ be expressed by $A(y, \vec{x}_n)$. Thus, say that we have established that

$$R(y, \vec{x}_n) \text{ is true iff } A(\widetilde{y}, \widetilde{x}_1, \ldots, \widetilde{x}_n) \text{ is true}$$

   then

$$R(y, \vec{x}_n) \text{ is arithmetical (I.H. on formulas)} \tag{1}$$

   Since we also have that (cf. proof of Lemma 8.3.4)

$$(\forall y)R(y, \vec{a}_n) \text{ is true iff } (\forall y)A(y, \widetilde{a}_1, \ldots, \widetilde{a}_n) \text{ is true}$$

   it follows from (1) that $(\forall y)R(y, \vec{x}_n)$ is also arithmetical. The remaining cases are as easy and are left to the reader.

   $\square$

12.8.2 Exercise. Argue the cases not covered in the proof of Theorem 12.8.1. □

12.8.3 Definition. (The Set CA) We denote by CA the set of all *closed formulas* over L_{ROB} that are true in the *standard interpretation* of Chap. 8.

CA is also called the *complete arithmetic* in the literature. □

12.8.1 A Bit of Arithmetisation and Tarski's Theorem

The incompleteness theorem that we will revisit in this chapter owes its name on the fact that *consistent* formal arithmetics —this is plural (cf. Definition 12.6.1)— cannot prove *all* the sentences of CA. These arithmetics are *incomplete* (in their goal to verify arithmetical truths syntactically).

12.8.4 Remark. Lemma 8.3.4 along with Theorem 12.8.1 concludes the proof of the fact that the concepts of *expressible* and *arithmetical* are equivalent.

By Theorem 12.8.1, for any formula $A(x_1, \ldots, x_n)$ over L_{ROB}, the informal relation R given by

$$R \overset{Def}{=} \{\vec{a}_n : A(\widetilde{a}_1, \ldots, \widetilde{a}_n) \text{ is true in the standard interpretation}\} \tag{1}$$

is arithmetical since it is expressible, noting that (1) is equivalent to (1′) below

$$R(\vec{a}_n) \equiv A(\widetilde{a}_1, \ldots, \widetilde{a}_n) \text{ is true in the standard interpretation} \tag{1′}$$

Therefore CA may be viewed in the metatheory (i.e., *informally*) as the set of all true *arithmetical* sentences over L_{ROB}. □

Tarski's theorem (Tarski 1955) says that CA is so complex that the informal predicate

$$F \in CA$$

is not arithmetical, hence also not expressible in L_{ROB}.

Pause. *How can we make the above predicate number-theoretic so that we can explore Tarski's theorem with the mathematical tools of computability?* ◄

We can do so by *Gödel numbering* all formulas of the language $L_{ROB} = L_{PA}$, *a process by which every string over L_{ROB} is assigned a unique "identity card"*: a number.

Our theory can talk and reason about numbers, so, by proxy it can talk and reason about formulas and *sets* of formulas (informal unary predicates).

The easiest way forward to obtain a Gödel numbering of all formulas of ROB is to *refine* somewhat and make more explicit the idea contained in Sect. 8.1 of Chap. 8. Recall that the L_{ROB} alphabet is obtained from Tentative Definition 0.2.2 (reproduced below) by tailoring the mathematical part to the needs of ROB:

$$\overbrace{\#, v, \neg, \vee, (,), =, \forall,}^{\text{logical part}} \quad \overbrace{f, p, c}^{\text{mathematical part}} \tag{*}$$

The ROB *mathematical* symbols

$$0, S, +, \times, <$$

are just user-friendly *abbreviations* (metatheoretical names) of the following strings

$$(c\#), (\# f\#), (\#\# f\#), (\#\# f\#\#), (\#\# p\#) \tag{**}$$

in that order (cf. discussion in Tentative Definition 0.2.2 and Remark 0.2.4).

As in Tentative Definition 0.2.2 the *actual ontology* of the *object variables* is the following *strings*, generated by the symbols "v", "(", ")", and "#":

$$(v\#), \ (v\#\#), \ (v\#\#\#), \dots$$

in short,

$$\text{the strings } (v\#^{n+1}), \text{ for } n = 0, 1, 2, \dots \tag{2}$$

where $\#^{n+1}$ denotes

$$\underbrace{\#\#\dots\#}_{n+1 \text{ times}}$$

 In other words, for $n \geq 0$, the variable name "v_n" is convenient *meta*notation *for* the string

$$(v\overbrace{\#\dots\#}^{n+1\,\#}), \text{ for } n \geq 0 \tag{3}$$

This metanotation is not the most flexible, and the practicing mathematician and logician will prefer to use an even simpler metanotation for object variables, namely, u, v, w, x, y, z *with or without primes, with or without subscripts.*

Thus, our terms and formulas are strings over L_{ROB} according to the standard definitions (Definitions 0.2.5 and 0.2.9) found in Sect. 0.2. We *view* each such *string* in any one of *two* ways, as convenience for the argument in hand dictates at the moment:

1. As a *string* over the alphabet (∗) on p. 350 or
2. As *the number* whose *11-adic notation is* the string in question. The 11-adic digits are the members of the sequence (∗) in *the order shown* where each digit's *position* in the sequence is its integer *value*. That is, the digit # has value 1 while

the digit p has value 10.[15] Thus we have obtained a 1-1 correspondence between strings A over ($*$) and the non zero natural numbers they denote. In this approach λ —the empty string— is "coded" as (is an alias for) 0.

This Gödel numbering *requires no computation* to "code" a string. It only requires us to have two *dual points of view* when we look at a string (over alphabet ($*$)): to *view* it as a *string* over a language of arithmetic, OR as a *number* just as we have learnt in elementary school to view the *string* of symbols "111" as the number "one hundred eleven".

This approach to Gödel numbering originated with (Quine 1946). To appreciate how complex alternative Gödel numberings can be see Gödel (1931), Shoenfield (1967), Hilbert and Bernays (1968), and Tourlakis (2003a).

We can, *in principle*, denote the number that the string A denotes by the notation $\langle \overline{A} \rangle_{11}$ of Definition 7.0.22.

We prefer however to denote this number by the notation $gn\{A\}$ —rather than $\langle \overline{A} \rangle_{11}$— and call it the (*informal*[16]) Gödel number of A.

We use "gn{A}" throughout rather than the normal round brackets-form "gn(A)" to avoid confusion when, for example, A is something like ")(((". That is, we use gn{)(((} rather than gn()(((.

12.8.5 Remark. (Worth Noting) Thus, in particular, given that any formula A (over L_{ROB}) has a *string view* and a *number view*, "$F \in CA$" can be viewed as saying, interchangeably, either that "the formula F is in CA (CA viewed as a set of sentences)" or "the Gödel number of the formula F is in CA (CA viewed as *a set of Gödel numbers* of sentences)".

One thing must be clear:

Syntactically, any expression (string) E over L_{ROB} is exactly the same string as gn(E) if both are displayed in 11-adic notation.

Which view we favour at any given time will be *implicit in the context of the discussion in hand*, or might even be emphasised by an explicit statement. □

Let us next adapt two lemmata that we proved before in Lemmata 7.0.28 and 7.0.29 to 11-adic notation (from 2-adic):

12.8.6 Lemma. *For any x, the smallest y such that $10x + 1 < 11^y$, is $|x| + 1$, where $|x|$ denotes the 11-adic length of x.*

---

[15] This is not any more "peculiar" than hexadecimal ("hex") —16-ary— notation in computing, where the sixteen *hexadecimal digits* are 0,1,2,3,4,5,6,7,8,9 that have as values themselves, and a, b, c, d, e, f that have as values 10, 11, 12, 13, 14, 15 in that order. The point is that the digits must be *symbols of length one*, so that, say, "11" will not be ambiguous: Its value in the familiar decimal notation is "17" since "11" in "hex" notation denotes $1 \times 16 + 1$. Of course, "eleven" in "hex" is represented by the digit "b".

[16] This number is computed outside any formal theory, in the realm of everyday informal mathematics.

Proof

- (Case $x > 0$:) First we have the obvious (we omit the subscript 11 of $\langle\ldots\rangle$)

$$\frac{11^{|x|} - 1}{10} = \underbrace{\langle 11 \ldots 1 \rangle}_{|x|}\,{}^{17} \leq x \leq \underbrace{\langle cc \ldots c \rangle}_{|x|}\,{}^{18} = 11\frac{11^{|x|} - 1}{10}$$

hence,

$$10x + 1 < 10x + 11 < 11^{|x|+1}$$

and

$$11^{|x|} \leq 10x + 1$$

Combining the above two we have

$$11^{|x|} \leq 10x + 1 < 11^{|x|+1} \tag{1}$$

- (Case $x = 0$:) (1) holds in this case too. Indeed, here $|x| = 0$ by definition since 0 is represented by the empty sequence λ. Thus, we have $10x + 1 = 1 = 11^0 = 11^{|x|} < 11^{|x|+1}$. Another way of writing this is $11^{|x|} \leq 10x + 1 < 11^{|x|+1}$, verifying the lemma in the $x = 0$ case. □

12.8.7 Remark. In the balance of this chapter we stay with 11-adic notation. In particular, $x * y$ will mean 11-adic concatenation $x *_{11} y$, and $|x|$ will mean 11-adic length. □

12.8.8 Lemma. *The predicates and functions below are in* $\mathcal{R}_{alt,*}$

- $10x + 1 < 11^y$
- $\lambda x.|x|$ *where* $|x|$ *denotes 11-adic length.*
- $\lambda x.11^{|x|}$
- $z = x * y$ *(see Remark 12.8.7 above)*

Proof

- $10x + 1 < 11^y$: $\lambda y.11^y \in \mathcal{PR} \subseteq \mathcal{R}_{alt}$.
- $\lambda x.|x|$: $|x| = \left((\tilde{\mu}y)10x + 1 < 11^y\right) \dot{-} 1$.
  Of course, the predicate $10x + 1 < 11^y$ is y-regular.

---

[17] We wrote "1" for each of the digits of the smallest x of given length $|x| = n$. But we could also use the symbol "#" for each such "1".

[18] We wrote $\langle cc \ldots c \rangle$ rather than $\langle 11, 11, \ldots, 11 \rangle$ avoiding the "two-symbol digit" "11".

One can also use bounded search, therefore establishing that $\lambda x.|x|$ is actually primitive recursive: $|x| = \left((\mu y)_{\le 10x+1} 10x + 1 < 11^y \right) \dot- 1$.

- $\lambda x.11^{|x|}$: Now trivial.
- $z = x * y$: $z = x11^{|y|} + y$.

$\square$

12.8.9 Remark.

(i) All $*_2$-specific results that can be translated to $*_{11}$ results, from Corollary 7.0.30 to the end of Chap. 7, are trivially valid for 11-adic notation. In particular xPy —in 11-adic notation— is in $\mathcal{R}_{alt,*}$, and the latter set is closed under part-of quantification in 11-adic notation.

(ii) It is conceivable that one is conditioned to assume that notation such as $15 * x * y * 6$ means to concatenate —in 11-adic notation— the (decimal) number 15 with $x * y$ and then append the 11-adic digit 6 at the end. Thus if we mean that 1 and 5 are 11-adic digits as opposed to decimal digits we ought to write $1 * 5 * x * y * 6$ or better still $\# * (* x * y *)$.

The "shorthand" $\#(x * y)$ for the above is *much less desirable* as we may mistake the brackets as *indicators of grouping* concatenation operations —not as *alphabet symbols* that participate in the concatenation— as in "do $x * y$ and then append $\#$ at the left".

(iii) Using the $gn\{\ldots\}$ notation we can also unambiguously write the above as $gn\{\ \#(\ \}* x * y * gn\{\ )\ \}$. $\square$

We now indicate how one can go about showing that we can translate the standard definitions of terms and formulas of first-order logic using recursive predicates on (Gödel) numbers.

12.8.10 Proposition. *The following predicates are in $\mathcal{R}_{alt,*}$:*

(1) *$tally(x)$ says that x is a tally of "1" or —same thing— "#" symbols.*
(2) *$var(x)$ says that x is the Gödel number of an object variable —i.e., x expressed in 11-adic has the form depicted in (2) on p. 350.*
(3) *$cons(x)$ says that x is the Gödel number of a constant.*
(4) *$Suc(x)$ says that x is the symbol S —that is, the string $(\#f\#)$ (cf. p. 350, $(**)$).*
(5) *$Plus(x)$ says that x is the symbol $+$ —that is, the string $(\#\#f\#)$*
(6) *$Times(x)$ says that x is the symbol $\times$— that is, the string $(\#\#f\#\#)$*
(7) *$Less(x)$ says that x is the symbol $<$ —that is, the string $(\#\#p\#)$ (cf. p. 350, $(**)$).*
(8) *$Term(x)$ says that x is the Gödel number of a term.*
(9) *$AF(x)$ says that x is the Gödel number of an atomic formula.*
(10) *$wff(x)$ says that x is the Gödel number of a first-order formula.*

Proof

(1) $tally(x)$: $tally(x) \equiv (\forall y)_{Px}(y \ne 0 \to 1Py)$.
(2) $var(x)$: Note that "#", "v", "(" and ")" in (2) on p. 350 are the 11-adic digits 1, 2, 5 and 6 respectively being at positions 1, 2, 5 and 6 in the alphabet $(*)$.

$$var(x) \equiv (\exists y)_{Px}\left(tally(y) \wedge x = gn\{(v\} * y * gn\{)\}\right)$$

 We will use *recursion*, naturally, throughout this and future proofs in this section in our definitions of the relevant predicates. The qualifier "naturally" is apt because in these proofs we often are tracking faithfully the recursion that we used in the *definitions* for the syntactic constructs we name here, e.g., *terms, formulas*, etc. (Cf. Definitions 0.2.5, 0.2.9).

The reader is invited to use the characteristic functions of the predicates we define here in order to convert the *informal recursions* that I display below — for *predicate definitions*— into course-of-values recursions for (characteristic) *function* definitions of said predicates.

(3) *cons*(*x*):

$$x = gn\{(c\#)\}$$

Note that the string "(*c*#)" —named in short "0"— is the only constant in *ROB* (p. 350, (∗∗)).

(4) *Suc*(*x*): $x = gn\{(\#f\#)\}$

 Since, *in decimal*, $gn\{(\#f\#)\} = 5 \cdot 11^4 + 1 \cdot 11^3 + 9 \cdot 11^2 + 1 \cdot 11^1 + 6 = 75642$, we could have defined *Suc*(*x*) by $x = 75642$ but that would have been extremely obscure.

(5) *Plus*(*x*): $x = gn\{(\#\#f\#)\}$
(6) *Times*(*x*): $x = gn\{(\#\#f\#\#)\}$
(7) *Less*(*x*): $x = gn\{(\#\#p\#)\}$
(8) *Term*(*x*) says that *x* is the Gödel number of a term.

$$Term(x) \equiv var(x) \vee cons(x) \vee (\exists u, v, w)_{Px}\Big($$

$$Suc(w) \wedge Term(u) \wedge x = w * u \vee$$

$$Plus(w) \wedge Term(u) \wedge Term(v) \wedge x = gn\{(\} * u * w * v * gn\{)\} \vee$$

$$Times(w) \wedge Term(u) \wedge Term(v) \wedge x = gn\{(\} * u * w * v * gn\{)\}\Big)$$

Note that for the cases of + and × above we used the *formal* notation, *with* brackets around the infix expressions (cf. Definition 12.6.1) "*u* ∗ *w* ∗ *v*", since our relations "*Term*(*x*)", "*wff*(*x*)", etc., speak about the *formal syntax*. Of course, writing "*Sx*" *is* formal.

(9) *AF*(*x*):

$$AF(x) \equiv (\exists u, v, w)_{Px}\Big(Term(u) \wedge Term(v) \wedge Less(w) \wedge$$

$$x = u * gn\{=\} * v \vee x = u * w * v\Big)$$

(10) $wff(x)$:

$$wff(x) \equiv AF(x) \vee (\exists u, z, w)_{Px} \Big($$

$$wff(u) \wedge x = gn\{(\neg\} * u * gn\{)\} \vee$$

$$wff(u) \wedge var(w) \wedge x = gn\{((\forall\} * w * gn\{)\} * u * gn\{)\} \vee$$

$$wff(u) \wedge wff(v) \wedge x = gn\{(\} * u * \vee * z * gn\{)\}\Big)$$

where again we used as 11-adic digits the symbols listed in the ROB alphabet (∗) rather than using their decimal (position) values.   □

12.8.11 Remark. (Tarski's Substitution Trick) We have briefly encountered in Chap. 9 Gödel's *substitution function*, $sub(x, y)$, which returns the Gödel number of the formula we obtain from A (where $gn\{A\} = x$), *after* a certain free variable of A, say v_0, is replaced by the term (that has Gödel number[19]) y. Thus, if we write $A(v_0)$ for A, to indicate that the only free variable of A is v_0, we have $sub(x, y) = gn\{A(y)\}$. In particular, if y is the term $\widetilde{b}$, then we have for the informal sub:[20]

$$sub(x, \widetilde{b}) = gn\{A(\widetilde{b})\}$$

while for the formal **sub** —which we depict here in bold to keep track of what is what[21]— we have:

$$\mathbf{sub}(\widetilde{x}, \widetilde{b}) = \ulcorner A(\widetilde{b}) \urcorner$$

since, if $x = gn\{A\}$, then $\widetilde{x} = \ulcorner A \urcorner$ (cf. p. 281).

Tarski proved that CA —viewed in informal mathematics as a set of (Gödel) numbers (cf. Remark 12.8.5) — is not expressible in the language L_{ROB}, that is, the predicate $x \in CA$ is not so expressible— equivalently, it is not *arithmetical*.

Well, *assume* for a moment that we *do* have **sub**. How would a proof of Tarski's theorem go?

---

[19] I must try to avoid repeating the "that has Gödel number". As we note in Remark 12.8.5, the number $gn\{y\}$ —expressed in 11-adic— and the string y are *identical strings*, so the quoted redundant qualifier-reminder might soon become annoying!

[20] In the literature one often introduces "s" by the definition $s(x, y) = sub(x, Num(y))$, where Num is that of Example 12.8.12 below. Then $s(x, b) = gn\{A(\widetilde{b})\}$.

[21] Note that the need to have a formal version is clear from the *proof* of the fix-point Lemma (Lemma 9.1.1), where we construct *inside the formal language* the formula $A(\mathbf{sub}(v_0, v_0))$.

Let us argue by contradiction, so, let us believe that some formula $T(v_0)$ over L_{ROB} expresses the set CA, that is, for all $a \in \mathbb{N}$,

$$a \in CA \text{ iff } T(\widetilde{a}) \text{ is true} \tag{1}$$

By Gödel's fixpoint lemma (Chap. 9), there is an $e \in \mathbb{N}$ such that PA proves

$$\widetilde{e} = \ulcorner \neg T(\widetilde{e}) \urcorner \tag{2}$$

and hence, informally, the following is true.

$$e = gn\{\neg T(\widetilde{e})\} \tag{2'}$$

Thus,

$\neg T(\widetilde{e})$ is true iff its (informal) Gödel number e is in CA iff $T(\widetilde{e})$ is true, by (1)

The equivalence (between the 1st and last formulas) above is a contradiction. Done!

Tarski's trick (below) sidesteps two complex requirements:

- Define a function that computes the Gödel number of a formula A *after* the *substitution* —*everywhere* into a variable v_0 of A— of a term t, and said function does so receiving as inputs the Gödel number of A *before* the substitution, and the Gödel number of the term t.
- Define the function *not only* informally, but also *formally*, so that writing $A(sub(x, y))$ is *syntactically permissible* in the language of formal arithmetic —that is, $sub(x, y)$ is a formal term of the language.

Tarski introduced another version of *sub* —we will name it *Sub* in what follows— by utilising instead the ubiquitous "*one-point-rule*", (3) below (cf. Theorem 0.2.94 and the many informal uses we have had, for example, in the proofs of Theorem 8.3.6, Proposition 12.7.2 and Theorem 12.8.1).

$$\vdash^{22} A[t] \equiv (\exists x)(x = t \wedge A) \tag{3}$$

where x is *one of* the free variables in A —without loss of generality we can take it to be v_0— x does not occur in t, and the substitution $A[x := t]$, whose result is denoted by $A[t]$, is *legal* (cf. 0.2). We use here the version (Theorem 0.2.94)(1)

$$\vdash A[t] \equiv (\forall x)(x = t \to A) \tag{4}$$

under the same restrictions as above.

---

[22] This is a theorem of pure logic, i.e., not requiring mathematical axioms (cf. Theorem 0.2.94).

Tarski used (4) with $\tilde{n}$ for t, in order to express the substitution $A[\tilde{n}]$ below by a simple equivalent formula *where the substitution is eliminated*:

$$\vdash A[\tilde{n}] \equiv (\forall v_0)(v_0 = \tilde{n} \to A) \tag{5}$$

This bypasses the need for using Gödel's *sub*.

A slight sleight of hand was employed next! Instead of computing $gn\{A[\tilde{n}]\}$ from $gn\{A\}$ and n —as a function of n and $z = gn\{A\}$,

compute the Gödel number of the *equivalent* formula $(\forall v_0)(v_0 = \tilde{n} \to A)$.

□

The computation of

$$gn\left\{\left((\forall v_0)\left((\neg v_0 = \tilde{n}) \vee A\right)\right)\right\}$$

where we have expressed here the defined symbol $\to$ in terms of the primitive symbols $\neg$ and $\vee$ of our alphabet and inserted all brackets the definition of formula requires (cf. Definition 0.2.9), is simple:

$$gn\left\{\left((\forall v_0)\left((\neg v_0 = \tilde{n}) \vee A\right)\right)\right\}$$
$$= gn\left\{((\forall v_0)((\neg v_0 = \right\} * gn\{\tilde{n}\} * gn\{)\vee\} * gn\{A\} * gn\{))\} \tag{6}$$

writing

$$k = gn\{((\forall v_0)((\neg v_0 =\}, \qquad r = gn\{)\vee\} \qquad \text{and} \qquad m = gn\{))\} \tag{7}$$

we need first an expression for $gn\{\tilde{n}\}$ *as a function of n*— that is, with n as a free (input) variable —to conclude. To avoid notation that may be subject to misinterpretation,[23] we will use "$Num(n)$" for the expression that, *for any natural number n*, computes $gn\{\tilde{n}\}$. This is achieved by the following primitive recursion.

$$Num(0) = gn\{(c\#)\} \quad \text{this is the Gödel number of the "formal 0"}$$
$$Num(n+1) = gn\{(\#f\#)\} * Num(n)$$

Thus,

$$\lambda n.Num(n) \in \mathcal{PR}$$

---

[23] For any specific value of n, $\tilde{n}$ is a specific string over our alphabet ($*$) therefore represents a *number* in 11-adic notation. This number *depends* on n.

Equipped with the above we have (from (6) and (7)) the following primitive recursive function *Sub* of arguments n and $z = gn\{A\}$, where A is *any* string over L_{ROB}— even a nonsensical non-formula —to ensure total-ness of *Sub*.

$$Sub \overset{Def}{=} \lambda zn.k * Num(n) * r * z * m \in \mathcal{PR} \tag{†}$$

12.8.12 Example. *Num* works as it should, namely, *for any fixed natural number* n, $Num(n) = gn\{\widetilde{n}\}$. We do induction on n (outside the formal system!):

For $n = 0$, $Num(0) = gn\{\mathbf{0}\}$ —where the bold $\mathbf{0}$, the formal zero, is an abbreviation of $(c\#)$— but by Definition 8.1.1, $\widetilde{\mathbf{0}} = \mathbf{0}$. Taking the I.H. for some fixed n, we have

$$Num(n + 1) = gn\{(\#f\#)\} * Num(n)$$

$$= gn\{(\#f\#)\} * gn\{\widetilde{n}\} \quad \text{by the I.H.}$$

$$= gn\{(\#f\#) * \widetilde{n}\}, \quad \text{``}gn\{\cdots\}\text{'' distributes over} * \text{due to Remark 12.8.5}$$

$$= gn\{\widetilde{n + 1}\}, \quad \text{definition of } \widetilde{\ }$$

$$\square$$

We have established that

12.8.13 Proposition. *There is a primitive recursive, and therefore* arithmetical, *function Sub given by* (†) *above such that*

$$gn\{(\forall v_0)(v_0 = \widetilde{n} \rightarrow A)\} = Sub(gn\{A\}, n) \tag{‡}$$

for all $n \in \mathbb{N}$ *and strings A over the alphabet* (*) *of p. 350, where we have used the abbreviation* "$\rightarrow$" *in* (‡). *In particular,* (‡) *is valid when A is a wff.*

For convenience we will denote $(\forall v_0)(v_0 = \widetilde{n} \rightarrow A)$ by the short name $\mathscr{A}(\widetilde{n})$ — note the calligraphic "$\mathscr{A}$" in the abbreviation vs. the use of "A" in its expansion. By Theorem 0.2.94 we have the *absolutely*[24] provable equivalence

$$\mathscr{A}(\widetilde{n}) \equiv A(\widetilde{n}) \tag{**}$$

hence the above is true in the metatheory by the soundness of (pure) predicate logic.

Moreover, Proposition 12.8.13 says that the following equality is true in the metatheory (which is where we "compute" the Gödel numbers of strings)

$$Sub(gn\{A\}, n) = gn\{\mathscr{A}(\widetilde{n})\} \tag{***}$$

---

[24] That is, without using mathematical axioms.

12.8.14 Definition. (Diagonal Function) The following is another version of Gödel's *diagonal function* (Smullyan 1992) that we encountered in the proof of the fixpoint lemma (Lemma 9.1.1) as $sub(x, x)$. We define here a *new* (and different) diagonal:

$$D = \lambda x.Sub(x, x)$$

where *Sub* is that of Proposition 12.8.13.   □

From the fact that *Sub* is arithmetical (Proposition 12.8.13), so is D by explicit transformation.

12.8.15 Theorem. (Tarski's Theorem) *The predicate* $x \in CA$ *—viewing* CA *as the set of* Gödel *numbers of all true sentences over* L_{ROB}*— is not arithmetical.*

Proof Arguing by contradiction, let $x \in CA$ be arithmetical. Then so is $x \notin CA$, and therefore so is $D(x) \notin CA$ by the fact that D is arithmetical (cf. ⧎-remark after Definition 12.8.14 and Proposition 12.7.2).

Let then $G(v_0)$ express $D(x) \notin CA$ (cf. Lemma 8.3.4 and Corollary 12.7.4) and also let $\tilde{g} = \ulcorner G(v_0) \urcorner$ —or, $g = gn\{G(v_0)\}$. Thus, to begin with, we have

$$\text{For all } a, \ D(a) \notin CA \text{ iff } G(\tilde{a}) \ is \ true \tag{1}$$

In particular we have

$$D(g) \notin CA \text{ iff } G(\tilde{g}) \ is \ true \text{ iff } \mathscr{G}(\tilde{g}) \ is \ true \tag{2}$$

Noting that $D(g) = gn\{\mathscr{G}(\tilde{g})\}$, the first formula in the iff-chain (2) says that

> The sentence $\mathscr{G}(\tilde{g})$ is *not* in CA, that is, it is *not true*

But then —in view of its rightmost "iff"— (2) is a contradiction.
This concludes our proof that CA is *not* arithmetical.   □

12.9 Yet Another Semantic Proof of Gödel's Incompleteness Theorem

As a warmup for what follows, and similarly (but not identically) to the proof in Chap. 8 this time our proof will be by "an effective cardinality argument", that is, if "enumerable" means "effectively enumerable", that is, r.e., then Gödel's theorem follows by noting

- The set of provable sentences is r.e. (cf. Corollary 8.1.11).
- The set of true (in the standard interpretation of arithmetic) sentences is not r.e.

12.9.1 Theorem. *Given a correct formal arithmetic $\mathcal{T}$ (correct extension of ROB; cf. Definition 12.6.1). Then there are infinitely many sentences that are true in the standard interpretation, yet not provable in $\mathcal{T}$.*

Proof By Tarski's theorem (Theorem 12.8.15) we have that CA is not r.e. (semi-recursive) because if it were, then it would be arithmetical by the Kleene normal form theorem and the results of Sect. 8.3.

Thus we verified the second bullet above. As noted above, the first bullet is also correct.

As the set of closed theorems —denoted extensionally by $\mathcal{T}$— of a correct formal arithmetic $\mathcal{T}$ is a subset of CA, the verified two bullets above show that

$$\mathcal{T} \subsetneq \text{CA}$$

So let $A \in \text{CA} - \mathcal{T}$. Then $\mathcal{T} \cup \{A\}$ is also a correct *formal arithmetic*, thus we have a $A' \in \text{CA} - \mathcal{T} \cup \{A\}$. Now add next A' as a mathematical axiom to form a new correct formal arithmetic $\mathcal{T} \cup \{A, A'\}$. Then we have a $A'' \in \text{CA} - \mathcal{T} \cup \{A, A'\}$, etc.

Continuing like this we can build an infinite sequence

$$A, A', A'', A''', \ldots$$

of formulas of CA that are not provable by $\mathcal{T}$. $\qquad\qquad\square$

 12.9.2 Remark. The incompleteness theorem hinges heavily on the presence of a recursive set of mathematical axioms. Does the set of axioms remain recursive as we make the indicated additions above? See Exercise 12.13.3. $\qquad\square$

12.10   Definability

The formal (syntactic) counterpart of *expressibility* is relation and function *definability* in a formal arithmetic.

 A word of caution. The term "definable" is not universally used, and other parts in the literature may use other terms such as "representable". Even the definitions themselves are often different, albeit, under some reasonable conditions they turn out to be equivalent.

12.10.1 Definition. (Formally Definable Relations)

Let the theory $\mathcal{T}$ over some language L_A of arithmetic ($\supseteq L_{ROB}$) be an extension of ROB. *At this point* we do not postulate either that $\mathcal{T}$ *is consistent* or that it is a formal arithmetic, *that is,* we do not require that its *mathematical axiom set is recursive*.

We say that a relation $R \subseteq \mathbb{N}^r$ is

1. *Weakly formally definable in* $\mathcal{T}$, or just *weakly definable* in $\mathcal{T}$, iff there is a formula $F(v_0, \ldots, v_{r-1})$[25] over $L_\mathcal{A}$ such that, for all $\vec{a}_r$ in $\mathbb{N}$,

$$R(\vec{a}_r) \text{ iff } \vdash_\mathcal{T} F(\widetilde{a}_1, \ldots, \widetilde{a}_r)$$

We say that F *weakly formally defines* R in $\mathcal{T}$, but often just say that it *weakly defines* R. The context will determine in which theory $\mathcal{T}$.

2. *Formally definable in* $\mathcal{T}$, or just *definable* in $\mathcal{T}$, iff there is a formula $F(v_0, \ldots, v_{r-1})$ over $L_\mathcal{A}$ such that, for all $\vec{a}_r$ in $\mathbb{N}$,

$$\text{If } R(\vec{a}_r), \text{ then } \vdash_\mathcal{T} F(\widetilde{a}_1, \ldots, \widetilde{a}_r)$$

and

$$\text{If } \neg R(\vec{a}_r), \text{ then } \vdash_\mathcal{T} \neg F(\widetilde{a}_1, \ldots, \widetilde{a}_r)$$

Note that the left "$\neg$" is informal (metamathematical), the right one is formal.

   We say that F *formally defines* R in $\mathcal{T}$, but often just say that it *defines* R. Again, the context will determine in which theory $\mathcal{T}$.

   □

It is clear that if R is definable in ROB and $\mathcal{T}$ extends ROB (over the same or over an extended language), then R is also definable in $\mathcal{T}$.

   The following is easy.

12.10.2 Proposition. *If the theory $\mathcal{T}$ is consistent, then definability implies weak definability.*

Proof   Let $\mathcal{T}$ be such a theory. Let R be definable by F. Thus, for all $\vec{a}_r$ in $\mathbb{N}$, we have

$$\text{If } R(\vec{a}_r), \text{ then } \vdash_\mathcal{T} F(\widetilde{a}_1, \ldots, \widetilde{a}_r) \tag{1}$$

and

$$\text{If } \neg R(\vec{a}_r), \text{ then } \vdash_\mathcal{T} \neg F(\widetilde{a}_1, \ldots, \widetilde{a}_r) \tag{2}$$

(1) is half of what we want. We also want the other half, namely,

$$\text{If } \vdash_\mathcal{T} F(\widetilde{a}_1, \ldots, \widetilde{a}_r), \text{ then } R(\vec{a}_r) \tag{3}$$

---

[25] Recall that variables enclosed in round brackets to the right of a formula denote *all* the free free variables of the formula.

Well, if we do not have (3) then we have its hypothesis,

$$\vdash_{\mathcal{T}} F(\widetilde{a}_1, \ldots, \widetilde{a}_r) \tag{4}$$

but also have that $\neg R(\vec{a}_r)$ is true. By (2) we can prove

$$\vdash_{\mathcal{T}} \neg F(\widetilde{a}_1, \ldots, \widetilde{a}_r)$$

which along with (4) implies that $\mathcal{T}$ is inconsistent. But this is contrary to our main assumption.   □

12.10.3 Corollary. *If the theory $\mathcal{T}$ is consistent, then definability also implies* $\neg R(\vec{a}_r)$ *iff* $\vdash_{\mathcal{T}} \neg F(\widetilde{a}_1, \ldots, \widetilde{a}_r)$.

Proof Exactly as before, we need to establish the $\leftarrow$ direction of "iff", since the $\rightarrow$ direction we have by definability. So let $\vdash_{\mathcal{T}} \neg F(\widetilde{a}_1, \ldots, \widetilde{a}_r)$. If, contrary to what we want, $R(\vec{a}_r)$ is true, then definability implies $\vdash_{\mathcal{T}} F(\widetilde{a}_1, \ldots, \widetilde{a}_r)$ and we have just contradicted consistency of $\mathcal{T}$.   □

 It is known that both ROB and PA are consistent (formal arithmetics). So in each of them the results Proposition 12.10.2 and Corollary 12.10.3 hold.

We have three versions of *total* function definability, of which two are equivalent under some easily met conditions.

12.10.4 Definition. (Formally Definable Functions) Let $\mathcal{T}$ name an extension of ROB over some language $L_A (\supseteq L_{ROB})$. We say that a *total* function $f : \mathbb{N}^r \to \mathbb{N}$ is

1. *Weakly formally definable in $\mathcal{T}$*—or just *weakly definable*— iff there is a formula $F(v_0, \ldots, v_{r-1}, v_r)$ that weakly defines the *graph* $f(\vec{x}_r) = y$ of f in the sense of Definition 12.10.1 (case 1).
2. *Formally definable in $\mathcal{T}$*—or just *definable*— iff there is a formula $F(v_0, \ldots, v_{r-1}, v_r)$ that *defines* the *graph* $f(\vec{x}_r) = y$ of f in the sense of Definition 12.10.1 (case 2), namely, for all $\vec{a}_r$ and b in $\mathbb{N}$,

$$\text{If } f(\vec{a}_r) = b, \text{ then } \vdash_{\mathcal{T}} F(\widetilde{a}_1, \ldots, \widetilde{a}_r, \widetilde{b})$$

and

$$\text{If } f(\vec{a}_r) \neq b, \text{ then } \vdash_{\mathcal{T}} \neg F(\widetilde{a}_1, \ldots, \widetilde{a}_r, \widetilde{b})$$

3. *Strongly formally definable in $\mathcal{T}$*—or just *strongly definable*— iff we have the conditions under 2. above satisfied, along with the *single-valued-ness conditions*, for each a_i,

$$\vdash_{\mathcal{T}} F(\widetilde{a}_1, \ldots, \widetilde{a}_r, y) \wedge F(\widetilde{a}_1, \ldots, \widetilde{a}_r, z) \to y = z$$

     □

12.10.5 Lemma. *The relation $x = y$ is ROB-definable.*

Proof We show that $v_0 = v_1$ defines $x = y$.

Let $a = b$ (be true). Thus $\widetilde{a}$ and $\widetilde{b}$ are *identical* strings over L_A. Hence $\vdash \widetilde{a}=\widetilde{b}$ by *substitution* in the logical axiom $x = x$, a process legitimised by the proof

$$\begin{array}{lll}(1) & z = z & \langle\text{axiom}\rangle \\ (2) & \widetilde{a} = \widetilde{a} & \langle(1) + \text{substitution (Proposition 0.2.85)}\rangle \\ (2') & \widetilde{a} = \widetilde{b} & \langle\text{rewrote (2): the strings } \widetilde{a} \text{ and } \widetilde{b} \text{ are identical}\rangle\end{array}$$

In the present sequence of lemmata regarding the power of ROB we abuse notation slightly and simply write "$\vdash$ …" (for "… is provable") rather than "$\vdash_{ROB}$ …" (for "… is provable in ROB").

Let next $a \neq b$. We use (*metamathematical*) induction on b to show $\vdash \neg\widetilde{a}=\widetilde{b}$.

Basis. If $b = 0$, then $a = c + 1$ for some c, under the assumption. We want $\vdash \neg\widetilde{c + 1}=\widetilde{0}$, i.e., $\vdash \neg S\widetilde{c}=\widetilde{0}$, which we have, by axiom "$S1$" of ROB (p. 345) and substitution of $\widetilde{c}$ into x (Proposition 0.2.85).

Induction step. We now go to the case $a \neq b + 1$. If $a = 0$, then we are back to what we have already argued (using the logical theorem $\vdash x = y \to y = x$). Let then $a = c + 1$. Thus, $c \neq b$, and hence (I.H.), $\vdash \neg\widetilde{c}=\widetilde{b}$. By axiom "$S2$" (and tautological implication) $\vdash \neg S\widetilde{c}= S\widetilde{b}$, i.e., $\vdash \neg\widetilde{c + 1}=\widetilde{b + 1}$.

$\square$

12.10.6 Lemma. $x < y$ *is ROB-definable.*

Proof Indeed, $v_0 < v_1$ defines the relation $x < y$. We prove so by induction on b, proving simultaneously

$$a < b \text{ implies } \vdash \widetilde{a}<\widetilde{b} \tag{i}$$

and

$$a \not< b \text{ implies } \vdash \neg\widetilde{a}<\widetilde{b} \tag{ii}$$

Worth noting. Induction performed to prove properties of ROB (what it can do, for example) is necessarily *external* to ROB, that is, it is done in the metatheory. ROB does *not* have the induction tool (axiom schema); this tool is in PA.

Basis. For $b = 0$, (i) is vacuously satisfied (because the left hand side of the implication, $a < 0$, is false). On the other hand, (ii) follows from axiom "1" and substitution (similarly with the technique applied in the proof of Lemma 12.10.5).

Induction step. Let $a < b + 1$. Thus, $a < b$ or $a = b$. One case yields (I.H.) $\vdash \widetilde{a}<\widetilde{b}$ and the other $\vdash \widetilde{a}=\widetilde{b}$ (Lemma 12.10.5).

By tautological implication, $\vdash \widetilde{a}<\widetilde{b} \vee \widetilde{a}=\widetilde{b}$ in either case.

Hence $\vdash \widetilde{a} < S\widetilde{b}$, by substitution, tautological implication, and axiom "< 2" of ROB (p. 345). That is, $\vdash \widetilde{a} < \widetilde{b+1}$.

Let next $a \not< b + 1$. Thus, $a \not< b$ and $a \neq b$. Thus we have both $\vdash \neg\widetilde{a} < \widetilde{b}$ (by I.H.) and $\vdash \neg\widetilde{a} = \widetilde{b}$ (by Lemma 12.10.5), thus, (tautological implication)

$$\vdash \neg(\widetilde{a}{<}\widetilde{b} \vee \widetilde{a}{=}\widetilde{b}) \tag{iii}$$

Via the equivalence theorem (Theorem 0.2.103), tautological equivalence, and axiom "< 2", (iii) yields $\vdash \neg\widetilde{a}{<} S\widetilde{b}$, that is, $\vdash \neg\widetilde{a}{<}\widetilde{b+1}$.   □

12.10.7 Definition. (Term Definability) Let $\mathcal{T}$ over some language L_A be an extension of ROB.

We say that a *total* function $\lambda \vec{x}_n . f(\vec{x}_n)$ is *formally term definable in* $\mathcal{T}$, or just *term definable* in $\mathcal{T}$, iff there is a term $t(v_0, \ldots, v_{n-1})$ over $L_{\mathcal{T}}$ such that, for all $\vec{a}_n$ and b in $\mathbb{N}$,

$$\text{If } f(\vec{a}_n) = b, \text{ then } \vdash_{\mathcal{T}} t(\widetilde{a}_1, \ldots, \widetilde{a}_n) = \widetilde{b}$$

We say that the t above *formally term defines* f in $\mathcal{T}$, but often just say that it *term defines* f. The context will determine in which theory $\mathcal{T}$.

In the presence of the terminology "term defines", we may also say that a formula F *"formula defines"* a function, if it does so in the sense of Definition 12.10.4, case 2.   □

12.10.8 Proposition. *If the total function f is term defined by t, then the formula $t = y$, where y does not occur in t, defines f in the sense of Definition 12.10.4, case 2.*

Proof So let $\mathcal{T}$ be an extension of ROB. Let also t be $t(v_0, \ldots, v_n)$ and y be distinct from the v_i listed as variables of t. The hypothesis says that we have, for all a_i and b,

$$\text{If } f(\vec{a}_n) = b, \text{ then } \vdash_{\mathcal{T}} t(\widetilde{a}_1, \ldots, \widetilde{a}_n) = \widetilde{b} \tag{1}$$

To prove the claim about the formula $t = y$ we must prove that

$$\text{If } f(\vec{a}_n) \neq b, \text{ then } \vdash_{\mathcal{T}} \neg t(\widetilde{a}_1, \ldots, \widetilde{a}_n) = \widetilde{b} \tag{2}$$

To prove (2) we do a formal proof by contradiction (within $\mathcal{T}$; cf. Theorem 0.2.93). So, we add $t(\widetilde{a}_1, \ldots, \widetilde{a}_n) = \widetilde{b}$ to $\mathcal{T}$ and go for a contradiction. The informal hypothesis of (2) says

$$f(\vec{a}_n) = c \text{ and } c \neq b, \text{ for some } c \tag{3}$$

By (1) we have

$$\vdash_{\mathcal{T}} t(\tilde{a}_1, \ldots, \tilde{a}_n) = \tilde{c} \tag{4}$$

and by the second conjunct of (3) and Lemma 12.10.5 we have (since this already is provable in ROB)

$$\vdash_{\mathcal{T}} \neg \tilde{b} = \tilde{c} \tag{5}$$

Now, (1) and (4) and transitivity of (formal) equality (Theorem 0.2.105) yield

$$\vdash_{\mathcal{T}} \tilde{b} = \tilde{c}$$

The above contradicts (5). □

We next explore the concepts of function definability, Definition 12.10.4, a bit further.

12.10.9 Proposition. *A total function f is strongly definable in $\mathcal{T}$ in the sense of Definition 12.10.4 (case 3) iff for some formula $F(\vec{x}_n, y)$, where the free variables x_i and y are distinct, we have, for all a_i and b,*

$$\textit{If } f(\vec{a}_n) = b, \textit{ then } \vdash_{\mathcal{T}} F(\tilde{a}_1, \ldots, \tilde{a}_n, y) \equiv y = \tilde{b}$$

Proof In this proof we will simplify the notation "$\vdash_{\mathcal{T}}$" writing just "$\vdash$".

(1) *If*-part. So let it be that

$$\boxed{f(\vec{a}_n) = b \text{ implies } \vdash F(\tilde{a}_1, \ldots, \tilde{a}_n, y) \equiv y = \tilde{b}.} \tag{†}$$

Assume $f(\vec{a}_n) = b$. By substitution in the formal side of (†) we get

$$\vdash F(\tilde{a}_1, \ldots, \tilde{a}_n, \tilde{b}) \equiv \tilde{b} = \tilde{b} \tag{*}$$

and by tautological implication from (*) and the logical theorem $\tilde{b} = \tilde{b}$ we get

$$\vdash F(\tilde{a}_1, \ldots, \tilde{a}_n, \tilde{b}) \tag{**}$$

Let next $f(\vec{a}_n) \neq b$. Then $f(\vec{a}_n) = c$ and $c \neq b$. From (†) we get

$$\vdash F(\tilde{a}_1, \ldots, \tilde{a}_n, y) \equiv y = \tilde{c}$$

and by substitution $\vdash F(\tilde{a}_1, \ldots, \tilde{a}_n, \tilde{b}) \equiv \tilde{b} = \tilde{c}$ hence, by tautological implication,

$$\vdash \neg F(\tilde{a}_1, \ldots, \tilde{a}_n, \tilde{b}) \equiv \neg \tilde{b} = \tilde{c} \tag{‡}$$

By $b \neq c$ and Lemma 12.10.5 we get $\vdash \neg \widetilde{b} = \widetilde{c}$, and then tautological implication and (‡) yield $\vdash \neg F(\widetilde{a_1}, \ldots, \widetilde{a_n}, \widetilde{b})$.

Thus, F *defines the graph of* f *in* $\mathcal{T}$ in the sense of Definition 12.10.4 (case 2). We need to show the single-valued-ness condition too,

$$\vdash F(\widetilde{a_1}, \ldots, \widetilde{a_r}, y) \wedge F(\widetilde{a_1}, \ldots, \widetilde{a_r}, z) \to y = z$$

in order to have strong definability. Well, say, $f(\vec{a}_n) = b$. By (†) we have (using y and z, distinct from the v_i and from each other)

$$\vdash F(\widetilde{a_1}, \ldots, \widetilde{a_r}, y) \equiv y = \widetilde{b}$$

and

$$\vdash F(\widetilde{a_1}, \ldots, \widetilde{a_r}, z) \equiv z = \widetilde{b}$$

By tautological implication from the two lines above we obtain

$$\vdash F(\widetilde{a_1}, \ldots, \widetilde{a_r}, y) \wedge F(\widetilde{a_1}, \ldots, \widetilde{a_r}, z) \to y = \widetilde{b} \wedge z = \widetilde{b} \qquad (\P)$$

But we can prove $\vdash y = \widetilde{b} \wedge z = \widetilde{b} \to y = z$ (Theorem 0.2.105 and Exercise 0.2.106) without using mathematical axioms from $\mathcal{T}$. One more tautological implication using this and ($\P$) proves the single-valued-ness condition.

(2) *Only if*-part. So let us assume, for some formula $H(v_0, \ldots, v_n, y)$ —where all free variables shown are distinct— for all choices of a_i and b below,

$$\boxed{f(\vec{a}_n) = b \text{ implies } \vdash H(\widetilde{a_1}, \ldots, \widetilde{a_n}, \widetilde{b}).} \qquad (i)$$

and

$$\boxed{f(\vec{a}_n) \neq b \text{ implies } \vdash \neg H(\widetilde{a_1}, \ldots, \widetilde{a_n}, \widetilde{b}).} \qquad (ii)$$

that is, we have the definability, case 2, of Definition 12.10.4. We also assume the single-valued-ness condition holds

$$\vdash H(\widetilde{a_1}, \ldots, \widetilde{a_r}, y) \wedge H(\widetilde{a_1}, \ldots, \widetilde{a_r}, z) \to y = z \qquad (iii)$$

where the v_i and y and z are distinct.

We want to prove (†).

$$\text{So, let } f(\vec{a}_n) = b \qquad (iv)$$

We will show that

$$\vdash H(\widetilde{a_1}, \ldots, \widetilde{a_n}, y) \equiv y = \widetilde{b} \qquad (v)$$

From the equality axiom we get

$$\vdash y = \widetilde{b} \rightarrow \Big(H(\widetilde{a}_1, \ldots, \widetilde{a}_n, y) \equiv H(\widetilde{a}_1, \ldots, \widetilde{a}_n, \widetilde{b}) \Big)$$

hence by (the conclusion of) (i) we get via tautological implication

$$\vdash y = \widetilde{b} \rightarrow H(\widetilde{a}_1, \ldots, \widetilde{a}_n, y) \qquad (vi)$$

If we prove the $\leftarrow$ direction of (vi), then we are done.
So, do substitution of $\widetilde{b}$ into z in (iii). We obtain

$$\vdash H(\widetilde{a}_1, \ldots, \widetilde{a}_r, y) \wedge H(\widetilde{a}_1, \ldots, \widetilde{a}_r, \widetilde{b}) \rightarrow y = \widetilde{b}$$

Now the conclusion of (i) and tautological implication yields

$$\vdash H(\widetilde{a}_1, \ldots, \widetilde{a}_r, y) \rightarrow y = \widetilde{b}$$

$\square$

12.10.10 Corollary. *Let f be total. The two conditions*

$$f(\vec{a}_n) = b \text{ implies } \Gamma \vdash H(\widetilde{a}_1, \ldots, \widetilde{a}_n, \widetilde{b}).$$

and

$$\Gamma \vdash H(\widetilde{a}_1, \ldots, \widetilde{a}_r, y) \wedge H(\widetilde{a}_1, \ldots, \widetilde{a}_r, z) \rightarrow y = z$$

are equivalent to

$$f(\vec{a}_n) = b \text{ implies } \Gamma \vdash F(\widetilde{a}_1, \ldots, \widetilde{a}_n, y) \equiv y = \widetilde{b}.$$

Proof A direct observation on the preceding proof. Exercise. $\square$

 In the context of ROB and its extensions, is strong definability really "strong"? To answer this we need a bit more work on the power of ROB.

12.10.11 Lemma. *If Γ is an extension of ROB, then*

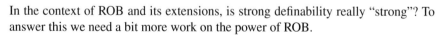
$$x = 0 \vee x = \widetilde{1} \vee \ldots \vee x = \widetilde{n} \vdash_\Gamma x < \widetilde{n+1}$$

Proof We do external (in the metatheory) induction on n.
  Basis. $n = 0$. The additional hypothesis $x = 0 \vee x = \widetilde{1} \vee \ldots \vee x = \widetilde{n}$ is just $x = 0$. By axiom "< 2" on p. 345, we have $\Gamma \vdash x < Sy \equiv x < y \vee x = y$ hence $\Gamma \vdash x < S0 \equiv x < 0 \vee x = 0$. By tautological implication from (the additional hypothesis) $x = 0$ we have $\Gamma \vdash x < S0$.

I.H. Assume then (fixing n)

$$\Gamma, x = 0 \lor x = \widetilde{1} \lor \ldots \lor x = \widetilde{n} \vdash x < \widetilde{n+1} \tag{1}$$

and prove the *Induction Step (I.S.)*

$$\Gamma, x = 0 \lor x = \widetilde{1} \lor \ldots \lor x = \widetilde{n} \lor x = \widetilde{n+1} \vdash x < \widetilde{n+2} \tag{2}$$

From $x = \widetilde{n+1} \vdash x = \widetilde{n+1}$ and (1) we get via proof by cases

$$\Gamma, x = 0 \lor x = \widetilde{1} \lor \ldots \lor x = \widetilde{n} \lor x = \widetilde{n+1} \vdash x < \widetilde{n+1} \lor x = \widetilde{n+1} \tag{3}$$

Again by axiom "< 2" on p. 345, we have

$$\Gamma \vdash x < \widetilde{Sn+1} \equiv x < \widetilde{n+1} \lor x = \widetilde{n+1}$$

$\square$

thus, (3) and tautological implication yield (2).

12.10.12 Lemma. *If Γ is an extension of ROB and A is a formula in which x occurs free —in which case we may write $A[x]$ to indicate our interest in x leaving open whether there are more free variables— then*

$$\Gamma, A[0], A[\widetilde{1}], \ldots, A[\widetilde{n-1}] \vdash (\forall x)(x < \widetilde{n} \rightarrow A[x])$$

or, using bounded quantification notation,

$$\Gamma, A[0], A[\widetilde{1}], \ldots, A[\widetilde{n-1}] \vdash (\forall x)_{<\widetilde{n}} A[x]$$

Proof We do metamathematical (external to Γ) induction on n.

Basis. $n = 0$, so the set of added (to Γ) hypotheses is empty and we want $\Gamma \vdash (\forall x)(x < 0 \rightarrow A[x])$. Well, why not see this as

$$\Gamma \vdash (\forall x)(\neg x < 0 \lor A[x]) \tag{1}$$

By axiom "< 1" on p. 345 we have $\Gamma \vdash \neg x < 0$ and hence (tautological implication) $\Gamma \vdash \neg x < 0 \lor A[x]$, which, via generalisation, yields (1).

I.H. Now *fix n* and assume

$$\Gamma, A[0], A[\widetilde{1}], \ldots, A[\widetilde{n-1}] \vdash (\forall x)(x < \widetilde{n} \rightarrow A[x]) \tag{2}$$

We will prove the *Induction Step (I.S.)* that

$$\Gamma, A[0], A[\widetilde{1}], \ldots, A[\widetilde{n}] \vdash (\forall x)(x < \widetilde{n+1} \rightarrow A[x]) \tag{3}$$

By Theorem 0.2.94, (2), and tautological implication

$$\Gamma, A[0], A[\widetilde{1}], \ldots, A[\widetilde{n}] \vdash (\forall x)(x < \widetilde{n} \to A[x]) \wedge (\forall x)(x = \widetilde{n} \to A[x])$$

hence

$$\Gamma, A[0], A[\widetilde{1}], \ldots, A[\widetilde{n}] \vdash (\forall x)\Big((x < \widetilde{n} \to A[x]) \wedge (x = \widetilde{n} \to A[x])\Big) \qquad (4)$$

by "$\forall$ over $\wedge$" distributive law. We have the tautology[26] (hence (logical) theorem)

$$(x < \widetilde{n} \to A[x]) \wedge (x = \widetilde{n} \to A[x]) \equiv (x < \widetilde{n} \vee x = \widetilde{n} \to A[x])$$

The above, via axiom "< 2" and tautological implication, yields

$$\Gamma \vdash (x < \widetilde{n} \to A[x]) \wedge (x = \widetilde{n} \to A[x]) \equiv (x < \widetilde{n+1} \to A[x])$$

The latter and the equivalence theorem (Theorem 0.2.103) applied to (4) proves (3).

$\square$

12.10.13 Theorem. *"Normal" definability Definition 12.10.4 case 2 is "strong".*

Proof Let f be defined by F in the sense of Definition 12.10.4 case 2. That is, we have

$$\text{If } f(\vec{a}_r) = b, \text{ then } \Gamma \vdash F(\widetilde{a}_1, \ldots, \widetilde{a}_r, \widetilde{b}) \qquad (1)$$

and

$$\text{If } f(\vec{a}_r) \neq b, \text{ then } \Gamma \vdash \neg F(\widetilde{a}_1, \ldots, \widetilde{a}_r, \widetilde{b}) \qquad (2)$$

We will show that f is strongly *defined* by H given by the definition

$$H(v_0, \ldots, v_{n-1}, y) \overset{Def}{\equiv} F(v_0, \ldots, v_{n-1}, y) \wedge (\forall z)_{<y} \neg F(v_0, \ldots, v_{n-1}, z) \qquad (3)$$

where the v_i and y are distinct. Defining H as above is a well known trick to obtain (from a given formula) a formula that is single-valued in one of its variables (here y). We do so by picking the smallest value of y that works —that value is unique. For an analogy, see Remark 2.4.24, second bullet.

We show that H strongly defines f. In so doing we will rely on Proposition 12.10.9.

$$\text{So let } f(a_1, \ldots, a_n) = b. \qquad (4)$$

---

[26] That we have nicknamed "proof by cases".

By (1) we have

$$\Gamma \vdash F(\tilde{a}_1, \ldots, \tilde{a}_n, \tilde{b}) \tag{$1'$}$$

but also —since $f(\vec{a}_n) \neq c$ for all $c < b$— (2) yields

$$\Gamma \vdash \neg F(\tilde{a}_1, \ldots, \tilde{a}_n, \tilde{c}), \text{ for } c = 0, 1, 2, \ldots b - 1 \tag{$2''$}$$

hence, by Lemma 12.10.12,

$$\Gamma \vdash (\forall z)\left(z < \tilde{b} \rightarrow \neg F(\tilde{a}_1, \ldots, \tilde{a}_n, z) \right) \tag{$2'$}$$

($1'$) and ($2'$) yield by tautological implication

$$\Gamma \vdash H(\tilde{a}_1, \ldots, \tilde{a}_n, \tilde{b}) \tag{5}$$

As in the proof of Proposition 12.10.9, by (5) and the equality axiom we get

$$\Gamma \vdash y = \tilde{b} \rightarrow H(\tilde{a}_1, \ldots, \tilde{a}_n, y) \tag{6}$$

We also need the converse of (6), that is, to prove that under assumption (4) we have

$$\Gamma \vdash F(\tilde{a}_1, \ldots, \tilde{a}_n, y) \wedge (\forall z)_{<y} \neg F(\tilde{a}_1, \ldots, \tilde{a}_n, z) \rightarrow y = \tilde{b} \tag{7}$$

Now, by a specialisation application to ($2'$) we get $\Gamma \vdash y < \tilde{b} \rightarrow \neg F(\tilde{a}_1, \ldots, \tilde{a}_r, y)$ therefore also

$$\Gamma \vdash F(\tilde{a}_1, \ldots, \tilde{a}_r, y) \rightarrow \neg y < \tilde{b} \tag{8}$$

Note also the substitution axiom application (hence absolute and thus also Γ-) theorem.

$$\Gamma \vdash (\forall z)_{<y} \neg F(\tilde{a}_1, \ldots, \tilde{a}_n, z) \rightarrow \left(\tilde{b} < y \rightarrow \neg F(\tilde{a}_1, \ldots, \tilde{a}_n, \tilde{b}) \right)$$

or

$$\Gamma \vdash (\forall z)_{<y} \neg F(\tilde{a}_1, \ldots, \tilde{a}_n, z) \rightarrow \left(F(\tilde{a}_1, \ldots, \tilde{a}_n, \tilde{b}) \rightarrow \neg \tilde{b} < y \right)$$

The latter simplifies to

$$\Gamma \vdash (\forall z)_{<y} \neg F(\tilde{a}_1, \ldots, \tilde{a}_n, z) \rightarrow \neg \tilde{b} < y \tag{9}$$

via ($1'$) and tautological implication. (8) and (9) tautologically imply

$$\Gamma \vdash H(\tilde{a}_1, \ldots, \tilde{a}_r, y) \rightarrow \neg y < \tilde{b} \wedge \neg \tilde{b} < y$$

Since $\neg y < \widetilde{b} \wedge \neg \widetilde{b} < y$ tautologically implies $y = \widetilde{b}$ via the trichotomy axiom of ROB (p. 345), we have got (7), and therefore, along with (6), we have

$$\Gamma \vdash H(\widetilde{a}_1, \ldots, \widetilde{a}_r, y) \equiv y = \widetilde{b}$$

The theorem is proved by an invocation of Proposition 12.10.9.   □

We next list (and prove some) specific function definability results in preparation for exploring the connection between computability and formal arithmetic.

12.10.14 Proposition. $\lambda xy.x + y$ *is term definable in any extension* Γ *of ROB by the term* $v_0 + v_1$.

Proof We show by (external) induction on b (and forall a, c) that

$$a + b = c \text{ implies } \Gamma \vdash \widetilde{a} + \widetilde{b} = \widetilde{c} \tag{1}$$

Basis. $b = 0$. Then $a = c$ and the symbols $\widetilde{a}$ and $\widetilde{c}$ denote the same term

$$\underbrace{S \ldots S}_{a \text{ times}} 0$$

Thus (p. 345, axiom "+1") we have —via substitution— $\Gamma \vdash \widetilde{a} + 0 = \widetilde{c}$. But $\widetilde{0}$ is 0 by definition.

Now *fix* b and take the *I.H.* that we have (1) for all a, c.

For the *I.S.* we work to establish

$$a + (b + 1) = c \text{ implies } \Gamma \vdash \widetilde{a} + \widetilde{b + 1} = \widetilde{c} \tag{2}$$

for all a, c.

The informal assumption implies that for some d, $c = d + 1$. But then $a + b = d$ and the I.H. yields

$$\Gamma \vdash \widetilde{a} + \widetilde{b} = \widetilde{d} \tag{2}$$

Substitution in the term Sx (cf. 0.2.81) and (2) yields $\Gamma \vdash S(\widetilde{a} + \widetilde{b}) = S\widetilde{d}$ and we are done by ROB axiom "+2" and transitivity of "=" seeing that $S\widetilde{b}$ is

$$\underbrace{S \ldots S}_{b+1 \text{ times}} 0$$

□

12.10.15 Proposition. $\lambda xy.x \times y$ *is term definable in a formal arithmetic* Γ *by the term* $v_0 \times v_1$.

Proof Exercise that imitates the proof of Proposition 12.10.14.   □

12.10.16 Proposition. *If R and Q are formally definable, then so are $\neg R$ and $R \vee Q$, as well as $R \rightarrow Q$, $R \equiv Q$ and $R \wedge Q$, where we suppressed mention of the arguments for simplicity of notation.*

That is, formally definable relations are closed under Boolean operations.

Proof We need only deal with the cases of $\neg$ and $\vee$.

[$\neg$] Let F define R. Thus

$$R \text{ implies } \Gamma \vdash F$$

and

$$\neg R \text{ implies } \Gamma \vdash \neg F$$

Trivially (by informal (on R) and formal (on F) double negation) $\neg R$ is defined by $\neg F$.

[$\vee$] Let R be defined as above and Q be defined by G. Noting that $\Gamma \vdash F$ and $\Gamma \vdash G$ yield by tautological implication that $\Gamma \vdash F \vee G$ —actually, only one of the two premises of this implication is needed to conclude this— while $\Gamma \vdash \neg F$ and $\Gamma \vdash \neg G$ and tautological implication yield $\Gamma \vdash \neg F \wedge \neg G$, we conclude that $F \vee G$ defines $R \vee Q$ since $\vdash \neg F \wedge \neg G \equiv \neg(F \vee G)$.

$\square$

12.10.17 Proposition. *If $R(\vec{x}_n)$ is formally definable in Γ, then so is*

$$R(x_1, \ldots, x_i, k, x_{i+2}, \ldots, x_n)$$

where k is a constant.

Proof Let $F(v_1, \ldots, v_i, v_{i+1}, v_{i+2}, \ldots, v_n)$ define R. Thus, for all a_j,

$$R(a_1, \ldots, a_i, a_{i+1}, a_{i+2}, \ldots, a_n) \text{ implies } \Gamma \vdash F(\tilde{a}_1, \ldots, \tilde{a}_i, \tilde{a}_{i+1}, \tilde{a}_{i+2}, \ldots, \tilde{a}_n)$$

and

$$\neg R(a_1, \ldots, a_i, a_{i+1}, a_{i+2}, \ldots, a_n) \text{ implies } \Gamma \vdash \neg F(\tilde{a}_1, \ldots, \tilde{a}_i, \tilde{a}_{i+1}, \tilde{a}_{i+2}, \ldots, \tilde{a}_n)$$

But then, for all a_j —with $j \neq i + 1$

$$R(a_1, \ldots, a_i, k, a_{i+2}, \ldots, a_n) \text{ implies } \Gamma \vdash F(\tilde{a}_1, \ldots, \tilde{a}_i, \tilde{k}, \tilde{a}_{i+2}, \ldots, \tilde{a}_n)$$

and

$$\neg R(a_1, \ldots, a_i, k, a_{i+2}, \ldots, a_n) \text{ implies } \Gamma \vdash \neg F(\tilde{a}_1, \ldots, \tilde{a}_i, \tilde{k}, \tilde{a}_{i+2}, \ldots, \tilde{a}_n)$$

That is, $F(v_1, \ldots, v_i, \widetilde{k}, v_{i+2}, \ldots, v_n)$ defines $R(x_1, \ldots, x_i, k, x_{i+2}, \ldots, x_n)$. □

12.10.18 Exercise.

- Prove that $\lambda x.x + 1$ is definable.
- Prove that $\lambda x.x$ is definable.
- Prove that $\lambda \vec{x}_n.x_i$ is definable. □

12.10.19 Proposition. *The function $\lambda xy.x \mathbin{\dot{-}} y$ is definable.*

Proof Since $x \mathbin{\dot{-}} y = z \equiv x < y \wedge z = 0 \vee \neg x < y \wedge x = y + z$ we are done by Lemmata 12.10.6, 12.10.5, Propositions 12.10.16 and 12.10.17. □

12.10.20 Proposition. *A relation $R(\vec{x}_n)$ is definable in an extension of ROB Γ iff its characteristic function is.*

Proof Given a relation $R(\vec{x}_n)$ and its characteristic function c_R which, as we know (Definition 2.2.1), satisfies

$$c_R(\vec{x}_n) = \begin{cases} 0 & \text{if } R(\vec{x}_n) \\ 1 & \text{if } \neg R(\vec{x}_n) \end{cases} \tag{1}$$

(1) means

$$c_R(\vec{x}_n) = y \equiv R(\vec{x}_n) \wedge y = 0 \vee \neg R(\vec{x}_n) \wedge y = 1 \tag{2}$$

where y is distinct from the x_i. We have two directions

- (*R is definable:*)   Say, $F(v_0, \ldots, v_n)$ defines $R(\vec{x}_n)$. We argue that

$$\boxed{H(v_0, \ldots, v_n, y), \text{ short for } F(v_0, \ldots, v_n) \wedge y = 0 \vee \neg F(v_0, \ldots, v_n) \wedge y = \widetilde{1}}$$

defines $c_R(\vec{x}_n) = y$. On the formal side in the box above we chose y to be distinct from the v_i.
So, let $c_R(\vec{a}_n) = 0$ be true. Then so is $R(\vec{a}_n)$ therefore

$$\Gamma \vdash F(\widetilde{a}_1, \ldots, \widetilde{a}_n) \tag{3}$$

By Proposition 12.10.9, we want

$$\Gamma \vdash H(\widetilde{a}_1, \ldots, \widetilde{a}_n, y) \equiv y = 0 \tag{4}$$

- For the $\leftarrow$ direction note that (by (3)) $\Gamma \vdash y = 0 \rightarrow F(\widetilde{a}_1, \ldots, \widetilde{a}_n)$, which via tautological implication along with the tautology $y = 0 \rightarrow y = 0$, yields $\Gamma \vdash y = 0 \rightarrow F(\widetilde{a}_1, \ldots, \widetilde{a}_n) \wedge y = 0$ and via one more tautological implication we get

$$\Gamma \vdash H(\tilde{a}_1, \ldots, \tilde{a}_n, y) \leftarrow y = 0$$

- For the $\rightarrow$ direction

  1. We note the tautology (hence Γ-theorem) $\Gamma \vdash F(\tilde{a}_1, \ldots, \tilde{a}_n) \wedge y = 0 \rightarrow y = 0$
  2. By tautological implication from (3) we have $\Gamma \vdash \neg F(\tilde{a}_1, \ldots, \tilde{a}_n) \wedge y = \tilde{1} \rightarrow y = 0$
  3. Proof by cases (i.e., tautological implication from the two conclusions, in 1. and 2. above) yields

$$\Gamma \vdash H(\tilde{a}_1, \ldots, \tilde{a}_n, y) \rightarrow y = 0$$

This proves (4)

There is one more case for $c_R(\vec{a}_n)$, where $c_R(\vec{a}_n) = 1$ is true. Then so is $\neg R(\vec{a}_n)$ therefore

$$\Gamma \vdash \neg F(\tilde{a}_1, \ldots, \tilde{a}_n) \tag{3'}$$

In this case we want

$$\Gamma \vdash H(\tilde{a}_1, \ldots, \tilde{a}_n, y) \equiv y = \tilde{1} \tag{4'}$$

As there are no new insights in this argument it is left as an exercise.

- (c_R *is definable*:)   Here we assume the definability of c_R, which entails the existence of a formula G such that

$$c_R(\vec{a}_n) = 0 \text{ implies } \Gamma \vdash G(\tilde{a}_1, \ldots, \tilde{a}_n, y) \equiv y = 0 \tag{i}$$

and

$$c_R(\vec{a}_n) = 1 \text{ implies } \Gamma \vdash G(\tilde{a}_1, \ldots, \tilde{a}_n, y) \equiv y = \tilde{1} \tag{ii}$$

We show now that $G(\tilde{a}_1, \ldots, \tilde{a}_n, 0)$ defines $R(\vec{x}_n)$. We start from noting

$$R(\vec{x}_n) \equiv c_R(\vec{x}_n) = 0$$

Let then $R(\vec{a}_n)$ be true. Then $c_R(\vec{a}_n) = 0$ is true, hence (i) holds. By substitution (Proposition 0.2.85) we get $\Gamma \vdash G(\tilde{a}_1, \ldots, \tilde{a}_n, 0) \equiv 0 = 0$ hence

$$\Gamma \vdash G(\tilde{a}_1, \ldots, \tilde{a}_n, 0)$$

by tautological implication since $\vdash 0 = 0$.

Let next $\neg R(\vec{a}_n)$ be true. Then $c_R(\vec{a}_n) = 1$ is true hence (ii) holds. By substitution,

$$\Gamma \vdash G(\tilde{a}_1, \dots, \tilde{a}_n, 0) \equiv 0 = \tilde{1}$$

and hence

$$\Gamma \vdash \neg G(\tilde{a}_1, \dots, \tilde{a}_n, 0) \equiv \neg 0 = \tilde{1}$$

Tautological implication from the substitution instance $\neg 0 = \tilde{1}$ of axiom $S1$, p. 345, now yields

$$\Gamma \vdash \neg G(\tilde{a}_1, \dots, \tilde{a}_n, 0)$$

$\square$

12.11 Gödel's First Incompleteness Theorem: Syntactic Versions

We saw that a number of functions and predicates —all were primitive recursive— are formally definable. Our first significant result here is that all *recursive functions and predicates* are definable. In view of Proposition 12.10.20 we need to show this only for functions. Our other —even more significant— results in this section are Gödel's First Incompleteness theorem and Church's negative solution to Hilbert's *Entscheidungsproblem*.

12.11.1 Theorem. *Every recursive function is definable in any extension of ROB* Γ.

Proof It suffices to show definability in ROB. The proof is by induction on the set of recursive functions $\mathcal{R}$, using the inductive definition of Corollary 7.0.37.

1. *Basis* —the initial functions (Definition 7.0.5).

   - For $\lambda x.0$ the definability follows from that of $x + y = z$, via Proposition 12.10.17.
   - For $\lambda x.x + 1$ and $\lambda \vec{x}_n.x_i$ see Exercise 12.10.8.
   - Proposition 12.10.14 settles definability of $\lambda xy.x + y$, while definability of $\lambda xy.x \times y$ was left as an exercise (Proposition 12.10.15).
   - Finally, definability of $\lambda xy.x \,\dot{-}\, y$ is proved in Proposition 12.10.19.

2. *Property (of function definability) is preserved by composition.* So let h of n variables, and $\lambda \vec{y}_m.g_i(\vec{y}_m)$, for $i = 1, \dots, n$, be all definable. Show that so is

$$f = \lambda \vec{y}_m.h\Big(g_1(\vec{y}_m), \dots, g_n(\vec{y}_m)\Big)$$

So, let h be defined by $H(v_1, \ldots, v_n, z)$ and each g_i by $G_i(u_1, \ldots, u_m, y_i)$, where the variables v_i, u_j, z, and y_k are distinct. Thus,

$$\boxed{g_i(\vec{a}_m) = b_i \text{ implies } \Gamma \vdash G_i(\tilde{a}_1, \ldots, \tilde{a}_m, y_i) \equiv y_i = \tilde{b}_i, \text{ for } 1 \leq i \leq n}$$

$$(1)$$

and

$$\boxed{h(b_1, \ldots, b_n) = c \text{ implies } \Gamma \vdash H(\tilde{b}_1, \ldots, \tilde{b}_n, z) \equiv z = \tilde{c}} \qquad (2)$$

Now, by the ubiquitous informal "one point rule" we have

$$f(\vec{a}_m) = c \equiv (\exists \vec{y}_n)\Big(g_1(a_1, \ldots, a_m) = y_1 \wedge \cdots \wedge g_n(a_1, \ldots, a_m) = y_n \wedge$$

$$h(\vec{y}_n) = c\Big)$$

which suggests that we should try F below as a candidate formula to define f.

$$F(\vec{u}_m, z) \stackrel{Def}{\equiv} (\exists \vec{y}_n)\Big(G_1(u_1, \ldots, u_m, y_1) \wedge \cdots \wedge G_n(u_1, \ldots, u_m, y_n) \wedge$$

$$H(\vec{y}_n, z)\Big)$$

So, let

$$f(\vec{a}_m) = c$$

This entails, for *some* $\vec{b}_n$, that $g_i(\vec{a}_m) = b_i$ $(1 \leq i \leq n)$ and $h(b_1, \ldots, b_n) = c$, hence also the right hand sides (of the "implies") in each box (1) and (2). The definition of F and substitution yields the absolute theorem:

$$\vdash F(\vec{\tilde{a}}_m, z) \equiv (\exists \vec{y}_n)\Big(G_1(\tilde{a}_1, \ldots, \tilde{a}_m, y_1) \wedge \cdots \wedge G_n(\tilde{a}_1, \ldots, \tilde{a}_m, y_n) \wedge H(\vec{y}_n, z)\Big)$$

Now the equivalence theorem 0.2.103 yields, via (1),

$$\Gamma^{27} \vdash F(\vec{\tilde{a}}_m, z) \equiv (\exists \vec{y}_n)\Big(y_1 = \tilde{b}_1 \wedge \cdots \wedge y_n = \tilde{b}_n \wedge H(\vec{y}_n, z)\Big)$$

On the other hand, by Theorem 0.2.94,

$$\vdash (\exists \vec{y}_n)\Big(y_1 = \tilde{b}_1 \wedge \cdots \wedge y_n = \tilde{b}_n \wedge H(\vec{y}_n, z)\Big) \equiv H(\tilde{b}_1, \ldots, \tilde{b}_n, z)$$

---

[27] Γ enters here due to its use in (1).

hence

$$\Gamma \vdash F(\widetilde{a}_1, \ldots, \widetilde{a}_m, z) \equiv H(\widetilde{b}_1, \ldots, \widetilde{b}_n, z)$$

by transitivity of $\equiv$. We conclude by (2): $\Gamma \vdash F(\vec{\widetilde{a}}_m, z) \equiv z = \widetilde{c}$.

3. *Property (of function definability) is preserved by* $(\widetilde{\mu} y)$. Thus, let a total function $\lambda y \vec{x}_n . f(y, \vec{x}_n)$ be definable by $F(y, v_0, \ldots, v_n, z)$, where all the y, v_i and z are distinct. Thus

$$f(b, \vec{\widetilde{a}}_n) = c \text{ implies } \Gamma \vdash F(\widetilde{b}, \widetilde{a}_1, \ldots, \widetilde{a}_n, z) \equiv z = \widetilde{c} \tag{3}$$

Moreover let

$$g = \lambda \vec{x}_n . (\widetilde{\mu} y) f(y, \vec{x}_n)$$

be total. We show that g is definable.

First off, informally,

$$g(\vec{a}_n) = b \equiv f(b, \vec{a}_n) = 0 \wedge (\forall z)(z < b \rightarrow \neg f(z, \vec{a}_n) = 0) \tag{4}$$

This suggests that we explore if indeed G given by (with distinct v_i and y —the latter as "output variable")

$$G(\vec{v}_n, y) \stackrel{Def}{\equiv} F(y, \vec{v}_n, 0) \wedge (\forall z)\Big(z < y \rightarrow \neg F(z, \vec{v}_n, 0)\Big)$$

works to define g.

So let $g(\vec{a}_n) = b$. That is, we have by (4)

- $f(b, \vec{a}_n) = 0$ and thus

$$\Gamma \vdash F(\widetilde{b}, \widetilde{a}_1, \ldots, \widetilde{a}_n, 0) \tag{5}$$

and $f(i, \vec{a}_n) \neq 0$, for $i = 0, \ldots, b - 1$, thus

$$\Gamma \vdash \neg F(\widetilde{i}, \widetilde{a}_1, \ldots, \widetilde{a}_n, 0), \text{ for } i = 0, \ldots b - 1$$

The above yields (cf. Lemma 12.10.12),

$$\Gamma \vdash (\forall z)\Big(z < \widetilde{b} \rightarrow \neg F(z, \widetilde{a}_1, \ldots, \widetilde{a}_n, 0)\Big) \tag{6}$$

Imitating the segment of the proof of Theorem 12.10.13, that follows the establishment of $(1')$ and $(2')$ there, we have here that (5) and (6) similarly yield

$$\Gamma \vdash F(y, \widetilde{a}_1, \ldots, \widetilde{a}_n, 0) \wedge (\forall z)\left(z < y \rightarrow \neg F(z, \widetilde{a}_1, \ldots, \widetilde{a}_n, 0)\right) \equiv y = \widetilde{b}$$

that is,

$$\Gamma \vdash G(\widetilde{a}_1, \ldots, \widetilde{a}_n, y) \equiv y = \widetilde{b}$$

$\square$

12.11.2 Corollary. *Every $P(\vec{x}_n) \in \mathcal{R}_*$ is definable in any extension of ROB.*

Proof   By Proposition 12.10.20. $\square$

We recall the concept of *weak (formal) definability* Definition 12.10.1(1), and Proposition 12.10.2 and Corollary 12.10.3 that in the present context yield

12.11.3 Proposition.   *If the extension $\mathcal{T}$ of ROB is consistent, then every recursive relation $R(\vec{x}_n)$ is weakly definable in $\mathcal{T}$. That is, there is a formula $F(\vec{x}_n)$ such that, for any $\vec{a}_n \in \mathbb{N}^n$, we have*

$$R(\vec{a}_n) \text{ iff } \vdash_{\mathcal{T}} F(\vec{\widetilde{a}}_n)$$

and

$$\neg R(\vec{a}_n) \text{ iff } \vdash_{\mathcal{T}} \neg F(\vec{\widetilde{a}}_n)$$

The next corollary is important in the provability theory of Peano arithmetic, in particular in the syntactic proof of the first incompleteness theorem, so I will call it a "theorem".

12.11.4 Theorem.   *If a relation has the form $(\exists y)P(y, \vec{x}_n)$ where $P(y, \vec{x}_n) \in \mathcal{R}_*$, then there is a formula $F(\vec{x}_n)$ over the language L_{ROB} such that the truth of $(\exists y)P(y, \vec{a}_n)$ for the natural numbers $\vec{a}_n$ implies the provability in ROB of $F(\vec{\widetilde{a}}_n)$.*

Proof   By Corollary 12.11.2, let $\mathscr{P}(y, \vec{x}_n)$ be a formula that defines (in ROB) the relation $P(y, \vec{x}_n)$. We can take $F(\vec{x}_n)$ to be the formula $(\exists y)\mathscr{P}(y, \vec{x}_n)$.

Indeed, let $a_1, \ldots, a_n$ be given natural numbers. If $(\exists y)P(y, \vec{a}_n)$ is true, then, for some b, $P(b, \vec{a}_n)$ is true. But then we have $\vdash_{ROB} \mathscr{P}(\widetilde{b}, \vec{\widetilde{a}}_n)$. Now, by Corollary 0.2.108 we have $\vdash_{ROB} (\exists y)\mathscr{P}(y, \vec{\widetilde{a}}_n)$. $\square$

12.11.5 Remark.   The above usually appears in the literature as "if the standard interpretation of a Σ_1 formula is true, then said formula is provable in any extension of ROB". We will talk about Σ_n relations and formulas in Sect. 13.11.2.

Neither the above theorem nor the definability results preceding it in this section require that the extensions of ROB considered have a recursive set of axioms. $\square$

12.11.1 More Arithmetisation

We continue in this section the arithmetisation of Sect. 12.8.1 as needed, among other things, to give a more complete proof of Lemma 8.1.8. For example, we need to show that the set of axioms is "recognisable" —technically, recursive. Let us approach this in a modular fashion:

12.11.6 Lemma. *The set of Boolean axioms in 0.2.28(1.–4.) is recursive.*

Proof In this proof, and in the interest of clarity and convenience, I will allow A, B, C —capital letters that we normally use as metavariables for strings or for formulas— to be also *natural number variables* so that I will avoid unnecessary acrobatics in notation.[28]

1. For $A \rightarrow A \vee B$, we first translate the schema using primitive symbols only, and fully bracket it:

$$\Big(\neg A \vee (A \vee B)\Big) \tag{1}$$

Let $A1(x)$ say "x is (when viewed in 11-adic) the Gödel number of (1)", that is, it *is* (1) (cf. Remark 12.8.5). The statement of the lemma means to show that the relation $A1(x)$ is *recursive*.
The following settles it

$$Ax1(x) \equiv (\exists A, B)_{Px}\Big(wff(A) \wedge wff(B) \wedge$$

$$x = gn\{(\neg\} * A * gn\{\vee(\} * A * gn\{\vee\} * B * gn\{))\}\Big)$$

2. Axiom 2 is $A \vee A \rightarrow A$, that is

$$\Big((\neg(A \vee A)) \vee A\Big) \tag{2}$$

Thus

$$Ax2(x) \equiv (\exists A)_{Px}\Big(wff(A) \wedge$$

$$x = gn\{((\neg(\} * A * gn\{\vee\} * A * gn\{)) \vee\} * A * gn\{)\}\Big)$$

---

[28] Like "x is the Gödel number for A", "y is the Gödel number for B", etc. Rather, we should remember that A *is* the Gödel number of A —it is just a point of view, string vs. number! So I am not letting names proliferate.

3. Axiom 3 is $A \vee B \rightarrow B \vee A$, that is,

$$\Big((\neg(A \vee B)) \vee (B \vee A)\Big) \tag{3}$$

Thus

$$Ax3(x) \equiv (\exists A, B)_{Px}\Big(wff(A) \wedge wff(B) \wedge x = gn\{((\neg(\} * A *$$

$$gn\{\vee\} * B * gn\{)) \vee (\} * B * gn\{\vee\} * A * gn\{))\}\Big)$$

4. Axiom 4 is $(A \rightarrow B) \rightarrow C \vee A \rightarrow C \vee B$, that is,

$$\Big((\neg((\neg A) \vee B)) \vee ((\neg(C \vee A)) \vee (C \vee B))\Big) \tag{4}$$

Thus

$$Ax4(x) \equiv (\exists A, B, C)_{Px}\Big(wff(A) \wedge wff(B) \wedge wff(C) \wedge$$

$$x = gn\{((\neg(\neg(\} * A * gn\{) \vee \} * B * gn\{)) \vee ((\neg(\} *$$

$$C * gn\{\vee\} * A * gn\{)) \vee (\} * C * gn\{\vee\} * B * gn\{)))\}\Big)$$

Let $Bool(x)$ denote the set of all Boolean axioms, that is, it is true of x iff x is a Boolean axiom. Then $Bool(x)$ is recursive since $Bool(x) \equiv Ax1(x) \vee Ax2(x) \vee Ax3(x) \vee Ax4(x)$.

$\square$

For the recursiveness of the set of of the first-order axioms we will need some more technical tools, to handle substitution $A[x := t]$.

12.11.7 Lemma. *The function $\lambda xy.Fr(x, y)$ is defined to output 0 if the variable with Gödel number y occurs free in the term or formula of Gödel number x. It outputs 1 otherwise.*

$Fr \in \mathcal{PR}$.

Proof Cf. Proposition 12.8.10.

$$Fr(x, y) = \begin{cases} 0 & \text{if } \Big(Term(x) \vee AF(x)\Big) \wedge var(y) \wedge yPx \\ 0 & \text{if } (\exists u)_{\leq x}\Big(wff(u) \wedge x = gn\{(\neg\} * u * gn\{)\} \wedge Fr(u, y) = 0\Big) \\ 0 & \text{if } (\exists u, v)_{\leq x}\Big(wff(u) \wedge wff(v) \wedge x = gn\{(\} * u * gn\{\vee\} * v * gn\{)\} \\ & \quad \wedge (Fr(u, y) = 0 \vee Fr(v, y) = 0)\Big) \\ 0 & \text{if } (\exists u, v)_{\leq x}\Big(wff(u) \wedge var(v) \wedge x = gn\{((\forall\} * v * gn\{)\} * u * gn\{)\} \\ & \quad \wedge Fr(u, y) = 0 \wedge \neg y = v\Big) \\ 1 & \text{otherwise} \end{cases}$$

$\square$

12.11.8 Lemma. *The relation* $closed(x)$ *that is true iff* x *is (the Gödel number of) a closed formula or term is primitive recursive.*

Proof $closed(x) \equiv (wff(x) \vee Term(x)) \wedge \neg(\exists y)_{Px} Fr(x, y) = 0.$ $\square$

12.11.9 Lemma. *The function* $\lambda xyz.legal(x, y, z)$ *returns 0 if the term of Gödel number* z *is substitutable in the variable of Gödel number* y *in the expression of Gödel number* x; *else it returns 1.*
  We have that $legal \in \mathcal{PR}.$

Proof

$$
legal(x, y, z) = \begin{cases}
0 & \text{if } Term(x) \vee AF(x) \\
0 & \text{if } (\exists u)_{\leq x}\Big(wff(u) \wedge x = gn\{(\neg\} * u * gn\{)\} \wedge legal(u, y, z) = 0\Big) \\
0 & \text{if } (\exists u, v)_{\leq x}\Big(wff(u) \wedge wff(v) \wedge x = gn\{(\} * u * gn\{\vee\} * v * gn\{)\} \\
& \qquad \wedge legal(u, y, z) = 0 \wedge legal(v, y, z) = 0\Big) \\
0 & \text{if } (\exists u, v)_{\leq x}\Big(wff(u) \wedge var(v) \wedge x = gn\{((\forall\} * v * gn\{)\} * u * gn\{)\} \\
& \qquad \wedge (Fr(z, v) = 1 \vee Fr(u, y) = 1)\Big) \\
1 & \text{otherwise}
\end{cases}
$$

$\square$

12.11.10 Lemma. *The function* $subst(x, y, z)$ *denotes the Gödel number a formula or term that we obtain as follows:*
  Start with a term or wff of Gödel number x. *If the variable of Gödel number* y *in said wff or term is everywhere replaced by a term of Gödel number* z, *then* $subst(x, y, z)$ *denotes the Gödel number of the resulting expression.*
  Note that we do the substitution with no regard for free variable capture.
  This $sub(x, y, z)$ *is primitive recursive.*

The note "*Note that we do the substitution with no regard for free variable capture*" is justified by the fact that we may check using Lemma 12.11.9 before we attempt the substitution.

Proof To keep the description simple we do two different *subst* functions, one for terms, *subst_T* and one for formulas, *subst_F* and then put them together.

1. For terms:
   $subst_T(x, y, z) =$

$$
\begin{cases}
x & \text{if } var(x) \wedge \neg y = x \\[4pt]
z & \text{if } var(x) \wedge y = x \wedge Term(z) \\[4pt]
x & \text{if } cons(x) \\[4pt]
u * subst_T(v, y, z) & \text{if } (\exists u, v)_{\leq x}(Suc(u) \wedge x = u * v \\
& \qquad \wedge\, Term(v)) \\[8pt]
gn\{(\} * subst_T(u, y, z) * & \\
\quad w * subst_T(v, y, z) * gn\{)\} & \text{if } (\exists u, v, w)_{\leq x}\Big((Plus(w) \vee \\
& \qquad Times(w)) \wedge x = gn\{(\} * u * w * v * gn\{)\} \\
& \qquad \wedge\, Term(u) \wedge Term(v)\Big) \\[8pt]
0 & \text{otherwise}
\end{cases}
$$

The 0 value for "otherwise" is apt since no expression has Gödel number 0.

2. For formulas:
   $subst_F(x, y, z) =$

$$
\begin{cases}
gn\{(\} * subst_F(u, y, z) * w * & \\
\quad subst_F(v, y, z) * gn\{)\} & \text{if } (\exists u, v, w)_{\leq x}\Big((Less(w) \vee w = \\
& \qquad gn\{=\}) \wedge Term(u) \wedge Term(v) \\
& \qquad \wedge\, x = gn\{(\} * u * w * v * gn\{)\}\Big) \\[8pt]
gn\{(\neg\} * subst_F(u, y, z) * gn\{)\} & \text{if } (\exists u)_{\leq x}\Big(wff(u) \wedge x = gn\{(\neg\} * u * gn\{)\}\Big) \\[8pt]
gn\{(\} * subst_F(u, y, z) * gn\{\vee\} * & \\
\quad subst_F(v, y, z) * gn\{)\} & \text{if } (\exists u, v)_{\leq x}\Big(wff(u) \wedge wff(v) \\
& \qquad \wedge\, x = gn\{(\} * u * gn\{\vee\} * v * gn\{)\}\Big) \\[8pt]
gn\{((\forall\} * v * gn\{)\} * & \\
\quad subst_F(u, y, z) * gn\{)\} & \text{if } (\exists u, v)_{\leq x}\Big(wff(u) \wedge var(v) \\
& \qquad \wedge\, x = gn\{((\forall\} * v * gn\{)\} * u * gn\{)\} \\
& \qquad\quad \wedge \neg v = y)\Big) \\[8pt]
x & \text{if } (\exists u, v)_{\leq x}\Big(wff(u) \wedge var(v) \\
& \qquad \wedge\, x = gn\{((\forall\} * v * gn\{)\} * u * gn\{)\} \\
& \qquad\quad \wedge v = y)\Big) \\[8pt]
0 & \text{otherwise}
\end{cases}
$$

$subst_F(x, y, z)$ effects unconditional substitution (oblivious to free variable "capture"; cf. discussion on p. 20). As such in its uses it will be accompanied by the function *legal*, which will test for substitution legitimacy. See proof of Lemma 12.11.11 below.

Finally,

$$subst(x, y, z) = \begin{cases} subst_T(x, y, z) & \text{if } Term(x) \\ subst_F(x, y, z) & \text{if } wff(x) \\ 0 & \text{otherwise} \end{cases}$$

□

We can now show the recursiveness of the first-order axioms, that for each schema, the set of its instances is recursive.

12.11.11 Lemma. *The set of first-order axioms is recursive.*

Proof

1. Axiom 5: $(\forall x)A[x] \to A[t]$. Let us rewrite (fully parenthesised) in terms of primitive connectives as

$$\Big((\neg(\forall x)A) \vee A[x := t] \Big) \tag{1}$$

Let $Ax5(x)$ mean "x is the Gödel number of (1) and t is substitutable in x". Thus

$$Ax5(x) \equiv (\exists A, v, z)_{\leq x}\Big(wff(A) \wedge var(v) \wedge Term(z) \wedge legal(A, v, z) = 0 \wedge$$

$$x = gn\{((\neg(\forall) * v * gn\,()\} * A * gn\,()\vee\} * subst(A, v, z) * gn\,()\}\Big)$$

2. Axiom 6: $y = y$. $Ax6(x) \equiv (\exists v)_{\leq x}(var(v) \wedge x = v * gn\{=\} * v)$
3. Axiom 7: $t = s \to (A[x := t] \equiv A[x := s])$. *Exercise:* Express $Ax7(x)$ that says "x is the Gödel number of an instance of the equality axiom" as a recursive relation of one variable (x).

Thus, "x is the Gödel number of an instance of a first-order axiom" —$FirstO(x)$— is recursive since $FirstO(x) \equiv Ax5(x) \vee Ax6(x) \vee Ax7(x)$. □

12.11.12 Lemma. *The set of mathematical axioms of each of ROB and PA are recursive.*

Proof We outline the proof, by highlighting three cases leaving the rest to the reader.

1. For ROB: We will call the axioms by their nicknames from p. 345. For readability I will write "S" for "$(\#f\#)$" and "$+$" for "$(\#\#f\#)$".

   a. Axiom **S2**: $Sx = Sy \to x = y$, that is, $((\neg Sx = Sy) \vee x = y)$ thus

   $$MathAxS2(z) \equiv (\exists x, y)_{\leq z}\Big(var(x) \wedge var(y) \wedge z = gn\{((\neg S\} * x *$$

   $$gn\{=\} * gn\{S\} * y * gn\,()\vee\} * x * gn\{=\} * y * gn\,()\}\Big)$$

b. Axiom $+\mathbf{2}$: $((x + Sy) = S(x + y))$.

$$MathAx + 2(z) \equiv (\exists x, y)_{\leq z}\Big(var(x) \wedge var(y) \wedge z = gn\,\{((\{\} * x * gn\,\{+S\}$$

$$* y * gn\,\{) = S(\{\} * x * gn\,\{+\} * y * gn\,\{))\}\Big)$$

2. For PA: We have all ROB axioms, plus, the induction schema, for any formula
A, and y fresh for A,

$$A[x := 0] \wedge (\forall y)(A[x := y] \rightarrow A[x := Sy]) \rightarrow A$$

or, eliminating the derived "$\rightarrow$" and putting on all required brackets,

$$\Big(\big((\neg A[x := 0]) \vee \big(\neg\big((\forall y)\big((\neg A[x := y]) \vee A[x := Sy]\big)\big)\big)\big) \vee A\Big)$$

Thus,

$$MathInd(z) \equiv (\exists x, y, A)_{\leq z}\Big(var(x) \wedge var(y) \wedge wff(A) \wedge \neg y\,P\,A\,\wedge$$

$$z = gn\,\{(((\neg\} * subst(A, x, Num(0)) * gn\,\{\} \vee (\neg((\forall\} * y * gn\,\{\}((\neg\}$$

$$* sub(A, x, Num(y)) * gn\,\{)\vee\} * subst(A, x, Num(y + 1)) * gn\,\{))))\vee\} * A * gn\,\{)\}\Big)$$

Thus, the *mathematical axioms* relation is

$$Math(z) \equiv MathAxS1(z) \vee MathAxS2(z) \vee MathAx + 1(z) \vee MathAx + 2(z)$$

$$\vee MathAx \times 1(z) \vee MathAx \times 2(z) \vee MathAx < 1(z) \vee MathAx < 2(z)$$

$$\vee MathAx < 3(z) \vee MathInd(z)$$

therefore, $Math(z)$ too is recursive.   □

Since the relation that says "z" is the Gödel number of an axiom, $Axiom(z)$, satisfies

$$Axiom(z) \equiv Bool(z) \vee FirstO(z) \vee Math(z)$$

thus is recursive.

12.11.13 Lemma. *The relations* $MP(A, B; C)$ *and* $AIntro(A; B)$ *saying respectively, "C is (the Gödel number of) a formula that follows from two formulas of Gödel numbers A and B via MP" and "B is the Gödel number of a formula that follows from the formula of Gödel number A via ∀-Intro" are recursive.*

Proof

$$MP(A, B; C) \equiv wff(A) \wedge wff(C) \wedge B = gn\{((\neg\} * A * gn\{)\vee\} * C * gn\{)\}$$

and

$$AIntro(A; B) \equiv (\exists X, Y, v)_{\leq B}\Big(wff(X) \wedge wff(Y) \wedge var(v) \wedge Fr(B, v) = 1 \wedge$$

$$A = gn\{((\neg\} * X * gn\{)\vee\} * Y * gn\{)\} \wedge$$

$$B = gn\{((\neg\} * X * gn\{) \vee ((\forall\} * v * gn\{)\} * Y * gn\{)\}\Big) \qquad \square$$

We now restate and prove Lemma 8.1.8 within our evolved notation and with the recursiveness of the axiom set and rules of inference, not just postulated but *proved*.

12.11.14 Proposition. *The predicate* $Proof(y)$ *that is true iff* y *prime-power codes a PA or ROB proof is recursive.*

Proof See also the proof of Theorem 4.3.2 for the notations $u \in v$ and $u <_w v$.

$$Proof(y) \equiv Seq(y) \wedge lh(y) \geq 1 \wedge (\forall A)_{\leq y}\Big(A \in y \to wff(A) \wedge$$

$$\Big\{Axiom(A) \vee$$

$$(\exists B, C)_{\leq y}\Big(B <_y A \wedge C <_y A \wedge MP(B, C; A)\vee$$

$$B <_y A \wedge AIntro(B; A)\Big)\Big\}\Big)$$

$$\square$$

12.11.15 Remark. Stated more carefully, Proposition 12.11.14 is *two* propositions, one for ROB and a *separate* one for PA.

More properly we should index "*Proof*" by *ROB* or *PA* depending on whether "*Axiom(x)*" is about ROB's axioms —say, $Axiom_{ROB}(x)$— or PA's axioms — $Axiom_{PA}(x)$. $\qquad \square$

We immediately have

12.11.16 Corollary. *The predicate* $Proof_T(y)$ *that is true iff* y *prime-power codes a proof written within a* recursive extension *theory* T *of ROB is recursive.*

By a recursive extension *we mean a theory over the language* L_{ROB} *that has a recursive set of axioms* $Math_T$ *that include the axioms of ROB.*

12.11.17 Corollary. (The Provability Predicate (in the Metatheory)) *For any recursive extension* T *of ROB we define* $\Theta_T(x)$ *in the metatheory by "x is the Gödel number of a theorem of* T*". This* $\Theta_T(x)$ *is semi-recursive.*

Proof Define $Bew_T(y, x)$ by

$$Bew_T(y, x) \equiv Proof_T(y) \wedge x \in y$$

Then $Bew_T \in \mathcal{R}_*$. Thus, $\Theta_T(x) \equiv (\exists y) Bew_T(y, x)$. We conclude by Theorem 6.3.14.   □

12.11.18 Remark. A function is *intentionally* a "rule" that says how to obtain outputs from inputs. *Extensionally* it is a (potentially) infinite *set* of input/output pairs.

Analogously, a theory, *intentionally*, is the set of its axioms —they, along with the rules of inference, are used to produce all the theory's theorems. *Extensionally* the theory is the set of *all* its theorems.

In general, the two sets —of axioms and theorems— are not the same, so one can be recursive, even if the other is not.

In what follows we will look at a formal arithmetic T extensionally, with no concern about its intentional nature (unless otherwise specified explicitly).

Thus a theory T is recursive, iff the set of its theorems —i.e., T— is.

As we have explained and practised several times already, the set of theorems is a set of strings which we painlessly arithmetised using the Quine/Smullyan approach utilising the duality of strings: as *strings* over the alphabet $(*)$ of p. 350 or as *numbers* written in 11-adic notation.

But is any *arithmetic recursive?*

It turns out that no consistent extension of ROB is, *even if we relax the requirement that the extension has a recursive set of axioms.*   □

12.11.19 Theorem. (Unsolvability of the *Entscheidungsproblem*) *Every consistent extension[29] of ROB, T, has an unsolvable Entscheidungsproblem; it is not recursive (extensionally).*

The above is Church's theorem (Church 1936a). Of course, the specification of T includes the special case of being ROB itself.

Proof The proof closely mirrors the proof we gave for Tarski's theorem (Theorem 12.8.15).

So let us argue by way of contradiction, and *assume* that T *is* recursive.
Thus also

---

[29] An inconsistent extension T consists of *all* wff over the language of T. But $wff(x)$ is recursive.

$$x \notin \mathcal{T} \text{ is recursive} \tag{1}$$

Since the diagonal function D of the proof of Theorem 12.8.15 is primitive recursive we have —by Grzegorczyk operations— that

$$D(x) \notin \mathcal{T} \text{ is recursive} \tag{2}$$

as well, and therefore, for some formula $F(x)$ in the language of $\mathcal{T}$ we have (Proposition 12.11.3)

$$D(a) \notin \mathcal{T} \text{ iff } \vdash_{\mathcal{T}} F(\tilde{a}), \text{ for all } a \tag{3}$$

Let $f = gn\{F(x)\}$ and thus

$$D(f) = gn\left\{\mathcal{F}(\tilde{f})\right\} \tag{4}$$

where $\vdash_{\mathcal{T}} F(x) \equiv \mathcal{F}(x)$ as in the proof of 12.8.15 or indeed from Theorem 0.2.94. Thus (3) specialises to

$$D(f) \notin \mathcal{T} \text{ iff } \vdash_{\mathcal{T}} F(\tilde{f}) \text{ iff } \vdash_{\mathcal{T}} \mathcal{F}(\tilde{f}) \tag{5}$$

(5) is a contradiction due to (4), since $D(f) \notin \mathcal{T}$ is the same as $\nvdash_{\mathcal{T}} \mathcal{F}(\tilde{f})$.   □
   Now we can obtain a few interesting corollaries:

12.11.20 Corollary. *Each of ROB and PA has a non recursive set of theorems.*

12.11.21 Corollary. Pure *first-order logic has a non recursive set of theorems, hence an unsolvable* Entscheidungsproblem, *provided its alphabet contains at least one constant and one unary function, at least two binary functions and one binary predicate.*

Proof   By the stipulation above, and since the symbols of an alphabet are *a priori* "meaningless", we may assume that our pure logic is over the alphabet of ROB

$$0, S, +, \times, < \tag{1}$$

We argue that if we could solve the decision problem for this pure logic, then we could also solve the decision problem for ROB:
   Indeed, let $\mathscr{A}$ be the conjunction of the universal closures of ROB's axioms.

 It helps that ROB has finitely many axioms!

   Let F be a formula over the alphabet (1). Then

Ded. theorem

$$\vdash_{ROB} F \quad \begin{array}{c} \Longrightarrow \\ \Longleftarrow \\ MP \end{array} \quad \vdash \mathscr{A} \to F \tag{2}$$

Thus if the *Entscheidungsproblem* for pure predicate calculus is solvable, then I can computably answer *yes/no* to "$\vdash_{ROB} F$?" by doing so for "$\vdash \mathscr{A} \to F$?"; impossible by Theorem 12.11.19. □

Below we note a much *weaker* form of Tarski's theorem. Why "weaker"? Because we already know (Definition 12.8.3 and Theorem 12.8.15) that CA is not arithmetical. But *not arithmetical* implies *not recursive*.

12.11.22 Definition. We call CA^{all} the set of *all formulas* over L_{ROB} that are true in the standard interpretation of said language (8.1). □

 Thus, both CA and CA^{all} contain $\widetilde{1} = \widetilde{1}$ but only the latter contains $\neg Sx = 0$ (the former will contain $(\forall x)\neg Sx = 0$, as will also the latter).

We have at once in the notation of Definition 0.2.83 (cf. also Proposition 0.2.84):

12.11.23 Lemma. *If $A \in CA^{all}$ then $A^u \in CA$. Conversely, if A^u for some A is in CA, then both A and A^u are in CA^{all}.*

12.11.24 Corollary. *CA^{all} is not recursive.*

Proof

1. CA^{all} is an extension of ROB: Indeed, all of ROB's theorems are in CA^{all} because all are true. (Why?).
2. CA^{all} is consistent: Indeed,

   - CA^{all} is *deductively closed*: Let $\vdash_{CA^{all}} A$. Then A is true in the standard interpretation by soundness. Hence $A \in CA^{all}$.
   - But then CA^{all} is consistent: Here is something it cannot prove: $0 = \widetilde{1}$, because otherwise $0 = \widetilde{1}$ would be in CA^{all} by the previous bullet; impossible as $0 = \widetilde{1}$ is false.

We are done by invoking Theorem 12.11.19. □

12.11.25 Corollary. *CA is not recursive.*

Proof If I can solve the decision problem $A \in CA$, then I can also solve that for $B \in CA^{all}$ contradicting Corollary 12.11.24:

Given B, there are two cases:

If B is closed, then answer true or false according as the call $B \in CA$ does. If B does have free variables, then answer true or false according as the call $B^u \in CA$ does. □

 Is there a recursive set of axioms for CA? No, because $x \in CA$ is not semi-recursive. Relevance? By Corollary 12.11.17, a recursive set of axioms would have implied the semi-recursiveness of CA.

But *why* is $CA \notin \mathcal{P}_*$?
Here is why:

1. Semi-recursive relations are arithmetical: Indeed, let $P(\vec{x}_n)$ be semi-recursive. Then, for some $e \in \mathbb{N}$,

$$P(\vec{x}_n) \equiv (\exists y)T^{(n)}(e, \vec{x}_n, y) \tag{1}$$

by Theorem 5.5.3. But $T^{(n)}(e, \vec{x}_n, y)$ is (primitive) recursive *hence arithmetical* by Theorem 12.7.3. Thus so is $P(\vec{x}_n)$ (Definition 12.7.1).
2. But by Tarski's theorem CA is not arithmetical (Theorem 12.8.15). So, CA cannot be semi-recursive either, by 1 above.

12.11.2 Gödel's Formulation of the Incompleteness Theorem

12.11.26 Definition. (Undecidable Sentences of Formal Arithmetic) A sentence A in the language of a formal arithmetic $\mathcal{T}$ is *undecidable* iff it satisfies *both* $\nvdash_{\mathcal{T}} A$ and $\nvdash_{\mathcal{T}} \neg A$.

A theory $\mathcal{T}$ that has no undecidable sentences is called *syntactically complete*, or *simply complete*. □

 A simply incomplete theory cannot certify either sentence A or $\neg A$. The concept is totally distinct from the undecidability of *sets* and *relations*.

12.11.27 Theorem. *A simply complete consistent formal arithmetic $\mathcal{T}$ is decidable (has a solvable* Entscheidungsproblem*).*

Proof We use the $Bew_{\mathcal{T}}$ of Corollary 12.11.17. Define f for all $x \in \mathbb{N}$ by

$$f(x) \overset{Def}{=} \begin{cases} (\mu y)\Big(Bew_{\mathcal{T}}(y, x) \vee Bew_{\mathcal{T}}\big(y, gn\,\{(\neg\} * x * gn\,\{)\}\big)\Big) & \text{if } wff(x) \wedge \\ & \hspace{3em} closed(x) \\ 0 & \text{otherwise} \end{cases}$$

$f \in \mathcal{P}$ and total, by assumption on $\mathcal{T}$, thus is recursive.

Hence the subset $\mathcal{S}$ of $\mathcal{T}$ that *includes only sentences* is recursive: $x \in \mathcal{S} \equiv Bew_{\mathcal{T}}(f(x), x) \wedge wff(x) \wedge closed(x)$.

We can also decide if *any* wff A is in $\mathcal{T}$, because $\vdash_{\mathcal{T}} A$ iff $\vdash_{\mathcal{S}} A^u$ (Proposition 0.2.84), that is, extensionally, $A \in \mathcal{T}$ iff $A^u \in \mathcal{S}$. This proves $\mathcal{T} \in \mathcal{R}_*$ □

 Compare the above proof with the 2nd (the mathematically exact one) proof of Theorem 6.1.7.

12.11.28 Corollary. (Gödel and Rosser Incompleteness) *If $\mathcal{T}$ is a consistent formal arithmetic, then it must be simply incomplete.*

Proof   By Theorems 12.11.19 and 12.11.27.   □

12.11.29 Definition. (ω-Consistency) A formal arithmetic $\mathcal{T}$ is ω-*consistent* iff, *for no formula A of its language*, can it prove all of

$$(\exists x)A[x], \text{ and, for all } n \geq 0, \ \neg A[\widetilde{n}] \tag{1}$$

Otherwise it is ω-*inconsistent*.   □

12.11.30 Remark. Since ω-consistency entails that $\mathcal{T}$ does not prove *everything*, it follows that $\mathcal{T}$ is consistent. It is known that the converse is not true; see Remark 12.11.40 item 4.   □

Here is a formulation of Gödel's first incompleteness theorem that uses ω-consistency (original version, Gödel 1931). The proof below *is not the original one*. It uses recursion theoretic techniques and offers additional insights on the formal arithmetic $\mathcal{T}$.

12.11.31 Theorem. (Gödel) *If a formal arithmetic $\mathcal{T}$ is ω-consistent, then it is simply incomplete.*

Proof   First a lemma within the proof: *Every semi-recursive set A is weakly definable.*

Proof of Lemma: By the projection theorem,

$$x \in A \equiv (\exists y)Q(y, x) \tag{1}$$

where Q is recursive. Let the formula $\mathbf{Q}(y, x)$ define Q in $\mathcal{T}$ (Definition 12.10.1), cf. Corollary 12.11.2.

Thus, let $a \in A$. By (1), for some b, $Q(b, a)$ is true, so, by definability, $\vdash_{\mathcal{T}} \mathbf{Q}(\widetilde{b}, \widetilde{a})$ hence

$$\text{If } a \in A, \text{ then } \vdash_{\mathcal{T}} (\exists y)\mathbf{Q}(y, \widetilde{a})$$

by Corollary 0.2.108.

For the converse assume $\vdash_{\mathcal{T}} (\exists y)\mathbf{Q}(y, \widetilde{a})$, and yet

$$\text{assume also that } a \in A \text{ is false.} \tag{3}$$

By ω-consistency, there must be some m for which

$$\nvdash_{\mathcal{T}} \neg\mathbf{Q}(\widetilde{m}, \widetilde{a}) \tag{4}$$

and thus $Q(m, a)$ is true, otherwise we have, by definability of Q, that $\vdash_{\mathcal{T}} \neg\mathbf{Q}(\widetilde{m}, \widetilde{a})$, which contradicts (4). So, (3) is invalid and we have proved $a \in A$.

That is, *we proved the lemma*:

$$a \in A \text{ iff } \vdash_\mathcal{T} (\exists y)\mathbf{Q}(y, \widetilde{a})$$

Now apply the lemma about the set A to the halting set K: As K is semi-recursive we have a formula $\mathbf{H}(x)$ such that

$$a \in K \text{ iff } \vdash_\mathcal{T} \mathbf{H}(\widetilde{a}) \tag{5}$$

that is

$$a \in K \text{ iff } gn\,\{\mathbf{H}(\widetilde{a})\} \in \mathcal{T} \text{ iff } s(gn\,\{\mathbf{H}\}, a) \in \mathcal{T} \tag{6}$$

where we wrote $s(z, w)$ for $subst(z, x, Num(w))$ since in this discussion the variable (of Gödel number) x stays fixed.

$\lambda xy.s(x, y)$ is primitive recursive just as $subst$ is (Lemma 12.11.10).

Define also the set B by

$$a \in B \text{ iff } gn\,\{(\neg\mathbf{H}(\widetilde{a}))\} \in \mathcal{T} \text{ iff } s\Big(gn\,\{(\neg\} * gn\,\{\mathbf{H}\} * gn\,\{)\}, a\Big) \in \mathcal{T} \tag{7}$$

By (7) we have $B \leq \mathcal{T}$ via the primitive recursive $\lambda a.s\Big(gn\,\{(\neg\} * gn\,\{\mathbf{H}\} * gn\,\{)\}, a\Big)$.

Thus

$$\boxed{B \in \mathcal{P}_*, \text{ since } \mathcal{T}\text{ —being a formal arithmetic— is.}}$$

Now comparing (6) and (7) we see that $K \cap B = \emptyset$ (consistency 12.11.30), hence $B \subseteq \overline{K}$. Thus, $B \subsetneq \overline{K}$.

So pick an $m \in \overline{K} - B$. Thus $m \notin K$. By (6) and (7) we have discovered an undecidable sentence: $\mathbf{H}(\widetilde{m})$ —and its companion $\neg\mathbf{H}(\widetilde{m})$.   □

12.11.32 Remark.   We have several observations.

1. We could also have used Tarski's "trick" and the function *Sub* in the above argument, instead of Gödel's s:

   For example, recalling our notation $\mathscr{H}(\widetilde{a})$ for $(\forall x)(x = \widetilde{a} \to \mathbf{H})$ and Theorem 0.2.94, the equivalence theorem 0.2.103 gives us the equivalent formulation of (5)

$$a \in K \text{ iff } \vdash_\mathcal{T} \mathscr{H}(\widetilde{a}) \tag{5'}$$

   Better still, using the primitive recursive *Sub* of Proposition 12.8.13 we state (5′) as

$$a \in K \text{ iff } Sub(gn\{\mathbf{H}\}, a) \in \mathcal{T} \tag{6'}$$

But then, we now do *also* have s!

2. In the proof above we obtain the byproduct, from (6), that $K \leq \mathcal{T}$.

As a consequence we have the startling results

3. **12.11.33 Corollary.** *Every ω-consistent formal arithmetic is creative.*

and

4. **12.11.34 Corollary.** *Any ω-consistent formal arithmetic is recursively isomorphic to K.*

and the (less startling) unsolvability of the Entscheidungsproblem under the stronger ω-consistency assumption.

5. **12.11.35 Corollary.** *As a creative theory is not recursive (extensionally), then every ω-consistent formal arithmetic has an unsolvable Entscheidungsproblem.*

6. **12.11.36 Corollary.** *Every ω-consistent formal arithmetic has infinitely many undecidable sentences, an infinite subsequence of which can be recursively enumerated.*

Proof Let $B = W_i$. At the beginning of the chapter we saw that $\overline{K}$ is productive, and that inside any productive set we can recursively generate an infinite sequence that avoids any given W_i that is a subset of said set. In our situation with $\overline{K} \supsetneq B = W_i$, let

$$f(0), f(1), f(2), \ldots, f(n), \ldots$$

be any such sequence in $\overline{K} - B$ enumerated by $f \in \mathcal{R}$.
Then all of

$$\mathbf{H}(\widetilde{f(n)}), \text{ for } n = 0, 1, 2, \ldots$$

are distinct undecidable statements.   □

12.11.3   Gödel's Incompleteness Theorem: Original Proof

We presented and proved the original formulation of the first incompleteness in the previous subsection, however the proof was not Gödel's, but was informed by advances in computability since the appearance of his theorem.

In this subsection we will imitate Gödel's original proof that is based in finding a sentence that says "I am not a theorem" and then proving it undecidable under the assumption of ω-consistency.

We revisit then the diagonalisation or fixpoint lemma of Carnap-Gödel. In this and the next subsections we will utilise bold face type for the names of formulas of the formal language. Moreover, if we are discussing an informal relation R and also introduce its formal counterpart, then the later will be denoted by the bold version of the same letter, **R**, and will have precisely the same variables as the informal R. We will not bold the names of formal variables.

12.11.37 Lemma. **(Fixpoint Lemma; Formal)** *For any formula* $\mathbf{F}(x)$ *over* $L_{\mathcal{T}}$ *of one variable* x *there is a* sentence $\mathbf{G}$ *such that we have* $\vdash_{\mathcal{T}} \mathbf{G} \equiv \mathbf{F}(\ulcorner\mathbf{G}\urcorner)$ *for any* (not necessarily recursive[30]) *extension* $\mathcal{T}$ *of ROB.* $\mathbf{G}$ *is called a fixpoint of* $\mathbf{F}$.

Proof It suffices to prove the lemma for ROB, since anything we can do in ROB that does not use the recursivess of its axiom-set we can also do in $\mathcal{T}$.

We will write "$\vdash$" below for "$\vdash_{ROB}$". The proof is simple in its core, essentially it is the one given in Chap. 9 (Lemma 9.1.1), but we will have to do some acrobatics since our "s" function of the proof of 12.11.31 is informal (lives in the metatheory of ROB).[31]

We define s in this lemma —in the metatheory of ROB— for all z and w, as

$$s(z, w) \overset{Def}{=} subst(z, x, Num(w)), \text{ keeping } x \text{ constant}[32]$$

Since $s \in \mathcal{R}$, so is the "diagonalisation" function d that we here define by

$$d = \lambda x.s(x, x)[33]$$

Let $\mathbf{D}(x, y)$ define d, that is, for any natural numbers a and b,

$$d(a) = b \text{ implies } \vdash \mathbf{D}(\tilde{a}, y) \equiv y = \tilde{b} \tag{1}$$

As in Lemma 9.1.1 we will want next to consider $\mathbf{F}(d(x))$. But s being informal, we consider instead

$$(\exists z)\Big(\mathbf{D}(x, z) \wedge \mathbf{F}(z)\Big) \tag{2}$$

Let us use the short (metamathematical) name $\mathbf{H}$[34] for the formula in (2), and let

$$\tilde{h} = \ulcorner\mathbf{H}(x)\urcorner \tag{3}$$

that is, informally, $h = gn\,\{\mathbf{H}(x)\}$.

By (3) and the equivalent explanatory statement immediately following it, the behaviour of *subst* (and thus of s and d) implies that we have $d(h) = s(h, h) =$

---

[30] I mean, not necessarily with a recursive set of *mathematical axioms*.

[31] See Tourlakis (2003a) for a proof that there is a term t of two variables such that it term-defines s in PA.

[32] That is, $s = \lambda zw.subst(z, x, Num(w))$.

[33] Lower case d so that we will not confuse it with the D of the proof of Theorem 12.11.19.

[34] Just as $\exists$ is a short metamathematical name for $\neg\forall\neg$.

$gn\left\{\mathbf{H}(\widetilde{h})\right\}$. Let us write

$$e \stackrel{Def}{=} gn\left\{\mathbf{H}(\widetilde{h})\right\} \text{ and hence also } \widetilde{e} = \ulcorner\mathbf{H}(\widetilde{h})\urcorner \tag{4}$$

Thus $d(h) = e$ hence, from (1), we have $\vdash \mathbf{D}(\widetilde{h}, y) \equiv y = \widetilde{e}$, and via Theorem 0.2.103 and (2) plus substitution

$$\vdash (\exists z)\left(\mathbf{D}(\widetilde{h}, z) \wedge \mathbf{F}(z)\right) \equiv (\exists z)\left(y = \widetilde{e} \wedge \mathbf{F}(z)\right) \tag{5}$$

or (Theorem 0.2.94) and definition of $\mathbf{H}$,

$$\vdash \mathbf{H}(\widetilde{h}) \equiv \mathbf{F}(\widetilde{e}) \tag{5'}$$

Expanding the abbreviation $\widetilde{e}$ from (4), (5') becomes $\vdash \mathbf{H}(\widetilde{h}) \equiv \mathbf{F}\left(\ulcorner\mathbf{H}(\widetilde{h})\urcorner\right)$. Thus $\mathbf{H}(\widetilde{h})$ is the "$\mathbf{G}$" we want. $\square$

12.11.38 Remark. From the proof we see that $\mathbf{G}$ is over the language of $\mathbf{F}$. $\square$

12.11.39 Theorem. (The 1st Incompleteness Theorem; Original Proof) *If $\mathcal{T}$ is an ω-consistent formal arithmetic, then it is incomplete.*

Proof Let

$$\Theta_{\mathcal{T}}(x) \stackrel{Def}{=} (\exists y)Bew_{\mathcal{T}}(y, x)$$

be the provability predicate for $\mathcal{T}$. Let $\mathbf{Bew}_{\mathcal{T}}$ over L_{ROB} *define* (Corollary 12.11.2) the recursive $Bew_{\mathcal{T}}$, that is,

$$Bew_{\mathcal{T}}(b, a) \text{ true } implies \ \vdash \mathbf{Bew}_{\mathcal{T}}(\widetilde{b}, \widetilde{a}), \text{ for all } a \text{ and } b \tag{1}$$

$$Bew_{\mathcal{T}}(b, a) \text{ false } implies \ \vdash \neg\mathbf{Bew}_{\mathcal{T}}(\widetilde{b}, \widetilde{a}), \text{ for all } a \text{ and } b \tag{1'}$$

where we write $\vdash$ for $\vdash_{PA}$ throughout this proof.[35]
    Let $\mathbf{G}$ —in the language of ROB (Remark 12.11.38 above)— be a fixpoint of $\neg(\exists y)\mathbf{Bew}_{\mathcal{T}}(y, x)$, that is,

$$\vdash \mathbf{G} \equiv \neg(\exists y)\mathbf{Bew}_{\mathcal{T}}(y, \ulcorner\mathbf{G}\urcorner) \tag{2}$$

We show that $\mathbf{G}$ is an undecidable sentence.

1. First we show that $\nvdash \mathbf{G}$. *Suppose instead* that $\vdash \mathbf{G}$. Then for some $b \in \mathbb{N}$, $Bew_{\mathcal{T}}(b, gn\{\mathbf{G}\})$ is true (b codes a proof of $\mathbf{G}$). Hence, by (1), we have

---

[35] In fact, $\vdash$ is for $\vdash_{ROB}$ here, but we mention the *a fortiori* $\vdash_{PA}$ for future reference in Sect. 12.12.

$\vdash \mathbf{Bew}_{\mathcal{T}}(\widetilde{b}, \ulcorner \mathbf{G} \urcorner)$ and therefore (Corollary 0.2.108)

$$\vdash (\exists y)\mathbf{Bew}_{\mathcal{T}}(y, \ulcorner \mathbf{G} \urcorner) \tag{3}$$

But we also have $\vdash \neg(\exists y)\mathbf{Bew}_{\mathcal{T}}(y, \ulcorner \mathbf{G} \urcorner)$ by (2) and Post, contradicting consistency of $\mathcal{T}$ (cf. Remark 12.11.30).

2. Now show that $\nvdash \neg\mathbf{G}$. *Suppose instead* that $\vdash \neg\mathbf{G}$. By (2) and Post this yields (3). Now, $Bew_{\mathcal{T}}(b, gn\,\{\mathbf{G}\})$ *cannot be false* for all $b \in \mathbb{N}$. Otherwise we have by $(1')$

$$\vdash \neg\mathbf{Bew}_{\mathcal{T}}(\widetilde{b}, \ulcorner \mathbf{G} \urcorner)$$

for *all* $b \in \mathbb{N}$. This and (3) contradict ω-consistency. So for some b we have $Bew_{\mathcal{T}}(b, gn\,\{\mathbf{G}\})$ is true, and thus $\vdash \mathbf{G}$. But this renders $\mathcal{T}$ inconsistent once more.

$\square$

12.11.40 Remark.

1. Think of $\mathcal{T}$ as PA or ROB. The right hand side (of $\equiv$) in (2) above says "$\mathbf{G}$ is not a theorem". By (2), $\mathbf{G}$ says the same thing.

   Thus it says "I am not a theorem".

   Having proved that this is indeed the case, and since PA and ROB are correct (cf. Remark 0.2.91), $\mathbf{G}$ is true in the standard interpretation of $L_{ROB} = L_{PA}$. But then $\neg\mathbf{G}$ must be false.

2. Clearly, the theory $PA + \neg\mathbf{G}^{36}$ is consistent since otherwise $\vdash_{PA} \mathbf{G}$ (cf. Theorem 0.2.93). Thus $PA + \neg\mathbf{G}$ is an example of a consistent but *incorrect* (cf. Remark 0.2.91) formal arithmetic. Why *formal* arithmetic? Because it includes the axioms of ROB (and of PA at that), and adding just one axiom —namely, $\neg\mathbf{G}$— to the axioms of PA the new set of mathematical axioms is still recursive, by closure of recursive relations under $\vee$.

3. With entirely analogous reasoning as above, $ROB + \mathbf{G}$ is a formal arithmetic. We can then apply Theorem 12.11.39 to find a new $\mathbf{G}'$.
   Pause. Why "new"?◄
   Then add $\mathbf{G}'$ to the axioms, find a new $\mathbf{G}''$, add it to the axioms, and so on.
   After each addition the new mathematical axiom set remains recursive, as we always add one member to a recursive set.
   Pause. In fact, even if we added an *infinite* recursive set of new mathematical axioms in each step, we would still obtain, in *each step* and after the addition, a recursive mathematical axiom set (again by closure of $\mathcal{R}_*$ under $\vee$).◄

---

[36] Recall the notation introduced in the remark below Metatheorem 0.2.51 where "Γ, A" and "$\Gamma + A$" both mean $\Gamma \cup \{A\}$ where Γ is some set of mathematical axioms (or even non-axiom hypotheses).

This means that ROB, and PA, are not just incomplete; they are *incompletable*, in that we cannot render them complete by adding some recursive set of axioms to the original mathematical axioms.

By the way, the fixation on recursive axiom sets is pragmatic: Unless we can *tell* (by some method or algorithm) whether a formula is a mathematical axiom or not we cannot check a proof for *syntactic correctness*, let alone *write one!*

4. Let us take $\mathcal{T} = PA$. The *consistent* formal arithmetic $PA + \neg\mathbf{G}$ is interesting. $PA + \neg\mathbf{G}$ proves (3) of the proof of Theorem 12.11.39 via (2), since the subtheory PA proves (2). On the other hand (being a ROB extension) $PA + \neg\mathbf{G}$ also proves $\neg\mathbf{Bew}_{\mathcal{T}}(b, \ulcorner\mathbf{G}\urcorner)$, for all $b \in \mathbb{N}$, because $Bew_{\mathcal{T}}(b, gn\,\{\mathbf{G}\})$ is false, for all $b \in \mathbb{N}$ (otherwise we would have $\vdash \mathbf{G}$, contradicting consistency of $PA + \neg\mathbf{G}$). Thus $PA + \neg\mathbf{G}$ is consistent but not ω-consistent.   □

12.11.4   Rosser's Incompleteness Theorem: Original Version

Rosser (1936) via a somewhat more complex proof of the first incompleteness theorem sharpened the requirement of ω-consistency to the *weaker* one of *consistency* —consistency does not imply ω-consistency: cf. Remark 12.11.40, 4. We present here a proof that does not rely on the unsolvability of the Entscheidungsproblem,[37] —cf. Corollary 12.11.28— but nevertheless, the proof in this subsection *is not Rosser's original proof*. The latter extended Gödel's proof idea via an appropriate *fixpoint*. We will present that proof too, but in the next subsection.

In the present subsection we will give a proof that looks back to the Rosser result through the lens of recursion theoretic techniques. To this end we will need some recursion theoretic lemmata.

12.11.41 Lemma. (Separation Lemma)   *Let A and B be any two* disjoint *semi-recursive subsets of* $\mathbb{N}$. *Then there is formula* $\mathbf{H}(x)$ *that "separates" them within ROB, that is, for all* $m \in \mathbb{N}$,

$$m \in A \text{ implies } \vdash_{ROB} \mathbf{H}(\widetilde{m})$$

and

$$m \in B \text{ implies } \vdash_{ROB} \neg\mathbf{H}(\widetilde{m}).$$

Proof   By Theorem 6.3.14, there are relations $R(x, y)$ and $Q(x, y)$ in $\mathcal{R}_*$ such that

$$m \in A \equiv (\exists x)R(x, m) \tag{1}$$

and

$$m \in B \equiv (\exists x) Q(x, m) \tag{2}$$

for all m.

We will assume that $R(x, y) \equiv T(e, y, x)$ and $Q(x, y) \equiv T(f, y, x)$

---

where T is the Kleene predicate and e, f are constants
By Corollary 12.11.2, let $\mathbf{R}$ and $\mathbf{Q}$ define R and Q respectively.
We claim that $\mathbf{H}(y)$ given by

$$\mathbf{H}(y) \overset{Def}{\equiv} (\exists x)\Big(\mathbf{R}(x, y) \wedge (\forall z)\big(z < x \to \neg\mathbf{Q}(z, y)\big)\Big) \tag{3}$$

works.

1. Let $m \in A$. Then $m \notin B$, hence (by (1) and (2)), $R(n + 1, m)$ holds for some n —why $n + 1$? Cf. Remark 5.5.4 and boxed statement above— while $Q(i, m)$, for all $i \geq 0$, is *false* —in particular, $\neg Q(i, m)$ is *true*, for $0 \leq i < n + 1$. Thus (Corollary 12.11.2),

$$\vdash \mathbf{R}(S\widetilde{n}, \widetilde{m}) \tag{4}$$

and $\vdash \neg\mathbf{Q}(\widetilde{i}, \widetilde{m})$, for $0 \leq i < n + 1$.[38] By Lemma 12.10.12, $\vdash z < S\widetilde{n} \to \neg\mathbf{Q}(z, \widetilde{m})$. Now, apply generalisation first, then use (4) and Post to obtain

$$\vdash \mathbf{R}(S\widetilde{n}, \widetilde{m}) \wedge (\forall z)\Big(z < S\widetilde{n} \to \neg\mathbf{Q}(z, \widetilde{m})\Big)$$

Thus (Corollary 0.2.108)

$$\vdash (\exists x)\Big(\mathbf{R}(x, \widetilde{m}) \wedge (\forall z)\big(z < x \to \neg\mathbf{Q}(z, \widetilde{m})\big)\Big)$$

i.e., by (3), $\vdash \mathbf{H}(\widetilde{m})$.

2. Let next $m \in B$. We want to show that $\vdash \neg\mathbf{H}(\widetilde{m})$, meaning

$$\vdash (\forall x)\Big(\neg\mathbf{R}(x, \widetilde{m}) \vee (\exists z)\big(z < x \wedge \mathbf{Q}(z, \widetilde{m})\big)\Big)$$

It suffices to prove instead the following (and then apply generalisation):

---

[38] Throughout this proof "$\vdash$" is short for "$\vdash_{ROB}$".

$$\neg \mathbf{R}(x, \widetilde{m}) \vee (\exists z)(z < x \wedge \mathbf{Q}(z, \widetilde{m})) \tag{5}$$

The assumption yields $m \notin A$, hence (by (1) and (2)), $Q(n, m)$ holds for some n, thus

$$\vdash \mathbf{Q}(\widetilde{n}, \widetilde{m}) \tag{6}$$

while $R(i, m)$, *for all $i \geq 0$, is false*, hence as above and by Lemma 12.10.12,

$$\vdash x < S\widetilde{n} \rightarrow \neg \mathbf{R}(x, \widetilde{m})$$

By Theorem 0.2.103 and ROB axiom "< 2" the latter yields,

$$\vdash (x <\widetilde{n} \vee x = \widetilde{n}) \rightarrow \neg \mathbf{R}(x, \widetilde{m}) \tag{7}$$

Now, (6) and Post derive $\vdash \widetilde{n} < x \rightarrow \widetilde{n} < x \wedge \mathbf{Q}(\widetilde{n}, \widetilde{m})$, therefore

$$\vdash \widetilde{n} < x \rightarrow (\exists z)(z < x \wedge \mathbf{Q}(z, \widetilde{m})) \tag{8}$$

by Theorem 0.2.107[39] and one more tautological implication.
Yet another tautological implication from (7) and (8) yields

$$\vdash (x <\widetilde{n} \vee x = \widetilde{n} \vee \widetilde{n} < x) \rightarrow \neg \mathbf{R}(x, \widetilde{m}) \vee (\exists z)(z < x \wedge \mathbf{Q}(z, \widetilde{m}))$$

from which MP and ROB axiom < 3 yield (5). $\square$

12.11.42 Remark. In view of Corollary 12.11.2, the Separation Lemma holds *in any extension of ROB*. $\square$

12.11.43 Definition. Two disjoint semi-recursive subsets of $\mathbb{N}$, A and B, are *recursively inseparable* iff there is *no* recursive set S such that $A \subseteq S$ and $S \subseteq \overline{B}$ (where $\overline{B} = \mathbb{N} - B$).

12.11.44 Lemma. $A = \{x : \phi_x(x) = 0\}$ *and* $B = \{x : \phi_x(x) = 1\}$ *are recursively inseparable.*

Proof Clearly A and B are disjoint semi-recursive sets. For example $x \in A \equiv (\exists y)(T(x, x, y) \wedge out(y, x, x) = 0)$ (cf. Theorem 5.5.3).
    Let S be recursive and suppose that it separates A and B, i.e.,

$$B \subseteq S \subseteq \overline{A} \tag{1}$$

Let $\phi_i = c_S$.

---

[39] Applied as $\vdash \widetilde{n} < x \wedge \mathbf{Q}(\widetilde{n}, \widetilde{m}) \rightarrow (\exists z)(z < x \wedge \mathbf{Q}(z, \widetilde{m}))$.

Assume that $i \in S$. Then $\phi_i(i) = 0$, hence $i \in A$ (definition of A). But also $i \in \overline{A}$, by (1). A contradiction.

Try $i \in \overline{S}$, then. Well, if so, then $\phi_i(i) = 1$, hence $i \in B$ (definition of B). By (1), $i \in S$ as well. Another contradiction.

Thus $i \notin S \cup \overline{S} = \mathbb{N}$, which is absurd. We conclude that no recursive set such as S exists. $\square$

12.11.45 Theorem. (Gödel-Rosser First Incompleteness Theorem)
Any consistent formal arithmetic $\mathcal{T}$ is incomplete.

Proof *In this proof $\vdash$ is short for $\vdash_{\mathcal{T}}$.*

By Lemma 12.11.41 we have a formula $\mathbf{H}(x)$ such that, for the A and B of Lemma 12.11.44, the following hold

$$a \in A \text{ implies } \vdash \mathbf{H}(\widetilde{a}) \tag{1}$$

and

$$b \in B \text{ implies } \vdash \neg\mathbf{H}(\widetilde{b}) \tag{2}$$

Let us define

$$S \overset{Def}{=} \{a : \vdash \mathbf{H}(\widetilde{a})\}$$

and

$$S' \overset{Def}{=} \{a : \vdash \neg\mathbf{H}(\widetilde{a})\}$$

Trivially, from (1) and (2),

$$A \subseteq S \text{ and } B \subseteq S' \tag{3}$$

Since

$$a \in S \text{ iff } gn\{\mathbf{H}(\widetilde{a})\} \in \mathcal{T} \text{ iff } s(gn\{\mathbf{H}\}, a) \in \mathcal{T}$$

we have $S \leq \mathcal{T} \in \mathcal{P}_*$ —the $\in$ from the semi-recursiveness of $\Theta_{\mathcal{T}}$— we obtain $S \in \mathcal{P}_*$. Similarly, $S' \in \mathcal{P}_*$.

Now we invoke consistency: We have

$$S \cap S' = \emptyset \tag{4}$$

since otherwise, for some a, $a \in S \cap S'$, hence we have both $\vdash \mathbf{H}(\widetilde{a})$ and $\vdash \neg\mathbf{H}(\widetilde{a})$; impossible.

We claim now that $S \cup S' \subsetneq \mathbb{N}$, hence there is an

$$e \in \mathbb{N} - (S \cup S') \tag{5}$$

If not so, $S' = \mathbb{N} - S = \overline{S}$, hence S is *recursive* and separates A and B —$A \subseteq S$ and $B \subseteq \overline{S}$. This contradicts Lemma 12.11.44.

(5) now quickly leads to $\nvdash \mathbf{H}(\widetilde{e})$ *and* $\nvdash \neg\mathbf{H}(\widetilde{e})$, that is, we found an undecidable sentence: $\mathbf{H}(\widetilde{e})$. □

12.11.46 Remark.

(I) It is clear that the set $\mathbb{N} - S \cup S'$ is infinite: If $M = \mathbb{N} - S \cup S'$ is finite, then $S \cup M$ is recursive, since it is semi-recursive (closure of $\mathcal{P}_*$ under $\vee$) and so is its complement S'. Thus $S \cup M$ recursively separates A and B. Absurd!
Thus *we have infinitely many undecidable sentences*, one for each distinct $e \in \mathbb{N} - S \cup S'$.

(II) For each $e \in \mathbb{N} - S \cup S'$, exactly one of $\mathbf{H}(\widetilde{e})$ and $\neg\mathbf{H}(\widetilde{e})$ is true in the standard interpretation.
Let us determine which one. To this end let us work in a *correct* (cf. Remark 0.2.91) formal arithmetic (e.g., one of ROB or PA will do).
$\mathbf{H}(y)$ was defined in the proof of Lemma 12.11.41 by (3), namely,

$$\mathbf{H}(y) \overset{Def}{\equiv} (\exists x)\Big(\mathbf{R}(x, y) \wedge (\forall z)\big(z < x \to \neg\mathbf{Q}(z, y)\big)\Big)$$

where $\mathbf{R}$ and $\mathbf{Q}$ were introduced as the formal counterparts —in the sense of Proposition 12.11.3— of the recursive R and Q in (1) and (2) below respectively.

$$m \in A \equiv (\exists x)R(x, m) \tag{1}$$

and

$$m \in B \equiv (\exists x)Q(x, m) \tag{2}$$

for all m. From the definition of $\mathbf{H}$ we immediately derive that

$$\vdash \neg\mathbf{H}(y) \equiv (\forall x)\Big(\neg\mathbf{R}(x, y) \vee (\exists z)\big(z < x \wedge \mathbf{Q}(z, y)\big)\Big) \tag{i}$$

Now, say $e \in \mathbb{N} - S \cup S'$. Then, in particular, $e \notin A$, hence by (1) $\neg(\exists x)R(x, e)$ is true, that is, $(\forall x)\neg R(x, e)$ is true. By Proposition 12.11.3 we have

$$\text{for all } m, \ \neg R(m, e) \text{ iff } \vdash \neg\mathbf{R}(\widetilde{m}, \widetilde{e}) \tag{3}$$

Since the left hand side of iff is true (for all m) in (3), so is the right hand side (for all m). By tautological implication the latter yields (for all m),

$$\vdash \left(\neg \mathbf{R}(\widetilde{m}, \widetilde{e}) \vee (\exists z)\left(z < \widetilde{m} \wedge \mathbf{Q}(z, \widetilde{e}) \right) \right) \tag{4}$$

To conclude note that, by correctness, $\left(\neg \mathbf{R}(\widetilde{m}, \widetilde{e}) \right)^{\mathbb{N}}$, that is,

$$\neg \mathbf{R}^{\mathbb{N}}(m, e)$$

is true for all m and the fixed e. So is, (trivially due to the "$\vee$") —for all m and the fixed e— the interpretation of (4), namely

$$\left(\neg \mathbf{R}^{\mathbb{N}}(m, e) \vee (\exists z)\left(z < m \wedge \mathbf{Q}^{\mathbb{N}}(z, e) \right) \right)$$

Hence, also

$$(\forall x)\left(\neg \mathbf{R}^{\mathbb{N}}(x, e) \vee (\exists z)\left(z < x \wedge \mathbf{Q}^{\mathbb{N}}(z, e) \right) \right)$$

is true. But this is the interpretation of $\neg \mathbf{H}(\widetilde{e})$.

(III) *ω-incompleteness of arithmetic.* Say $e \in \mathbb{N} - S \cup S'$. Then, in particular, $e \notin A$, hence by (1) in the proof of Lemma 12.11.41, $\neg(\exists x)R(x, e)$ is true, that is, $(\forall x)\neg R(x, e)$ is true. By Proposition 12.11.3,

$$\text{for all } m \geq 0, \ \neg R(m, e) \text{ iff } \vdash \neg \mathbf{R}(\widetilde{m}, \widetilde{e})$$

Since we do have, for all $m \geq 0$, that $\neg R(m, e)$ is true, tautological implication and the above yield,

$$\text{for all } m \geq 0, \ \vdash \neg \mathbf{R}(\widetilde{m}, \widetilde{e}) \vee (\exists z)\left(z < \widetilde{m} \wedge \mathbf{Q}(z, \widetilde{e}) \right) \tag{4}$$

Let us use the shorthand $\mathbf{F}(x, y)$ for $\neg \mathbf{R}(x, y) \vee (\exists z)\left(z < x \wedge \mathbf{Q}(z, y) \right)$ thus, (4) translates to

$$\text{for } m \geq 0, \ \vdash \mathbf{F}(\widetilde{m}, \widetilde{e})$$

However

$$\nvdash (\forall x)\mathbf{F}(x, \widetilde{e})$$

Why? Because it is the $\neg \mathbf{H}(\widetilde{e})$ of the proof of Theorem 12.11.45.
Thus a formal arithmetic is *not* closed under the so-called *ω-rule*

$$\frac{A(\widetilde{0}), A(\widetilde{1}), A(\widetilde{2}), A(\widetilde{3}), \dots, A(\widetilde{k}), \dots}{(\forall x)A(x)}$$

Every such theory is what we call *ω-incomplete.*   □

12.11.47 Theorem. (Yet Another Proof of Church's Theorem) *Every* consistent extension T of ***ROB*** *is not* recursive.

Proof The proof takes a different turn than that of the proof of Theorem 12.11.45. We reuse said proof here, noting (1), (2) and definition (*ibid*)

$$S \stackrel{Def}{=} \{a : \vdash \mathbf{H}(\widetilde{a})\}$$

We next note (*ibid*) that

$$S \leq T \tag{*}$$

By (2) (*ibid*) it is $B \cap S = \emptyset$, hence $S \subseteq \overline{B}$. That is, S separates A and B of Lemma 12.11.41, thus $S \notin \mathcal{R}_*$. But then, by (*), $T \notin \mathcal{R}_*$ as well. $\qquad\square$

 The above proof —unlike that of Theorem 12.11.45— did not assume or use that T has a recursive set of axioms. This result (Church 1936a,b) was published some five years after Gödel published his *incompleteness* results, and shows that the *decision problem* for *any consistent theory* — *not just for one that is recursively axiomatized*— that contains ROB is recursively unsolvable.

Church's theorem directly contradicted Hilbert's belief that the *Entscheidungsproblem* of any axiomatic theory ought to be solvable by "mechanical means".

On the other hand, Gödel's incompleteness theorems had already shown the untenability of Hilbert's hope to address the consistency of axiomatic theories by "finitary means".[40]

12.11.5 Rosser's Incompleteness Theorem: Original Proof

The proof here is similar to that of Theorem 12.11.39, starting with a more complex provability predicate in order to circumvent the need of postulating ω-consistency. Namely, instead of $(\exists y) Bew_T(y, x)$ we will use

$$(\exists y)\Big(Bew_T(y, x) \wedge (\forall z)\big(z < y \rightarrow \neg Bew_T(z, gn \{(\neg\} * x * gn \{\})\})\big)\Big) \tag{1}$$

intuitively saying that some y verifies the provability of x, but no proof-code $z < y$ verifies the refutation of "x" (i.e., the provability of its negation).

---

[40] The idea is that "finitary means" are syntactic means describable by finitary processes over a finite alphabet. *BUT*: As such they can be formalised within PA, say via arithmetisation if needed. However, the consistency of PA is not provable from within PA (by Gödel's Second Incompleteness Theorem, whose proof we will *sketch* shortly). Hence not provable by "finitary means". For a complete self-contained proof see one or both of Hilbert and Bernays (1968), Tourlakis (2003a).

It will be clear that the proof details track very closely the details of the proof of the *Separation Lemma* (Lemma 12.11.41). The essence of the proof is also very similar to that of the proof of Theorem 12.11.45 this given via Lemma 12.11.41:

The latter lemma finds a formula $\mathbf{H}(y)$ that separates two disjoint semi-recursive sets A and B. Then an e is shown to exist that is outside two disjoint r.e. supersets of A and B, and $\mathbf{H}(\tilde{e})$ furnishes an example of an undecidable sentence.

Intuitively, an example of such sets A and B (other than those chosen for Theorem 12.11.45) might be to take A to be the r.e. set of all (Gödel numbers of) provable sentences ψ (in, say, ROB) and B taken to be the r.e. set of all *disprovable* or *refutable* sentences ψ (i.e., such that $\vdash_{ROB} \neg\psi$) and find a sentence $\mathbf{R}$ that is in $\overline{A \cup B}$. This is the idea in the proof below (Rosser 1936).

12.11.48 Theorem. *If $\mathcal{T}$ is a consistent formal arithmetic, then it is incomplete.*

Proof Let, as before, $\mathbf{Bew}_{\mathcal{T}}$ weakly define (Proposition 12.11.3) $Bew_{\mathcal{T}}$, that is:

$$\text{For all } a \text{ and } b, \, Bew_{\mathcal{T}}(a, b) \text{ iff } \vdash \mathbf{Bew}_{\mathcal{T}}(a, b) \tag{2}$$

and

$$\text{For all } a \text{ and } b, \, \neg Bew_{\mathcal{T}}(a, b) \text{ iff } \vdash \neg\mathbf{Bew}_{\mathcal{T}}(a, b) \tag{2'}$$

To follow the outlined plan that we spoke of in the preamble of this subsection we will also need to talk of *disprovable* sentences ψ, hence, by (2) and (2') we have also, for all a and b,

$$Bew_{\mathcal{T}}\big(a, gn\{(\neg\} * b * gn\{)\}\big) \text{ iff } \vdash \mathbf{Bew}_{\mathcal{T}}\big(\tilde{a}, \ulcorner(\neg\urcorner * b * \ulcorner)\urcorner\big) \tag{3}$$

and, for all a, b,

$$\neg Bew_{\mathcal{T}}\big(a, gn\{(\neg\} * b * gn\{)\}\big) \text{ iff } \vdash \neg\mathbf{Bew}_{\mathcal{T}}\big(\tilde{a}, \ulcorner(\neg\urcorner * \tilde{b} * \ulcorner)\urcorner\big) \tag{3'}$$

 Dis*provable* vs. un*decidable* ψ; a reminder: The former means $\vdash \neg\psi$, the latter means that we have *neither* $\vdash \neg\psi$ nor $\vdash \psi$. Of course, un*provable* ψ means $\nvdash \psi$.

By Lemma 12.11.37 let $\mathbf{R}$ be a fixpoint of

$$\neg(\exists y)\Big(\mathbf{Bew}_{\mathcal{T}}(y, x) \wedge (\forall z)(z < y \to \neg\mathbf{Bew}_{\mathcal{T}}(z, \ulcorner(\neg\urcorner * x * \ulcorner)\urcorner))\Big)$$

that is,

$$\vdash \mathbf{R} \equiv$$
$$\neg(\exists y)\Big(\mathbf{Bew}_{\mathcal{T}}(y, \ulcorner\mathbf{R}\urcorner) \wedge (\forall z)\big(z < y \to \neg\mathbf{Bew}_{\mathcal{T}}(z, \ulcorner\neg\mathbf{R}\urcorner)\big)\Big) \tag{4}$$

where throughout the proof "$\vdash$" is short for "$\vdash_\mathcal{T}$" Incidentally, in (4) and below I am not using the outermost brackets in "$\neg\mathbf{R}$". We show that $\mathbf{R}$ is an undecidable sentence.

1. First we show that $\nvdash \mathbf{R}$. *Suppose instead* that $\vdash \mathbf{R}$. Then for some $b \in \mathbb{N}$, $Bew_\mathcal{T}(b, gn\,\{\mathbf{R}\})$ is true (b codes a proof of $\mathbf{R}$) and hence, by (2),

$$\vdash \mathbf{Bew}_\mathcal{T}(\widetilde{b}, \ulcorner\mathbf{R}\urcorner) \tag{5}$$

By consistency, we cannot have $\vdash \neg\mathbf{R}$ too, so, for *all* k, $Bew_\mathcal{T}(\widetilde{k}, gn\,\{\neg\mathbf{R}\})$ is *false*, and in particular it is so for $k = 0, 1, \ldots b - 1$.
Therefore $\vdash \neg\mathbf{Bew}_\mathcal{T}(\widetilde{k}, \ulcorner\neg\mathbf{R}\urcorner)$ by (3'), for for $k = 0, 1, \ldots b - 1$. By Lemma 12.10.12 followed by generalisation we have

$$\vdash (\forall z)(z < \widetilde{b} \to \neg\mathbf{Bew}_\mathcal{T}(z, \ulcorner\neg\mathbf{R}\urcorner))$$

thus, using (5), $\vdash \mathbf{Bew}_\mathcal{T}(\widetilde{b}, \ulcorner\mathbf{R}\urcorner) \wedge (\forall z)(z < \widetilde{b} \to \neg\mathbf{Bew}_\mathcal{T}(z, \ulcorner\neg\mathbf{R}\urcorner))$ and hence (Corollary 0.2.108)

$$\vdash (\exists y)\Big(\mathbf{Bew}_\mathcal{T}(y, \ulcorner\mathbf{R}\urcorner) \wedge (\forall z)(z < y \to \neg\mathbf{Bew}_\mathcal{T}(z, \ulcorner\neg\mathbf{R}\urcorner))\Big) \tag{6}$$

(4) and (6) via Post yield $\vdash \neg\mathbf{R}$, contradicting consistency.

2. Now show that $\nvdash \neg\mathbf{R}$. *Suppose instead* that $\vdash \neg\mathbf{R}$. Thus $Bew_\mathcal{T}(b, gn\,\{\neg\mathbf{R}\})$ is true for some $b \in \mathbb{N}$. Hence, by (3),

$$\vdash \mathbf{Bew}_\mathcal{T}(\widetilde{b}, \ulcorner\neg\mathbf{R}\urcorner) \tag{7}$$

Now, we cannot have $\vdash \mathbf{R}$ due to consistency, so $Bew_\mathcal{T}(b, gn\,\{\mathbf{R}\})$ is *false* for *all* $k \in \mathbb{N}$, in particular it is so for $k = 0, 1, 2, \ldots, b$. By (2') and Lemma 12.10.12,

$$\vdash y < S\widetilde{b} \to \neg\mathbf{Bew}_\mathcal{T}(y, \ulcorner\mathbf{R}\urcorner)$$

which by ROB axiom $< \mathbf{2}$ (p. 345) and Theorem 0.2.103 can be rewritten as

$$\vdash (y < \widetilde{b} \vee y = \widetilde{b}) \to \neg\mathbf{Bew}_\mathcal{T}(y, \ulcorner\mathbf{R}\urcorner) \tag{8}$$

By Post and (7), we have also $\widetilde{b} < y \to \widetilde{b} < y \wedge \mathbf{Bew}_\mathcal{T}(\widetilde{b}, \ulcorner\neg\mathbf{R}\urcorner)$. Combining with Theorem 0.2.107: $\widetilde{b} < y \wedge \mathbf{Bew}_\mathcal{T}(\widetilde{b}, \ulcorner\neg\mathbf{R}\urcorner) \to (\exists z)(z < y \wedge \mathbf{Bew}_\mathcal{T}(z, \ulcorner\neg\mathbf{R}\urcorner))$ and Post, we get

$$\vdash \widetilde{b} < y \to (\exists z)(z < y \wedge \mathbf{Bew}_\mathcal{T}(z, \ulcorner\neg\mathbf{R}\urcorner)) \tag{9}$$

Proof by cases using (8), (9) and the trichotomy axiom yields

$$\vdash \neg\mathbf{Bew}_\mathcal{T}(y, \ulcorner\mathbf{R}\urcorner) \vee (\exists z)(z < y \wedge \mathbf{Bew}_\mathcal{T}(z, \ulcorner\neg\mathbf{R}\urcorner))$$

One generalisation later we have

$$\vdash (\forall y)\Big(\neg\mathbf{Bew}_T(y,\ulcorner\mathbf{R}\urcorner) \vee (\exists z)(z < y \wedge \mathbf{Bew}_T(z,\ulcorner\neg\mathbf{R}\urcorner))\Big)$$

Hence, by $\vdash (\forall x)A \equiv \neg(\exists x)\neg A$ and several easy applications of Theorem 0.2.103 we get

$$\vdash \neg(\exists y)\Big(\mathbf{Bew}_T(y,\ulcorner\mathbf{R}\urcorner) \wedge (\forall z)(z < y \rightarrow \neg\mathbf{Bew}_T(z,\ulcorner\neg\mathbf{R}\urcorner))\Big)$$

By (4) and Post we have $\vdash \mathbf{R}$, contradicting consistency.   □

12.12 The Second Incompleteness Theorem: Briefly

Having come this far on the phenomenon of *incompletableness*, it would be a shame not to talk about the second incompleteness theorem of Gödel. It is notable that Gödel did not provide the complete self-contained proof in Gödel (1931); but then how could he in the confines of a single paper! The machinery to get to the final proof can fill half a book! (Cf. Hilbert and Bernays 1968, Tourlakis 2003a. The first published complete proof appeared in the first of these two references.)

The technical "machinery" towards a proof aims at proving first the three so-called *derivability conditions* of Löb —in short "DC"— from which the second incompleteness theorem follows routinely, if not trivially.

The DC are derivable from the *Ableitungsforderungen* (derivability conditions) of Hilbert and Bernays (1968) (cf. also Tourlakis 2003a).

The first one, DC1, we have already proved in the first part of the proof of Theorem 12.11.39 and only relies on the metatheoretical properties of ROB and of its consistent recursively axiomatised extensions —e.g., their ability to formally define all recursive relations. DC1 is a direct special case of Theorem 12.11.4 —which can also be stated as

12.12.1 Theorem. *If a formally Σ_1[41] sentence is true in the standard interpretation over the language L_T of a consistent recursively axiomatised extension T of PA, then it is provable in T.*

On the other hand DC2 and DC3 each require a formidable amount of perseverance (and page space), in particular, they require the development of primitive recursive functions and their properties *formally within PA*.[42] This development (Hilbert and Bernays 1968; Tourlakis 2003a) redoes Chap. 2 formally, all the way to and

---

[41] "Formally Σ_1" is denoted by the bold $\mathbf{\Sigma}_1$ symbol.

[42] ROB does not support formal induction, that is why this work *must* be done within PA.

including demonstrating that we can code sequences, an important tool towards defining **Bew**(y, x) *within PA*,[43] whose first variable is the "code" of a proof of the formula coded by the second variable.

PA as is given contains (axioms defining) only *three* primitive recursive functions (as function symbols): S, $+$ and $\times$. One introduces several more primitive recursive functions formally —in principle we introduce *all of them*— by *definitions*. Each such function **h** of n variables is introduced by a derivation, just as in Definition 2.1.1, although the "f_i" here are already introduced (in PA) *names* —written in bold, $\mathbf{f}_i$; one of these is our **h**— of formal primitive recursive functions. This creates an interesting situation: By the time we are done[44] we have built a theory $\mathcal{C}$ ($\mathcal{C}$ for "coding") over a language $L_\mathcal{C} \supseteq L_{PA}$. It turns out that this theory $\mathcal{C}$ is a "*conservative extension*" of PA, that is,

any B over L_{PA} that $\mathcal{C}$ proves, PA proves it as well

That is, $\mathcal{C}$ *has no more proving power than PA as long as one stays in the* L_{PA} *language.*

This is a bit annoying since now if $\mathcal{T}$ is an extension of PA it is so by virtue of adding its symbols and axioms to L_{PA} and PA respectively. It is not necessarily the case that $\mathcal{C}$ is contained in $\mathcal{T}$ or that $L_\mathcal{C} \subseteq L_\mathcal{T}$! But we can force this to be the case, essentially, by adding all the machinery of $\mathcal{C}$ (all the definitions of primitive recursive functions) to $\mathcal{T}$ and $L_\mathcal{T}$. Then we obtain a *conservative extension* $\mathcal{T}'$ of $\mathcal{T}$ that does contain all the coding tools; all of $\mathcal{C}$.

Since $\mathcal{T}'$ is conservative over $\mathcal{T}$ we may work with the former and pretend that we are working with the latter; after all, *defined symbols* can be "eliminated" from a formula (cf. Tourlakis 2003a). E.g., we have that $\mathcal{T}$ is consistent iff $\mathcal{T}'$ is. One direction is trivial. In the other, say $\mathcal{T}$ is consistent. Did our extending of $\mathcal{T}$ introduce axioms that along with the other axioms can derive a contradiction? *No:* Say, $\nvdash_\mathcal{T} (\forall x)\neg x = x$. But then $\nvdash_{\mathcal{T}'} (\forall x)\neg x = x$ as well, for in the contrary case conservatism entails that $\vdash_\mathcal{T} (\forall x)\neg x = x$ contradicting the assumption.

Therefore, without loss of generality, we can safely pretend that $\mathcal{C}$ is *part of* PA, and identify the extension $\mathcal{T}$ of PA with $\mathcal{T}'$, so it contains all the coding apparatus too. That is, in the end we work as if $\mathcal{T}$ extends PA *and as if all the coding apparatus resides in PA*.

Having defined all $f \in \mathcal{PR}$ within PA (or $\mathcal{C}$) has a useful consequence: If **f** is a formally defined $\mathcal{PR}$ function in PA, and if f is defined *informally* mimicking faithfully the derivation of the former, then one can prove (Tourlakis 2003a):

---

[43] **Bew**(y, x) is a meta name for a long formula in the language of a recursively axiomatised extension of PA, whose only potentially non-primitive recursive subformula part says "x is an axiom".

[44] Philosophically speaking; there are infinitely many formal primitive recursive functions, one for each informal $f \in \mathcal{PR}_*$.

$$\mathbf{f}^{\mathbb{N}} = f$$

and moreover $\mathbf{f}^{\mathbb{N}}$ is term-defined by $\mathbf{f}(\vec{x})$:

For all $\vec{a}$ and b, $\mathbf{f}^{\mathbb{N}}(\vec{a}) = b$ implies [45] $\vdash_{PA} \mathbf{f}(\widetilde{\vec{a}}) = \widetilde{b}$

It should come as no surprise that we also need versions of **Bew** within PA (and its recursively axiomatised extensions) towards proving the second incompleteness theorem. After all, the proof hinges on the *first part* —the one that does not rely on ω-consistency— of the proof of Theorem 12.11.39. In turn, the latter theorem relies on the *definability* (Corollary 12.11.2) of the recursive $Bew(u, v)$ in (3) below by some *formula* **Bew**(u, v) in the familiar (Definition 12.10.1) sense of (2) and (2′) below.

$$Bew(a, gn\{A\}) \text{ implies } \vdash_{\mathcal{T}} \mathbf{Bew}(\widetilde{a}, \ulcorner A \urcorner) \tag{2}$$

and

$$\neg Bew(a, gn\{A\}) \text{ implies } \vdash_{\mathcal{T}} \neg\mathbf{Bew}(\widetilde{a}, \ulcorner A \urcorner) \tag{2′}$$

The idea is to introduce the formula (abbreviation) **Bew** in PA as the *formal counterpart* of (3) below (cf. Proposition 12.11.14 and Corollary 12.11.17).

$$Bew(y, x) \overset{Def}{\equiv} Seq(y) \wedge lh(y) \geq 1 \wedge (\forall A)_{\leq y}\Bigg(A \in y \rightarrow wff(A) \wedge$$

$$\Bigg\{ Axiom(A) \vee$$

$$(\exists B, C)_{\leq y}\Big(B <_y A \wedge C <_y A \wedge MP(B, C; A)\vee$$

$$B <_y A \wedge AIntro(B; A)\Big)\Bigg\} \Bigg) \wedge x \in y \tag{3}$$

By "formal counterpart" we mean to replace all primitive recursive predicates and functions (e.g., $Seq, lh, \in, <_y, MP$) by their formal primitive recursive counterparts —denoted in bold face— and to do the same for the recursive $Axiom(x)$. One obtains (Tourlakis 2003a)

---

[45] The "implies" is sharpened to an "iff" in a consistent theory, as we already know.

$$\mathbf{Bew}(y, x) \overset{Def}{\equiv} \mathbf{Seq}(y) \wedge \mathbf{lh}(y) \geq \widetilde{1} \wedge (\forall A)_{\leq y}\Big(A \in y \rightarrow \mathbf{wff}(A) \wedge$$

$$\Big\{ \mathbf{Axiom}(A) \vee$$

$$(\exists B, C)_{\leq y}\Big(B <_y A \wedge C <_y A \wedge \mathbf{MP}(B, C; A)\vee$$

$$B <_y A \wedge \mathbf{AIntro}(B; A)\Big)\Big\}\Big) \wedge x \in y \qquad (3')$$

Finally one introduces the *metasymbol* $\Theta(x) \overset{Def}{\equiv} (\exists y)\mathbf{Bew}(y, x)$.

Since $\big(\mathbf{Bew}(\widetilde{b}, \widetilde{a})\big)^{\mathbb{N}} = {}^{46}\mathbf{Bew}^{\mathbb{N}}(b, a) = Bew(b, a)$ (cf. the table and discussion following on p. 270, and also Corollary 8.1.6) we obtain (2) and (2′) by Theorem 12.12.1 because each of $\mathbf{Bew}(\widetilde{b}, \widetilde{a})$ and $\neg\mathbf{Bew}(\widetilde{b}, \widetilde{a})$ —being (formally) recursive— is Σ_1.

12.12.2 Remark.

1) The proof of Theorem 12.11.39 goes through with the **Bew** introduced in PA via (3′).

2) In (3) and in Proposition 12.11.14 the coding tools $Seq, lh, (z)_i$ refer to prime-power coding (cf. Definitions 2.4.2 and 2.4.3). In Chap. 7 we introduced the Smullyan-Bennett (Smullyan 1961; Bennett 1962) dyadic coding to obtain a characterisation of $\mathcal{R}$ without the *a priori* knowledge that it is closed under primitive recursion.

   In the present chapter we introduced yet another coding based on m-adic representation of numbers ($m = 11$). Our goal has been to build and employ the most readily accessible tools —while, arguably, the Chinese remainder theorem is not that "accessible". In Tourlakis (2003a) however, coding of sequences within PA *is* done throughout using Gödel's β-function that is based on the Chinese remainder theorem.

   As long as the coding of sequences is primitive recursive no one will notice the difference at the macro level, but definitions and proof details (in PA), of course, will be markedly different. In the end, the prime power coding proponents will notice the need for some bootstrapping —say, via β-function coding— if they want to prove that *arbitrary primitive recursive functions* can *be introduced in PA*. After that is done, prime power coding *can* be introduced in PA by primitive recursive definitions —mimicking formally Definitions 2.4.2 and 2.4.3— and used. Since we are omitting these proofs anyway, we will not elaborate further on (3) or (3′). $\qquad\qquad\qquad\qquad\qquad\qquad\qquad\qquad\qquad\qquad\qquad\square$

---

[46] Equality as strings.

To appreciate the need for a formal theory of $\mathcal{PR}$ functions, consider DC2: It is derived from a proof of

$$\vdash_{PA} \textbf{Bew}(u, \ulcorner A \urcorner) \rightarrow \Big(\textbf{Bew}(v, \ulcorner A \rightarrow B \urcorner) \rightarrow \textbf{Bew}(u * v * \langle \ulcorner B \urcorner \rangle, \ulcorner B \urcorner) \Big)^{47} \qquad (1)$$

where **Bew** is that of $(3')$ and we are using the notation of Definitions 2.4.2 and 2.4.3 "$\langle \ulcorner B \urcorner \rangle$" to denote the code of a one-element sequence whose only element is $\ulcorner B \urcorner$.

Note that u and v in (1) are *distinct* free *variables*, which enables the steps following $(1')$ below.

Now pure logic yields (Theorem 0.2.107)

$$\vdash_{PA} \textbf{Bew}(u * v * \langle \ulcorner B \urcorner \rangle, \ulcorner B \urcorner) \rightarrow (\exists y)\textbf{Bew}(y, \ulcorner B \urcorner) \qquad (4)$$

Next, (1) and (4) and Post yield

$$\vdash_{PA} \textbf{Bew}(u, \ulcorner A \urcorner) \rightarrow \Big(\textbf{Bew}(v, \ulcorner A \rightarrow B \urcorner) \rightarrow (\exists y)\textbf{Bew}(y, \ulcorner B \urcorner) \Big) \qquad (1')$$

By $\exists$-intro Theorem 0.2.109, $(1')$ yields

$$\vdash_{PA} (\exists y)\textbf{Bew}(y, \ulcorner A \urcorner) \rightarrow \Big(\textbf{Bew}(v, \ulcorner A \rightarrow B \urcorner) \rightarrow (\exists y)\textbf{Bew}(y, \ulcorner B \urcorner) \Big)$$

which by Post is an equivalent assertion to

$$\vdash_{PA} \textbf{Bew}(v, \ulcorner A \rightarrow B \urcorner) \rightarrow \Big((\exists y)\textbf{Bew}(y, \ulcorner A \urcorner) \rightarrow (\exists y)\textbf{Bew}(y, \ulcorner B \urcorner) \Big)$$

On more application of Theorem 0.2.109 now yields

$$\vdash_{PA} (\exists y)\textbf{Bew}(y, \ulcorner A \rightarrow B \urcorner) \rightarrow \Big((\exists y)\textbf{Bew}(y, \ulcorner A \urcorner) \rightarrow (\exists y)\textbf{Bew}(y, \ulcorner B \urcorner) \Big)$$

This is DC2 as stated below.

 12.12.3 Remark. Since all coding tools reside in $\mathcal{C}$ (PA), the proofs of formulas that contain **Bew** or Θ, as expressed symbolically via "$\vdash$", will be carried out in PA ($\mathcal{C}$) therefore, unless otherwise indicated, "$\vdash$" will stand for "$\vdash_{PA}$". On the other hand, if we study the theorems of a theory $\mathcal{T}$, then **Bew** and Θ help us *express* provability in $\mathcal{T}$ so they will be subscripted by $\mathcal{T}$ —after all, the subformula **Axiom**(A) of

---

[47] Intuitively, if we have a proof coded by u for A and one coded by v for $A \rightarrow B$, then $u * v * \langle \ulcorner B \urcorner \rangle$ codes a proof for B.

Bew$_\mathcal{T}$ says that A is an axiom, which if it is neither logical nor in PA, then it is an axiom of $\mathcal{T}$.

But remember that **Bew** and Θ reside in the language L_{PA} (L_C). □

The following specialised result, also stated without proof (cf. Hilbert and Bernays 1968, Tourlakis 2003a), is responsible for proving DC3:

12.12.4 Theorem. *For any* Σ_1 *formula A over* L_{PA}, $A(x_1, \ldots, x_n)$ *(cf. Remark 0.2.14), we have* $\vdash_{PA} A(x_1, \ldots, x_n) \to \Theta(\{A(x_1, \ldots, x_n)\})$.

Notation used above:

1. A Σ_1 relation R of the metatheory is a semi-recursive one, specifically, one that is expressible as $(\exists y)Q$ where Q is recursive (informally). A bold Σ_1 formula A over the language L_{PA} is a *formula* of the form $(\exists y)\mathbf{B}$, where $\mathbf{B}$ is provably recursive in PA.
2. $\Theta_\mathcal{T}$ is $(\exists y)\mathbf{Bew}_\mathcal{T}(y, x)$, where $\mathbf{Bew}_\mathcal{T}(y, x)$ is provably recursive in PA —hence it is Σ_1. But then $\Theta_\mathcal{T}$ is also Σ_1, since the latter set is closed under $(\exists w)$.
3. $\{\mathbf{E}(x_1, \ldots, x_n)\}$ for any term or formula $\mathbf{E}$ of the indicated free variables is a formal primitive recursive term (albeit the notation here of said term is metatheoretical) associated with $\mathbf{E}$. It was introduced in Hilbert and Bernays (1968) (and later used in Tourlakis 2003a) and has the property that for any $a_i \in \mathbb{N}$, for $i = 1, \ldots, n$, we have $\vdash_{PA} \{\mathbf{E}(\tilde{a}_1, \ldots, \tilde{a}_n)\} = \ulcorner \mathbf{E}(\tilde{a}_1, \ldots, \tilde{a}_n) \urcorner$.

If we believe Theorem 12.12.4, then DC3 follows at once as $\Theta_\mathcal{T}(\ulcorner A \urcorner)$ has *no free variables* and, since it stands for $(\exists y)\mathbf{Bew}_\mathcal{T}(y, \ulcorner A \urcorner)$, it is Σ_1 over L_{PA}.

Therefore $\vdash_{PA} \{\Theta_\mathcal{T}(\ulcorner A \urcorner)\} = \ulcorner \Theta_\mathcal{T}(\ulcorner A \urcorner) \urcorner$ and we next invoke Theorem 12.12.4.

Thus, notwithstanding careless statements often made in the literature[48] that the proof of the Second incompleteness theorem is obtained "by formalising the proof of the First within PA" it is not that "simple" —after all, the proof of the *First theorem has nothing to owe to DC2 or DC3*.

12.12.5 Definition. (The Derivability Conditions of Löb) Let $\mathcal{T}$ be a formal arithmetic extending PA, and let A, B be sentences over $L_\mathcal{T}$. The three *derivability conditions* are:

DC1. If $\vdash_\mathcal{T} A$, then $\vdash_{PA} \Theta_\mathcal{T}(\ulcorner A \urcorner)$

DC2. $\vdash_{PA} \Theta_\mathcal{T}(\ulcorner A \to B \urcorner) \to \left(\Theta_\mathcal{T}(\ulcorner A \urcorner) \to \Theta_\mathcal{T}(\ulcorner B \urcorner) \right)$

DC3. $\vdash_{PA} \Theta_\mathcal{T}(\ulcorner A \urcorner) \to \Theta_\mathcal{T}\left(\ulcorner \Theta_\mathcal{T}(\ulcorner A \urcorner) \urcorner \right)$

□

What does it mean that PA —indeed any consistent formal arithmetic $\mathcal{T}$ that extends PA— cannot prove its own consistency? How is this consistency *formulated*

---

[48] Smoryński (1978) laments, essentially, that "In expositions one often replaces precise statements by imprecise ones, or by precise but false ones" and identifies one of the latter kind as "the second incompleteness theorem is proved by formalising the proof of the first one".

for $\mathcal{T}$ to ponder it? But, of course, as $\neg\Theta_{\mathcal{T}}(\mathbf{X})$, where we choose $\mathbf{X}$ to be disprovable —that is, $\neg\mathbf{X}$ is provable— in $\mathcal{T}$. Then we have consistency if we cannot also prove $\mathbf{X}$.

The literature seems to favour "$0 = \tilde{1}$" as $\mathbf{X}$, but $(\forall x)\neg x = x$ would do as well, and so would $\neg 0 = 0$! Incidentally, $0 = \tilde{1}$ *is* disprovable in any consistent formal arithmetic since $\vdash_{ROB} \neg 0 = \tilde{1}$ (by the first ROB axiom).

Thus we will follow standard practise and define "$Con_{\mathcal{T}}$" below, the formal statement within $\mathcal{T}$ that "says" that $\mathcal{T}$ is *consistent*. So, we define (metatheoretical abbreviation):

$$Con_{\mathcal{T}} \stackrel{Def}{\equiv} \neg\Theta_{\mathcal{T}}(\ulcorner 0 = \tilde{1}\urcorner) \qquad\qquad (\dagger)$$

We will stay with our simplifying view where we identify PA with the theory where all the coding tools were constructed, $\mathcal{C}$. Thus, if $\mathcal{T}$ extends PA, then all coding tools are in $\mathcal{T}$ as well.

The actual picture is, to repeat our earlier remarks, that $\mathcal{C}$ extends PA conservatively, $\mathcal{T}$ extends PA, and $\mathcal{T}'$ extends $\mathcal{T}$ conservatively by importing all the $\mathcal{C}$ tools. Then we take the simplifying assumptions (by conservativeness) that $PA = \mathcal{C}$ and $\mathcal{T} = \mathcal{T}'$.

We now state and prove (suppressing some 100 pages needed to completely establish DC2 and DC3):

12.12.6 Theorem. (Second Incompleteness Theorem) *Any consistent formal arithmetic $\mathcal{T}$ that extends PA (it may be $\mathcal{T} = PA$, of course) cannot prove its own consistency.*

Proof   We will omit the subscripts $\mathcal{T}$ or PA from everything except from $Con_{\mathcal{T}}$.

Now the following is a theorem of pure logic (tautology)

$$\neg\Theta(\ulcorner\mathbf{G}\urcorner) \to \Theta(\ulcorner\mathbf{G}\urcorner) \to 0 = \tilde{1} \qquad\qquad (1)$$

where $\mathbf{G}$ (over L_{PA}) is the fixpoint sentence in Theorem 12.11.39. *Ibid*, Theorem 0.2.103 and (1) entail

$$\vdash \mathbf{G} \to \Theta(\ulcorner\mathbf{G}\urcorner) \to 0 = \tilde{1} \qquad\qquad (1')$$

After applying DC1 and DC2 in that order we have from $(1')$

$$\vdash \Theta(\ulcorner\mathbf{G}\urcorner) \to \Theta\left(\ulcorner\Theta(\ulcorner\mathbf{G}\urcorner)\urcorner\right) \to \Theta(\ulcorner 0 = \tilde{1}\urcorner)$$

Applying DC3 plus Post to the above yield (after utilising the abbreviation $(\dagger)$)

$$\vdash \Theta(\ulcorner\mathbf{G}\urcorner) \to \Theta(\ulcorner\mathbf{G}\urcorner) \to \neg Con_{\mathcal{T}}$$

which by tautological implication and Theorem 12.11.39 simplifies to

$$\vdash \neg \mathbf{G} \to \neg Con_{\mathcal{T}}$$

or, contrapositively,

$$\vdash Con_{\mathcal{T}} \to \mathbf{G} \tag{2}$$

Recall that $\vdash$ means $\vdash_{PA}$. From (2) and the fact that $\mathcal{T}$ extends PA we have

$$\vdash_{\mathcal{T}} Con_{\mathcal{T}} \to \mathbf{G}$$

So, if $\vdash_{\mathcal{T}} Con_{\mathcal{T}}$, then $\vdash_{\mathcal{T}} \mathbf{G}$ contradicting part 1 of the proof of Theorem 12.11.39.

$\square$

12.12.7 Remark. It is noteworthy that we can easily also prove the converse of (2) above:

From $\vdash_{PA} 0 = \tilde{1} \to \mathbf{G}$ we get by DC1 followed by DC2 and MP

$$\vdash_{PA} \Theta(\ulcorner 0 = \tilde{1} \urcorner) \to \Theta(\ulcorner \mathbf{G} \urcorner)$$

That is, via Theorem 12.11.39, $\vdash_{PA} \neg Con_{\mathcal{T}} \to \neg \mathbf{G}$, hence $\vdash_{PA} \mathbf{G} \to Con_{\mathcal{T}}$, and thus also $\vdash_{\mathcal{T}} \mathbf{G} \to Con_{\mathcal{T}}$. **Worth recording**: $\vdash_{PA} \mathbf{G} \equiv Con_{\mathcal{T}}$ $\square$

The following result due to Löb is intimately connected with the second incompleteness theorem.

12.12.8 Theorem. (Löb's Theorem) *For any sentence A of a consistent formal arithmetic $\mathcal{T}$, we have $\vdash_{\mathcal{T}} \Theta(\ulcorner A \urcorner) \to A$ iff $\vdash_{\mathcal{T}} A$.*

Proof The *if* direction is by a trivial tautological implication from A. For the *only if* direction the standard proof proceeds contrapositively: So, assume that $\nvdash_{\mathcal{T}} A$. Then the theory $\mathcal{T} + \neg A$ is a recursively axiomatised consistent (cf. Theorem 0.2.93) extension of PA. Now $\mathcal{T} + \neg A \nvdash A$ because of consisitency and $\mathcal{T} + \neg A \vdash \neg A$. Thus the statement $Con_{\mathcal{T} + \neg A}$ can be taken to be $\neg\Theta(\ulcorner A \urcorner)$. By Theorem 12.12.6 we thus have

$$\mathcal{T} + \neg A \nvdash \neg\Theta(\ulcorner A \urcorner)$$

from which $\mathcal{T} \nvdash \neg A \to \neg\Theta(\ulcorner A \urcorner)$ follows.

Pause. Deduction theorem? MP? Other?◄

Hence also $\mathcal{T} \nvdash \Theta(\ulcorner A \urcorner) \to A$. $\square$

12.12.9 Remark. Conversely, one can prove Theorem 12.12.6 from Theorem 12.12.8.

We start with $\nvdash_{\mathcal{T}} 0 = \tilde{1}$, where $\mathcal{T}$ is a consistent formal arithmetic.

Thus, by Löb's theorem,

$$\nvdash_{\mathcal{T}} \Theta(\ulcorner 0 = \tilde{1} \urcorner) \to 0 = \tilde{1} \tag{1}$$

We get at once

$$\nvdash_{\mathcal{T}} \underbrace{\neg\Theta(\ulcorner 0 = \tilde{1}\urcorner)}_{Con_{\mathcal{T}}}$$

since otherwise tautological implication would prove $\vdash_{\mathcal{T}} \neg\Theta(\ulcorner 0 = \tilde{1}\urcorner) \lor 0 = \tilde{1}$, contradicting (1).   □

There is an easy direct proof of Theorem 12.12.8. "Easy" once the DC1–DC2 are proved, that is!

12.12.10 Theorem. (Löb's Theorem —Direct) *For any sentence A of a consistent formal arithmetic* $\mathcal{T}$, *we have* $\vdash_{\mathcal{T}} \Theta(\ulcorner A\urcorner) \to A$ *iff* $\vdash_{\mathcal{T}} A$.

Proof Again, only the left to right direction is nontrivial. First, assume

$$\vdash_{\mathcal{T}} \Theta(\ulcorner A\urcorner) \to A \qquad (1)$$

Now apply the fixpoint theorem in $\mathcal{T}$ to the formula $\Theta(x) \to A$ of one variable x. We obtain a sentence B such that

$$\vdash_{\mathcal{T}} B \equiv \Theta(\ulcorner B\urcorner) \to A \qquad (2)$$

From (2) we have $\vdash_{\mathcal{T}} B \to \Theta(\ulcorner B\urcorner) \to A$. Apply to this DC1 followed by DC2 (MP is also used): $\vdash_{\mathcal{T}} \Theta(\ulcorner B\urcorner) \to \Theta(\ulcorner\Theta(\ulcorner B\urcorner)\urcorner) \to \Theta(\ulcorner A\urcorner)$.

The above via DC3 and tautological implication leads to

$$\vdash_{\mathcal{T}} \Theta(\ulcorner B\urcorner) \to \Theta(\ulcorner A\urcorner)$$

which along with (1) plus Post yields

$$\vdash_{\mathcal{T}} \Theta(\ulcorner B\urcorner) \to A \qquad (3)$$

By (2) I have $\vdash_{\mathcal{T}} B$ and hence $\vdash_{\mathcal{T}} \Theta(\ulcorner B\urcorner)$ by DC1. Therefore $\vdash_{\mathcal{T}} A$ by (3).   □

12.13 Exercises

1. Complete the proof of Proposition 12.2.3
2. Show that $\{x : \phi_x \in \mathcal{R}\} \sim \{x : W_x$ is infinite$\}$.
   Hint. Prove $\{x : \phi_x \in \mathcal{R}\} \leq_1 \{x : W_x$ is infinite$\}$ and $\{x : W_x$ is infinite$\} \leq_1 \{x : \phi_x \in \mathcal{R}\}$.
3. Answer the question raised in Remark 12.9.2.
4. Prove Corollary 12.10.10.
5. Prove Proposition 12.10.15.

6. Prove Lemma 12.11.11.

7. Prove Lemma 12.11.12.

8. Prove that every infinite primitive recursive sets contains a recursive but not primitive recursive subset.

9. Let $A \leq_m B$ via f and $A \notin \mathcal{R}_*$. Prove that $f[A] \notin \mathcal{R}_*$.

10. Let finite sets be prime power coded. Let A be r.e. and $K \leq_m A$ via $f \in \mathcal{R}$. Show that an $h \in \mathcal{R}$ exists, such that

$$W_{h(x)} = \begin{cases} f^{-1}[D_x] & \text{if } D_x \cap A = \emptyset \\ \mathbb{N} & \text{if } D_x \cap A \neq \emptyset \end{cases}$$

Hint. Given x, program a $\lambda x.g(x)$, which enumerates $f^{-1}[D_x]$ or $\mathbb{N}$ depending on the two cases in the definition above. (If now we take $t \in \mathcal{R}$ such that $\mathrm{ran}(\phi_x) = W_{t(x)}$, then h can be taken to be $\lambda x.tg(x)$). Further hints: There is an $m \in \mathcal{R}$ such that $W_{m(x)} = f^{-1}[D_x]$. Then the program for $g(x)$ uses $m(x)$ and enumerates $f^{-1}[D_x]$ and A in parallel, as long as no enumerated A-member is in D_x. The first time this happens $g(x)$ exits from running $m(x)$ and starts enumerating $\mathbb{N}$ instead. Etc.

11. Let finite sets be prime power coded. Let A be r.e. and $K \leq_m A$ via $f \in \mathcal{R}$. Construct a g such that for all x

- $\emptyset \neq D_x \subsetneq A \rightarrow g(x) \in A - D_x$
- $\emptyset \neq D_x \subsetneq \overline{A} \rightarrow g(x) \in \overline{A} - D_x$

Hint. Rogers 1967 Note that $f[K]$ is infinite. It is also r.e. (Why?). Thus let $t \in \mathcal{R}$ enumerate $f[K]$ ($\mathrm{ran}(t) = f[K]$). Let h be that in Exercise 12.13. Now let

$$g(x) \simeq \text{if } fh(x) \notin D_x \text{ then } fh(x) \text{ else } t\Big((\mu y)t(y) \notin D_x\Big)$$

12. Let A be r.e. Show that there is a $g \in \mathcal{R}$ such that

$$W_{g(x,y)} = \text{if } y \in A \text{ then } W_x \text{ else } \emptyset$$

13. Suppose $\emptyset \in \mathcal{C} \subsetneq \mathcal{P}$. Prove that $\{x : \phi_x \in \mathcal{C}\}$ is productive.

14. Suppose $\emptyset \neq A \neq \mathbb{N}$ and that A is r.e. Show that A is creative.

15. Is $\{x : W_x = \emptyset\}$ creative? How about productive?

16. Same question as in the Exercise above, but for $\{x : W_x \neq \emptyset\}$.

17. Is $\{x : 42 \in W_x\}$ creative? If yes, give a proof.

18. Show that $\{x : W_x \neq \emptyset\}$ is 1-complete.

Chapter 13
Relativised Computability

Overview

We have had the briefest glimpse of computing *relative* to a given function f (cf. Definition 9.5.1) —or *in* a function f— that is, having the computation rely on obtaining from time to time values $f(a)$ for various a using an *external agent*, the computation being totally uninterested in how the agent computes, or obtains in some perhaps *non effective* (not "computable") way, said values.

The details of how the agent obtains answers being unknown and irrelevant suggested the nickname "*oracle* for f" for the agent. So we have a "secret" process —*that is not part of our explicit computation*— which when our computation requires a value $f(a)$ of the function the said secret process furnishes that value if (the function call "$f(a)$" returns) $f(a)\downarrow$, or, if $f(a)\uparrow$, returns nothing, in which case, by convention, the explicit part of the computation diverges as well because the latter is "stuck" waiting for an answer that will not be forthcoming.

This processes that we described in intuitive terms is *analogous* —but not identical— to a program, say, written in FORTRAN, calling a "library function", say, $\cos x$, without the user being at all interested in *how* a value

$$\cos 5.5$$

might be obtained.

We stress the word "analogous" —as opposed to "identical"— as in this example there *is* an actual computation that obtains $\cos 5.5$, albeit it is hidden and irrelevant as far as the user is concerned.

 Externally, the FORTRAN analogy is apt. Internally though the "library functions" such as "$\cos x$" *are* actually computed, while in our mathematical model for

© Springer Nature Switzerland AG 2022

G. Tourlakis, *Computability*, https://doi.org/10.1007/978-3-030-83202-5_13

computation in f it is profitable to deliberately choose non computable f, e.g., such as the characteristic function of K.[1]

The behaviour according to which a computation, every now and then, consults an oracle for f —i.e., calls $f(a)$ but we just want the final answer, not the details— can be achieved mathematically either by allowing a URM instruction of the type $X \leftarrow f(Y)$ involving said function, or do as in Definition 9.5.1, where f is adopted as one of the initial functions.

This is just the idea. There are several ways to implement oracle computations and their generalisations —such as computing with a mix of numerical and function inputs— of which we will visit a few. We will visit a variety of topics including the recursion theorems (first and second), Turing degrees, the priority method, and the arithmetical hierarchy.

13.1   Functionals

We will take the avenue of defining, as the basic concept, that of *computable functionals*.

Functionals are "higher type" (type-2) functions —that have as inputs both members of $\mathbb{N}$ but also entire partial functions $\alpha : \mathbb{N}^n \to \mathbb{N}$.

It is common in the literature (e.g., Hinman 1978; Odifreddi 1989) to consider only *one-argument* α since the n-argument case can be trivially accommodated by, say, prime power coding vector inputs. Thus in all that follows we will have $n = 1$, that is, $\alpha : \mathbb{N} \to \mathbb{N}$.

We will be dealing with objects of three types in this chapter: numbers (type-0), number-theoretic functions (type-1) and functionals (type-2). A call to a functional thus looks like $\mathscr{F}(x, \alpha)$ where a specific number is inputed to the type-0 variable x and a specific function is inputed to the type-1 variable α.

How does one input a potentially (extensionally) *infinite* f into a variable α (say, case where f is total)? The theory of computable functionals is interested in the computation itself, and not in how the infinite input was inputed.

The type-1 inputs are assumed to be present *extensionally* (sets of input/output pairs for each function input) before the computation on a given hybrid input starts.

It turns out that (terminating) computations are finite, as we will soon see.

Pause. What other than *finite* might a "terminating" computation be?

---

[1] If $f \in \mathcal{P}$ then this idea of oracle computation furnishes nothing beyond $\mathcal{P}$, as we saw in Lemma 9.5.2

It could be an infinite path the nodes of which are calls labelled by *ordinals* some of which might be infinite (forcing an infinite path length).[2] This happens in computation of functionals that are of type higher than 2.◄

How does *the computation* get, on an as needed basis, number pairs (x, y) from a type-1 input α? They are provided by an *oracle*, of course, which means, we don't care how the (x, y) *is* fetched *by* the computation, and certainly it is *not* computed *by the computation*. The transaction with the oracle, say (in URM terms) $X \leftarrow \alpha(Y)$ counts only as *one* "step" —if successful— just like $X \leftarrow X + 1$ does. Of course, if $\alpha(Y) \uparrow$ the computation cannot proceed and *hangs* (diverges).

 Incidentally,

(1) the term "computation" above is used in the dynamic (intuitive) sense, which so far we witnessed as a *linear* one. That is, by *computation* we understand the step-by-step *process* that we encountered several times in the earlier chapters of this volume.
(2) In the present chapter we will also term "computations" certain *tuples* that *syntactically represent* the relation "the program of index e with input $\vec{x}, \vec{\alpha}$ outputs the number y", and
(3) We will find it convenient to view the *dynamic* computations of functionals as trees rather than linear structures.[3]

From the concept of computation of a functional we will derive, in particular, the important "computable in α" concept, where α is some *fixed* number-theoretic function $\alpha : \mathbb{N} \to \mathbb{N}$. That is, we will be looking at (the special) cases of computation of functionals such as $\lambda x.\mathscr{F}(x, \alpha)$ to obtain the mathematisation of the concept of *relative* (to a function) computation, once we have figured out the case of how one computes $\lambda x\alpha.\mathscr{F}(x, \alpha)$.

13.1.1 Definition. (Type-2 Functionals) Let us denote by $\mathcal{P}(\mathbb{N} : \mathbb{N})$ the set of all partial one-argument number-theoretic functions, while, as usual, $\mathbb{N}^{\mathbb{N}}$ [4] denotes the set of unary total number-theoretic functions.

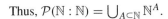 Thus, $\mathcal{P}(\mathbb{N} : \mathbb{N}) = \bigcup_{A \subseteq \mathbb{N}} \mathbb{N}^{A}$.

We will write $\alpha, \beta, \gamma, \delta$ with or without primes or subscripts to denote members of $\mathbb{N}^{\mathbb{N}}$ and $\mathcal{P}(\mathbb{N} : \mathbb{N})$.

---

[2] We will not use ordinals in this volume. These are sets that denote not only finite *numbers* —0, 1000, 10^{350000}— but also infinite *numbers* —ω, $\omega + 5$. The reader who might be unfamiliar with ordinals and may want to find out more about them may refer to any recent set theory text, e.g., Levy (1979), Tourlakis (2003b).

[3] But any tree can be linearised, just as we linearised the Ackermann computation that naturally presents itself as a tree of recursive calls; cf. Sect. 4.3.

[4] Also denoted by $^{\mathbb{N}}\mathbb{N}$ in parts of the literature.

A *functional* of *rank* (k, l) (in this chapter[5]) is a function

$$\mathcal{F} : \mathbb{N}^k \times \mathcal{P}(\mathbb{N} : \mathbb{N})^l \to \mathbb{N}$$

We may write $\mathcal{F}(\vec{x}_k, \vec{\alpha}_l)$ to indicate an arbitrary invocation (call) of $\mathcal{F}$ with arguments $\vec{x}_k, \vec{\alpha}_l$. Of course, we may also write

$$\mathcal{F} = \lambda \vec{x}_k \vec{\alpha}_l . \mathcal{F}(\vec{x}_k, \vec{\alpha}_l)$$

in which case the $\vec{x}_k$ and $\vec{\alpha}_l$ are variables. We shall use script style capital F, G, H with or without subscripts or primes for functionals.

A *restricted functional* of *rank* (k, l) is a function $\boldsymbol{F}$ whose *domain* is a subset of $\mathbb{N}^k \times (\mathbb{N}^{\mathbb{N}})^l$.

That is, *these functionals are undefined if called with nontotal type-1 inputs, regardless of the type-0 input used in a call*. Of course, total type-1 inputs alone will not guarantee an output if the computation gets into an infinite loop.

We may write $\boldsymbol{F}(\vec{x}_k, \vec{\alpha}_l)$ to indicate an arbitrary invocation (call) of $\boldsymbol{F}$ with arguments $\vec{x}_k, \vec{\alpha}_l$. Of course, we may also write

$$\boldsymbol{F} = \lambda \vec{x}_k \vec{\alpha}_l . \boldsymbol{F}(\vec{x}_k, \vec{\alpha}_l)$$

We shall use bold style capital F, G, H with or without primes or subscripts for *restricted* functionals.

An *unrestricted functional* $\mathcal{F}$ is called *total* if it is *everywhere defined* —for all values of its inputs— *whenever all* its type-1 input variables contain *total* inputs.

Thus, a total unrestricted functional may be *undefined* for some non total α-inputs. One such is $\lambda x \alpha . \alpha(x)$.

A *restricted* functional $\boldsymbol{F}$ is *total* if it is defined for all inputs: type-0 *and* type-1. There is no change of meaning to the symbols $\downarrow, \uparrow$, dom, ran. The terminology

$$\mathcal{F}(\vec{x}, \vec{\alpha}) \; converges, \text{ for } \mathcal{F}(\vec{x}, \vec{\alpha}) \downarrow$$

and

$$\mathcal{F}(\vec{x}, \vec{\alpha}) \; diverges, \text{ for } \mathcal{F}(\vec{x}, \vec{\alpha}) \uparrow$$

are still available.

Until now we used the notations $f(\vec{x}) \simeq y$ and $f(\vec{x}) \simeq g(\vec{y})$ in the context of *single-valued* functions. Section 13.6, which is about computing *partial multiple*

---

[5] There are even higher type computable functionals that we will not deal with in this volume; cf. e.g., Moldestad (1977), Hinman (1978), Fenstad (1980).

valued functions —or *p.m.v. functions* (cf. Definition 0.4.1)— will require us to use a richer notation (Fenstad 1980), which we introduce here, at the outset of the chapter.

1. $\mathscr{F}(\vec{x}, \vec{\alpha}) \simeq y$ means that y is *an* output of $\mathscr{F}(\vec{x}, \vec{\alpha})$, but there may be other outputs too. The relation is false exactly if *no* such y exists ($\mathscr{F}(\vec{x}, \vec{\alpha}) \uparrow$).

2. $\mathscr{F}(\vec{x}, \vec{\alpha}) \subseteq \mathscr{G}(\vec{y}, \vec{\beta})$ means $(\forall z)\Big(\mathscr{F}(\vec{x}, \vec{\alpha}) \simeq z \to \mathscr{G}(\vec{y}, \vec{\beta}) \simeq z\Big)$, which is the same as $\mathscr{F}(\vec{x}, \vec{\alpha}) \simeq z \to \mathscr{G}(\vec{y}, \vec{\beta}) \simeq z$ as we know from the behaviour of "$\forall$" detailed in Sect. 0.2. In particular, if $\mathscr{F}(\vec{x}, \vec{\alpha}) \uparrow$, then the previous implication is true.

3. $\mathscr{F}(\vec{x}, \vec{\alpha}) \simeq \mathscr{G}(\vec{y}, \vec{\beta})$ means $\mathscr{F}(\vec{x}, \vec{\alpha}) \subseteq \mathscr{G}(\vec{y}, \vec{\beta})$ and $\mathscr{F}(\vec{x}, \vec{\alpha}) \supseteq \mathscr{G}(\vec{y}, \vec{\beta})$, or

$$\mathscr{F}(\vec{x}, \vec{\alpha}) \simeq z \equiv \mathscr{G}(\vec{y}, \vec{\beta}) \simeq z$$

In particular $\mathscr{F}(\vec{x}, \vec{\alpha}) \uparrow$ and $\mathscr{G}(\vec{y}, \vec{\beta}) \uparrow$ imply $\mathscr{F}(\vec{x}, \vec{\alpha}) \simeq \mathscr{G}(\vec{y}, \vec{\beta})$.

4. $\mathscr{F} \subseteq \mathscr{G}$ is understood in the set-theoretic sense viewing each of $\mathscr{F}$ and $\mathscr{G}$ extensionally, as sets of input/output tuples $(\vec{x}, \vec{\alpha}, y)$.

5. $\mathscr{F} = \mathscr{G}$ is understood in the set-theoretic sense viewing each of $\mathscr{F}$ and $\mathscr{G}$ extensionally, as sets of input/output tuples $(\vec{x}, \vec{\alpha}, y)$.

6. $\mathscr{F}(\vec{x}, \vec{\alpha}) = z$ means: there exists a *unique* z such that $\mathscr{F}(\vec{x}, \vec{\alpha}) \simeq z$. $\quad\square$

13.1.2 Definition. (Type-2 Relations) A *type-2 relation* of rank (k, l) is a set $\mathbb{R} \subseteq \mathbb{N}^k \times \mathcal{P}(\mathbb{N} : \mathbb{N})^l$. As usual, we write $\mathbb{R}(\vec{x}_k, \alpha_l)$ for the statement $(\vec{x}_k, \alpha_l) \in \mathbb{R}$.

Thus $\mathbb{R}(\vec{x}_k, \alpha_l)$ is *one* of true or false; it *never* is undefined.

We will normally use the capital letters P, Q, R with or without subscripts or primes but in "blackboard bold" font (like $\mathbb{P}$) to denote type-2 relations. $\quad\square$

13.1.3 Definition. (Substitutions) $\mathscr{F}(\mathscr{G}(\vec{y}, \vec{\beta}), \vec{x}, \vec{\alpha})$ means that

$$\mathscr{F}(\mathscr{G}(\vec{y}, \vec{\beta}), \vec{x}, \vec{\alpha}) \simeq z \text{ iff } (\exists u)\Big(\mathscr{G}(\vec{y}, \vec{\beta}) \simeq u \wedge \mathscr{F}(u, \vec{x}, \vec{\alpha}) \simeq z\Big)$$

$\mathscr{F}(\vec{x}, \vec{\alpha}, \lambda y.\mathscr{G}(y, \vec{z}, \vec{\beta}))$ means that

$$\mathscr{F}(\vec{x}, \vec{\alpha}, \lambda y.\mathscr{G}(y, \vec{z}, \vec{\beta})) \simeq w \text{ iff } (\exists \gamma)\Big(\lambda y.\mathscr{G}(y, \vec{z}, \vec{\beta}) = \gamma \wedge \mathscr{F}(\vec{x}, \vec{\alpha}, \gamma) \simeq w\Big)$$

Both of the above are semantic instances of the "one-point rule" Theorem 0.2.94 as are the following two.

$$\mathbb{P}(\mathscr{G}(\vec{y}, \vec{\beta}), \vec{x}, \vec{\alpha}) \text{ means } (\exists u)\Big(\mathscr{G}(\vec{y}, \vec{\beta}) \simeq u \wedge \mathbb{P}(u, \vec{x}, \vec{\alpha})\Big)$$

$$\mathbb{P}(\vec{x}, \vec{\alpha}, \lambda y.\mathscr{G}(y, \vec{z}, \vec{\beta})) \text{ means } (\exists \gamma)\Big(\lambda y.\mathscr{G}(y, \vec{z}, \vec{\beta}) = \gamma \wedge \mathbb{P}(\vec{x}, \vec{\alpha}, \gamma)\Big) \quad\square$$

13.2   Partial Recursive and Recursive Functionals Following Kleene

This section deals with *unrestricted single-valued* functionals. How do we compute a functional? We may describe this using URMs with instructions of the kind $X \leftarrow \alpha(X)$, or, alternatively, by an inductive definition, where —rather than fixing an arbitrary function f and making it *initial*, as in Definition 9.5.1— we take care of all possible type-1 inputs of the initial functionals by adding an "application" (or "evaluation") *functional Ap* as *initial*. Ap behaves as $Ap(x, \alpha) = \alpha(x)$, for all x and α.

These approaches will get us along for a while, but at some point we will need indexing ("ϕ index" analogs) towards obtaining fundamental results such as the S-m-n and the second recursion theorems. But if so we might as well build the indexing up in front as Kleene (1959) originally did and as is the norm in the modern literature that extends computation to functional types higher than 2 (Fenstad 1980; Hinman 1978; Moldestad 1977; Moschovakis 1969).

13.3   Indexing

In this section we present the "indexing approach" to computability of functionals, building the indexing up in front. In essence, we build inductively a set of numbers that codify instructions for computing functionals. We met the concept and approach already in Chap. 10 where we defined indices for, essentially, $\mathcal{PR}$-derivations. We will work analogously here.

In the definition and in much of what follows in this chapter we will use the prime power coding of Definition 2.4.2 $\langle \cdots \rangle$, that is:

For the empty sequence λ we set (by definition) $\langle \lambda \rangle = 1$.

Moreover, $\langle x_0, \ldots x_{n-1} \rangle = \prod_{i=0}^{n-1} p_i^{x_i+1}$, where p_i is the i-th prime ($p_0 = 2$). On the other hand "$(\ldots)$" denotes (set-theoretic) ordered tuples (finite sequences). We will soon rediscover that the computation theories introduced in this chapter all include $\mathcal{PR}$ as a subclass and thus can compute their (primitive recursive) indices. This fact is not needed in all its gory details ahead of time.

13.3.1 Definition. (Computations; the *Syntax*) This definition by induction gives the syntax of what we will call, following the nomenclature in the literature, *computations*. Intuitively, one understands the concept of computation "dynamically" as a *process* that unfolds in a finite amount of time as a finite sequence of "steps". The technical term applied to the tuples below is thus a "static" version of "computation". However, this static concept leads to the dynamic concept by the same name in a straightforward way as we will see later in this chapter.

A *computation* here is a *tuple* $(e, \vec{x}, \vec{\alpha}, y)$ that encodes a *derivation* of a computable functional for a *given* hybrid *input* $\vec{x}, \vec{\alpha}$. The tuple also encapsulates the output, y, for the given input, *if indeed an output exists*. Crucial is the role of the *index e* as it packs the information —the "instructions"— of *how* to verify computationally, intuitively speaking, that y is indeed the output on input $(\vec{x}_k, \vec{\alpha}_l)$.

Thus "$(e, \vec{x}_k, \vec{\alpha}_l, y)$" —adding the lengths k and l of the vectors $\vec{x}$ and $\vec{\alpha}$ in the notation— is stripped down notation for $\phi_e^{(k,l)}(\vec{x}_k, \vec{\alpha}_l) \simeq y$ in Rogers' notation, or, what is the same, $\{e\}^{(k,l)}(\vec{x}_k, \vec{\alpha}_l) \simeq y$ in Kleene's notation (this being the preferred notation in much of the literature). The stripping down omits the redundant symbols "$\phi, \{\}, \simeq, (k, l)$".

Back to syntax: Let us define "k" and "l" up in front as $k = \text{length}(\vec{x})$ and $l = \text{length}(\vec{\alpha})$. Now that this has been stated we will be writing $\vec{x}$ for $\vec{x}_k$ and $\vec{\alpha}$ for $\vec{\alpha}_l$ in the rest of this definition.

The set Ω of *computations* on input $(\vec{x}, \vec{\alpha}) \in \mathbb{N}^k \times \mathcal{P}(\mathbb{N} : \mathbb{N})^l$, for all $k + l \geq 1$,[6] $\vec{x}, \vec{\alpha}, y, z, u, v, e, f, g, c$ is inductively defined as the *smallest* set of tuples $(e, \vec{x}, \vec{\alpha}, y)$ satisfying I–VIII below:

Basis (Initial) cases: I–VI

I.	$(\langle 1, k, l, i\rangle, \vec{x}, \vec{\alpha}, x_i) \in \Omega$	**for** $1 \leq i \leq k$
II.	$(\langle 2, k, l, i\rangle, \vec{x}, \vec{\alpha}, x_i + 1) \in \Omega$	**for** $1 \leq i \leq k$
III.	$(\langle 3, k, l, c\rangle, \vec{x}, \vec{\alpha}, c) \in \Omega$	**for** $c \in \mathbb{N}$
IV.	$(\langle 4, k + 2, l\rangle, a, b, \vec{x}, \vec{\alpha}, Sb_1(a, b)) \in \Omega$	**for** $Sb_1(a, b)$ defined in Theorem 13.3.18. **Cf.** also Theorem 10.2.1.
V.	$(\langle 5, k + 4, l\rangle, z, y, u, v, \vec{x}, \vec{\alpha}, z) \in \Omega$	**if** $u = v$
	$(\langle 5, k + 4, l\rangle, z, y, u, v, \vec{x}, \vec{\alpha}, y) \in \Omega$	**if** $u \neq v$
VI.	$(\langle 6, k, l, i, j\rangle, \vec{x}, \vec{\alpha}, y) \in \Omega$	**if** $\alpha_j(x_i) = y$

Inductive cases: VII–VIII

VII.	$(\langle 7, k, l, f, \vec{g}_m\rangle, \vec{x}, \vec{\alpha}, y) \in \Omega$	**if** $(f, \vec{z}_m, \vec{\alpha}, y) \in \Omega$
	and$(g_i, \vec{x}, \vec{\alpha}, z_i) \in \Omega$	**for** $i = 1, \ldots, m$
VIII.	$(\langle 8, k + 1, l\rangle, e, \vec{x}, \vec{\alpha}, y) \in \Omega$	**if** $(e, \vec{x}, \vec{\alpha}, y) \in \Omega$

Paraphrased in the style of Sect. 0.6, $\Omega = \text{Cl}(\mathcal{I}, \mathcal{O})$, where the *initial* objects form the set $\mathcal{I}$ of all tuples from I–VI —for all $k + l \geq 1$, $\vec{x}, \vec{\alpha}, y, z, u, v, c$— while we have just *two types* of operations on tuples in $\mathcal{O}$ (from VII and VIII): *composition operations* (VII) and *the universal functional operation* (VIII).

Both operations are finitary, that is, have a finite number of arguments. The indices $\langle 7, k, l, f, \vec{g}_m\rangle$ and $\langle 8, k + 1, l\rangle$ determine the number of inputs for the respective operations: $m + 1$ for the former and one for the latter. We have one operation for each input arity (case VII).

---

[6] As was the case with number-theoretic functions, we prefer *not* to allow functionals to have no inputs at all.

1. From the tuples $(f, \vec{z}_m, \vec{\alpha}, y)$ and $(g_i, \vec{x}, \vec{\alpha}, z_i)$, for $i = 1, \ldots, m = (f)_1$, we can generate the tuple $(\langle 7, k, l, f, \vec{g}_m \rangle, \vec{x}, \vec{\alpha}, y)$.
2. From the tuple $(e, \vec{x}, \vec{\alpha}, y)$ we can generate the tuple $(\langle 8, k+1, l \rangle, e, \vec{x}, \vec{\alpha}, y)$. □

13.3.2 Remark. We note that a correctly formed index e satisfies $Seq(e)$ and moreover

(1) $(e)_0 \in \{1, 2, 3, 4, 5, 6, 7, 8\}$, indicating the case (I)–(VIII) relevant to e according to the Definition 13.3.1.
(2) $(e)_1$ is the length of $\vec{x}$.
(3) $(e)_2$ is the length of $\vec{\alpha}$.
(4) e has additional components $(e)_i$, $i > 2$ only as needed to specify the relevant *initial* or *operation* cases.   □

We turn to semantics, that is, what these indices do:

13.3.3 Definition. (Computations; the *Semantics*) The symbol

$$\{e\}^{(k,l)}(\vec{x}_k, \vec{\alpha}_l) \simeq y \tag{1}$$

says exactly

$$(e, \vec{x}_k, \vec{\alpha}_l, y) \in \Omega$$

In words, *the functional* $\{e\}^{(k,l)}$ *of rank* (k, l) *given by the "program" e has as output the number y if the input is* $\vec{x}_k, \vec{\alpha}_l$.

This interpretation *makes no claim of uniqueness* or *single-valued-ness* of the y variable, but see the following proposition.

We also have the notations below signalling that we will usually write $\{e\}(\vec{x}_k, \alpha_l)$ without the superscript (k, l) letting the context fend for itself.

$$\{e\}(\vec{x}_k, \vec{\alpha}_l) \downarrow \text{ iff } (\exists y)\Big((e, \vec{x}_k, \vec{\alpha}_l, y) \in \Omega\Big) \text{ iff } (\exists y)\{e\}(\vec{x}_k, \vec{\alpha}_l) \simeq y$$

and

$$\{e\}(\vec{x}_k, \vec{\alpha}_l) \uparrow \text{ iff } \neg(\exists y)\Big((e, \vec{x}_k, \vec{\alpha}_l, y) \in \Omega\Big) \tag{2}$$

□

What is it that encourages the view of an index as a program? Well, for example, $\langle 1, k, l, i \rangle$ can be viewed as instructing a "computer" to compute the output x_i for any input $\vec{x}, \vec{\alpha}$. $\langle 6, k, l, i, j \rangle$ instructs a computer to consult an oracle (for α_j) and output the oracle's answer (if an answer *was* indeed offered), namely $\alpha_j(x_i)$.

The instruction $\langle 7, k, l, f, \vec{g}_m \rangle$ is more complex: It instructs the computer to use each of the instructions g_i on input $\vec{x}, \vec{\alpha}$ to (recursively) compute the z_i and then to use the instruction f on input $\vec{z}_m, \vec{\alpha}$ to (recursively) compute and output y. Why "recursively"? Because our "computer" can discharge the call to $\langle 7, k, l, f, \vec{g}_m \rangle$ with arguments $(\vec{x}, \vec{\alpha})$ as long as it knows how to discharge the *"simpler"* calls g_i on input $\vec{x}, \vec{\alpha}$ and f on input $\vec{z}_m, \vec{\alpha}$, and does so *following the inductive Definition* 13.3.1.

Why "simpler calls"? For one thing, Theorem 0.6.10 tells us that in a derivation of $(\langle 7, k, l, f, \vec{g}_m \rangle, \vec{x}, \vec{\alpha}, y)$ the tuples $(g_i, \vec{x}, \vec{\alpha}, z_i)$ and $(f, \vec{z}_m, \vec{\alpha}, y)$ occur *earlier*. This "earlier" is reasonably viewed as tantamount to "simpler"; for example, if "simplicity", *intuitively*, is gauged by run time, then no procedure that a program calls runs longer than the program itself. Another way to gauge relative simplicity of *computations* (in the sense of Definition 13.3.1) is contained in the definition below.

13.3.4 Definition. (Subcomputations and Computation Lengths) We define *immediate subcomputations* —in short *i.s.*— and *computation lengths* for each tuple in Definition 13.3.1.

I.–VI. Each of the computations $(e, \vec{x}, \vec{\alpha}, y)$ in cases I–VI of Definition 13.3.1 has *no* i.s.

  Their *length* $\|e, \vec{x}, \vec{\alpha}, y\|$ is equal to zero. In the case of VI, the length is 0 precisely when $(e, \vec{x}, \vec{\alpha}, y) = (\langle 6, k, l, j, i \rangle, \vec{x}, \vec{\alpha}, \alpha_j(x_i))$ *and* $\alpha_j(x_i) \downarrow$, else $\|e, \vec{x}, \vec{\alpha}, y)\| = \infty$.

Here "∞" is an attempt to avoid a knowledge of ordinals (which would allow us to use ω, the first infinite ordinal). The advantage of the ordinal ω is that it can engage with natural numbers in some arithmetic, and also satisfies $n < \omega$ for all $n \in \mathbb{N}$, as well as $n + \omega = \omega$.

Let us *postulate* the above property for "∞" (as is usually done) and leave ordinals out of this definition. Incidentally, the rudiments of arithmetic involving ∞[7] above will be needed to complete the inductive definition of $\| \ldots \|$ in the inductive cases below.

VII. If there exist y and $z_i, i = 1, \ldots, m$ such that $(f, \vec{z}, \vec{\alpha}, y) \in \Omega$ and $(e_i, \vec{x}, \vec{\alpha}, z_i) \in \Omega$, $i = 1, \ldots, m$, then *the* i.s. of

$$(\langle 7, k, l, f, e_1, \ldots, e_m \rangle, \vec{x}, \vec{\alpha}, y)$$

are $(f, \vec{z}, \vec{\alpha}, y)$ and $(e_i, \vec{x}, \vec{\alpha}, z_i)$, for $i = 1, 2, \ldots, m$. Moreover,

$$\| \langle 7, k, l, f, e_1, \ldots, e_m \rangle, \vec{x}, \vec{\alpha}, y \| =$$

$$\max \left(\| f, \vec{z}, \vec{\alpha}, y \|, \| e_1, \vec{x}, \vec{\alpha}, z_i \|, \ldots, \| e_m, \vec{x}, \vec{\alpha}, z_m \| \right) + 1$$

VIII. If there exists a y such that $(e, \vec{x}, \vec{\alpha}, y) \in \Omega$, then *the* i.s. of

---

[7] The mathematically rigorous use of the symbol ∞ is very limited, normally in something like (the interval) $(0, +\infty)$, $\lim_{x \to \infty} f(x) = a$ or $\int_0^\infty g(x)dx$.

$$(\langle 8, k+1, l \rangle, e, \vec{x}, \vec{\alpha}, y)$$

is $(e, \vec{x}, \vec{\alpha}, y)$.

Moreover, $\| \langle 8, k+1, l \rangle, e, \vec{x}, \vec{\alpha}, y \| = \| e, \vec{x}, \vec{\alpha}, y \| + 1$.

A computation $c \in \Omega$ is a *subcomputation* of $d \in \Omega$ just in case there is finite sequence $c = d_1, d_2, \dots, d_n = d$ such that d_i is an i.s. of d_{i+1}, for $i = 1, \dots, n-1$. $\square$

13.3.5 Remark. We used "the" in the two cases above that do have i.s. (VII and VIII) because the indices $\langle 7, k, l, f, e_1, \dots, e_m \rangle$ and $\langle 8, k+1, l \rangle$ uniquely determine the i.s. that there are, when they (it) exist(s). $\square$

13.3.6 Proposition. *The relation* $(e, \vec{x}_k, \alpha_l, y) \in \Omega$ *is single-valued in* y.

Proof Induction on Ω: For the Basis we sample two cases:

1. *Case I*: Let $(\langle 1, k, l, i \rangle, \vec{x}_k, \vec{\alpha}_l, x_i) \in \Omega$. Clearly the "$y$" (output) part is x_i, that is uniquely determined from $\vec{x}_k$ and $\langle 1, k, l, i \rangle$.
2. *Case VI*. Suppose $(\langle 6, k, l, i, j \rangle, \vec{x}, \vec{\alpha}, y) \in \Omega$ and $(\langle 6, k, l, i, j \rangle, \vec{x}, \vec{\alpha}, z) \in \Omega$. The only way to get these tuples (for index $\langle 6, k, l, i, j \rangle$) is to have $\alpha_j(x_i) = y$ and $\alpha_j(x_i) = z$. Since α_j is a *single-valued* function, we have $y = z$.

The single-valued-ness propagates with the two operation in $\mathcal{O}$. We sample operation VIII, assuming as I.H. that the *input* tuples in the operation are single-valued with respect to their last component.

Let

$$(\langle 8, k+1, l \rangle, e, \vec{x}, \vec{\alpha}, y) \in \Omega \text{ and } (\langle 8, k+1, l \rangle, e, \vec{x}, \vec{\alpha}, z) \in \Omega$$

By this particular operation (on tuples) (Definition 13.3.1) we get the above results if $(e, \vec{x}, \vec{\alpha}, y) \in \Omega$ and $(e, \vec{x}, \vec{\alpha}, z) \in \Omega$. But this implies $y = z$ by the I.H. $\square$

13.3.7 Remark. Because of the above, one often uses *"truncated computations"* — that is, computations with the output part omitted.

For example, we may write "$(e, \vec{x}, \vec{\alpha})$ is the i.s. of $(\langle 8, k+1, l \rangle, e, \vec{x}, \vec{\alpha})$", and also "$\| \langle 8, k+1, l \rangle, e, \vec{x}, \vec{\alpha} \| = \| e, \vec{x}, \vec{\alpha} \| + 1$", "$\| \langle 1, k, l, i \rangle, \vec{x}, \vec{\alpha} \| = 0$". $\square$

13.3.8 Definition. (The Set of Partial Recursive Functionals) In view of Proposition 13.3.6, we may define a single-valued functional $\{e\}$ of rank (k, l) extensionally (by its graph) by

$$\{e\}(\vec{x}, \vec{\alpha}) \simeq y \text{ iff } (e, \vec{x}, \vec{\alpha}, y) \in \Omega$$

We next define the set of *partial recursive functionals* (*partial computable functionals*):

It is the set of all $\mathcal{F}$ of rank (k, l) —for all $k + l \geq 1$— such that $\mathcal{F} = \{e\}^{(k,l)}$, for some $e \in \mathbb{N}$.

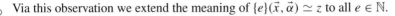

In particular, if e is not a correctly formed index, as defined inductively in Definition 13.3.1, we have —due to (2) p. 422— $\{e\} = \emptyset$, the everywhere undefined (*computable*) functional.

Via this observation we extend the meaning of $\{e\}(\vec{x}, \vec{\alpha}) \simeq z$ to all $e \in \mathbb{N}$.

The set of *recursive functionals* is formed by collecting *all* total partial recursive functionals.

We will *not* introduce special symbols for the *class* of recursive and partial recursive functionals. However, if we let $\mathcal{P}^{(k,l)}$ (resp. $\mathcal{R}^{(k,l)}$) be the set of all partial recursive (resp. *total* recursive) functionals of rank (k, l) then the class of *all* such, for all ranks, is $\bigcup_{k+l\geq 1} \mathcal{P}^{(k,l)}$ (resp. $\bigcup_{k+l\geq 1} \mathcal{R}^{(k,l)}$). □

13.3.9 Example. It is clear that each of the indices e defined in (I)–(VIII) in Definition 13.3.1 satisfies $Seq(e)$ (cf. Example 2.3.10(8)). Now $Seq(0)$ is false, thus, $\{0\} = \emptyset$, the empty functional. □

The following binds things well with our intuitive understanding, and issues the first few straightforward properties of the partial recursive functionals.

13.3.10 Proposition.

   I. $\{\langle 1, k, l, i \rangle\}(\vec{x}, \vec{\alpha}) \simeq x_i$
  II. $\{\langle 2, k, l, i \rangle\}(\vec{x}, \vec{\alpha}) \simeq x_i + 1$
 III. $\{\langle 3, k, l, c \rangle\}(\vec{x}, \vec{\alpha}) \simeq c$
 IV. $\{\langle 4, k+2, l \rangle\}(a, b, \vec{x}, \vec{\alpha}) \simeq Sb_1(a, b)$
  V. $\{\langle 5, k+4, l \rangle\}(p, q, a, b, \vec{x}, \vec{\alpha}) \simeq p \; if \; a = b$
 VI. $\{\langle 5, k+4, l \rangle\}(p, q, a, b, \vec{x}, \vec{\alpha}) \simeq q \; if \; a \neq b$
 VII. $\{\langle 6, k, l, i, j \rangle\}(\vec{x}, \vec{\alpha}) \simeq \alpha_j(x_i)$
VIII. $\{\langle 7, k, l, f, \vec{e}_m \rangle\}(\vec{x}, \vec{\alpha}) \simeq \{f\}\Big(\{e_1\}(\vec{x}, \vec{\alpha}), \ldots, \{e_m\}(\vec{x}, \vec{\alpha}), \vec{\alpha}\Big)$
 IX. $\{\langle 8, k+1, l \rangle\}(e, \vec{x}, \vec{\alpha}) \simeq \{e\}(\vec{x}, \vec{\alpha})$

Proof Cases I–VI are immediate.
  For

Case VII.
  We want

$$\{\langle 7, k, l, f, \vec{e}_m \rangle\}(\vec{x}, \vec{\alpha}) \simeq y \text{ iff } \{f\}\Big(\{e_1\}(\vec{x}, \vec{\alpha}), \ldots, \{e_m\}(\vec{x}, \vec{\alpha}), \vec{\alpha}\Big) \simeq y \quad (1)$$

- $\leftarrow$ direction of (1). Immediate from case VII of Definition 13.3.1, and from Definition 13.3.3.
- $\rightarrow$ direction. We are given $\{\langle 7, k, l, f, \vec{e}_m \rangle\}(\vec{x}, \vec{\alpha}) \simeq y$, that is, $(\langle 7, k, l, f, \vec{e}_m \rangle, \vec{x}, \vec{\alpha}, y) \in \Omega$. Induction on derivation length:
  We inspect a derivation of $(\langle 7, k, l, f, \vec{e}_m \rangle, \vec{x}, \vec{\alpha}, y)$ (this computation tuple cannot be the first element of a derivation (why?)).
  The structure of the index $c = \langle 7, k, l, f, \vec{e}_m \rangle$ implies that we must have —
  for some $z_i, i = 1, \ldots, m$— $(f, \vec{z}_m, \alpha, y)$ and $(e_i, \vec{x}, \alpha, z_i)$ appear *before* c in

the derivation and are in Ω. By Definition 13.3.3 and the I.H. (on derivation length) we have that

$$\{f\}(\vec{z}_m, \vec{\alpha}) \simeq y$$
$$\{e_1\}(\vec{x}, \vec{\alpha}) \simeq z_1$$
$$\{e_2\}(\vec{x}, \vec{\alpha}) \simeq z_2$$

$$\vdots$$

$$\{e_m\}(\vec{x}, \vec{\alpha}) \simeq z_m$$

(2)

We can summarise (2) by one true formula:

$$(\exists \vec{z}_m)\Big(\{f\}(\vec{z}_m, \vec{\alpha}) \simeq y \wedge \{e_1\}(\vec{x}, \vec{\alpha}) \simeq z_1 \wedge \ldots \wedge \{e_m\}(\vec{x}, \vec{\alpha}) \simeq z_m\Big)$$

which, by Definition 13.1.3, says

$$\{f\}(\{e_1\}(\vec{x}, \vec{\alpha}), \{e_2\}(\vec{x}, \vec{\alpha}), \ldots, \{e_m\}(\vec{x}, \vec{\alpha}), \vec{\alpha}) \simeq y$$

- (Case VIII.) We want

$$\{\langle 8, k+1, l\rangle\}(e, \vec{x}, \vec{\alpha}) \simeq y \text{ iff } \{e\}(\vec{x}, \vec{\alpha}) \simeq y$$

This case being much less involved is left as an exercise. Again the $\leftarrow$ direction is immediate from Definition 13.3.1.

$\square$

13.3.11 Remark.

- The $\rightarrow$ direction can also be argued by *induction on computation lengths*. Indeed, looking at case (VII) once more, let

$$(\langle 7, k, l, f, \vec{e}_m\rangle, \vec{x}, \vec{\alpha}, y) \in \Omega$$

By Definition 13.3.4, each of $\|f, \vec{z}_m, \vec{\alpha}, y\|$ and $\|e_i, \vec{x}, \vec{\alpha}, z_i\|$, for $i = 1, \ldots, m$, is

$$< \|\langle 7, k, l, f, \vec{e}_m\rangle, \vec{x}, \vec{\alpha}, y\|$$

By the obvious I.H. on computation lengths we have that all the statements in (2) of the preceding proof are true, and may conclude the rest of the proof exactly as before.
- We can even do CVI on the POset of all subcomputations ordered by the subcomputation relation, since the latter relation has clearly MC. Indeed, in any set of computations we can find a minimal element by creating a finite (why

finite?) descending chain of subcomputations (ordered by the subcomputation relation) starting from any element in the set. □

In more familiar (from earlier parts of the book) terminology we restate Proposition 13.3.10 as two corollaries.

13.3.12 Corollary. *The following functionals are partial computable.*

I. *Identity,* $\mathcal{U}_i^{(k,l)}$. *For all* $\vec{x}$ *and* $\vec{\alpha}$, $\mathcal{U}_i^{(k,l)}(\vec{x}, \vec{\alpha}) \simeq x_i$.

II. *Successor,* $\mathcal{S}_i^{(k,l)}$. *For all* $\vec{x}$ *and* $\vec{\alpha}$, $\mathcal{S}_i^{(k,l)}(\vec{x}, \vec{\alpha}) \simeq x_i + 1$.

III. *Constant (for each c),* $\mathcal{C}_c^{(k,l)}$. *For all* $\vec{x}$ *and* $\vec{\alpha}$, $\mathcal{C}_c^{(k,l)}(\vec{x}, \vec{\alpha}) \simeq c$.

IV. $\mathcal{S}_1^{(k+2,l)}$. *For all* $a, b, \vec{x}$ *and* $\vec{\alpha}$, $\mathcal{S}_1^{(k+2,l)}(a, b, \vec{x}, \vec{\alpha}) \simeq Sb_1(a, b)$.

V. *If-then else, or definition by cases,* $\mathcal{DC}^{(k+4,l)}$. $\mathcal{DC}^{(k+4,l)}(p, q, a, b, \vec{x}, \vec{\alpha}) \simeq p$ *if* $a = b$ $\mathcal{DC}^{(k+4,l)}(p, q, a, b, \vec{x}, \vec{\alpha}) \simeq q$ *if* $a \neq b$

VI. *Oracle call or function-input application,* $\mathcal{AP}_{i,j}^{(k,l)}$. $\mathcal{AP}_{i,j}^{(k,l)}(\vec{x}, \vec{\alpha}) \simeq \alpha_j(x_i)$

 13.3.13 Remark. Note the all the functionals I–VI above are *total* (cf. Definition 13.1.1), thus *recursive*. □

13.3.14 Example. Every partial recursive functional has infinitely many distinct indices. For example, if $\{a\} = \mathcal{F}$, then since $\mathcal{F} = \mathcal{U}_1^{(1,l)}\mathcal{F}$ we also have $\{\langle 7, k, l, \langle 1, 1, l, 1 \rangle, a \rangle\} = \mathcal{F}$. But $\langle 7, k, l, \langle 1, 1, l, 1 \rangle, a \rangle > a$. □

13.3.15 Corollary.

VII. *If the functionals* $\mathcal{F}^{(m,l)}$ *and* $\mathcal{E}_i^{(k,l)}$, *for* $i = 1, 2, \ldots, m$, *are partial recursive, then so is the one below obtained by* composition:

$$\lambda \vec{x}\vec{\alpha}.\mathcal{F}^{(m,l)}\left(\mathcal{E}_1^{(k,l)}(\vec{x}, \vec{\alpha}), \mathcal{E}_2^{(k,l)}(\vec{x}, \vec{\alpha}), \ldots, \mathcal{E}_m^{(k,l)}(\vec{x}, \vec{\alpha}), \vec{\alpha}\right)$$

VIII. *There is a* universal functional *of rank* $(k + 1, l)$ *—*$\mathcal{UN}^{(k+1,l)}$*— for all computable functionals* $\{e\}^{(k,l)}$ *of rank* (k, l). *That is,* $\mathcal{UN}^{(k+1,l)}(e, \vec{x}, \vec{\alpha}) \simeq \{e\}(\vec{x}, \vec{\alpha})$, *for all* $e, \vec{x}, \vec{\alpha}$.

In view of the above corollaries, in particular items I–III and VII, we have from Exercise 2.1.10:

13.3.16 Proposition. *The set of partial recursive functionals is closed under Grzegorczyk substitution into type-0 inputs, as long as the type-1 variable list is the same in the substitution parts:* $\lambda \vec{x}\vec{z}\vec{\alpha}.\mathcal{F}(\mathcal{G}(\vec{z}, \vec{\alpha}), \vec{x}, \vec{\alpha})$.

 13.3.17 Remark. If $\lambda y\vec{x}\vec{\alpha}.\mathcal{F}(y, \vec{x}, \vec{\alpha})$ and $\lambda \vec{z}\vec{\beta}.\mathcal{G}(\vec{z}, \vec{\beta})$ are partial recursive, is this true also for $\lambda \vec{x}\vec{z}\vec{\alpha}\vec{\beta}.\mathcal{F}(\mathcal{G}(\vec{z}, \vec{\alpha}), \vec{x}, \vec{\beta})$?

We will answer this question affirmatively in Sect. 13.4. A partial answer is found in the remark following Proposition 13.3.29. □

The following technical results bootstrap the theory.

13.3.18 Theorem. (S-m-n Theorem; Gödel, Kleene) *Let* $k = lh(\vec{x})$ *and* $l = lh(\vec{\alpha})$. *There is an* index $Sb_1(e, y)$ *that depends on e and y and, for all e and y, we have*

$$\{Sb_1(e, y)\}(\vec{x}, \vec{\alpha}) \simeq z \text{ iff } \{e\}(y, \vec{x}, \vec{\alpha}) \simeq z$$

Moreover, $\|e, y, \vec{x}, \vec{\alpha}, z\| < \|Sb_1(e, y), \vec{x}, \vec{\alpha}, z\|$.

Proof Note that

$$Sb_1(e, y) = \langle 7, k, l, e, \langle 3, k, l, y \rangle, \langle 1, k, l, 1 \rangle, \langle 1, k, l, 2 \rangle, \ldots, \langle 1, k, l, k \rangle \rangle$$

works. The inequality is from Definition 13.3.4, case VII —using i.s. $(e, y, \vec{x}, \vec{\alpha}, z)$.

$\square$

It is clear that $Sb_1(e, y)$ is strictly increasing with respect to e and y.

Using the approach from Theorem 10.2.1 we can define S-m-n functions Sb_n exactly as in Definition 10.2.2 and obtain

13.3.19 Corollary. *For each* $n \geq 1$, *there is a rank-*$(n + 1, 0)$ $Sb_n \in \mathcal{PR}$ *and we have that, for all* $m \geq 1$, *the following:*

$$\{Sb_n(a, \vec{y}_n)\}(\vec{x}_m, \vec{\alpha}) \simeq \{a\}(\vec{y}_n, \vec{x}_m, \vec{\alpha})$$

Moreover, $\|e, \vec{y}_n, \vec{x}, \vec{\alpha}, z\| < \|Sb_n(e, \vec{y}_n), \vec{x}, \vec{\alpha}, z\|$.

13.3.20 Remark. As we noted already in Proposition 13.3.16, the presence of I. in Corollary 13.3.12 enables Grzegorczyk operations on type-0 inputs, thus, a relaxed version of the S-m-n theorem (as already noted in Theorem 5.7.1) is

There is a primitive recursive h of n variables, such that, for any $\mathcal{F}$ *of rank* $(n + m, l)$ *and any* $\vec{y}_n$ *and* $\vec{x}_m$, *we have*

$$\{h(\vec{y}_n)\}(\vec{x}_m, \vec{\alpha}) \simeq \mathcal{F}(\vec{x}_m, \vec{y}_n, \vec{\alpha}) \tag{1}$$

Pause. Does it matter if the order of type-0 variables on the right hand side of (1) is any permutation of the x_i, y_r in $\vec{x}_m$, $\vec{y}_n$? ◀ $\square$

13.3.21 Theorem. (The 2nd Recursion Theorem)

(Gödel, Carnap, Kleene.) *For any partial recursive* $\lambda e \vec{x} \vec{\alpha}.\mathcal{F}(e, \vec{x}, \vec{\alpha})$ *there is an index* $\overline{e}$ *such*

$$\mathcal{F}(\overline{e}, \vec{x}, \vec{\alpha}) \simeq z \text{ iff } \{\overline{e}\}(y, \vec{x}, \vec{\alpha}) \simeq z$$

Moreover, if we let f be an index for $\mathcal{F}$, *then* $\|\overline{e}, \vec{x}, \vec{\alpha}, z\| > \|f, \overline{e}, \vec{x}, \vec{\alpha}, z\|$.

Proof Via Gödel's "substitution trick". Let $\{f\} = \mathcal{F}$ and consider the partial recursive functional $\lambda z \vec{x} \vec{\alpha}.\{f\}(\mathcal{S}_1^{(k+1,l)}(z, z, \vec{x}, \vec{\alpha}), \vec{x}, \vec{\alpha})$. This functional has the typical call below (presented as full composition):

$$\{f\}(\mathscr{S}_1^{(k+1,l)}(z, z, \vec{x}, \vec{\alpha}), \mathscr{U}_2^{k+1}(z, \vec{x}, \alpha), \ldots, \mathscr{U}_{k+1}^{k+1}(z, \vec{x}, \alpha), \vec{\alpha}) \tag{1}$$

Let s_1 be an index of $\mathscr{S}_1^{(k+1,l)}$. Then a "natural"[8] index b of the functional (whose typical call is displayed) in (1) is

$$b = \langle 7, k+1, f, s_1, \langle 1, k+1, l, 2 \rangle, \ldots, \langle 1, k+1, l, k+1 \rangle \rangle \tag{2}$$

Then

$$\{Sb_1(b, z)\}(\vec{x}, \vec{\alpha}) \overset{\text{Theorem 13.3.18}}{\simeq} \{b\}(z, \vec{x}, \vec{\alpha}) \simeq \{f\}(Sb_1(z, z), \vec{x}, \vec{\alpha}) \simeq \mathscr{F}(Sb_1(z, z), \vec{x}, \vec{\alpha})$$

Take $\bar{e} = Sb_1(b, b)$. Finally, note the inequalities

$$\|\bar{e}, \vec{x}, \vec{\alpha}, z\| \overset{\text{Theorem 13.3.18}}{>} \|b, b, \vec{x}, \vec{\alpha}, z\| \overset{(2)}{>} \|f, \bar{e}, \vec{x}, \vec{\alpha}, z\| \qquad \square$$

We have a number of applications, using $\mathscr{DC}$ (cf. Corollary 13.3.12, V) in conjunction with the recursion theorem.

13.3.22 Theorem. (Closure Under (μy)) *Let* $\lambda \vec{x} \vec{\alpha}.\mathscr{F}(\vec{x}, \vec{\alpha})$ *be partial computable. Then so is* $\mathscr{H}$ *given by*

$$\mathscr{H}(y, \vec{x}, \vec{\alpha}) \simeq \begin{cases} y & \text{if } \mathscr{F}(y, \vec{x}, \vec{\alpha}) \simeq 0 \\ \mathscr{H}(y+1, \vec{x}, \vec{\alpha}) & \text{othw} \end{cases} \tag{1}$$

Proof We note that $\mathscr{H}$ is uniquely determined as follows: Suppose that

$$\mathscr{H}(m, \vec{x}, \vec{\alpha}) = \mathscr{H}(m+1, \vec{x}, \vec{\alpha}) = \mathscr{H}(m+2, \vec{x}, \vec{\alpha}) = \ldots$$

$$= \mathscr{H}(k, \vec{x}, \vec{\alpha}) = k = \begin{cases} \min\{y \geq m : \mathscr{F}(y, \vec{x}, \vec{\alpha}) \simeq 0 \wedge \\ \qquad (\forall z)_{m \leq z < k} \mathscr{F}(z, \vec{x}, \vec{\alpha}) > 0\} \\ \uparrow \text{ othw} \end{cases}$$

In particular

$$\mathscr{H}(0, \vec{x}, \vec{\alpha}) = (\mu y)\mathscr{F}(y, \vec{x}, \vec{\alpha}) \tag{2}$$

As for the partial recursiveness of $\mathscr{H}$ we argue as in Sect. 9.5: Define

$$\mathscr{G}(z, y, \vec{x}, \vec{\alpha}) \simeq \begin{cases} y & \text{if } \mathscr{F}(y, \vec{x}, \vec{\alpha}) \simeq 0 \\ \{z\}(y+1, \vec{x}, \vec{\alpha}) & \text{othw} \end{cases}$$

---

[8] Computed faithfully to the displayed composition.

By composition and the recursiveness of $\mathscr{DC}$, $\mathscr{G}$ is partial recursive. By Theorem 13.3.21 there is an e such that $\{e\}(y, \vec{x}, \vec{\alpha}) \simeq \mathscr{G}(e, y, \vec{x}, \vec{\alpha})$, thus $\{e\} = \mathscr{H}$. By (2) we are done.   □

13.3.23 Proposition. (The Predecessor) $\mathscr{P}_i^{(k,l)} = \lambda \vec{x}\vec{\alpha}.x_i \,\dot{-}\, 1$, *is recursive.*

Proof We want $x_i \,\dot{-}\, 1 \simeq (\mu y)(y + 1 = x_i)$. Thus define

$$\mathscr{H}(y, \vec{x}, \vec{\alpha}) = \begin{cases} 0 & \text{if } x_i = y + 1 \\ \mathscr{H}(y + 1, \vec{x}, \vec{\alpha}) + 1 & \text{othw} \end{cases} \tag{†}$$

The definition of $\mathscr{H}$ above is a standard variant of that in (1) of the previous proof. With the help of $\mathscr{DC}$ we can dress (†) up to not only follow our rules of formation but also to be seen that it does:

$$\mathscr{H}(y, \vec{x}, \vec{\alpha}) = \mathscr{DC}^{(k+4,l)}\Big(0, \mathscr{H}(y + 1, \vec{x}, \vec{\alpha}) + 1, x_i, y + 1, \vec{x}, \vec{\alpha}\Big)$$

By the recursion theorem, we obtain an e such that $\{e\} = \mathscr{H}$. Then

$$\mathscr{H}(0, \vec{x}, \vec{\alpha}) = \begin{cases} x_i \,\dot{-}\, 1 & \text{if } x_i > 0 \\ \uparrow & \text{othw} \end{cases}$$

To conclude, mindful of Proposition 13.3.16, we set

$$\mathscr{P}_i^{(k,l)}(\vec{x}, \vec{\alpha}) \stackrel{Def}{=} \mathscr{DC}^{(k+4,l)}\Big(0, \mathscr{H}(0, \vec{x}, \vec{\alpha}), x_i, 0, \vec{x}, \vec{\alpha}\Big)$$

It is clear that $\mathscr{P}_i^{(k,l)}$ is total.   □

13.3.24 Theorem. (Closure Under Primitive Recursion) *The set of partial recursive functionals is closed under primitive recursion:*

$$\mathscr{F}(0, \vec{x}, \vec{\alpha}) = \mathscr{H}(\vec{x}, \vec{\alpha})$$
$$\mathscr{F}(z + 1, \vec{x}, \vec{\alpha}) = \mathscr{G}(z, \vec{x}, \vec{\alpha}, \mathscr{F}(z, \vec{x}, \vec{\alpha})) \tag{1}$$

That is, if $\mathscr{H}$ and $\mathscr{G}$ are partial recursive then so is $\mathscr{F}$.

We shall call z the iteration variable *and* $\mathscr{G}$ the iteration functional, *while $\mathscr{H}$ is the* Basis *functional.*

Proof We rewrite (1) as one equation using $\mathscr{DC}$ and $\mathscr{P}_1^{(k+1,l)}$ of Proposition 13.3.23:

$$\mathscr{F}(z, \vec{x}, \vec{\alpha}) = \mathscr{DC}\Big(\mathscr{H}(\vec{x}, \vec{\alpha}), \mathscr{G}(z \,\dot{-}\, 1, \vec{x}, \vec{\alpha}, \mathscr{F}(z, \vec{x}, \vec{\alpha})), z, 0, \vec{x}, \vec{\alpha}\Big) \tag{2}$$

where we wrote above $z \,\dot{-}\, 1$ rather than $\mathscr{P}_1^{(k+1,l)}(z, \vec{x}, \vec{\alpha})$ for readability.

By Theorem 0.4.43, we have a unique $\mathscr{F}$ that "solves" equations (1). By (2), the recursion theorem guarantees the existence of an e such that $\{e\} = \mathscr{F}$. □

13.3.25 Corollary. *The set of (total) recursive functionals is closed under primitive recursion.*

Proof It is clear from the above proof and and an induction on the iteration variable that if $\mathscr{H}$ and $\mathscr{G}$ are total, then so is $\mathscr{F}$. □

Incidentally,

13.3.26 Proposition. *The class of recursive functionals is closed under composition.*

Proof Because the class of all partial recursive functionals is, and composition preserves totalness. □

One can now see immediately that all the tools from Sect. 2.1 have become immediately available, in particular the coding tools of Definition 2.4.2. We state the obvious, without proofs:

13.3.27 Corollary. $\mathcal{PR}$ —a class of $(k, 0)$ functionals, for all $k \geq 1$— is a subset of the the class of recursive functionals.

We have more primitive recursive functionals than the above:

13.3.28 Definition. (Primitive Recursive Functionals) The class of the *primitive recursive* functionals of rank (k, l) —$k + l \geq 1$— is the *smallest* class that includes the initial functionals (Corollary 13.3.12, I–VI and is closed under *composition* (equivalently,[9] *Grzegorczyk operations on type-0 variables*) and *primitive recursion*. □

It is immediate that all primitive recursive functionals are total (cf. Definition 13.1.1) and therefore they form a subset of the recursive functionals, since the latter class contains the same initial functions and is closed under the same operations.

For Definition 13.3.28 it suffices to include as initial functionals only I–III and VI.

The following is useful.

13.3.29 Proposition. (Type-1 Argument List Expansion) If $\lambda \vec{x} \vec{\alpha}.\mathscr{F}(\vec{x}, \vec{\alpha})$ is primitive recursive then so is $\lambda \vec{x} \vec{y} \vec{\alpha} \vec{\beta}.\mathscr{G}(\vec{x}, \vec{y}, \vec{\alpha}, \vec{\beta}) = \lambda \vec{x} \vec{y} \vec{\alpha} \vec{\beta}.\mathscr{F}(\vec{x}, \vec{\alpha})$.

Proof The case for type-0 argument list expansion follows by Grzegorczyk operations (Proposition 13.3.16) that holds for all partial recursive functionals. To show that we can do $\lambda \vec{x} \vec{\alpha} \vec{\beta}.\mathscr{F}(\vec{x}, \vec{\alpha})$ in the primitive recursive case we do induction on the inductive definition of such functionals (Definition 13.3.28).

---

[9] In the presence of I–III.

1. *The claim checks for the initial functionals.*
   We sample case VI: Then $\mathscr{F}(\vec{x}, \vec{\alpha}) \simeq \{\langle 6, k, l, i, j\rangle\}(\vec{x}, \vec{\alpha}) \simeq \alpha_j(i)$, where
   $k = length(\vec{x})$, $l = length(\vec{\alpha})$. Clearly, $\alpha_j(i) \simeq \{\langle 6, k, l + l', i, j\rangle\}(\vec{x}, \vec{\alpha}, \vec{\beta}) \simeq$
   $\lambda \vec{x} \vec{\alpha} \vec{\beta}.\mathscr{F}(\vec{x}, \vec{\alpha}, \vec{\beta})$, where $l' = length(\beta)$.

2. *The claim propagates with Grzegorczyk substitution.* Suppose $\mathscr{H}(\vec{x}, \vec{z}, \vec{\alpha}) \simeq$
   $\mathscr{F}(\mathscr{G}(\vec{z}, \vec{\alpha}), \vec{x}, \vec{\alpha})$ and for the primitive recursive $\mathscr{F}$ and $\mathscr{G}$ we have

   $$\mathscr{F}(u, \vec{x}, \vec{\alpha}) \simeq \mathscr{F}(u, \vec{x}, \vec{\alpha}, \vec{\beta}) \simeq w \tag{1}$$

   and

   $$\mathscr{G}(\vec{z}, \vec{\alpha}) \simeq \mathscr{G}(\vec{z}, \vec{\alpha}, \vec{\beta}) \simeq v \tag{2}$$

   Thus,

   $$\mathscr{H}(\vec{x}, \vec{z}, \vec{\alpha}, \vec{\beta}) \simeq w$$

   iff

   $$(\exists u)\Big(\mathscr{F}(u, \vec{x}, \vec{\alpha}, \vec{\beta}) \simeq w \wedge \mathscr{G}(\vec{z}, \vec{\alpha}, \vec{\beta}) \simeq u\Big)$$

   iff, by (1) and (2)

   $$(\exists u)\Big(\mathscr{F}(u, \vec{x}, \vec{\alpha}) \simeq w \wedge \mathscr{G}(\vec{z}, \vec{\alpha}) \simeq u\Big)$$

   iff

   $$\mathscr{H}(\vec{x}, \vec{z}, \vec{\alpha}) \simeq w$$

3. *The claim propagates with primitive recursion.* This case is left to the reader as it
   is entirely similar to the above. Let $\mathscr{H} = prim(\mathscr{F}, \mathscr{G})$ and accept the I.H. that

   $$\mathscr{F}(\vec{y}, \vec{\alpha}) \simeq \mathscr{F}(\vec{y}, \vec{\alpha}, \vec{\beta}) \simeq u \tag{1}$$

   and

   $$\mathscr{G}(x, \vec{y}, w, \vec{\alpha}) \simeq \mathscr{G}(x, \vec{y}, w, \vec{\alpha}, \vec{\beta}) \simeq v \tag{2}$$

   Next compare $\mathscr{H}(x, \vec{y}, \vec{\alpha})$ with $\mathscr{H}(x, \vec{y}, \vec{\alpha}, \vec{\beta})$. $\square$

13.3.30 Remark. From the above follows immediately that for *primitive recursive
functionals* we can do[10] substitutions and primitive recursions where the two component functionals in each case have different type-1 lists of inputs, $\vec{\alpha}$ and $\vec{\beta}$. Using

---

[10] "Can do" is a colloquialism for "if we do so, then we get a primitive recursive functional".

the above proposition we can equalise the two lists forming the list $\vec{\alpha}, \vec{\beta}$. Then, e.g., a substitution like $\mathscr{F}(\mathscr{G}(\vec{z}, \vec{\alpha}), \vec{x}, \vec{\beta})$ is the same as $\mathscr{F}(\mathscr{G}(\vec{z}, \vec{\alpha}, \vec{\beta}), \vec{x}, \vec{\alpha}, \vec{\beta})$. But the latter does not lead outside the class of primitive recursive functionals, hence nor does the former. □

13.3.31 Definition. (Partial Characteristic Functional) A *partial characteristic functional* $\mathscr{K}_{\mathbb{P}}$ of a predicate $\mathbb{P}$ is defined as

$$\mathscr{K}_{\mathbb{P}}(\vec{x}, \vec{\alpha}) \simeq \begin{cases} 0 & \text{if } \mathbb{P}(\vec{x}, \vec{\alpha}) \\ \neq 0 \text{ or } \uparrow & \text{othw} \end{cases}$$

In short, the defining property of a partial characteristic functional $\mathscr{K}_{\mathbb{P}}$ of $\mathbb{P}$ is $\mathbb{P}(\vec{x}, \vec{\alpha}) \equiv \mathscr{K}_{\mathbb{P}}(\vec{x}, \vec{\alpha}) \simeq 0$. □

 13.3.32 Remark. Partial characteristic functionals of a given relation are not unique. For example, see ($\ddagger$) in Example 13.3.35, item (5), where one can use —instead of $\mathscr{DC}(0, 42, y, \alpha(x), x, \alpha)$— $\mathscr{DC}(0, c, y, \alpha(x), x, \alpha)$, for *any* $c \geq 1$, as a partial characteristic functional for $y = \alpha(x)$. □

13.3.33 Definition. [Semi-Recursiveness and (Primitive) Recursiveness] A relation $\mathbb{P}$ of rank (k, l) is called *semi-recursive* iff it is the *domain of a partial recursive functional* $\mathscr{F}$ of rank (k, l).

An index e for $\mathscr{F}$ is by definition a *semi-recursive index*, or simply *semi-index* or just *index* of $\mathbb{P}$.

A relation $\mathbb{P}$ is called *(primitive) recursive* iff it has a *(primitive) recursive partial characteristic functional* $\mathscr{K}_{\mathbb{P}}$. □

13.3.34 Proposition. *A relation $\mathbb{P}$ of rank (k, l) is* semi-recursive *iff its partial characteristic functional $\mathscr{K}_{\mathbb{P}}$ is partial recursive.*

Proof

1. (*If* part.) Let $\mathscr{K}_{\mathbb{P}}^{(k,l)}$ be partial recursive. We have noted earlier that $\{0\}^{(k,l)} = \emptyset$, the empty function.
   Thus if

$$\mathscr{G} \overset{Def}{=} \lambda \vec{x}\vec{\alpha}.\mathscr{DC}^{(k+4,l)}\left(0, \{0\}(0), \mathscr{K}_{\mathbb{P}}^{(k,l)}(\vec{x}, \vec{\alpha}), 0, \vec{x}, \vec{\alpha}\right)$$

   then $\mathbb{P}(\vec{x}, \vec{\alpha}) \equiv \mathscr{G}(\vec{x}, \vec{\alpha}) \downarrow (\simeq 0)$ and $\mathscr{G}$ is partial recursive.
2. (*Only If* part.) Let $\mathbb{P}(\vec{x}, \vec{\alpha}) \equiv \mathscr{F}(\vec{x}, \vec{\alpha}) \downarrow$ for some partial recursive $\mathscr{F}$.
   Note that $\lambda \vec{x}\vec{\alpha}.\mathscr{C}_0(\mathscr{F}(\vec{x}, \vec{\alpha}), \vec{x}, \vec{\alpha})$ is a partial characteristic function of $\mathbb{P}$ *and is* partial recursive. □

 13.3.35 Example. This is an ad hoc collection of some useful primitive recursive functionals and primitive and semi-recursive relations.

(1) Firstly, by Proposition 13.3.29 we can add "don't care" type-1 arguments[11] to any function in $\mathcal{PR}$ to obtain a primitive recursive functional.

For example, from $\lambda xy.x \doteq y$ of $\mathcal{PR}$ we get the primitive recursive functional $\lambda xy\vec{\alpha}.x \doteq y$. Similarly from $\lambda xy.|x-y| \in \mathcal{PR}$[12] we get the primitive recursive functional $\lambda xy\vec{\alpha}.|x-y|$.

(2) If $\lambda y\vec{x}\vec{\alpha}.\mathscr{F}(y,\vec{x},\vec{\alpha})$ is primitive recursive then the primitive recursion below shows that so is $\lambda z\vec{x}\vec{\alpha}.\sum_{i<z}\mathscr{F}(i,\vec{x},\vec{\alpha})$.

$$\sum_{i<0}\mathscr{F}(i,\vec{x},\vec{\alpha}) \simeq 0$$

$$\sum_{i<z+1}\mathscr{F}(i,\vec{x},\vec{\alpha}) \simeq \mathscr{F}(z,\vec{x},\vec{\alpha}) + \sum_{i<z}\mathscr{F}(i,\vec{x},\vec{\alpha})$$

Similarly, $\lambda z\vec{x}\vec{\alpha}.\prod_{i<z}\mathscr{F}(i,\vec{x},\vec{\alpha})$ is primitive recursive as the following shows:

$$\prod_{i<0}\mathscr{F}(i,\vec{x},\vec{\alpha}) \simeq 1$$

$$\prod_{i<z+1}\mathscr{F}(i,\vec{x},\vec{\alpha}) \simeq \mathscr{F}(z,\vec{x},\vec{\alpha}) \times \prod_{i<z}\mathscr{F}(i,\vec{x},\vec{\alpha})$$

(3) If $\mathbb{P}(y,\vec{x},\vec{\alpha})$ is a primitive recursive relation[13] and $\lambda\vec{y}\vec{\beta}.\mathscr{F}(\vec{y},\vec{\beta})$ is a primitive recursive functional, then $\mathbb{P}(\mathscr{F}(\vec{y},\vec{\beta}),\vec{x},\vec{\alpha})$ is a primitive recursive predicate. Indeed, let $\mathscr{K}$ be a primitive recursive partial characteristic functional of $\mathbb{P}$. By Definition 13.3.33 we have

$$\mathbb{P}(z,\vec{x},\vec{\alpha}) \equiv \mathscr{K}(z,\vec{x},\vec{\alpha}) \simeq 0 \tag{¶}$$

Thus $\lambda\vec{x}\vec{y}\vec{\alpha}\vec{\beta}.\mathscr{K}(\mathscr{F}(\vec{y},\vec{\beta}),\vec{x},\vec{\alpha})$ is primitive recursive by Grzegorczyk substitution. We claim that

$$\mathbb{P}(\mathscr{F}(\vec{y},\vec{\beta}),\vec{x},\vec{\alpha}) \equiv \mathscr{K}(\mathscr{F}(\vec{y},\vec{\beta}),\vec{x},\vec{\alpha}) \simeq 0$$

To verify, note that

$$\mathbb{P}(\mathscr{F}(\vec{y},\vec{\beta}),\vec{x},\vec{\alpha}) \equiv (\exists u)\Big(\mathscr{F}(\vec{y},\vec{\beta}) \simeq u \wedge \mathbb{P}(u,\vec{x},\vec{\alpha})\Big)$$

and

$$\mathscr{K}(\mathscr{F}(\vec{y},\vec{\beta}),\vec{x},\vec{\alpha}) \simeq 0 \equiv (\exists u)\Big(\mathscr{F}(\vec{y},\vec{\beta}) \simeq u \wedge \mathscr{K}(u,\vec{x},\vec{\alpha}) \simeq 0\Big)$$

---

[11] Some authors, unkindly, call such variables or arguments "dummy".

[12] $|x-y| = (x \doteq y) + (y \doteq x)$.

[13] Going forward we will do as the literature does and —when working in computability— we shall use the term "*predicate*" interchangeably with "*relation*".

We are done because the right hand sides (of $\equiv$) in the last two displayed statements above are logically equivalent due to ($\P$).

(4) The class of primitive recursive predicates is closed under the Boolean operation $\wedge$.

So let $\mathbb{P}(\vec{x}, \vec{\alpha})$ and $\mathbb{Q}(\vec{y}, \vec{\beta})$ be primitive recursive, that is, we have $\mathscr{K}_P$ and $\mathscr{K}_Q$, primitive recursive, such that

$$\mathbb{P}(\vec{x}, \vec{\alpha}) \equiv \mathscr{K}_P(\vec{x}, \vec{\alpha}) \simeq 0 \qquad (i)$$

and

$$\mathbb{Q}(\vec{y}, \vec{\beta}) \equiv \mathscr{K}_Q(\vec{y}, \vec{\beta}) \simeq 0 \qquad (ii)$$

Then

$$\mathbb{P}(\vec{x}, \vec{\alpha}) \wedge \mathbb{Q}(\vec{y}, \vec{\beta}) \equiv \mathscr{K}_P(\vec{x}, \vec{\alpha}) + \mathscr{K}_Q(\vec{y}, \vec{\beta}) \simeq 0 \qquad (iii)$$

Indeed, if the left hand side (of $\equiv$) in (iii) is true, then each of $\mathbb{P}(\vec{x}, \vec{\alpha})$ and $\mathbb{Q}(\vec{y}, \vec{\beta})$ are true thus each of the right hand sides (of $\equiv$) in (i) and (ii) are true, and thus so is $\mathscr{K}_P(\vec{x}, \vec{\alpha}) + \mathscr{K}_Q(\vec{y}, \vec{\beta}) \simeq 0$.

For the converse, the truth of the last statement above implies the truth of each right hand side in (i) and (ii), which yields the truth of $\mathbb{P}(\vec{x}, \vec{\alpha}) \wedge \mathbb{Q}(\vec{y}, \vec{\beta})$.

Pause. Why not use $\times$ between $\mathscr{K}_P$ and $\mathscr{K}_Q$ to settle primitive recursiveness of "$\vee$"?

Well, suppose we claim (iv) below.

Say, $\mathbb{P}(\vec{x}, \vec{\alpha}) \vee \mathbb{Q}(\vec{y}, \vec{\beta})$ is true due to $\mathbb{P}(\vec{x}, \vec{\alpha})$ being true. Assume, however, that $\mathbb{Q}(\vec{y}, \vec{\beta}))$ is false, manifested as $\mathscr{K}_Q(\vec{y}, \vec{\beta}) \uparrow$ (cf. Definition 13.3.31). Then the right hand side of (iv) is undefined; it is *not* $\simeq 0$ and thus our attempt to show closure under $\vee$ fails.

$$\mathbb{P}(\vec{x}, \vec{\alpha}) \vee \mathbb{Q}(\vec{y}, \vec{\beta}) \equiv \mathscr{K}_P(\vec{x}, \vec{\alpha}) \times \mathscr{K}_Q(\vec{y}, \vec{\beta}) \simeq 0 \qquad (iv)$$

$\blacktriangleleft$

(5) The relation $y = \alpha(x)$ is primitive recursive. In fact, in the equivalence below

$$y = \alpha(x) \equiv |y - \alpha(x)| \simeq 0 \qquad (\dagger)$$

the right hand side of $\equiv$ contains a primitive recursive partial characteristic functional of the left hand side relation by (1) and (3).

Alternatively,

$$\mathscr{DC}(0, 42, y, \alpha(x), x, \alpha) \qquad (\ddagger)$$

is another primitive recursive partial characteristic functional for $y = \alpha(x)$.

Thus we conclude, once again, that $y = \alpha(x)$ is primitive recursive because it has primitive recursive partial characteristic functionals —e.g., (†) and (‡).

(6) The class of primitive recursive predicates is closed under bounded universal quantification (this is the natural extension of closure under $\wedge$).
So let

$$\mathbb{P}(y, \vec{x}, \vec{\alpha}) \equiv \mathscr{K}(y, \vec{x}, \vec{\alpha}) \simeq 0 \tag{*}$$

where $\mathscr{K}$ is primitive recursive. Then

$$(\forall y)_{<z}\mathbb{P}(y, \vec{x}, \vec{\alpha}) \equiv \left(\sum_{y<z} \mathscr{K}(y, \vec{x}, \vec{\alpha}) \right) \simeq 0$$

Indeed, if $\mathbb{P}(y, \vec{x}, \vec{\alpha})$ is true for $y < z$ then, by (*), it is $\left(\sum_{y<z} \mathscr{K}(y, \vec{x}, \vec{\alpha}) \right) \simeq 0$.
Conversely, assuming the latter we have $\mathscr{K}(y, \vec{x}, \vec{\alpha}) \downarrow\simeq 0$, for $y < z$, hence $\mathbb{P}(y, \vec{x}, \vec{\alpha})$ is true, for all $y < z$, by (*).

(7) We will use this example in Sect. 13.4. Recall Definition 11.3.6 and the notations $[u]$ and alternatively ξ_u or θ_u for finite functions of code u. The predicate $\lambda u\alpha.[u] \subseteq \alpha$ is primitive recursive.

Some authors (Davis 1958b; Odifreddi 1989) write $\alpha \mid u$ for $[u] \subseteq \alpha$.

Note that

$$[u] \subseteq \alpha \equiv u \geq 1 \wedge (\forall y)_{\leq u}\left(p_y \mid u \to (u)_y = \alpha(y) \right) \tag{**}$$

To conclude, we observe that $p_y \mid u \to (u)_y = z$ is primitive recursive (rank $(3, 0)$) and hence so is $p_y \mid u \to (u)_y = \alpha(y)$ —rank$(2, 1)$— by item 3.
By items (4) and (6) the primitive recursiveness of (**) follows.   □

13.3.1   Functional Substitution

Suppose $\mathscr{F}$ of rank $(k, l + 1)$ and $\lambda y.\mathscr{G}(y, \vec{x}, \vec{\alpha})$ are partial recursive.
How about $\lambda \vec{x}\vec{\alpha}.\mathscr{F}(\vec{x}, \vec{\alpha}, \lambda y.\mathscr{G}(y, \vec{x}, \vec{\alpha}))$?

Intuitively, this should be partial recursive too, as any call to the oracle of the last type-1 argument β would become a call to the partial computable $\lambda y.\mathscr{G}(y, \vec{x}, \vec{\alpha})$ and we could in principle obtain the answer computationally with no need for an oracle for this specific "β".

Our theory of computable partial functionals supports this rough analysis.

13.3.36 Theorem. *There is a rank* $(2, 0)$ *recursive function(al) h such that if we have two partial recursive $\lambda \vec{x}\vec{\alpha}\beta.\{f\}(\vec{x}, \vec{\alpha}, \beta)$ and $\lambda y.\{e\}(y, \vec{\alpha})$, then*

$$\{f\}\left(\vec{x}, \vec{\alpha}, \lambda y.\{e\}(y, \vec{\alpha})\right) \simeq z \text{ iff } \{h(f, e)\}(\vec{x}, \vec{\alpha}) \simeq z \tag{‡}$$

Proof The idea is to define a rank $(3, 0)$ recursive $\lambda f e t.g(f, e, t)$ for all f, e, t and then freeze t by the recursion theorem to obtain h. The construction of g is done by cases over Definition 13.3.1.

I. $f = \langle 1, k, l + 1, i \rangle$. Set $g(f, e, t) = \langle 1, (f)_1, (f)_2 \dot- 1, i \rangle$.

II. $f = \langle 2, k, l + 1, i \rangle$. Set $g(f, e, t) = \langle 2, (f)_1, (f)_2 \dot- 1, i \rangle$.

III. $f = \langle 3, k, l + 1, c \rangle$. Set $g(f, e, t) = \langle 3, (f)_1, (f)_2 \dot- 1, c \rangle$.

IV. $f = \langle 4, k + 2, l + 1 \rangle$. Set $g(f, e, t) = \langle 4, (f)_1 + 2, (f)_2 \dot- 1 \rangle$.

V. $f = \langle 5, k + 4, l + 1 \rangle$. Set $g(f, e, t) = \langle 5, (f)_1 + 4, (f)_2 \dot- 1 \rangle$.

VI. $f = \langle 6, k, l + 1, i, j \rangle$. Set $g(f, e, t) = \langle 6, (f)_1, (f)_2 \dot- 1, i, j \rangle$, if $j \le l$. Else $g(f, e, t)$ is essentially e, that is,
$g(f, e, t) = \langle 7, (f)_1, (f)_2 \dot- 1, e, \langle 1, (f)_1, (f)_2 \dot- 1, 1 \rangle \rangle$.
In terms of functionals, $\{g(f, e, t)\}(\vec{x}, \vec{\alpha}) \simeq \{e\}(x_1, \vec{\alpha})$.

VII. $f = \langle 7, k, l + 1, a, q_1, \dots, q_r \rangle$. Then set

$$g(f, e, t) = \langle 7, (f)_1, (f)_2 \dot- 1, g(a, e, t), g(q_1, e, t), \dots, g(q_r, e, t) \rangle$$

VIII. $f = \langle 8, k + 1, l + 1 \rangle$. We obtain first —in the "obvious way"[14]— a primitive recursive $d(f, t)$ such that

$$\{d(f, t)\}(f, e, r, \vec{x}, \vec{\alpha}) \simeq \{\{t\}(r, e)\}(\vec{x}, \vec{\alpha})$$

Then set

$$g(f, e, t) = Sb_2(d(f, t), f, e) \tag{0}$$

In some detail we construct the function d as follows by calling $\{\{t\}(r, e)\}$ recursively using Definition 13.3.1:

- By VIII:

$$\{\langle 8, (f)_1 + 1, (f)_2 \dot- 1 \rangle\}(\{t\}(r, e), \vec{x}, \vec{\alpha}) \simeq z \text{ iff } \{\{t\}(r, e)\}(\vec{x}, \vec{\alpha}) \simeq z \tag{1}$$

Using truncated computations we record

$$\|\{t\}(r, e), \vec{x}, \vec{\alpha}\| < \|\langle 8, (f)_1 + 1, (f)_2 \dot- 1 \rangle, \{t\}(r, e), \vec{x}, \vec{\alpha}\| \tag{1'}$$

- By VII:

$$\{\langle 8, k + 1, l \rangle\}(\{t\}(r, e), \vec{x}, \vec{\alpha}) \simeq z \text{ iff}$$

$$\{\langle 7, k + 3, l, \langle 8, k + 1, l \rangle, \langle 7, k + 3, l, t, \langle 1, k + 3, l, 3 \rangle, \langle 1, k + 3, l, 2 \rangle \rangle,$$

$$\langle 1, k + 3, l, 4 \rangle, \dots, \langle 1, k + 3, l, k + 3 \rangle \rangle\} \Big(f, e, r, \vec{x}, \vec{\alpha} \Big) \simeq z \tag{2}$$

---

[14] That is, by unpacking the program $\{\{t\}(r, e)\}$.

In the interest of clarity we used "k, l" and "$\ldots$-notation" as if k, l were constants. In reality they are functions of the independent variable "f".
We can make the notation in the above bullet precise by using the primitive recursive rank-$(2, 0)$ functional $\lambda x y . x * y$ of Definition 2.4.3. We define for any primitive recursive rank-$(m+1, 0)$ function $\lambda \vec{y}_m i . q(\vec{y}_m, i)$ the function $\lambda \vec{y}_m z. \underset{i<z}{*} q(\vec{y}_m, i)$ by the primitive recursion below.

$$\underset{i<0}{*} q(\vec{y}_m, i) = 1$$

$$\underset{i<z+1}{*} q(\vec{y}_m, i) = \left(\underset{i<z}{*} q(\vec{y}_m, i) \right) * \langle q(\vec{y}_m, z) \rangle \tag{3}$$

Using (3) we can express $\langle \langle 1, k + 3, l, 4 \rangle, \ldots, \langle 1, k + 3, l, k + 3 \rangle \rangle$ primitive recursively as

$$\underset{i<(f)_1}{*} \langle 1, (f)_1 + 3, (f)_2 \doteq 1, 4 + i \rangle$$

and rewrite (2) as

$$\{ \langle 8, (f)_1 + 1, (f)_2 \doteq 1 \rangle \}(\{t\}(r, e), \vec{x}, \vec{\alpha}) \simeq z \text{ iff}$$

$$\Big\{ \langle 7, (f)_1 + 3, (f)_2 \doteq 1, \langle 8, (f)_1 + 1, (f)_2 \doteq 1 \rangle,$$

$$\langle 7, (f)_1 + 3, (f)_2 \doteq 1, t, \langle 1, (f)_1 + 3, (f)_2 \doteq 1, 3 \rangle, \tag{2'}$$

$$\langle 1, (f)_1 + 3, (f)_2 \doteq 1, 2 \rangle \rangle \rangle$$

$$* \Big(\underset{i<(f)_1}{*} \langle 1, (f)_1 + 3, (f)_2 \doteq 1, 4 + i \rangle \Big) \Big\}(f, e, r, \vec{x}, \vec{\alpha}) \simeq z$$

Thus the $d(f, t)$ mentioned at the outset of case VIII is

$$d(f, t) = \langle 7, (f)_1 + 1, (f)_2 \doteq 1, \langle 8, (f)_1 + 1, (f)_2 \doteq 1 \rangle,$$

$$\langle 7, (f)_1 + 3, (f)_2 \doteq 1, t, \langle 1, (f)_1 + 3, (f)_2 \doteq 1, 3 \rangle,$$

$$\langle 1, (f)_1 + 3, (f)_2 \doteq 1, 2 \rangle \rangle \rangle * \tag{4}$$

$$\Big(\underset{i<(f)_1}{*} \langle 1, (f)_1 + 3, (f)_2, 4 + i \rangle \Big)$$

and we can finally write (2) as

$$\{ \langle 8, (f)_1 + 1, (f)_2 \doteq 1 \rangle \}(\{t\}(r, e), \vec{x}, \vec{\alpha}) \simeq z \text{ iff } \{d(f, t)\}(f, e, r, \vec{x}, \vec{\alpha}) \simeq z \tag{5}$$

thus we also have

$$\|\langle 8, (f)_1 + 1, (f)_2 \dot- 1\rangle, \{t\}(r, e), \vec{x}, \vec{\alpha}\| < \|d(f, t), f, e, r, \vec{x}, \vec{\alpha}\|$$

$$\overset{(0)}{<} \|Sb_2(d(f, t), f, e), r, \vec{x}, \vec{\alpha}\| = \|g(f, e, t), r, \vec{x}, \vec{\alpha}\| \tag{5'}$$

If f does not fit any cases among I–VIII, then set $g(f, e, t) = 0$.
By the recursion theorem, there is a $\bar{t}$ such that $g(f, e, \bar{t}) = \{\bar{t}\}(f, e)$.
$\lambda fe.g(f, e, \bar{t})$ is total, hence recursive.
Let us define

$$h(f, e) \overset{Def}{=} g(f, e, \bar{t})$$

We next show ($\ddagger$) in the theorem statement.

1. $\rightarrow$ direction. *Set* $\beta = \lambda y.\{e\}(y, \vec{\alpha})$ for clarity of notation and *assume*

$$\{f\}(\vec{x}, \vec{\alpha}, \beta) \simeq z \tag{6}$$

We do induction along subcomputations of $(f, \vec{x}, \vec{\alpha}, \beta, z)$, that is, along Definitions 13.3.1 and 13.3.4. We sample a few cases among I–VIII.

- (Case I.) For $f = \langle 1, k, l + 1, i\rangle$ we have $z = x_i$. We also have $h(f, e) = g(f, e, \bar{t}) = \langle 1, (f)_1, (f)_2 \dot- 1, i\rangle$. Thus $\{h(f, e)\}(\vec{x}, \vec{\alpha}) \simeq x_i = z$.
- (Case VI.) $f = \langle 6, k, l + 1, i, j\rangle$.

  - Subcase where $j \leq l$.
    Then $h(f, e) = g(f, e, \bar{t}) = \langle 6, (f)_1, (f)_2 \dot- 1, i, j\rangle$.
    Thus $\{f\}(\vec{x}, \vec{\alpha}, \beta) \simeq \alpha_j(x_i)$ and

    $$\{h(f, e)\}(\vec{x}, \vec{\alpha}) = \{\langle 6, (f)_1, (f)_2 \dot- 1, j, i\rangle\}(\vec{x}, \alpha)$$

    $$\simeq \alpha_j(x_i)$$

    as well.
  - Subcase where $j = l + 1$ and $\{f\}(\vec{x}, \vec{\alpha}, \beta) \simeq \{e\}(x_1, \vec{\alpha})$.
    Also

    $$\{h(f, e)\}(\vec{x}, \vec{\alpha}) = \{\langle 7, (f)_1, (f)_2 \dot- 1, e, \langle 1, (f)_1, (f)_2 \dot- 1, 1\rangle\rangle\}(\vec{x}, \vec{\alpha})$$

    $$\simeq \{e\}(x_1, \vec{\alpha})$$

  - (Case VII.) Here we have $f = \langle 7, k, l + 1, a, q_1, \ldots, q_r\rangle$ and $\{f\}(\vec{x}, \vec{\alpha}, \beta) \simeq z$. By Proposition 13.3.10, case VII, we have

    $$\{f\}(\vec{x}, \vec{\alpha}, \beta) \simeq \{a\}\Big(\{q_1\}(\vec{x}, \vec{\alpha}, \beta), \ldots, \{q_r\}(\vec{x}, \vec{\alpha}, \beta), \vec{\alpha}, \beta\Big)$$

that is, for some z_i,

$$\{a\}(\vec{z}_r, \vec{\alpha}, \beta) \simeq z$$

$$\{q_1\}(\vec{x}, \vec{\alpha}, \beta) \simeq z_1$$

$$\{q_2\}(\vec{x}, \vec{\alpha}, \beta) \simeq z_2 \tag{7}$$

$$\vdots$$

$$\{q_r\}(\vec{x}, \vec{\alpha}, \beta) \simeq z_r$$

the z being the one in the hypothesis (6). By the induction hypothesis on subcomputations we have that

$$\{h(a, e)\}(\vec{z}_r, \vec{\alpha}) \simeq z$$

$$\{h(q_1, e)\}(\vec{x}, \vec{\alpha}) \simeq z_1$$

$$\{h(q_2, e)\}(\vec{x}, \vec{\alpha}) \simeq z_2 \tag{8}$$

$$\vdots$$

$$\{h(q_r, e)\}(\vec{x}, \vec{\alpha}) \simeq z_r$$

By (8) we have (cf. proof of Proposition 13.3.10)

$$\{h(a, e)\}(\{h(q_1, e)\}(\vec{x}, \vec{\alpha}), \ldots, \{h(q_r, e)\}(\vec{x}, \vec{\alpha}), \vec{\alpha}) \simeq z \tag{8'}$$

The *obvious* index for the composition denoted above is

$$\langle 7, (f)_1, (f)_2 \dotminus 1, h(a, e), h(q_1, e), \ldots, h(q_r, e) \rangle = h(f, e)$$

by case VII in the construction of h (p.437).

Thus (8′) directly implies $\{h(f, e)\}(\vec{x}, \vec{\alpha}) \simeq z$ as was to be proved.

– (Case VIII.) $f = \langle 8, k+1, l+1 \rangle$. What can we directly conclude from $\{f\}(r, \vec{x}, \vec{\alpha}, \beta) \simeq z$? That we must have $\{r\}(\vec{x}, \vec{\alpha}, \beta) \simeq z$ by Proposition 13.3.10, item VIII. By the I.H. on subcomputations we obtain

$$\{h(r, e)\}(\vec{x}, \vec{\alpha}) \simeq z \tag{9}$$

By (0), (1) and (5) (remembering $\bar{t}$) we have

$$\{h(f, e)\}(r, \vec{x}, \vec{\alpha}) \simeq \{Sb_2(d(f, \bar{t}), f, e)\}(r, \vec{x}, \vec{\alpha})$$

$$\simeq \{d(f, \bar{t})\}(f, e, r, \vec{x}, \vec{\alpha}) \simeq \{\{\bar{t}\}(r, e)\}(\vec{x}, \vec{\alpha}) \simeq z$$

Thus we have shown that the hypothesis of Case VIII —$\{f\}(r, \vec{x}, \vec{\alpha}, \beta) \simeq z$— implies $\{h(f, e)\}(r, \vec{x}, \vec{\alpha}) \simeq z$, concluding the induction.

2. $\leftarrow$ direction. Again *set* $\beta = \lambda y.\{e\}(y, \vec{\alpha})$ and *assume*

$$\{h(f, e)\}(\vec{x}, \vec{\alpha}) \simeq z \qquad (\P)$$

The task is to prove $\{f\}(\vec{x}, \vec{\alpha}, \beta) \simeq z$.

We do induction on the computation length $\|h(f, e), \vec{x}, \vec{\alpha}, z\|$ of $(h(f, e), \vec{x}, \vec{\alpha}, z)$.

- (Cases I.–VI.) These are rather straightforward. We sample the most interesting of those:
- (Subcase VI.)

  - (Subsubcase) $f = \langle 6, (f)_1, (f)_2, i, j \rangle$ and $j < (f)_2$.
    Then $h(f, e) = g(f, e, \bar{t}) = \langle 6, (f)_1, (f)_2 \dot{-} 1, i, j \rangle$ and the assumption yields
    $\{\langle 6, (f)_1, (f)_2, i, j \rangle\}(\vec{x}, \vec{\alpha}) \simeq z$ by $j < (f)_2$.
  - (Subsubcase) $f = \langle 6, (f)_1, (f)_2, i, j \rangle$ and $j = (f)_2$.
    Then $h(f, e) = \langle 7, (f)_1, (f)_2 \dot{-} 1, e, \langle 1, (f)_1, (f)_2 \dot{-} 1, 1 \rangle \rangle$, that is, $\{h(f, e)\}(\vec{x}, \vec{\alpha}) \simeq z \simeq \{e\}(x_1, \vec{\alpha})$. But then $\{f\}(\vec{x}, \vec{\alpha}, \beta) \simeq z$.

- (Case VII.) Left to the reader.
- (Case VIII.) $f = \langle 8, (f)_1 + 1, (f)_2 \rangle$. So, for all $r, \vec{x}, \vec{\alpha}$ and the given β we have

$$\{f\}(r, \vec{x}, \vec{\alpha}, \beta) \simeq \{r\}(\vec{x}, \vec{\alpha}, \beta) \qquad (*)$$

The assumption ($\P$) in this context (of inputs) is $\{h(f, e)\}(r, \vec{x}, \vec{\alpha}) \simeq z$

As before, by (1) and (5) we have

$$\{d(f, \bar{t})\}(f, e, r, \vec{x}, \vec{\alpha}) \simeq \{\{\bar{t}\}(r, e)\}(\vec{x}, \vec{\alpha}) \simeq \{h(r, e)\}(\vec{x}, \vec{\alpha})$$

By (1′) and (5′) we have

$$\|h(r, e), \vec{x}, \vec{\alpha}\| < \|d(f, \bar{t}), f, e, r, \vec{x}, \vec{\alpha}\| < \|h(f, e), r, \vec{x}, \vec{\alpha}\|$$

Thus the I.H. on lengths $\|h(\cdot, \cdot), \ldots\|$ applies and we have

$$\{h(r, e)\}(\vec{x}, \vec{\alpha}) \simeq z \text{ implies } \{r\}(\vec{x}, \vec{\alpha}, \beta) \simeq z$$

thus, by $(*)$, we proved $\{f\}(r, \vec{x}, \vec{\alpha}, \beta) \simeq z$. $\qquad\square$

13.3.37 Corollary. *There is a rank* $(1 + n, 0)$ *recursive function(al) h such that if we have two partial recursive* $\lambda \vec{x}\vec{\alpha}\beta.\{f\}(\vec{x}, \vec{\alpha}, \beta)$ *and* $\lambda y.\{e\}(y, \vec{x}, \vec{\alpha})$, *then*

$$\{f\}\Big(\vec{x}, \vec{\alpha}, \lambda y.\{e\}(y, \vec{x}, \vec{\alpha})\Big) \simeq z \;\; iff \; \{h(f, e)\}(\vec{x}, \vec{\alpha}) \simeq z$$

Proof Using Remark 13.3.20 we have a primitive recursive function k of rank $(n, 0)$ (n is the length of $\vec{x}$) such that $\{k(\vec{x})\}(y, \vec{\alpha}) \simeq \{e\}(y, \vec{x}, \vec{\alpha})$. This trivially modifies only step VI. in the previous proof. We use the "$k(\vec{x})$" as the previous "e" throughout. This changes the rank of our previous h to $(1 + n, 0)$. $\square$

13.4 Normal Form Theorems for Computable Functionals

In this subsection, following Kleene (1952); cf. also Odifreddi (1989), Sasso (1971) we will generalise and adapt *the normal form theorem* of Theorem 5.5.3 to the computability theory of partial recursive functionals.

13.4.1 Arithmetisation Again

How do we "compute" a yes/no answer to the question "$\{e\}(\vec{a}, \vec{\alpha}) = b$?" for arbitrary $e, \vec{a}, \vec{\alpha}, b$? Just as we did in the case of the Ackermann function (Sect. 4.3), we think of the question "$\{e\}(\vec{a}, \vec{\alpha}) = b$?" as a "*call*" to a function

$$Eval(e, \vec{a}, \vec{\alpha}, b)$$

which is true iff

$$\{e\}(\vec{a}, \vec{\alpha}) = b \tag{1}$$

is true.

We can "program" a *verifier* for $Eval$ by the following informal "recursive" program:

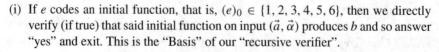

This "recursively programmed" verifier will verify that (1) holds iff it does; it will give no answer otherwise. Why do I call it "recursive"? Because as a recursive procedure $f(x)$ that we program in, say, a programming course will in general call itself —$f(y)$— for "smaller" (than x) appropriate inputs y, so will our verifier for $Eval(e, \vec{a}, \vec{\alpha}, b)$ call $Eval(e', \vec{a}', \vec{\alpha}', b')$, $Eval(e'', \vec{a}'', \vec{\alpha}'', b'')$, etc., for various "smaller" or "earlier" in a certain sense (read on!) input tuples $(e', \vec{a}', \vec{\alpha}', b')$, $(e'', \vec{a}'', \vec{\alpha}'', b'')$, etc.

(i) If e codes an initial function, that is, $(e)_0 \in \{1, 2, 3, 4, 5, 6\}$, then we directly verify (if true) that said initial function on input $(\vec{a}, \vec{\alpha})$ produces b and so answer "yes" and exit. This is the "Basis" of our "recursive verifier".

(ii) If $(e)_0 \in \{7, 8\}$, then we have two cases shown in tree form below. In each case the *root call* is $Eval(e, \vec{a}, \vec{\alpha}, b)$ and the "children" —e.g., the $Eval(g_i, \vec{a}_m, \vec{\alpha}_l, z_i)$, for appropriate i, in the case of Fig. 13.1— represent the recursive calls *needed* according to the inductive definition of *computation tuples* Definition 13.3.1 (see also Proposition 13.3.10 and Corollary 13.3.12) in order to eventually verify (yes) the value of the root —if true.

Thus, the computation of $Eval(e, \vec{a}_n, \vec{\alpha}_l, b)$ can be arranged in a *tree* —a *computation tree*— with its root (1st "call") labelled $Eval(e, \vec{a}_n, \vec{\alpha}_l, b)$ and all the other nodes labelled by the relevant calls. Clearly, the name "$Eval$" does not add *any* information to the call "$Eval(e, \vec{a}_n, \vec{\alpha}_l, b)$" and we might as well choose any other name, for example "$George$", and still carry out all the calls perfectly well.

That is, all we need to know for each call are the relevant arguments, $e, \vec{a}_n, \vec{\alpha}_l, b$. We will thus ignore the name "$Eval$" and label the tree nodes by just the call-arguments *coded as single numbers*, that is,

$$\langle e, \langle \vec{a}_n \rangle, \langle \vec{s}_l \rangle, b \rangle$$

will label the tree nodes, where we replaced the vector of partial functions $\vec{\alpha}_l$ by a vector of *numbers* $\vec{s}_l$ such that *each $[s_j]$ is a finite subfunction of α_j.*

The reason we can do the latter is because as the computation is finite in length (there are finitely many calls, leading to a finite tree) we have only finitely many input-output pairs for each α_j that we obtained from the "oracle" during the entire computation.

Next, our purpose is to have a predicate $Tree(y)$ that says that y codes a computation tree.

A computation tree is ordered from left to right —it is *ordered*— as Figs. 13.1 and 13.2 indicate, thus we can code it from left to right, *recursively*, as a root, followed by the *trees* hanging from the root (called *subtrees* of the root[15]) *from left to right.*

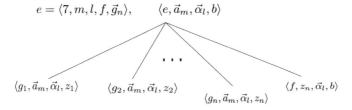

Fig. 13.1 Composition

---

[15] In Figs. 13.1 and 13.2 we only show the roots of the subtrees of the root.

$$e = \langle 8, n+1, l \rangle, \qquad \langle e, f, \vec{a}_n, \vec{\alpha}_l, b \rangle$$

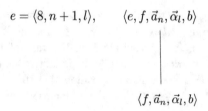

$$\langle f, \vec{a}_n, \vec{\alpha}_l, b \rangle$$

Fig. 13.2 Universal function

13.4.1 Definition. (Coding Trees) Let $T, T_1, T_2, \ldots$ name *ordered* trees where we let $t, t_1, t_2, \ldots$ name their numerical codes respectively.

We code an ordered tree T as follows:

(1) If T has just a single node coded v, then $t = \langle v \rangle$.
(2) Let T be the tree denoted by the ordered tuple $v, T_1, T_2, \ldots, T_r$, where v codes its root and $T_1, T_2, \ldots, T_r$ are the subtrees of the root from left to right. Let also $t_i, i = 1, \ldots, r$, be all the *codes* of the subtrees of the root. Then we let $t = \langle v, t_1, t_2, \ldots, t_r \rangle$ and hence $lh(t) = r + 1$.

□

13.4.2 Remark. Thus a *coded computation tree* is either a single node t, where $((t)_0)_0 \in \{1, 2, 3, 4, 5, 6\}$, or is coded as in Definition 13.4.1, namely, $\langle v, t_1, \ldots, t_r \rangle$. Here v is one of

(1) $v = \langle \langle 7, m, l, f, \vec{g}_n \rangle, \langle \vec{a}_m \rangle \langle \vec{s}_l \rangle, b \rangle$ ($r = n + 1$ in this case; cf. Fig. 13.1).
(2) $v = \langle \langle 8, n, f \rangle, \langle \vec{a}_n \rangle, \langle \vec{s}_l \rangle, b \rangle$ (here $n > 0$ and $r = 1$, see Fig. 13.2).

and each of the sub tree codes t_i is, recursively, a coded computation tree with root as indicated in the aforementioned two figures on p. 443.

Of course, the root of T_i is $(t_i)_0$. If $y = \langle v, t_1, \ldots, t_r \rangle$, then we can gather the top level information regarding the computation easily:

1. $(y)_0 = v$ is the root call, and $((y)_0)_0$ tells us the *type* of the main call (the *index*).
2. $((y)_i)_0$ tells us the type of call at the second level (just after the main call) for $i = 1, \ldots, lh(y) \dot{-} 1$.
3. In what follows we simplify expressions such as $((z)_i)_j$ writing instead $(z)_{i,j}$ and $(((z)_i)_j)_k$ writing instead $(z)_{i,j,k}$. Thus,
4. $(y)_0$ is the main call with index $(y)_{0,0}$, the vector of type-0 variables is coded as $(y)_{0,1}$ while the vector of type-1 variables (*finite approximations thereof*) are coded as $(y)_{0,2}$.

   The *expected output* b of the *root tuple* $\{e\}(\vec{x}_m, \vec{\alpha}_l) \simeq b$, which our "recursive algorithm" sets out to verify (if true) —the "b" (Figs. 13.1, 13.2)— is $(y)_{0,3}$.   □

We now embark on a *course-of-values recursive* definition of (the characteristic function $\lambda y.c_{Tree}(y)$ of) $Tree(y)$. We bypass the notational demands of the characteristic functions by presenting the definition directly in terms of relations, in logical notation.

The definition will be split into five parts,

1. Part $P(y)$ for the "preamble" that is applicable and common to all following parts.
2. Part $I(y)$ for the initial functions.
3. Part $C(y)$ for composition.
4. Part $U(y)$ for the universal functional application.
5. A part $O(y)$ for incorporating the variables $\vec{\alpha}_l$ about which we rely on the oracle will be added as an *after thought* externally to $Tree(y)$ to ensure that this predicate is *absolutely* primitive recursive.[16]

Parts 3 and 4 prepare the arguments for the recursive calls of $Tree$ but do not make these calls.

(I) The $P(y)$:

$$P(y) \equiv Seq(y) \wedge lh(y) \geq 1 \wedge (\forall i)_{<lh(y)} Seq((y)_i) \wedge lh((y)_0) = 4 \wedge$$
$$Seq((y)_{0,0}) \wedge Seq((y)_{0,1}) \wedge Seq((y)_{0,2})$$

(II) The $I(y)$:

$$I(y) \equiv lh(y) = 1 \wedge \Big($$

$$\{\mathscr{U}_i^{n,l}\} \quad (\exists p, q, a, b, c, x, s, n, l, i, j)_{<y} \Big[Seq(x) \wedge Seq(s) \wedge$$

$$lh(x) = n \wedge lh(s) = l \wedge$$

$$\Big\{ i < n \wedge (y)_0 = \langle\langle 1, n, l, i + 1\rangle\rangle * x * s * \langle(x)_i\rangle \vee$$

$$\{\mathscr{S}_i^{n,l}\} \qquad i < n \wedge (y)_0 = \langle\langle 2, n, l, i + 1\rangle\rangle * x * s * \langle(x)_i + 1\rangle \vee$$

$$\{\mathscr{C}_c^{n,l}\} \qquad (y)_0 = \langle\langle 3, n, l, i + 1\rangle\rangle * x * s * \langle c\rangle \vee$$

$$\{\mathscr{S}_1^{(n+2,l)}\} \qquad (y)_0 = \langle\langle 4, n + 2, l\rangle, a, b\rangle * x * s * \langle Sb_1(a, b)\rangle \vee$$

$$\{\mathscr{D}\mathscr{C}^{(n+4,l)}\} \qquad (y)_0 = \langle\langle 5, n + 4, l\rangle, p, q, a, b\rangle * x * s * \langle p\rangle \wedge a = b \vee$$

$$\{\mathscr{D}\mathscr{C}^{(n+4,l)}\} \qquad (y)_0 = \langle\langle 5, n + 4, l\rangle, p, q, a, b\rangle * x * s * \langle q\rangle \wedge a \neq b \vee$$

$$\{\mathscr{A}\mathscr{P}_{i,j}^{(n,l)}\} \qquad P_{(x)_i} \mid (s)_j \wedge$$

$$(y)_0 = \langle\langle 6, n, l, i + 1, j + 1\rangle\rangle * x * s * \langle(s)_{j,(x)_i}\rangle \Big\} \Big]$$

$$\Big)$$

---

[16] That is, no type-1 variables involved.

(III) <u>The $C(y)$:</u>   We have $y = \langle v, t_1, \dots, t_{n+1} \rangle$, where (cf. Fig. 13.1) $v = \langle \langle 7, m, l, f \rangle * g \rangle * a * s * \langle b \rangle$, and we have represented the vectors $\vec{g}_n = g_1, \dots, g_n$ by $g = \langle \vec{g}_n \rangle$, $\vec{a}_m$ by $a = \langle \vec{a}_m \rangle$, and $\vec{z}_n$ (outputs of the g_i) by $z = \langle \vec{z}_n \rangle$, s being the code of the vector of approximations for $\vec{\alpha}_l$. Finally, we have set $t = \langle \vec{t}_{n+1} \rangle$ for the tree codes of the $n + 1$ subtrees (Fig. 13.1) so that $y = \langle v \rangle * t$. Note that as before, $v = (y)_0$,

In particular, the intuitive picture $\langle g_i, \vec{a}_m, \vec{s}_l, z_i \rangle$ for the recursive calls to the g_i is captured by (i-th call) $\langle (g)_i \rangle * a * s * \langle (z)_i \rangle$ (root of t_i, for $1 \le i \le n$).

$$\{\text{Fig. 13.1}\}\ C(y) \equiv (\exists a, z, f, g, m, n, s, l, b)_{<y}\Big\{lh(y) = n + 2 \wedge Seq(a) \wedge$$
$$Seq(z) \wedge Seq(f) \wedge Seq(g) \wedge n = lh(z) \wedge n = lh(g) \wedge$$
$$Seq(s) \wedge lh(s) = l \wedge m = lh(a) \wedge (f)_1 = n \wedge$$
$$(\forall i)_{<n}\big(Seq((g)_i) \wedge ((g)_i)_1 = m\big) \wedge$$

$\{\text{Root (main) call}\}\quad (y)_0 = \langle \langle 7, m, l, f \rangle * g \rangle * a * s * \langle b \rangle \wedge$

$\{\text{Root of } t_n\}\qquad\quad ((y)_{n+1})_0 = \langle f \rangle * z * s * \langle b \rangle \wedge$

$\{\text{Root of } t_i,\ i < n\}\ \ (\forall i)_{<n}((y)_{i+1})_0 = \langle (g)_i \rangle * a * s * \langle (z)_i \rangle\Big\}$

(IV) <u>The $U(u)$:</u> $y = \langle v, t_0 \rangle$, where (cf. Fig. 13.2) $v = \langle \langle 8, n + 1, l \rangle, f \rangle * a * s * \langle b \rangle$ and the root of t_0 is $(t_0)_0 = (y)_{1,0} = \langle f \rangle * a * s * \langle b \rangle$, where as above we set $a = \langle \vec{a}_n \rangle$ and $s = \langle s_l \rangle$.

Thus,

$$\{\text{Fig. 13.2}\}\ U(y) \equiv (\exists n, f, a, b, s, l)_{<y}\Big\{lh(y) = 2 \wedge Seq(f) \wedge$$
$$Seq((y)_1) \wedge Seq(s) \wedge lh(s) = l \wedge$$
$$(f)_1 = n \wedge n > 0 \wedge$$
$$Seq(a) \wedge lh(a) = n \wedge$$

$\{\text{The root call}\}\qquad\ (y)_0 = \langle \langle 8, n + 1, l \rangle, f \rangle * a * s * \langle b \rangle \wedge$

$\{\text{The recursive call}\}\ (y)_{1,0} = \langle f \rangle * a * s * \langle b \rangle\Big\}$

We can now conclude the course-of-values recursive definition of $Tree(y)$.

$$Tree(y) \equiv P(y) \wedge \Big(I(y) \vee C(y) \vee U(y)\Big) \wedge (\forall i)_{<lh(y)}(i > 0 \rightarrow Tree((y)_i))$$

It is clear by the course-of-values recursion above —that we expressed in a more direct way via the predicate $Tree$ rather than the characteristic function c_{Tree}— that $Tree(y)$ is absolutely primitive recursive (is in $\mathcal{PR}_*$).

We can now define the Kleene T-predicate for functionals.

13.4.3 Definition. (The Kleene Predicate of Rank (n, l))

For any given n and l we define the *Kleene predicate* of rank (n, l) by

$$T^{(n,l)}(z, \vec{x}_n, \vec{\alpha}_l, y) \overset{Def}{\equiv} Tree(y) \wedge (y)_{0,0} = z \wedge (y)_{0,1} = \langle \vec{x}_n \rangle \wedge lh\big((y)_{0,2}\big) = l \wedge$$

$$(\forall i)_{<l}[(y)_{0,2,i}] \subseteq \alpha_{i+1}$$

Since $(y)_0 = \langle z, \langle \vec{x}_n \rangle , \langle \vec{s}_l \rangle , b \rangle$, where $\langle \vec{s}_l \rangle$ is a vector of approximations to $\vec{\alpha}_l$ during the computation and b is the output of the computation,

Pause. How do we *know* that there *is* an output? ◄

we have that $b = (y)_{0,3}$. We encounter such a function —$\lambda \vec{x}_n \vec{\alpha}_l.(y)_{0,3}$— in the literature under the name "*decoding function*" of the computation *code y*, and it is (often) given the short name *d*. *We will too adopt this name.*   □

13.4.4 Theorem. (Kleene Normal Form Theorem —Form I) *There is an absolute primitive recursive function d of one argument, and, for any given n and l, a primitive recursive predicate* $T^{(n,l)}(z, \vec{x}_n, \vec{\alpha}_l, y)$ *as defined above such that*

(a) $\{z\}^{(n,l)}(\vec{x}_n, \vec{\alpha}_l) \downarrow \equiv (\exists y) T^{(n,l)}(z, \vec{x}_n, \vec{\alpha}_l, y)$
(b) $\{z\}^{(n,l)}(\vec{x}_n, \vec{\alpha}_l) \simeq u \equiv (\exists y)\big(T^{(n,l)}(z, \vec{x}_n, \vec{\alpha}_l, y) \wedge d(y) = u\big)$

Proof The primitive recursiveness of $T^{(n,l)}$ follows from that of $Tree(y)$ and $\lambda s\alpha.[s] \subseteq \alpha$ (cf. Example 13.3.35, (3), (4), (6), (7)).

All the proof work for (a)–(b) is already done.   □

13.4.5 Theorem. (Kleene Normal Form Theorem —Form II) *For each* $l \geq 0$, *there exists an absolute primitive recursive predicate* $T_l(z, x, s, y)$ *of rank* $(4, 0)$ *such that for each functional of rank* (n, l) *we have*

(i) $\{z\}^{(n,l)}(\vec{x}_n, \vec{\alpha}_l) \downarrow \equiv (\exists y)(\exists \vec{s}_l)_{<y}\Big(T_l\big(z, \langle \vec{x}_n \rangle , \langle \vec{s}_l \rangle , y\big) \wedge \bigwedge_{1 \leq j \leq l}[s_j] \subseteq \alpha_j\Big).$
(ii)

$$\{z\}^{(n,l)}(\vec{x}_n, \vec{\alpha}_l) \simeq u \equiv (\exists y)(\exists \vec{s}_l)_{<y}\Big(T_l\big(z, \langle \vec{x}_n \rangle , \langle \vec{s}_l \rangle , y\big) \wedge d(y) = u$$

$$\wedge \bigwedge_{1 \leq j \leq l} [s_j] \subseteq \alpha_j\Big)$$

where the empty conjunction (case of no oracle variables) when $l = 0$ *is by definition true.*

Proof

(i) Define T_l as

$$\boxed{T_l(z, x, s, y) \overset{Def}{\equiv} Tree(y) \wedge (y)_{0,0} = z \wedge (y)_{0,1} = x \wedge (y)_{0,2} = s \wedge lh(s) = l}$$

The claims (i) and (ii) follow directly from the definition of $Tree(y)$ (cf. also Theorem 13.4.4).   □

13.4.6 Long Remark. We list a few consequences of the Normal Form theorems for functionals.

(I) Normal form theorem for semi-recursive predicates.

13.4.7 Corollary. (Normal Form for Semi-Recursive Relations. I) $\mathbb{P}(\vec{x}_n, \vec{\alpha}_l)$ *is semi-recursive iff we have*

$$\mathbb{P}(\vec{x}_n, \vec{\alpha}_l) \equiv (\exists y) T^{(n,l)}\big(z, \vec{x}_n, \vec{\alpha}_l, y\big) \tag{1}$$

where z is a semi-index of $\mathbb{P}$.

Proof *If part.* Immediate by Theorem 13.4.4, item (a), since the assumption entails $\mathbb{P}(\vec{x}_n, \vec{\alpha}_l) \equiv \{z\}^{(n,l)}(\vec{x}_n, \vec{\alpha}_l) \downarrow$.
Only if part. Assume (1). The right hand side of it says $\{z\}^{(n,l)}(\vec{x}_n, \vec{\alpha}_l) \downarrow$ by Theorem 13.4.4, item (a). $\square$

13.4.8 Corollary. (Normal Form for Semi-Recursive Relations. II)
    $\mathbb{P}(\vec{x}_n, \vec{\alpha}_l)$ *is semi-recursive iff we have*

$$\mathbb{P}(\vec{x}_n, \vec{\alpha}_l) \equiv (\exists y)(\exists \vec{s}_l)_{<y}\left(T_l\big(z, \langle \vec{x}_n \rangle, \langle \vec{s}_l \rangle, y\big) \wedge \bigwedge_{1 \le j \le l} [s_j] \subseteq \alpha_j \right) \tag{2}$$

where z is a semi-index of $\mathbb{P}$.

Proof Because the right hand side of $\equiv$ in (2) says $\{z\}^{(n,l)}(\vec{x}_n, \vec{\alpha}_l) \downarrow$ by Theorem 13.4.5, item (i). $\square$

(II) Partial recursive functionals and semi-recursive relations are closed under type-1 *argument list expansion.*

13.4.9 Corollary. *If $\lambda \vec{x}_n \vec{\alpha}_l . \mathscr{F}(\vec{x}_n, \vec{\alpha}_l)$ is partial recursive, then so is $\mathscr{G} = \lambda \vec{x}_n \vec{\alpha}_l \vec{\beta}_k . \mathscr{F}(\vec{x}_n, \vec{\alpha}_l)$ and we have*

$$\mathscr{F}(\vec{x}_n, \vec{\alpha}_l) \simeq u \equiv \mathscr{G}(\vec{x}_n, \vec{\alpha}_l, \vec{\beta}_k) \simeq u$$

Proof Let $\{f\} = \lambda \vec{x}_n \vec{\alpha}_l . \mathscr{F}(\vec{x}_n, \vec{\alpha}_l)$. Then

$$\mathscr{F}(\vec{x}_n, \vec{\alpha}_l) \simeq u \overset{\text{Theorem 13.4.4,(b)}}{\equiv} (\exists y)\Big(T^{(n,l)}(f, \vec{x}_n, \vec{\alpha}_l, y) \wedge d(y) = u \Big) \tag{i}$$

The predicate quantified by $(\exists y)$ is primitive recursive thus it supports expansion of its type-1 variables list (cf. Proposition 13.3.29), say, by $\vec{\beta}_k$. Of course, the result regarding the introduction of dummy arguments to primitive recursive *functionals* transfers to primitive recursive *relations* by using the (primitive recursive) partial characteristic functionals of the latter.

Hence

$$\lambda u \vec{x}_n \vec{\alpha}_l \vec{\beta}_k . \mathscr{F}(\vec{x}_n, \vec{\alpha}_l) \simeq u \overset{(i)}{\Leftrightarrow} \lambda u \vec{x}_n \vec{\alpha}_l \vec{\beta}_k . (\exists y) \left(T^{(n,l)}(f, \vec{x}_n, \vec{\alpha}_l, y) \wedge d(y) = u \right)$$

$$\overset{13.3.29}{\Leftrightarrow} \lambda u \vec{x}_n \vec{\alpha}_l . (\exists y) \left(T^{(n,l)}(f, \vec{x}_n, \vec{\alpha}_l, y) \wedge d(y) = u \right)$$

$$\overset{(i)}{\Leftrightarrow} \lambda u \vec{x} \vec{\alpha} . \mathscr{F}(\vec{x}_n, \vec{\alpha}_l) \simeq u$$

$\square$

13.4.10 Corollary. *Suppose that* $\lambda z \vec{x} \vec{\alpha} . \mathscr{F}(z, \vec{x}, \vec{\alpha})$ *and* $\lambda \vec{y} \vec{\beta} . \mathscr{G}(\vec{y}, \vec{\beta})$ *are partial recursive.*
Then so is $\lambda \vec{x} \vec{y} \vec{\alpha} \vec{\beta} . \mathscr{F}(\mathscr{G}(\vec{y}, \vec{\beta}), \vec{x}, \vec{\alpha})$.

Proof Using the previous corollary and Grzegorczyk operations (the latter for type-0 arguments). $\square$

(III)

13.4.11 Corollary. *If* $\mathscr{F}$ *is partial recursive of rank* (n, l), *then there is a partial recursive* ψ *of rank* $(n + l, 0)$ *such that*

$$\mathscr{F}(\vec{x}_n, \vec{\alpha}_l) \simeq u \equiv (\exists \vec{s}_l) \left(\psi(\vec{x}_n, \vec{s}_l) \simeq u \wedge \bigwedge_{1 \le j \le l} [s_j] \subseteq \alpha_j \right)$$

Proof Using Theorem 13.4.5, (ii) and removing the bound on $(\exists \vec{s}_l)$ we express two equivalences (conjunctionally).

$$\{z\}^{(n,l)}(\vec{x}_n, \vec{\alpha}_l) \simeq u \Leftrightarrow (\exists y)(\exists \vec{s}_l) \left(T_l(z, \langle \vec{x}_n \rangle, \langle \vec{s}_l \rangle, y) \wedge d(y) = u \wedge \right.$$

$$\left. \bigwedge_{1 \le j \le l} [s_j] \subseteq \alpha_j \right)$$

$$\Leftrightarrow (\exists \vec{s}_l) \left((\exists y) \left[T_l(z, \langle \vec{x}_n \rangle, \langle \vec{s}_l \rangle, y) \wedge d(y) = u \right] \wedge \right.$$

$$\left. \bigwedge_{1 \le j \le l} [s_j] \subseteq \alpha_j \right)$$

The relation

$$Q(\vec{x}_n, \vec{s}_l, u) \overset{Def}{\equiv} (\exists y) \left[T_l(z, \langle \vec{x}_n \rangle, \langle \vec{s}_l \rangle, y) \wedge d(y) = u \right] \tag{$\ddagger$}$$

is *absolutely* semi-recursive and is *single-valued in* u by the semantics of the right hand side of its definition. Thus, we can define a *function* ψ by

$$\psi(\vec{x}_n, \vec{s}_l) \simeq u \overset{Def}{\equiv} Q(\vec{x}_n, \vec{s}_l, u) \tag{¶}$$

We have that $\psi \in \mathcal{P}$ by the Graph Theorem (Theorem 6.6.1).

Thus we have proved the corollary: that if $\lambda \vec{x}_n \vec{\alpha}_l . \mathcal{F}(\vec{x}_n, \vec{\alpha}_l)$, of index z, is partial recursive, then there is a $\psi \in \mathcal{P}$ such that

$$\mathcal{F}(\vec{x}_n, \vec{\alpha}_l) \simeq u \equiv (\exists \vec{s}_l) \left(\psi(\vec{x}_n, \vec{s}_l) \simeq u \wedge \bigwedge_{1 \le j \le l} [s_j] \subseteq \alpha_j \right)$$

What can we say if we want to view z as a variable? Well, then we obtain a single-valued in u predicate $Q'(z, \vec{x}_n, \vec{s}_l, u)$ from ($\ddagger$), instead of Q. Thus ($\P$) becomes

$$\psi'(z, \vec{x}_n, \vec{s}_l) \simeq u \overset{Def}{\equiv} Q'(z, \vec{x}_n, \vec{s}_l, u)$$

and we have, by S-m-n, a $\sigma \in \mathcal{R}$ such that $\{\sigma(z)\}(\vec{x}_n, \vec{s}_l) \simeq u \equiv Q'(z, \vec{x}_n, \vec{s}_l, u)$. We conclude that, for any z we have

$$\{z\}(\vec{x}_n, \vec{\alpha}_l) \simeq u \equiv (\exists \vec{s}_l) \left(\{\sigma(z)\}(\vec{x}_n, \vec{s}_l) \simeq u \wedge \bigwedge_{1 \le j \le l} [s_j] \subseteq \alpha_j \right) \qquad \square$$

(IV) Partial recursive functionals are *continuous*, also called *compact* (when confusion with the topological notion is possible).

13.4.12 Corollary. (Computable Functionals are Compact) *Partial recursive functionals $\mathcal{F}$ of rank (n, l) are* continuous *or* compact, *that is, if $\mathcal{F}(\vec{x}_n, \vec{\alpha}_l) \simeq u$, then for some appropriate* finite *sub functions ξ_i of the α_i ($1 \le i \le l$) we have $\mathcal{F}(\vec{x}_n, \vec{\xi}_l) \simeq u$.*

Proof So, let $\mathcal{F} = \{e\}^{(n,l)}$. By Theorem 13.4.5, (*ii*) we have

$$\{e\}^{(n,l)}(\vec{x}_n, \vec{\alpha}_l) \simeq u \equiv (\exists y)(\exists \vec{s}_l)_{<y} \left(T_l(e, \langle \vec{x}_n \rangle, \langle \vec{s}_l \rangle, y) \wedge \right.$$
$$\left. d(y) = u \wedge \bigwedge_{1 \le j \le l} [s_j] \subseteq \alpha_j \right) \tag{$\dagger$}$$

Fix an array of s_i, $i = 1, \ldots, l$, that work in ($\dagger$). Then, trivially,

$$\{e\}^{(n,l)}(\vec{x}_n, [\vec{s}_l]) \simeq u \equiv (\exists y) \left(T_l(e, \langle \vec{x}_n \rangle, \langle \vec{s}_l \rangle, y) \wedge \right.$$
$$\left. d(y) = u \wedge \bigwedge_{1 \le j \le l} [s_j] \subseteq [s_j] \right) \tag{$\ddagger$}$$

Note that the last conjunct "$\bigwedge_{1 \le j \le l} \ldots$" of ($\ddagger$) is true trivially, while the two conjuncts to its left are true by the italicised sentence above. Thus, $\{e\}^{(n,l)}(\vec{x}_n, [\vec{s}_l]) \simeq u$ is true. $\square$

(V) Partial recursive functionals are *consistent*, also known as *monotone*.

13.4.13 Corollary. (Monotonicity of Computable Functionals) *Partial recursive functionals $\mathscr{F}$ of rank (n, l) are* monotone *or* consistent: *That is, if $\mathscr{F}(\vec{x}_n, \vec{\alpha}_l) \simeq u$, and $\alpha_i \subseteq \beta_i$, for $i = 1, \ldots, l$, then we also have $\mathscr{F}(\vec{x}_n, \vec{\beta}_l) \simeq u$.*

Proof   Let $\mathscr{F} = \{e\}^{(n,l)}$. By Theorem 13.4.5, (ii) we have

$$\{e\}^{(n,l)}(\vec{x}_n, \vec{\alpha}_l) \simeq u \equiv (\exists y)(\exists \vec{s}_l)_{<y}\Big(T_l(e, \langle \vec{x}_n \rangle, \langle \vec{s}_l \rangle, y) \wedge$$
$$d(y) = u \wedge \bigwedge_{1 \leq j \leq l} [s_j] \subseteq \alpha_j \Big) \tag{i}$$

Fix an array of s_i, $i = 1, \ldots, l$, that work in (i). Then, by (i), we have

$$\{e\}^{(n,l)}(\vec{x}_n, \vec{\beta}_l) \simeq u \equiv (\exists y)\Big(T_l(e, \langle \vec{x}_n \rangle, \langle \vec{s}_l \rangle, y) \wedge$$
$$d(y) = u \wedge \bigwedge_{1 \leq j \leq l} [s_j] \subseteq \beta_j \Big) \tag{ii}$$

Note that the last conjunct (of three) in $\Big(\ldots \Big)$ in (ii) above is true by $\alpha_j \subseteq \beta_j$, for all applicable j (by (i)). The first two conjuncts are true by (i) and the choice of the s_j.

   (ii) Now yields that $\{e\}^{(n,l)}(\vec{x}_n, \vec{\beta}_l) \simeq u$ is true. $\square$

13.5   Partial Recursive and Recursive Functionals: Index Independent Approach

A theory of computability with oracles, which handle the type-1 partial inputs for functional calls such as $\mathscr{F}(\vec{x}_n, \vec{\alpha}_l)$, need not be necessarily built on an *a priori* concept of "program" or "index" (cf. Davis 1958b) of a computational process. The starting point of view here is that since *dynamic* (say, pencil and paper) *terminating* computations of the kind of functionals that we allow are finite, then, necessarily, the oracle(s) must have dealt with only finitely many input-output pairs of the type-1 arguments α_j.

Thus in trying to discharge a call $\mathscr{F}(\vec{x}_n, \vec{\alpha}_l)$ —for a non restricted functional $\mathscr{F}$ and inputs as shown— we need *only have* at our disposal the *finite* sub functions $[s_j] = \{\ldots, (x, y), \ldots\} \subseteq \alpha_j$ that are actually *used* (*extensionally*), for $j = 1, \ldots, l$. A sub function of α_j is "the actually used one" means that each of its pairs (x, y) are precisely the ones we get by asking the oracle (during the computation) "$\alpha_j(x) =$?" and getting the answer "y".

13.5.1 Definition. (Davis Computability Davis (1958b)) An *index-independent* computable functional $\lambda \vec{x}_n \vec{\alpha}_l . \mathscr{F}(\vec{x}_n, \vec{\alpha}_l)$ is one for which there is a rank $(n + l, 0)$ partial recursive function ψ that satisfies, for all $\vec{x}_n, \vec{\alpha}_l, u,$

$$\mathscr{F}(\vec{x}_n, \vec{\alpha}_l) \simeq u \equiv (\exists \vec{s}_l) \left(\psi(\vec{x}_n, \vec{s}_l) \simeq u \wedge \bigwedge_{1 \leq j \leq l} [s_j] \subseteq \alpha_j \right) \tag{1}$$

Functionals computable in an index-independent way were introduced by Davis (1958b). We will call them *Davis-computable* or simply *D-computable*. $\qquad\square$

We can immediately state

13.5.2 Proposition. *Every partial recursive functional in the sense of Definitions 13.3.1 and 13.3.8 is also Davis-computable.*

Proof Corollary 13.4.11. $\qquad\square$

In particular, the D-computable functionals class contains $\mathcal{P}$ as a subset.

We cannot prove the converse of the above proposition with the tools of Kleene-computability of unrestricted *functionals*. We essentially need a "selection function" like the one for computability of *functions* (no type-1 inputs; cf. Theorem 6.9.9) to *select the u* that satisfies the relation on the right hand side of (1). Our tools derived from Definitions 13.3.1 and 13.3.8 cannot reach that far. We *will* have a computable selection operation in the next section within Moscovakis's *Search Computability* —but we achieve this axiomatically. Or, we can limit our study to *restricted* computable functionals, Sect. 13.8, where again we can *prove* the selection theorem. See also the approach in Tourlakis (1986, 1996).

13.5.3 Proposition. *Every D-computable partial functional is compact.*

Proof Assume $\mathscr{F}(\vec{x}_n, \vec{\alpha}_l) \simeq u$. By Definition 13.5.1 this means

$$(\exists \vec{s}_l) \left(\psi(\vec{x}_n, \vec{s}_l) \simeq u \wedge \bigwedge_{1 \leq j \leq l} [s_j] \subseteq \alpha_j \right)$$

But then, for the selected (by $\exists$) s_j we have

$$\left(\psi(\vec{x}_n, \vec{s}_l) \simeq u \wedge \bigwedge_{1 \leq j \leq l} [s_j] \subseteq [s_j] \right)$$

and thus (logic) $(\exists \vec{s}_l) \left(\psi(\vec{x}_n, \vec{s}_l) \simeq u \wedge \bigwedge_{1 \leq j \leq l} [s_j] \subseteq [s_j] \right)$ is true.

This says $\mathscr{F}(\vec{x}_n, [\vec{s}_l]) \simeq u$. $\qquad\square$

13.5.4 Proposition. *Every D-computable partial functional is monotone.*

Proof   Assume $\mathscr{F}(\vec{x}_n, \vec{\alpha}_l) \simeq u$. By Definition 13.5.1 this means

$$(\exists \vec{s}_l) \left(\psi(\vec{x}_n, \vec{s}_l) \simeq u \wedge \bigwedge_{1 \leq j \leq l} [s_j] \subseteq \alpha_j \right) \tag{†}$$

Let us fix an array of s_i that work in (†). Say now $\alpha_1 \subseteq \beta$. Then $[s_1] \subseteq \beta$ and hence

$$(\exists \vec{s}_l) \left(\psi(\vec{x}_n, \vec{s}_l) \simeq u \wedge [s_1] \subseteq \beta \wedge \bigwedge_{2 \leq j \leq l} [s_j] \subseteq \alpha_j \right) \text{ is true} \tag{‡}$$

(‡) says $\mathscr{F}(\vec{x}_n, \beta, \alpha_2, \ldots, \alpha_l) \simeq u$.   □

In the next little while we will focus attention to D-computable functions of rank $(0,1)$.

13.5.5 Definition. (Recursive Operators)   The D-computable functionals

$$\mathscr{G} : \mathcal{P}(\mathbb{N} : \mathbb{N}) \rightarrow \mathbb{N}$$

naturally give rise to *recursive operators* (Davis 1958b; Rogers 1967), from $\mathcal{P}(\mathbb{N} : \mathbb{N}) \rightarrow \mathcal{P}(\mathbb{N} : \mathbb{N})$ respectively, defined, for all x, α, by

$$\mathscr{G}(\alpha) \stackrel{Def}{=} \lambda x.\mathscr{G}(\alpha)(x)$$

where, of course, (Definition 13.5.1) for some $\psi \in \mathcal{P}$,

$$\mathscr{G}(\alpha)(x) \simeq u \equiv (\exists s)(\psi(s, x) \simeq u \wedge [s] \subseteq \alpha) \qquad \square$$

A recursive operator $\mathscr{G}$ is *total* in the natural sense that, for any α, we do have a function $\lambda x.\mathscr{G}(\alpha)(x)$ even though for some α this function may be $\emptyset$.

We have at once from Propositions 13.5.3 and 13.5.4

13.5.6 Proposition.   *Every recursive operator $\mathscr{F}$ is compact, that is, $\mathscr{F}(\alpha)(x) \simeq u$ implies $\mathscr{F}([s])(x) \simeq u$, for some $s \in \mathbb{N}$ such that $[s] \subseteq \alpha$.*

13.5.7 Proposition.   *Every recursive operator $\mathscr{F}$ is monotone, that is, $\mathscr{F}(\alpha)(x) \simeq u$ and $\alpha \subseteq \beta$ imply $\mathscr{F}(\beta)(x) \simeq u$.*

In the literature, recursive operators usually are denoted by the Greek capital letters Ψ and Φ. We prefer to continue using the *same notation as for unrestricted functionals* that give rise to these operators.

13.5.8 Theorem. (Detecting Recursive Operators)   *A mapping $\mathscr{G} : \mathcal{P}(\mathbb{N} : \mathbb{N}) \rightarrow \mathcal{P}(\mathbb{N} : \mathbb{N})$ is a recursive operator iff*

1. $\mathscr{G}$ *is compact and monotone.*
2. *The function g of two arguments given by*

$$g(y, x) \overset{Def}{=} \begin{cases} \mathscr{G}([y])(x) & \text{if y codes (cf. Definition 11.3.6) a function } [y] \\ \uparrow & \text{if not} \end{cases}$$

is partial recursive.

Proof

- (*Only If* part): Let $\mathscr{G}$ be a recursive operator. Then it is compact and monotone by Propositions 13.5.6 and 13.5.7. Next let ψ be as in Definition 13.5.5:

$$\mathscr{G}(\alpha)(x) \simeq u \equiv (\exists s)(\psi(s, x) \simeq u \wedge [s] \subseteq \alpha) \tag{\dagger}$$

Replacing the left hand side above by $\mathscr{G}([y], x)$, but using $g(y, x)$ instead, we obtain

$$g(y, x) \simeq u \equiv (\exists s)(\psi(s, x) \simeq u \wedge [s] \subseteq [y])$$

We have enough tools to see at once that the right hand side above is semi-recursive, and hence so is the left hand side. We have got 2. above via the graph theorem (Theorem 6.6.1).

- (*If* part): Here we assume 1. and 2. and prove —for an appropriate "ψ"— the last formula of Definition 13.5.5, which is the same as (†) above. By 1. (compactness part) $\mathscr{G}(\alpha)(x) \simeq u$ implies $(\exists s)([s] \subseteq \alpha \wedge \mathscr{G}([s])(x) \simeq u)$, which becomes an *equivalence* by 1. (second conjunct and monotonicity part). Thus

$$\mathscr{G}(\alpha)(x) \simeq u \Leftrightarrow (\exists s)([s] \subseteq \alpha \wedge \mathscr{G}([s])(x) \simeq u)$$

$$\overset{using\ 2.}{\Leftrightarrow} (\exists s)([s] \subseteq \alpha \wedge g(s, x) \simeq u)$$

The last equivalence yields (†) using g for "ψ".   □

In our first application of topological methods we give below a characterisation of compact and monotone operators.

13.5.9 Theorem. (Uspenskii (1955), Nerode (1957)) *An operator $\lambda\alpha.\mathscr{F}(\alpha)$ is compact and monotone iff it is (topologically) continuous.*[17]

---

[17] Notwithstanding the parenthetical qualifier "topologically", the use of the term "continuous" in two different ways in the same sentence would have been awkward at the very least. Hence we used the alternative term "compact" on the "computation side".

Proof

- (*If* case.) Suppose $\mathscr{F} : \mathcal{P}(\mathbb{N} : \mathbb{N}) \to \mathcal{P}(\mathbb{N} : \mathbb{N})$ is continuous. By Proposition 0.7.5, $F^{-1}[\widehat{\theta}]$ is open for every (finite, as the notation suggests) θ. Let $\mathscr{F}(\alpha) = \beta$ and fix attention on some x where $\beta(x) \downarrow \simeq y$, for some y. Let $\theta = \{(x, y)\}$. Then

$$\mathscr{F}^{-1}[\widehat{\theta}] \text{ is open} \tag{i}$$

and $\beta \supseteq \theta$ hence

$$\beta \in \widehat{\theta}$$

Thus

$$\mathscr{F}^{-1}[\{\mathscr{F}(\alpha)\}] = \mathscr{F}^{-1}[\{\beta\}] \subseteq \mathscr{F}^{-1}[\widehat{\theta}]$$

from which we get $\alpha \in \mathscr{F}^{-1}[\widehat{\theta}]$. Apply now Proposition 0.7.9, 1. and 2., to obtain a ξ such that

$$\widehat{\xi} \subseteq \mathscr{F}^{-1}[\widehat{\theta}]$$

and

$$\xi \subseteq \alpha \text{ and } \xi \in \mathscr{F}^{-1}[\widehat{\theta}] \text{ (this conjunction is "1.")} \tag{ii}$$

and

$$\alpha \in F^{-1}[\widehat{\theta}] \wedge \alpha \subseteq \beta \text{ implies } \beta \in \mathscr{F}^{-1}[\widehat{\theta}] \text{ (this conjunction is "2.")} \tag{iii}$$

Now, (ii) implies $\mathscr{F}(\xi) \in \widehat{\theta}$ hence $\mathscr{F}(\xi)(x) \simeq y$, settling compactness.
On the other hand, (iii) translates as $\mathscr{F}(\alpha) \in \widehat{\theta} \wedge \alpha \subseteq \beta$ implies $\mathscr{F}(\beta) \in \widehat{\theta}$. In short, $\mathscr{F}(\alpha)(x) \simeq y$ and $\alpha \subseteq \beta$ imply $\mathscr{F}(\beta)(x) \simeq y$ settling monotonicity of $\mathscr{F}$.

- (*Only if* case.) Now we start with the assumption that $\mathscr{F}$ is compact *and* monotone and seek to prove that for any θ, $\mathscr{F}^{-1}[\widehat{\theta}]$ is open. By Proposition 0.7.9 we must prove

$$\text{If } \alpha \in \mathscr{F}^{-1}[\widehat{\theta}], \text{ then } (\exists \xi)(\xi \in \mathscr{F}^{-1}[\widehat{\theta}] \wedge \xi \subseteq \alpha) \tag{1}$$

and

$$\text{If } \alpha \in \mathscr{F}^{-1}[\widehat{\theta}] \wedge \alpha \subseteq \beta, \text{ then } \beta \in \mathscr{F}^{-1}[\widehat{\theta}] \tag{2}$$

For (1): The assumption in (1) yields $\mathscr{F}(\alpha) \in \widehat{\theta}$ hence (cf. Definition 0.7.8) $\theta \subseteq \mathscr{F}(\alpha)$.

Let $\theta = \{(x_0, y_0), (x_1, y_1), \ldots, (x_n, y_n)\}$. So $\mathscr{F}(\alpha)(x_i) \simeq y_i$, for $i = 0, 1, \ldots, n$. By compactness, for each $i = 0, 1, \ldots, n$, we have a $\sigma_i \subseteq \alpha$ such that

$$\mathscr{F}(\sigma_i)(x_i) \simeq y_i, \text{ for } i = 0, 1, \ldots, n$$

Thus the union $\xi = \bigcup_{0 \leq i \leq n} \sigma_i$ *is a subset of* α, so it is a *function* (single-valued). It is a finite function because all the σ_i are. By monotonicity, $\mathscr{F}(\xi)(x_i) \simeq y_i$, for $i = 0, 1, \ldots, n$. But this says $\mathscr{F}(\xi) \supseteq \theta$ or (cf. Definition 0.7.8) $\mathscr{F}(\xi) \in \widehat{\theta}$ or $\xi \in \mathscr{F}^{-1}[\widehat{\theta}]$. We proved (1).

For (2): Assume hypothesis. The part $\alpha \in \mathscr{F}^{-1}[\widehat{\theta}]$ means $\mathscr{F}(\alpha) \in \widehat{\theta}$, hence is the same as $\mathscr{F}(\alpha) \supseteq \theta$. Bringing in $\alpha \subseteq \beta$ and monotonicity we get $\mathscr{F}(\beta) \supseteq \mathscr{F}(\alpha) \supseteq \theta$. Hence $\mathscr{F}(\beta) \in \widehat{\theta}$. Thus $\beta \in \mathscr{F}^{-1}[\widehat{\theta}]$. This settles (2).

□

13.5.10 Definition. (Modulus of Continuity) For an operator $\mathscr{F}$ we call *modulus of continuity* the function $\lambda z.\left(\lambda x. f(z, x)\right)$ given by $f(z, x) \simeq \mathscr{F}([z])(x)$.

A continuous operator $\mathscr{F}$ is called *effectively continuous* if the function f is partial recursive. □

By Theorem 13.5.8, given a recursive operator $\mathscr{F}$, its modulus f is precisely defined as

$$f(z, x) \overset{Def}{=} \begin{cases} \mathscr{F}([z])(x) & \text{if } z \text{ codes (cf. Definition 11.3.6) a function } [z] \\ \uparrow & \text{if not} \end{cases} \tag{1}$$

We saw that f is partial recursive since $f(z, x) \simeq u$ is r.e. Thus the function we called "modulus of continuity", above is partial recursive in the case of a recursive functional.

Combining the preceding two theorems we have

13.5.11 Corollary. (Uspenskii (1955), Nerode (1957)) *The recursive operators are effectively continuous.*

Conversely, effectively continuous operators $\mathscr{F} : \mathcal{P}(\mathbb{N} : \mathbb{N}) \to \mathcal{P}(\mathbb{N} : \mathbb{N})$ *are recursive.*

Proof A recursive operator is compact and monotone. By Theorem 13.5.9 it is also continuous, topologically speaking. The *effectiveness* of continuity is the result of Definition 13.5.10 and the remarks following it about f ((1) above).

For the converse, topological continuity of $\mathscr{F} : \mathcal{P}(\mathbb{N} : \mathbb{N}) \to \mathcal{P}(\mathbb{N} : \mathbb{N})$ implies the compactness and monotonicity of $\mathscr{F}$ (Theorem 13.5.9). Adding "effective" to continuity we get the "f" in (1) above and we are done by Theorem 13.5.8. □

13.5.1 Effective Operations on $\mathcal{P}$

We connect the material of this section with Corollary 6.10.6 and Theorem 11.3.10. This section provides a very brief look at *effective operations*, that is, functionals that compute on the *indices* of their one-argument type-1 inputs (from $\mathcal{P}$). These functionals behave as they please (meaning we do not care how) if the set of type-1 inputs contains at least one function whose index (if it exists) *was not passed* to the computing agent of such a functional. Rogers (1967), Odifreddi (1989, 1999) cover effective operations extensively.

13.5.12 Definition. A recursive $\lambda x. f(x)$ is called *extensional* if, for all e and e', $\{e\} = \{e'\}$ implies $\{f(e)\} = \{f(e')\}$, or, in Rogers' notation, $\phi_e = \phi_{e'}$ implies $\phi_{f(e)} = \phi_{f(e')}$. □

Extensional functions arise, for example, as follows. Given a recursive operator $\mathcal{H}$. What is $\mathcal{H}(\{e\})$?

Well, $\mathcal{H}(\{e\})(x) \simeq u$ iff, for some $\psi \in \mathcal{P}$, $(\exists s)([s] \subseteq \{e\} \wedge \psi(s, x) \simeq u)$. We already know that the predicate $\mathbb{P}(e, x, u) \equiv (\exists s)([s] \subseteq \{e\} \wedge \psi(s, x) \simeq u)$ is semi-recursive, and single-valued in u by the semantics of $\mathcal{H}(\{e\})(x) \simeq u$. So, for some g (function) we have $g(e, x) \simeq u \Leftrightarrow^{18} P(e, x, u) \Leftrightarrow \mathcal{H}(\{e\})(x) \simeq u$, thus $g \in \mathcal{P}$ by Theorem 6.6.1. By S-m-n applied to g, and switching to Rogers' notation, let $f \in \mathcal{R}$ such that

$$\phi_{f(e)}(x) \simeq u \equiv \mathcal{H}(\phi_e)(x) \simeq u$$

This f is extensional: Say, $\phi_e = \phi_{e'}$. Then

$$\phi_{f(e)} = \mathcal{H}(\phi_e) = \mathcal{H}(\phi_{e'}) = \phi_{f(e')}$$

13.5.13 Definition. (Effective Operations) An *effective operation* on the set of partial recursive functions $\mathcal{P}$ is a functional $\mathcal{F}$ *induced* by an *extensional* recursive f via the definition, for all $e \in \mathbb{N}$,

$$\mathcal{F}(\phi_e) = \phi_{f(e)}$$

By the definition of *extensional function* and the discussion above, $\mathcal{F}$ restricted on $\mathcal{P}^{(1)}$[19] is single-valued.[20] □

---

[18] Conjunctional equivalence.

[19] $\mathcal{P}^{(1)}$ is the set of all one-argument functions in $\mathcal{P}$.

[20] In some terminology this functional is described as a "well-defined" (concept) on $\mathcal{P}$. Yet, there is nothing "ill defined" about the partial multi-valued functionals of the next section.

The following theorem of Myhill and Shepherdson (1955), Uspenskii (1955) characterises effective operations.

13.5.14 Theorem. *The effective operations on $\mathcal{P}$ are precisely the restrictions on $\mathcal{P}$ of the effectively continuous operators $\mathcal{H} : P(\mathbb{N} : \mathbb{N}) \to P(\mathbb{N} : \mathbb{N})$.*

Proof

- Suppose $\mathcal{H} : P(\mathbb{N} : \mathbb{N}) \to P(\mathbb{N} : \mathbb{N})$ is effectively continuous. By Corollary 13.5.11 $\mathcal{H}$ is a recursive operator hence quoting the argument found in the ②-enclosed passage following Definition 13.5.12 we get an $f \in \mathcal{R}$ such that

$$\phi_{f(e)}(x) \simeq u \equiv \mathcal{H}(\phi_e)(x) \simeq u$$

  So $\mathcal{H}$ is an effective operation (Definition 13.5.13).
- Conversely, let $\mathcal{F}$, with left and right fields equal to $\mathcal{P}$, satisfy

$$\text{for some extensional } g \text{ and all } e, \ \mathcal{F}(\phi_e)(x) \simeq u \equiv \phi_{g(e)}(x) \simeq u \tag{1}$$

The right hand side of (1) is trivially semi-recursive (the graph $\phi_z(x) \simeq u$ is, and then we substitute the recursive function call $g(e)$ into z). Thus so is the left hand side. Therefore, for each x and u, if we set

$$\mathcal{C}_{x,u} = \{\phi_e : \mathcal{F}(\phi_e)(x) \simeq u\} \tag{2}$$

then $\{e : \phi_e \in \mathcal{C}_{x,u}\}$ is semi-recursive. By Corollary 6.10.6 and (2) we have

$$\mathcal{F}(\phi_e)(x) \simeq u \equiv (\exists\theta)(\theta \subseteq \phi_e \wedge \mathcal{F}(\theta)(x) \simeq u) \tag{3}$$

We are looking for a recursive operator $\mathcal{G}$ that *extends* $\mathcal{F}$ and *agrees* with $\mathcal{F}$ on the set $\{\phi_e : e \in \mathbb{N}\}$ (we know that $\theta \in \{\phi_e : e \in \mathbb{N}\}$, Proposition 11.3.7).
By (3) and the requirement for $\mathcal{G}$ to be compact and continuous, the agreement requirement leads us to define, for all α and x,

$$\mathcal{G}(\alpha)(x) \simeq u \equiv (\exists s)([s] \subseteq \alpha \wedge \mathcal{F}([s])(x) \simeq u) \tag{4}$$

The $\to$ direction in the equivalence of (4) yields compactness of $\mathcal{G}$ while the $\leftarrow$ direction yields monotonicity. This establishes *topological continuity* of $\mathcal{G}$. The *modulus of continuity*

$$f(s, x) \simeq \mathcal{G}([s])(x) \simeq \mathcal{F}([s])(x)$$

is in $\mathcal{P}$ by our work following Definition 13.5.10. Thus by Corollary 13.5.11 $\mathcal{G}$ is a recursive operator.   □

13.6   Search Computability of Moschovakis

The *search computable functionals* introduce *nondeterminism* in (dynamic) computations so that in, say, an attempt to determine if $\mathbb{P}(\vec{x}, \vec{\alpha}) \vee \mathbb{Q}(\vec{y}, \vec{\alpha})$ is true we are not "stuck" as was the case in our attempt in Example 13.3.35, (4) if $\mathbb{P}(\vec{x}, \vec{\alpha})$ is true but $\mathscr{K}_Q(\vec{y}, \vec{\alpha}) \uparrow$.

In Kleene-style computation of *non restricted* functionals we must discharge *all* calls that present themselves in the course of a computation; we cannot ignore any one of them. Thus we cannot avoid the call $\mathscr{K}_Q(\vec{y}, \vec{\alpha})$ in our example, which results in $\left(0 \times \mathscr{K}_Q(\vec{y}, \vec{\alpha})\right) \uparrow$ because the computation must *wait* the outcome of the call $\mathscr{K}_Q(\vec{y}, \vec{\alpha})$ *before* it has a chance to multiply the result by zero: there *will* be *no* outcome though so we are stuck: we wait forever!

In *nondeterministic* computation *we* can steer the computation to take a path that *works* and ignore a path —that starts with a call, for example, to $\mathscr{K}_Q(\vec{y}, \vec{\alpha})$— which does not. Would that distort the result in our example? Of course not, since for the truth of $\mathbb{P}(\vec{x}, \vec{\alpha}) \vee \mathbb{Q}(\vec{y}, \vec{\alpha})$ it suffices to verify just the truth of $\mathbb{P}(\vec{x}, \vec{\alpha})$.

The emphasis on "we" is meant to distinguish *nondeterminism* from *parallelism*. The former is a mode of computation where there are many computation paths, but we can "guess" which is a good one and we take it so that we can conclude the computation.

The latter has no guessing. *All* possible paths are traversed (tried) in parallel.

Restricted computable functionals (as we shall see) can *simulate parallelism* — as does absolute computability that uses no oracles. Poor person's parallelism in either case is obtained by *dovetailing* computations— cf. Theorem 6.9.9 —just as it is done in practical computing in, say, the context of so-called "coroutines". These are several routines (programs) that *appear* to be running in parallel because the operating system gives each routine the opportunities to run for a few steps, then wait, then run some more steps, etc. Each coroutine gets successive *"time slices"* to fit its computation.

In the case of $\mathbb{P}(\vec{x}, \vec{\alpha}) \vee \mathbb{Q}(\vec{y}, \vec{\alpha})$ or $\mathscr{K}_P(\vec{x}, \vec{\alpha}) \times \mathscr{K}_Q(\vec{y}, \vec{\alpha}) \simeq 0$ dovetailing would have one coroutine for each call $\mathscr{K}_P(\vec{x}, \vec{\alpha})$ and $\mathscr{K}_Q(\vec{y}, \vec{\alpha})$. So, a few steps for $\mathscr{K}_P(\vec{x}, \vec{\alpha})$, then a few $\mathscr{K}_Q(\vec{y}, \vec{\alpha})$, and repeat until *any one* ends with a 0 returned value.

Why can't we dovetail in Kleene's theory of non restricted computable functionals?

Because we do not have the selection theorem which is a manifestation of dovetailing (see also *definition by positive cases*, Theorem 6.8.1). The selection function (of the selection theorem [or axiom]) will select ("computationally") a member of a nonempty semi-recursive set $\{y : f(y, \vec{x}) \simeq 0\}$. Unbounded search will *not* find such a member in general. For example, $(\mu y) f(y, \vec{x})$ will diverge if $f(0, \vec{x}) \uparrow$ even if it is, say, $f(1, \vec{x}) \simeq 0$.

Dovetailing (the method we used to prove selection in Theorem 6.9.9) allows us to "coexist" with the fact that $f(0, \vec{x}) \uparrow$ since we are running *all* computations $f(i, \vec{x})$ —$i \geq 0$— (in principle) "simultaneously" looking for $f(i, \vec{x}) \simeq 0$.

$$\begin{array}{ll}
\text{do } 1 & \text{step with} \qquad\quad i = 0 \\
\text{do } 2 & \text{steps with each of } i = 0, 1 \\
\text{do } 3 & \text{steps with each of } i = 0, 1, 2 \\
\vdots \;\; \vdots & \quad\vdots \;\; \vdots \qquad\qquad\quad \vdots \\
\text{do } k + 1 & \text{steps with each of } i = 0, 1, \ldots, k \\
\vdots \;\; \vdots & \quad\vdots \;\; \vdots \qquad\qquad\quad \vdots
\end{array}$$

Absolute computability as we saw (and computability of *restricted* functionals, as we shall see) *can* prove the selection theorem.

For computability of *non* restricted functionals Moschovakis (1969) introduced selection *axiomatically*, as part of the Ω set Definition 13.3.1. Our exposition here does not have the full generality of said reference. We confine attention to unrestricted functionals that accept as inputs *numbers* and *partial functions* $\alpha : \mathbb{N} \to \mathbb{N}$ (Definition 13.1.1). The α in principle *can* be p.m.v. but even if they are not, the theory will contain p.m.v. computable functions that result from an application of the selection operator.

The form of axiomatic selection is in the first instance defined in terms of computable functions and then extended to predicates (thought of as partial characteristic functionals):

If $\lambda y \vec{x} \vec{\alpha}.\mathscr{G}(y, \vec{x}, \vec{\alpha})$ is partial recursive, then so is $\lambda \vec{x} \vec{\alpha}.(\nu y)\mathscr{G}(y, \vec{x}, \vec{\alpha})$, where (νy) returns *all* y such that $\mathscr{G}(y, \vec{x}, \vec{\alpha}) \simeq 0$.

Thus, application of (νy) may lead to *partial multiple-valued functionals* or *p.m.v.*, even when applied to an ordinary (single-valued) functional $\mathscr{G}$.

Why multiple valued? The reason is that while the operator (μy) has a *computational definition* —that is, "search along $y = 0, 1, 2, \ldots$ until *the first $y = k$ is reached* that satisfies a computationally stated condition"— where the italicised "reached" implies that no $y < k$ led to an infinite loop (cf. Definition 1.1.25), we have no such *a priori computational* definition for a single-valued selection operator (νy) that is similar to a "*choice function*" of set theory.

One way to avoid this difficulty is to select *all* elements that work, analogously to the multiple-valued behaviour of $\bigcup$ of set theory. For example $\bigcup$ applied to the "single object" $\{\{a\}, \{b\}, \{c\}, \ldots\}$ returns the set of objects $\{a, b, c, \ldots\}$. Admittedly this is a somewhat fuzzy analogy, but the approach works; cf. Definition 13.6.21 below.

13.6.1 Definition. (Ω^{sel} **and the Selection Tuple**) We expand Ω of Definition 13.3.1 into a set of tuples Ω^{sel}: For all $k + l \geq 1, \vec{x}, \vec{\alpha}, y, z, u, v, e, f, g, c$ it is inductively defined as the *smallest* set of tuples $(e, \vec{x}, \vec{\alpha}, y)$ satisfying I–VIII from Definition 13.3.1 *plus* clause IX below:

The selection axiom: IX

$$IX. \quad (\langle 9, k, l, e \rangle, \vec{x}, \vec{\alpha}, y) \in \Omega \qquad if \, (e, y, \vec{x}, \vec{\alpha}, 0) \in \Omega$$

□

The semantics of the Ω^{sel} tuples I–IX is as in I-VIII (13.3.3) plus the semantics for IX.

13.6.2 Definition. (Computations) The symbol

$$\{e\}_v^{(k,l)}(\vec{x}_k, \vec{\alpha}_l) \simeq y \tag{1}$$

says exactly

$$(e, \vec{x}_k, \vec{\alpha}_l, y) \in \Omega^{sel}$$

In words, *the functional* $\{e\}_v^{(k,l)}$ *of rank* (k, l) *given by the "program"* e has among its outputs *the number* y if the *input* is $\vec{x}_k, \vec{\alpha}_l$.

The subscript v indicates that the set of tuples that mean $\{e\}_v^{(k,l)}(\vec{x}_k, \vec{\alpha}_l) \simeq y$ is the set of those in Ω^{sel} where *selection* is supported axiomatically via an operation v (in the notation in Moschovakis (1969)), in the computability so founded in the style of Sect. 13.2.

This interpretation *makes no claim of uniqueness* or *single-valued-ness* of the y variable, in fact in the case of *search computability* of this section this is *not* generally the case.

We will allow the *possibility* of p.m.v. type-1 inputs if only for the sake of allowing the substitution theorem (counterpart of Theorem 13.3.36) of a computable functional (extensionally) into a type-1 variable.

We also have the notations below signalling that we will usually write $\{e\}_v(\vec{x}_k, \alpha_l)$ without the superscript (k, l) —but retaining the selector symbol "v" as a subscript— letting the context fend for itself.

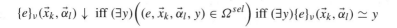

$$\{e\}_v(\vec{x}_k, \vec{\alpha}_l) \downarrow \text{ iff } (\exists y)\Big((e, \vec{x}_k, \vec{\alpha}_l, y) \in \Omega^{sel}\Big) \text{ iff } (\exists y)\{e\}_v(\vec{x}_k, \vec{\alpha}_l) \simeq y$$

and

$$\{e\}_v(\vec{x}_k, \vec{\alpha}_l) \uparrow \text{ iff } \neg(\exists y)\Big((e, \vec{x}_k, \vec{\alpha}_l, y) \in \Omega^{sel}\Big) \tag{2}$$

□

13.6.3 Definition. (Subcomputations and Computation Lengths) We expand Definition 13.3.4 to include the case of clause IX.

Immediate subcomputations for case IX (Definition 13.6.1) —in short *i.s.*— and *computation lengths* for each tuple in Definition 13.3.1.

IX. If $(e, y, \vec{x}, \vec{\alpha}, 0) \in \Omega^{sel}$, then the i.s. of $(\langle 9, k, l, e \rangle, \vec{x}, \vec{\alpha}, y)$ is $(e, y, \vec{x}, \vec{\alpha}, 0)$. Moreover, $\| \langle 9, k, l, e \rangle, \vec{x}, \vec{\alpha}, y \| = \| e, y, \vec{x}, \vec{\alpha}, 0 \| + 1$.

As before, a computation $c \in \Omega^{sel}$ is a *subcomputation* of $d \in \Omega^{sel}$ just in case there is finite sequence $c = d_1, d_2, \ldots, d_n = d$ such that d_i is an i.s. of d_{i+1}, for $i = 1, \ldots, n-1$.   □

13.6.4 Remark. Clause IX causes multi-valued-ness in y since several y (for fixed $e, \vec{x}, \vec{\alpha}$) might satisfy $(e, y, \vec{x}, \vec{\alpha}, 0) \in \Omega^{sel}$.

Thus we will *not* be able to use *truncated computations* in search computability.

□

13.6.5 Definition. (The Set of Partial Search Computable Functionals) We may define a p.m.v. functional $\{e\}_v$ of rank (k, l) extensionally (by its graph) by

$$\{e\}_v(\vec{x}, \vec{\alpha}) \simeq y \text{ iff } (e, \vec{x}, \vec{\alpha}, y) \in \Omega^{sel}$$

We next define the set of *partial recursive functionals* (*partial computable functionals*):

It is the set of all $\mathscr{F}$ of rank (k, l) —for all $k + l \geq 1$— such that $\mathscr{F} = \{e\}_v^{(k,l)}$, for some $e \in \mathbb{N}$.

Thus, if e is not an index, as defined inductively in Definitions 13.3.1 and 13.6.1, we have —due to (2), p. 461— $\{e\}_v = \emptyset$, the everywhere undefined (computable) functional.

Via this observation we extend the meaning of $\{e\}_v(\vec{x}, \vec{\alpha}) \simeq z$ to all $e \in \mathbb{N}$.

We will *not* introduce special symbols for the *class* of these functionals. However, if we let $\mathcal{P}_v^{(k,l)}$ (resp. $\mathcal{R}_v^{(k,l)}$) be the set of all *partial* (resp. *total*) search computable functionals of rank (k, l) then the class of *all* such, for all ranks, is $\bigcup_{k+l \geq 1} \mathcal{P}_v^{(k,l)}$ (resp. $\bigcup_{k+l \geq 1} \mathcal{R}_v^{(k,l)}$).   □

13.6.6 Example. It is clear that each of the indices e as given by Definitions 13.3.1 and 13.6.1 satisfies $Seq(e)$ (cf. Example 2.3.10(8)). Now $Seq(0)$ is false, thus, $\{0\}_v = \emptyset$, the empty functional.   □

The following is consistent with our intuitive understanding and issues the first few straightforward properties of the partial search computable p.m.v. functionals.

13.6.7 Proposition.

I. $\{\langle 1, k, l, i \rangle\}_v(\vec{x}, \vec{\alpha}) \simeq x_i$
II. $\{\langle 2, k, l, i \rangle\}_v(\vec{x}, \vec{\alpha}) \simeq x_i + 1$
III. $\{\langle 3, k, l, c \rangle\}_v(\vec{x}, \vec{\alpha}) \simeq c$
IV. $\{\langle 4, k+2, l \rangle\}_v(a, b, \vec{x}, \vec{\alpha}) \simeq Sb_1(a, b)$
V. $\{\langle 5, k+4, l \rangle\}_v(p, q, a, b, \vec{x}, \vec{\alpha}) \simeq p \text{ if } a = b \; \{\langle 5, k+4, l \rangle\}_v(p, q, a, b, \vec{x}, \vec{\alpha}) \simeq q \text{ if } a \neq b$
VI. $\{\langle 6, k, l, i, j \rangle\}_v(\vec{x}, \vec{\alpha}) \simeq \alpha_j(x_i)$

VII. $\{\langle 7, k, l, f, \vec{e}_m \rangle\}_\nu(\vec{x}, \vec{\alpha}) \simeq \{f\}\Big(\{e_1\}_\nu(\vec{x}, \vec{\alpha}), \dots, \{e_m\}_\nu(\vec{x}, \vec{\alpha}), \vec{\alpha}\Big)$

VIII. $\{\langle 8, k+1, l \rangle\}_\nu(e, \vec{x}, \vec{\alpha}) \simeq \{e\}_\nu(\vec{x}, \vec{\alpha})$

IX. $\{\langle 9, k, l, e \rangle\}_\nu(\vec{x}, \vec{\alpha}) \simeq y$ iff $\{e\}_\nu(y, \vec{x}, \vec{\alpha}) \simeq 0$

Proof   Cases I–VI are immediate.

For

Cases VII, VIII.

The argument is the same as in the proof of Proposition 13.3.10, this time remembering to carry along the subscript ν.

Case IX.

We want

$$\{\langle 9, k, l, e \rangle\}_\nu(\vec{x}, \vec{\alpha}) \simeq y \text{ iff } \{e\}_\nu(y, \vec{x}, \vec{\alpha}) \simeq 0$$

The $\leftarrow$ direction is immediate by Definition 13.6.1.

For the $\rightarrow$ direction, assume that

$$\{\langle 9, k, l, e \rangle\}_\nu(\vec{x}, \vec{\alpha}) \simeq y \tag{†}$$

is true. That is $(\langle 9, k, l, e \rangle, \vec{x}, \vec{\alpha}, y) \in \Omega^{sel}$.

Thus (Definition 13.6.3), $\|e, y, \vec{x}, \vec{\alpha}, 0\| < \|\langle 9, k, l, e \rangle, \vec{x}, \vec{\alpha}, y\|$ and (by the obvious I.H. on computation lengths) we have $\{e\}_\nu(y, \vec{x}, \vec{\alpha}) \simeq 0$ is true.   □

13.6.8 Definition. (The Selection Operation ν)   We introduce the operator ν and rewrite the equivalence

$$\{\langle 9, k, l, e \rangle\}_\nu(\vec{x}, \vec{\alpha}) \simeq y \text{ iff } \{e\}_\nu(y, \vec{x}, \vec{\alpha}) \simeq 0$$

that we proved above as the *definition* of (νy):

$$(\nu y)\Big(\{e\}_\nu(y, \vec{x}, \vec{\alpha}) \simeq 0\Big) \simeq u \stackrel{Def}{\equiv} \{\langle 9, k, l, e \rangle\}_\nu(\vec{x}, \vec{\alpha}) \simeq u \tag{1}$$

that is, *all* the u-values that make $\{e\}_\nu(u, \vec{x}, \vec{\alpha}) \simeq 0$ true are returned by the call $\{\langle 9, k, l, e \rangle\}_\nu(\vec{x}, \vec{\alpha})$ and therefore by $(\nu y)\Big(\{e\}_\nu(y, \vec{x}, \vec{\alpha}) \simeq 0\Big)$.

In analogy with the use of μ we may also write the right hand side of (1) above as $(\nu y)\Big(\{e\}_\nu(y, \vec{x}, \vec{\alpha})\Big)$, omitting the "$\simeq 0$" part.   □

13.6.9 Definition. (Semi-Recursive Relations)   Once again, semi-recursive relations in search computability are defined unremarkably:

$\mathbb{P}(\vec{x}, \vec{\alpha})$ is *semi-recursive* iff for some search computable $\mathscr{F}$ we have $\mathbb{P}(\vec{x}, \vec{\alpha}) \equiv \mathscr{F}(\vec{x}, \vec{\alpha}) \downarrow$.   □

As in the case of Kleene computability we have via the $\mathscr{UC}$ function,

13.6.10 Proposition. $\mathbb{P}$ *is semi-recursive in search computability iff it has a search computable partial characteristic function.*

The selection theorem is valid in search computability and takes the form

13.6.11 Theorem. (Selection Theorem)

(1) $(vy)\Big(\{e\}_v(y,\vec{x},\vec{\alpha}) \simeq 0\Big) \downarrow$ *iff* $(\exists y)\{e\}_v(y,\vec{x},\vec{\alpha}) \simeq 0.$

(2) $(vy)\Big(\{e\}_v(y,\vec{x},\vec{\alpha}) \simeq 0\Big) \downarrow$ *implies* $\{e\}_v\Big((vy)\{e\}_v(y,\vec{x},\vec{\alpha}) \simeq 0, \vec{x}, \vec{\alpha}\Big) \simeq 0.$

Proof

Going forward we omit the subscript v from our computable functionals "$\{e\}$" unless we need to distinguish them from other functionals that are computable in some other theory.

(1) We proved in Proposition 13.6.7 $\{\langle 9, k+1, l, e\rangle\}(\vec{x},\vec{\alpha}) \simeq y \equiv \{e\}(y,\vec{x},\vec{\alpha}) \simeq 0$
By the equivalence theorem in logic, we get $(\exists y)\{\langle 9, k+1, l, e\rangle\}(\vec{x},\vec{\alpha}) \simeq y \equiv (\exists y)\{e\}(y,\vec{x},\vec{\alpha}) \simeq 0$ which says

$$\{\langle 9, k+1, l, e\rangle\}(\vec{x},\vec{\alpha}) \downarrow \equiv (\exists y)\{e\}(y,\vec{x},\vec{\alpha}) \simeq 0$$

and, by Definition 13.6.8, we can rewrite the above as

$$(vy)\Big(\{e\}(y,\vec{x},\vec{\alpha}) \simeq 0\Big) \downarrow \equiv (\exists y)\{e\}(y,\vec{x},\vec{\alpha}) \simeq 0$$

(2) Let $(vy)\Big(\{e\}_v(y,\vec{x},\vec{\alpha}) \simeq 0\Big) \downarrow$. By Definition 13.6.8, the assumption says $(\exists y)\{\langle 9, k+1, l, e\rangle\}(\vec{x},\vec{\alpha}) \simeq y$. By Proposition 13.6.7 we have

$$\{\langle 9, k+1, l, e\rangle\}(\vec{x},\vec{\alpha}) \simeq y \equiv \{e\}(y,\vec{x},\vec{\alpha}) \simeq 0$$

hence

$$\{\langle 9, k+1, l, e\rangle\}(\vec{x},\vec{\alpha}) \simeq y \to \{e\}(y,\vec{x},\vec{\alpha}) \simeq 0$$

and by logic

$$\{\langle 9, k+1, l, e\rangle\}(\vec{x},\vec{\alpha}) \simeq y \to \{\langle 9, k+1, l, e\rangle\}(\vec{x},\vec{\alpha}) \simeq y \wedge \{e\}(y,\vec{x},\vec{\alpha}) \simeq 0$$

By $\exists$-monotonicity (0.2.99)

$$(\exists y)\{\langle 9, k+1, l, e\rangle\}(\vec{x},\vec{\alpha}) \simeq y \to (\exists y)\Big(\{\langle 9, k+1, l, e\rangle\}(\vec{x},\vec{\alpha}) \simeq y \wedge$$
$$\{e\}(y,\vec{x},\vec{\alpha}) \simeq 0\Big)$$

By assumption and modus ponens,

$$(\exists y)\Big(\{\langle 9, k+1, l, e\rangle\}(\vec{x}, \vec{\alpha}) \simeq y \wedge \{e\}(y, \vec{x}, \vec{\alpha}) \simeq 0\Big)$$

The above says (cf. Definition 13.1.3) $\{e\}(\{\langle 9, k+1, l, e\rangle\}(\vec{x}, \vec{\alpha}), \vec{x}, \vec{\alpha}) \simeq 0$, that is (13.6.8), $\{e\}\Big((\nu y)\big[\{e\}(y, \vec{x}, \vec{\alpha}) \simeq 0\big], \vec{x}, \vec{\alpha}\Big) \simeq 0.$ □

13.6.12 Definition. (ν on Relations) $(\nu y)\mathbb{P}(y, \vec{x}, \vec{\alpha})$ means

$$(\nu y)\mathscr{K}_P(y, \vec{x}, \vec{\alpha}) \simeq 0$$

where $\mathscr{K}_P$ is a partial characteristic functional of $\mathbb{P}$. □

The S-m-n and the (second) recursion theorems hold in search computability with the same statements and same proofs. We state them below for the record:

13.6.13 Theorem. (S-m-n Theorem; Gödel, Kleene, Moschovakis)
Let $k = lh(\vec{x})$ and $l = lh(\vec{\alpha})$. There is an index $Sb_1(e, y)$ that depends on e and y and, for all e and y, we have

$$\{Sb_1(e, y)\}(\vec{x}, \vec{\alpha}) \simeq z \text{ iff } \{e\}(y, \vec{x}, \vec{\alpha}) \simeq z$$

Moreover, $\|e, y, \vec{x}, \vec{\alpha}, z\| < \|Sb_1(e, y), \vec{x}, \vec{\alpha}, z\|$.

It is clear that $Sb_1(e, y)$ is strictly increasing with respect to e and y.

Using the approach from Theorem 10.2.1 we can define S-m-n functions Sb_n exactly as in Definition 10.2.2 and obtain

13.6.14 Corollary. *For each $n \geq 1$, there is a rank-$(n + 1, 0)$ $Sb_n \in \mathcal{PR}$ and we have that, for all $m \geq 1$, the following:*

$$\{Sb_n(a, \vec{y}_n)\}(\vec{x}_m, \vec{\alpha}) \simeq \{a\}(\vec{y}_n, \vec{x}_m, \vec{\alpha})$$

Moreover, $\|e, \vec{y}_n, \vec{x}, \vec{\alpha}, z\| < \|Sb_n(e, \vec{y}_n), \vec{x}, \vec{\alpha}, z\|$.

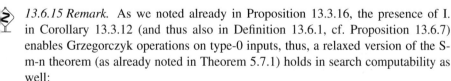

13.6.15 Remark. As we noted already in Proposition 13.3.16, the presence of I. in Corollary 13.3.12 (and thus also in Definition 13.6.1, cf. Proposition 13.6.7) enables Grzegorczyk operations on type-0 inputs, thus, a relaxed version of the S-m-n theorem (as already noted in Theorem 5.7.1) holds in search computability as well:

There is a primitive recursive h of n variables, such that, for any $\mathscr{F}$ of rank $(n + m, l)$ and any $\vec{y}_n$ and $\vec{x}_m$, we have

$$\{h(\vec{y}_n)\}(\vec{x}_m, \vec{\alpha}) \simeq \mathscr{F}(\vec{x}_m, \vec{y}_n, \vec{\alpha}) \tag{1}$$

□

13.6.16 Theorem. (The 2nd Recursion Theorem) Gödel, Carnap, Kleene, Moschovakis. *For any partial recursive $\lambda e\vec{x}\vec{\alpha}.\mathscr{F}(e, \vec{x}, \vec{\alpha})$ there is an index $\overline{e}$ such*

$$\mathscr{F}(\overline{e}, \vec{x}, \vec{\alpha}) \simeq z \text{ iff } \{\overline{e}\}(y, \vec{x}, \vec{\alpha}) \simeq z$$

Moreover, if we let f be an index for $\mathscr{F}$, then $\|\overline{e}, \vec{x}, \vec{\alpha}, z\| > \|f, \overline{e}, \vec{x}, \vec{\alpha}, z\|$.

It is trivial that any partial computable functional $\mathscr{F}$ according to Kleene (via Definition 13.3.1) is also search computable since the clauses used to prove the computability of these functionals are also clauses of Definition 13.6.1. In particular, all primitive recursive functionals are search computable.

 13.6.17 Remark. The class of Kleene-computable functionals is a subset of the class of search computable functionals.   □

In particular, all the results from Corollary 13.3.12 to Proposition 13.3.34 hold for search computable functionals, including, to apply some emphasis, the S-m-n theorem 13.3.18 and the recursion theorem 13.3.21 —restated as Theorem 13.6.13 and the recursion theorem 13.6.16— with all the applications enumerated for the latter, including closure of the class of search computable functionals under unbounded search.

The remark above can be sharpened to

13.6.18 Theorem. *There is a recursive function τ (functional of rank $(1,0)$) such that, for any Kleene-computable partial functional $\{f\}^{(k,l)}$ of rank (k, l), for all $\vec{x}, \vec{\alpha}, u$, we have*

$$\{f\}^{(k,l)}(\vec{x}, \vec{\alpha}) \simeq u \text{ iff } \{\tau(f)\}_v^{(k,l)}(\vec{x}, \vec{\alpha}) \simeq u \qquad (\dagger)$$

Proof The proof is similar to but simpler than the one for Theorem 13.3.36. The idea is to define a rank $(2, 0)$ recursive $\lambda ft.g(f, t)$ for all f, t and then freeze t by the recursion theorem to obtain τ. The construction of g is done by cases over Definition 13.6.1, that are common with the cases of Definition 13.3.1, p. 420.

For the cases I–VI we define $g(f, t) = f$, where f stands for the index of the corresponding computation tuple.

For VII, we start with $f = \langle 7, k, l, a, q_1, \ldots, q_r \rangle$. Then set

$$g(f, t) = \langle 7, (f)_1, (f)_2, g(a, t), g(q_1, t), \ldots, g(q_r, t) \rangle$$

For VIII, where $f = \langle 8, k + 1, l \rangle$, the construction is more complicated, as it was in the proof of Theorem 13.3.36:

1. We obtain first —in the "obvious way"[21]— a primitive recursive $d(f, t)$ such that

---

[21] That is, by unpacking the program $\{\{t\}_v(f)\}_v$.

$$\{d(f, t)\}_v(f, r, \vec{x}, \vec{\alpha}) \simeq \{\{t\}_v(r)\}_v(\vec{x}, \vec{\alpha}) \tag{$\ddagger$}$$

Then set

$$g(f, t) = Sb_1(d(f, t), f) \tag{0}$$

In detail we construct the function d as follows by recursively unpacking $\{\{t\}_v(r)\}_v$ below, using Definition 13.6.1, case VIII (same as Definition 13.3.1, case VIII):

- By VIII:

$$\{\langle 8, (f)_1 + 1, (f)_2 \rangle\}_v(\{t\}_v(r), \vec{x}, \vec{\alpha}) \simeq z \text{ iff } \{\{t\}_v(r)\}_v(\vec{x}, \vec{\alpha}) \simeq z \tag{1}$$

  We also record

$$\|\{t\}_v(r), \vec{x}, \vec{\alpha}, z\| < \|\langle 8, (f)_1 + 1, (f)_2 \rangle, \{t\}_v(r), \vec{x}, \vec{\alpha}, z\| \tag{1$'$}$$

- By VII:

$$\{\langle 8, k + 1, l \rangle\}_v(\{t\}_v(r), \vec{x}, \vec{\alpha}) \simeq z \text{ iff}$$
$$\{\langle 7, k + 2, l, \langle 8, k + 1, l \rangle, \langle 7, k + 2, l, t, \langle 1, k + 2, l, 2 \rangle \rangle ,$$
$$\langle 1, k + 2, l, 3 \rangle, \dots, \langle 1, k + 2, l, k + 2 \rangle \rangle\}_v \Big(f, r, \vec{x}, \vec{\alpha} \Big) \simeq z \tag{2}$$

See also the $\langle\!\!\langle\,2\,\rangle\!\!\rangle$-delimited paragraph regarding k and l in the proof of Theorem 13.3.36.

Thus the $d(f, t)$ mentioned at the outset of case VIII is

$$d(f, t) = \langle 7, k + 2, l, \langle 8, k + 1, l \rangle, \langle 7, k + 2, l, t, \langle 1, k + 2, l, 2 \rangle \rangle ,$$
$$\langle 1, k + 2, l, 3 \rangle, \dots, \langle 1, k + 2, l, k + 2 \rangle \rangle \tag{3}$$

and we can finally write (2) as

$$\{\langle 8, (f)_1 + 1, (f)_2 \rangle\}_v(\{t\}_v(r), \vec{x}, \vec{\alpha}) \simeq z \text{ iff } \{d(f, t)\}_v(f, r, \vec{x}, \vec{\alpha}) \simeq z$$
$$\text{iff } \{g(f, t)\}_v(r, \vec{x}, \vec{\alpha}) \simeq z \tag{4}$$

and we also have

$$\|\langle 8, (f)_1 + 1, (f)_2 \rangle, \{t\}_v(r), \vec{x}, \vec{\alpha}, z\| < \|d(f, t), f, r, \vec{x}, \vec{\alpha}, z\|$$
$$\overset{by\ (0)}{<} \|Sb_1(d(f, t), f), r, \vec{x}, \vec{\alpha}, z\| = \|g(f, t), r, \vec{x}, \vec{\alpha}, z\| \tag{5}$$

If f does not fit any cases among I–VIII, then set $g(f, t) = 0$.

By the recursion theorem in search computability, there is a $\bar{t}$ such that $g(f, \bar{t}) = \{\bar{t}\}_v(f)$, for all f. Note that $\lambda f.g(f, \bar{t})$ is total, hence recursive.

Let us define

$$\tau(f) \overset{Def}{=} g(f, \bar{t})$$

Verification of (†) of the theorem.

1. $\rightarrow$ direction. This is analogous to the same direction in the related proof of Theorem 13.3.36, and it is by induction on the inductively defined set Ω of Definition 13.3.1. As such it is left mostly as an exercise. We sample the interesting case here:

   Case VIII. $f = \langle 8, k+1, l \rangle$.

   What can we directly conclude from $\{f\}(r, \vec{x}, \vec{\alpha}) \simeq z$? We conclude that we must have $\{r\}(\vec{x}, \vec{\alpha}) \simeq z$ by the technique of Proposition 13.3.10, case VII, and Proposition 13.6.7, case IX.

   By the I.H. on subcomputations we obtain

   $$\{\tau(r)\}_v(\vec{x}, \vec{\alpha}) \simeq z \tag{6}$$

   By (0), (‡) and definition of τ, we have

   $$\{\tau(f)\}_v(r, \vec{x}, \vec{\alpha}) \simeq \{g(f, \bar{t})\}_v(r, \vec{x}, \vec{\alpha}) \simeq \{Sb_1(d(f, \bar{t}), f)\}_v(r, \vec{x}, \vec{\alpha})$$
   $$\simeq \{d(f, \bar{t})\}_v(f, r, \vec{x}, \vec{\alpha})$$
   $$\simeq \{\{\bar{t}\}_v(r)\}_v(\vec{x}, \vec{\alpha}) \tag{7}$$
   $$\simeq \{\tau(r)\}_v(\vec{x}, \vec{\alpha})$$
   $$\simeq z \text{ (by (6))}$$

   Thus we have shown that the hypothesis of Case VIII —$\{f\}(r, \vec{x}, \vec{\alpha}) \simeq z$— implies $\{\tau(f)\}_v(r, \vec{x}, \vec{\alpha}) \simeq z$, concluding the induction.

2. $\leftarrow$ direction. Induction on computation lengths. Only the case VIII is nontrivial. So consider the case (VIII) where $f = \langle 8, k+1, l \rangle$ and $\{f\}(r, \vec{x}, \vec{\alpha}) \simeq \{r\}(\vec{x}, \vec{\alpha})$. Let then $\{\tau(f)\}_v(r, \vec{x}, \vec{\alpha}) \simeq z$. Thus also, by (7),

   $$\{\tau(r)\}_v(\vec{x}, \vec{\alpha}) \simeq z \tag{8}$$

   We need to show that $(f, r, \vec{x}, \vec{\alpha}, z) \in \Omega$ or $\{f\}(r, \vec{x}, \vec{\alpha}) \simeq z$. From (5) and (1′) we have

   $$\| \overbrace{\{\bar{t}\}_v(r)}^{\tau(r)}, \vec{x}, \vec{\alpha}, z \| < \| \overbrace{g(f, \bar{t})}^{\tau(f)}, r, \vec{x}, \vec{\alpha}, z \|$$

By the I.H. on computation lengths and (8) we have $\{r\}(\vec{x}, \vec{\alpha}) \simeq z$. Hence also $\{f\}(r, \vec{x}, \vec{\alpha}) \simeq z$ by rule VIII of Ω, Definition 13.3.1.

□

The substitution into a type-1 variable theorem is valid in search computability with a proof only trivially augmented vis a vis the one for Kleene computability that we saw as Theorem 13.3.36.

13.6.19 Theorem. *There is a rank* $(2, 0)$ *recursive function(al) h such that if we have two search computable p.m.v. functionals* $\lambda \vec{x} \vec{\alpha} \beta.\{f\}(\vec{x}, \vec{\alpha}, \beta)$ *and* $\lambda y.\{e\}(y, \vec{\alpha})$, *then*

$$\{f\}\left(\vec{x}, \vec{\alpha}, \lambda y.\{e\}(y, \vec{\alpha})\right) \simeq z \text{ iff } \{h(f, e)\}(\vec{x}, \vec{\alpha}) \simeq z \tag{‡}$$

Proof The definition of $g(f, e, t)$ is as in the proof of Theorem 13.3.36 but here we add a case for $f = \langle 9, k, l, q \rangle$: $g(f, e, t) = \langle 9, k, l, g(q, e, t) \rangle$. The complicated segment in the proof that the construction works is, as before, the case of formation rule VIII but it does not differ from that in Theorem 13.3.36. The details are left as an exercise.

□

13.6.20 Corollary. *There is a rank* $(1 + n, 0)$ *recursive function(al) h such that if we have two search computable p.m.v. functionals* $\lambda \vec{x} \vec{\alpha} \beta.\{f\}(\vec{x}, \vec{\alpha}, \beta)$ *and* $\lambda y.\{e\}(y, \vec{x}, \vec{\alpha})$, *then*

$$\{f\}\left(\vec{x}, \vec{\alpha}, \lambda y.\{e\}(y, \vec{x}, \vec{\alpha})\right) \simeq z \text{ iff } \{h(f, e)\}(\vec{x}, \vec{\alpha}) \simeq z$$

Proof As in Corollary 13.3.37.

□

13.6.21 Definition. (Definition by Positive Cases) (Cf. also Theorem 6.8.1) We say that the p.m.v. functional $\mathscr{F}$ is defined by the positive cases $\mathscr{G}_i(\vec{x}, \vec{\alpha}) \simeq 0$ from the p.m.v. functionals $\mathscr{H}_i$ —for $i = 1, 2, \ldots, k$— iff

$$\mathscr{F}(\vec{x}, \vec{\alpha}) \simeq u \equiv$$

$$\left(\mathscr{H}_1(\vec{x}, \vec{\alpha}) \simeq u \wedge \mathscr{G}_1(\vec{x}, \vec{\alpha}) \simeq 0 \right) \vee \ldots \vee \left(\mathscr{H}_k(\vec{x}, \vec{\alpha}) \simeq u \wedge \mathscr{G}_k(\vec{x}, \vec{\alpha}) \simeq 0 \right) \tag{1}$$

□

13.6.22 Remark. The "normal" way to write the definition in a user-friendly way is as in Theorem 6.8.1

$$\mathscr{F}(\vec{x}, \vec{\alpha}) \simeq \begin{cases} \mathscr{H}_1(\vec{x}, \vec{\alpha}) & \text{if } \mathscr{G}_1(\vec{x}, \vec{\alpha}) \simeq 0 \\ \vdots & \vdots \\ \mathscr{H}_k(\vec{x}, \vec{\alpha}) & \text{if } \mathscr{G}_k(\vec{x}, \vec{\alpha}) \simeq 0 \end{cases} \tag{2}$$

Since we only have the *positive* cases —$\mathscr{G}_i(\vec{x}, \vec{\alpha}) \simeq 0$— drive the outcomes, the "otherwise", i.e., when none of the positive cases is true, is left undefined. This was explicit in Theorem 6.8.1 with a "↑ otherwise" case.

The reader should *not* take (2) to mean (computationally) that we evaluate the $\mathcal{G}_i(\vec{x}, \vec{\alpha}) \simeq 0$ in turn ($i = 1, \dots, k$) until we find one that works. With partial functions this will *not* work as we can get stuck in a call $\mathcal{G}_i(\vec{x}, \vec{\alpha})$, and yet some $\mathcal{G}_j(\vec{x}, \vec{\alpha}) \simeq 0$, for $j > i$, is true. One must imagine that we *nondeterministically pick* a case (or more cases, if we allow $\mathcal{F}$ to be multiple-valued) that is true. Indeed that is what the proof below does with the help of ν.   □

13.6.23 Theorem. *If in (1), Definition 13.6.21, all the $\mathcal{G}_i$ and $\mathcal{H}_i$ are search computable p.m.v. functionals, then $\mathcal{F}$ is also a search computable p.m.v. functional.*

Proof The relation

$$R(u, \vec{h}_k, \vec{g}_k) \overset{Def}{\equiv} \left(h_1 = u \wedge g_1 = 0\right) \vee \dots \vee \left(h_k = u \wedge g_k = 0\right)$$

is absolutely primitive recursive, hence for some (absolute) primitive recursive r, we have

$$r(u, \vec{h}_k, \vec{g}_k) = 0 \equiv \left(h_1 = u \wedge g_1 = 0\right) \vee \dots \vee \left(h_k = u \wedge g_k = 0\right)$$

Thus, by closure under composition, $\lambda u \vec{x} \vec{\alpha}. r\left(u, \vec{\mathcal{H}_k}(\vec{x}, \vec{\alpha}), \vec{\mathcal{G}_k}(\vec{x}, \vec{\alpha})\right)$ is search computable and so is, by Theorem 13.6.11,

$$\mathcal{F} = \lambda \vec{x} \vec{\alpha}.(\nu u)\left(r\left(u, \vec{\mathcal{H}_k}(\vec{x}, \vec{\alpha}), \vec{\mathcal{G}_k}(\vec{x}, \vec{\alpha})\right) \simeq 0\right)$$ □

13.6.24 Corollary. *If $\mathcal{F}$ is given by semi-recursive cases $\mathbb{P}_i(\vec{x}, \vec{\alpha})$ and search computable p.m.v. $\mathcal{H}_i$ as follows*

$$\mathcal{F}(\vec{x}, \vec{\alpha}) \simeq \begin{cases} \mathcal{H}_1(\vec{x}, \vec{\alpha}) & \text{if } \mathbb{P}_1(\vec{x}, \vec{\alpha}) \\ \vdots & \vdots \\ \mathcal{H}_k(\vec{x}, \vec{\alpha}) & \text{if } \mathbb{P}_k(\vec{x}, \vec{\alpha}) \end{cases}$$

then it is search computable.

Proof This reduces to the previous proof by using the search computable partial characteristic functionals of the $\mathbb{P}_i$ relations in the cases.   □

 13.6.25 Example. In search computability, semi-recursive relations are closed under $\vee$ (compare and contrast with Example 13.3.35, (4)). We show that if $\mathbb{P}(\vec{x}, \alpha)$ and $\mathbb{Q}(\vec{x}, \alpha)$ are search computable, then so is $\mathbb{P}(\vec{x}, \alpha) \vee \mathbb{Q}(\vec{x}, \alpha)$.

Indeed, the following is a search computable partial characteristic function of the disjunction.

$$\mathcal{F}(\vec{x}, \vec{\alpha}) \simeq \begin{cases} 0 & \text{if } \mathbb{P}(\vec{x}, \vec{\alpha}) \\ 0 & \text{if } \mathbb{Q}(\vec{x}, \vec{\alpha}) \end{cases}$$

□

13.6.1 The Normal Form Theorem, Again

In this subsection we present the normal form theorem for search computable functionals. The main novelty is the use of selection, v, and an extra segment in the definition of $Tree(y)$ that starts on p.445) that we will call $S(u)$ to take care of the selection clause IX in Ω^{sel} (Definition 13.6.1). We add this sub predicate $S(y)$ in the definition of $Tree$ which we will now call $Tree^{sel}$. See figure below.

$$e = \langle 9, n, l, f \rangle, \qquad \langle e, \vec{a}_n, \vec{\alpha}_l, b \rangle$$

$$\langle f, b, \vec{a}_n, \vec{\alpha}_l, 0 \rangle$$

Case for IX

The $S(u)$: Similarly to $U(y)$ (for the Ω-case VIII) we have $y = \langle v, t_0 \rangle$ — one subtree of code t_0— where (cf. the figure above) $v = \langle \langle 9, n, l, f \rangle \rangle * a * s * \langle b \rangle$ and the root of t_0 is $(t_0)_0 = (y)_{1,0} = \langle f, b \rangle * a * s * \langle 0 \rangle$, where as above we set $a = \langle \vec{a}_n \rangle$ and $s = \langle s_l \rangle$.

Thus,

$$
\begin{aligned}
\{\text{Fig. above}\}\ S(y) \equiv (\exists n, f, a, b, s, l)_{<y} \Big\{ lh(y) &= 2 \wedge Seq(f) \wedge \\
Seq((y)_1) &\wedge Seq(s) \wedge lh(s) = l\ \wedge \\
(f)_1 &= n + 1 \wedge n > 0\ \wedge \\
Seq(a) &\wedge lh(a) = n\ \wedge \\
\{\text{The root call}\} \qquad (y)_0 &= \langle \langle 9, n, l, f \rangle \rangle * a * s * \langle b \rangle\ \wedge \\
\{\text{The recursive call}\} \quad (y)_{1,0} &= \langle f, b \rangle * a * s * \langle 0 \rangle \Big\}
\end{aligned}
$$

We can now conclude the course-of-values recursive definition of the $Tree$ predicate, $Tree^{sel}(y)$, for search computability (dynamic) computations.

$$Tree^{sel}(y) \equiv P(y) \wedge \Big(I(y) \vee C(y) \vee U(y) \vee S(y) \Big) \wedge (\forall i)_{<lh(y)}(i > 0 \rightarrow Tree^{sel}((y)_i)$$

Just as in the case of $Tree$, it is clear by the course-of-values recursion above, that $Tree^{sel}(y)$ is absolutely primitive recursive (is in $\mathcal{PR}_*$).

The Kleene predicates (form I and form II) for search computability with type-0 and type-1 inputs are defined as in Definition 13.4.3 and Theorem 13.4.5 and we will denote them by $T^{sel}_{(k,l)}$ and T^{sel}_l respectively. They are both easily seen to be primitive recursive.

13.6.26 Definition. (Kleene Predicate for Search Computability)

For any given n and l we define the *Kleene predicate* of rank (n, l) for search computability by

$$T^{sel}_{(n,l)}(z, \vec{x}_n, \vec{\alpha}_l, y) \overset{Def}{\equiv} Tree^{sel}(y) \wedge (y)_{0,0} = z \wedge (y)_{0,1} = \langle \vec{x}_n \rangle \wedge lh\big((y)_{0,2}\big) = l \wedge$$

$$(\forall i)_{<lh((y)_{0,2})}[(y)_{0,2,i}] \subseteq \alpha_{i+1}$$

Since $(y)_0 = \langle z, \langle \vec{x}_n \rangle, \langle \vec{s}_l \rangle, b \rangle$, where $\langle \vec{s}_l \rangle$ is a vector of approximations to $\vec{\alpha}_l$ during the computation and b is the output of the computation; we again have that $b = (y)_{0,3}$. We thus define $d = \lambda \vec{x}_n \vec{\alpha}_l.(y)_{0,3}$, the *decoding function* of the computation y. □

The respective Normal Form theorems are hence (with proofs as before omitted):

13.6.27 Theorem. (Kleene Normal Form I; Search Computability) *There is an absolute primitive recursive function d of one argument, and, for any given n and l, a primitive recursive predicate $T^{sel}_{(n,l)}(z, \vec{x}_n, \vec{\alpha}_l, y)$ as defined above such that*

(a) $\{z\}^{(n,l)}_v(\vec{x}_n, \vec{\alpha}_l) \downarrow \equiv (\exists y) T^{sel}_{(n,l)}(z, \vec{x}_n, \vec{\alpha}_l, y)$

(b) $\{z\}^{(n,l)}_v(\vec{x}_n, \vec{\alpha}_l) \simeq u \equiv (\exists y)\Big(T^{sel}_{(n,l)}(z, \vec{x}_n, \vec{\alpha}_l, y) \wedge d(y) = u\Big)$

(c) $\{z\}^{(n,l)}_v(\vec{x}_n, \vec{\alpha}_l) \simeq d\Big((vy) T^{sel}_{(n,l)}(z, \vec{x}_n, \vec{\alpha}_l, y)\Big)$

Proof The primitive recursiveness of $T^{sel}_{(n,l)}$ follows from that of $Tree^{sel}(y)$ and $\lambda s \alpha.[s] \subseteq \alpha$ (cf. Example 13.3.35, (3), (4), (6), (7)).

All the proof work for (a)–(b) is already done. (c) follows from (a) and Theorem 13.6.11, (1), (vy) picking *all* y that work in $(\exists y) T^{sel}_{(n,l)}(z, \vec{x}_n, \vec{\alpha}_l, y)$ iff any such exist. d decodes each of the selected y to get the (possibly) multiple outputs of the left hand side. Of course, $T^{sel}_{(n,l)}$ has a primitive recursive partial characteristic function (this is total for total α_j) $\mathcal{K}_T$, which is also search computable. □

13.6.28 Corollary. (Relational Normal Form I; Search Computability) $\mathbb{P}(\vec{x}_n, \vec{\alpha}_l)$ *is semi-recursive in search computability iff we have*

$$\mathbb{P}(\vec{x}_n, \vec{\alpha}_l) \equiv (\exists y) T^{sel}_{(n,l)}\big(z, \vec{x}_n, \vec{\alpha}_l, y\big) \tag{1}$$

where z is a semi-index of $\mathbb{P}$.

Proof As in Corollary 13.4.7. □

We have the extremely useful two corollaries below.

13.6.29 Corollary. (Type-1 Argument List Expansion) *Corollary 13.4.9 holds in search computability.*

13.6.30 Corollary. *Corollary 13.4.10 holds in search computability.*

13.6.31 Theorem. (Closure Properties of Semi-Recursive Relations) *Semi-recursive relations in search computability are closed under*

$$\wedge, \vee, (\exists y), (\exists y)_{<z}, (\forall y)_{<z}$$

Proof

(1) Let $\mathbb{P}(\vec{x}, \vec{\alpha})$ and $\mathbb{Q}(\vec{y}, \vec{\beta})$ be semi-recursive. Let $\mathscr{K}_P$ and $\mathscr{K}_Q$ be partial characteristic functionals for them. Then $\mathbb{P}(\vec{x}, \vec{\alpha}) \wedge \mathbb{Q}(\vec{y}, \vec{\beta})$ has $\mathscr{K}_P(\vec{x}, \vec{\alpha}) + \mathscr{K}_Q(\vec{y}, \vec{\beta})$ as a partial characteristic functional. Now invoke Corollary 13.6.30.

(2) With $\mathbb{P}$ and $\mathbb{Q}$ and $\mathscr{K}_P$, $\mathscr{K}_Q$ given as above we define

$$\mathscr{K}_{P \vee Q}(\vec{x}, \vec{y}, \vec{\alpha}, \vec{\beta}) = \begin{cases} 0 & \text{if } \mathscr{K}_P(\vec{x}, \vec{\alpha}) \\ 0 & \text{if } \mathscr{K}_Q(\vec{y}, \vec{\beta}) \end{cases}$$

We conclude by Example 13.6.25 and Corollary 13.6.30.

(3) Let i be a semi-index of $\mathbb{P}$ in search computability. By Corollary 13.6.28 we have $\mathbb{P}(y, \vec{x}_n, \vec{\alpha}_l) \equiv (\exists w) T^{sel}_{(n+1,l)}(i, y, \vec{x}_n, \vec{\alpha}_l, w)$ hence (using $\Leftrightarrow$ conjunctionally)

$$(\exists y)\mathbb{P}(y, \vec{x}_n, \vec{\alpha}_l) \Leftrightarrow (\exists y)(\exists w) T^{sel}_{(n+1,l)}(i, y, \vec{x}_n, \vec{\alpha}_l, w)$$
$$\Leftrightarrow (\exists u) T^{sel}_{(n+1,l)}(i, (u)_0, \vec{x}_n, \vec{\alpha}_l, (u)_1)$$
$$\Leftrightarrow \left((\nu u) T^{sel}_{(n+1,l)}(i, (u)_0, \vec{x}_n, \vec{\alpha}_l, (u)_1) \right) \downarrow$$

(4) Let $\mathbb{P}$ be semi-recursive in search computability. Since

$$(\exists y)_{<z}\mathbb{P}(y, \vec{x}, \vec{\alpha}) \equiv (\exists y)\left(y < z \wedge \mathbb{P}(y, \vec{x}, \vec{\alpha}) \right)$$

We are done since semi-recursive relations are closed under $\wedge$ (same proof as that in Example 13.3.35,(4)) and $(\exists y)$.

(5) Let i be a semi-index of $\mathbb{P}$ in search computability. We have

$$(\forall y)_{<z}\mathbb{P}(y, \vec{x}, \vec{\alpha}) \Leftrightarrow (\forall y)_{<z}(\exists w) T^{sel}_{(n+1,l)}(i, y, \vec{x}, \vec{\alpha}, w)$$
$$\Leftrightarrow (\exists u)(\forall y)_{<z} T^{sel}_{(n+1,l)}(i, y, \vec{x}, \vec{\alpha}, (u)_y)$$
$$\Leftrightarrow \left((\nu u)(\forall y)_{<z} T^{sel}_{(n+1,l)}(i, y, \vec{x}, \vec{\alpha}, (u)_y) \right) \downarrow$$

the last equivalence from the fact that after the "(νu)" we have a primitive recursive, hence semi-recursive relation (Example 13.3.35,(6)) and by Theorem 13.6.11, (1).

□

13.6.32 Corollary. (Kleene Normal Form II; Search Computability) *For each* $l \geq 0$, *there exists an absolute* primitive recursive *predicate* T_l^{sel} *of rank* $(4, 0)$ *such that for each search computable functional of rank* (n, l) *we have*

(i) $\{z\}_v^{(n,l)}(\vec{x}_n, \vec{\alpha}_l) \downarrow \equiv (\exists y)(\exists \vec{s}_l)_{<y}\left(T_l^{sel}(z, \langle \vec{x}_n \rangle, \langle \vec{s}_l \rangle, y) \wedge \bigwedge_{1 \leq j \leq l}[s_j] \subseteq \alpha_j\right).$

(ii)

$$\{z\}_v^{(n,l)}(\vec{x}_n, \vec{\alpha}_l) \simeq u \equiv (\exists y)(\exists \vec{s}_l)_{<y}\left(T_l^{sel}(z, \langle \vec{x}_n \rangle, \langle \vec{s}_l \rangle, y) \wedge d(y) = u\right.$$

$$\left. \wedge \bigwedge_{1 \leq j \leq l}[s_j] \subseteq \alpha_j\right)$$

where the empty conjunction (case of no oracle variables) —when $l = 0$— is true.

(iii) $\{z\}_v^{(n,l)}(\vec{x}_n, \vec{\alpha}_l) \simeq d\left((vy)(\exists \vec{s}_l)_{<y}\left(T_l^{sel}(z, \langle \vec{x}_n \rangle, \langle \vec{s}_l \rangle, y) \wedge \bigwedge_{1 \leq j \leq l}[s_j] \subseteq \alpha_j\right)\right).$

Proof Define T_l^{sel} as

$$T_l^{sel}(z, x, s, y) \overset{Def}{\equiv} Tree^{sel}(y) \wedge$$

$$(y)_{0,0} = z \wedge (y)_{0,1} = x \wedge (y)_{0,2} = s \wedge lh(s) = l$$

The claims (*i*) and (*ii*) follow directly from the definition of $Tree^{sel}(y)$ (cf. also Theorem 13.6.27). (*iii*) follows from (*i*) and Theorem 13.6.11, (1), via Theorem 13.6.31, (4) that guarantees the semi-recursiveness of the part following "(vy)".   □

13.6.33 Corollary. (Compactness and Monotonicity) *Search computable p.m.v. functionals are compact and monotone.*

Proof Exercise.   □

13.6.34 Corollary. (Relational Normal Form II; Search Computability) $\mathbb{P}(\vec{x}_n, \vec{\alpha}_l)$ *is semi-recursive iff we have*

$$\mathbb{P}(\vec{x}_n, \vec{\alpha}_l) \equiv (\exists y)(\exists \vec{s}_l)_{<y}\left(T_l(z, \langle \vec{x}_n \rangle, \langle \vec{s}_l \rangle, y) \wedge \bigwedge_{1 \leq j \leq l}[s_j] \subseteq \alpha_j\right) \qquad (2)$$

where z is a semi-index of $\mathbb{P}$.

Proof (2) is true because the right hand side of $\equiv$ above says $\{z\}^{(n,l)}(\vec{x}_n, \vec{\alpha}_l) \downarrow$ by Corollary 13.6.32, item (*i*).   □

13.6.35 Theorem. (The Graph Theorem) $\mathscr{F}$ *is search computable iff its graph,* $\mathscr{F}(\vec{x}, \vec{\alpha}) \simeq u$, *is semi-recursive.*

Proof

1. *(If.)* Say $\mathscr{F}(\vec{x}, \vec{\alpha}) \simeq u$ is semi-recursive. Then

$$\mathscr{F}(\vec{x}, \vec{\alpha}) \simeq (vu)(\mathscr{F}(\vec{x}, \vec{\alpha}) \simeq u)$$

Pause. Would $(\mu u)(\mathscr{F}(\vec{x}, \vec{\alpha}) \simeq u)$ also work? ◄
No, not in general. While $\lambda \vec{x} \vec{\alpha}(\mu u)(\mathscr{F}(\vec{x}, \vec{\alpha}) \simeq u)$ is partial recursive, recall that μu may fail to find the u we want. Say we are working with a partial characteristic functional $\mathscr{K}$ such that $\mathscr{K}(\vec{x}, \vec{\alpha}, u) \simeq 0 \equiv \mathscr{F}(\vec{x}, \vec{\alpha}) \simeq u$. But it may happent that for some output values u we have that $(\exists z < u)(\mathscr{K}(\vec{x}, \vec{\alpha}, z) \uparrow)$.
2. *(Only if.)* Say $\mathscr{F} = \{z\}_v$. By Theorem 13.6.27, (b) and the fact that we have a primitive recursive (hence semi-recursive) relation existentially quantified to the right hand side of $\equiv$. Now apply Theorem 13.6.31, (3). □

13.6.36 Theorem. *Davis-computable functionals (13.5.1) are search computable. Conversely, a single-valued search computable functional is D-computable.*

Proof → direction. Say $\mathscr{F}$ is Davis-computable. Then, for some rank $(n + l, 0)$ partial recursive function ψ we have that $\mathscr{F}$ satisfies

$$\mathscr{F}(\vec{x}_n, \vec{\alpha}_l) \simeq u \equiv (\exists \vec{s}_l) \left(\psi(\vec{x}_n, \vec{s}_l) \simeq u \wedge \bigwedge_{1 \leq j \leq l} [s_j] \subseteq \alpha_j \right) \tag{1}$$

The predicate following "$(\exists \vec{s}_l)$" in (1) is semi-recursive in the Kleene sense since $[s_j] \subseteq \alpha_j$ is primitive recursive, $\psi(\vec{x}_n, \vec{s}_l) \simeq u$ is absolutely semi-recursive and semi-recursive relations (in Kleene computability) are closed under $\wedge$. Thus the said predicate is also semi-recursive in search computability and so is the quantified part by Theorem 13.6.31, (3). We conclude via the graph theorem 13.6.35.
← direction. Cf. Proposition 13.5.2. □

Thus Davis-computability includes Kleene computable functionals but is a sub theory of *search computability*.

Next let us revisit Theorem 13.6.19 via the Normal Form theorem.

13.6.37 Theorem. *There is a rank $(2, 0)$ recursive function(al) h such that if we have two search computable p.m.v. functionals $\lambda \vec{x} \vec{\alpha} \beta.\{f\}(\vec{x}, \vec{\alpha}, \beta)$ and $\lambda y.\{e\}(y, \vec{x}, \vec{\alpha})$, then*

$$\{f\}\left(\vec{x}, \vec{\alpha}, \lambda y.\{e\}(y, \vec{x}, \vec{\alpha})\right) \simeq z \text{ iff } \{h(f, e)\}(\vec{x}, \vec{\alpha}) \simeq z \tag{‡}$$

Proof   By Corollary 13.6.32, (iii), we have

$$\{f\}(\vec{x}_n, \vec{\alpha}_l, \beta) \simeq d\Bigg((\nu y)(\exists s, \vec{s}_l)_{<y}\Big(T_l^{sel}\big(f, \langle \vec{x}_n \rangle, \langle \vec{s}_l, s \rangle, y\big) \wedge$$

$$[s] \subseteq \beta \wedge \tag{13.1}$$

$$\bigwedge_{1 \leq j \leq l} [s_j] \subseteq \alpha_j\Big)\Bigg)$$

Replacing β by $\lambda y.\{e\}(y, \vec{x}_n, \vec{\alpha}_l)$ we obtain

$$\{f\}(\vec{x}_n, \vec{\alpha}_l, \{e\}(y, \vec{x}_n, \vec{\alpha}_l)) \simeq d\Bigg((\nu y)(\exists \vec{s}_l)_{<y}\Big[\overbrace{T_l^{sel}\big(f, \langle \vec{x}_n \rangle, \langle \vec{s}_l \rangle, y\big)}^{prim.\ rec.} \wedge$$

$$\Big\{ \underbrace{[s] \subseteq \lambda y.\{e\}(y, \vec{x}_n, \vec{\alpha}_l)}_{semi-rec.} \Big\} \wedge$$

$$\bigwedge_{1 \leq j \leq l} [s_j] \subseteq \alpha_j\Big]\Bigg)$$

$$\underbrace{}_{prim.\ rec.}$$

$$\overset{S-m-n}{\simeq} \{h(f, e)\}(\vec{x}_n, \vec{\alpha}_l)$$

where the middle conjunct in the $\{\dots\}$-brackets above is semi-recursive by Example 13.3.35, (7) as it expands to $s \geq 1 \wedge (\forall y)_{\leq u}\Big(p_y \mid s \rightarrow (s)_y = \{e\}(y, \vec{x}_n, \vec{\alpha}_l)\Big)$. $\square$

The following consequence of the normal form theorems has no counterpart in absolute (oracle-less) computability.

13.6.38 Proposition. (($\exists \alpha$) **Closure**)   *Semi-recursive relations are closed under type-1 existential quantification.*

Proof   Let e be a semi-index of $\mathbb{P}(\vec{x}, \beta, \vec{\alpha})$. By the Corollary 13.6.34 we have

$$\mathbb{P}(\vec{x}_n, \beta, \vec{\alpha}_l) \equiv (\exists y)(\exists s, \vec{s}_l)\Big(T_l^{sel}(e, \langle \vec{x}_n \rangle, \langle s, \vec{s}_l \rangle, y) \wedge [s] \subseteq \beta \wedge \bigwedge_{1 \leq j \leq l} [s_j] \subseteq \alpha_j\Big)$$

We claim that

$$(\exists \beta)\mathbb{P}(\vec{x}_n, \beta, \vec{\alpha}_l) \equiv (\exists y)(\exists s, \vec{s}_l)\Big(T_l^{sel}(e, \langle \vec{x}_n \rangle, \langle s, \vec{s}_l \rangle, y) \wedge$$

$$\bigwedge_{1 \leq j \leq l} [s_j] \subseteq \alpha_j\Big) \tag{1}$$

Indeed, assume the left hand side of $\equiv$ in (1). Then fix a β that verifies $\mathbb{P}(\vec{x}_n, \beta, \vec{\alpha}_l)$ and thus by compactness, we have an s such that $[s] \subseteq \beta$ such that

$$\mathbb{P}(\vec{x}_n, [s], \vec{\alpha}_l) \equiv (\exists y)(\exists s, \vec{s}_l)\Big(T_l^{sel}(e, \langle \vec{x}_n \rangle, \langle s, \vec{s}_l \rangle, y) \wedge$$

$$[s] \subseteq [s] \tag{2}$$

$$\bigwedge_{1 \le j \le l} [s_j] \subseteq \alpha_j\Big)$$

It follows that the right hand side of $\equiv$ in (2), and hence in (1), are true.

Conversely, assume the right hand side of (1). Fix an s that works. Thus the right hand side says the same thing as the right hand side of (2). Therefore, $\mathbb{P}(\vec{x}_n, [s], \vec{\alpha}_l)$ is true and hence there is a $\beta = [s]$ such that $\mathbb{P}(\vec{x}_n, \beta, \vec{\alpha}_l)$ is true.

At the end of all this we also note that the right hand side of (1) is semi-recursive (Theorem 13.6.31) and we are done. $\square$

We conclude this section with another "staple ingredient" of computability, that if a predicate and its negation are both semi-recursive then the predicate is recursive. But first, let us recall

13.6.39 Definition. (Recursive Predicates in Search Computability) Cf. Definition 13.3.33. A relation $\mathbb{P}(\vec{x}, \vec{\alpha})$ is *recursive* iff it has a recursive *partial characteristic functional* $\mathcal{H}_P$, that is, this functional is search computable and *total* in the sense of Definition 13.1.1. $\square$

13.6.40 Theorem. *If $\mathbb{P}(\vec{x}, \vec{\alpha})$ and $\neg \mathbb{P}(\vec{x}, \vec{\alpha})$ are both semi-recursive, then $\mathbb{P}(\vec{x}, \vec{\alpha})$ is recursive.*

Proof We will define a provably recursive partial characteristic functional of $\mathbb{P}(\vec{x}, \vec{\alpha})$.

By assumption we have i and j such that (Corollary 13.6.28)

$$\mathbb{P}(\vec{x}_n, \vec{\alpha}_l) \equiv (\exists y)T_{(n,l)}^{sel}(i, \vec{x}_n, \vec{\alpha}_l, y)$$

$$\neg \mathbb{P}(\vec{x}_n, \vec{\alpha}_l) \equiv (\exists y)T_{(n,l)}^{sel}(j, \vec{x}_n, \vec{\alpha}_l, y)$$

Define

$$\mathcal{H} = \lambda \vec{x}\vec{\alpha}.(\nu y)\Big(T_{(n,l)}^{sel}(i, \vec{x}_n, \vec{\alpha}_l, y) \vee T_{(n,l)}^{sel}(j, \vec{x}_n, \vec{\alpha}_l, y)\Big) \tag{1}$$

$\mathcal{H}$ is search computable and is *total* for total $\vec{\alpha}$ since by the semantics of the disjunction in (1) a y-value searched for by (νy) always exists. So $\mathcal{H}$ is recursive. Now note

$$\mathbb{P}(\vec{x}_n, \vec{\alpha}_l) \equiv T_{(n,l)}^{sel}\Big(i, \vec{x}, \vec{\alpha}, \mathcal{H}(\vec{x}, \vec{\alpha})\Big)$$

But the right hand side above is recursive. $\square$

13.7   The First Recursion Theorem: Fixpoints of Computable Functionals

The first recursion theorem or fixpoint theorem states the existence of a *least computable fixed point* of a function-equation —with "unknown" α— such as

$$\alpha(x) \simeq \mathscr{F}(\vec{x}, \alpha), \text{ for all } x$$

under some reasonable restrictions for $\mathscr{F}$. We will look at the theorem for various types of computable functionals. Makes sense to start from the most general type of such functionals and proceed to special cases where it also holds. Note that the first recursion theorem, unlike the second (9.2.1) does not refer to indices.

13.7.1   Weakly Computable Functionals Version

13.7.1 Definition. (Weakly Computable Functionals) A *weakly partial recursive functional*

$$\mathscr{F} : \mathbb{N} \times \mathcal{P}(\mathbb{N} : \mathbb{N}) \to \mathbb{N}$$

is one that on computable type-1 inputs of the form $\{e\}$ it rather *computes* on the "program" e of $\{e\}$ and on the type-0 input x. It behaves as it pleases (i.e., we do *not* care how) on non computable type-1 inputs. Thus, for some $f \in \mathbb{N}$, the behaviour of $\mathscr{F}$ is defined by

$$\mathscr{F}(x, \{e\}) \simeq u \equiv \{f\}(e, x) \simeq u$$

The concept can be easily extended to functionals of many variables, that is, to functionals $\mathscr{G} : \mathbb{N}^k \times \left(\mathcal{P}(\mathbb{N} : \mathbb{N})\right)^l \to \mathbb{N}$; but we will not do so.   □

 13.7.2 Remark. Every partial recursive functional in each of Kleene's and Moschovakis's sense is also weakly partial recursive. For example, using Theorem 13.3.36 (Kleene computability) or Theorem 13.6.19 (Moschovakis computability), for a partial recursive functional $\lambda x \alpha.\{e\}(x, \alpha)$ we get a recursive h such that

$$\{e\}(x, \lambda y.\{f\}(y)) \simeq u \equiv \{h(e, f)\}(x) \simeq u \tag{1}$$

Applying the universal function theorem (say, with index r) followed by a computation of an index k for the substitution $\{r\}(h(e, f), x)$ we can rewrite the right hand side of (1) as $\{k\}(e, f, x)$ thus (1) becomes

$$\{e\}(x, \lambda y.\{f\}(y)) \simeq u \equiv \{k\}(e, f, x) \simeq u \tag{1'}$$

We note that the functional $\mathscr{E} = \{e\}$ is weakly partial recursive according to the definition.[22]

The converse of this remark is not true. See Remark 13.7.7.   □

13.7.3 Theorem. (The First Recursion Theorem) (For weakly computable functionals) *For every* monotone *weakly partial recursive functional* $\lambda x \alpha. \mathscr{F}(x, \alpha)$, *the equation*

$$\alpha(x) \simeq \mathscr{F}(x, \alpha)$$

has a least *partial (absolutely) recursive solution* α *called a* fixed point *or* fixpoint, *"least" meaning that if* β *is another fixpoint, then* $\alpha \subseteq \beta$.

Proof   The least solution is found via a sequence of approximations.

$$\text{Set } \alpha_0 = \emptyset$$

$$\text{Set } \alpha_{n+1}(x) \simeq \mathscr{F}(x, \alpha_n)$$

Next we show that $\alpha_n \subseteq \alpha_{n+1}$ for all $n \geq 0$.
     For $n = 0$ (*Basis*) this is trivial.
     Taking the I.H. $\alpha_n \subseteq \alpha_{n+1}$ for a fixed n we next examine the $n + 1$ case: By *assumed monotonicity* we have that $\alpha_{n+1}(x) \simeq \mathscr{F}(x, \alpha_n) \downarrow \simeq y$ implies

$$\mathscr{F}(x, \alpha_n) \simeq \mathscr{F}(x, \alpha_{n+1}) \downarrow \simeq \alpha_{n+2}(x) \simeq y$$

Thus, $\alpha_{n+1} \subseteq \alpha_{n+2}$. Thus define

$$\alpha \stackrel{Def}{=} \bigcup_n \alpha_n \tag{1}$$

By the above sub result, α is single-valued. Moreover,

$$\alpha(x) \simeq u \stackrel{(1)}{\Rightarrow} (\text{for some } n)\, \alpha_{n+1}(x) \simeq u \Rightarrow \mathscr{F}(x, \alpha_n) \simeq u \stackrel{mon}{\Rightarrow} \mathscr{F}(x, \alpha) \simeq u$$

Having established the α of (1) as a fixpoint of $\mathscr{F}$, assume that β is also a fixpoint. We show that $\alpha \subseteq \beta$ establishing the constructed α as *the least* such.

---

[22] Were we to ignore the (fixed) index of $\mathscr{E}$ that caused (1'), then correspondingly we would rewrite (1') as

$$\mathscr{E}(x, \lambda y.\{f\}(y)) \simeq u \equiv \{k'\}(f, x) \simeq u \tag{1''}$$

(1'') totally fits the form of Definition 13.7.1.

Idea. Why not prove that $\alpha_n \subseteq \beta$, for all $n \geq 0$. Then we would immediately conclude by (1).◀

Let's do so: (*Basis*). That $\alpha_0 \subseteq \beta$ is trivial.

Assume (I.H.) that for a fixed n it is $\alpha_n \subseteq \beta$.

The case for $\alpha_{n+1} \subseteq \beta$.

We have that

$$\alpha_{n+1}(x) \simeq u \Rightarrow \mathscr{F}(x, \alpha_n) \simeq u \stackrel{I.H. + mon}{\Rightarrow} \mathscr{F}(x, \beta) \simeq u$$

By (1) and $(\forall n)(\alpha_n \subseteq \beta)$ proved, we have $\alpha = \bigcup_{n \geq 0} \alpha_n \subseteq \beta$.

Finally we show that $\alpha \in \mathcal{P}$. By Definition 13.7.1 there is an $f \in \mathbb{N}$ such that

$$\mathscr{F}(x, \lambda y.\{e\}(y)) \simeq \{f\}(e, x) \stackrel{\text{S-m-n}}{\simeq} h(e)(x)^{23}$$

We will find a recursive g such that, for all n, $\{g(n)\} = \alpha_n$.

$$\begin{aligned} g(0) &= z \\ g(n+1) &= h(g(n)) \end{aligned}$$

where we fixed an index z of $\emptyset$. Indeed, by the definition above, $\alpha_0 = \{g(0)\}$. Suppose $\{g(n)\} = \alpha_n$, for a fixed n. Then

$$\{g(n+1)\}(x) \simeq \{h(g(n))\}(x) \simeq \mathscr{F}(x, \lambda y.\{g(n)\}) \stackrel{I.H.}{\simeq} \mathscr{F}(x, \alpha_n) \simeq \alpha_{n+1}(x)$$

With this out of the way, we have

$$u \simeq \alpha(x) \equiv (\exists n)\alpha_n(x) \simeq u \equiv (\exists n)\{g(n)\}(x) \simeq u$$

Thus $u \simeq \alpha(x)$ is semi-recursive and hence α is in $\mathcal{P}$ by Theorem 6.6.1.   □

13.7.2 Kleene Version

13.7.4 Theorem. (The First Recursion Theorem (Kleene 1952)) *For every partial recursive functional $\lambda x\alpha.\mathscr{F}(x, \alpha)$ (unrestricted, in the Kleene sense Sect. 13.2) the equation*

$$\alpha(x) \simeq \mathscr{F}(x, \alpha)$$

---

[23] Since f is fixed we do not show it as an argument of h.

has a least *partial (absolutely) recursive solution* α, *called a* fixpoint *or* fixed point, *meaning that if* β *is another fixpoint, then* $\alpha \subseteq \beta$.

Proof The proof is essentially the same as that in the weakly computable case (Theorem 13.7.3) with minor variations. For example here monotonicity is not assumed, but is provable (Corollary 13.4.13).

The least solution is found via a sequence of approximations.

$$\text{Set } \alpha_0 \quad = \emptyset$$
$$\text{Set } \alpha_{n+1}(x) \simeq \mathscr{F}(x, \alpha_n)$$

That $\alpha_n \subseteq \alpha_{n+1}$ for all $n \geq 0$, is proved exactly as in Theorem 13.7.3.

Thus, we define

$$\alpha \overset{Def}{=} \bigcup_n \alpha_n \tag{1}$$

By the above sub result, α is a function (single-valued). Moreover,

$$\alpha(x) \simeq u \overset{(1)}{\Rightarrow} \text{(for some } n) \, \alpha_{n+1}(x) \simeq u \Rightarrow \mathscr{F}(x, \alpha_n) \simeq u \overset{\text{Corollary } 13.4.13}{\simeq} \mathscr{F}(x, \alpha)$$

Having established that α as a fixpoint of $\mathscr{F}$, assume that β is also a fixpoint. We show that $\alpha \subseteq \beta$ establishing the constructed α as being *the least*.

We also prove that $\alpha_n \subseteq \beta$, for all $n \geq 0$, exactly as in Theorem 13.7.3.

Finally we show that $\alpha \in \mathcal{P}$. By Theorem 13.3.36 there is a recursive $\lambda x.h(x)$ such that

$$\{h(e)\}(x) \simeq \mathscr{F}(x, \lambda y.\{e\}(y))^{24}$$

We will find a recursive g such that, for all n, $\{g(n)\} = \alpha_n$.

$$g(0) \quad = z$$
$$g(n+1) = h(g(n))$$

where we fixed an index z of $\emptyset$. Indeed, by the definition above, $\alpha_0 = \{g(0)\}$. Suppose $\{g(n)\} = \alpha_n$, for a fixed n. Then

$$\{g(n+1)\}(x) \simeq \{h(g(n))\}(x) \simeq \mathscr{F}(x, \lambda y.\{g(n)\}) \overset{I.H.}{\simeq} \mathscr{F}(x, \alpha_n) \simeq \alpha_{n+1}(x)$$

With this out of the way, we have

---

[24] The "h" function in Theorem 13.3.36 has two arguments one being an index of $\mathscr{F}$. The latter being a fixed functional, we take a fixed index for it so it does not enter as an argument in h.

$$u \simeq \alpha(x) \equiv (\exists n)\alpha_n(x) \simeq u \equiv (\exists n)\{g(n)\}(x) \simeq u$$

Thus $u \simeq \alpha(x)$ is semi-recursive and thus α is in $\mathcal{P}$ by Theorem 6.6.1.   □

We have a straightforward corollary:

13.7.5 Corollary. *If $\mathcal{G}$ is a recursive operator, then the equation*

$$\alpha \simeq \mathcal{G}(\alpha)$$

has a least partial computable (in the sense of Kleene) fixpoint α.

Proof The proof is immediate: By Theorem 13.5.14, $\mathcal{G}$ satisfies, for some *extensional $f \in \mathcal{R}$* and all unary ϕ_e,

$$\mathcal{G}(\phi_e) \simeq \phi_{f(e)} \tag{1}$$

Now the proposed fixpoint —and its minimality proof— are as in the proof of Theorem 13.7.4: We have $\alpha(x) \simeq \bigcup_{n \geq 0} \alpha_n$, where $\alpha_0 = \emptyset$ and $\alpha_{n+1} \simeq \mathcal{G}(\alpha_n)$.

For the partial recursiveness of α we recursively compute indices for the α_n and define g as before, using the extensional f:

$$\begin{aligned} g(0) &= z \\ g(n+1) &= f(g(n)) \end{aligned}$$

where we fixed an index z of $\emptyset$. Indeed, by the definition above, $\alpha_0 = \{g(0)\}$. Suppose $\{g(n)\} = \alpha_n$, for a fixed n. Then

$$\{g(n+1)\}(x) \simeq \{f(g(n))\}(x) \overset{(1)}{\simeq} \mathcal{G}(\lambda y.\{g(n)\})(x) \overset{I.H.}{\simeq} \mathcal{G}(\alpha_n)(x) \simeq \alpha_{n+1}(x)$$

With this out of the way, we have

$$u \simeq \alpha(x) \equiv (\exists n)\alpha_n(x) \simeq u \equiv (\exists n)\{g(n)\}(x) \simeq u$$

Thus $u \simeq \alpha(x)$ is semi-recursive and hence α is partial computable in the absolute sense (no oracles) by Theorem 6.6.1.   □

13.7.3   Moschovakis Version

13.7.6 Theorem. (The First Recursion Theorem) Moschovakis (1969) *For every search computable p.m.v. functional $\lambda x \alpha.\mathscr{F}(x, \alpha)$ the equation*

$$\alpha(x) \simeq \mathscr{F}(x, \alpha)$$

has a least *search (absolutely) computable p.m.v. solution* α, *called a* fixpoint *or* fixed point, *meaning that if* β *is another fixpoint, then* $\alpha \subseteq \beta$.

Proof   The proof is essentially the same as that in the Kleene case (Theorem 13.7.4) with minor variations.

The least solution is found via a sequence of approximations.

$$\text{Set } \alpha_0 \quad = \emptyset$$
$$\text{Set } \alpha_{n+1}(x) \simeq \mathscr{F}(x, \alpha_n)$$

We are reminded that the $\simeq$-notation is consistent with p.m.v. functions and functionals, in particular

1. $\mathscr{F}(\vec{x}, \vec{\alpha}) \simeq y$ means that y is an —not necessarily *the*— output of $\mathscr{F}(\vec{x}, \vec{\alpha})$.
2. $\mathscr{F}(\vec{x}, \vec{\alpha}) \subseteq \mathscr{G}(\vec{y}, \vec{\beta})$ means $(\forall z)\Big(\mathscr{F}(\vec{x}, \vec{\alpha}) \simeq z \rightarrow \mathscr{G}(\vec{y}, \vec{\beta}) \simeq z\Big)$, which is the same as $\mathscr{F}(\vec{x}, \vec{\alpha}) \simeq z \rightarrow \mathscr{G}(\vec{y}, \vec{\beta}) \simeq z$ as we know from the behaviour of "$\forall$" detailed in Sect. 0.2. In particular, if $\mathscr{F}(\vec{x}, \vec{\alpha}) \uparrow$, then the previous implication is true.
3. $\mathscr{F}(\vec{x}, \vec{\alpha}) \simeq \mathscr{G}(\vec{y}, \vec{\beta})$ means $\mathscr{F}(\vec{x}, \vec{\alpha}) \subseteq \mathscr{G}(\vec{y}, \vec{\beta})$ and $\mathscr{F}(\vec{x}, \vec{\alpha}) \supseteq \mathscr{G}(\vec{y}, \vec{\beta})$, or

$$\mathscr{F}(\vec{x}, \vec{\alpha}) \simeq z \equiv \mathscr{G}(\vec{y}, \vec{\beta}) \simeq z$$

In particular $\mathscr{F}(\vec{x}, \vec{\alpha}) \uparrow$ and $\mathscr{G}(\vec{y}, \vec{\beta}) \uparrow$ imply $\mathscr{F}(\vec{x}, \vec{\alpha}) \simeq \mathscr{G}(\vec{y}, \vec{\beta})$.

Next we show that $\alpha_n \subseteq \alpha_{n+1}$ for all $n \geq 0$.

For $n = 0$ (*Basis*) this is trivial.

Taking as I.H. $\alpha_n \subseteq \alpha_{n+1}$ for a fixed n we next examine the $n+1$ case: By *monotonicity* (a direct consequence of Corollary 13.6.33 as for Kleene computability) we have that $\alpha_{n+1}(x) \simeq \mathscr{F}(x, \alpha_n) \downarrow\simeq y$ implies

$$\mathscr{F}(x, \alpha_n) \simeq \mathscr{F}(x, \alpha_{n+1}) \downarrow\simeq \alpha_{n+2}(x) \simeq y$$

Thus, define

$$\alpha \stackrel{Def}{=} \bigcup_n \alpha_n, \text{ equivalently } \alpha(x) \simeq u \equiv (\exists n)\alpha_n(x) \simeq u \tag{1}$$

We note that

$$\alpha(x) \simeq u \stackrel{(1)}{\Rightarrow} \text{ (for some } n) \; \alpha_{n+1}(x) \simeq u \Rightarrow \mathscr{F}(x, \alpha_n) \simeq u \stackrel{\text{monotonicity}}{\simeq} \mathscr{F}(x, \alpha)$$

Having established that α as a fixpoint of $\mathscr{F}$, assume that β is also a fixpoint. We show that $\alpha \subseteq \beta$ establishing the constructed α as being *the least*.

We first prove that $\alpha_n \subseteq \beta$, for all $n \geq 0$.
(*Basis*). That $\alpha_0 \subseteq \beta$ is trivial.
Assume (I.H.) that for a fixed n it is $\alpha_n \subseteq \beta$.
The case for $\alpha_{n+1} \subseteq \beta$.
We have that

$$\alpha_{n+1}(x) \simeq u \Rightarrow \mathscr{F}(x, \alpha_n) \simeq u \overset{I.H. + mon}{\Rightarrow} \mathscr{F}(x, \beta) \simeq u$$

By (1) and $(\forall n)(\alpha_n \subseteq \beta)$ proved, we have $\alpha = \bigcup_{n \geq 0} \alpha_n \subseteq \beta$.
   Finally we show that $\alpha \in \mathcal{P}$. By Theorem 13.6.19 there is a recursive $\lambda x.h(x)$
such that

$$\{h(e)\}(x) \simeq \mathscr{F}(x, \lambda y.\{e\}(y))^{25}$$

We will find a(n absolute) recursive g such that, for all n, $\{g(n)\} = \alpha_n$.

$$g(0) \quad = z$$
$$g(n+1) = h(g(n))$$

where we fixed an index z of $\emptyset$. Indeed, by the definition above, $\alpha_0 = \{g(0)\}$.
Suppose $\{g(n)\} = \alpha_n$, for a fixed n. Then

$$\{g(n+1)\}(x) \simeq \{h(g(n))\}(x) \simeq \mathscr{F}(x, \lambda y.\{g(n)\}) \overset{I.H.}{\simeq} \mathscr{F}(x, \alpha_n) \simeq \alpha_{n+1}(x)$$

With this out of the way, we have

$$u \simeq \alpha(x) \equiv (\exists n)\alpha_n(x) \simeq u \equiv (\exists n)\{g(n)\}(x) \simeq u$$

Thus $u \simeq \alpha(x)$ is semi-recursive and thus α is search computable by Theorem 13.6.35.   □

13.7.7 Remark. The converse of Remark 13.7.2 is not true. Pick an uncomputable
total function α of one argument. Now, we saw in Chap. 0 that the set of all functions
in $\mathbb{N}^{\mathbb{N}}$ is *uncountable*. Since computable functions of one argument form a *countable*
set —due to the onto mapping $e \mapsto \{e\}$ from $\mathbb{N}$ to $\mathcal{P}^{(1)}{}^{26}$— there *is* an uncomputable
$\alpha \in \mathbb{N}^{\mathbb{N}}$. Then $\beta = \lambda x.\alpha(x) + 1$ is also uncomputable (Why?).
   Consider now the functional $\mathscr{F}$ given by

$$\mathscr{F}(x, \delta) \simeq \begin{cases} 0 & \text{if } \delta = \phi_e, \text{ for some } e \in \mathbb{N} \\ \beta(x) & \text{othw} \end{cases}$$

We have a few observations:

---

[25] The "h" function in Theorem 13.6.19 has two arguments one being an index of $\mathscr{F}$. The latter
being a fixed functional, we take a fixed index for it so it does not enter as an argument in h.
[26] In general for any class of functions $\mathcal{C}$, $\mathcal{C}^{(n)}$ means the subset of n-argument functions in $\mathcal{C}$.

1. $\mathscr{F}$ is weakly partial recursive since, for all x and all e in $\mathbb{N}$, $\mathscr{F}(x, \phi_e) \simeq 0$.
2. It is not monotone:

   - $\mathscr{F}(x, \phi_z) = 0$, where z is a program code for $\emptyset$, but
   - even though $\emptyset \subseteq \alpha$, for any $\alpha \in \mathbb{N}^{\mathbb{N}} - \mathcal{P}$, we have $\mathscr{F}(x, \alpha) \simeq \beta(x) > 0$ (for all x). We need $\simeq 0$ here!

3. $\mathscr{F}$ is *total* (defined everywhere) and does not have a *least* fixed point. It has *two total* fixpoints

   - $\lambda y.0$ yields $\mathscr{F}(x, \lambda y.0) \simeq 0$,
     and
   - β yields $\mathscr{F}(x, \beta) \simeq \beta(x)$

   These two are incomparable, hence none of the two is least.
4. Thus: $\mathscr{F}$

   a. is not partial recursive (otherwise it would have been monotone), which justifies the opening statement of this remark.
   b. fails the first recursion theorem, which aptly requires monotonicity, and
   c. *non monotone* weakly partial recursive functionals exist. □

13.7.4 The Second vs. the First Recursion Theorem

The second recursion theorem implies the first Rogers (1967). Since the second recursion theorem is applicable to non-monotone weak partial computable functionals to obtain a fixpoint (computable function of rank $(k, 0)$) but the first is not according the remark above, one is justified to nickname the second the "strong" recursion theorem and the first the "weak" recursion theorem.

   Below we prove the claimed implication. Our proof is a simplification and adaptation (to Kleene partial computable functionals) of the proof in Moschovakis (1971) (that was also reproduced in Fenstad (1980)). These two proofs were given in an *abstract* setting (not over the integers) for, essentially, *weakly* partial recursive functionals where compactness (essentially[27]) and the property (6) in the proof below were adopted *axiomatically*.

13.7.8 Theorem. *Suppose that* $\lambda x \alpha . \mathscr{F}(x, \alpha)$ *is partial recursive in the Kleene sense. Then for an* $f \in \mathcal{P}$ *obtained by the functional substitution Theorem 13.3.36*

$$f(e, x) \simeq \mathscr{F}(x, \{e\})$$

---

[27] The concept of the sub-functions g_i in (Moschovakis 1971, Def.7, p.208) where the term "weakly computable" is not used, and the identical concept of the sub-functions g_i in (Fenstad 1980, Def.2.1.2, p.44) where the term "weakly (Θ-)computable" *is* used come close to the finite sub functions of the compactness phenomenon. In abstract recursion theory setting ordinary finiteness is replaced by a different concept that we will not get into.

an application of the second recursion theorem obtains an $\bar{e}$ such that $f(\bar{e}, x) \simeq \{\bar{e}\}(x)$, for all x. Then $\{\bar{e}\}$ is the least *fixpoint of $\mathscr{F}$.*

Proof By the second recursion theorem

$$\{\bar{e}\}(x) \simeq f(\bar{e}, x) \simeq \mathscr{F}(x, \{\bar{e}\})$$

thus $\{\bar{e}\}$ is indeed *a* fixpont of $\mathscr{F}$. To show that $\{\bar{e}\}$ is the *least* fixpoint it suffices to show that $\{\bar{e}\} \subseteq \beta$ where β is *any* fixpoint of $\mathscr{F}$.

To this end, let

$$\{\bar{e}\}(x) \downarrow \simeq z \tag{1}$$

hence also

$$\mathscr{F}(x, \{\bar{e}\}) \downarrow \simeq z \tag{2}$$

and moreover, by (1) and Theorem 13.3.21,

$$\|\bar{e}, x, z\| > \|\hat{f}, \bar{e}, x, z\|, \text{ where } \hat{f} \text{ is some fixed index of } f \tag{3}$$

Towards $\{\bar{e}\} \subseteq \beta$ take as I.H. on computation lengths that

$$\text{if } \|\bar{e}, p, q\| < \|\bar{e}, x, z\|, \text{ then } \beta(p) \simeq q \tag{4}$$

By (2) and compactness, there is a u such that $[u] \subseteq \{\bar{e}\}$ satisfying

$$\mathscr{F}(x, [u]) \simeq z \tag{5}$$

We choose a $[u]$ that is $\subseteq$-minimal with the stated property.

  Then if $[u](p) \simeq q$ was used in the computation (2), so was $\{\bar{e}\}(p) \simeq q$.

Thus whenever $[u](p) \simeq q$, then —due to the way $[u] \subseteq \{\bar{e}\}$ enters in the computation (2) (cf. Theorem 13.4.5 and Example 13.3.35, (7))— the computation will *call* the tuple $(\bar{e}, p, q)$ —that is, it will "evaluate" $\{\bar{e}\}(p) \simeq q$— *before* it computes and returns the value z in (2) and hence (1).

This means that if $[u](p) \simeq q$, then

$$\|\bar{e}, p, q\| \overset{\text{previous paragraph}}{<} \|\hat{f}, \bar{e}, x, z\| \overset{(3)}{<} \|\bar{e}, x, z\| \tag{6}$$

By the I.H. (4), $\beta(p) \simeq q$ for all p, q satisfying (6). Therefore $[u] \subseteq \beta$. We now have the following conjunctional iff-chain that leads to the end of proof:

$$\{\bar{e}\}(x) \simeq z \text{ iff } \mathscr{F}(x, \{\bar{e}\}) \simeq z \overset{(5)}{\text{ iff }} \mathscr{F}(x, [u]) \simeq z \overset{Monoton.}{\text{ implies }} \mathscr{F}(x, \beta) \simeq \beta(x) \simeq z$$

$\square$

13.7.5 A Connection with Program Semantics

Suppose that we formulated and proved the first recursion theorem (say, in Kleene computability) for the case where α has n number variables, $\vec{x}_n$. Then the equation

$$\alpha(\vec{x}_n) \simeq \mathscr{F}(\vec{x}_n, \alpha) \tag{†}$$

also has a computable least fixpoint (Exercise 13.12.17). We can use this fact to define *semantics* for "recursive programs" of the type

$$P : \qquad \alpha(\vec{x}) ::= \mathscr{F}(\vec{x}, \alpha) \tag{‡}$$

where we named the program "P" (the label to the left) and "::=" says "is defined as". $\mathscr{F}$ may be a monotone weakly partial recursive functional, or a partial recursive functional in the sense, say, of Kleene.

Is the "label" P redundant? No. P is the *program* for obtaining a fixpoint α of $\mathscr{F}$, say, the *least* one. On the other hand, $\mathscr{F}$ is a computable "rule" that will obtain various functions β for various inputs γ —that is, for all $\vec{x}$, it will obtain $\beta(x) \simeq \mathscr{F}(\vec{x}, \gamma)$. So P and $\mathscr{F}$ are two different objects.

So what does the *program* (‡) *mean*, what does it *compute*? One approach is of course, to find the least fixpoint of (†) and say, *this is what P means*. The other is to try to compute what the program does, with this or that method, being mindful of the meaning of ::= ("is defined as").

In this connection, here is an example from Moris (1968).

13.7.9 Example. (Morris) Let

$$\mathscr{F}(x, y, \alpha) \simeq if\ x = 0\ then\ 1\ else\ \alpha(x \,\dot{-}\, 1, \alpha(x, y)) \tag{0}$$

Consider the program below

$$Q : \qquad \alpha(x, y) ::= \mathscr{F}(x, y, \alpha) \tag{1}$$

and let's see what program Q computes under fixpoint semantics.

Clearly $\mathscr{F}$ is partial recursive in the Kleene sense (cf. Exercise 13.12.16)

We follow the process of generating the sequence

$$\emptyset = \alpha_0 \subseteq \alpha_1 \subseteq \alpha_2 \subseteq \ldots$$

Note that, for all x and y, we set $\alpha_0(x, y) \uparrow$ and

$$\alpha_{n+1}(x, y) \simeq \mathscr{F}(x, y, \alpha_n) \simeq if\ x = 0\ then\ 1\ else\ \alpha_n(x \,\dot{-}\, 1, \alpha_n(x, y))$$

Thus

$$\alpha_1(x, y) \simeq \text{if } x = 0 \text{ then } 1 \text{ else } \uparrow$$
$$\alpha_2(x, y) \simeq \text{if } x = 0 \text{ then } 1 \text{ else } \alpha_1(x \mathbin{\dot-} 1, \alpha_1(x, y))$$
$$ \simeq \text{if } x = 0 \text{ then } 1 \text{ else if } x = 1 \text{ then } 1 \text{ else } \uparrow$$
$$ \simeq \text{if } x \in \{0, 1\} \text{ then } 1 \text{ else } \uparrow$$

$$\vdots \qquad\qquad \vdots$$

$$\alpha_n(x, y) \simeq \text{if } x \in \{0, 1, \dots, n\} \text{ then } 1 \text{ else } \uparrow \quad \Leftarrow I.H. \text{ on } n$$
$$\alpha_{n+1}(x, y) \simeq \text{if } x = 0 \text{ then } 1 \text{ else } \alpha_n(x \mathbin{\dot-} 1, \alpha_n(x, y))$$
$$\phantom{\alpha_{n+1}(x, y)} \simeq \text{if } x = 0 \text{ then } 1 \text{ else if } x \mathbin{\dot-} 1 \in \{0, 1, \dots, n\} \text{ then } 1 \text{ else } \uparrow$$
$$\phantom{\alpha_{n+1}(x, y)} \simeq \text{if } x \in \{0, 1, \dots, n, n+1\} \text{ then } 1 \text{ else } \uparrow$$

$$\vdots \qquad\qquad \vdots$$

The above contains an induction that proves

$$\alpha_n(x, y) \simeq \text{if } x \in \{0, 1, \dots, n\} \text{ then } 1 \text{ else } \uparrow \tag{2}$$

Thus, for the *least fixpoint* $\alpha_\infty = \bigcup_{n \geq 0}$ we have

$$\alpha_\infty(x, y) \simeq \text{if } x \geq 0 \text{ then } 1 \text{ else } \uparrow \tag{3}$$

The above condition is always true, thus

$$\alpha_\infty = \lambda xy.1 \tag{4}$$

Suppose now we wanted to implement program Q with *call-by-value* calls.

A function call $f(a, b, c)$ is called-by-value if we evaluate *all* inputs —a, b, c— *first*, and then pass their values to f.

Let us apply call-by-value to (1). Say, we want the value of $\alpha(1, 0)$ and are going to get it by evaluating the right hand side of (1), $\mathscr{F}(1, 0, \alpha)$, applying call-by-value to all calls.

Well, we have a call to $\alpha(0, \alpha(1, 0))$ but we cannot evaluate this before we evaluate the right input.

Attempting this leads to a call $\alpha(0, \alpha(0, \alpha(1, 0)))$ and it is obvious we are going to loop forever. That is, $\alpha(1, 0) \uparrow$ if we are using call-by-value.

It can be shown that $\alpha(x, y)$ computed by call-by-value is $\alpha(x, y) \simeq \text{if } x = 0 \text{ then } 1 \text{ else } \uparrow$ (cf. Exercise 13.12.18.) Thus, this solution is *not* a fixpoint, since $\alpha \subsetneq \alpha_\infty$.

There is another computation rule for calls, called *call-by-name* or the *leftmost rule*. In this rule evaluation of the arguments *proceeds left to right*. We *do not* have to evaluate *all* arguments *first*.

Thus $\alpha(1, 0)$ leads to the call $\alpha(0, \alpha(1, 0))$ and this reurns 1 and we terminate the process. More generally,

$$\alpha(x, y) \stackrel{(2)}{\simeq} if \ x = 0 \ then \ 1 \ else \ \alpha(x \dotdiv 1, \alpha(x, y)) \tag{5}$$

leads to a terminating sequence of calls

$$\alpha(x, y), \ \alpha(x \dotdiv 1, \alpha(x, y)), \quad \alpha(x \dotdiv 2, \alpha(x \dotdiv 1, \alpha(x, y))), \quad \dots, \quad \alpha(0, \dots) \simeq 1$$

evaluating *depth-first* on the first argument (cf. (5)) until it becomes 0. This shows the call-by-name computed α is the same as α_∞. This is *true in general*, not just a coincidence: call-by-name computes the least fixpoint (Manna 1974).[28]

The fixpoint approach to program semantics complements the call-by-name computational approach in that it lends itself in proving *properties of programs* like ($\ddagger$). To simplify the exposition, assume our program is

$$P: \qquad \alpha ::= \mathscr{G}(\alpha) \tag{6}$$

where α is unary, and $\mathscr{G}$ is a recursive operator.

The process of building the least fixpoint α_∞ is building the sequence

$$\emptyset = \alpha_0 \subseteq \alpha_1 \subseteq \alpha_2 \subseteq \dots$$

(cf. proof of Theorem 13.7.4 and its corollary).

Naturally, one wishes to prove that the *least fixpoint* α_∞ has a property $\mathscr{C}$ (i.e., $\alpha_\infty \in \mathscr{C}$) by doing *induction* —called *computational induction*— on n for the statement $\alpha_n \in \mathscr{C}$ to prove $(\forall n)\alpha_n \in \mathscr{C}$, *hoping to conclude from this that* $\alpha_\infty \in \mathscr{C}$ as well.

This narrows the range of "properties" we may use to what is called "*admissible properties*" Manna (1974). These are $\mathscr{C}$ such that, for any recursive operator $\mathscr{G}$,

$$\text{If, for all } n \geq 0, \mathscr{G}^n(\emptyset) \in \mathscr{C}, \text{ then } \bigcup_{n \geq 0} \mathscr{G}^n(\emptyset) \in \mathscr{C} \tag{A}$$

Of course, $\bigcup_{n \geq 0} \mathscr{G}^n(\emptyset)$ is α_∞, the least fixpoint solution for (6). Examples of admissible properties include those for which $\{x : \phi_e \in \mathscr{C}\}$ is semi-recursive (cf. Theorem 6.10.3).

The *fixpoint theory* of *program semantics* originated with Scott (1970). Other references are Manna (1974), Bird (1976). $\qquad\qquad\qquad\qquad\qquad\qquad \square$

---

[28] Manna (1974) assumes the generally accepted "$0 \cdot \uparrow \ \simeq \uparrow$" in the proof. Interestingly, if the convention changes to "$0 \cdot \ \uparrow \simeq 0$" loc. cit. offers a counterexample: *With this convention*, the *leftmost rule* does *not in general* compute the least fixpoint. See Exercise 13.12.23.

13.8 Partial Recursive and Recursive *Restricted* Functionals

We develop here, in the style of Kleene, the elementary theory of computable restricted functionals (cf. Definition 13.1.1) to the extent needed for the last sections of this chapter. The approach follows Kleene's index-driven approach, but this theory is much more user friendly in view of the fact that the type-1 inputs are total. In particular we get all basic facts of absolute computability (oracle-less), including the *selection theorem* with no need for a clause IX in the definition of "Ω", which, incidentally, we will call Ω^r ("r" for restricted) in this section.

For the record, we repeat the definition of Ω (as Ω^r) here:

13.8.1 Definition. (Computations; the *Syntax*) This definition gives the syntax of (static) *computations* each such *stating* "$\{e\}(\vec{x}_n, \vec{\alpha}_l) \simeq y$" saying "program e with input $\vec{x}, \vec{\alpha}$ has output y" —abstracted as a tuple "$(e, \vec{x}_n, \vec{\alpha}_l, y)$".

The definition is identical in its clauses I-VIII to Definition 13.3.1, but the present one restricts the type-1 inputs to *total* functions $\beta : \mathbb{N} \to \mathbb{N}$.

This concept for static computation leads to the dynamic concept of computation just as we showed in the cases of Ω and Ω^{sel}.

The syntax of Ω^r tuples: Let us define the lengths of type-0 and type-1 inputs "k" and "l" up in front as $k = \text{length}(\vec{x})$ and $l = \text{length}(\vec{\alpha})$. Now that this has been stated we will be writing $\vec{x}$ for $\vec{x}_k$ and $\vec{\alpha}$ for $\vec{\alpha}_l$ in the rest of this definition.

The set Ω^r of *computations* on input $(\vec{x}, \vec{\alpha}) \in \mathbb{N}^k \times (\mathbb{N}^{\mathbb{N}})^l$, for all $k + l \geq 1$,[29] $\vec{x}, \vec{\alpha}, y, z, u, v, e, f, g, c$ is inductively defined as the *smallest* set of tuples $(e, \vec{x}, \vec{\alpha}, y)$ satisfying I–VIII below:

Basis (Initial) cases: I–VII

I. $(\langle 1, k, l, i \rangle, \vec{x}, \vec{\alpha}, x_i) \in \Omega$ **for** $1 \leq i \leq k$

II. $(\langle 2, k, l, i \rangle, \vec{x}, \vec{\alpha}, x_i + 1) \in \Omega$ **for** $1 \leq i \leq k$

III. $(\langle 3, k, l, c \rangle, \vec{x}, \vec{\alpha}, c) \in \Omega$ **for** $c \in \mathbb{N}$

IV. $(\langle 4, k+2, l \rangle, a, b, \vec{x}, \vec{\alpha}, Sb_1(a, b)) \in \Omega$ **for** $Sb_1(a, b)$ defined in Theorem 13.3.18. **Cf.** also Theorem 10.2.1.

V. $(\langle 5, k+4, l \rangle, z, y, u, v, \vec{x}, \vec{\alpha}, z) \in \Omega$ **if** $u = v$

 $(\langle 5, k+4, l \rangle, z, y, u, v, \vec{x}, \vec{\alpha}, y) \in \Omega$ **if** $u \neq v$

VI. $(\langle 6, k, l, i, j \rangle, \vec{x}, \vec{\alpha}, y) \in \Omega$ **if** $\alpha_j(x_i) = y$

Inductive cases: VII–VIII

VII. $(\langle 7, k, l, f, \vec{g}_m \rangle, \vec{x}, \vec{\alpha}, y) \in \Omega$ **if** $(f, \vec{z}_m, \vec{\alpha}, y) \in \Omega$
 and $(g_i, \vec{x}, \vec{\alpha}, z_i) \in \Omega$ **for** $i = 1, \ldots, m$

VIII. $(\langle 8, k+1, l \rangle, e, \vec{x}, \vec{\alpha}, y) \in \Omega$ **if** $(e, \vec{x}, \vec{\alpha}, y) \in \Omega$

---

[29] Again we prefer *not* to allow functionals to have no inputs at all.

Paraphrased in the style of Sect. 0.6, $\Omega^r = \text{Cl}(\mathcal{I}, \mathcal{O})$, where the *initial* objects form the set $\mathcal{I}$ of all tuples from I–VI —for all $k + l \geq 1, \vec{x}, \vec{\alpha}, y, z, u, v, c$— while we have just *two types* of operations on tuples in $\mathcal{O}^r$ (from VII and VIII): *composition operations* (VII) and *the universal operation* (VIII).

Both operations are finitary, that is, have a finite number of arguments. The indices $\langle 7, k, l, f, \vec{g}_m \rangle$ and $\langle 8, k + 1, l \rangle$ determine the number of inputs for the respective operations: $m + 1$ for the former and one for the latter. We have one operation for each input arity (case VII).

1. From the tuples $(f, \vec{z}_m, \vec{\alpha}, y)$ and $(g_i, \vec{x}, \vec{\alpha}, z_i)$, for $i = 1, \ldots, m = (f)_1$, we can generate the tuple $(\langle 7, k, l, f, \vec{g}_m \rangle, \vec{x}, \vec{\alpha}, y)$.
2. From the tuple $(e, \vec{x}, \vec{\alpha}, y)$ we can generate the tuple $(\langle 8, k+1, l \rangle, e, \vec{x}, \vec{\alpha}, y)$. $\square$

We reproduce much of the *definitions* and *results* from Sect. 13.3 the latter without proof if the earlier proofs require no changes. We will also introduce and work with *characteristic functionals* in place of *partial characteristic functionals* that we employed instead because of the presence of partial type-1 inputs (until now).

13.8.2 Definition. (Computations; *Semantics*) The symbol

$$\{e\}^{(k,l)}(\vec{x}_k, \vec{\alpha}_l) \simeq y \tag{1}$$

says exactly

$$(e, \vec{x}_k, \vec{\alpha}_l, y) \in \Omega$$

In words, *the functional* $\{e\}^{(k,l)}$ *of rank* (k, l) *given by the "program" e has as output the number y if the input is* $\vec{x}_k, \vec{\alpha}_l$.

Similarly to Definition 13.3.3, this interpretation *makes no claim of uniqueness* or *single-valued-ness* of the y-variable thus, *so far* in our development, the functionals $\{e\}^{(k,l)}$ in this definition are p.m.v. (Definition 13.1.1); but see the following proposition.

We also have the notations below signalling that we will usually write $\{e\}(\vec{x}_k, \alpha_l)$ without the superscript (k, l) letting the context fend for itself.

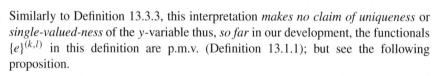

$$\{e\}(\vec{x}_k, \vec{\alpha}_l) \downarrow \text{ iff } (\exists y)\Big((e, \vec{x}_k, \vec{\alpha}_l, y) \in \Omega\Big) \text{ iff } (\exists y)\{e\}(\vec{x}_k, \vec{\alpha}_l) \simeq y$$

and

$$\{e\}(\vec{x}_k, \vec{\alpha}_l) \uparrow \text{ iff } \neg(\exists y)\Big((e, \vec{x}_k, \vec{\alpha}_l, y) \in \Omega\Big) \tag{2}$$

$\square$

13.8.3 Definition. (Subcomputations and Computation Lengths) We define *immediate subcomputations* —in short *i.s.*— and *computation lengths* for each tuple in Definition 13.8.1.

- **(I.–VI.)** Each of the computations $(e, \vec{x}, \vec{\alpha}, y)$ in cases I–VI of Definition 13.8.1 has *no* i.s.
  Their *length* $\|e, \vec{x}, \vec{\alpha}, y\|$ is equal to zero. In the case of VI, the length is 0 precisely when $(e, \vec{x}, \vec{\alpha}, y) = (\langle 6, k, l, j, i \rangle, \vec{x}, \vec{\alpha}, \alpha_j(x_i))$ *and* $\alpha_j(x_i) \downarrow$, which is always the case in the computability of restricted functionals.
- **(VII.)** If there exist y and $z_i, i = 1, \ldots, m$ such that $(f, \vec{z}, \vec{\alpha}, y) \in \Omega^r$ and $(e_i, \vec{x}, \vec{\alpha}, z_i) \in \Omega^r$, $i = 1, \ldots, m$, then *the* i.s. of $(\langle 7, k, l, f, e_1, \ldots, e_m \rangle, \vec{x}, \vec{\alpha}, y)$ are $(f, \vec{z}, \vec{\alpha}, y)$ and $(e_i, \vec{x}, \vec{\alpha}, z_i)$, for $i = 1, 2, \ldots, m$.
  Moreover,

$$\| \langle 7, k, l, f, e_1, \ldots, e_m \rangle, \vec{x}, \vec{\alpha}, y \| =$$

$$\max\left(\|f, \vec{z}, \vec{\alpha}, y\|, \|e_1, \vec{x}, \vec{\alpha}, z_1\|, \ldots, \|e_m, \vec{x}, \vec{\alpha}, z_m\| \right) + 1$$

- **(VIII.)** If there exists a y such that $(e, \vec{x}, \vec{\alpha}, y) \in \Omega^r$, then *the* i.s. of $(\langle 8, k+1, l \rangle, e, \vec{x}, \vec{\alpha}, y)$ is $(e, \vec{x}, \vec{\alpha}, y)$.
  Moreover, $\| \langle 8, k+1, l \rangle, e, \vec{x}, \vec{\alpha}, y \| = \|e, \vec{x}, \vec{\alpha}, y\| + 1$.
- **(Otherwise.)** If $(e, \vec{x}, \vec{\alpha}, y) \notin \Omega^r$, then we write $\|e, \vec{x}, \vec{\alpha}, y\| = \infty$.

A computation $c \in \Omega^r$ is a *subcomputation* of $d \in \Omega^r$ just in case there is finite sequence $c = d_1, d_2, \ldots, d_n = d$ such that d_i is an i.s. of d_{i+1}, for $i = 1, \ldots, n-1$.   □

13.8.4 Remark. We used "the" in the two cases above that do have i.s. (VII and VIII) because the indices $\langle 7, k, l, f, e_1, \ldots, e_m \rangle$ and $\langle 8, k+1, l \rangle$ uniquely determine the i.s. we will have, when they (it) exist(s).   □

13.8.5 Proposition. *The relation* $(e, \vec{x}_k, \alpha_l, y) \in \Omega^r$ *is single-valued in* y.

Proof See proof of Proposition 13.3.6.   □

13.8.6 Remark. Because of the above, one often uses "truncated computations" — that is, computations with the output part omitted.
   For example, "$(e, \vec{x}, \vec{\alpha})$ is the i.s. of $(\langle 8, k+1, l \rangle, e, \vec{x}, \vec{\alpha})$", "$\| \langle 8, k+1, l \rangle, e, \vec{x}, \vec{\alpha} \| = \|e, \vec{x}, \vec{\alpha}\| + 1$", "$\| \langle 1, k, l, i \rangle, \vec{x}, \vec{\alpha} \| = 0$", etc.   □

13.8.7 Definition. (The Set of Partial Recursive Functionals) In view of Proposition 13.8.5, we may define a single-valued functional $\{e\}$ of rank (k, l) extensionally (by its graph) via

$$\{e\}(\vec{x}, \vec{\alpha}) \simeq y \text{ iff } (e, \vec{x}, \vec{\alpha}, y) \in \Omega^r$$

We next define the set of *partial recursive restricted functionals* (*partial computable functionals*):

It is the set of all F of rank (k, l) —for all $k + l \geq 1$— such that $F = \{e\}^{(k,l)}$, for some $e \in \mathbb{N}$.

Thus, if e is not an index, as defined inductively in Definition 13.8.1, then we have —due to (2) p. 491— $\{e\} = \emptyset$, the everywhere undefined (computable) functional.

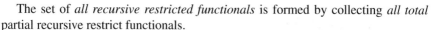

Via this observation we extend the meaning of $\{e\}(\vec{x}, \vec{\alpha}) \simeq z$ to all $e \in \mathbb{N}$.

The set of *all recursive restricted functionals* is formed by collecting *all total* partial recursive restrict functionals.

We will *not* introduce special symbols for the *class* of recursive and partial recursive restricted functionals. However, if we let $P^{(k,l)}$ (resp. $R^{(k,l)}$) be the set of all partial recursive (resp. *total* recursive) restricted functionals of rank (k, l) then the class of *all* such, for all ranks, is $\bigcup_{k+l \geq 1} P^{(k,l)}$ (resp. $\bigcup_{k+l \geq 1} R^{(k,l)}$). $\qquad \square$

13.8.8 Example. It is clear that each of the indices e satisfies $Seq(e)$ (cf. Example 2.3.10(8)). Now $Seq(0)$ is false, thus, $\{0\} = \emptyset$, the empty functional. $\qquad \square$

The following restates Proposition 13.3.10 for the case of partial recursive restricted functionals.

13.8.9 Proposition.

    I. $\{\langle 1, k, l, i \rangle\}(\vec{x}, \vec{\alpha}) \simeq x_i$
    II. $\{\langle 2, k, l, i \rangle\}(\vec{x}, \vec{\alpha}) \simeq x_i + 1$
    III. $\{\langle 3, k, l, c \rangle\}(\vec{x}, \vec{\alpha}) \simeq c$
    IV. $\{\langle 4, k + 2, l \rangle\}(a, b, \vec{x}, \vec{\alpha}) \simeq Sb_1(a, b)$
    V. $\{\langle 5, k + 4, l \rangle\}(p, q, a, b, \vec{x}, \vec{\alpha}) \simeq p$ if $a = b$
       $\{\langle 5, k + 4, l \rangle\}(p, q, a, b, \vec{x}, \vec{\alpha}) \simeq q$ if $a \neq b$
    VI. $\{\langle 6, k, l, i, j \rangle\}(\vec{x}, \vec{\alpha}) \simeq \alpha_j(x_i)$
    VII. $\{\langle 7, k, l, f, \vec{e}_m \rangle\}(\vec{x}, \vec{\alpha}) \simeq \{f\}\Big(\{e_1\}(\vec{x}, \vec{\alpha}), \ldots, \{e_m\}(\vec{x}, \vec{\alpha}), \vec{\alpha}\Big)$
    VIII. $\{\langle 8, k + 1, l \rangle\}(e, \vec{x}, \vec{\alpha}) \simeq \{e\}(\vec{x}, \vec{\alpha})$

Proof Cases I–VI are immediate.

    For

Case VIII.

    We want

$$\{\langle 8, k + 1, l \rangle\}(e, \vec{x}, \vec{\alpha}) \simeq y \text{ iff } \{e\}(\vec{x}, \vec{\alpha}) \simeq y$$

The $\leftarrow$ direction is by Definition 13.8.1, case VIII.

    We argue the $\rightarrow$ direction via derivations: So let $\{\langle 8, k + 1, l \rangle\}(e, \vec{x}, \vec{\alpha}) \simeq y$. By Definition 13.8.7 we have $(\langle 8, k + 1, l \rangle, e, \vec{x}, \vec{\alpha}, y) \in \Omega^r$. This item is derived in Ω^r (Theorem 0.6.10) either because it is *initial* (I–VI) —but it is not— or it is the result of a rule. Only rule VIII applies, so $(e, \vec{x}, \vec{\alpha}, y)$ occurs in the derivation before $(\langle 8, k + 1, l \rangle, e, \vec{x}, \vec{\alpha}, y)$. But all items in such derivation are in Ω^r, hence $(e, \vec{x}, \vec{\alpha}, y) \in \Omega^r$, that is, $\{e\}(\vec{x}, \vec{\alpha}) \simeq y$.

$\square$

In more familiar (from earlier parts of the book) terminology we restate Proposition 13.8.9 as two corollaries.

13.8.10 Corollary. *The following restricted functionals are (total) recursive.*

I. *Identity,* $U_i^{(k,l)}$. *For all* $\vec{x}$ *and* $\vec{\alpha}$, $U_i^{(k,l)}(\vec{x}, \vec{\alpha}) \simeq x_i$.

II. *Successor,* $S_i^{(k,l)}$. *For all* $\vec{x}$ *and* $\vec{\alpha}$, $S_i^{(k,l)}(\vec{x}, \vec{\alpha}) \simeq x_i + 1$.

III. *Constant (for each c),* $C_c^{(k,l)}$. *For all* $\vec{x}$ *and* $\vec{\alpha}$, $C_c^{(k,l)}(\vec{x}, \vec{\alpha}) \simeq c$.

IV. $S_1^{(k+2,l)}$. *For all* $a, b, \vec{x}$ *and* $\vec{\alpha}$, $S_1^{(k+2,l)}(a, b, \vec{x}, \vec{\alpha}) \simeq Sb_1(a, b)$.

V. *If-then else, or definition by cases,* $DC^{(k+4,l)}$. $DC^{(k+4,l)}(p, q, a, b, \vec{x}, \vec{\alpha}) \simeq p$ *if* $a = b$. $DC^{(k+4,l)}(p, q, a, b, \vec{x}, \vec{\alpha}) \simeq q$ *if* $a \neq b$

VI. *Oracle call or function-input application,* $AP_{i,j}^{(k,l)}$. $AP_{i,j}^{(k,l)}(\vec{x}, \vec{\alpha}) \simeq \alpha_j(x_i)$

13.8.11 Example. Every partial recursive functional has infinitely many distinct indices. For example, if $\{a\} = F$, then since $F = U_1^{(1,l)}F$ we also have $\{\langle 7, k, l, \langle 1, 1, l, 1 \rangle, a \rangle\} = F$. But $\langle 7, k, l, \langle 1, 1, l, 1 \rangle, a \rangle > a$. $\square$

13.8.12 Corollary.

VII. *If the functionals* $F^{(m,l)}$ *and* $E_i^{(k,l)}$, *for* $i = 1, 2, \ldots, m$, *are partial recursive, then so is the one below obtained by* composition:

$$\lambda \vec{x}\vec{\alpha}.F^{(m,l)}\left(E_1^{(k,l)}(\vec{x}, \vec{\alpha}), E_2^{(k,l)}(\vec{x}, \vec{\alpha}), \ldots, E_m^{(k,l)}(\vec{x}, \vec{\alpha}), \vec{\alpha} \right)$$

VIII. *There is a* universal functional *of rank* $(k + 1, l)$ —$UN^{(k+1,l)}$— *for all computable functionals* $\{e\}^{(k,l)}$ *of rank* (k, l). *That is,* $UN^{(k+1,l)}(e, \vec{x}, \vec{\alpha}) \simeq \{e\}(\vec{x}, \vec{\alpha})$, *for all* $e, \vec{x}, \vec{\alpha}$.

In view of the above corollaries, in particular items I–III and VII, we have from Exercise 2.1.10:

13.8.13 Proposition. *The set of partial recursive functionals is closed under Grzegorczyk substitution into type-0 inputs, as long as the type-1 variable list is the same in the substitution parts:* $\lambda \vec{x}\vec{z}\vec{\alpha}.F(G(\vec{z}, \vec{\alpha}), \vec{x}, \vec{\alpha})$.

13.8.14 Remark. If $\lambda y\vec{x}\vec{\alpha}.F(y, \vec{x}, \vec{\alpha})$ and $\lambda \vec{z}\vec{\beta}.G(\vec{z}, \vec{\beta})$ are partial recursive, is this true also for $\lambda \vec{x}\vec{z}\vec{\alpha}\vec{\beta}.F(G(\vec{z}, \vec{\alpha}), \vec{x}, \vec{\beta})$?

We will answer this question in Sect. 13.8.2. A partial answer is found in the remark following Proposition 13.8.26. $\square$

The following technical results bootstrap the theory.

13.8.15 Theorem. (S-m-n Theorem; Gödel, Kleene) *There is an* index $Sb_1(e, y)$ *that depends on* e *and* y *and, for all* e *and* y, *we have*

$$\{Sb_1(e, y)\}(\vec{x}, \vec{\alpha}) \simeq z \;\; iff \; \{e\}(y, \vec{x}, \vec{\alpha}) \simeq z$$

Moreover, $\|e, y, \vec{x}, \vec{\alpha}, z\| < \|Sb_1(e, y), \vec{x}, \vec{\alpha}, z\|$.

Proof Note that

$$Sb_1(e, y) = \langle 7, k+1, l, e, \langle 3, k, l, y \rangle, \langle 1, k, l, 1 \rangle, \langle 1, k, l, 2 \rangle, \ldots, \langle 1, k, l, m \rangle \rangle$$

works. The inequality is from Definition 13.8.3, case VII. □

It is clear that $Sb_1(e, y)$ is strictly increasing with respect to e and y.

Using the approach from Theorem 10.2.1 we can define S-m-n functions Sb_n exactly as in Definition 10.2.2 and obtain

13.8.16 Corollary. *For each $n \geq 1$, there is a rank-$(n + 1, 0)$ $Sb_n \in \mathcal{PR}$ and we have that, for all $m \geq 1$, the following:*

$$\{Sb_n(a, \vec{y}_n)\}(\vec{x}_m, \vec{\alpha}) \simeq \{a\}(\vec{y}_n, \vec{x}_m, \vec{\alpha})$$

Moreover, $\|e, \vec{y}_n, \vec{x}, \vec{\alpha}, z\| < \|Sb_n(e, \vec{y}_n), \vec{x}, \vec{\alpha}, z\|$.

13.8.17 Remark. As we noted already in Proposition 13.8.13, the presence of I. in Corollary 13.8.10 enables Grzegorczyk operations on type-0 inputs, thus, a relaxed version of the S-m-n theorem (as already noted in Theorem 5.7.1) is

*There is a primitive recursive h of n variables, such that, for any **F** of rank $(n + m, l)$ and any $\vec{y}_n$ and $\vec{x}_m$, we have*

$$\{h(\vec{y}_n)\}(\vec{x}_m, \vec{\alpha}) \simeq F(\vec{x}_m, \vec{y}_n, \vec{\alpha}) \tag{1}$$

Pause. Does it matter if the order of type-0 variables on the right hand side of (1) is any permutation of the x_i, y_r in $\vec{x}_m, \vec{y}_n$? ◄ □

13.8.18 Theorem. (The 2nd Recursion Theorem; Gödel, Carnap, Kleene) *For any partial recursive $\lambda e \vec{x} \vec{\alpha}.F(e, \vec{x}, \vec{\alpha})$ there is an index $\overline{e}$ such*

$$F(\overline{e}, \vec{x}, \vec{\alpha}) \simeq z \;\; iff \; \{\overline{e}\}(y, \vec{x}, \vec{\alpha}) \simeq z$$

Proof Via Gödel's "substitution trick". Let $\{a\} = F$ and consider the partial recursive functional $\lambda z \vec{x} \vec{\alpha}.F(S_1^{(k+1,l)}(z, z, \vec{x}, \vec{\alpha}), \vec{x}, \vec{\alpha})$ (cf. Proposition 13.8.13). Let b be an index of the latter. Then

$$\{Sb_1(b, z)\}(\vec{x}, \vec{\alpha}) \simeq \{b\}(z, \vec{x}, \vec{\alpha}) \simeq F(Sb_1(z, z), \vec{x}, \vec{\alpha})$$

Let $\overline{e} = Sb_1(b, b)$. □

We have a number of applications, using **DC** (cf. Corollary 13.8.10, V) in conjunction with the recursion theorem.

13.8.19 Theorem. ((μ) on Restricted Partial Computable Functionals) *Let* $\lambda \vec{x} \vec{\alpha}.F(\vec{x}, \vec{\alpha})$ *be partial computable. Then so is* H *given by*

$$H(y, \vec{x}, \vec{\alpha}) \simeq \begin{cases} y & \text{if } F(y, \vec{x}, \vec{\alpha}) \simeq 0 \\ H(y+1, \vec{x}, \vec{\alpha}) & \text{othw} \end{cases} \tag{1}$$

Proof We note that H is uniquely determined as follows: Suppose that

$$H(m, \vec{x}, \vec{\alpha}) = H(m+1, \vec{x}, \vec{\alpha}) = H(m+2, \vec{x}, \vec{\alpha}) = \dots$$

$$= H(k, \vec{x}, \vec{\alpha}) = k = \begin{cases} \min\{y \geq m : F(y, \vec{x}, \vec{\alpha}) \simeq 0 \wedge \\ \qquad\qquad (\forall z)_{m \leq z < k} F(z, \vec{x}, \vec{\alpha}) > 0\} \\ \uparrow \text{othw} \end{cases}$$

In particular

$$H(0, \vec{x}, \vec{\alpha}) = (\mu y) F(y, \vec{x}, \vec{\alpha}) \tag{2}$$

As for the partial recursiveness of H we argue as in Sect. 9.5: Define

$$G(z, y, \vec{x}, \vec{\alpha}) \simeq \begin{cases} y & \text{if } F(y, \vec{x}, \vec{\alpha}) \simeq 0 \\ \{z\}(y+1, \vec{x}, \vec{\alpha}) & \text{othw} \end{cases}$$

By composition and the recursiveness of DC, G is partial recursive. By Theorem 13.8.18 there is an e such that $\{e\}(y, \vec{x}, \vec{\alpha}) \simeq G(e, y, \vec{x}, \vec{\alpha})$, thus $\{e\} = H$. By (2) we are done.   □

13.8.20 Proposition. (The Predecessor) $P_i^{(k,l)} = \lambda \vec{x} \vec{\alpha}.x_i \dot{-} 1$, *is recursive.*

Proof We want $x_i \dot{-} 1 \simeq (\mu y)(y+1 = x_i)$. Thus define

$$H(y, \vec{x}, \vec{\alpha}) = \begin{cases} 0 & \text{if } x_i = y+1 \\ H(y, \vec{x}, \vec{\alpha}) + 1 & \text{othw} \end{cases} \tag{$\dagger$}$$

The definition of H above is a standard variant of that in (1) of the previous proof. With the help of DC we can dress ($\dagger$) up to not only follow our rules of formation but also to be seen that it does:

$$H(y, \vec{x}, \vec{\alpha}) = DC^{(k+4,l)}\Big(0, H(y, \vec{x}, \vec{\alpha})+1, x_i, y+1, \vec{x}, \vec{\alpha}\Big)$$

By the recursion theorem, we obtain an e such that $\{e\} = H$. Then

$$H(0, \vec{x}, \vec{\alpha}) = \begin{cases} x_i \dot{-} 1 & \text{if } x_i > 0 \\ \uparrow & \text{othw} \end{cases}$$

To conclude, mindful of (Proposition 13.8.13), we set

$$P_i^{(k,l)}(\vec{x}, \vec{\alpha}) \stackrel{Def}{=} DC^{(k+4,l)}\left(0, H(0, \vec{x}, \vec{\alpha}), x_i, 0, \vec{x}, \vec{\alpha}\right)$$

It is clear that $P_i^{(k,l)}$ is total. □

13.8.21 Theorem. (Closure Under Primitive Recursion) *The set of partial recursive restricted functionals is closed under primitive recursion:*

$$F(0, \vec{x}, \vec{\alpha}) = H(\vec{x}, \vec{\alpha})$$
$$F(z + 1, \vec{x}, \vec{\alpha}) = G(z, \vec{x}, \vec{\alpha}, F(z, \vec{x}, \vec{\alpha})) \tag{1}$$

That is, if H and G are partial recursive then so is F.

We shall call z the iteration variable *and G the* iteration functional, *while H is the* Basis *functional.*

Proof We rewrite (1) as one equation using DC and $P_1^{(k+1,l)}$ of Proposition 13.8.20:

$$F(z, \vec{x}, \vec{\alpha}) = DC\left(H(\vec{x}, \vec{\alpha}), G(z \dot{-} 1, \vec{x}, \vec{\alpha}, F(z, \vec{x}, \vec{\alpha})), z, 0, \vec{x}, \vec{\alpha}\right) \tag{2}$$

where we wrote above $z \dot{-} 1$ rather than $P_1^{(k+1,l)}(z, \vec{x}, \vec{\alpha})$ for readability.

By Theorem 0.4.43, we *have* a *unique F* that "solves" equations (1). By (2), the recursion theorem guarantees the existence of an e such that $\{e\} = F$.

□

13.8.22 Corollary. *The set of (total) recursive restricted functionals is closed under primitive recursion.*

Proof It is clear from the above proof and Theorem 0.4.43 that if H and G are total, then so is F (induction on the iteration variable). □

Also,

13.8.23 Proposition. *The class of recursive restricted functionals is closed under composition.*

Proof Because the class of all partial recursive restricted functionals is, and composition preserves totalness. □

One can now see immediately that all the tools from Sect. 2.1 are immediately available, in particular the coding tools of Definition 2.4.2. We state the obvious, without proofs:

13.8.24 Corollary. $\mathcal{PR}$ —*a class of* $(k, 0)$ *functionals, for all* $k \geq 1$— *is a subset of the the class of recursive restricted functionals.*

We have more primitive recursive functionals than the above:

13.8.25 Definition. (Primitive Recursive Restricted Functionals) The class of the *primitive recursive* functionals of rank (k, l) —$k + l \geq 1$— is the *smallest* class that includes the initial functionals (13.8.10, I–VI) and is closed under *composition* (equivalently,[30] *Grzegorczyk operations on type-0 variables*) and *primitive recursion.* □

 It is immediate that *all primitive recursive restricted functionals are total* (cf. Definition 13.1.1) and therefore they form a subset of the recursive restricted functionals, since the latter class contains the same initial functions and is closed under the same operations.

For Definition 13.8.25 it suffices to include as initial functionals only I–III and VI.

The following is quite useful.

13.8.26 Proposition. (Type-1 Argument List Expansion) *If* $\lambda \vec{x}\vec{\alpha}.F(\vec{x}, \vec{\alpha})$ *is a primitive recursive restricted functional, then so is* $\lambda \vec{x}\vec{y}\vec{\alpha}\vec{\beta}.G(\vec{x}, \vec{y}, \vec{\alpha}, \vec{\beta}) = \lambda \vec{x}\vec{y}\vec{\alpha}\vec{\beta}.F(\vec{x}, \vec{\alpha})$.

Proof The case for type-0 argument list expansion follows by Grzegorczyk operations (Proposition 13.8.13) that holds for all partial recursive functionals. To show that we can do $\lambda \vec{x}\vec{\alpha}\vec{\beta}.F(\vec{x}, \vec{\alpha})$ in the primitive recursive case we do induction on the inductive definition of such functionals (Definition 13.8.25).

1. *The claim checks for the initial functionals.*
   We sample case VI: Then $F(\vec{x}, \vec{\alpha}) \simeq \{\langle 6, k, l, i, j \rangle\}(\vec{x}, \vec{\alpha}) \simeq \alpha_j(i)$, where $k = length(\vec{x})$, $l = length(\vec{\alpha})$. Clearly, $\alpha_j(i) \simeq \{\langle 6, k, l + l', i, j \rangle\}(\vec{x}, \vec{\alpha}, \vec{\beta}) \simeq \lambda \vec{x}\vec{\alpha}\vec{\beta}.F(\vec{x}, \vec{\alpha}, \vec{\beta})$, where $l' = length(\vec{\beta})$.

2. *The claim propagates with Grzegorczyk substitution.* Suppose $H(\vec{x}, \vec{z}, \vec{\alpha}) \simeq F(G(\vec{z}, \vec{\alpha}), \vec{x}, \vec{\alpha})$ and for the primitive recursive F and G we have

$$F(u, \vec{x}, \vec{\alpha}) \simeq F(u, \vec{x}, \vec{\alpha}, \vec{\beta}) \simeq w \tag{1}$$

   and

$$G(\vec{z}, \vec{\alpha}) \simeq G(\vec{z}, \vec{\alpha}, \vec{\beta}) \simeq v \tag{2}$$

---

[30] In the presence of I–III.

Thus,

$$H(\vec{x}, \vec{z}, \vec{\alpha}, \vec{\beta}) \simeq w$$

iff

$$(\exists u)\left(F(u, \vec{x}, \vec{\alpha}, \vec{\beta}) \simeq w \wedge G(\vec{z}, \vec{\alpha}, \vec{\beta}) \simeq u \right)$$

iff, by (1) and (2)

$$(\exists u)\left(F(u, \vec{x}, \vec{\alpha}) \simeq w \wedge G(\vec{z}, \vec{\alpha}) \simeq u \right)$$

iff

$$H(\vec{x}, \vec{z}, \vec{\alpha}) \simeq w$$

3. *The claim propagates with primitive recursion.* This case is left to the reader as it is entirely similar to the above. Let $H = prim(F, G)$ and accept the I.H. that

$$F(\vec{y}, \vec{\alpha}) \simeq F(\vec{y}, \vec{\alpha}, \vec{\beta}) \simeq u \tag{1}$$

and

$$G(x, \vec{y}, w, \vec{\alpha}) \simeq G(x, \vec{y}, w, \vec{\alpha}, \vec{\beta}) \simeq v \tag{2}$$

Next compare $H(x, \vec{y}, \vec{\alpha})$ with $H(x, \vec{y}, \vec{\alpha}, \vec{\beta})$. $\qquad\qquad\square$

13.8.27 Remark. From the above follows immediately that for *primitive recursive restricted functionals* we can do[31] substitutions and primitive recursions where the two component restricted functionals in each case have different type-1 lists of inputs, α and β. Using the above proposition we can equalise the two lists forming the list $\vec{\alpha}, \vec{\beta}$. Then, e.g., a substitution like $F(G(\vec{z}, \vec{\alpha}), \vec{x}, \vec{\beta})$ is the same as $F(G(\vec{z}, \vec{\alpha}, \vec{\beta}), \vec{x}, \vec{\alpha}, \vec{\beta})$. But the latter does not lead outside the class of primitive recursive functionals, hence nor does the former. $\qquad\square$

13.8.28 Definition. (Characteristic Functional) A *characteristic functional* $C_{\mathbb{P}}$ of a predicate $\mathbb{P}$ is defined as

$$C_{\mathbb{P}}(\vec{x}, \vec{\alpha}) \simeq \begin{cases} 0 & \text{if } \mathbb{P}(\vec{x}, \vec{\alpha}) \\ 1 & \text{othw} \end{cases}$$

$\square$

---

[31] "Can do" is a colloquialism for "if we do so, then we get a primitive recursive restricted functional".

13.8.29 Definition. (Semi-Recursiveness and (Primitive) Recursiveness) A relation $\mathbb{P}$ of rank (k, l) is called *semi-recursive* in the Kleene theory of computable restricted functionals iff it is the *domain of a partial recursive restricted functional* F of rank (k, l).

An index e for F is by definition a *semi-recursive index*, or simply *semi-index* or just *index* of $\mathbb{P}$.

A relation $\mathbb{P}$ is called *(primitive) recursive* iff it has a *(primitive) recursive characteristic restricted functional* $C_\mathbb{P}$.   □

13.8.30 Proposition. *A relation $\mathbb{P}$ of rank (k, l) is* semi-recursive *iff for some partial restricted functional F we have*

$$\mathbb{P}(\vec{x}, \vec{\alpha}) \equiv F(\vec{x}, \vec{\alpha}) \simeq 0$$

Proof

If part. Let F be partial recursive as above. We will adjust it (and rename G) so that when $F(\vec{x}, \vec{\alpha}) \not\simeq 0$ the call is undefined.

We have noted earlier that $\{0\}^{(k,l)} = \emptyset$, the empty function.

Thus if

$$G \overset{Def}{=} \lambda \vec{x}\vec{\alpha}.DC^{(k+4,l)}\left(0, \{0\}(0), F(\vec{x}, \vec{\alpha}), 0, \vec{x}, \vec{\alpha}\right)$$

then $\mathbb{P}(\vec{x}, \vec{\alpha}) \equiv G(\vec{x}, \vec{\alpha}) \downarrow (\simeq 0)$ and G is partial recursive.

Only If part. Let $\mathbb{P}(\vec{x}, \vec{\alpha}) \equiv H(\vec{x}, \vec{\alpha}) \downarrow$ for some partial recursive H.

Note that $F = \lambda \vec{x}\vec{\alpha}.C_0(H(\vec{x}, \vec{\alpha}), \vec{x}, \vec{\alpha})$ is a partial recursive restricted functional that fits Definition 13.8.29.   □

13.8.31 Example. (Example 13.3.35 Revisited) This is a much more satisfactory version of Example 13.3.35 as primitive recursiveness is easily established in all cases.

(1) Firstly, by Proposition 13.8.26 we can add "don't care" or "dummy" type-1 arguments to any function in $\mathcal{PR}$ to obtain a primitive recursive functional.

For example, from $\lambda xy.x \dotdiv y$ of $\mathcal{PR}$ we get the primitive recursive functional $\lambda xy\vec{\alpha}.x \dotdiv y$. Similarly from $\lambda xy.|x - y| \in \mathcal{PR}$ we get the primitive recursive functional $\lambda xy\vec{\alpha}.|x - y|$.

(2) If $\lambda y\vec{x}\vec{\alpha}.F(y, \vec{x}, \vec{\alpha})$ is primitive recursive then the primitive recursion below shows that so is $\lambda z\vec{x}\vec{\alpha}. \sum_{i<z} F(i, \vec{x}, \vec{\alpha})$.

$$\sum_{i<0} F(i, \vec{x}, \vec{\alpha}) \simeq 0$$

$$\sum_{i<z+1} F(i, \vec{x}, \vec{\alpha}) \simeq F(z, \vec{x}, \vec{\alpha}) + \sum_{i<z} F(i, \vec{x}, \vec{\alpha})$$

Similarly, $\lambda z \vec{x} \vec{\alpha} . \prod_{i<z} F(i, \vec{x}, \vec{\alpha})$ is primitive recursive as the following shows:

$$\prod_{i<0} F(i, \vec{x}, \vec{\alpha}) \simeq 1$$

$$\prod_{i<z+1} F(i, \vec{x}, \vec{\alpha}) \simeq F(z, \vec{x}, \vec{\alpha}) \times \prod_{i<z} F(i, \vec{x}, \vec{\alpha})$$

(3) If $\mathbb{P}(y, \vec{x}, \vec{\alpha})$ is a primitive predicate and $\lambda \vec{y} \vec{\beta} . F(\vec{y}, \vec{\beta})$ is a primitive recursive restricted functional, then

$$\mathbb{P}(F(\vec{y}, \vec{\beta}), \vec{x}, \vec{\alpha}) \tag{1}$$

is a primitive recursive predicate.

Indeed, let $C_\mathbb{P}$ be the primitive recursive characteristic functional of $\mathbb{P}$.
Then $\lambda \vec{x} \vec{y} \vec{\alpha} \vec{\beta} . C_\mathbb{P}(F(\vec{y}, \vec{\beta}), \vec{x}, \vec{\alpha})$ is primitive recursive by assumption and Remark 13.8.27, and is also the characteristic functional of (1).

(4) The class of primitive recursive predicates is closed under Boolean operations. So let $\mathbb{P}(\vec{x}, \vec{\alpha})$ and $\mathbb{Q}(\vec{y}, \vec{\beta})$ be primitive recursive, that is, we have primitive recursive characteristic functions $C_\mathbb{P}$ and $C_\mathbb{Q}$ for them respectively.

$\neg$ case: Since in the context of restricted functionals "total" means *defined for all inputs, type-0 and type-1 alike*, we readily have (as in the absolute case of Chap. 2) that $C_{\neg\mathbb{P}} = \lambda \vec{x} \vec{\alpha} . 1 \dotminus C_\mathbb{P}(\vec{x}, \vec{\alpha})$, a primitive recursive restricted functional.

$\vee$ case: Note that $C_{\mathbb{P}\vee\mathbb{Q}} = \lambda \vec{x} \vec{y} \vec{\alpha} \vec{\beta} . C_\mathbb{P}(\vec{x}, \vec{\alpha}) \times C_\mathbb{Q}(\vec{y}, \vec{\beta})$, a primitive recursive restricted functional (cf. also Remark 13.8.27).

$\wedge, \rightarrow, \equiv$: These follow from the above two as they are derived connectives.

(5) The relation $y = \alpha(x)$ is primitive recursive. In fact, in the equivalence below

$$y = \alpha(x) \equiv |y - \alpha(x)| \simeq 0$$

the right hand side of $\equiv$ is obtained by substituting (3 above) in the primitive recursive predicate $y = z$ the primitive recursive $\lambda x \alpha . \alpha(x)$ (1).
Alternatively,

$$DC(0, 1, y, \alpha(x), x, \alpha)$$

is the primitive recursive characteristic functional for $y = \alpha(x)$.

(6) The class of primitive recursive predicates is closed under bounded quantification.

Exactly as in the absolute case from Chap. 2.
So let $C_\mathbb{P}$ be the characteristic functional of $\mathbb{P}(y, \vec{x}, \vec{\alpha})$. Let us also call $C_{\exists\mathbb{P}}$ the characteristic functional of $(\exists)_{<z}\mathbb{P}(y, \vec{x}, \vec{\alpha})$.

Then

$$C_{\exists \mathbb{P}}(0, \vec{x}, \vec{\alpha}) \simeq 1$$

$$C_{\exists \mathbb{P}}(z + 1, \vec{x}, \vec{\alpha}) \simeq C_{\mathbb{P}}(z, \vec{x}, \vec{\alpha}) \times C_{\exists \mathbb{P}}(z, \vec{x}, \vec{\alpha})$$

Alternatively,

$$(\exists)_{<z}\mathbb{P}(y, \vec{x}, \vec{\alpha}) \equiv \prod_{i<z} C_{\mathbb{P}}(i, \vec{x}, \vec{\alpha}) \simeq 0$$

and we conclude by substitution into $w = 0$ (3 and 2 above) of the primitive recursive $\lambda z \vec{x} \vec{\alpha}. \prod_{i<z} C_{\mathbb{P}}(i, \vec{x}, \vec{\alpha})$.
The case for $(\forall y)_{<z}\mathbb{P}$ follows from

$$(\forall y)_{<z}\mathbb{P} \equiv \neg(\exists y)_{<z}\neg\mathbb{P}$$

(7) Recall Definition 11.3.6 and the notations $[u]$ and alternatively ξ_u or θ_u for finite functions of code u. The predicate $\lambda u\alpha.[u] \subseteq \alpha$ is primitive recursive since, clearly, the right hand side of

$$[u] \subseteq \alpha \equiv u \geq 1 \wedge (\forall y)_{\leq u}\left(p_y \mid u \rightarrow (u)_y = \alpha(y)\right) \qquad (**)$$

is by previous observations in this example.

(8)

> All the foregoing hold if we replace "primitive recursive" by "recursive".

□

13.8.1   Course-of-Values Recursion

The following definition strongly echoes Definition 2.4.11.

13.8.32 Definition. (Course-of-Values Recursion) We say that F, of rank $(n + 1, l)$, is defined from two total functionals —namely, the *basis* functional $\lambda\vec{y}_n\vec{\alpha}_l.B(\vec{y}_n, \vec{\alpha}_l)$ and the *iterator* functional $\lambda x \vec{y}_n z \vec{\alpha}_l.G(x, \vec{y}_n, z, \vec{\alpha}_l)$— by *course-of-values recursion* if, for all x, $\vec{y}_n$, $\vec{\alpha}_l$, the following equations hold:

$$\begin{cases} F(0, \vec{y}_n, \vec{\alpha}_l) & \simeq B(\vec{y}_n, \vec{\alpha}_l) \\ F(x + 1, \vec{y}_n, \vec{\alpha}_l) & \simeq G(x, \vec{y}_n, \overline{F}(x, \vec{y}_n, \vec{\alpha}_l), \vec{\alpha}_l) \end{cases} \qquad (1)$$

where $\lambda x \vec{y}_n \vec{\alpha}_l . \overline{F}(x, \vec{y}_n, \vec{\alpha}_l)$ is the *history function* of F at the point $(x, \vec{y}_n, \vec{\alpha}_l)$, that is,

$$\langle F(0, \vec{y}_n, \vec{\alpha}_l), F(1, \vec{y}_n, \vec{\alpha}_l), \ldots, F(x, \vec{y}_n, \vec{\alpha}_l) \rangle \qquad \square$$

With an identical proof as in Theorem 2.4.13 —except for the obvious trivial deviations— we obtain

13.8.33 Theorem. *The class of all primitive recursive restricted functionals of ranks (k, l), for all $k + l \geq 1$, is closed under course-of-values recursion in the sense of the above definition.*

13.8.2   Normal Form Theorems; **One Last Time**

This subsection quickly derives two versions of the Normal Form Theorem, and concentrates on applications such as the Selection Theorem and the Graph Theorem. How "quickly"? We will totally rely on the construction of "$Tree(y)$" of Sect. 13.4.1. Starting on p. 445 we showed that the $Tree(y)$ predicate is primitive recursive. Here we will not add to that construction and discussion except for the occasional obvious and very helpful observation: With *restricted functionals* "total" really means *total*: That is, *defined* for *all* type-0 *and* type-1 arguments. We display the definition of $Tree(y)$ below. The definition will be split into five parts,

1. Part $P(y)$ for the "preamble" that is applicable and common to all following parts.
2. Part $I(y)$ for the initial functions.
3. Part $C(y)$ for composition.
4. Part $U(y)$ for the universal functional application.
5. A part $O(y)$ for incorporating the variables $\vec{\alpha}_l$ about which we rely on the oracle will be added as an *after thought* externally to $Tree(y)$ to ensure that this predicate is *absolutely* primitive recursive.

(I) The $P(y)$:

$$P(y) \equiv Seq(y) \wedge lh(y) \geq 1 \wedge (\forall i)_{<lh(y)} Seq((y)_i) \wedge lh\big((y)_0\big) = 4 \wedge$$
$$Seq\big((y)_{0,0}\big) \wedge Seq\big((y)_{0,1}\big) \wedge Seq\big((y)_{0,2}\big)$$

(II) <u>The $I(y)$</u>:

$$I(y) \equiv lh(y) = 1 \wedge \Big($$

$$\{U_i^{n,l}\}(\exists p, q, a, b, c, x, s, n, l, i, j)_{<y}\Big[Seq(x) \wedge Seq(s) \wedge$$

$$lh(x) = n \wedge lh(s) = l \wedge$$

$$\Big\{i < n \wedge (y)_0 = \langle\langle 1, n, l, i+1\rangle\rangle * x * s * \langle(x)_i\rangle \vee$$

$$\{S_i^{n,l}\} \qquad i < n \wedge (y)_0 = \langle\langle 1, n, l, i+1\rangle\rangle * x * s * \langle(x)_i + 1\rangle \vee$$

$$\{C_c^{n,l}\} \qquad (y)_0 = \langle\langle 1, n, l, i+1\rangle\rangle * x * s * \langle c\rangle \vee$$

$$\{S_1^{(n+2,l)}\} \qquad (y)_0 = \langle\langle 4, n+2, l\rangle, a, b\rangle * x * s * \langle Sb_1(a, b)\rangle \vee$$

$$\{DC^{(n+4,l)}\} \qquad (y)_0 = \langle\langle 5, n+4, l\rangle, p, q, a, b\rangle * x * s * \langle p\rangle \wedge a = b \vee$$

$$\{DC^{(n+4,l)}\} \qquad (y)_0 = \langle\langle 5, n+4, l\rangle, p, q, a, b\rangle * x * s * \langle q\rangle \wedge a \neq b \vee$$

$$\{AP_{i,j}^{(n,l)}\} \qquad p_{(x)_i} \mid (s)_j \wedge$$

$$(y)_0 = \langle\langle 6, n, l, i+1, j+1\rangle\rangle * x * s * \langle(s)_{j,(x)_i}\rangle\Big\}\Big]$$

$$\Big)$$

(III) <u>The $C(y)$</u>:   We have $y = \langle v, t_1, \ldots, t_{n+1}\rangle$, where (cf. Fig. 13.1) $v = \langle\langle 7, m, l, f\rangle * g\rangle * a * s * \langle b\rangle$, and we have represented the vectors $\vec{g}_n = g_1, \ldots, g_n$ by $g = \langle\vec{g}_n\rangle$, $\vec{a}_m$ by $a = \langle\vec{a}_m\rangle$, and $\vec{z}_n$ (outputs of the g_i) by $z = \langle\vec{z}_n\rangle$, s being the code of the vector of approximations for $\vec{\alpha}_l$. Finally, we have set $t = \langle\vec{t}_{n+1}\rangle$ for the tree codes of the $n + 1$ subtrees (Fig. 13.1) so that $y = \langle v\rangle * t$. Note that as before, $v = (y)_0$.

In particular, the intuitive picture $\langle g_i, \vec{a}_m, \vec{s}_l, z_i\rangle$ for the recursive calls to the g_i is captured by (i-th call) $\langle(g)_i\rangle * a * s * \langle(z)_i\rangle$ (root of t_i, for $1 \leq i \leq n$).

$$\{\text{Fig. 13.1}\}\ C(y) \equiv (\exists a, z, f, g, m, n, s, l, b)_{<y}\Big\{lh(y) = n + 2 \wedge Seq(a) \wedge$$
$$Seq(z) \wedge Seq(f) \wedge Seq(g) \wedge n = lh(z) \wedge n = lh(g) \wedge$$
$$Seq(s) \wedge lh(s) = l \wedge m = lh(a) \wedge (f)_1 = n \wedge$$
$$(\forall i)_{<n}\big(Seq((g)_i) \wedge ((g)_i)_1 = m\big) \wedge$$

$\{\text{Root (main) call}\}\quad (y)_0 = \langle\langle 7, m, l, f\rangle * g\rangle * a * s * \langle b\rangle \wedge$

$\{\text{Root of } t_n\}\qquad ((y)_{n+1})_0 = \langle f\rangle * z * s * \langle b\rangle \wedge$

$\{\text{Root of } t_i,\ i < n\}\quad (\forall i)_{<n}((y)_{i+1})_0 = \langle(g)_i\rangle * a * s * \langle(z)_i\rangle\Big\}$

(IV) <u>The $U(u)$</u>: $y = \langle v, t_0\rangle$, where (cf. Fig. 13.2) $v = \langle\langle 8, n+1, l\rangle, f\rangle * a * s * \langle b\rangle$ and the root of t_0 is $(t_0)_0 = (y)_{1,0} = \langle f\rangle * a * s * \langle b\rangle$, where as above we set $a = \langle\vec{a}_n\rangle$ and $s = \langle s_l\rangle$.

Thus,

$$\{\text{Fig. 13.2}\}\ U(y) \equiv (\exists n, f, a, b, s, l)_{<y}\Big\{ lh(y) = 2 \wedge Seq(f) \wedge$$
$$Seq((y)_1) \wedge Seq(s) \wedge lh(s) = l \wedge$$
$$(f)_1 = n \wedge n > 0 \wedge$$
$$Seq(a) \wedge lh(a) = n \wedge$$

$$\{\text{The root call}\}\qquad (y)_0 = \langle\langle 8, n+1, l\rangle, f\rangle * a * s * \langle b\rangle \wedge$$
$$\{\text{The recursive call}\}\ (y)_{1,0} = \langle f \rangle * a * s * \langle b \rangle \Big\}$$

We can now conclude the course-of-values recursive definition of $Tree(y)$.

$$Tree(y) \equiv P(y) \wedge \Big(I(y) \vee C(y) \vee U(y) \Big) \wedge (\forall i)_{<lh(y)}(i > 0 \rightarrow Tree((y)_i))$$

It is clear by the course-of-values recursion above that $Tree(y)$ is absolutely primitive recursive (is in $\mathcal{PR}_*$).

We can now define the Kleene $\mathbb{T}$-predicate for restricted functionals, using $\mathbb{T}$ to distinguish the predicate for such functionals from the ones we derived for Kleene computability and Moschovakis computability of unrestricted functionals.

13.8.34 Definition. (The Kleene Predicate of Rank (n, l)) For any given n and l we define the *Kleene predicate* of rank (n, l) by

$$\mathbb{T}^{(n,l)}(z, \vec{x}_n, \vec{\alpha}_l, y) \stackrel{Def}{\equiv} Tree(y) \wedge (y)_{0,0} = z \wedge (y)_{0,1} = \langle \vec{x}_n \rangle \wedge lh((y)_{0,2}) = l \wedge$$
$$(\forall i)_{<lh((y)_{0,2})}[(y)_{0,2,i}] \subseteq \alpha_{i+1}$$

Since $(y)_0 = \langle z, \langle \vec{x}_n \rangle, \langle \vec{s}_l \rangle, b \rangle$, where $\langle s_l \rangle$ is a vector of the sub functions of the $\vec{\alpha}_l$ *that were* actually *used* during the computation and b is the output of the computation, we have that $b = (y)_{0,3}$. We will call this "*decoding function*" of the computation by the short name D (note capitalisation in this section) in the context of computation with *unrestricted functionals*.   □

13.8.35 Theorem. (Kleene Normal Form Theorem I) *There is an absolute primitive recursive function D of one argument, and, for any given n and l, a primitive recursive predicate* $\mathbb{T}^{(n,l)}(z, \vec{x}_n, \vec{\alpha}_l, y)$ *as defined above such that*

(a) $\{z\}^{(n,l)}(\vec{x}_n, \vec{\alpha}_l) \downarrow \equiv (\exists y)\mathbb{T}^{(n,l)}(z, \vec{x}_n, \vec{\alpha}_l, y)$

(b) $\{z\}^{(n,l)}(\vec{x}_n, \vec{\alpha}_l) \simeq u \equiv (\exists y)\Big(\mathbb{T}^{(n,l)}(z, \vec{x}_n, \vec{\alpha}_l, y) \wedge D(y) = u \Big)$

(c) $\{z\}^{(n,l)}(\vec{x}_n, \vec{\alpha}_l) \simeq D\Big((\mu y)\mathbb{T}^{(n,l)}(z, \vec{x}_n, \vec{\alpha}_l, y) \Big)$

Proof The primitive recursiveness of $\mathbb{T}^{(n,l)}$ follows from that of $Tree(y)$ and $\lambda s\alpha.[s] \subseteq \alpha$ (cf. Example 13.8.31, (3), (4), (6), (7)).

All the proof work for (a)–(b) is already done.

For (c) we translate (a) applying (μy) to $\mathbb{T}^{(n,l)}(z, \vec{x}_n, \vec{\alpha}_l, y)$ to obtain the least y that the existential statement promises (when the left hand side is defined) since $C_\mathbb{T}$

is "*really*" *defined* for *all* inputs and the unbounded search does not suffer from any $w < y$ on which $C_{\mathbb{T}}(z, \vec{x}_n, \vec{\alpha}_l, w)$ may be *undefined*; there are *no* such w.

13.8.36 Remark. The finite sub functions $[(y)_{0,2,j-1}] - j = 1, \ldots, l-$ of the α_j actually *used* in the computation coded by y are also sub functions of the finite (given extensionally below)

$$\alpha_j \upharpoonright y \stackrel{Def}{=} \{(x, \alpha_j(x)) : x < y\}$$

since no query to the oracle can *use* a higher input (for α_j) than y because the code y contains *all* numbers r *referred* to (*used* in) the computation and as such $r < y$.

We can code $\alpha_j \upharpoonright y$ by a single number; that number is the *history* of (the *total*) α_j at $y \doteq 1$, denoted in this chapter by an *overline* as

$$\overline{\alpha_j}(y) = \langle \alpha_j(0), \ldots, \alpha_j(y \doteq 1) \rangle$$

The "y" in "$\overline{\alpha_j}(y)$" is the *length* of the history:

$$y = lh\left(\overline{\alpha_j}(y)\right), \text{ cf. Definition 2.4.2}$$

Thus, for $j < l$, $[(y)_{0,2,j}] \subseteq \alpha_{j+1}$ in Definition 13.8.34 is equivalent to $[(y)_{0,2,j}] \subseteq \alpha_{j+1} \upharpoonright y$, where y is the *computation* ("*Tree*") *code*, that is,

$$(y)_{0,2,j} \geq 1 \wedge (\forall x)_{<y}\left(p_x | (y)_{0,2,j} \rightarrow ((y)_{0,2,j})_x = \left(\overline{\alpha_{j+1}}(y)\right)_x\right)$$

The above comments readily lead to the *Normal Form II* below.   □

We translate Definition 13.8.34 and define $\mathbb{T}_l$ as

13.8.37 Definition. (The Kleene Predicate of Rank $(4, 0)$) For any given l we define the *Kleene predicate* of rank $(4, 0)$ by

$$\mathbb{T}_l(z, x, w, y) \stackrel{Def}{\Leftrightarrow} Tree(y) \wedge (y)_{0,0} = z \wedge lh((y)_{0,2}) = l \wedge Seq(w) \wedge lh(w) = l \wedge$$

$$(\forall v)_{<y}(v \in w \rightarrow Seq(v)) \wedge$$

$$(\forall j)_{<lh((y)_{0,2})}[(y)_{0,2,j}] \subseteq [(w)_j]$$

$$\Leftrightarrow Tree(y) \wedge (y)_{0,0} = z \wedge lh((y)_{0,2}) = l \wedge Seq(w) \wedge lh(w) = l \wedge$$

$$(\forall v)_{<y}(v \in w \rightarrow Seq(v)) \wedge$$

$$(\forall j)_{<lh((y)_{0,2})}\Big\{(y)_{0,2,j} \geq 1 \wedge$$

$$(\forall x)_{<y}\left(p_x | (y)_{0,2,j} \rightarrow ((y)_{0,2,j})_x = (w)_{j,x}\right)\Big\}$$

Here, the input x in $\mathbb{T}_l$ will receive input $\langle \vec{x}_n \rangle$ while w will receive input $\langle \overline{\alpha_1}(y), \ldots, \overline{\alpha_l}(y) \rangle$. By the preceding remark if $w = \langle \overline{\alpha_1}(y), \ldots, \overline{\alpha_l}(y) \rangle$, then, for $0 \leq j < l$, $(w)_j = \overline{\alpha_{j+1}}(y)$ and if $x < y$, then $(w)_{j,x} = \left(\overline{\alpha_{j+1}}(y) \right)_x = \alpha_{j+1}(x)$. $\square$

13.8.38 Theorem. (Kleene Normal Form Theorem II) *For each $l \geq 0$, there exists an absolute* primitive recursive *predicate* $\mathbb{T}_l(z, x, w, y)$ *of rank $(4, 0)$ such that for each functional of rank (n, l) we have*

(*i*) $\{z\}^{(n,l)}(\vec{x}_n, \vec{\alpha}_l) \downarrow \equiv (\exists y) \mathbb{T}_l\big(z, \langle \vec{x}_n \rangle, \langle \overline{\alpha_1}(y), \ldots, \overline{\alpha_l}(y) \rangle, y\big)$

(*ii*)

$$\{z\}^{(n,l)}(\vec{x}_n, \vec{\alpha}_l) \simeq u \equiv (\exists y)\Big(\mathbb{T}_l\big(z, \langle \vec{x}_n \rangle, \langle \overline{\alpha_1}(y), \ldots, \overline{\alpha_l}(y) \rangle, y\big) \wedge D(y) = u \Big)$$

(*iii*) $\{z\}^{(n,l)}(\vec{x}_n, \vec{\alpha}_l) \simeq D\Big((\mu y) \mathbb{T}_l\big(z, \langle \vec{x}_n \rangle, \langle \overline{\alpha_1}(y), \ldots, \overline{\alpha_l}(y) \rangle, y\big) \Big)$

Proof

(*i*) By Theorem 13.8.35, (*a*), Remark 13.8.36 and by comparing Definitions 13.8.34 and 13.8.37.
(*ii*) No special comment needed.
(*iii*) No special comment needed, except, once again we note that (μy) applies here to a *really* total[32] functional —$C_{\mathbb{T}_l}\big(z, \langle \vec{x}_n \rangle, \langle \overline{\alpha_1}(y), \ldots, \overline{\alpha_l}(y) \rangle, y\big)$— thus the minimum y is indeed computed. $\square$

13.8.39 Long Remark. (Corollaries of the Normal Form Theorems) We list a few consequences of the Normal Form theorems for restricted functionals.

(I) Normal form theorem for semi-recursive predicates.

13.8.40 Corollary. (Semi-Recursive Relations, Normal Form I) $\mathbb{P}(\vec{x}_n, \vec{\alpha}_l)$ *is semi-recursive iff we have*

$$\mathbb{P}(\vec{x}_n, \vec{\alpha}_l) \equiv (\exists y) \mathbb{T}^{(n,l)}\big(z, \vec{x}_n, \vec{\alpha}_l, y\big) \tag{1}$$

where z is a semi-index of $\mathbb{P}$.

Proof *If part.* Immediate by Theorem 13.8.35, item (*a*), since the assumption entails $\mathbb{P}(\vec{x}_n, \vec{\alpha}_l) \equiv \{z\}^{(n,l)}(\vec{x}_n, \vec{\alpha}_l) \downarrow$.

Only if part. Assume (1). The right hand side of it says $\{z\}^{(n,l)}(\vec{x}_n, \vec{\alpha}_l) \downarrow$ by Theorem 13.8.35, item (*a*). $\square$

13.8.41 Corollary. (Semi-Recursive Relations, Normal Form II)
$\mathbb{P}(\vec{x}_n, \vec{\alpha}_l)$ *is semi-recursive iff we have*

---

[32] Defined for *all* type-0 and type-1 arguments.

$$\mathbb{P}(\vec{x}_n, \vec{\alpha}_l) \equiv (\exists y)\mathbb{T}_l\big(z, \langle \vec{x}_n \rangle, \langle \overline{\alpha_1}(y), \dots, \overline{\alpha_l}(y)\rangle, y\big) \tag{2}$$

where z is a semi-index of $\mathbb{P}$.

Proof Because the right hand side of $\equiv$ in (i) says $\{z\}^{(n,l)}(\vec{x}_n, \vec{\alpha}_l) \downarrow$ by Theorem 13.8.38, item (i). □

(II) Partial recursive restricted functionals and semi-recursive relations are closed under type-1 *argument list expansion*.

13.8.42 Corollary. *If $\lambda \vec{x}_n \vec{\alpha}_l.F(\vec{x}_n, \vec{\alpha}_l)$ is partial recursive, then so is $G = \lambda \vec{x}_n \vec{\alpha}_l \vec{\beta}_k.F(\vec{x}_n, \vec{\alpha}_l)$ and we have*

$$\boldsymbol{F}(\vec{x}_n, \vec{\alpha}_l) \simeq u \equiv \boldsymbol{G}(\vec{x}_n, \vec{\alpha}_l, \vec{\beta}_k) \simeq u$$

Proof Let $\{f\} = \lambda \vec{x}_n \vec{\alpha}_l.F(\vec{x}_n, \vec{\alpha}_l)$. Then

$$\boldsymbol{F}(\vec{x}_n, \vec{\alpha}_l) \simeq u \overset{\text{Theorem 13.8.35,(b)}}{\equiv} (\exists y)\Big(\mathbb{T}^{(n,l)}(f, \vec{x}_n, \vec{\alpha}_l, y) \wedge D(y) = u\Big) \tag{i}$$

The predicate quantified by $(\exists y)$ is primitive recursive thus it supports expansion of its type-1 variables list (cf. Proposition 13.8.26), say, by $\vec{\beta}_k$.

Hence

$$\lambda u \vec{x}_n \vec{\alpha}_l \vec{\beta}_k.F(\vec{x}_n, \vec{\alpha}_l) \simeq u \overset{(i)}{\Leftrightarrow} \lambda u \vec{x}_n \vec{\alpha}_l \vec{\beta}_k.(\exists y)\Big(\mathbb{T}^{(n,l)}(f, \vec{x}_n, \vec{\alpha}_l, y) \wedge D(y) = u\Big)$$

$$\overset{13.8.26}{\Leftrightarrow} \lambda u \vec{x}_n \vec{\alpha}_l.(\exists y)\Big(\mathbb{T}^{(n,l)}(f, \vec{x}_n, \vec{\alpha}_l, y) \wedge D(y) = u\Big)$$

$$\overset{(i)}{\Leftrightarrow} \lambda u \vec{x} \vec{\alpha}.F(\vec{x}_n, \vec{\alpha}_l) \simeq u \qquad \Box$$

13.8.43 Corollary. *Suppose that $\lambda z \vec{x} \vec{\alpha}.F(z, \vec{x}, \vec{\alpha})$ and $\lambda \vec{y} \vec{\beta}.G(\vec{y}, \vec{\beta})$ are partial recursive.*
Then so is $\lambda \vec{x} \vec{y} \vec{\alpha} \vec{\beta}.F(G(\vec{y}, \vec{\beta}), \vec{x}, \vec{\alpha})$.

Proof Using the previous corollary and Grzegorczyk operations (the latter for type-0 arguments). □

(III) *Projection Theorem.*

13.8.44 Corollary. *A relation $\mathbb{P}(\vec{x}_n, \vec{\alpha}_l)$ is semi-recursive iff, for some primitive recursive $\mathbb{Q}(y, \vec{x}_n, w)$, we have*

$$\mathbb{P}(\vec{x}_n, \vec{\alpha}_l) \equiv (\exists y)\mathbb{Q}(y, \vec{x}_n, \langle \overline{\alpha_1}(y), \dots, \overline{\alpha_l}(y)\rangle) \tag{1}$$

Proof

- *If* direction. Say we have (1). We are done by noting

$$\mathbb{P}(\vec{x}_n, \vec{\alpha}_l) \Leftrightarrow (\exists y)\mathbb{Q}(y, \vec{x}_n, \langle \overline{\alpha_1}(y), \ldots, \overline{\alpha_l}(y)\rangle)$$

$$(\mu y)\mathbb{Q}(y, \vec{x}_n, \langle \overline{\alpha_1}(y), \ldots, \overline{\alpha_l}(y)\rangle) \downarrow$$

Recall that $C_\mathbb{Q}$ is *really* total.

- *Only if* direction. Let p be a semi-index of $\mathbb{P}(\vec{x}_n, \vec{\alpha}_l)$.
  Take $\mathbb{Q}(y, \vec{x}_n, w) \equiv \mathbb{T}_l(f, \langle \vec{x}_n\rangle, w, y)$. $\square$

(IV) The "semi-recursive *and* co-semi-recursive theorem".

13.8.45 Theorem. *If $\mathbb{Q}(\vec{x}, \vec{\alpha})$ and $\neg\mathbb{Q}(\vec{x}, \vec{\alpha})$ are both semi-recursive, then $\mathbb{Q}(\vec{x}, \vec{\alpha})$ is recursive.*
And conversely.

For a "property" $\mathscr{P}$ of predicates we say that "a predicate $\mathbb{R}$ is co-$\mathscr{P}$" iff $\neg\mathbb{Q}$ is $\mathscr{P}$ (that is, *has* property $\mathscr{P}$).

Proof   For the forward direction assume the assumption on $\mathbb{Q}(\vec{x}, \vec{\alpha})$.
Let e and f be semi-indices of $\mathbb{Q}(\vec{x}, \vec{\alpha})$ and $\neg\mathbb{Q}(\vec{x}, \vec{\alpha})$ respectively. Define, for all $\vec{x}, \vec{\alpha}$,

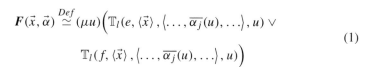

$$F(\vec{x}, \vec{\alpha}) \overset{Def}{\simeq} (\mu u)\Big(\mathbb{T}_l(e, \langle \vec{x}\rangle, \langle \ldots, \overline{\alpha_j}(u), \ldots\rangle, u) \vee \tag{1}$$

$$\mathbb{T}_l(f, \langle \vec{x}\rangle, \langle \ldots, \overline{\alpha_j}(u), \ldots\rangle, u)\Big)$$

By closure under (μu), F is partial recursive. Since finding the smallest u that works is guaranteed by the choice of e, f, F is total, hence recursive. But now

$$\mathbb{Q}(\vec{x}, \vec{\alpha}) \equiv \mathbb{T}_l(e, \langle \vec{x}\rangle, \langle \ldots, \overline{\alpha_j}(F(\vec{x}, \vec{\alpha})), \ldots\rangle, F(\vec{x}, \vec{\alpha}))$$

hence we are done by Grzegorczyk substitution.
For the converse, pick a fresh (new) z. Then $\mathbb{Q}(\vec{x}, \vec{\alpha}) \equiv (\exists z)\mathbb{Q}(\vec{x}, \vec{\alpha})$, hence $\mathbb{Q}$ is semi-recursive by the projection theorem. Same argument for $\neg\mathbb{Q}$. $\square$

(V) The *Selection Theorem* for Kleene computability of restricted functionals.

13.8.46 Corollary. (Selection Theorem) *For any $n+l \geq 1$ and any semi-recursive $\mathbb{P}(y, \vec{x}_n, \vec{\alpha}_l)$, there is a partial computable restricted functional $Sel_\mathbb{P}^{(n,l)}$ of rank (n, l) such that for we have*

(1) $Sel_\mathbb{P}^{(n,l)}(\vec{x}_n, \vec{\alpha}_l) \downarrow$ *iff* $(\exists y)\mathbb{P}(y, \vec{x}_n, \vec{\alpha}_l)$.
(2) $Sel_\mathbb{P}^{(n,l)}(\vec{x}_n, \vec{\alpha}_l) \downarrow$ *implies* $\mathbb{P}(Sel_\mathbb{P}^{(n,l)}(\vec{x}_n, \vec{\alpha}_l), \vec{x}_n, \vec{\alpha}_l)$.

Proof   By the projection theorem there is a primitive recursive $\mathbb{Q}$ such that

$$\mathbb{P}(y, \vec{x}_n, \vec{\alpha}_l) \equiv (\exists u)\mathbb{Q}(u, \langle y, \vec{x}_n \rangle, \langle \overline{\alpha_1}(u), \ldots, \overline{\alpha_l}(u) \rangle)$$

hence also

$$(\exists y)\mathbb{P}(y, \vec{x}_n, \vec{\alpha}_l) \equiv (\exists y)(\exists u)\mathbb{Q}(u, \langle y, \vec{x}_n \rangle, \langle \overline{\alpha_1}(u), \ldots, \overline{\alpha_l}(u) \rangle) \qquad (\dagger)$$

We compute the selection functional $\boldsymbol{Sel}_{\mathbb{P}}^{(n,l)}(\vec{x}_n, \vec{\alpha}_l)$ by selecting a y and u that *work* —that is, they make $\mathbb{Q}(u, \langle y, \vec{x}_n \rangle, \langle \overline{\alpha_1}(u), \ldots, \overline{\alpha_l}(u) \rangle)$ true.

In principle, we use *"parallelism"*:

> **do**   "simultaneously" , for **each** $u = 0, 1, 2, 3, \ldots$,
> **do**   "sequentially" , for **each** $y = 0, 1, 2, 3, \ldots$,
> **until**
> $$\left\{ \begin{array}{l} \\ \\ \mathbb{Q}(u, \langle y, \vec{x}_n \rangle, \langle \overline{\alpha_1}(u), \ldots, \overline{\alpha_l}(u) \rangle) \text{ is true} \end{array} \right.$$
> **return** y
> $$\Big\}$$

In so doing we will *not* be met by undefinedness, since $\boldsymbol{C}_{\mathbb{Q}}$ is "truly" total.

In practice we *dovetail* to *simulate* parallelism: Look for the first —if any— $w = 0, 1, 2, \ldots$, where $w = \langle u, y \rangle$, such that

$$\mathbb{Q}((w)_0, \langle (w)_1, \vec{x}_n \rangle, \langle \overline{\alpha_1}((w)_0), \ldots, \overline{\alpha_l}((w)_0) \rangle)$$

Return $(w)_1$ as the value of $\boldsymbol{Sel}_{\mathbb{P}}^{(n,l)}(\vec{x}_n, \vec{\alpha}_l)$ *(if defined)*.

Thus

$$\boldsymbol{Sel}_{\mathbb{P}}^{(n,l)}(\vec{x}_n, \vec{\alpha}_l) \stackrel{Def}{\simeq}$$

$$\left((\mu w)\mathbb{Q}\Big((w)_0, \langle (w)_1, \vec{x}_n \rangle, \langle \overline{\alpha_1}((w)_0), \ldots, \overline{\alpha_l}((w)_0) \rangle \Big) \right)_1$$

We immediately see that (2) is proved. As for (1), we note that the search succeeds iff $(w)_1$ is indeed a number (not "↑") and verifies (†) in the sense that $w = \langle u, y \rangle$ that is $u = (w)_0$ and $y = (w)_1$.   □

(VI) The Graph Theorem.

13.8.47 Corollary. *The restricted functional $\boldsymbol{F}$ of rank (n, l) is partial recursive iff its graph is semi-recursive.*

Proof

- *If* direction. Suppose that the predicate $\mathbb{G}(\vec{x}_n, \vec{\alpha}_l, u) \equiv F(\vec{x}_n, \vec{\alpha}_l) \simeq u$ is semi-recursive. Then $F(\vec{x}_n, \vec{\alpha}_l) \simeq \mathbf{Sel}_{\mathbb{G}}^{(n,l)}(\vec{x}_n, \vec{\alpha}_l)$.
- *Only if* direction. Let z be an index of F. We conclude using the Normal form theorem I, (b), or II, (ii), noting that the right hand side of $\equiv$ in either case is primitive recursive. Thus we next apply the projection theorem. □

(VII) Closure Properties of semi-recursive relations in Kleene computability for restricted functionals.

13.8.48 Corollary. *The semi-recursive relations in Kleene computability for restricted functionals are closed under* $\vee, \wedge, (\exists w), (\exists w)_{<z}, (\forall w)_{<z}$.

Proof Given semi-recursive $\mathbb{P}(\vec{x}_n, \vec{\alpha}_l), \mathbb{Q}(\vec{y}_m, \vec{\beta}_k), \mathbb{R}(w, \vec{z}_s, \vec{\gamma}_r)$. We have (13.8.44) primitive recursive predicates $\mathbb{O}, \mathbb{T}$ and $\mathbb{S}$ such that

$$\mathbb{P}(\vec{x}_n, \vec{\alpha}_l) \equiv (\exists u)\mathbb{O}(u, \vec{x}_n, \langle\overline{\alpha_1}(u), \dots, \overline{\alpha_l}(u)\rangle) \tag{i}$$

$$\mathbb{Q}(\vec{y}_m, \vec{\beta}_k) \equiv (\exists u)\mathbb{T}(u, \vec{y}_m, \langle\overline{\beta_1}(u), \dots, \overline{\beta_k}(u)\rangle) \tag{ii}$$

$$\mathbb{R}(w, \vec{z}_s, \vec{\gamma}_r) \equiv (\exists u)\mathbb{S}(u, w, \vec{z}_s, \langle\overline{\gamma_1}(u), \dots, \overline{\gamma_r}(u)\rangle) \tag{iii}$$

Cf. Theorem 6.5.2.

a. OR case.

$$\mathbb{P}(\vec{x}_n, \vec{\alpha}_l) \vee \mathbb{Q}(\vec{y}_m, \vec{\beta}_k) \overset{(i)+(ii)}{\Leftrightarrow} (\exists u)\mathbb{O}(u, \vec{x}_n, \langle\overline{\alpha_1}(u), \dots, \overline{\alpha_l}(u)\rangle) \vee$$
$$(\exists u)\mathbb{T}(u, \vec{y}_m, \langle\overline{\beta_1}(u), \dots, \overline{\beta_k}(u)\rangle)$$
$$\overset{logic}{\Leftrightarrow} (\exists u)\Big(\mathbb{O}(u, \vec{x}_n, \langle\overline{\alpha_1}(u), \dots, \overline{\alpha_l}(u)\rangle) \vee$$
$$\mathbb{T}(u, \vec{y}_m, \langle\overline{\beta_1}(u), \dots, \overline{\beta_k}(u)\rangle)\Big)$$

We conclude by Corollary 13.8.44.

b. AND case.

$$\mathbb{P}(\vec{x}_n, \vec{\alpha}_l) \wedge \mathbb{Q}(\vec{y}_m, \vec{\beta}_k) \overset{(i)+(ii)}{\Leftrightarrow} (\exists u)\mathbb{O}(u, \vec{x}_n, \langle\overline{\alpha_1}(u), \dots, \overline{\alpha_l}(u)\rangle) \wedge$$
$$(\exists u)\mathbb{T}(u, \vec{y}_m, \langle\overline{\beta_1}(u), \dots, \overline{\beta_k}(u)\rangle)$$
$$\overset{coding}{\Leftrightarrow} (\exists u)\Big(\mathbb{O}((u)_0, \vec{x}_n, \langle\overline{\alpha_1}((u)_0), \dots, \overline{\alpha_l}((u)_0)\rangle) \wedge$$
$$\mathbb{T}((u)_1, \vec{y}_m, \langle\overline{\beta_1}((u)_1), \dots, \overline{\beta_k}((u)_1)\rangle)\Big)$$

We conclude by Corollary 13.8.44 since primitive recursive predicates are closed under $\wedge$ (Example 13.8.31).

c. $(\exists w)$ case.

$$(\exists w)\mathbb{R}(w, \vec{z}_s, \vec{\gamma}_r) \overset{(iii)}{\Leftrightarrow} (\exists w)(\exists u)\mathbb{S}(u, w, \vec{z}_s, \langle \overline{\gamma_1}(u), \ldots, \overline{\gamma_r}(u) \rangle)$$

$$\overset{coding}{\Leftrightarrow} (\exists v)\mathbb{S}((v)_1, (v)_0, \vec{z}_s, \langle \overline{\gamma_1}((v)_1), \ldots, \overline{\gamma_r}((v)_1) \rangle)$$

We are done by Corollary 13.8.44.

d. $(\exists w)_{<v}$ case. $(\exists w)_{<v}\mathbb{R}(w, \vec{z}_s, \vec{\gamma}_r) \equiv (\exists w)\Big(w < v \wedge \mathbb{R}(w, \vec{z}_s, \vec{\gamma}_r)\Big)$. We are done by cases (b) and (c).

e. $(\forall w)_{<v}$ case. By (iii), we have

$$(\forall w)_{<v}\mathbb{R}(w, \vec{z}_s, \vec{\gamma}_r) \equiv (\forall w)_{<v}(\exists u)\mathbb{S}(u, w, \vec{z}_s, \langle \overline{\gamma_1}(u), \ldots, \overline{\gamma_r}(u) \rangle) \qquad (\ddagger)$$

In "$(\forall w)_{<v}(\exists u)$" we claim that for each of the v values of w below v there is a u that works. Now these "u" in general *depend* on the w so we *cannot* assume that *one u* works for *all w* and bring $(\exists u)$ up in front (to the left of "$(\forall w)_{<v}$"). However there is a t bigger than all the "u" that depend on all the $w < v$; for example, I can take

$$t = \max\{\text{all the said } u\} + 1$$

Then I can rewrite $(\ddagger)$ as

$$(\forall w)_{<v}\mathbb{R}(w, \vec{z}_s, \vec{\gamma}_r) \equiv (\exists t) \overbrace{(\forall w)_{<v}(\exists u)_{<t}\mathbb{S}(u, w, \vec{z}_s, \langle \overline{\gamma_1}(u), \ldots, \overline{\gamma_r}(u) \rangle)}^{\text{prim. recursive}}$$

The above clearly rests the case by Corollary 13.8.44.   □

(VIII) Substituting a partial recursive (restricted) functional into a type-0 argument of a semi-recursive relation.

13.8.49 Corollary. *If $\mathbb{P}(u, \vec{x}, \vec{\alpha})$ is semi-recursive and F is partial recursive, then $\mathbb{P}(F(\vec{y}, \vec{\beta}), \vec{x}, \vec{\alpha})$ is semi-recursive.*

Proof

$$\mathbb{P}(F(\vec{y}, \vec{\beta}), \vec{x}, \vec{\alpha}) \equiv (\exists u)\Big(\mathbb{P}(u, \vec{x}, \vec{\alpha}) \wedge F(\vec{y}, \vec{\beta}) \simeq u\Big)$$

We are done by closure properties of semi-recursive relations (Corollary 13.8.48) and the graph theorem (Corollary 13.8.47).   □

(IX) Definition by positive (semi-recursive) cases.

13.8.50 Corollary. *The restricted functional F given by (1) below, where the F_i are partial recursive while the $\mathbb{P}_i$ are semi-recursive for each input are mutually exclusive, then F is also partial recursive.*

$$F(\vec{x}, \vec{\alpha}) \simeq \begin{cases} F_1(\vec{x}, \vec{\alpha}) & \text{if } \mathbb{P}_1(\vec{x}, \vec{\alpha}) \\ \vdots & \vdots \\ F_i(\vec{x}, \vec{\alpha}) & \text{if } \mathbb{P}_i(\vec{x}, \vec{\alpha}) \\ \vdots & \vdots \\ F_k(\vec{x}, \vec{\alpha}) & \text{if } \mathbb{P}_k(\vec{x}, \vec{\alpha}) \end{cases} \tag{1}$$

The semantics of the definition by cases (1) are meant to be *nondeterministic*: For any given input $\vec{x}, \vec{\alpha}$ the $\mathbb{P}_i(\vec{x}, \vec{\alpha})$ that *is* true —if any— is *nondeterministically selected* ("guessed" by the "user") and *verified* and the corresponding $F_i(\vec{x}, \vec{\alpha})$ is then called and the value of this call —if any— is returned (as the value of $F(\vec{x}, \vec{\alpha})$).

Proof The proof is exactly as in Theorems 6.8.1 and 13.6.23 based on closure properties of the semi-recursive relations and on the graph theorem. Exercise.

(X) Substitution of a total restricted computable functional into a type-1 variable.

13.8.51 Theorem. *There is a rank $(2,0)$ primitive recursive function(al) h such that, for any restricted computable functionals $\lambda \vec{x} \vec{\alpha} \beta . \{f\}(\vec{x}, \vec{\alpha}, \beta)$ and $\lambda y \vec{x} \vec{\alpha} . \{e\}(y, \vec{x}, \vec{\alpha})$ —the latter being* total— *we have*

$$\{f\}\Big(\vec{x}, \vec{\alpha}, \lambda y.\{e\}(y, \vec{x}, \vec{\alpha})\Big) \simeq z \text{ iff } \{h(f, e)\}(\vec{x}, \vec{\alpha}) \simeq z \tag{‡}$$

Proof By Theorem 13.8.38, (iii), we have

$$\{f\}^{(n,l)}(\vec{x}_n, \vec{\alpha}_l, \beta) \simeq D\Big((\mu y)\mathbb{T}_l\left(f, \langle \vec{x}_n \rangle, \langle \overline{\alpha_1}(y), \ldots, \overline{\alpha_l}(y) \rangle * \langle \overline{\beta}(y) \rangle, y\right)\Big)$$

$$D\Big((\mu y)\Big(\mathbb{T}_l\left(f, \langle \vec{x}_n \rangle, \langle \overline{\alpha_1}(y), \ldots, \overline{\alpha_l}(y) \rangle *\right. \tag{1}$$
$$\left.\langle \overline{\{e\}}(y, \vec{x}_n, \vec{\alpha}_l) \rangle\right), y\Big)\Big)$$

where $\overline{\beta}(y)$ is $\overline{\{e\}}(y, \vec{x}_n, \vec{\alpha}_l) = \langle \{e\}(0, \vec{x}_n, \vec{\alpha}_l), \ldots, \{e\}(y, \vec{x}_n, \vec{\alpha}_l) \rangle$. Applying S-m-n to (1) we obtain an $h \in \mathcal{R}$ such that

$$\{f\}(\vec{x}_n, \vec{\alpha}_l, \{e\}(y, \vec{x}_n, \vec{\alpha}_l)) \simeq \{h(f, e)\}(\vec{x}_n, \vec{\alpha}_l) \qquad \square$$

13.9   Computing Relative to a *Fixed Total* Type-1 Input

We remain within Kleene computability of restricted functionals. An important special case is when we have just *one* total type-1 input α that we keep fixed and proceed to study the properties of the Kleene-computable $\lambda \vec{x}_n . F(\vec{x}_n, \alpha)$.

13.9.1 Definition. (Computing Relative to a *Total Function* or a *Set*)

1. A function f is partial *computable in* or *relative to* the *total* function α of one argument iff, for some e and all x, $f(x) \simeq \{e\}(x, \alpha)$.

   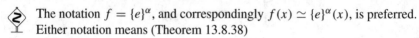 The notation $f = \{e\}^{\alpha}$, and correspondingly $f(x) \simeq \{e\}^{\alpha}(x)$, is preferred. Either notation means (Theorem 13.8.38)

$$f(x) \simeq y \equiv (\exists u)\Big(\mathbb{T}_1(e, \langle x \rangle, \langle \overline{\alpha}(u) \rangle, u) \wedge D(u) \simeq y\Big)$$

   This "*reducibility*" —this function is computable, if such and such function is— is called *Turing reducibility* and we introduce a symbol commonly used in the literature that is the same (but has a different subscript) as the symbol for *strong reducibility* (Definition 12.2.1).

$$f \leq_T \alpha$$

   read "f is Turing reducible to α".

   A *total* function g is *recursive in* (or *recursive relative to*) α iff it is partial recursive in α.

2. If $\alpha = c_A$ for some $A \subseteq \mathbb{N}$ we prefer to write

$$f = \{e\}^A \text{ rather than } f = \lambda x.\{e\}(x, c_A)$$

   that is,

$$f(x) \simeq y \Leftrightarrow \{e\}^A(x) \simeq y \Leftrightarrow (\exists u)\Big(\mathbb{T}_1(e, \langle x \rangle, \langle \overline{c_A}(u) \rangle, u) \wedge D(u) \simeq y\Big)$$

   or

$$f = \{e\}^A = \lambda x.D\Big((\mu u)\mathbb{T}_1(e, \langle x \rangle, \langle \overline{c_A}(u) \rangle, u)\Big)$$

   We can also write in short $f \leq_T A$.

3. A set $A \subseteq \mathbb{N}$ is *Turing reducible*, or *recursive in* a set $B \subseteq \mathbb{N}$ iff, for some e, $c_A = \{e\}^B = \lambda x.\{e\}^B(x)$. Or, $c_A = \lambda x.D\Big((\mu u)\mathbb{T}_1(e, \langle x \rangle, \langle \overline{c_B}(u) \rangle, u)\Big)$.
   We write $A \leq_T B$ to indicate this situation. □

13.9.2 Proposition. $\leq_T$ *is transitive.*

Proof Say $A \leq_T B \leq_T C$. Thus, for all x,

$$c_A = \{f\}^B, \text{ for some } f \tag{1}$$

and

$$c_B = \{e\}^C, \text{ for some } e \tag{2}$$

By Theorem 13.8.51, there is an h that is primitive recursive and

$$c_A(x) \simeq \{h(f, e)\}^C(x) \simeq D((\mu u)\mathbb{T}_1(h(f, e), \langle x \rangle, \langle \overline{c_C}(u) \rangle, u)$$

for all x. Thus, $A \leq_T C$.   □

13.9.3 Exercise. Show that $\leq_T$ on sets is reflexive.

Hint. Express $x \in A$ via the characteristic function.   □

13.9.4 Exercise. Show that if $A \leq_T B$ and B is recursive, then so is A.   □

13.9.5 Remark. Turing reducibility is more intuitively pleasing than strong reducibility. For example, with the latter we have $\overline{A} \leq_T A$ and hence *conversely as well* since $c_{\overline{A}}(x) \simeq 1 \div c_A(x)$.

On the other hand (cf. Remark 12.2.9) $\overline{K} \not\leq_m K$ even though, *intuitively*, "if I can *solve* $x \in \overline{K}$, then I ought to be able to *solve* $x \in K$ and *vice versa* by changing *yes* to *no* answers and *vice versa*". Thus the idea in quotation marks is captured well for the intuitive "solve" by $\leq_T$ but not by $\leq_m$.

Another issue in favour of $\leq_T$ is that all recursive sets are reducible to each other, but not so with $\leq_m$ (cf. Remark 12.2.9). See Exercise 13.12.7.   □

We next introduce some notation and concepts that are especially useful in Sects. 13.10 and 13.11.2.

13.9.6 Long Definition.

(a) A set A is *B-semi-recursive* (*B-r.e.*) iff $A = \text{dom}(\{e\}^B)$ or $x \in A \equiv \{e\}^B(x) \downarrow$, for some e. That is,

$$x \in A \equiv (\exists y)\mathbb{T}_1(e, \langle x \rangle, \langle \overline{c_B}(y) \rangle, y)$$

(b) $W_e^B \overset{Def}{=} \text{dom}(\{e\}^B) = \left\{ x : (\exists y)\mathbb{T}_1(e, \langle x \rangle, \langle \overline{c_B}(y) \rangle, y) \right\}$.

(c) σ, τ, ρ with or without subscripts or primes denote *finite* strings over the alphabet $\{0, 1\}$. If $\sigma = \sigma_0\sigma_1 \cdots \sigma_n$, then

- We view σ as the characteristic function —given as an array of output values for $i = 0, 1, 2, \ldots, n$— of the finite set A:

$$x \in A \equiv \sigma(x) = 0 \equiv \sigma_x = 0$$

- For the above we define "*length*" of σ —$lh(\sigma)$— by $lh(\sigma) = n + 1$.
- If B is a subset of $\mathbb{N}$, then $\sigma \subseteq B$ means

$$\text{If } \sigma(x) \downarrow \text{, then } \sigma(x) = c_B(x)$$

- We recall that for a total α, $\alpha \upharpoonright u$ means $\{(x, \alpha(x)) \ : \ x \ < \ u\}$ (cf. Remark 13.8.36) and we extend this notation to two cases

  a. $A \upharpoonright u$ for $c_A \upharpoonright u$
  and
  b. $\sigma \upharpoonright u$ for $\tau = \{(x, \sigma(x)) : x < u\}$
  Clearly the finite functions $A \upharpoonright u$ and $\sigma \upharpoonright u$ each have length u.

(d) It is intuitively immediate that in $\mathbb{T}^{(k,l)}(e, \ldots, y)$ and $\mathbb{T}_l(e, \ldots, y)$ the code y of the *computation tree* is a *strictly increasing function* of the number of *steps* taken by the *dynamic computation* as such number is gauged by the number of *calls* to the computation tuples on the roots of subtrees, as the latter are recursively defined in Figs. 13.1 and 13.2.
Thus *intuitively* we may let y gauge (a strictly increasing function of) *steps* themselves. This motivates the definitions/notation below.

- $\{e\}_s^A(x) \simeq y$ means that *all* of (a)–(c) hold:

  a. $\{e\}^A(x) \simeq y$ in $< s$ (dynamic) computation "*steps*",
  b. $s > 0$ and $e, x, y < s$,[33]
  and
  c. The largest number w used in an oracle query ("$w \in A$?") during the computation is also $< s$.

  Note that

$$(\exists u)_{<s}\left(\mathbb{T}_1(e, \langle x \rangle, \langle \overline{c_A}(u) \rangle, u) \wedge D(u) \simeq y\right) \tag{1}$$

  satisfies all of (a)–(c) above, hence is a mathematical depiction of "$\{e\}_s^A(x) \simeq y$".
  Note that the query with the *largest* (input) *number* (for c_A) is $< u < s$ (cf. Remark 13.8.36).

 Thus, by (1), $\{e\}_s^A(x) \simeq y$ is primitive recursive in A.

- $W_{e,s}^A$ means $(\exists u)_{<s}\mathbb{T}_1(e, \langle x \rangle, \langle \overline{c_A}(u) \rangle, u)$. Thus it is primitive recursive in A.
  It follows that $x \in W_e^A \equiv (\exists s)x \in W_{e,s}^A$ is semi-recursive in A by the projection theorem. In set notation we have $W_e^A = \bigcup_{s \in \mathbb{N}} W_{e,s}^A$.

---

[33] This is sloppy but well established in use notation for the exact "$e < s \wedge x < s \wedge y < s$".

- The number mentioned in (c) is important enough (e.g., in Sect. 13.10) to be also associated with a function (as its output) called the *use function u* given by

$$u(A, e, x, s) \overset{Def}{=} \begin{cases} (\mu u)_{<s} \mathbb{T}_1(e, \langle x \rangle, \langle \overline{c_A}(u) \rangle, u) & \text{if } \{e\}_s^A(x) \downarrow \\ 0 & \text{if } \{e\}_s^A(x) \uparrow \end{cases} \tag{2}$$

By (1), the symbol "$\{e\}_s^A(x) \downarrow$" means "$(\exists u)_{<s} \mathbb{T}_1(e, \langle x \rangle, \langle \overline{c_A}(u) \rangle, u)$", or $x \in W_{e,s}^A$, thus the conditions in (2) are primitive recursive in A.
Therefore, $\lambda e x s A.u(A, e, x, s)$ is primitive recursive in A.

- Referring to (2) note that a *convergent* computation $\{e\}_s^A(x) \simeq y$ *uses a finite subfunction* $\sigma = A \restriction u$ of c_A, whose domain is an initial segment of $\mathbb{N}$ (i.e., $\subseteq \{0, 1, 2, \ldots u - 1\}$). In fact, (2) uses $\sigma = A \restriction u(A, e, x, s)$.

We write —by abuse of notation since we do not use nontotal function oracles— the suggestive

$$\{e\}_s^\sigma(x) \simeq y \text{ if } \{e\}_s^A(x) \simeq y, \text{ for some } A \supseteq \sigma, \text{ where } lh(\sigma) \geq u(A, e, x, s)$$

Correspondingly we set

$$\{e\}^\sigma(x) \simeq y \overset{Def}{\equiv} (\exists s)\{e\}_s^\sigma(x) \simeq y$$

$\square$

After such a long definition it is only fair that we should establish a few easy results:

13.9.7 Proposition. (Semi-Recursive and Co-Semi-Recursive Sets) *If B is W_e^A and $\overline{B} = W_f^A$, then B is recursive in A —that is, $B \leq_T A$.*
And conversely.

Proof By Theorem 13.8.45. $\square$

13.9.8 Proposition. (Relativised $\overline{K}$) *The set $\overline{K}^A$, that is, the predicate $x \in \overline{K}^A$ or $x \notin W_x^A$ is not semi-recursive in A.*

Proof Well, if the opposite is true, then for some e,

$$x \in W_e^A \equiv x \notin W_x^A$$

Setting $x = e$ we get the contradiction $e \in W_e^A \equiv e \notin W_e^A$. $\square$

13.9.9 Corollary. (Relativised Halting Problem) K^A *is not recursive in A.*

Proof By Proposition 13.9.8 we have that $\overline{K}^A$ is not semi-recursive in A hence K^A cannot be recursive in A: $K^A \not\leq_T A$. $\square$

The following is often called "the use principle" (e.g., Soare (1987)) as it encapsulates the properties of the use function.

The trivial (in terms of its proof) proposition that follows is worth recording.

13.9.10 Proposition. *If* $\{e\}_s^A(x) \simeq y$, *then* $(\forall t \geq s)\Big(\{e\}_t^A(x) \simeq y \wedge u(A, e, x, s) = u(A, e, x, t)\Big)$.

Proof Let

$$\{e\}_s^A(x) \simeq y \tag{i}$$

and $t \geq s$.

Note that (i) translates to (1) of Long Definition 13.9.6 (p. 516). Since (1) trivially remains true if s is replaced by t, we have

$$\{e\}_t^A(x) \simeq y \tag{ii}$$

Turning attention to (2) of Long Definition 13.9.6, we have

$$u(A, e, x, s) = (\mu u)_{<s} \mathbb{T}_1(e, \langle x \rangle, \langle \overline{c_A}(u) \rangle, u) \stackrel{(i)+(ii)}{\simeq}$$

$$(\mu u)_{<t} \mathbb{T}_1(e, \langle x \rangle, \langle \overline{c_A}(u) \rangle, u) = u(A, e, x, t)$$

□

13.9.11 Proposition. *The following hold*

(1) $\{e\}^A(x) \simeq y$ *implies* $(\exists s)(\exists \sigma \subseteq A)\{e\}_s^\sigma \simeq y$.
(2) $\{e\}_s^\sigma(x) \simeq y$ *implies* $(\forall t \geq s)(\forall \tau \supseteq \sigma)\{e\}_t^\tau \simeq y$.
(3) $\{e\}^\sigma(x) \simeq y$ *implies* $(\forall A \supseteq \sigma)\{e\}^A \simeq y$.

Proof

(1) This follows from the normal form theorem as we discussed in the Long Definition 13.9.6 (cf. (1), p. 516):

$$\{e\}^\sigma(x) \simeq y \equiv (\exists s)(\exists u)_{<s}\Big(\mathbb{T}_1(e, \langle x \rangle, \langle \overline{c_A}(u) \rangle, u) \wedge D(u) \simeq y\Big)$$

The finite sub function $\sigma = A \restriction u$ works.

(2) If

$$\{e\}^\sigma(x) \simeq y \equiv (\exists s)(\exists u)_{<s}\Big(\mathbb{T}_1(e, \langle x \rangle, \langle \overline{c_A}(u) \rangle, u) \wedge D(u) \simeq y\Big) \tag{†}$$

is true, then any $t > s$ works along with any $\tau \supseteq \sigma$: For, say, we have $t \geq s$ and $\tau \supseteq \sigma$. Then, firstly, we have that

$$(\exists u)_{<t}\Big(\mathbb{T}_1(e, \langle x \rangle, \langle \overline{c_A}(u) \rangle, u) \wedge D(u) \simeq y\Big)$$

is true, using the same $u < s$ and the same $\sigma = A \upharpoonright u$ as in (†). That is, $\{e\}_t^{\sigma}(x) \simeq y$. Then, if $\tau \supseteq \sigma$, we also have $\{e\}_t^{\tau}(x) \simeq y$ since only the part $\sigma \subseteq \tau$ of τ is used in the computation.

(3) By Long Definition 13.9.6, $\{e\}^{\sigma}(x) \simeq y$ if $\{e\}^{B}(x) \simeq y$ for some $B \supseteq \sigma$ where $lh(\sigma) \geq u(B, e, x, s)$. Thus

$$\{e\}^{\sigma}(x) \simeq y \equiv (\exists u)\Big(\mathbb{T}_1(e, \langle x \rangle, \langle \overline{c_B}(u) \rangle, u) \wedge D(u) \simeq y\Big) \qquad (\ddagger)$$

Now any $A \supseteq \sigma$ will be such that $\sigma = A \upharpoonright u = B \upharpoonright u$. Hence we have (‡) with B replaced by A. That is, $\{e\}^A(x) \simeq y$.

$\square$

13.10 Turing Degrees and Post's Problem

13.10.1 Definition. (Turing Equivalence; Turing Degrees) We define the relation $A \equiv_T B$ to mean $(A \leq_T B) \wedge (B \leq_T A)$, pronounced "$A$ and B are *Turing equivalent* sets". Since $\leq_T$ is reflexive —Exercise 13.9.3— and transitive it follows that $\equiv_T$ is, trivially, an *equivalence relation*. The *equivalence classes* $[X]_{\equiv_T}$ are called *Turing degrees* or *degrees of unsolvability*.

1. We denote Turing degrees by short names such as a, b, c, d with or without primes or subscripts, typically choosing a for $[A]_{\equiv_T}$. On occasion, if we want to name a degree by a set A *in* the degree we will use the preceding symbol for equivalence class or, preferably, Rogers' notation $d_T(A)$.
2. The set of all Turing degrees is denoted by D.
3. We write $A <_T B$ for $A \leq_T B$ and $B \not\leq_T A$ (equivalently, for $A \leq_T B$ and $A \not\equiv_T B$).
4. We define $a \leq b$ to mean $A \leq_T B$ where $A \in a$ and $B \in b$. On the other hand, $a < b$ means $a \leq b$ and $a \neq b$.

$\square$

13.10.2 Proposition. ($\leq$ on Turing Degrees Is an Order) *The introduction of the relation $\leq$ on Turing degrees is well-defined, that is, independent of choice of A and B above.*

Moreover, $\leq$ is an order on all Turing degrees.

Proof Let $\{A, A_1\} \subseteq a$ and $\{B, B_1\} \subseteq b$. If $A \leq_T B$, then also $A_1 \leq_T B_1$ since (conjunctionally)

$$A_1 \leq_T A \leq_T B \leq_T B_1$$

Now apply transitivity.

For $\leq$ on degrees now:

- (Reflexivity) $a \leq a$, since $A \leq_T A$ has already been noted.
- (Antisymmetry) Let $(a \leq b) \wedge (b \leq a)$. Thus there are A and B in a and b respectively, such that $A \leq_T B \leq_T A$. By Definition 13.10.1, $A \equiv_T B$, therefore $a = b$.
- (Transitivity) Let (conjunctionally) $a \leq b \leq c$. Then for three representative sets in these three degrees we have $A \leq_T B \leq_T C$. By transitivity of $\leq_T$ we have $A \leq_T C$ hence $a \leq c$.

$\square$

13.10.3 Proposition. *For any sets A and B, $A \leq_m B$ implies $A \leq_T B$.*

Proof By assumption, $x \in A \equiv f(x) \in B$, for some $f \in \mathcal{R}$. The right hand side of the equivalence is recursive in B, hence so is the left hand side. $\square$

13.10.4 Proposition. *If A is recursive and B is any set then $A \leq_T B$.*

Proof $F = \lambda x \alpha . c_A(x)$ is recursive by type-1 argument list expansion (Corollary 13.8.42). By definition of F, $F(x, c_B) \simeq c_A(x)$, for all x. But $\lambda x . F(x, c_B)$ is recursive in B. $\square$

13.10.5 Corollary. *$A \leq_T \emptyset$ iff A is recursive.*

Proof The *If* is by the preceding proposition.

For the *Only if* the assumption means that, for some e,

$$c_A(x) \simeq D\Big((\mu y)\mathbb{T}_1(e, \langle x \rangle, \langle \overline{c_\emptyset}(y) \rangle, y)\Big) \tag{†}$$

Now, referring to Definition 13.8.37, $\mathbb{T}_1(e, z, w, y)$ is a conjunction, the last conjunct of which is

$$(\forall j)_{<lh((y)_{0,2})}\Big\{(y)_{0,2,j} \geq 1 \wedge (\forall x)_{<y}\Big(p_x|(y)_{0,2,j} \to ((y)_{0,2,j})_x = \big(w\big)_{j,x}\Big)\Big\} \tag{1}$$

where in our case here $lh((y)_{0,2} = 1$ and $w = \langle \overline{c_\emptyset}(y) \rangle$, thus $(w)_0 = \overline{c_\emptyset}(y)$ and, if $x < y$, then $(w)_{0,x} = \big(\overline{c_\emptyset}(y)\big)_x = c_\emptyset(x) = 1$.

Hence (1) simplifies to

$$\Big\{(y)_{0,2,0} \geq 1 \wedge (\forall x)_{<y}\Big(p_x|(y)_{0,2,0} \to ((y)_{0,2,0})_x = 1\Big)\Big\} \tag{2}$$

which is absolutely primitive recursive. All other conjuncts being trivially absolutely primitive recursive (Definition 13.8.37) we have that $\mathbb{T}_1(e, \langle x \rangle, \langle \overline{c_\emptyset}(y) \rangle, y)$ is absolutely primitive recursive, hence c_A is recursive by (†). $\square$

 13.10.6 Remark. Thus, since $\emptyset \leq_T A$ trivially, $\mathbf{0} = \mathbf{d}_T(\emptyset)$ is the degree of all recursive sets. $\square$

13.10.7 Definition. If $R \subseteq \mathbb{N}^k$ and $A \subseteq \mathbb{N}$, then we write $R \leq_m A$ meaning, for some $\lambda \vec{x}. f(\vec{x}) \in \mathcal{R}$, we have

$$R(\vec{x}) \equiv f(\vec{x}) \in A \square$$

13.10.8 Remark The properties below carry unchanged for the case of strong reducibility of a relation of a vector (type-0) input, absolute or relativised.

- If $R \leq A$ —where $\leq$ stands for $\leq_m$— and A is recursive, then so is R.
- If $R \leq A$ —where $\leq$ stands for $\leq_m$— and A is semi-recursive, then so is R.
- If $R \leq A$ —where $\leq$ stands for $\leq_m$— and A is recursive in B, then so is R.
- If $R \leq A$ —where $\leq$ stands for $\leq_m$— and A is semi-recursive in B, then so is R.
  The proofs are left as easy exercises. $\square$

13.10.9 Definition. (The Jump Operator) The set K^A is denoted by A' and called *the jump of* A.
   Inductively, we define, for all $n \geq 0$,

$$A^{(0)} \stackrel{Def}{=} A$$
$$A^{(n+1)} \stackrel{Def}{=} \left(A^{(n)} \right)' = K^{A^{(n)}}$$

0 denotes the degree of $\emptyset$, that is, the set of all recursive sets (cf. Exercise 13.12.14)
$\mathbf{0}^{(n)}$ is the degree of $\emptyset^{(n)}$.
   As "(n)" in "$A^{(n)}$" means "n primes", we write A'' for $(A')'$, A''' for $((A')')'$, etc. $\square$

13.10.10 Proposition. *For any* $R \subseteq \mathbb{N}^k$ *and* $A \subseteq \mathbb{N}$, $R \leq_m A'$ *iff* R *is semi-recursive in* A.

Proof

- (*If* case, cf. Example 12.3.3.) Say R is semi-recursive in A. Thus, for some e,

$$R(\vec{x}) \equiv \langle \vec{x} \rangle \in W_e^A \tag{1}$$

Define (by positive cases)

$$\psi(x, y) \simeq y \text{ if } x \in W_e^A \tag{1'}$$

hence (by S-m-n, Theorem 13.8.15 and Remark 13.8.17) there is *a recursive h* such that $\{h(x)\}(y) \simeq y$ if $x \in W_e^A$.
In particular,

$$\{h(x)\}^A(h(x)) \downarrow \text{ iff } x \in W_e^A \tag{1''}$$

Pause. Why was the "if" of $(1')$ changed to "iff" in $(1'')$?◄
The left hand side of $(1'')$ is "$h(x) \in K^A$", that is, $h(x) \in A'$.

We proved $x \in W_e^A$ iff $h(x) \in A'$. Thus

$$R(\vec{x}) \overset{(1)}{\Leftrightarrow} \langle \vec{x} \rangle \in W_e^A \overset{(1'')}{\Leftrightarrow} h(\langle \vec{x} \rangle) \in A'$$

This says that $R \leq A'$.

- (Only if case.) It is given that $R(\vec{x}) \equiv f(\vec{x}) \in A'$, for some $f \in \mathcal{R}$. Thus

$$R(\vec{x}) \equiv \{f(\vec{x})\}^A \Big(f(\vec{x}) \Big) \downarrow$$

which says that R is semi-recursive in A.

$\square$

13.10.11 Corollary.

(1) A' is semi-recursive in A.
(2) $A <_T A'$
(3) $a < b < c$ implies $a < c$
(4) In the ascending sequence of Turing degrees $\mathbf{0} < \mathbf{0}' < \mathbf{0}'' < \mathbf{0}''' < \cdots$ all degrees are distinct.

Proof

(1) $x \in A'$ says $\{x\}^A(x) \downarrow$.
(2) $A \leq_T A$ implies $A \leq A'$ (Proposition 13.10.10) and hence $A \leq_T A'$. But (Corollary 13.9.9) $A' \nleq_T A$, thus $A <_T A'$.
(3) Assumption implies $a \leq b \leq c$. By Proposition 13.10.2 we have $a \leq c$. Can it be $a = c$? If so, $a \leq b \leq a$, hence $a = b$ by antisymmetry of $\leq$. A contradiction to $a < b$.
(4) If $\mathbf{0}^{(k)} = \mathbf{0}^{(n)}$, for some $k < n$, note also

$$\mathbf{0}^{(k)} < \mathbf{0}^{(k+1)} < \cdots < \mathbf{0}^{(n)}$$

hence $\mathbf{0}^{(k)} < \mathbf{0}^{(n)}$ by transitivity ((3) above). A contradiction.

$\square$

13.10.12 Definition. (*T*-Complete Sets) A set A is *T-complete* iff it fulfils *both* requirements below:

1. It is semi-recursive.
2. If B is semi-recursive, then $B \leq_T A$.

$\square$

13.10.13 Proposition. $K = \emptyset'$ is *T-complete*.

Proof Let A be semi-recursive. Then $A \leq_1 K$ by 1-completeness of K (Example 12.3.3). Thus $A \leq_T K$ as well.

$\square$

13.10.14 Definition. A *recursive sequence of functions* $\left(\lambda x. f(n, x)\right)_{n\geq 0}$ is one that $\lambda x n. f(n, x)$ is recursive. Its *limit* $g(x)$ is defined "*point-wise*"[34] by

$$\text{For each } x, \text{ we have } \lim_{n\to\infty} f(n, x) = g(x)$$

meaning that, for each x, $f(n, x) = g(x)$ a.e. with respect to n (cf. Definition 4.1.10) $\qquad\qquad\square$

13.10.15 Lemma. (Shoenfield's Limit Lemma (Shoenfield 1971)) *A set A satisfies $A \leq_T K$ iff there is a recursive sequence* $\left(\lambda x. f(n, x)\right)_{n\geq 0}$ *satisfying*

$$\lim_{n\to\infty} f(n, x) = c_A(x)$$

for each x.

Proof (1) Say that the sequence $\left(\lambda x. f(n, x)\right)_{n\geq 0}$ exists as stated.

Then $c_A(x) = 0 \equiv (\exists m)(\forall n)(n > m \to f(n, x) = 0)$, thus c_A is Σ_2. But also $c_A(x) = 1 \equiv (\exists m)(\forall n)(n > m \to f(n, x) = 1)$, thus —given that $f(n, x) = 0 \equiv \neg(f(n, x) = 1)$— it is also $c_A(x) = 0 \equiv (\forall m)(\exists n)(n > m \wedge f(n, x) = 0)$. Hence c_A is also Π_2. Thus c_A is Δ_2, therefore we have $A \leq_T \emptyset'$ by Post's theorem on the arithmetical hierarchy.

(2) Conversely, assume $A \leq_T K$ and construct the sequence $\left(\lambda x. f(n, x)\right)_{n\geq 0}$. Let then $e \in \mathbb{N}$ such that $c_A(x) \simeq \{e\}^K(x)$, for all x. Using the standard notation of Definition 13.9.6 let us set $f(n, x) \simeq \{e\}_n^{K\upharpoonright u(K,e,x,n)}(x) \simeq \{e\}_n^{K\upharpoonright n}(x)$. Then f is recursive (cf. comments on (2) on p. 517) and $\lim_{n\to\infty} f(n, x) = c_A(x)$, for all x. $\qquad\square$

13.10.16 Remark One can recast part (1) of the proof in a manner that does not use *Post's hierarchy theorem*: Assume without loss of generality that $f(0, x) = 1$, for all x[35] and that the range of f is a subset of $\{0, 1\}$.[36] Define B by $\langle x, n\rangle \in B \equiv |\{s : f(s, x) \neq f(s + 1, x)\}| \geq n$. This is semi-recursive since a verifier for $\langle x, n\rangle \in B$ is: Start at $s = 0$ and increment by 1, keeping a running tally of the alternations in the f-value, until the cardinality condition is first met. Thus B is strongly, and hence Turing, reducible to K —by the latter's 1- and m-completeness— that is, B is K-computable ($B \leq_T K$). Having a decider (*in K*) for B we then obtain one for A: To decide $x \in A$ compute the smallest n such that $\langle x, n\rangle \notin B$. This is a total (by assumption the limit exists for all x) and computable (*in K*) operation. Clearly, $n - 1$ is the number of alternations of f (on the given x and all s) between the values

---

[34] For each input "point" separately.

[35] If not, modify f to behave so. This does not affect the limit for any x.

[36] If not, use $1 \dotminus (1 \dotminus f)$ instead of f.

0 and 1. As $f(0, x) = 1$ for all x, we have $x \in A$ iff n is odd (on the convention that $c_A(x) = 0$ iff $x \in A$).   □

Relativising the above argument we obtain

13.10.17 Corollary. *A set A satisfies $A \leq_T K^B$ iff there is a recursive in B sequence $\left(\lambda x. f(n, x)\right)_{n \geq 0}$ such that $\lim_{n \to \infty} f(n, x) = c_A(x)$ for all x.*

13.10.18 Definition. (Post's Problem) Post asked: *Is there an r.e. degree —$d_T(A)$ where A is r.e.— that is strictly between $\mathbf{0}$ and $\mathbf{0}'$?* Equivalently, are there any r.e. degrees other than $\mathbf{0}$ and $\mathbf{0}'$?   □

13.10.1   Sacks' Version of the Finite Injury Priority Method

The priority method used to construct r.e sets was independently discovered by Muchnik (1956) and Friedberg (1957) toward answering (affirmatively) Post's question contained in Definition 13.10.18. To this end they constructed two r.e. sets A and B with *incomparable* degrees, that is, $A \not\leq_T B$ and $B \not\leq_T A$. As a corollary, one answers Post's question as follows:

13.10.19 Corollary. *There is an r.e. degree x such that $\mathbf{0} < x < \mathbf{0}'$.*

Proof Indeed none of the sets constructed by Friedberg-Muchnik can be in $\mathbf{0}'$. For example, say $A \in \mathbf{0}'$. Then

$$B \overset{13.10.13}{\underset{\leq_T}{}} K \overset{\equiv_T}{\underset{thus \ \leq_T}{}} A$$

a contradiction (by transitivity of $\leq_T$).

Similarly it cannot be that any one of them is recursive (cf. Proposition 13.10.4). Thus, for example, take $x = d_T(A)$. We have $\mathbf{0} < d_T(A) < \mathbf{0}'$.   □

The Sacks *agreement* method is a generalisation and improvement of the method of Muchnik and Friedberg that leads to shorter proofs and deeper results. We will introduce this method below to give a solution to *Post's Problem* by showing directly, as Sacks did, the existence of a *simple set* (Definition 12.5.1) A such that $\mathbf{0} < d_T(A) < \mathbf{0}'$.

Comparing roughly the two approaches, in the original proof (Muchnik (1956); Friedberg (1957)) we have a construction of *two* r.e. sets A and B by stages, such that, for all $e \in \mathbb{N}$, $c_A \neq \{e\}^B$ and $c_B \neq \{e\}^A$ (Muchnik 1956; Friedberg 1957; Shoenfield 1967; Rogers 1967; Tourlakis 1984), that is, $A \not\leq_T B$ and $B \not\leq_T A$.

Sacks on the other hand approaches the problem by constructing *one simple* set (Definition 12.5.1) A that satisfies $C \not\leq_T A$ for some given fixed *r.e. non recursive* set C.

As A is constructed by stages by a recursive procedure that at stage s adds a number to A_s to form A_{s+1} we want at the limit ($s \to \infty$) to have

$$c_C \neq \{e\}^A \tag{1}$$

In the rest of this sub section we will write S for c_S and $S(x)$ for $c_S(x)$ identifying characteristic function with the set. We will continue with our convention that $c_S(x) = S(x) = 0$ means $x \in S$, while $c_S(x) = S(x) = 1$ means $x \notin S$.

The Friedberg-Muchnik strategy would translate to maintaining *disagreement* as much as possible between $C_s{}^{37}$ and $\{e\}^{A_s}$ at each stage s. Counterintuitively, Sacks rather tries to preserve *agreement* (via his *restraint function* $r(e, s)$, see the proof below). This works because at the limit *if* $C = \{e\}^A$ holds we get a contradiction. The construction of the *simple* set is an elaboration of that in Theorem 12.5.2 that takes into account the requirement (1). In Friedberg-Muchnik jargon, the *requirements* to add elements into the A_s constructed below are the *positive* requirements, P_e. The requirement to prove (1) is the *negative* one: N_e.

We say that a requirement is *injured* at a stage if at said stage we backtrack over an action we did earlier for this requirement. We assign *priorities* to requirements so the ones of *low priority* are allowed to be injured. The priority method in either the original Friedberg-Muchnik form or in the Sacks proof *injure* requirements finitely many times, hence the term "*finite injury* priority method". There is no obvious activity regarding priorities in the Sacks approach.

13.10.20 Theorem. (Sacks 1963) *For all r.e. non recursive C we can build a simple set A such that $C \nleq_T A$.*

Proof (Soare (1976), Soare (1987), Odifreddi (1989)) We construct A by stages. At the end of stage s we have A_s. At stage 0 we have $A_0 = \emptyset$.

The *positive* requirement builds the simple set A by a variation of the proof of Theorem 12.5.2.

$$P_e : \quad W_e \text{ infinite implies } W_e \cap A \neq \emptyset$$

The negative requirement aims to achieve at the limit

$$N_e : \quad C \neq \{e\}^A, \text{ that is, } C \nleq_T A$$

We fix a recursive enumeration of C as a sequence of finite subsets induced by the assumed recursive enumeration of the *members* of C (it is r.e.) Cf. Exercise 13.12.25.

$$C_0 = \emptyset \subseteq C_1 \subseteq C_2 \subseteq C_3 \subseteq \dots$$

---

[37] C is given outright, but being r.e., there is a recursive enumeration of it $C^0, C^1, C^2, \dots$. Exercise 13.12.25.

The construction of A is by stages. At stage 0 we set $A_0 = \emptyset$. Having got A_s, here is how we proceed in

Stage $s + 1$:

Define for all e,

1. the *length of agreement*:

$$l(e, s) = \max \left\{ x : (\forall y)_{<x} \{e\}_s^{A_s}(y) \downarrow \simeq C_s(y) \right\}$$

2. the *restraint*:

$$r(e, s) = \max \{ u(A_s, e, x, s) : x \leq l(e, s) \}$$

Comment: The restraint restrains additions of elements to A_s (towards forming A_{s+1}) with a view of avoiding injury to the negative requirement N_e. The definition of $r(e, s)$ restrains the construction steps in order to *preserve* not only the *agreement* ($x < l(e, s)$) but also to *preserve* the *first point* of *disagreement*, $x = l(e, s)$. Thus the ingenious idea to preserve agreement is not paradoxical. Some disagreements are respected too when possible.

Now, *for each $e \leq s$, if $W_{e,s} \cap A_s = \emptyset$ and*

$$(\exists x)\left(x \in W_{e,s} \wedge x > 2e \wedge (\forall i)_{\leq e} r(i, s) < x \right) \tag{1}$$

we add to A_s the minimum x that works in (1) towards the formation of A_{s+1} which is the so augmented A_s, after all additions.[38] One often says that such P_e "*receive attention*" at step $s + 1$.[39]

We set

$$A = \bigcup_{s \geq 0} A_s \tag{2}$$

which is trivial to verify that is r.e. since all the A_s are finite the procedure is recursive and we can code the A_s with any one of the codings available to us.

The part $(\forall e)_{\leq i} r(e, s) < x$ in (1) restrains the x we pick to include in A_{s+1} so that N_e is not injured. Injury thus happens if an $x \leq r(e, s)$ is put in $A_{s+1} - A_s$. These "injurious" elements, collected *for all e*, form an r.e. set (cf. Exercise 13.12.26.)

---

[38] We can compute this x since all the conjuncts in (1) are recursive. Cf. also the proof of Theorem 12.5.2.

[39] One may also say that once condition (1) and $W_{e,s} \cap A_s = \emptyset$ are detected at step $s + 1$ that each corresponding P_e "*requires attention*" at step $s + 1$.

3. the *injury set*

$$I_e = \{x : (\exists s)(x \in A_{s+1} - A_s \wedge x \leq r(e,s)\}$$

I_e is finite. In fact it has at most e members since N_e is injured at most once by each P_i, $i < e$. The P_i are never injured in the given process.

The proof concludes via three in-proof lemmata.

13.10.21 Lemma. $(\forall e)C \neq \{e\}^A$.

Proof We argue by contradiction so let, for some e,

$$C = \{e\}^A \tag{3}$$

Thus,

$$\lim_{s \to \infty} l(e,s) = \infty \tag{4}$$

We will prove that $C \in \mathcal{R}$, contradicting one of the assumptions of the theorem.

Firstly, we argue that $l(e,s) \leq l(e,t)$ for $t > s$.

Suppose not, so for some $t > s$, we have $l(e,s) > l(e,t)$ —that is, $l(e,t)$ is a point in the open interval $(0, l(e,s)) \cap \mathbb{N}$— *the set of points of agreement* has shrunk. This means that *it was C_s that changed* since, by Proposition 13.9.11(2) and the preceding inequality, we have

$$\overbrace{\{e\}_s^{A_s}(l(e,t)) \downarrow}^{agrees\ with\ C_s(l(e,t))} \simeq \overbrace{\{e\}_t^{A_s}(l(e,t))}^{disagrees\ with\ C_t(l(e,t))} \tag{5}$$

that is, $\{e\}_s^{A_s}(l(e,t)) \downarrow$ will not change at any stage after s. But a $C_s(u)$ can change only *once*: From u not in, to u in: From $C_s(u) \simeq 1$ to $C_t(u) \simeq 0$, $t > s$ (C is a given set; at no stage can points be taken away from it).

Diagrammatically, (5) says that from stage t onwards the *disagreement* $C_t(l(e,t)) \not\simeq \{e\}_t^{A_t}(l(e,t))$ is *preserved ad infinitum* contradicting hypothesis (3).

$$\text{So it must be } l(e,s) \leq l(e,t) \text{ after all} \tag{6}$$

We can now show how to compute $C(u)$, for any u. Take an $s > s'$, where s' is a state at which, and onwards, N_e does not get injured.

By (4), we will let the s we just chose also satisfy $l(e,s) > u$.
Thus

$$(\forall t \geq s)\left(C_t(u) \overset{l(e,t) \geq l(e,s) > u}{\simeq} \{e\}_t^{A_t}(u) \downarrow \overset{\text{Proposition 13.9.11}}{\simeq} \{e\}^A(u) \overset{(3)}{\simeq} C(u)\right)$$

So, to compute $C(u)$ we compute $C_s(u)$, that is, $\{e\}_s^{A_s}(u)$. This is a *primitive recursive in A_s* calculation (cf. Long Definition 13.9.6, item (d)). The finiteness of A_s (hence recursiveness) makes it an *absolutely* recursive calculation. Having obtained $C(u)$ (absolutely) recursively we have just contradicted one of the hypotheses of the theorem.   □

13.10.22 Lemma. *For all e, there is an $a \in \mathbb{N}$ such that $\lim_{s \to \infty} r(e, s) = a$.*

Proof By the preceding lemma, let u be the *smallest* such that $C(u) \not\simeq \{e\}^A(u)$. Pick a stage s' such that, for $s \geq s'$,

- N_e is not injured at stage s
- $(\forall x)_{<u}\{e\}_s^{A_s}(x) \simeq \{e\}^A(x)$
- $(\forall x)_{\leq u}C_s(x) \simeq C(x)$
  We have two cases,

1. $(\forall s \geq s')\{e\}_s^{A_s}(u) \uparrow$. No change to $\{e\}_s^{A_s}(u) \uparrow$ in perpetuity, hence by definition of $r(e, s)$ (as a function of $u(A, e, x, s)$ cf. Long Definition 13.9.6(d)) $s \geq s'$ implies $r(e, s) = r(e, s')$. That is, $\lim_{s \to \infty} r(e, s) = r(e, s')$ in this case.

2. $(\exists s \geq s')\{e\}_s^{A_s}(u) \downarrow$. Let t be the smallest $s \geq s'$ such that $\{e\}_t^{A_t}(u) \downarrow$. By Proposition 13.9.11(2), for all $t' \geq t$,

$$\{e\}_t^{A_t}(u) \simeq \{e\}_{t'}^{A_{t'}}(u) \simeq \{e\}^A(u) \not\simeq C(u) \tag{1}$$

By (1) and bullet one we have that disagreement is preserved and so is $\{e\}_t^{A_t}(u)$. That is, $r(e, s)$ has not changed since the value $r(e, t)$ (cf. definition of $r(e, s)$ and $u(A, e, x, s)$). Thus, $\lim_{s \to \infty} r(e, s) = r(e, t)$ in this case.
                                                                                      □

13.10.23 Lemma. *P_e is satisfied, for all e.*

Proof Set $r(i) = \lim_{s \to \infty} r(i, s)$, for all $i \leq e$ (cf. when and how P_e receives attention in the proof of the theorem, p.526). Then set $R(e) = \max(r(i) : i \leq e)$. Also, since all $r(i, s)$ —for $i \leq e$— converge as s goes to infinity, *and are finitely many*, let s' be a stage such that all $r(i, s)$ achieved their limit values:

$$(\forall s \geq s')(\forall i \leq e)r(i, s) = r(i, s') \tag{1}$$

Now if W_e is infinite the following is true

$$(\exists x)(x \in W_e \land x > 2e \land x > R(e)) \tag{2}$$

We have two cases

1. $W_{e,s'} \cap A_{s'} \neq \emptyset$ (cf. (1) on p.526). Thus $W_e \cap A \neq \emptyset$.
2. Alternatively, there is a stage $s > s'$ such that an x satisfying (2) above still did not get into A. But then P_e (cf. p. 526) will require *and* receive attention at s,

since $x > R(e)$ implies the third conjunct of (1) (in loc. cit.) by (1) in the present proof. Thus we will have $x \in W_{e,s+1} \cap A_{s+1} \neq \emptyset$, hence $W_e \cap \overline{A} \neq \emptyset$.

The second conjunct in (2) takes care of the infinity of $\overline{A}$ as in the proof of Theorem 12.5.2. □

This is the end of the global proof of Sacks' theorem. □

13.10.24 Corollary. (Sacks' Answer to Post's Problem) *There is an r.e. non recursive set A with degree strictly between $\mathbf{0}$ and $\mathbf{0}'$.*

Proof The set A constructed in the proof of Theorem 13.10.20 is *simple*, hence not recursive. Thus $\mathbf{0} < d_T(A)$. Next, $A \notin \mathbf{0}'$, otherwise we have

$$C \overset{13.10.13}{\leq_T} K \underset{thus}{\overset{\equiv_T}{\leq_T}} A$$

Therefore, $d_T(A) < \mathbf{0}'$. □

13.11 The Arithmetical Hierarchy

We have met the *Arithmetical relations* (arithmétic in Smullyan (1992)) in Definition 8.3.3, then revisited them (in Sect. 12.7) in the form of the arithmetical relations —lower case, *a* designating the omission of the initial relation $z = x^y$— and now we revisit them one last time in an investigation that arranges said relations into a hierarchy of relation sets according to the definitional complexity of their members, that ends up being also a *true complexity*, in the computational sense, that ranks them according to *increasing levels of unsolvability*.

So far we know that recursive relations are solvable, e.g., $\lambda nxz.z = A_n(x)$,[40] while the r.e. or semi-recursive set of relations includes, e.g., the *unsolvable* but *verifiable* K. $\overline{K}$ —or $x \notin K$— must be placed at a subsequent level, beyond the level for r.e. relations as it is even more (computationally) complex than K, not being verifiable.

Interestingly we have a *hierarchy* of infinitely many levels where at each higher level we have more complex ("*more unsolvable*") relations than in the levels before it.

All the levels taken together (the set union of all sets of relations at each level) constitute the set of all *Arithmetical* or *arithmetical* relations that we met before, but now these relations are viewed under the (complexity) microscope as it were.

Because of this aim, it will be purposeless to split hair as to whether or not this or that relation is in the *zeroth set* of the hierarchy, so we will take said set of relations to include *all* recursive relations.

---

[40] *Ackermann* function graph (Chap. 4).

The *arithmetical hierarchy* is due to Kleene (1943), Mostowski (1947).

13.11.1 Definition. The set of all *arithmetical relations* (or *arithmetical predicates*) is the least set that includes $\mathcal{R}_*$ and is closed under $(\exists x)$ and $(\forall x)$. We will denote this set by Δ.

A *function* (no type-1 variables) is in Δ iff its graph is.   □

We sort arithmetical relations into a *hierarchy* as follows.

13.11.2 Definition. (The Arithmetical Hierarchy)   We define by induction on $n \in \mathbb{N}$ the sets:

$$\Sigma_0 = \Pi_0 = \mathcal{R}_*$$

$$\Sigma_{n+1} = \{(\exists x)P : P \in \Pi_n\}$$

$$\Pi_{n+1} = \{(\forall x)P : P \in \Sigma_n\}$$

The variable "x" above is generic.

We also define, for all n, $\Delta_n = \Sigma_n \cap \Pi_n$ and also set $\Delta_\infty = \bigcup_{n \geq 0}(\Sigma_n \cup \Pi_n)$.

13.11.3 Remark. Intuitively, the arithmetical hierarchy is composed of all relations of the form $(Q_1 x_1)(Q_2 x_2) \ldots (Q_n x_n)R$, where $R \in \mathcal{R}_*$ and Q_i is one of $\exists$ or $\forall$, for $i = 1, \ldots, n$. If $n = 0$ there is no quantifier prefix. Since $(\exists x)(\exists y)$ and $(\forall x)(\forall y)$ can be "collapsed" into a single $\exists$ and single $\forall$ respectively,

Pause. Use a pairing function to code x and y by a single number and prove this.◄

we can think of the prefix as a sequence of alternating quantifiers with no adjacent duplicates (for the "no adjacent duplicates" part see also Exercise 0.8.1). The relation is placed in a Π-set (respectively, Σ-set) iff the leftmost quantifier is a "$\forall$" (respectively, "$\exists$").

The definitional or static complexity of an arithmetical relation is the length of the quantifier prefix that we obtain once we convert any given such relation to the above "normal form" (a prefix acting on a recursive relation). See also Exercise 13.12.27.

Important. Usually we think of a set as a "property" of the members it contains. Thus, membership in a set like Π_n, say, $P \in \Pi_n$ is attaching to P the property Π_n. Therefore

$$\text{"}P \text{ is } \Sigma_n\text{" (resp. "}P \text{ is } \Pi_m\text{") means } P \in \Sigma_n \text{ (resp. } P \in \Pi_m)$$

□

13.11.4 Lemma. $R \in \Sigma_n$ *iff* $(\neg R) \in \Pi_n$. $R \in \Pi_n$ *iff* $(\neg R) \in \Sigma_n$.

Proof We handle both equivalences simultaneously. For $n = 0$ this is so by closure properties of $\mathcal{R}_*$.

Assuming the claim for n, we have (recall that "iff" is used conjunctionally)

$$R \in \Sigma_{n+1} \text{ iff } R \equiv (\exists y)Q \text{ and } Q \in \Pi_n$$
$$\text{iff (I.H.) } R \equiv \neg(\forall y)\neg Q \text{ and } (\neg Q) \in \Sigma_n$$
$$\text{iff } \neg R \equiv (\forall y)\neg Q \text{ and } (\neg Q) \in \Sigma_n$$
$$\text{iff } \neg R \in \Pi_{n+1}$$

and

$$R \in \Pi_{n+1} \text{ iff } R \equiv (\forall y)Q \text{ and } Q \in \Sigma_n$$
$$\text{iff (I.H.) } R \equiv \neg(\exists y)\neg Q \text{ and } (\neg Q) \in \Pi_n$$
$$\text{iff } \neg R \equiv (\exists y)\neg Q \text{ and } (\neg Q) \in \Pi_n$$
$$\text{iff } \neg R \in \Sigma_{n+1}$$

$\square$

It is immediate that $\Delta_\infty \subseteq \Delta$. The following lemma has the converse inclusion as a corollary.

13.11.5 Lemma. *Among the following numbered closure properties we have those indicated in each bulleted case below.*

(1) *Substitution of a variable by a recursive function call*[41]
(2) $\vee, \wedge$
(3) $(\exists y)_{<z}, (\forall y)_{<z}$
(4) $\neg$
(5) $(\exists y)$
(6) $(\forall y)$

- $\Delta_\infty :$ *Is closed under* (1)–(6).
- $\Delta_n(n \geq 0) :$ *Is closed under* (1)–(4).
- $\Sigma_n(n \geq 0) :$ *Is closed under* (1)–(3), *and, if* $n > 0$, *also* (5).
- $\Pi_n(n \geq 0) :$ *Is closed under* (1)–(3), *and, if* $n > 0$, *also* (6).

Proof (1) follows from the corresponding closure of $\mathcal{R}_*$. The rest follow at once from Definition 13.11.2 and the techniques of Theorem 6.5.2 ((4) also uses Lemma 13.11.4). $\square$

13.11.6 Corollary. $\Delta = \Delta_\infty$, *hence, Definition* 13.11.2 *classifies* all *arithmetical predicates according to* definitional complexity .

Proof The direction $\Delta_\infty \subseteq \Delta$ already being noted, we establish the equally obvious $\Delta \subseteq \Delta_\infty$ direction.

---

[41] Since U_i^n, $\lambda x.0$ and $\lambda x.x + 1$ are recursive, this allows the full range of Grzegorczyk substitutions, i.e., additionally to function substitution, also expansion of the variable list, permutation of the variable list, identification of variables, substitution of constants into variables.

We do induction of the formation of Δ. For the initial relations $\mathcal{R}_*$, they all are in Δ_∞ (empty quantifier prefix). Let then $(\exists x)P \in \Delta$. By the I.H. $P \in \Delta_\infty$. By Definition 13.11.2, say P is Π_n then $(\exists x)P \in \Sigma_{n+1} \subseteq \Delta_\infty$. If on the other hand P is Σ_n, then $(\exists x)P \in \Sigma_n$ by Lemma 13.11.5. $\qquad\square$

13.11.7 Proposition. $\Sigma_n \cup \Pi_n \subseteq \Delta_{n+1}$, for $n \geq 0$.

Proof Induction on n.

Basis. For $n = 0$, $\Sigma_n \cup \Pi_n = \mathcal{R}_*$. On the other hand, $\Sigma_1 = \mathcal{P}_*$ (why?), while (by Lemma 13.11.4) $\Pi_1 = \{Q : (\neg Q) \in \mathcal{P}_*\}$ —the *co-r.e.* predicates. Thus $\Delta_1 = \Sigma_1 \cap \Pi_1 = \mathcal{R}_*$ by Theorem 6.1.7.

We consider now the $(n + 1)$-case (under the obvious I.H.). Let $R(\vec{x}_r) \in \Sigma_{n+1}$. Then, by Lemma 13.11.5(1), so is $\lambda z \vec{x}_r . R(\vec{x}_r)$, where z is not among the $\vec{x}_r$. Now, $R \equiv (\forall z)R$, hence $(\forall z)R \in \Pi_{n+2}$, that is, $R \in \Pi_{n+2}$.

Next, let $R(\vec{x}_r) \equiv (\exists z)Q(z, \vec{x}_r)$, where $Q \in \Pi_n$. By the I.H., $Q \in \Delta_{n+1}$, hence $Q \in \Pi_{n+1}$. Thus, $R \in \Sigma_{n+2}$. We conclude the induction step in this case: $R(\vec{x}_r) \in \Sigma_{n+2} \cap \Pi_{n+2} = \Delta_{n+2}$.

The argument if we start with an $R(\vec{x}_r) \in \Pi_{n+1}$ is entirely analogous and is omitted. $\qquad\square$

13.11.8 Corollary. $\Sigma_n \subseteq \Sigma_{n+1}$ and $\Pi_n \subseteq \Pi_{n+1}$, for $n \geq 0$.

13.11.9 Corollary. $\Delta_n \subseteq \Delta_{n+1}$, for $n \geq 0$.

13.11.10 Lemma. *If $R(\vec{x})$ is Σ_n or Π_n then its characteristic function c_R is Δ_{n+1}.*

Proof Note that $c_{R(\vec{x})} \simeq u \equiv \big((R(\vec{x}) \wedge u = 0) \vee (\neg R(\vec{x}) \wedge u = 1)\big)$. Each disjunct in the right hand side of the equivalence is Δ_{n+1} (Proposition 13.11.7). But Δ_{n+1} is closed under $\vee$. $\qquad\square$

We next sharpen the inclusions above to proper inclusions.

13.11.11 Definition. (Kleene) For $n \geq 1$ we define a sequence of relations of two arguments $\lambda x y . E_n(x, y)$ by induction:

$$E_1(x, y) \equiv (\exists z)T(x, y, z)$$

$$E_{n+1}(x, y) \equiv (\exists z)\neg E_n(x, \langle z \rangle * y)$$

where "$\langle \rangle$" and "$*$" are as in Definitions 2.4.2 and 2.4.1, and T is the Kleene predicate, say, as defined in Definition 5.5.1. $\qquad\square$

Of course, $\langle z \rangle = 2^{z+1}$.

13.11.12 Lemma. $E_n \in \Sigma_n$ and $\neg E_n \in \Pi_n$, for $n \geq 1$.

Proof Induction on n. For $n = 1$, E_1 is Σ_1 and $\neg E_1$ is Π_1 since T is (primitive) recursive.

Taking the obvious I.H. for fixed n, we look at $n + 1$. By the I.H. $\neg E_n$ is Π_n, so E_{n+1} is Σ_{n+1} by Lemma 13.11.5(1) and the second defining equation in Definition 13.11.11. But then $\neg E_{n+1}$ is Π_{n+1} by Lemma 13.11.4. $\qquad\square$

The following is an enumeration and representation theorem where the E_n play a role analogous to that of the Kleene predicate for the enumeration and representation of the functions in $\mathcal{P}$.

13.11.13 Theorem. (Enumeration or Indexing Theorem (Kleene))

(1) $R(\vec{x}_r) \in \Sigma_{n+1}$ iff $R(\vec{x}_r) \equiv E_{n+1}(i, \langle \vec{x}_r \rangle)$, for some i.
(2) $R(\vec{x}_r) \in \Pi_{n+1}$ iff $R(\vec{x}_r) \equiv \neg E_{n+1}(i, \langle \vec{x}_r \rangle)$, for some i.

Proof The *if*-part is Lemma 13.11.12 (using Lemma 13.11.5(1)).

We prove the *only if*-part ((1) and (2) simultaneously) by induction on n.

Basis. $n = 0$. Say, $R(\vec{x}_r)$ is Σ_1. Then so is $R((u)_0, \dots, (u)_{r-1})$ (Lemma 13.11.5(1)). Thus, for some i (semi-index), $R((u)_0, \dots, (u)_{r-1}) \equiv (\exists z)T(i, u, z)$, hence $R(\vec{x}_r) \equiv (\exists z)T(i, \langle \vec{x}_r \rangle, z) \equiv E_1(i, \langle \vec{x}_r \rangle)$.

If $R(\vec{x}_r)$ is Π_1, then $(\neg R(\vec{x}_r))$ is Σ_1. Thus, for some e, $\neg R(\vec{x}_r) \equiv E_1(e, \langle \vec{x}_r \rangle)$, hence $R(\vec{x}_r) \equiv \neg E_1(e, \langle \vec{x}_r \rangle)$.

The induction step. (From the obvious I.H.) Let $R(\vec{x}_r) \in \Sigma_{n+2}$. Then $R(\vec{x}_r) \equiv (\exists z)Q(z, \vec{x}_r)$, where $Q \in \Pi_{n+1}$, hence (I.H.)

$$Q(z, \vec{x}_r) \equiv \neg E_{n+1}(e, \langle z, \vec{x}_r \rangle), \text{ for some } e$$

Since $\langle z, \vec{x}_r \rangle = \langle z \rangle * \langle \vec{x}_r \rangle$, the last equivalence implies

$$R(\vec{x}_r) \equiv (\exists z)\neg E_{n+1}(e, \langle z \rangle * \langle \vec{x}_r \rangle)$$
$$\equiv E_{n+2}(e, \langle \vec{x}_r \rangle) \qquad \text{(by the definition of } E_{n+2})$$

An entirely analogous argument applies to the case of $R(\vec{x}_r)$ being Π_{n+2}. $\square$

13.11.14 Theorem. (Hierarchy Theorem (Kleene, Mostowski))

1. $\Sigma_{n+1} - \Pi_{n+1} \neq \emptyset$,
2. $\Pi_{n+1} - \Sigma_{n+1} \neq \emptyset$.

Moreover, all the inclusions in 13.11.7 $(n > 0)$, 13.11.8 $(n \geq 0)$ *and* 13.11.9 $(n > 0)$ *are* proper.

Proof

(1) $E_{n+1} \in \Sigma_{n+1} - \Pi_{n+1}$. See Lemma 13.11.12). If also $E_{n+1} \in \Pi_{n+1}$, then (proceeding by diagonalisation), for some e, we have (Lemma 13.11.5(1) and Theorem 13.11.13)

$$E_{n+1}(x, \langle x \rangle) \equiv \neg E_{n+1}(e, \langle x \rangle)$$

Letting $x = e$ in the above we get a contradiction.

(2) Claim (2) is proved as in (1), but use $\neg E_{n+1}$ as the "counterexample".

- For Proposition 13.11.7 with $n > 0$: Define $R(x, y) \equiv E_n(x, \langle x \rangle) \vee \neg E_n(y, \langle y \rangle)$. Since $E_n \in \Sigma_n$ and $(\neg E_n) \in \Pi_n$, Lemma 13.11.5(1), Proposition 13.11.7 and closure of Δ_{n+1} under $\vee$ yield $R \in \Delta_{n+1}$. Let x_0 and y_0 be such that $E_n(x_0, \langle x_0 \rangle)$ is false and $E_n(y_0, \langle y_0 \rangle)$ is true (if no such x_0, y_0 exist, then E_n is recursive (why?), not so for $n > 0$ (why?)).

Now, $R \notin \Sigma_n \cup \Pi_n$, for, otherwise, say it is in Σ_n. Then so is $R(x_0, y)$ (that is, $\neg E_n(y, \langle y \rangle)$) which is absurd. Similarly, we are led to the absurd conclusion that $R(x, y_0) \in \Pi_n$, if $R \in \Pi_n$.

- For Corollary 13.11.8 with $n \geq 0$: $K \in \Sigma_1 - \Sigma_0$. For $n > 0$, $\Sigma_n = \Sigma_{n+1}$ implies $\Pi_n = \Pi_{n+1}$ (why?), hence $\Sigma_n \cup \Pi_n = \Sigma_{n+1} \cup \Pi_{n+1} \supseteq \Sigma_{n+1} \cap \Pi_{n+1} = \Delta_{n+1}$, hence $\Sigma_n \cup \Pi_n = \Delta_{n+1}$ by Proposition 13.11.7, contrary to what we have just established above. Similarly for the "dual" Π_n vs. Π_{n+1}.
- For Corollary 13.11.9 with $n > 0$: If $\Delta_{n+1} = \Delta_n$, then $\Delta_{n+1} = \Sigma_n \cap \Pi_n \subseteq \Sigma_n \cup \Pi_n$, which has been established as an absurdity for $n > 0$.   □

So *the hierarchy is proper*, that is, there is no n after which any of Σ_n, Π_n, Δ_n stay the same forever. But still we have not connected definitional with computational complexity. We will do this shortly.

But first we present a tool here that helps to place sets (relations) *accurately* in the hierarchy. By "accurately" we mean placement in the lowest possible level and class (of type Σ, Π or Δ).

The tool in mind is (strong) reducibility introduced in Definition 13.10.7. This tool is not omni-applicable, but works if we apply it to select cases.

13.11.15 Definition. (Completeness) A set A is called Σ_n- (resp. Π_n-) *complete* iff *every* relation $R(\vec{x})$ that is Σ_n (resp. Π_n) is strongly reducible to A, that is (Definition 13.10.7), $R(\vec{x}) \leq_m A$.

Intuitively, A is "hardest" among relations in Σ_n (resp. Π_n).   □

13.11.16 Example. Consider the set of total computable functions, given by their indices, $Tot = \{e : \{e\} \text{ is total}\}$.

Thus, $e \in Tot \equiv (\forall x)(\exists y)T(e, x, y)$, which is Π_2.

Now take any relation $R(\vec{x})$ in Π_2. Thus (Definition 13.11.2), for some *recursive* P we have $R(\vec{x}) \equiv (\forall u)(\exists w)P(u, w, \vec{x})$. Thus the function

$$f \overset{Def}{=} \lambda u\vec{x}.(\mu w)P(u, w, \vec{x})$$

is partial recursive and total: it is *recursive*. By the S-m-n theorem there is a recursive $h(\vec{x})$ such that $\{h(\vec{x})\} = \lambda u.f(u, \vec{x})$. But $R(\vec{x}) \equiv (\forall u)f(u, \vec{x}) \downarrow \equiv \{h(\vec{x})\} \in Tot$ and we proved $R(\vec{x}) \leq Tot$. Thus placing Tot in Π_2 is optimal.   □

We next establish a strong correlation between the *definitional complexity* and the *degree of unsolvability* of the arithmetical relations, the latter gauged by T-degree bounds on the degrees of the predicates in the hierarchy (Post's theorem below).

The following pair of theorems is collectively known as *Post's theorem* and together establish the connection between *definitional complexity* and (un)*computational complexity*.

13.11.17 Theorem. (Post's Theorem) *For* $n \geq 0$,

(1) R *is* Σ_{n+1} *iff* R *is* A-r.e. *for some* set $A \in \Sigma_n$ *(or in* Π_n*)*.
(2) R *is* Δ_{n+1} *iff* R *is* A-recursive *for some* $A \in \Sigma_n$ *(or in* Π_n*)*.

Proof We prove each direction of iff for both (1) and (2).

- $\rightarrow$ direction.

    (1) We have $R(\vec{x}) \equiv (\exists z)Q(z,\vec{x})$, where Q is Π_n. *Define* $A = \{\langle z,\vec{x}\rangle : \neg Q(z,\vec{x})\}$. Thus A is Σ_n. But $R(\vec{x}) \equiv (\exists z)\langle z,\vec{x}\rangle \notin A$, so it is A-r.e. since the quantified part is A-recursive.
    (2) Here both R and $\neg R$ are in Σ_{n+1}.[42] By (1), R is A-r.e. and $\neg R$ is B-r.e., for A and B in Σ_n. Trivially each of R and $\neg R$ is $A \oplus B$-r.e. where

    $$A \oplus B = \{2x : x \in A\} \cup \{2x+1 : x \in B\}$$

    (the *join*, "$\oplus$" was first defined and used on p. 336). Thus R is recursive in A by Proposition 13.9.7 (via Theorem 13.8.45).

- $\leftarrow$ direction.

    (1) Let R be A-r.e. for some *set* $A \in \Sigma_n$ (or in Π_n). We work with the Σ_n assumption leaving the alternative Π_n to the reader. By Theorem 13.8.38, if e is a semi-index (relative to A) for R, then we have (conjunctionally)

    $$\begin{aligned} R(\vec{x}) &\Leftrightarrow \{e\}^A(\vec{x}) \downarrow \\ &\Leftrightarrow (\exists y)\mathbb{T}_1(e,\langle\vec{x}\rangle,\langle\overline{c_A}(y)\rangle,y) \\ &\Leftrightarrow (\exists y)(\exists u)\Big(u = \langle\overline{c_A}(y)\rangle \wedge \mathbb{T}_1(e,\langle\vec{x}\rangle,\langle\overline{c_A}(y)\rangle,y)\Big) \end{aligned}$$

    The unquantified part of the last formula in the equivalence chain above is in Δ_{n+1} by the presence of $u = \langle\overline{c_A}(y)\rangle$ (cf. Lemma 13.11.10 and Definition 13.11.2) hence in Σ_{n+1}. But Σ_{n+1} is closed under $\exists$.
    (2) Let R be A-*recursive* for some *set* $A \in \Sigma_n$ (or in Π_n). Thus R *and* $\neg R$ are A-semi-recursive. By part (1) in this direction, R is Σ_{n+1} and $\neg R$ is also Σ_{n+1}, hence R is also Π_{n+1}. Thus R is Δ_{n+1}. □

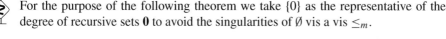

For the purpose of the following theorem we take $\{0\}$ as the representative of the degree of recursive sets **0** to avoid the singularities of $\emptyset$ vis a vis $\leq_m$.

13.11.18 Theorem. (Post's Theorem) *For* $n \geq 0$,

(1) R *is* Σ_n *iff* $R \leq_m \emptyset^{(n)}$.
(2) R *is* Σ_{n+1} *iff* R *is* $\emptyset^{(n)}$-*r.e.*
(3) R *is* Δ_{n+1} *iff* $R \leq_T \emptyset^{(n)}$.

---

[42] $R \in \Sigma_{n+1} \cap \Pi_{n+1}$ yields $R \in \Pi_{n+1}$ hence $\neg R \in \Sigma_{n+1}$.

Proof We prove all of (1)–(3) simultaneously by induction on n. Note that since $S \leq_m S$ for nonempty sets, (1) implies also that $\emptyset^{(n)} \in \Sigma_n$ for $n \geq 0$ (as we cautioned above, $\mathbf{0} = \boldsymbol{d}_T(\{0\})$).

Note now that (1) implies (2) and (3): Indeed, say R is Σ_{n+1}. Then, by (1), $R \leq_m \emptyset^{(n+1)}$. Now $\emptyset^{(n+1)} = (\emptyset^{(n)})' = K^{\emptyset^{(n)}}$ hence R is $\emptyset^{(n)}$-r.e.

Conversely (the right to left direction in (2)) let R be $\emptyset^{(n)}$-r.e. Since $\emptyset^{(n)}$ is Σ_n, Theorem 13.11.17 establishes R is Σ_{n+1}.

We leave (1)$\rightarrow$(3) to the reader (Exercise 13.12.29).

Next note that (1) is true for $n = 0$ as long as we do not try $S = \emptyset$[43] due to the awkwardness of $\leq_m$. Indeed, $\Sigma_0 = \mathcal{R}_*$ and a set is recursive iff it is many-one-reducible to a recursive set.

Assume all the cases for (1)–(3) for a fixed n (The I.H.)

Suffices to to prove the $n + 1$ case for (1). This follows from (2) at n (I.H.) by Proposition 13.10.10, which in our case offers $R \leq_m \emptyset^{(n+1)}$ iff R is $\emptyset^{(n)}$-r.e. $\qquad\square$

13.11.19 Corollary. *For $n \geq 0$, R is Π_n iff $R \leq_m \overline{\emptyset^{(n)}}$.*

13.11.20 Example. Let $A = \{x : W_x \text{ is finite}\}$. Then

$$x \in A \equiv (\exists u)(\forall y)\Big(y > u \rightarrow \neg(\exists z)T(x, y, z)\Big)$$

The predicate in the large brackets being Π_1, A is Σ_2. Now neither A nor $\overline{A}$ are semi-recursive, hence we have a very accurate placement for A in the hierarchy: $A \in \Sigma_2 - \Sigma_1 \cup \Pi_1$. We note that $A \leq_m \emptyset^{(2)}$ by Theorem 13.11.18. By Exercise 13.11.2 we note that $\{x : \phi_x \in \mathcal{R}\} \sim \{x : W_x \text{ is infinite}\}$ and that the left of the two these sets —which we earlier called Tot— is Π_2-complete (13.11.16). Thus $\overline{\emptyset^{(2)}} \leq_1 Tot$ (1-completeness and m-completeness coincide). Hence $\overline{\emptyset^{(2)}} \equiv_1 Tot$ and thus (Myhill's theorem) $\overline{\emptyset^{(2)}} \sim Tot$. Therefore

$$\{x : W_x \text{ is infinite}\} \sim Tot \sim \overline{\emptyset^{(2)}}$$

and

$$\{x : W_x \text{ is finite}\} \sim \overline{Tot} \sim \emptyset^{(2)}$$

$\qquad\square$

13.11.21 Example. Let $B = \{x : W_x = K\}$. B is neither r.e. nor co-r.e. as seen via an application of Theorem 11.3.5. We have

$$x \in B \equiv (\forall y)\Big((\neg y \in W_x \vee y \in K) \wedge (\neg y \in K \vee y \in W_x)\Big)$$

---

[43] Of course, the theorem for $S = \emptyset$ can be established directly and trivially.

Now, each of $y \in W_x$ and $y \in K$ are semi-recursive, hence Σ_1, thus (due to the negations) each conjunct in the big brackets is ($\Sigma_1 \cup \Pi_1$ thus) Δ_2 hence also Π_2. Thus B is Π_2 by Lemma 13.11.5(6). By the opening remark about B, we have the accurate placement B is in $\Pi_2 - \Sigma_1 \cup \Pi_1$. The reader may want to explore additional placement examples in Rogers (1967). □

13.11.22 Theorem. (Another Look at Tarski's Theorem) *The set of all true arithmetical sentences, CA, is not arithmetical.*

Proof We assume the arithmetisation toolbox of Chap. 12 and in particular the symbols $gn\{\ldots\}$ and $\ulcorner \ldots \urcorner$ for informal (outside PA) and formal (within PA) Gödel numbers of string expressions respectively.

Arguing by contradiction, suppose that CA *is* arithmetical, say,

$$(x^{44} \in CA) \in \Sigma_m \cup \Pi_m$$

for some m.

Pick a relation $R(x) \in \Delta_{m+1} - \Sigma_m \cup \Pi_m$ (cf. Theorem 13.11.14: there is no "last level" in the hierarchy) and let the formula $G(u)$, of one variable, in the language of, say, ROB (Definition 12.6.1) *express R* over said language (cf. Definition 8.3.1), that is,

$$(\forall a \in \mathbb{N}): R(a) \text{ is true iff } G(\widetilde{a}) \text{ is true in the } standard \; model \; of \; arithmetic^{45}$$

 The reader who may be starting the exploration of this volume with the present chapter may be wondering "what do we need a formal arithmetic like ROB or PA for?"

Well, only for their alphabet here —which we use to write down formulas (as finite strings) among other things. Note that we need CA to be a set of *numbers* in order to say things about it like "it is not r.e." or, worse, "it is not arithmetical". The next best thing to having numbers in CA is to have strings over a finite alphabet. Such strings are in a natural 1-1 correspondence with numbers. But how can an *informal* relation —an extensional object[46]— get a *representing* unique number? It does so indirectly! For each such relation the concept of *expressibility* assigns to it a *formula counterpart* that is true at $\widetilde{a}$ in the standard model iff the relation is true at a. The formula counterpart[47] is a *finite string over a finite alphabet* of, say, b elements. *So a formula can be viewed as a number written in base-b notation (approximately*

---

[44] Of course, the variable x varies over closed arithmetical formulas indirectly by actually varying over their informal Gödel numbers.

[45] That is, variables and constants $\widetilde{a}$ get their values from $\mathbb{N}$ —$\widetilde{a}$ necessarily getting the value a— and $<$ is *translated* as less than, S as $\lambda x.x + 1$, $+$ as plus, $\times$ as times.

[46] A set of numbers.

[47] This assignment is not unique but we can work with it!

speaking) and it *names* the informal relation. Thus if $R(x)$ is expressed by a formula $G(x)$ then the *string* $G(x)$ is viewed as a *number* that "codes" not only $G(x)$ but also codes the *relation*. If that number is informally called g, then the formal number (in the, say, ROB alphabet) is

$$\underbrace{SS \cdots S}_{g\ S\ symbols}\ 0$$

denoted by $\widetilde{g}$, and is called the (*formal*) Gödel number (of $G(x)$). We write $\widetilde{g} = \ulcorner G(x) \urcorner$. We also write $g = gn\ \{G(x)\}$ to indicate that $g \in \mathbb{N}$ is the *informal* Gödel number of $G(x)$.

All this is elaborated in great detail in Chap. 12.

The formula

$$\mathscr{G}(y) \overset{Def}{\equiv} (\exists u)\big(y = u \wedge G(u)\big)$$

also expresses $R(x)$ by Theorem 0.2.94 and we prefer $\mathscr{G}(y)$ over the original $G(y)$ since an informal *and* a formal Gödel number for $\mathscr{G}(\widetilde{n})$ —$n \in \mathbb{N}$— can be easily computed (Chap. 12). Specifically, $\mathscr{G}(\widetilde{n})$ has an informal Gödel number that is a *recursive* function of n, or $\lambda n.gn\ \{\mathscr{G}(\widetilde{n})\} \in \mathcal{R}$.

We can now conclude our argument:

for all $n \in \mathbb{N}$, $R(n)$ is true iff $gn\ \{\mathscr{G}(\widetilde{n})\} \in CA$

Thus $R(x) \in \Sigma_m \cup \Pi_m$ by the recursiveness of $\lambda n.gn\ \{\mathscr{G}(\widetilde{n})\}$ and the assumed placement of CA in the hierarchy, a contradiction.   □

13.12   Exercises

1. Prove that the set of all indices of *computations* defined in Definition 13.3.1 is recursive. Can you improve "recursive"?
2. Complete the$z$ proof of Proposition 13.3.29.
3. Complete the proof of Theorem 13.6.18.
4. Complete the proof of Theorem 13.6.19.
5. Prove that the search computable functionals are compact and monotone.
6. For any α define the *finite* sub-function $\lambda xu\alpha.\ \langle \alpha \mid u \rangle\ (x)$ (Davis (1958a)) by letting, extensionally,

$$\langle \alpha \mid u \rangle \overset{Def}{=} \{(x, y) \in \alpha : x \le u \wedge y \le u\}$$

Prove that $\lambda xu\alpha.\ \langle \alpha \mid u \rangle\ (x)$ is neither Kleene- nor Moschovakis-computable. Is it weakly partial computable? Why?

7. Prove that if A and B are any recursive sets, then $A \leq_T B$.
8. A function α is recursive in a function β —$\alpha \leq_T \beta$— means that $\alpha = \{e\}^\beta$ for some $e \in \mathbb{N}$.

   Prove that there are total α and β satisfying $\alpha \leq_T \beta$ but there is no partial recursive $\mathscr{F}$ in the sense of Kleene such that $\alpha = \lambda x.\mathscr{F}(x, \beta)$.

   Hint. Let **Ind** be the set of indices from Exercise 13.12 in this section. Define the set A by

$$x \in A \overset{Def}{\equiv} x \in \textbf{Ind} \wedge \{x\}(x, c_A) \simeq 1$$

   Trivially, $A \leq_T \textbf{Ind}$. However you should obtain a contradiction from the assumption $c_A = \{e\}(x, c_{\textbf{Ind}})$, for some e.
9. Prove that a Kleene-computable partial functional $\mathscr{F} : \mathbb{N} \times \mathcal{P}(\mathbb{N} : \mathbb{N}) \to \mathbb{N}$ is (topologically) *partial continuous* (Example 0.7.10). Recall that $\mathbb{N}$ is equipped with the discrete topology.
10. Prove that a functional $\mathscr{F} : \mathbb{N} \times \mathcal{P}(\mathbb{N} : \mathbb{N}) \to \mathbb{N}$ is monotone and compact iff it is (topologically) *partial continuous*.
11. Prove that partial recursive *restricted* functionals are closed under a *restricted* type-1 substitution: Given two partial recursive (restricted) functionals $\boldsymbol{F}$ (rank $(n, l+1)$) and $\boldsymbol{G}$ (rank (n, l)) *there is* a partial recursive restricted functional $\boldsymbol{H}$ (rank (n, l)) such that for all $(\vec{x}, \vec{\alpha})$ where $\lambda y.\boldsymbol{G}(y, \vec{x}, \vec{\alpha})$ is *total* we have

$$\boldsymbol{H}(\vec{x}, \vec{\alpha}) \simeq u \text{ iff } \boldsymbol{F}\big(\vec{x}, \vec{\alpha}, \lambda y.\boldsymbol{G}(y, \vec{x}, \vec{\alpha})\big) \simeq u \tag{1}$$

   Hint. Use the Normal Form theorem.

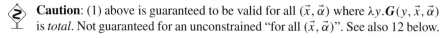 **Caution**: (1) above is guaranteed to be valid for all $(\vec{x}, \vec{\alpha})$ where $\lambda y.\boldsymbol{G}(y, \vec{x}, \vec{\alpha})$ is *total*. Not guaranteed for an unconstrained "for all $(\vec{x}, \vec{\alpha})$". See also 12 below.

12. Prove that substitution of type-1 inputs (in the case of restricted partial recursive functionals) does not hold in general. That is, Given two partial recursive (restricted) functionals $\boldsymbol{F}$ (rank $(n, l+1)$) and $\boldsymbol{G}$ (rank (n, l)) we do *not* in general have a restricted partial recursive $\boldsymbol{H}$ for which (1) in 11 works *for all* $(\vec{x}, \vec{\alpha})$.

    Hint. Take $l = 0$ and then define $\boldsymbol{F}(\vec{x}, \beta) \simeq 0$ for all arguments. Take $\boldsymbol{G}(x, z) \simeq 0 \equiv x \notin W_{x,z}$ (cf. $W_{e,s}$ notation in Long Definition 13.9.6). Then investigate the statement $\boldsymbol{F}(\vec{x}, \lambda z.\boldsymbol{G}(x, z)) \downarrow$ and draw a contradiction.
13. While non monotone *weakly partial computable* functionals $\mathscr{F}(x, \alpha)$ may fail to have a least (computable or not) fixpoint (Remark 13.7.7), they always have a *computable fixpoint*.

    Hint. Use the *second* recursion theorem and imitate the technique of Sect. 9.5.
14. Prove that **0** is the degree of all recursive sets.
15. Prove that the jump of A, A', is semi-recursive in A.
16. Prove that the $\mathcal{F}$ defined in (1) in Example 13.7.9 is partial recursive in the Kleene sense.

17. Adapt the one-variable (type-0) case to the n variable case and prove that

$$\alpha(\vec{x}_n) \simeq \mathscr{F}(\vec{x}_n, \alpha)$$

has a computable *least fixpoint* in the Kleene and Moschovakis computabilities.

18. Prove that the call-by-value approach for

$$Q: \qquad \alpha(x, y) ::= \mathscr{F}(x, y, \alpha) \tag{1}$$

yields the function $\alpha(x, y) \simeq$ *if* $x = 0$ *then* 1 *else* $\uparrow$.

19. Let $\mathscr{F}$ be given by $\mathscr{F}(x, \alpha) \simeq$ *if* $x = 0$ *then* 1 *else* $\alpha(x + 1)$. Find the least fixpoint of $\mathscr{F}$. Does it have any recursive fixpoints?

20. (McCarthy's "91 function") Show that $\lambda x.if\ x > 100\ then\ x \mathbin{\dot-} 10\ else\ 91$ is the only fixed point of $\mathscr{F} = \lambda x \alpha.if\ x > 100\ then\ x \mathbin{\dot-} 10\ else\ \alpha(\alpha(x + 11))$.

21. Prove that $\mathscr{C} \subseteq \mathcal{P}^{(1)}$ is *completely r.e.* iff for some partial recursive functional $\mathscr{F}$ of rank (0, 1) we have $\mathrm{dom}(\mathscr{F}) = \mathscr{C}$.

22. Prove that a class of r.e. *sets* $\mathscr{S}$ is completely r.e iff for some partial recursive functional $\mathscr{G}$ of rank (0, 1) we have $\mathscr{G}(\alpha) \simeq 0$ if $\mathrm{dom}(\alpha) \in \mathscr{S}$ and $\mathscr{G}(\alpha) \uparrow$ if $\mathrm{dom}(\alpha) \notin \mathscr{S}$

23. Manna (1974) Adopt here the convention $0 \cdot \uparrow \simeq 0$. Verify that the *call-by-name* rule does *not* compute the *least fixpoint* of the following program

$$P: \qquad \alpha(x, y) ::= \mathscr{F}(x, y, \alpha)$$

where $\mathscr{F}(x, y, \alpha) \simeq$ *if* $x = 0$ *then* 0 *else* $\alpha(x + 1, \alpha(x, y)) \cdot \alpha(x \mathbin{\dot-} 1, \alpha(x, y))$.

24. Prove Corollary 13.8.50.

25. Use your favourite coding for finite sets and show that from a recursive enumeration f of a set $C - \mathrm{ran}(f) = C -$ you can obtain a recursive enumeration of (the codes of) the sets

$$\emptyset, \{f(0)\}, \{f(0), f(1)\}, \{f(0), f(1), f(2)\}, \dots, \{f(0), f(1), \dots, f(n)\}, \dots$$

Thus C is enumerated by a recursive $\subseteq$-increasing sequence of finite subsets C_n.

26. Prove that the predicate $\lambda x e.x \in I_e$ is r.e., where I_e is defined on p. 527.

27. Prove using pure logic that any arithmetical relation R is absolutely (using no mathematical axioms) provably equivalent to one of the form given in Remark 13.11.3.

28. Prove the omitted cases "(or in Π_n)" in the proof of Theorem 13.11.17.

29. Fill the gaps in the proof of Theorem 13.11.18.

30. Prove Corollary 13.11.19.

31. Prove $\{x : \phi_x \in \mathcal{R}\} \leq_m \overline{\emptyset^{(2)}}$.

32. Prove that neither $A = \{x : W_x \text{ is finite}\}$ nor $\overline{A} = \{x : W_x \text{ is infinite}\}$ are r.e.

33. Locate $\{x : W_x \text{ is simple}\}$ in the arithmetical hierarchy.

34. Locate $\{x : W_x = \{0\}\}$ in the arithmetical hierarchy.

35. Locate $\{x : W_x = \emptyset\}$ in the arithmetical hierarchy.

36. Locate $\{(x, y) : \phi_x = \phi_y\}$ in the arithmetical hierarchy.

Chapter 14
Complexity of $\mathcal{P}$ Functions

Overview

Our investigations so far were aimed mainly towards understanding *limitations of computing* and *logic*, exploring *uncomputability*, *undecidability* and *unprovability* (in ROB, Definition 12.6.1) and its recursive extensions, most notable of which being PA), with short studies of what *is* computable —e.g., in the early chapters on primitive recursive functions and loop programs— doing the latter task (so far) mostly for "tooling" reasons, for example, obtaining *coding* tools. Our last excursion into the unsolvable was the Kleene-Mostowski *Arithmetical Hierarchy* of *increasingly undecidable* relations and sets.

There is another kind of limitation, this one found among the *computable* functions. In "real life" the programmer is interested not only in finding a *correct* computational solution to a problem, but also one that is "*efficient*" meaning it spends, relatively to other solutions, fewer computation resources such as computer *memory* or computer (execution) *time*.

Thus, the theoretician classifies problems according to their *degree of unsolvability* —a measure of problem complexity (the higher it is in the Kleene-Mostowski hierarchy the more uslovabhe or complex it is). The serious programming practitioner[1] on the other hand would rather classify *solvable*, *decidable*, problems according to the amount of computational resources they consume —their *computational complexity*.

Mathematicians and computer scientists have contributed to the body of knowledge known as computational complexity at least since the mid 1960s. We divide our brief review of this vast area into a chapter (the present one) on the complexity theory of the set $\mathcal{P}$ of *partial recursive* functions and into a chapter (the next one)

---

[1] Computer programming can be quite a mathematically challenging activity. See, for example the series of volumes titled "the Art of Computer Programming" (with each volume having an appropriate subtitle that fits the specific topics in it) starting with Vol. 1, "fundamental algorithms" Knuth (1973).

© Springer Nature Switzerland AG 2022

G. Tourlakis, *Computability*, https://doi.org/10.1007/978-3-030-83202-5_14

on the complexity of the *primitive recursive* functions. In both cases —right at the outset in the case of $\mathcal{PR}$— we adopt a hierarchy point of view.

In the present chapter we follow the axiomatic approach of Blum (1967). The cited work is a marvelous mathematical feat where *just two* "intuitively *obvious* and acceptable" axioms lead to deep and at times unexpected (by intuition) results.

We already saw in the case of the arithmetical hierarchy that the classification of relations by complexity was at first "*static*" or *definitional* but Post's theorem demonstrated the tight connection between static and dynamic (or *computational*) complexity.

This chapter is solely about the computational, or dynamic, complexity of the functions is $\mathcal{P}$. A gauge for such complexity, *intuitively*, may be the amount of "steps" or "time" a computation takes to conclude (halt), or may be the amount of memory, e.g., *maximum total size of stored data* —we often say "*space*"— in all the variables of a URM that provides instructions for the computation.

Chapter 15 on the other hand starts by classifying the $\mathcal{PR}$ functions in a provably *infinite hierarchy* —actually we introduce several different hierarchies that we compare and prove equivalent in a very strong sense— according to *static* or *definitional* complexity and concludes by showing via the "*Ritchie-Cobham property*" that the dynamic or *computational* complexity of the functions —say, computed by a URM— at any level of the hierarchy are accurately predicted by their static complexity!

This static complexity of a function could be the size of a (URM or Loop) program that computes it (e.g., the program length, as a string of symbols) or more usually some structural attributes of program complexity, e.g., depth of nesting of the Loop-end instruction, or equivalently, but more mathematically the static complexity of a *function* (rather than a *program*) may be the *maximum depth of nesting over all primitive recursions* needed in a given primitive recursive derivation of said function. Or, it could even be the *size* of the function's output (as is the case with the Grzegorczyk hierarchy).

This discussion will make more sense in Chap. 15. In particular, "the *maximum depth of nesting over all primitive recursions*" will be carefully and simply defined.

14.1   Blum's Axioms

Blum (1967) introduces the concept of (dynamic) complexity Φ_i (note the capital Φ) of the arbitrary $\phi_i \in \mathcal{P}$ without concretely specifying it —as time, space or something else— but rather by two "defining properties", that is, two *axioms*. This is entirely analogous to, say, defining the concept of partial *order* in an abstract way *with no reference to any familiar set* by two axioms: *irreflexivity* and *transitivity*. Then one has a wide field of applicability, such as order on $\mathbb{N}$, on $\mathbb{R}$ (*the reals*) and many others.

The results in this chapter are not specific to any particular model of computation any more that are particular to any specific complexity measure Φ. They are valid for any model equivalent to the URM (which in particular includes Turing machines and the foundation of computability as it is done in Chap. 13, omitting type-1 inputs in the latter case). Each such model of computation leads —via arithmetisation or other techniques— to, in general, *distinct enumerations* ϕ_i —or $\{i\}$— of the *unary* functions of $\mathcal{P}$. Such enumerations constitute, in the terminology of Rogers (1967), an *acceptable numbering* of the (unary) partial recursive functions.

We will not define the general term "acceptable numbering" here, and we will not issue reminders, when speaking of the enumeration ϕ_i in the rest of the chapter that we assume that we *have* fixed such a numbering —say the one for the URMs (cf. Definition 5.4.5). Blum's results stay true f*or any acceptable numbering*.

14.1.1 Definition. (Blum (1967) Abstract Complexity Measure)

A *complexity measure* for the partial recursive (unary) functions $(\phi_i)_{i \geq 0}$ is a sequence of partial (unary) functions $(\Phi_i)_{i \geq 0}$ that satisfy the two axioms below:

Ax 1. For all i, x, $\mathrm{dom}(\phi_i) = \mathrm{dom}(\Phi_i)$.
Ax 2. The predicate $\lambda i x y.\Phi_i(x) \simeq y$ is recursive. □

The definition is parsimonious: It does not say what computability attributes the *functions* Φ_i have (are $\lambda x.\Phi_i(x)$ or $\lambda i x.\Phi_i(x)$ computable?). This information is actually coded into the axiom 2. Cf. Proposition 14.1.7.

14.1.2 Remark Intuitively obvious remarks:

- The first axiom is trivial in concrete circumstances and is the very minimum we can expect of a "reasonable" measure. We can measure complexity of $\phi_i(x)$ iff $\phi_i(x) \downarrow$ since otherwise there is nothing to measure: If $\phi_i(x) \uparrow$, then the complexity is infinite, if we are thinking of URM steps (say), and that is not a number nor can we verify $\phi_i(x) \uparrow$ in general (we *can* verify $\phi_i(x) \downarrow$).
- The second axiom is eminently reasonable too: Imagine that the resource we are gauging with Φ_i is "time" (or number of "steps" of a URM, say). So, let i, x, y be given.

  Then, intuitively, equipped with a stopwatch I do

1. I set it to go off at "time" y.
2. I start my computation $\phi_i(x)$ (on the i-th URM, with input x) and start the stopwatch at the same time.
   Decisions:
3. If the time the stopwatch goes off coincides with the time that the computation ends (that is, the **stop** instruction of the URM is reached at that very time), then I exit my process and say "yes".

4. If at the time the stopwatch goes off the URM is still going in an *essential manner*,[2] then I exit my process and say "no".
5. If at any time in the process the URM reaches **stop** but *the stopwatch has not gone off*, then I exit my process and say "no".

By Church's thesis, $\lambda ixy.\Phi_x(x) \simeq y$ computed in outline above is recursive.   □

No part of the above concrete discussion *proves any* of the axioms. However it provides, in the sense of logic, a model for the axioms (a concrete interpretation where the axioms are true), hence the two axioms are *consistent* (Definition 0.2.89 and Theorem 0.2.92). *We do not expect any contradictions derived from them.*

14.1.3 Example. Here are two mathematical models for the axioms A 1. and A 2.

(a) Let us define, for all i, x,

$$\Psi_i(x) \simeq (\mu y)T(i, x, y)$$

where T is the Kleene predicate introduced in Definition 5.5.1 (cf. also Theorem 5.5.3). Recall that this implementation of the T-predicate, the variable y indicates the number of *steps* of *the URM* found in position i (in the canonical enumeration) when it computes with input x.

   Trivially, $\phi_i(x) \downarrow$ iff $\Psi_i(x) \downarrow$ (see Theorem 5.5.3), thus Ψ_i makes Ax 1. true. Next,

$$\Psi_i(x) \le u \equiv (\exists y)_{\le u} T(i, x, y)$$

which is in $\mathcal{PR}_*$. Moreover

$$\Psi_i(x) \ge u \equiv \neg(\exists y)_{<u} T(i, x, y)$$

Thus,

$$\Psi_i(x) \simeq u \equiv \Psi_i(x) \le u \wedge \Psi_i(x) \ge u$$

proving that Ψ_i satisfies Ax 2.

(b) Any type of Kleene predicate T leads to a complexity measure just similar to Ψ_i above. For example, if we define, for all i, x,

$$\Psi_i'(x) \simeq (\mu y)T^{(1,0)}(i, x, , y) \tag{1}$$

---

[2] That is, it has not reached **stop** yet. Recall that if a URM reaches **stop** it will go on forever in an *inessential manner*: It will continue "executing" the **stop** instruction *without* changing any variable.

where $T^{(1,0)}$ is that of Theorem 13.4.4 with the $\vec{\alpha}$-part empty —which accounts for the two consecutive commas in (1)— one shows entirely analogously to (a) above that Ψ_i' satisfies Ax 1. and Ax 2. □

14.1.4 Proposition. *Any complexity measure Φ_i as defined in Definition 14.1.1 satisfies*

1. $\lambda i x y. \Phi_i(x) \leq y$ *is recursive.*
2. $\lambda i x y. \Phi_i(x) \geq y$ *is recursive.*

Proof

1. $\Phi_i(x) \leq y \equiv (\exists w)_{\leq y} \Phi_i(x) \simeq w$.
2. $\Phi_i(x) \geq y \equiv \neg(\exists w)_{<y} \Phi_i(x) \simeq w$. □

14.1.5 Corollary. *Any complexity measure Φ_i as defined in Definition 14.1.1 satisfies $\lambda i x y. \Phi_i(x) > y$ is recursive.*

Proof $\Phi_i(x) > y \equiv \Phi_i(x) \geq y + 1$. □

14.1.6 Example. If we define $\Psi_i = \phi_i$, then Ψ_i satisfies A 1. but fails Ax 2 because $\lambda i x y. y = \phi_i(x)$ is not recursive (cf. Exercise 6.11.1). □

14.1.7 Proposition. *In any complexity measure (Definition 14.1.1), $\lambda i x. \Phi_i(x)$ is partial recursive.*

Proof By Ax 2. and the graph theorem 6.6.1. □

14.1.8 Corollary. *There is an $h \in \mathcal{R}$ such that, for all i, x, $\Phi_i(x) \simeq \phi_{h(i)}(x)$, equivalently, $\Phi_i = \phi_{h(i)}$ for all $i \geq 0$.*

Proof Apply the S-m-n theorem to $\lambda i x. \Phi_i(x)$. □

The laconic axiom system of two (Definition 14.1.1) has considerable power in yielding a number of nontrivial intuition-consistent results as well as one *highly nontrivial*[3] (in both statement and proof) result that challenges our intuition. The Blum speed-up theorem.

The first is an easy diagonalisation that we will not use in what follows but is a good warmup for the key lemma below (Theorem 14.1.11) for the speed-up theorem.

14.1.9 Theorem. *For every complexity measure Φ and every chosen recursive function $\lambda x. g(x)$, we can construct an $f \in \mathcal{R}$ such that, for any i, if $f = \phi_i$, then $g(x) < \Phi_i(x)$, for all $x \geq i$.*

---

[3] Early on in my studies of mathematics and its *language* I learnt of the versatility of the word "nontrivial" especially when it is ornamented by appropriate qualifiers. So, "nontrivial" simply means not easy, or worth stating. But "highly nontrivial" means very-very hard (to prove, if it refers to a theorem).

We have a priori arbitrarily chosen a *level of computational "difficulty"*, g. We may choose an enormously "big" g —e.g., it could be $A_x(A_x(x))$, where A is the Ackermann function of Chap. 4. Then we show how to find a function f, which *no matter how* we program it (via a URM i), such a program will take *more* than $g(x)$ "steps" to terminate on *almost all inputs* x, indeed on all $x \geq i$.

Proof   We want to build an f that for $i \leq x$ cannot be computed within $\leq g(x)$ steps. We need to meet two requirements:

(1) The f that we build is recursive.
(2) *No* ϕ-indices i that satisfy

$$i \leq x \land \Phi_i(x) \leq g(x)$$

are indices of f.

We say that in our construction we *cancel* all such indices (as candidates for $f = \phi_i$).

Let us thus set

$$I(x) \overset{\text{Def}}{=} \{i : i \leq x \land \Phi_i(x) \leq g(x)\} \tag{1}$$

hence the predicate $i \in I(x)$ of two arguments is recursive, being equivalent to $i \leq x \land \Phi_i(x) \leq g(x)$ (Proposition 14.1.4).
   So is the predicate $I(x) \neq \emptyset$ since $I(x) \neq \emptyset \equiv (\exists i)_{\leq x} \Phi_i(x) \leq g(x)$. Define now f, for all x, by:

$$f(x) \overset{\text{Def}}{=} \begin{cases} 1 + \sum_{i \in I(x)} \phi_i(x) & \text{if } I(x) \neq \emptyset \\ 1 & \text{otherwise} \end{cases} \tag{2}$$

We next show that f is recursive:

- Let us put the bounded summation in (2) in recognisable form. Define

$$h(i, x) \overset{\text{Def}}{=} \text{if } i \in I(x) \text{ then } \phi_i(x) \text{ else } 0$$

Now if the condition is true, then $\phi_i(x) \downarrow$ due to $\Phi_i(x) \leq g(x)$ and thus $\Phi_i(x) \downarrow$. Therefore h is total, and trivially in $\mathcal{P}$. All told, $h \in \mathcal{R}$.
   Thus

$$f(x) \overset{\text{Def}}{=} \begin{cases} 1 + \sum_{i \leq x} h(i, x) & \text{if } I(x) \neq \emptyset \\ 1 & \text{otherwise} \end{cases}$$

and $f \in \mathcal{P}$ via definition by recursive cases.
- $f(x)$ is one of $1 + \sum_{i \in I(x)} \phi_i(x)$ which is defined since all $i \in I(x)$ and thus $\Phi(x) \leq g(x)$, or $f(x) = 1$. So $f \in \mathcal{R}$.

Finally, the diagonalisation (2) ensures that, for each x, all indices in $I(x)$ ((1) above) are *cancelled*, so, if $f = \phi_j$ then $x \geq j$ implies $\Phi_j(x) > g(x)$, *otherwise* we have $x \geq j \wedge \Phi_j(x) \leq g(x)$ and hence $j \in I(x)$. The definition (2) then yields $f(x) \simeq 1 + \ldots \phi_j(x) + \ldots > \phi_j(x) = f(x)$. A contradiction (totalness is important for the contradiction). $\qquad\square$

 The reader will note that we actually *constructed* an f with the stated properties.

14.1.10 Corollary. *Let Φ be a complexity measure. There is no $\lambda x.g(x) \in \mathcal{R}$ such that, for every recursive ϕ_i, we have $\Phi_i \leq g(x)$ a.e.*

Proof See Exercise 14.14.1. $\qquad\square$

 The corollary says that there is no upper bound on the complexities of the recursive functions.

It is noteworthy that there are also *arbitrarily hard to compute* —in the sense of Theorem 14.1.9— 0-1 valued recursive functions, that is, arbitrarily hard to compute *recursive predicates*. The following and its proof are due to Blum (1967). This is the key lemma for his speed-up theorem.

14.1.11 Theorem. (Key Lemma) *For every complexity measure Φ and every chosen recursive function $\lambda x.g(x)$, we can* construct *a 0-1 valued $f \in \mathcal{R}$ such that, for any i, if $f = \phi_i$, then $g(x) < \Phi_i(x)$ a.e. with respect to x*[4]

Proof Again, we define our *new f* by *cancelling uncancelled indices* found in the recursive set

$$I(x) = \{x : i \leq x \wedge \Phi_i(x) \leq g(x)\} \tag{1}$$

Since f must be 0-1 valued we cannot cancel indices by the simple trick of defining $f(x)$ as $1 + \sum_{i \in I(x)} \phi_i(x)$ when $I(x)$ has uncancelled indices. We will do this the hard way, recording in a stack (Definition 2.4.9) all the indices we have cancelled. Every time a new index is cancelled we add it to the top of the stack. Therefore we define f, for all x, by:

$$f(x) \stackrel{Def}{=} 1 \dot{-} \phi_i(x) \text{ where } i \text{ is } least \text{ } uncancelled \text{ in } I(x); \text{ } now \text{ cancel } i \text{ } and$$

add to the stack. *Case applies* when $I(x)$ has uncancelled indices.

$$1 \qquad\qquad Case \text{ } applies \text{ when } I(x) \text{ has } no \text{ uncancelled indices.}$$

Intuitively, because of $1 \dot{-} \phi_i(x)$, $f(x)$ is forced not to have as an index *any* cancelled index.

---

[4] "a.e." was defined in Definition 4.1.10.

Here is why this works.[5]

First, f is recursive since, in outline, for any x we already have the stack up to date, so consulting it (scanning its entire length) we can find the least i in the recursive $I(x)$ that is *not cancelled* (i.e., is not anywhere in the stack). If none found, then we output 1 and exit. If we found such an i, then since it implies $\Phi_i(x) \le g(x)$, we conclude that $\phi_i(x) \downarrow$, so we can compute $f(x) = 1 \dotminus \phi_i(x)$ in a finite number of steps. This precludes this i as an index for f. So we also put i in the stack to flag this cancellation. The calculation of $f(x)$ is complete in this case. Exit.

These observations make f more than partial recursive; it is *total* too as we exit successfully in each case. Trivially it is 0-1 valued.

Next we establish the main (lower bound) claim by contradiction. So let $f = \phi_m$, for some m, provide a counterexample to the lower bound claim. That is,

$$\text{Suppose that } g(x) < \Phi_m(x) \text{ a.e. is false} \tag{2}$$

$$\text{Note that by } f = \phi_m, m \text{ is never cancelled} \tag{3}$$

Then there is an infinite sequence of inputs x_i, all larger than m,

$$m < x_1 < x_2 < x_3 < x_4 < \cdots$$

such that, for $j \ge 1$,

$$\Phi_m(x_j) \le g(x_j) \tag{4}$$

Thus

x_1:   When computing $f(x_1)$ there is an uncancelled index, m. By (3) we pick a different index j_1 to cancel. Clearly $j_1 < m$ (m is not smallest uncancelled else it would have been chosen to cancel).

x_2:   When computing $f(x_2)$ there is an uncancelled index, m. By (3) we pick a different index j_2 (j_1 is *already* cancelled) to cancel. Clearly $j_1 < m$, and by the parenthetical remark, $j_1 < j_2 < m$.

x_3:   When computing $f(x_3)$ there is an uncancelled index, m. By (3) we pick a different index j_3 (j_1 *and* j_2 are *already* cancelled) to cancel. Clearly $j_3 < m$, and by the parenthetical remark, $j_1 < j_2 < j_3 < m$.

Continuing like this we obtain an infinite sequence of $j_i, i \ge 1$

$$j_1 < j_2 < j_3 < j_4 \cdots < m$$

But having an infinite number of (distinct) natural numbers below m is impossible. So (2) is *not* false after all.   □

14.1.12 Corollary. *Let Φ be a complexity measure. Then there is no $b \in \mathcal{R}$ such that for all i, $\Phi_i(x) \leq b(x, \phi_i(x))$ a.e. with respect to x.*

Proof We are using the f of Theorem 14.1.11. For the bound g used in the key lemma we take $g = \lambda x. \max(b(x, 0), b(x, 1))$. Then

$$b(x, \phi_i(x)) \leq \max(b(x, 0), b(x, 1)) = g(x) < \Phi_i(x), \text{ a.e. with respect to } x. \quad □$$

An interpretation of the above corollary is that we cannot use the outputs of a program to predict the program's complexity. This is intuitively evident especially when the outputs are 0 or 1.

On the other hand, in the converse direction we can in principle use the number of "steps" of a program, to bound the possible outputs. See Exercise 14.4.3.

14.1.13 Proposition. *For every complexity measure Φ, there is a two-argument $g \in \mathcal{R}$ such that $\phi_i(x) \leq g(x, \Phi_i(x))$.*

Proof Blum (1967) Define

$$b(i, x, y) \stackrel{Def}{\simeq} \begin{cases} \phi_i(x) & \text{if } \Phi_i(x) \simeq y \\ 0 & \text{othw} \end{cases} \tag{1}$$

This is a definition by recursive cases and the part-functions are in $\mathcal{P}$ so $b \in \mathcal{P}$. Moreover, $\phi_i(x) \downarrow$, trivially, if $\Phi_i(x) \simeq y$, hence $\Phi_i(x) \downarrow$ (Ax 1.) holds.

Thus $b \in \mathcal{R}$.

Next, $\phi_i(x) \downarrow$ implies (Ax 1.) that $\Phi_i(x) \downarrow$ hence

$$\Phi_i(x) \simeq y, \text{ for some } y \tag{2}$$

But then $b(i, x, y) \simeq \phi_i(x)$, or —using (2)— $\phi_i(x) \simeq b(i, x, \Phi_i(x))$.

Take $g \stackrel{Def}{=} \lambda xy. \max_{i \leq x} b(i, x, y)$.   □

The moral is that recursive functions $\phi_i(x)$ that generate huge outputs will necessarily be "hard to compute" (that is, will have huge $\Phi_i(x)$).

The next is rather intuitively obvious: We can write arbitrarily inefficient (resource gobblers) programs. It is surprising that the Blum theory of complexity can attest to that:

14.1.14 Proposition. *For any complexity measure Φ, any recursive f and any resource (lower) bound $g \in \mathcal{R}$, there is an m such that $f = \phi_m$ and $(\forall x)\Phi_m(x) > g(x)$, where "$\phi$" is our standard ϕ of Definition 5.4.2.*

Proof Define

$$\psi(i, x) \simeq \begin{cases} f(x) & \text{if } \Phi_i(x) > g(x) \\ 1 \dot{-} \phi_i(x) & \text{othw} \end{cases} \tag{1}$$

By the (second) recursion theorem there is an m such that $\phi_m(x) \simeq \psi(m, x)$, for all x. Thus, from (1),

$$\phi_m(x) \simeq \begin{cases} f(x) & \text{if } \Phi_m(x) > g(x) \\ 1 \dot{-} \phi_m(x) & \text{othw} \end{cases}$$

The otherwise above cannot hold, since $\phi_m(x) \simeq 1 \dot{-} \phi_m(x)$ is untenable, seeing that the "othw" case —$\Phi_m(x) \leq g(x)$— implies $\phi_m(x) \downarrow$. Thus $f = \phi_m$ and $\Phi_m(x) > g(x)$ holds for all x (as the top is the only case we have). $\square$

The preceding discussion established the reasonableness of any complexity measure Φ. What was proved so far was intuitively expected. The next result shows something that seriously challenges our intuition.

14.2   The Speed-up Theorem

14.2.1 Theorem. (The Speed-up Theorem) *Blum (1967) Let Φ be a complexity measure and ϕ that of Definition 5.4.2. Choose a recursive two-argument r.*

Then there exists a 0-1 valued f such that, for any i such that $f = \phi_i$, there is a j such that $\Phi_i(x) > r(x, \Phi_j(x))$ a.e.

Proof Blum (1967) The proof uses theorem 14.1.11, with a modification. Rather than building an f that for all i, if $f = \phi_i$, then $\Phi_i(x) > g(x)$ a.e., we choose f so that $\Phi_i(x) > g(x - i)$ a.e. —where g *will be specified later in the proof.*

Recall that defining $f(x)$ involves finding the least uncancelled i_0 among $\{0, 1, \ldots, x\}$ for which $\Phi_{i_0}(x) \leq g(x - i_0)$ and ensuring $f(x) \not\simeq \phi_{i_0}(x)$ by defining $f(x) \simeq 1 \dot{-} \phi_{i_0}(x)$. If there is no such i_0 (details in the proof of Theorem 14.1.11), then we set $f(x) \simeq 1$.

So to "standard-compute" $f(x)$[6] we do: Find i_0 as described. Compute and output $1 \dot{-} \phi_{i_0}(x)$; Exit.

Otherwise, if no such i_0 exists, output 1. Exit.

A shortcut in the computation that achieves speed-up is, essentially, to shorten the search for a least uncancelled $i_0 \leq x$. The idea (Blum) is for any selected u we need only scan $\{u, u + 1, \ldots, x\}$ to look for uncancelled i_0 entries as long as x is *large enough.* "Large enough" is gauged by another new parameter, v, agreeing that "large enough" means $x \geq v$.

---

[6] Compute in the "obvious" way according to the definition of f.

It turns out that this shortcut is sufficient to provide enough speed-up over any other standard computation that uses a smaller u and hence searches more entries.

What about *nonstandard* computations that might not be sped up?

This concern is eliminated by *choosing* a g so that, *for each k, there is a j* satisfying

(1) $f = \phi_j$
(2) $g(x \,\dot{-}\, k) \geq r(x, \Phi_j(x))$ a.e. with respect to x and if $\phi_k = f$, then $\Phi_k(x) >$
   $g(x \,\dot{-}\, k)$ a.e. by the construction of f.
       (2) now concludes the proof.

We next make the above sketch mathematically precise (in particular, where u, v and g are coming *from* and *how* exactly).

The mathematical proof starts by repeating the construction of Theorem 14.1.11 "effectively" or "uniformly"[7] on a ϕ-index l of a possibly nontotal (and unspecified *as yet*) function g.

Thus the presence of g in our informal discussion is replaced below by ϕ_l, l being a variable. We define

$$\psi(u, v, l, x) \simeq \begin{cases} \textbf{if } v < u & \textbf{then return } \psi(u, u, l, x) \\ & \textbf{otherwise } (it\ is\ v \geq u) \\ \textbf{case 1}: \ x < v & : \ \text{Find the } least \text{ uncancelled } i_0 \\ \text{in } \{0, \ldots, x\} & \text{satisfying } \Phi_{i_0}(x) \leq \phi_l(x \,\dot{-}\, i_0) \\ & \textbf{return } 1 \,\dot{-}\, \phi_{i_0}(x) \text{ and } \textbf{cancel } i_0 \\ & \text{if no such } i_0 \text{ exists } \textbf{return } 1 \\ & \textbf{End case 1.} \\ \textbf{case 2}: \ x \geq v & : \ (We\ can\ compute\ faster\ now.) \\ & \text{Find the } least \text{ uncancelled } i_0 \\ \text{in } \{u, u+1, \ldots, x\} & \text{satisfying } \Phi_{i_0}(x) \leq \phi_l(x \,\dot{-}\, i_0) \\ & \textbf{return } 1 \,\dot{-}\, \phi_{i_0}(x) \text{ and } \textbf{cancel } i_0 \\ & \text{if no such } i_0 \text{ exists } \textbf{return } 1 \\ & \textbf{End case 2.} \end{cases}$$

Using inequality under **case** 1 implies probing $\phi_l(x) \simeq \phi_l(x \,\dot{-}\, 0), \phi_l(x \,\dot{-}\, 1), \ldots, \phi_l(x \,\dot{-}\, x) \simeq \phi_l(0)$. If all these are defined then so is $\psi(u, v, l, x)$.

Similarly, using inequality under **case** 2 implies probing $\phi_l(x \,\dot{-}\, u), \ldots, \phi_l(x \,\dot{-}\, x) \simeq \phi_l(0)$. If all these are defined then so is $\psi(u, v, l, x)$.

---

[7] I.e., the construction yields a computable function of l, among other possible arguments.

After providing the mathematical details of the construction from the proof of Theorem 14.1.11 (augmented by Exercise 14.4.2) an application of the second recursion theorem 9.2.1 shows that $\psi \in \mathcal{P}$ using the technique from Sect. 9.5.

Thus, by S-m-n theorem, we have a $t \in \mathcal{R}$ such that

$$\psi(u, v, l, x) \simeq \phi_{t(u,v,l)}(x), \text{ for all } u, v, l, x$$

$\square$

We conclude the proof via three in-proof lemmata:

14.2.2 Lemma. *If $\phi_l \in \mathcal{R}$ and we set $f = \phi_{t(0,0,l)}$, then $f \in \mathcal{R}$ and is 0-1 valued. Moreover, $f = \phi_i$ implies $\Phi_i(x) > \phi_l(x \div l)$ a.e. with respect to x.*

Proof of Lemma By Theorem 14.1.11 and the definition of ψ. Only condition $v \geq u$, **case** 2, is applicable. $\square$

14.2.3 Lemma. *If $\phi_l \in \mathcal{R}$, then, for each u, there is a v such that $\phi_{t(u,v,l)} = \phi_{t(0,0,l)}$.*

Proof of Lemma There are only finitely many i_0 in $\{0, \ldots, u \div 1\}$ that are ultimately cancelled. Choose v such that the *last* (in the cancellation order) i_0 that was cancelled was *during the computation of* $\phi_{t(0,0,l)}(v \div 1)$.

By the definition of $\phi_{t(u,v,l)}(x)$, we compute this value when $x < v$ following the same steps as when we compute $\phi_{t(0,0,l)}(x)$. For $x \geq v$ the computation of $\phi_{t(u,v,l)}(x)$ still gives the same result as that of $\phi_{t(0,0,l)}(x)$ since the latter would first cancel the various "i_0" in the range $\{0, \ldots, u - 1\}$ and then proceed according to the computation of $\phi_{t(0,0,l)}(x)$. Note that the above discusses only the case $u \leq v$ since the first case reduces to the former via the call $\psi(u, u, l, x)$. $\square$

Finally, we select the g function of our informal discussion, that is, we select l.

14.2.4 Lemma. *There is a $\phi_l \in \mathcal{R}$ such that, for all u, v, $\phi_l(x \div u + 1^8) \geq r(x, \Phi_{t(u,v,l)}(x))$ a.e.*

Proof of Lemma Define p by the primitive recursion

$$p(n, x) \simeq 0$$

$$p(n, x + 1) \simeq \max_{\substack{0 \leq u \leq x \\ 0 \leq v \leq x}} r\big(x + u, \Phi_{t(u,v,n)}(x + u)\big)$$

Thus $p \in \mathcal{P}$. By the second recursion theorem there is an l such that $\phi_l(x) \simeq p(l, x)$.

First we show that $(\forall x)\phi_l(x) \downarrow$ by induction on x.

---

[8] Evaluation of the argument is meant to be left to right, so $x \div u + 1 = (x \div u) + 1$.

- (*Basis*) $\phi_l(0) \downarrow\simeq 0$.
- Fix k and take the I.H. that $\phi_l(x) \downarrow$, for $x \leq k$.

    We will show that $\phi_{t(u,v,l)}(x+u) \downarrow,$[9] for $x \leq k$, and $u \leq x$ and $v \leq x$: Indeed, $v \leq x$ implies $x+u \geq v,$[10] hence to compute $\phi_{t(u,v,l)}(x+u)$ we compute under **case** 2 (among other things) $\phi_l(u) \simeq \phi_l(x+u \dotminus x), \ldots, \phi_l(x+u \dotminus u) \simeq \phi_l(x),$[11] all of which converge by the I.H. and hence $\phi_{t(u,v,l)}(x+u) \downarrow$.

    Hence $\Phi_{t(u,v,l)}(x+u) \downarrow$, for $x \leq k$ (Ax 1). Thus $\phi_l(x+1) \downarrow$ by the second equation in the primitive recursive definition of p (its right hand side is defined). The induction is concluded.

    Finally, if $x \geq \max(2u, u+v)$, then $v \leq x \dotminus u$ and $u \leq x \dotminus u$. Thus, looking at the second equation of the definition of p, we have

$$\phi_l(x \dotminus u + 1) \geq r(x, \Phi_{t(u,v,l)}(x)) \ a.e.$$

in fact, the previous inequality holds for $x \geq \max(2u, u+v)$. $\square$

The *global proof of the speed-up theorem* will be concluded once we verify (1) and (2) of the informal preamble to the proof (p. 553). Pick a k, then set $u = k+1$ and select v as in the proof of Lemma 14.2.3. Then $\phi_l(x \dotminus k) \simeq \phi_l(x \dotminus u + 1) \geq r(x, \Phi_{t(u,v,l)}(x))$ a.e. by the above paragraph. Now take $j = t(u,v,l)$. But $f = \phi_{t(0,0,l)} = \phi_{t(u,v,l)}$. $\square$

This theorem is highly nontrivial and it is remarkable that it is the result of just two "obvious" axioms. As remarkable is what it says: For any complexity measure Φ, there are 0-1 valued recursive functions of one variable that do not have a "*best program*". *Any* program of such function can be *nontrivially* improved in terms of its Φ-complexity ad infinitum. This result suggests that a more profitable tool for gauging the complexity of functions of $\mathcal{P}$ is not via some "intrinsic" or "optimal" complexity attributes —that the speed-up theorem advises us not to look for *in general*.

Hence one turns to *upper bounds* of the Φ-complexity of functions, using which one can study hierarchies of increasingly complex functions in terms of upper bounds on their computational complexity —analogously to the classification of the arithmetical relations in the arithmetical hierarchy.

---

[9] Which via Ax 1 will imply $\Phi_{t(u,v,l)}(x+u) \downarrow$.

[10] Only the case $v \geq u$ is considered since (definition of ψ) the alternative case $v < u$ reduces to the former.

[11] Cf. ②-remark below the definition of ψ.

14.3   Complexity Classes

14.3.1 Definition. Let Φ be a complexity measure for $(\phi_i)_{i \geq 0}$ and let $t \in \mathcal{R}$. Then the *complexity class*, C_t^Φ, is defined by

$$C_t^\Phi \overset{Def}{=} \{\phi_i : \Phi_i(x) \leq t(x) \text{ a.e. with respect to } x\}$$

$\square$

The following is immediate by either Theorem 14.1.9 or Theorem 14.1.11.

14.3.2 Proposition. *There is an infinite sequence of recursive functions* $(t_i)_{i \geq 0}$ *such that* $C_{t_i}^\Phi \subsetneq C_{t_{i+1}}^\Phi$, *for* $i \geq 0$.

Can we make the sequence $(t_i)_{i \geq 0}$, "effective"? That is, is there a recursive f such that $C_t \subsetneq C_{ft}$, for all sufficiently "large" t? Here we wrote ft for $\lambda x . f(t(x))$.

The answer, due to Trakhtenbrot (1967) and independently to Borodin (1969, 1970) is negative and constitutes the *Gap theorem*.

14.3.3 Theorem. (Gap Theorem) *For any* Φ, *and any recursive* h *and* f *such that* $(\forall x) f(x) \geq x$, *there is a* $t \in R$ *such that* $(\forall x) t(x) \geq h(x)$, *and*

$$C_t^\Phi = C_{ft}^\Phi$$

 Noting that we said for each of h and f "any", t and f can be chosen to be arbitrarily *large* so that the "gap" from the upper bound t to the upper bound ft can be arbitrarily large, and yet no new functions were added to C_t.

By $(\forall x) t(x) \geq h(x)$, changing h we have another gap further up. So we have infinitely many gaps.

Together these two observations, intuitively, indicate that "tight" bounds are "sparse" in the class of recursive functions.

Proof We are looking for a $t \in \mathcal{R}$ such that

(1) $t(x) \geq h(x)$, for $x \geq 0$.
(2) For all i, the set $\{x : t(x) \leq \Phi_i(x) \wedge \Phi_i(x) \leq f(t(x))\}$ is finite.

Thus we define

$$t(x) \simeq (\mu y)\Big(y \leq h(x) \wedge (\forall i)_{\leq x}(\Phi_i(x) \leq f(y) \to \Phi_i(x) \leq y)\Big) \tag{1}$$

(μy) applies to a recursive relation (by Ax 2 and choice of f) thus $t \in \mathcal{P}$. Now for each $i \leq x$ we have two cases to consider.

- $\Phi_i(x) \uparrow$, thus $\Phi_i(x) \leq f(y)$ is false. Thus the implication in (1) holds, and we have $t(x) \downarrow$.
- $\Phi_i(x) \downarrow$ thus the right hand side of the implication is true a.e. with respect to y. In this case too, $t(x) \downarrow$.

Thus t is recursive.

Now assume $\Phi_i(x) \leq f(t(x))$ a.e. (this entails that the x of interest satisfy $x \geq i$). By (1), $\Phi_i(x) \leq t(x)$ a.e. hence $C_t^{\Phi} \supseteq C_{ft}^{\Phi}$. The other direction, $C_t^{\Phi} \subseteq C_{fot}^{\Phi}$, being trivial we are done. □

14.3.4 Definition. (Measured Sets) Blum (1967) A sequence of one argument functions $(\gamma_i)_{i \geq 0}$ from $\mathcal{P}$ is called a *measured set* iff $\lambda ixy.y \simeq \gamma_i(x)$ is recursive.

Note that the sequence Φ_i, for any complexity measure, is a measured set, but a "general" measured set $(\gamma_i)_{i \geq 0}$ may fail Ax 1.

14.3.5 Lemma. *If $(\gamma_i)_{i \geq 0}$ is a measured set, then there is a $\sigma \in \mathcal{R}$ such that $\gamma_i = \phi_{\sigma(x)}$, for all $i \geq 0$.*

Proof Exactly like the proof of Corollory 14.1.8. □

14.3.6 Lemma. *Blum (1967) There is an $h \in \mathcal{R}$ such that*

$$dom(\phi_i) = dom(\phi_{h(i)})$$

and for every j such that $\phi_j = \phi_{h(i)}$,

$$\Phi_j(x) > \phi_i(x) \ a.e.$$

This is Theorem 7 in Blum (1967), which is a "uniform" version of Theorem 14.1.11, that is, here we construct an "f" in terms of an *index recursively computed* from an *index of the bounding function* $g = \phi_i$. The added complication in this "uniform" version's proof (compared to that of Theorem 14.1.11) is that g is not assumed total this time.

Proof Refer to the proof of Theorem 14.1.11. The process —apart from the need to have a stack whose role is to record all indices cancelled before the "current" step, and hence also to record the index *we cancel at the current step*— is fairly straightforward: Say, we want to define how $f(x)$ is computed. What we did in the earlier proof was, essentially, to search the set $\{\phi_0(x), \phi_1(x), \ldots, \phi_x(x)\}$ for those entries $\phi_i(x)$ that satisfy $\Phi_i(x) \leq g(x)$ and we made sure that $f(x) \not\simeq \phi_i(x)$ for such entries —we cancelled those entries by cancelling (i.e., not using for f) their indices. Note that *we do not cancel* all *such entries* at once, but we cancel the *smallest index* in the set of uncancelled indices.

How do we know which ones are cancelled and which are not? We maintain a stack that essentially (this is nuanced below) contains what *indices* we cancelled *so far*, that is, we need to compute $f(0), \ldots, f(x-1)$ in order to compute $f(x)$. This will not work with a nontotal $g = \phi_i$ without applying a dovetailing technique. Note that the $I(y)$ in the proof of Theorem 14.1.11 was recursive. Here

$$I(y) = \{j : j \leq y \wedge (\exists z)(z = \overbrace{\phi_i}^{\text{the "}g\text{"}}(y) \wedge \Phi_j(y) \leq z)\}$$

is only semi-recursive, and $z = \phi_i(x)$ itself is not recursive 6.11.1. Thus if $y < x$, then, if $\phi_i(y) \uparrow$, we get $f(y) \uparrow$ and we will never reach $f(x)$ to compute it: it will also be $f(x) \uparrow$.

The idea to circumvent the difficulty is to use dovetailing, that is, enumerate all pairs of numbers (x, y) *coded* as $\langle x, y \rangle$ and define "f" as $\psi(i, x)$ below to emphasise effective dependence on the i of "g".

We define ψ *and* the associated stack C (holds cancelled indices) by stages $s \geq 0$. We want, among other things, to achieve $\psi(i, x) \downarrow$ iff $\phi_i(x) \downarrow$ (to satisfy claim one of the lemma).

The stack is being built independently of the value of $\psi(i, x)$, but synchronously with the computation of ψ.[12] The entries we "push" in the stack are not just indices but are *coded pairs* $\langle j, x \rangle$ of an index j and argument x such that we have set $\psi(i, x) \not\simeq \phi_j(x)$. The argument i of ψ is tracking the index of $g = \phi_i$ to ensure the uniformity we spoke of.

(I) Building the stack.

> **At stage s** : Let $s = \langle x, y \rangle$, that is, $x = (s)_0$ and $y = (s)_1$.
>
> if $\Phi_i(x) \not\simeq y$ then **goto** next stage, $s + 1$ **Comment**. $\phi_i(x) \downarrow$ *uncertain*.
>
> if $\Phi_i(x) \simeq y$ then (in particular (by Ax 1) $\phi_i(x) \downarrow$) **do**
>
> (*a*) **Call** $\phi_i(x)$ **note** *result*.
>
> (*b*) **In** $\{\phi_0(x), \ldots, \phi_x(x)\}$ **get** *the first uncancelled* $\phi_j(x)$; j *is* **least**;
>
> (*not cancelled* **in previous stages** $s' < s$) *that is*, $\langle j, x \rangle \notin C$
>
> which also satisfies $\Phi_j(x) \leq \phi_i(x)$; **push** entry $\langle j, x \rangle$ to the stack C
>
> **goto** stage $s + 1$ **Comment**. (I)(*b*) *is part of* (II)(*b*) *below*.

(II) Computing $\psi(i, x)$.

> **At stage s** : Let $s = \langle x, y \rangle$, that is, $x = (s)_0$ and $y = (s)_1$.
>
> We attempt to define $\psi(i, x)$ —i is that of $g = \phi_i$.
>
> if $\Phi_i(x) \not\simeq y$ then **goto** next stage, $s + 1$ **Comment**. "Do nothing".
>
> if $\Phi_i(x) \simeq y$ then **do** (I)(*a*) as above, and do an *augmented* step (*b*)
>
> [see (I)(*b*) above, **building the stack**] :

---

[12] The stack being infinite, since we have infinitely many stages, we do not expect to build it *first* and *then* build ψ.

Augmented (b) $\Big[$ **In** $\{\phi_0(x), \ldots, \phi_x(x)\}$ **get first** *uncancelled* $\phi_j(x)$

(i.e., the $\langle j, x \rangle$ was not put in C in **previous stages** *$s' < s$)*

Note that j is **least** *uncancelled among $j \leq x$;*

such that $\phi_j(x)$ **also satisfies** $\Phi_j(x) \leq \phi_i(x)$; *and* **do**

$\Big\{$**push** entry $\langle j, x \rangle$ to the stack C;

set $\psi(i, x) \simeq 1 \dotdiv \phi_j(x)$;

goto stage $s + 1$; $\Big\}$

else, *if no such $\langle j, x \rangle$ found, then* **do**

set $\psi(i, x) \simeq 1$;

goto stage $s + 1$; $\Big\}$

$\Big]$

We need next to verify the two claims of the lemma. But first, ψ is partial recursive, as the construction above indicates. While it can be made rigorous (Exercise 14.4.5), there is enough detail to allow Church's Thesis to settle the claim. Then by S-m-n, there is an $h \in \mathcal{R}$ such that $\psi(i, x) \simeq \phi_{h(i)}(x)$, for all i and x.

Claim 1. If $\phi_i(x) \uparrow$ then $\Phi_i(x) \simeq y$ is never true, so the definition of $\psi(i, x)$ is indefinitely postponed to the next stage. Thus $\psi(i, x) \uparrow$. If $\phi_i(x) \downarrow \simeq q$, then the construction (at i, x) will succeed to return a value for $\psi(i, x)$: either some $1 \dotdiv \phi_k(x)$ or the value 1.

All told, $\mathrm{dom}(\phi_i) = \mathrm{dom}(\phi_{h(i)})$.

Claim 2. Let

$$\phi_j = \phi_{h(i)} \tag{1}$$

Thus, j is never cancelled in the construction. Arguing by contradiction let instead be the case that on the domain of ϕ_j the following claim is false:

$$\Phi_j(x) > \phi_i(x), \text{ a.e. with respect to } x \text{ in } \mathrm{dom}(\phi_j) \text{ is } false \tag{2}$$

For (2) we consider only the case of infinite $\mathrm{dom}(\phi_j)$ since otherwise the claim is trivially true.[13] Note that

$$\mathrm{dom}(\Phi_j) \overset{Ax1}{=} \mathrm{dom}(\phi_j) \overset{(1)}{=} \mathrm{dom}(\phi_{h(i)}) \overset{Claim\ 1}{=} \mathrm{dom}(\phi_i) \overset{Ax1}{=} \mathrm{dom}(\Phi_i)$$

As in the proof of Theorem 14.1.11, the falsehood of the inequality in (2) implies the existence of an infinite sequence of inputs x_i in $\mathrm{dom}(\phi_j)$, all larger than j,

$$j < x_1 < x_2 < x_3 < x_4 < \cdots$$

such that, for $r \geq 1$,

$$\Phi_j(x_r) \leq \phi_i(x_r) \tag{3}$$

Now,

x_1:   since j is never cancelled and we have $\Phi_j(x_1) \leq \phi_i(x_1)$ it must be (cf. proof of Theorem 14.1.11) that some $m_1 < j$ was cancelled at this step.

x_2:   since j is never cancelled and m_1 is, *and* we have $\Phi_j(x_2) \leq \phi_i(x_2)$ it must be that some $m_1 < m_2 < j$ was cancelled at this step.

x_3:   since j is never cancelled and m_1, m_2 are, *and* we have $\Phi_j(x_3) \leq \phi_i(x_3)$ it must be that some $m_1 < m_2 < m_3 < j$ was cancelled at this step.

x_4:   And so on, we build an infinite sequence $m_1 < m_2 < m_3 < m_4 < \cdots$ sandwiched between m_1 at the left and j at the right. This is absurd.

Thus the inequality in (2) is actually *true*.

□

14.3.7 Theorem. (Compression Theorem) *Blum (1967) For any complexity measure Φ and any measured set $(\gamma_i)_{i\geq 0}$, and with σ and h as in lemmata 14.3.5 and 14.3.6, there is a $g \in \mathcal{R}$ such that for all i and j,*

(1) $\phi_j = \phi_{h(\sigma(i))}$ *implies* $\Phi_j(x) > \gamma_i(x)$ *a.e.*
(2) $\Phi_{h(\sigma(i))}(x) < g(x, \gamma_i(x))$ *a.e.*

Proof

(1) This is Lemma 14.3.6 using here $\phi_{\sigma(i)} = \gamma_i$ (Lemma 14.3.5) in place of "ϕ_i" (the bounding function "g").

---

[13] Recall that in general Definition 4.1.10 defines "$P(\vec{x})$ a.e." by "$\{\vec{x} : \neg P(\vec{x})\}$ is finite". This is especially noteworthy here as ϕ_i and hence $\phi_{h(i)}$ may not be total.

(2) For (2) —$\Phi_{h(\sigma(i))}(x) < g(x, \gamma_i(x))$ a.e.— define g by

$$g(x, y) \overset{Def}{\simeq} 1 + \max\{\Phi_{h(\sigma(m))}(x) : m \leq x \wedge \gamma_m(x) \simeq y\}$$

Clearly $g \in \mathcal{R}$ as we argue below (using the conjunctional implication):

$$\gamma_m(x) \simeq y \Rightarrow \gamma_m(x) \downarrow \overset{14.3.5}{\Rightarrow} \phi_{\sigma(m)}(x) \downarrow \overset{14.3.6}{\Rightarrow} \phi_{h(\sigma(m))}(x) \downarrow \overset{Ax1}{\Rightarrow} \Phi_{h(\sigma(m))}(x) \downarrow$$

Thus, $\Phi_{h(\sigma(i))}(x) < g(x, \gamma_i(x))$, for any $x \geq i$ such that $\gamma_i(x) \downarrow$. $\qquad\square$

14.3.8 Corollary. *With Φ and the γ_i as in Theorem 14.3.7, we have*

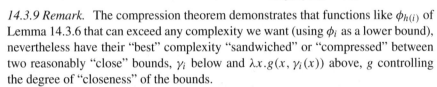

$$\gamma_i \in \mathcal{R} \rightarrow C_{\gamma_i}^{\Phi} \subsetneq C_{\lambda x.g(x,\gamma(x))}^{\Phi}$$

Proof In fact, under the assumptions, $\phi_{h(\sigma(i))} \in C_{\lambda x.g(x,\gamma(x))}^{\Phi} - C_{\gamma_i}^{\Phi}$. $\qquad\square$

14.3.9 Remark. The compression theorem demonstrates that functions like $\phi_{h(i)}$ of Lemma 14.3.6 that can exceed any complexity we want (using ϕ_i as a lower bound), nevertheless have their "best" complexity "sandwiched" or "compressed" between two reasonably "close" bounds, γ_i below and $\lambda x.g(x, \gamma_i(x))$ above, g controlling the degree of "closeness" of the bounds.

Moreover, the corollary translates in plain language that for any recursive "run time" γ_i from a measured set we can effectively "name" ($\gamma_i \mapsto \lambda x.g(x, \gamma_i(x))$) a more inclusive complexity class. Thus some names like t in Theorem 14.3.3 are "bad" while some others, like a recursive γ_i are good. $\qquad\square$

14.4 Exercises

1. Prove Corollary 14.1.10.
2. Convert the proof of Theorem 14.1.11 to a rigorous mathematical one explicitly using a stack.
3. A one-input URM, when presented with input x reaches **stop** in y steps. Find a *good* upper bound for the output of this URM as a function of x, y and any other parameters of the URM that you consider to be relevant. "Good", roughly, means an upper bound that is not larger than it needs to be. Some people call such bound "tight". For example, we know from elementary computing that we can search for an element x in an ordered array of numbers of length n in no more than n steps. Binary search can do it in about $\log_2 n$ steps. This is better or tighter than the former.

4. Let ϕ be one of the indexings (of unary functions in $\mathcal{P}$) we saw in this volume, and let Φ be a given complexity measure.

   Let $\phi_{i_0} = \lambda x.x$ Define $\widehat{\Phi}$ by

   $$\widehat{\Phi}_i(x) \simeq \text{ if } i = i_0 \text{ then } \Phi_{i_0} \text{ else } \max(g(x), \Phi_i(x))$$

   where $g \in \mathcal{R}$.
   Prove

   - $\widehat{\Phi}_{i_0} = \Phi_{i_0}$.
   - $\widehat{\Phi}$ is a complexity measure.
   - Choose g to be "horrendously" bigger than Φ_{i_0} (e.g., involve the latter as an argument of the Ackermann function to define g). Show that no matter what index m we choose for $\lambda x.x + 1$ we will have $\widehat{\Phi}_m(x) \geq g(x) > \widehat{\Phi}_{i_0}(x)$. Thus the measure $\widehat{\Phi}$ assigns a tremendously higher complexity to the computation of $\lambda x.x + 1$ than to that of $\lambda x.x$ despite the fact that reasonably straightforward programs for $\lambda x.x$ and $\lambda x.x + 1$ have almost identical definitional complexity.

5. Fill in the mathematical details missing in the proof of Lemma 14.3.6.
   Hint. Almost everything is found in the construction of C and ψ.

6. Consider the truncated computations of Remark 13.3.7, omitting the type-1 inputs. Is $\lambda x.\|i, x\|$ a complexity measure for the unary functions of $\mathcal{P}$? Justify.

7. Prove that for any complexity measure Φ there is a recursive g such that $\Phi_i(x) \simeq (\mu y)g(i, x, y)$.

8. (*Combining lemma* Borodin (1969, 1970)) For any complexity measure Φ and $c \in \mathcal{R}$, such that $\phi_i(x) \downarrow \wedge \phi_j(x) \downarrow \rightarrow \phi_{c(i,j)}(x) \downarrow$ there is an $h \in \mathcal{R}$ such that $\Phi_{c(i,j)}(x) \leq h(x, \Phi_i(x), \Phi_j(x))$ a.e.

9. Prove that there is a $c \in \mathcal{R}$ such that, for any i, j, $\phi_i(x) \downarrow \wedge \phi_j(x) \downarrow \equiv \phi_{c(i,j)}(x) \downarrow$

10. Consistently with the general understanding of what $Q(f(\vec{x}), \vec{y})$ means, namely "$(\exists u)(u = f(\vec{x}) \wedge Q(u, \vec{y}))$", we understand $\Phi_i(x) \geq y$ to mean $(\exists u)(u \simeq \Phi_i(x) \wedge u \geq y)$ or also $\neg(\exists w)_{<y}\Phi_i(x) \simeq w$. The latter is better as it is recursive, of course (the former form only prove semi-recursiveness).

    But what if we extend this meaning as we did for "Kleene equality"? Namely, imagine that $\Phi_i(x) \geq y$ means $\Phi_i(x) \uparrow \vee \Phi(x) \downarrow \wedge \Phi_i(x) \geq y$.
    Prove that under this interpretation "$\Phi_i(x) \geq y$" is not recursive.
    Is it r.e.?

11. Prove that $(\gamma_i)_{i \geq 0}$ is a measured set iff for some $g \in \mathcal{R}$ we have $\gamma_i = \lambda x.(\mu y)g(i, x, y)$.

Chapter 15
Complexity of $\mathcal{PR}$ Functions

Overview

The literature refers to the complexity theory of the partial recursive functions as *"high level complexity"* due to its generality and the mostly theoretical and far removed from practice essence of it. The present chapter is about more down to earth functions. We will be looking into the complexity of $\mathcal{PR}$ functions using a hierarchy approach. Much like in the arithmetical hierarchy, *in the first instance*, we build our hierarchies according to the *definitional* or *static* complexity of function derivations, or (loop) programs. As we develop the theme the reader will note that the measure of static complexity or complexity of the definition is not always the same (some times it is nesting level of primitive recursion, other times it is size of function output that *is determined at definition time*). In the end of all this we will have various hierarchies —called *sub-recursive hierarchies* in the literature— to *compare*, and a powerful tool, the Ritchie-Cobham property (theorem) that allows all these static hierarchies to be viewed as *complexity classes named* by time-bounding functions (of URM or Loop-program computations).

So, how much down to earth are the $\mathcal{PR}$ functions? Much more *than* the partial recursive and recursive functions; however, absolutely speaking, functions of interest to the practitioner are found only in the first couple of levels of the various hierarchies. For example, *level two* of the loop program hierarchy already contains the function we mentioned in the Preface, essentially,

$$\left. 2^{\cdot^{\cdot^{2^{2^x}}}} \right\} 10^{350000} \text{ 2s}$$

which has impossibly large output even for $x = 0$. Totally impractical! The following function, where the "ladder" height is a function of the input x (ladder of 2s height is equal to the argument x)

© Springer Nature Switzerland AG 2022

G. Tourlakis, *Computability*, https://doi.org/10.1007/978-3-030-83202-5_15

$$\left. 2^{\cdot^{\cdot^{2^x}}} \right\} x \text{ 2's}$$

cannot be computed if we only allow loop nesting depth (or level) two (that is, one loop inside the other) in one or more places in an (attempted) program; it *requires* nesting level *three* and is even more astronomical than the previous "ladder" function.

Impressive as this discussion may be, the dull truth is that, *in practice*, we compute *finite* functions because we only have *finite* (computing) resources such as CPU time and/or memory space.

Still we do *not* compute by *brute force table look-up*. Rather, we *pretend* that our programs will *run for any input*,[1] thus we write *general algorithms* (programs) with that in mind. It is then clear that theories like the complexity of primitive recursive functions can enrich our *programming practice*.

For example we learn that if we program with loop programs that never nest loops more than two levels, then all the functions $\lambda \vec{x}. f(\vec{x})$ that we can so compute have outputs that are *majorised* (or bounded; see Sect. 4.2) by

$$\left. 2^{\cdot^{\cdot^{2^{\max(\vec{x})}}}} \right\} c \text{ 2s}$$

where c depends on f.

15.1 The Axt, Loop-Program and Grzegorczyk Hierarchies

15.1.1 Definition. (Axt (1965), Heinermann (1961)) We build a hierarchy $\mathcal{K} = (K_n)_{n \geq 0}$ by induction on the level n, and at each level K_n is defined as a closure (cf. Definition 0.6.2).

We set $K_0 = \mathrm{Cl}(\{\lambda x.x + 1, \lambda x.x\}, \mathcal{O})$ where $\mathcal{O}$ contains only Grzegorczyk substitution.

Having defined K_n, we define K_{n+1}: First, let

$$R_{n+1} \stackrel{Def}{=} \{f : f = prim(h, g) \wedge \{h, g\} \subseteq K_n\}$$

Then set

$$K_{n+1} \stackrel{Def}{=} \mathrm{Cl}(K_n \cup R_{n+1}, \mathcal{O})$$

---

[1] They will not, due to computing time and memory finiteness.

If $f \in K_n$ then we say its *level* is $\leq n$. If $f \in K_{n+1} - K_n$ then we say its *level* is $= n + 1$.   □

The definition clearly assumes that *it is primitive recursion that adds to the (definitional) complexity,* namely, the *nesting levels* of it: K_n already has functions $f' = prim(h', g')$ with h', g' in K_{n-1}. Going to K_{n+1} we are adding another primitive recursion on top of $f' = prim(h', g')$.

15.1.2 Proposition.

1. For $n \geq 0$, $K_n \subseteq K_{n+1}$.
2. $\mathcal{PR} = \bigcup_{n \geq 0} K_n$

Proof Easy exercise. For item 2, $\subseteq$ direction, do induction over $\mathcal{PR}$ since it is a closure.   □

15.1.3 Lemma. $A_n \in K_n$, *for $n \geq 0$, where $\lambda n x . A_n(x)$ is the Ackermann function.*

Proof Induction on n. For $n = 0$ we have $A_0 = \lambda x . x + 2$ which is obtained from $\lambda x . x$ by Grzegorczyk operations. Thus $A_0 \in K_0$. Fix now n and assume (I.H.) $A_n \in K_n$.
   For K_{n+1}:

$$A_{n+1}(0) = 2$$

$$A_{n+1}(x + 1) = A_n(A_{n+1}(x))$$

By the I.H. and the definition of K_{n+1} the above primitive recursion places $A_{n+1} \in K_{n+1}$.   □

15.1.4 Lemma. (Majorising Lemma for $\mathcal{K}$) *For $n \geq 0$, $\lambda \vec{x} . f(\vec{x}) \in K_n$ implies that for some m depending on f we have $f(\vec{x}) \leq A_n^m(|\vec{x}|)$, for all $\vec{x}$.*

As this is related to Theorem 4.2.1 we use "$|\vec{x}|$" for "$max(\vec{x})$".

Proof This is a trivial corollary of the proof of Theorem 4.2.1: Grzegorczyk substitution does not raise the index n of the (bounding) Ackerman function, but *one* primitive recursion raises it *by one.*
   So fix n. If $f = prim(h, g)$ is in K_{n+1} because it is in R_{n+1} of Definition 15.1.1, then the I.H. and the proof of Theorem 4.2.1 say that f is majorised by A_{n+1}^m (some m) since (I.H.) h and g are majorised by A_{n+1}^k and A_n^r for some k and r. If f got into K_{n+1} after a finite number of substitutions *after* we first obtained $prim(h, g)$ —i.e., f is in $Cl(K_n \cup R_{n+1}, \mathcal{O})$— then the bounding function A_{n+1}^m will at most change to $A_{n+1}^{m'}$ from A_{n+1}^m.   □

15.1.5 Theorem. *The hierarchy $\mathcal{K} = (K_n)_{n \geq 0}$ is proper or nontrivial, that is, $K_n \subsetneq K_{n+1}$, for all n.*

Proof Suffices to find for each n some f in $K_{n+1} - K_n$. Well, $A_{n+1} \in K_{n+1} - K_n$, the positive part by Lemma 15.1.3: $A_{n+1} \in K_{n+1}$; and the negative part by Lemma

15.1.4: if $A_{n+1} \in K_n$ then $A_{n+1}(x) \leq A_n^m(x)$ for some m and all x. This is not so (Lemma 4.1.12.) □

If we replace *primitive recursion* by *simultaneous* (primitive) *recursion* then we obtain an easier to work with hierarchy.

15.1.6 Definition. We build a hierarchy $\mathcal{K}^{sim} = (K_n^{sim})_{n \geq 0}$ by induction on the level n, and at each level K_n^{sim} is defined as a closure.

We set $K_0^{sim} = \text{Cl}(\{\lambda x.x + 1, \lambda x.x\}, \mathcal{O}) = K_0$ where $\mathcal{O}$ contains only Grzegorczyk substitution.

Having defined K_n^{sim}, we define K_{n+1}^{sim}: First, let

$$R_{n+1}^{sim} \stackrel{Def}{=} \{f_j : j \leq r \wedge \vec{f_r} = simprim(\vec{h_r}, \vec{g_r}) \wedge \{\vec{h_r}, \vec{g_r}\} \subseteq K_n^{sim}\}$$

Then set

$$K_{n+1}^{sim} \stackrel{Def}{=} \text{Cl}(K_n^{sim} \cup R_{n+1}^{sim}, \mathcal{O})$$

If $f \in K_n^{sim}$ then we say its *level* is $\leq n$. If $f \in K_{n+1}^{sim} - K_n^{sim}$ then we say its *level* is $= n + 1$. □

15.1.7 Lemma. *For all* $n \geq 0$, $K_n \subseteq K_n^{sim}$.

Proof This is trivial since all other things being equal, primitive recursion is a special case of simultaneous recursion. □

15.1.8 Lemma. (Majorising Lemma for $\mathcal{K}^{sim}$) *For* $n \geq 0$, $\lambda \vec{x}.f(\vec{x}) \in K_n^{sim}$ *implies that for some* m *depending on* f *we have* $f(\vec{x}) \leq A_n^m(|\vec{x}|)$, *for all* $\vec{x}$.

Proof Exercise 15.4.2. □

15.1.9 Proposition.

1. *For* $n \geq 0$, $K_n^{sim} \subseteq K_{n+1}^{sim}$.
2. $\mathcal{PR} = \bigcup_{n \geq 0} K_n^{sim}$

Proof 1. is immediate from the definition. For 2. the $\supseteq$ is trivial while the $\subseteq$ follows from Proposition 15.1.2(2) and Lemma 15.1.7. See Exercise 15.4.3 □

15.1.10 Theorem. *The hierarchy* $\mathcal{K}^{sim} = (K_n^{sim})_{n \geq 0}$ *is proper, that is,* $K_n^{sim} \subsetneq K_{n+1}^n$, *for all* n.

Proof By Lemma 15.1.7, $A_{n+1} \in K_{n+1}^{sim}$. By Lemma 15.1.8 $A_{n+1} \notin K_n^{sim}$. □

At this point we pause to offer some examples of familiar functions and place them in the hierarchies we have so far, and to develop some tools that will help us do some clever coding (needed to go from simultaneous to single primitive recursion).

15.1.11 Proposition. (Examples and Tools) *The following table lists some simple $\mathcal{PR}$ functions and their placement in the $\mathcal{K}$ hierarchy.*

Function	Upper bound of level
(1) $\lambda xy.x + y$	1
(2) $\lambda xy.x \dot{-} 1$	1
(3) $\lambda xy.x(1 \dot{-} y)$ *(Restricted if-then-else)*	1
(4) $\lambda xy.x \times y$	2
(5) $\lambda x.2^x$	2
(6) $\lambda xy.x \dot{-} y$	2
(7) $\lambda xy.\|x - y\|$	2
(8) $\lambda xy.\max(x, y)$	2
(9) $\lambda x.\left\lfloor \dfrac{x(x + 1)}{2} \right\rfloor$	2
(10) $\lambda xy.rem(x, y)$ *(i.e., the remainder of the division x/y)*	2
(11) $\lambda x.\left\lfloor \dfrac{x}{2} \right\rfloor$	2
(12) $\lambda x.2^{\lfloor \log_2 x \rfloor}$ *(where we adopt the convention $\log_2 0 = 0$)*	2
(13) $\lambda x.\lfloor \log_2 x \rfloor$	3

Proof Schwichtenberg (1969) attributes the table to Heinermann (1961). The only nontrivial entries are (10)–(13). We follow Schwichtenberg (1969) for these four. We also do (9) and leave *all else* to the reader (Exercise 15.4.4).

(9)   Since $\left\lfloor \frac{x(x + 1)}{2} \right\rfloor = \sum_{j \le x} j$ we note

$$\sum_{j \le 0} j = 0$$

$$\sum_{j \le x+1} j = x + 1 + \sum_{j \le x} j$$

But $\lambda x.x + 1 \in K_0$ and the $+$-function is of level at most 1.

(10)   Define h as below

$$h(0, y) = y \dot{-} 1$$

$$h(x + 1, y) = h(x, y) \dot{-} 1 + (y \dot{-} 1) \cdot (1 \dot{-} h(x, y))$$

Whatever h does, it is of level ≤ 2 by (1), (2) and (3) in the table.

Now, h is periodic. It starts at a "high value" of $y \dot{-} 1$ and decreases linearly (by 1) to zero, at which point the next output is again $y \dot{-} 1$ and the cycle repeats. But then, $rem(x, y) = y \dot{-} 1 \dot{-} h(x, y)$ and substitution does not increase the level.

(11) Note that

$$\left\lfloor \frac{x}{2} \right\rfloor = \overbrace{rem\left(\left\lfloor \frac{x(x+1)}{2} \right\rfloor, x\right)}^{term\ 1} + \overbrace{rem\left(\left\lfloor \frac{x(x \dotdiv 1)}{2} \right\rfloor, x \dotdiv 1\right)}^{term\ 2}$$

This is so because, as it is easy to verify, *term* 1 $= if\ 2|x\ then\ \lfloor x/2 \rfloor\ else\ 0$ and *term* 2 $= if\ 2|x\ then\ 0\ else\ \lfloor x/2 \rfloor$. The rest is that substitution does not raise the level.

(12) Define g by

$$g(0) = 0$$

$$g(x+1) = if\ g(x) > 1\ then\ g(x) \dotdiv 1\ else\ x + 1$$

It is easy, once the statement below is discovered, to prove by induction on k that $g(2^k) = 2^k$ (all k), *and* moreover, $2^k \leq x < 2^{k+1}$ implies $g(x) + x = 2^{k+1}$ (all k).

Now, how does one discover this compound statement? By looking at a well-plotted graph of g.

The level of g is ≤ 2, since we have the full *if-then-else* in K_1: $ifte = \lambda xyz.if\ x = 0\ then\ y\ else\ z$ in $K_1{}^2$ and $g(x) > 1$ needs $x > 1$, which is equivalent to $1 \dotdiv (x \dotdiv 1) = 0$.

Thus

$$2^{\lfloor \log_2(x) \rfloor} = \left\lfloor \frac{g(x) + x}{2} \right\rfloor \quad for\ x \geq 1$$

hence —taking the stated convention into account,

$$2^{\lfloor \log_2(x) \rfloor} = 1 \dotdiv x + \left\lfloor \frac{g(x) + x}{2} \right\rfloor (1 \dotdiv (1 \dotdiv x))$$

$\square$

(13) $\lfloor \log_2 x \rfloor = \sum_{i \leq x} i(1 \dotdiv |2^{\lfloor \log_2(x) \rfloor} - 2^i|)$.

For the level *bound* note that if $g \in K_n$ then $\lambda x\vec{y}. \sum_{i \leq x} g(i, \vec{y}) \in K_{n+1}$ since

$$\sum_{i \leq 0} g(i, \vec{y}) = g(0, \vec{y})$$

$$\sum_{i \leq x+1} g(i, \vec{y}) = g(x+1, \vec{y}) + \sum_{i \leq x} g(i, \vec{y})$$

---

[2] $ifte(0, x, y) = y$ and $ifte(x+1, y, z) = z$.

Thus level of log is ≤ 3 since the definining sum applies to level ≤ 2 functions.

15.1.12 Definition. If $\mathcal{C}$ is any subclass of $\mathcal{PR}$ we denote by $\mathcal{C}_*$ the class of the *corresponding predicates*:

$$\mathcal{C}_* \overset{Def}{=} \{f(\vec{x}) = 0 : f \in \mathcal{C}\}$$

$\square$

We have the following easy lemmata:

15.1.13 Lemma. $K_{n,*}$ *and* $K_{n,*}^{sim}$ *are closed under Boolean operations (all* $n \geq 1$*).*

Proof Exercise. $\square$

15.1.14 Lemma. K_n *and* K_n^{sim} *are closed under definition by cases (all* $n \geq 1$*).*

Proof Because $ifte$ defined in the proof of Proposition 15.1.11 is in K_1. Now see the proof of Theorem 2.3.6. $\square$

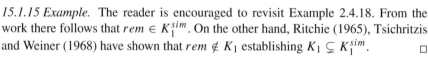

15.1.15 Example. The reader is encouraged to revisit Example 2.4.18. From the work there follows that $rem \in K_1^{sim}$. On the other hand, Ritchie (1965), Tsichritzis and Weiner (1968) have shown that $rem \notin K_1$ establishing $K_1 \subsetneq K_1^{sim}$. $\square$

15.1.16 Proposition. (A Characterisation of K_0) $K_0 = K_0^{sim}$ *contains only functions of the form* $\lambda\vec{x}, k$ *and* $\lambda\vec{x}.x_i + k$.[3]

Proof Induction on K_0 as a *closure*. For the initial functions we are done; they have the right form.

So let f and g have the right form and show that the form is preserved by substitution. We are only looking at the "interesting" case of substitution $h = \lambda\vec{z}\vec{y}.f(g(\vec{z}), \vec{y})$.

Case 1. $f(w, \vec{y}) = k$ (all $w, \vec{y}$)). Then h has the right form.
Case 2. $f(w, \vec{y}) = y_i + k$ (all $w, \vec{y}$)). Then $f(g(\vec{z}), \vec{y}) = y_i + k$ (all $\vec{z}, \vec{y}$)).
Case 3. $f(w, \vec{y}) = w + k$ (all $w, \vec{y}$)). Then $f(g(\vec{z}), \vec{y}) = g(\vec{z}) + k$ (all $\vec{z}$)) and the form is preserved as we see after replacing $g(\vec{z})$ in the last right hand side.

$\square$

15.1.17 Theorem. (Ritchie (1965), Tsichritzis and Weiner (1968)) K_1 *is the closure of the functions (1)–(3) in Proposition 15.1.11 under substitution.*

Proof Define

$$\mathcal{C} \overset{Def}{=} Cl\Big(\{\lambda xy.x + y, \lambda x.x \dot{-} 1, \lambda xy.x(1 \dot{-} y)\}, \{\text{substitution}\}\Big)$$

---

[3] This result goes at least as far back as Ritchie (1965); Tsichritzis and Weiner (1968).

It suffices —by Proposition 15.1.11, (1)–(3), that settle at once the $\supseteq$ direction— to only show $K_1 \subseteq C$.

We do induction on the definition of K_1 as a closure (Definition 15.1.1). For the nontrivial step let $\lambda y \vec{x}. f(y, \vec{x}) \in K_1 - K_0$. If f is the result of a primitive recursion on K_0 functions that obtained f', followed by a *finite* sequence of substitutions to yield f, these substitutions (a) did not raise the level of f', (b) C is closed under substitution. Thus we assume without loss of generality that f is given by the recursion below —from $\{h, g\} \subseteq K_0$— and no substitutions followed.

$$f(0, \vec{x}) = h(\vec{x})$$

$$f(y + 1, \vec{x}) = g(y, \vec{x}, f(y, \vec{x}))$$

We consider the following exhaustive cases.

1. $g(y, \vec{x}, z) = k$ (constant), for all $y, \vec{x}, z$.
   Then $f(y, \vec{x}) = if\ y = 0\ then\ h(\vec{x})\ else\ k$, thus $f \in C$ since $h \in K_0 \subseteq C$ and $if\ te \in C$ (Exercise 15.4.6).
2. $g(y, \vec{x}, z) = y + k$ or $= x_i + k$, for all $y, \vec{x}, z$.

   Then $f(y, \vec{x}) = if\ y = 0\ then\ h(\vec{x})\ else\ \begin{cases} y \mathrel{\dot{-}} 1 + k \\ \quad\ or \\ x_i + k \end{cases}$

   which is clearly in C.
   Finally,
3. $g(y, \vec{x}, z) = z + k$. Then $f(y, \vec{x}) = h(\vec{x}) + ky$. This is in C since

$$ky = \overbrace{y + \cdots + y}^{k\ summands}$$

   is obtained from $\lambda x y . x + y$ by substitution.

$\square$

We just saw that K_j, $j = 0, 1$, contain trivial functions and in the absence of the power of primitive recursion (in essence, that operation was traded off for a handful of initial functions as the above theorem demonstrates) the corresponding predicates (Definition 15.1.12) are trivial too.

We will see soon that the Grzegorczyk "small classes" $\mathcal{E}^0, \mathcal{E}^1$ and, the addition by Warkentin (1971), $\mathcal{E}^{-1}$ have enormously rich structure. Even their corresponding *relation* (predicate) *sets* are impressive: One can already place a version of the Kleene T-predicate in $\mathcal{E}_*^{-1}$. But we will postpone this until after we introduce the loop programmable functions hierarchy which is essentially $\mathcal{K}^{sim}$ from a "programming" perspective.

15.1.18 *Example.* We revisit Example 2.4.19 here to add a comment. The simultaneous recursion (ibid) immediately demonstrates that $\lambda x. \lfloor x/2 \rfloor \in K_1^{sim}$. A characterisation of K_1 by Tsichritzis and Weiner (1968) —and its corollary: if $f \in K_1$, then, for some constant B depending on f, the restriction of f on the set $\{\vec{x} : \min(\vec{x}) \geq B\}$ is linear— imply that neither *rem* nor $\lambda x. \lfloor x/2 \rfloor$ are in K_1.

□

15.1.19 Definition. (A Hierarchy of Loop Programs) L_0 is the set of all loop programs that contain *no loop-end instruction*.

Having defined L_n, we define L_{n+1} as a closure:

$$\mathrm{Cl}\Big(L_n \cup \{\textbf{Loop } X;\ P;\ \textbf{end} \,\Big|\, P \in L_n\}, \mathcal{O}\Big)^4$$

where $\mathcal{O}$ contains just the operation for *program concatenation* (cf. Definition 3.1.1), namely, $(P, Q) \mapsto P;\ Q$. □

15.1.20 Remark. It is clear that $L_n \subsetneqq L_{n+1}$ since L_{n+1} contains, but L_n does not, programs with nesting level of **Loop**-end instruction equal to $n + 1$, that is, L_0 programs have zero nesting (of **Loop**-end instruction), and if P has at least one occurrence of nesting level n, but none higher, then **Loop** $X;\ P$ **end** contains a loop of nesting level $n + 1$; the outermost.

Comparing with Definition 3.1.1 it is immediate that $L = \bigcup_{n \geq 0} L_n$. □

We can now separate the functions in $\mathcal{L} = \mathcal{PR}$ into a hierarchy of *functions*.

15.1.21 Definition. For $n \geq 0$,

$$\mathcal{L}_n \stackrel{Def}{=} \{P_Y^{\vec{X}} : P \in L_n, \text{ where } Y \text{ and the } X_i \text{ are all in } P\}$$ □

15.1.22 Theorem. ($\mathcal{K}^{sim}$ **vs.** $\mathcal{L}$) $K_n^{sim} = \mathcal{L}_n$, $n \geq 0$.

Proof An adaptation of the Theorems 3.2.1 and 3.3.2, from the $\mathcal{PR}$ vs. $\mathcal{L}$ case to the K_n^{sim} vs. $\mathcal{L}_n$ case. Exercise 15.4.7.

With Theorem 15.1.22 settled we directly get the hierarchy corollary.

15.1.23 Corollary. $\mathcal{L} = (\mathcal{L}_n)_{n \geq 0}$ *is a proper hierarchy, that is,*

1. $\mathcal{L}_n \subsetneqq \mathcal{L}_{n+1}$ *(all n)*
2. $\bigcup_{n \geq 0} \mathcal{L}_n = \mathcal{L}$.

15.1.1 The Grzegorczyk Hierarchy

This static hierarchy does not let primitive recursion to force functions into the next level. It does this by restricting *acceptable primitive recursions* to be those that produce functions that are *majorised* by functions earlier derived. Hence the concept of *bounded* or more frequently called *limited recursion*.

---

[4] We deviated from our norm and used here the notation $\{x \mid \ldots\}$ rather than $\{x : \ldots\}$ in the interest of more visual clarity.

15.1.24 Definition. Grzegorczyk (1953) Given functions h, g, b, we say that f is defined by *limited recursion* or *bounded recursion from them* provided the two equations and one inequality below are satisfied for all $y, \vec{x}$.

$$f(0, \vec{x}) = h(\vec{x})$$
$$f(y + 1, \vec{x}) = g(y, \vec{x}, f(y, \vec{x}))$$
$$f(y, \vec{x}) \leq b(y, \vec{x})$$
□

We can now define the classes $\mathcal{E}^n$ of the Grzegorczyk hierarchy.

15.1.25 Definition. (The Grzegorczyk Hierarchy) We select a sequence of bounding functions, $(g_n)_{n \geq 0}$ (using the same letter as in Grzegorczyk 1953) by

$$g_0 = \lambda x . x + 1$$
$$g_1 = \lambda x y . x + y$$
$$g_2 = \lambda x y . x y$$

and, for $n \geq 2$,

$$g_{n+1} = \lambda x y . A_n \big(\max(x, y) \big)$$

where $\lambda n x . A_n(x)$ is the Ackermann function *version* we used in Chap. 4.

The hierarchy $(\mathcal{E}^n)_{n \geq 0}$ is defined as follows: $\mathcal{E}^n$ is the *closure* of

$$\{ \lambda x . x + 1, \lambda x . x, g_n \}$$

under (Grzegorczyk) *substitution* and *bounded primitive recursion*.   □

15.1.26 Remark.

(1) Note that closure of $\mathcal{E}^n$ under limited (primitive) recursion means that if f is produced from h, g as $f = prim(h, g)$ and is majorised by b, then it is in $\mathcal{E}^n$, *if all h, g, b are.*

(2) The version of the Ackermann function and the few "initial" small bounding functions are as follows (still using "g" here for the original bounding functions, as in Grzegorczyk 1953): $g_0(x, y) = y + 1$, $g_1(x, y) = x + y$, $g_2(x, y) = (x + 1)(y + 1)$ and, for $n \geq 0$,

$$g_{n+1}(0, y) = g_n(y + 1, y + 1)$$
$$g_{n+1}(x + 1, y) = g_{n+1}(x, g_{n+1}(x, y))$$

Tourlakis (1984) contains a direct proof from first principles that g_{n+1} and A_n have the same order of magnitude. Here we will verify this indirectly by virtue of the comparison of the $\mathcal{E}^{n+1}$ with the $K_n^{sim} = \mathcal{L}_n$, for $n \geq 0$. The original $\mathcal{E}^{n+1}$ and the ones based on A_n are the same.

(3) There has been a lot of interest in $\mathcal{E}^0$ and its predicate counterpart, $\mathcal{E}^0_*$ due to its unexpected wealth of nontrivial (albeit "small") functions. It turns out as the reader will be able to discover that the version of the Kleene predicate $T^{(n,0)}$ introduced in Theorem 13.4.4, with no type-1 inputs, while pronounced "*just primitive recursive*" on p.447 can be actually proved —with no changes in its definition!— to be in $\mathcal{E}^0_*$, in fact, a tad lower. (Exercise 15.4.28). □

The following theorem is due to Warkentin (1971) and facilitates the study of the low Grzegorczyk classes. Among the listed results only (f) is nontrivial and not in Grzegorczyk (1953).

15.1.27 Theorem. *Let us call* $\mathcal{M}$ *any class of primitive recursive functions that contains* $\lambda x.x$ *and is closed under substitution and limited recursion. The smallest such class (closure) we will denote, following Warkentin (1971), by* $\mathcal{E}^{-1}$.

(a) $\mathcal{M}$ *contains* $\lambda x.x \doteq 1, \lambda xy.x \doteq y, \lambda xy.x(1 \doteq y)$.
(b) $\mathcal{M}_*$ *is closed under Boolean operations and bounded quantification (bounds* $< z$ *and* $\leq z$*).*
(c) $\lambda xy.x \leq y, \lambda xy.x < y, \lambda xy.x = y, \lambda xy.x \neq y$ *are in* $\mathcal{M}_*$.
(d) $\mathcal{M}$ *is closed under* $(\overset{\circ}{\mu}y)_{<x}$ *and* $(\overset{\circ}{\mu}y)_{\leq x}$, *where unsuccessful search is designated by returning 0.*
(e) $\mathcal{M}$ *is closed under definition by cases, provided the defined function* $\lambda \vec{x}_n.f(\vec{x}_n)$ *is majorised for all* $\vec{x}$ *by some constant, or by* x_i *(for some* $1 \leq i \leq n$*).*
(f) *If* f *is given by*

$$f(0, \vec{x}) = h(\vec{x})$$

$$f(y + 1, \vec{x}) = g(y, \vec{x}, f(y, \vec{x}))$$

and we have that the graphs of h *and* g *are in* $\mathcal{M}_*$, *then so is the graph of* f *provided* f *is increasing with respect to* y *(for any fixed* $\vec{x}$*).*

Proof

(a) • $\lambda x.x \doteq 1$:

$$0 \doteq 1 = 0$$

$$(x + 1) \doteq 1 = x$$

$$x \doteq 1 \leq x$$

• $\lambda xy.x \doteq y$:

$$x \doteq 0 = x$$

$$x \doteq (y + 1) = (x \doteq y) \doteq 1$$

$$x \doteq y \leq x$$

- $\lambda xy.x(1 \dotminus y)$:

$$x(1 \dotminus 0) = x$$
$$x(1 \dotminus (y+)) = 0$$
$$x(1 \dotminus y) \leq x$$

(b) • ($\neg$): Say $P(\vec{x}) \in \mathcal{M}_*$. Then (Definition 15.1.12), for some $p \in \mathcal{M}$, $P(\vec{x}) \equiv p(\vec{x}) = 0$. Then $\neg P(\vec{x}) \equiv (1 \dotminus p(\vec{x})) = 0$. But $\lambda \vec{x}.1 \dotminus p(\vec{x}) \in \mathcal{M}$ (use the substitution $x \leftarrow 1$ in $\lambda xy.x(1 \dotminus y)$ to obtain $\lambda y.1 \dotminus y$).

   • ($\vee$): Say $P(\vec{x})$ and $Q(\vec{y})$ in $\mathcal{M}_*$. Let p be as above, and similarly let q be so for Q. Then

$$P(\vec{x}) \vee Q(\vec{y}) \Leftrightarrow p(\vec{x}) \cdot q(\vec{y}) = 0 \Leftrightarrow p(\vec{x}) \cdot (1 \dotminus (1 \dotminus q(\vec{y}))) = 0$$

   The last equivalence is the operative one, since the "full"multiplication is not postulated as a member of $\mathcal{M}$, while the function in the third bullet of (a) *is* in $\mathcal{M}$.

   • $((\exists y)_{<z}$ and $(\exists y)_{\leq z})$: Let $R(y, \vec{x})$ be in $\mathcal{M}_*$, that is, for some $r \in \mathcal{M}$, it is, $R(y, \vec{x}) \equiv r(y, \vec{x}) = 0$. We are looking for a $g \in \mathcal{M}$ such that

$$(\exists y)_{<z} R(y, \vec{x}) \equiv g(z, \vec{x}) = 0$$

   Note that $(\exists y)_{<z+1} R(y, \vec{x}) \equiv R(z, \vec{x}) \vee (\exists y)_{<z} R(y, \vec{x})$, which leads to the limited recursion

$$g(0, \vec{x}) = 1 \quad \textbf{Comment.} \ (\exists y)_{<0} R(y, \vec{x}) \text{ is false}$$
$$g(z + 1, \vec{x}) = (1 \dotminus (1 \dotminus r(z, \vec{x}))(1 \dotminus (1 \dotminus g(z, \vec{x}))$$
$$g(z, \vec{x}) \leq 1$$

   Also note that $(\exists y)_{\leq z} R(y, \vec{x}) \equiv R(z, \vec{x}) \vee (\exists y)_{<z} R(y, \vec{x})$.

   • $((\forall y)_{<z}$ and $(\forall y)_{\leq z})$: Previous bullet and closure under $\neg$.

   Since all of $\equiv, \rightarrow, \wedge$ are defined in terms of $\neg, \vee$ we have closure of $\mathcal{M}_*$ under *all* Boolean operators (connectives).

(c) • $(\lambda xy.x \leq y:) \ x \leq y \equiv x \dotminus y = 0$.

   • $(\lambda xy.x < y:) \ x < y \equiv \neg(y \leq x)$.

   Thus we have at once,
   $x = y \equiv x \leq y \wedge y \leq x$, hence is in $\mathcal{M}_*$ and so is $x \neq y \equiv \neg x = y$ by closure under negation.

(d) $((\overset{\circ}{\mu}y)_{<x}$ and $(\overset{\circ}{\mu}y)_{\leq x})$: Let us set $g = \lambda z \vec{y}.(\overset{\circ}{\mu}x)_{<z} f(x, \vec{y})$, where $f \in \mathcal{M}$. Then

$$g(0, \vec{y}) = 0$$

$$g(z + 1, \vec{x}) = \text{if } \neg(\exists x)_{<z} f(x, \vec{y}) = 0 \wedge f(z, \vec{y}) = 0$$

$$\text{then } z \text{ else } g(z, \vec{y}) \tag{1}$$

$$g(z, \vec{y}) \leq z$$

We have to show how to rewrite the above limited recursion without using if-then-else, which is not postulated to be in $\mathcal{M}$ (in fact, it is *not* in the special cases $\mathcal{E}^{-1} \cup \mathcal{E}^0$ of $\mathcal{M}$).

Noting that $g(z, \vec{y}) \leq z$, the *iterator* function H of the primitive recursion part of (1) is given by

$$H(z, u, \vec{x}, w) \stackrel{Def}{=} \text{if } u = 0 \text{ then } z$$

$$\left[\text{else if } w \leq z \text{ then } w \text{ else } 0 \right]$$

The variable u receives the condition $\neg(\exists x)_{<z} f(x, \vec{y}) = 0 \wedge f(z, \vec{y}) = 0$ in the form of a function call $F(z, \vec{y})$ where $F \in \mathcal{M}$ is chosen to satisfy

$$\neg(\exists x)_{<z} f(x, \vec{y}) = 0 \wedge f(z, \vec{y}) = 0 \equiv F(z, \vec{y}) = 0$$

The variable w receives the "recursive call" in (1). We now rewrite the definition of H as a limited recursion:

$$H(z, 0, \vec{x}, w) = z$$

$$H(z, u + 1, \vec{x}, w) = \left(1 \div (w \div z)\right) w$$

$$H(z, u, \vec{x}, w) \leq z$$

Thus $H \in \mathcal{M}$. We are done.

Regarding the $\leq z$ bound note that

$$(\overset{\circ}{\mu}x)_{\leq z} f(x, \vec{y}) = \text{if } \neg(\exists x)_{<z} f(x, \vec{y}) = 0 \wedge f(z, \vec{y}) = 0 \text{ then } z \text{ else } (\overset{\circ}{\mu}x)_{<z} f(x, \vec{y})$$

where we handle the if-then-else as above.

(e) Let

$$f(\vec{x}) = \begin{cases} f_1(\vec{x}) & \text{if } R_1(\vec{x}) \\ \vdots & \vdots \\ f_r(\vec{x}) & \text{if } R_r(\vec{x}) \end{cases}$$

The proof that $f \in \mathcal{M}$ under the stated restrictions is reminiscent of that for Theorem 6.6.1: First,

$$y = f(\vec{x}) \equiv y = f_1(\vec{x}) \wedge R_1(\vec{x}) \vee y = f_2(\vec{x}) \wedge R_2(\vec{x}) \vee \cdots \vee y = f_r(\vec{x}) \wedge R_r(\vec{x})$$

Thus $y = f(\vec{x}) \in \mathcal{M}_*$ by (b). But $f(\vec{x}) = (\overset{\circ}{\mu}y) \le x_i{}^5(y = f(\vec{x}))$.

(f) Warkentin (1971) We define two functions — f_1 and f_2 — related to f, that will aid in establishing that the graph of the latter is in $\mathcal{M}_*$.

$$\text{For all } \vec{x} \text{ and } y,\ f_1(y, \vec{x}) \overset{Def}{=} 1 \dot{-} (1 \dot{-} f(y, \vec{x})) \tag{2}$$

Note that

$$f_1(0, \vec{x}) = \text{if } h(\vec{x}) = 0 \text{ then } 0 \text{ else } 1$$

$$f_1(y + 1, \vec{x}) = \text{if } f_1(y, \vec{x}) = 0 \text{ then } \Big[$$

$$\text{if } g(y, \vec{x}, 0) = 0 \text{ then } 0 \text{ else } 1 \Big]$$

$$\text{else if } g(y, \vec{x}, f(y, \vec{x})) = 0 \text{ then } 0 \text{ else } 1$$

$$f_1(y, \vec{x}) \le 1$$

To reformulate the above (second equation) totally in terms of f_1 note that if $f_1(y, \vec{x}) = 1$ we have $f(y, \vec{x}) > 0$ by (2), hence $f(y + 1, \vec{x}) = g(y, \vec{x}, f(y, \vec{x})) > 0$ by the assumption on f. Thus the last else returns 1 (unconditionally) and the second equation simplifies to

$$f_1(y + 1, \vec{x}) = \text{if } f_1(y, \vec{x}) = 0 \text{ then } \Big[$$

$$\text{if } g(y, \vec{x}, 0) = 0 \text{ then } 0 \text{ else } 1 \Big]$$

$$\text{else } 1$$

or simpler still,

$$f_1(y + 1, \vec{x}) = \text{if } f_1(y, \vec{x}) = 0 \wedge g(y, \vec{x}, 0) = 0 \text{ then } 0 \text{ else } 1$$

Since $f_1(y, \vec{x}) \le 1$ for all $y, \vec{x}$ this proves that $f_1 \in \mathcal{M}$.
Now define

$$f_2(y, \vec{x}, z) \overset{Def}{=} \begin{cases} f(y, \vec{x}) & \text{if } f(y, \vec{x}) \le z \\ 0 & \text{othw} \end{cases}$$

---

[5] Or $\le k$ as the case (for f) may be.

We can obtain f_2 by limited recursion within $\mathcal{M}$:

$$f_2(0, \vec{x}, z) = (\overset{\circ}{\mu}u)_{\leq z}(u = h(\vec{x}))$$

$$f_2(y + 1, \vec{x}, z) = (\overset{\circ}{\mu}u)_{\leq z}\Big(u = g(y, \vec{x}, f_2(y, \vec{x}, z)) \wedge f_2(y, \vec{x}, z) \neq 0$$

$$\vee\; f_1(y, \vec{x}) = 0 \wedge u = g(y, \vec{x}, 0)\Big)$$

Note that the iterator is in $\mathcal{M}$ by (b) and (c).

Since $f_2(y, \vec{x}, z) \leq z$, the above proves that $f_2 \in \mathcal{M}$. $\qquad\square$

There are several corollaries:

15.1.28 Corollary. *For $n \geq 0$, $\mathcal{E}^n$ is closed under $(\overset{\circ}{\mu}y)_{<z}$ and $(\overset{\circ}{\mu}y)_{\leq z}$, as well as under definition by cases. The latter is unrestricted for $n \geq 1$ but for $n = 0$ requires a restriction similar to that in (e) above: The resulting function must be bounded by $x_i + k$ for some x_i among its arguments, and some k.*

"Must" is not an overkill (cf. Exercise 15.4.11).

15.1.29 Corollary. *The graphs $z = x + y$ and $z = xy$ are in $\mathcal{M}$, hence in $\mathcal{E}^{-1}$.*

Proof Direct application of Theorem 15.1.27.(f). Why is the graph of the iterator $\lambda y.y + 1$ of $\lambda xy.x + y$ in $\mathcal{M}_*$? Note $z = y + 1 \equiv z \div 1 = y \wedge z > 0$. $\qquad\square$

15.1.30 Corollary. *$\lambda xz.z = A_n(x)$ is in $\mathcal{M}_*$ (hence in $\mathcal{E}^n_*$ for all $n \geq 0$).*

Proof Direct application of Theorem 15.1.27.(f) the fact that in the primitive recursion below (for a fixed n)

$$A_{n+1}(0) = 2$$

$$A_{n+1}(x + 1) = A_n(A_{n+1}(x))$$

A_{n+1} is increasing along x, and we also take the I.H. that $z = A_n(x)$ is in $\mathcal{M}_*$ for the n we fixed. $\qquad\square$

15.1.31 Corollary. *$\mathcal{E}^{-1} \subseteq \mathcal{E}^n$, $n \geq 0$.*

Proof By the definition of $\mathcal{E}^{-1}$ as a closure (in Theorem 15.1.27). All the $\mathcal{E}^n$, $n \geq 0$, include the initial function of $\mathcal{E}^{-1}$ and are closed under the same operations. $\qquad\square$

15.1.32 Corollary. *$\mathcal{E}^n \subseteq \mathcal{E}^{n+1}$, $n \geq -1$.*

Proof For $n \geq 3$, $A_{n-1} \in \mathcal{E}^{n+1}$ since by Corollary 15.1.30 $z = A_{n-1}(x) \in \mathcal{E}^{-1}_*$ and hence in $\mathcal{E}^{n+1}_*$. But then $A_{n-1} \in \mathcal{E}^{n+1}$ by

$$A_{n-1}(x) = (\overset{\circ}{\mu}z)_{\leq A_n(x)}z = A_{n-1}(x)$$

Thus $\mathcal{E}^{n+1}$ includes the initial functions of $\mathcal{E}^n$ and is closed under the same operations.

For $n < 3$, $\mathcal{E}^{-1} \subseteq \mathcal{E}^0$ is direct from Corollary 15.1.31 and for

$$\mathcal{E}^0 \subseteq \mathcal{E}^1 \subseteq \mathcal{E}^2 \subseteq \mathcal{E}^3$$

the argument is the same as for the general case, using the bounding (initial) functions $\lambda xy.y+1$, $\lambda xy.x+y$, $\lambda xy.xy$, knowing that $z = y+1$, $z = x+y$, $z = xy$ are all in $\mathcal{E}_*^{-1}$. □

15.1.33 Example. (Part of Example 2.3.10 Revisited)

1. We already saw that $z = x + y$ and $z = xy$ are in $\mathcal{M}_*$. Now, $x \mid y$ means $(\exists z)_{\leq y} y = xz$. Hence $x \mid y$ is in $\mathcal{M}_*$.

2. $z = x^y$ is also in $\mathcal{M}_*$: $\lambda xy.x^y$ is increasing in y and is given by the primitive recursion $x^0 = 1$ and $x^{y+1} = x \cdot x^y$. Since the iterator's graph $w = x \cdot z$ is in $\mathcal{M}_*$, we are done by Theorem 15.1.27.(f).

3. Where is $\lfloor x/y \rfloor$ located in the Grzegorczyk classes? We defined it in Example 2.3.10 as $(\overset{\circ}{\mu} z)_{\leq x} (z + 1)y > x$, using here $\overset{\circ}{\mu}$ instead of μ —this only changes the "artificially defined value" at $y = 0$ as 0 instead of $x + 1$. Let us write this definition differently:

$$z = \lfloor x/y \rfloor \equiv z = 0 \wedge y = 0 \vee (zy \leq x \wedge zy + y > x)$$

Now, $zy \leq x \equiv (\exists w)_{\leq x} w = zy$, thus it is in $\mathcal{M}_*$. Now, first note that $x + y \leq v \equiv (\exists u)_{\leq v}(x + y = u)$, hence it is $\mathcal{M}_*$. Thus

$$zy + y > x \equiv \neg(zy + y \leq x) \equiv \neg(\exists u)_{\leq x}(u = zy \wedge u + y \leq x)$$

hence it is in $\mathcal{M}_*$, and so is $z = \lfloor x/y \rfloor$.

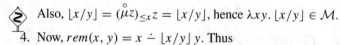 Also, $\lfloor x/y \rfloor = (\overset{\circ}{\mu} z)_{\leq x} z = \lfloor x/y \rfloor$, hence $\lambda xy.\lfloor x/y \rfloor \in \mathcal{M}$.

4. Now, $rem(x, y) = x \,\dot{-}\, \lfloor x/y \rfloor y$. Thus

$$z = rem(x, y) \equiv (\exists u, v)_{\leq x}(z = x \,\dot{-}\, u \wedge u = vy \wedge v = \lfloor x/y \rfloor)$$

Therefore $z = rem(x, y) \in \mathcal{M}_*$.

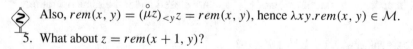 Also, $rem(x, y) = (\overset{\circ}{\mu} z)_{\leq y} z = rem(x, y)$, hence $\lambda xy.rem(x, y) \in \mathcal{M}$.

5. What about $z = rem(x + 1, y)$?

$$z = rem(x + 1, y) \equiv (\exists u)_{\leq x}(u = rem(x, y) \wedge [u + 1 = y \wedge z = 0 \vee$$
$$u + 1 < y \wedge z = u + 1])$$

Also, $rem(x+1, y) = (\overset{\circ}{\mu}z)_{<y}z = rem(x+1, y)$, hence $\lambda xy.rem(x+1, y) \in \mathcal{M}$.

6. $Pr(x)$ has been introduced in Example 2.3.10. Here we are nuancing it a bit. $Pr(x) \equiv x > 1 \wedge (\forall y)_{<x}(rem(x, y) = 0 \rightarrow y = 1)$. This is clearly in $\mathcal{M}_*$ (cf. Theorem 15.1.27).

   We will also need $Pr(x + 1)$ (this is in $\mathcal{E}_*^0$ as follows from the above by substitution of the call $x + 1$ in the argument of Pr. In $\mathcal{E}^{-1}$ we are not allowed such a call. The output is too "big"!

   So, proceed this way: $Pr(x + 1) \equiv x > 0 \wedge (\forall y)_{\leq x}(rem(x + 1, y) = 0 \rightarrow y = 1)$.

7. $\pi(x)$ (the number of primes $\leq x$) is in $\mathcal{M}$. Note that $\pi(0) = 0$, and $\pi(x+1) =$ if $Pr(x + 1)$ then $\pi(x) + 1$ else $\pi(x)$.

   Note that π is increasing and that the graph of the iterator —$w = if\ Pr(x + 1)\ then\ z + 1\ else\ z$— is in $\mathcal{M}_*$. Indeed, said graph is equivalent to

   $$w = z + 1 \wedge Pr(x + 1) \vee w = z \wedge \neg Pr(x + 1)$$

   Thus $w = \pi(x)$ is in $\mathcal{M}_*$.

 Since $\pi(x) \leq x$, for all x, $\pi(x) = (\overset{\circ}{\mu}w)_{\leq x}w = \pi(x)$, thus $\pi \in \mathcal{M}$.

8. The predicate $y = p_n$ (the graph of the "nth prime" function): Note that $y = p_n \equiv Pr(y) \wedge \pi(y) = n + 1$.[6] Hence $\lambda ny.y = p_n$ is in $\mathcal{M}_*$.

9. $z = p_n^x$ and $z = p_n^{x+1}$ are in $\mathcal{M}_*$. Indeed, $z = p_n^x \equiv (\exists u)_{\leq z}(u = p_n \wedge z = u^x)$. As for $z = p_n^{x+1}$, we have $z = p_n^{x+1} \equiv (\exists u)_{\leq z}(u = x + 1 \wedge z = p_n^u)$.

10. $\lambda nx. \exp(n, x)$ (the exponent of p_n in the prime factorization of x). $\exp(n, x) = (\overset{\circ}{\mu}y)_{<x}\neg(p_n^{y+1}|x) = (\overset{\circ}{\mu}y)_{<x}\neg\left((\exists w)_{\leq x}w = p_n^{y+1} \wedge w|x\right)$ (cf. Example 2.3.10.(7)). Thus $\lambda nx. \exp(n, x) \in \mathcal{M}$.

 Also $\lambda nx. \exp(n, x) \overset{\cdot}{-} 1 \in \mathcal{M}$. As earlier, we denote this function by $\lambda nx.(x)_n$.

11. $Seq(x)$ (says that x's prime number factorization contains 2, and has no gaps: *no* prime, between 2 and the largest prime in the factorisation, is *missing*.) $Seq(x) \equiv x > 1 \wedge (\forall y)_{\leq x}(\forall z)_{\leq x}(Pr(y) \wedge Pr(z) \wedge y < z \wedge z|x \rightarrow y|x)$. Thus, $Seq(x) \in \mathcal{M}_*$.

12. Our next tool (reintroduced; cf. Definition 2.4.2) is the *length function*. For numbers z that satisfy Seq it returns the number of distinct primes in their prime decomposition. It also returns a value for all *other* z (it is total) but we have no interest in those. Recall the definition: $lh(z)$ (pronounced "length of z") is shorthand for $(\overset{\circ}{\mu}y)_{<z}\neg(p_y|z)$. Since $p_y|z$ is in $\mathcal{M}_*$ due to its equivalence to $(\exists w)_{\leq z}(w = p_y \wedge w|z)$, we have $\lambda z.lh(z) \in \mathcal{M}$.

13. $\lambda nx. = \lfloor \log_{p_n} x \rfloor$ is in $\mathcal{M}$ (using the convention $\lfloor \log_{p_n} 0 \rfloor = 0$). Indeed,

---

[6] Substitution of the call $\pi(y)$ into z of $z = n + 1$.

$$\lfloor \log_{p_n} x \rfloor = (\overset{\circ}{\mu} y)_{\leq x} p_n^{y+1} > x = (\overset{\circ}{\mu} y)_{\leq x} \left(\neg (\exists u)_{\leq x} u = p_n^{y+1} \right)$$

See also 9. above.

14. Final tool: $\lambda x. \lfloor \sqrt{x} \rfloor \in \mathcal{M}$ (hence also in $\mathcal{E}^{-1}$).

$$\lfloor \sqrt{x} \rfloor = (\overset{\circ}{\mu} z)_{\leq x} (z+1)^2 > x = (\overset{\circ}{\mu} z)_{\leq x} \left(\neg (\exists w)_{\leq x} w = (z+1)(z+1) \right). \qquad \square$$

The Grzegorczyk classes ever since Grzegorczyk (1953) was published have been studied by many researchers. The low classes $\mathcal{E}^0, \mathcal{E}^1, \mathcal{E}^2$ have enormous expressive power (the reader will be asked to place the Kleene predicate of Chap. 13 in $\mathcal{E}^{-1}$, which will immediately render the equivalence problem of the functions in this class non semi-recursive).

Ritchie (1963) studied $\mathcal{E}^2$ extensively (we will see an important role it plays in the simulation of URMs by simultaneous recursions) while $\mathcal{E}^3$ coincides with the well-known class of *elementary functions* directly defined by Kalmár as

15.1.34 Definition. (Kalmár (1943)) The class of the *elementary functions* $\mathcal{E}$ is the closure of the set of initial functions $\{\lambda xy.x + y, \lambda xy.x \mathbin{\dot{-}} y\}$ under Grzegorczyk *substitution, summation* ($\sum_{i \leq z}$) and product ($\prod_{i \leq z}$). $\qquad \square$

The elementary functions contain $\lambda x.2^x$ and hence also "monster functions" like the one on p.564. Intuitively they are viewed as the boundary class between "practical" and "impractical" computing.

15.1.35 Lemma. *For* $n \geq 2$, $\lambda \vec{x}. f(\vec{x}) \in \mathcal{E}^{n+1}$ *implies that for some* k *(depending on* f, *we have* $f(\vec{x}) \leq A_n^k(\max(\vec{x}))$, *for all* $\vec{x}$.

Proof Induction on the formation of $\mathcal{E}^{n+1}$.

The initial functions are $\lambda xy.A_n(\max(x, y))$, $\lambda x.x$ and $\lambda x.x + 1$. The claim trivially follows for them (recall the growth properties of A_n, Lemmata 4.1.4–4.1.12). The property trivially *propagates* with *substitution* as this operation *does not raise* n[7] (cf. Theorem 4.2.1 and Remark 4.2.2).

Even more trivial is that the property propagates with limited recursion: Indeed, the I.H. posits that if $f = prim(h, g)$ and

$$f(\vec{x}) \leq b(\vec{x}) \tag{1}$$

for all $\vec{x}$, then, in particular, $b(\vec{x}) \leq A_n^k(\vec{x})$ for some k and all $\vec{x}$. By (1) $f(\vec{x}) \leq A_n^k(\vec{x})$ for the same k and all $\vec{x}$. $\qquad \square$

15.1.36 Remark Exercises 15.4.11, 15.4.12, 15.4.13 imply that $\lambda xy.x + y \in \mathcal{E}^1 - \mathcal{E}^0$, $\lambda xy.xy \in \mathcal{E}^2 - \mathcal{E}^1$, $\lambda x.2^x \in \mathcal{E}^3 - \mathcal{E}^2$, hence $\mathcal{E}^0 \subsetneq \mathcal{E}^1 \subsetneq \mathcal{E}^2 \subsetneq \mathcal{E}^3$.

In fact, it is also $\mathcal{E}^{-1} \subsetneq \mathcal{E}^0$ since $\lambda x.x + 1 \notin \mathcal{E}^{-1}$. $\qquad \square$

---

[7] That is, if $f = \lambda \vec{x} \vec{y}. g(h(\vec{x}), \vec{y})$ with g and h in $\mathcal{E}^{n+1}$, then (I.H.) g is, say, bounded by A_n^l and h is bounde by A_n^r. Then f is bounded by A_n^{r+l}.

15.1.37 Theorem. $(\mathcal{E}^n)_{n \geq -1}$ *is a proper hierarchy of* $\mathcal{PR}$, *that is,*

- $\mathcal{E}^n \subsetneq \mathcal{E}^{n+1}$, *for* $n \geq -1$
  and
- $\mathcal{PR} = \bigcup_{n \geq -1} \mathcal{E}^n$

Proof

- Firstly, the non-strict inclusion has been proved in Corollary 15.1.32. For $n \leq 2$, $\mathcal{E}^n \subsetneq \mathcal{E}^{n+1}$ by Remark 15.1.36. On the hand, for $n \geq 3$, we have $A_n \in \mathcal{E}^{n+1} - \mathcal{E}^n$. Note that $A_n \in \mathcal{E}^n$ entails $A_n(x) \leq A_{n-1}^k(x)$ for some k and all x, contradicting Lemma 4.1.12.
- The $\supseteq$ is trivial.
  Say then that $f \in \mathcal{PR}$ and let $d_1, d_2, \ldots, d_r, f$ be a primitive recursive derivation. By induction on length of derivation. Let n and k be such that all the d_i and f are majorised by A_n^k. But then all primitive recursions in the derivation are valid as limited recursions in $\mathcal{E}^{n+1}$, hence $f \in \mathcal{E}^{n+1}$. This proves $\subseteq$.
  How or where was the I.H. used? (cf. Exercise 15.4.18).   □

15.2 The Ritchie-Cobham Property and Hierarchy Comparison

The comparison of the hierarchies involves proving that the *dynamic complexity* of the functions in various classes $(K_n, K_n^{sim}, \mathcal{L}_n, \mathcal{E}^{n+1})$ goes hand in hand with their *definitional complexity* —one complexity predicts the other. This result is the *"Richie[8]-Cobham property"* which is the central tool in this section (along with some good old fashioned programming considerations on URMs and via loop programs). As we will simplify our "programming tasks" by using loop programs *in lieu* of URMs we will start by making explicit the implementation of the **Loop** X-**end** instruction that respects the stated loop-program semantics, in particular the part where *changing the loop variable in the body of the loop does not change the number of iterations*. We need this elaboration as we shall be looking into loop-program run times, thus we need to know what exactly is going on in the looping mechanism.

We imagine that we have *numbered* all instructions in our loop-programs in the style of URMs.

Now, each of several **Loop** X-**end** instruction-pairs —I emphasise: with the *same* variable X— is assigned a *distinct* "*hidden*" variable H_X^m assigned to the m-th such loop encountered from top to bottom of the overall program. We translate so ("macro expand") each **Loop** X-**end** pair —which may occur multiple times.

---

[8] D. Ritchie.

A program segment

$$L : \textbf{Loop } X$$

$$\vdots$$

$$R : \textbf{ end}$$

is *implemented* as follows (e.g., on a URM or "real life compiler"), where for simplicity of notation we call the hidden variable assigned just "H".

 Why "hidden"? This is an often used nomenclature for compiler-generated "internal" (to the compiler) variables that *are not accessible to the user* (programmer).

$$L : \textbf{Loop } X \textbf{ replace by} : \begin{cases} L : \ H \leftarrow X \\ L+1 : \ \text{if } H = 0 \text{ goto } R + 2 \text{ else goto } L + 2 \end{cases}$$

$$\vdots$$

$$R : \textbf{ end} \quad \textbf{replace by} : \begin{cases} R : \ H \leftarrow H \mathbin{\dot-} 1 \\ R+1 : \ \text{goto } L \end{cases}$$

 With the above *clarification* of what **Loop** X-**end** does exactly, we note that thus a URM can simulate a given loop program *without any run time loss*.

But what about the fact that the URM does not have the instruction $X \leftarrow Y$ (as primitive) that loop programs do have, and thus the former spends as much time as $O(y)$, where y is the contents of Y?[9]

The simple (and also theoretically correct) answer is "OK; let's retroactively fit the URM with such instructions. This does not change the computability theory we developed so far, since the instruction can be simulated anyway".

We next refer back to the *simulation of a URM* by a *simultaneous recursion*, Sect. 2.4.4, specifically, Theorem 2.4.20. We quote the concluding sentence in the proof of the latter theorem:

> Since the iterator functions only utilise the functions $\lambda z.a$, $\lambda z.z + 1$, $\lambda z.z \mathbin{\dot-} 1$, $\lambda z.z$, and predicates $\lambda z.z = 0$, and $\lambda z.z > 0$ —all in $\mathcal{PR}$ and $\mathcal{PR}_*$— it follows that all the simulating functions are in $\mathcal{PR}$.

 As a result of our work in Section 1 of this chapter, the part in the quote above "—all in $\mathcal{PR}$ and $\mathcal{PR}_*$— it follows that all the simulating functions are in $\mathcal{PR}$" can be

---

[9] The reader recalls that the simulation is rather "expensive". Cf. Example 1.1.14 and Exercise 1.1.15.

replaced by "—all in K_1^{sim} and $K_{1,*}^{sim}$— *it follows that all the simulating functions are in K_2^{sim}*".

Thus,

15.2.1 Lemma. *All the simulating functions in Theorem 2.4.20, i.e., the X_i and IC are in $K_2^{sim} = \mathcal{L}_2$.*

In fact, all simulating functions of a URM are in $\mathcal{E}^2$. We prove first

15.2.2 Lemma. *All $\mathcal{E}^n$, $n \geq 2$, are closed under limited simultaneous recursion.*

Proof Let

$$\text{for } i = 1, \ldots, k \begin{cases} f_i(0, \vec{y}) & = h_i(\vec{y}) \\ f_i(x+1, \vec{y}) & = g_i(x, \vec{y}, f_1(x, \vec{y}), \ldots, f_k(x, \vec{y})) \\ f_i(x, \vec{y}) & \leq b_i(x, \vec{y}) \end{cases} \tag{1}$$

where the h_i, g_i, b_i are in $\mathcal{E}^n$.

We convert the simultaneous recursion to a simple (single) limited recursion using the coding of Definition 2.4.35 and the projections from Example 2.4.37, expressed as substitutions using only K and L (Exercise 2.5.9). We use the quadratic (non onto) pairing function $J = \lambda xy.(x+y)^2 + x$ to define $[\![\vec{x}_k]\!]^{(k)}$ and the projections Π_i^k, $i = 1, \ldots, k$. All these projections are in $\mathcal{E}^2$ either directly via, for example, $Kx^{10} = (\overset{\circ}{\mu}x)_{\leq z}(\exists y)_{\leq z}z = J(x, y)$, or in terms of $\lfloor \sqrt{z} \rfloor$ (cf. Example 2.4.32). That $\lambda \vec{x}_k . [\![\vec{x}_k]\!]^{(k)}$ is in $\mathcal{E}^2$ is also immediate via a finite number of substitution instances of $J(x, y)$ (cf. Definition 2.4.35).

So, following the process of Theorem 2.4.17, we set

$$F(x, \vec{y}) \overset{Def}{=} [\![f_1(x, \vec{y}), \ldots, f_k(x, \vec{y})]\!]^{(k)}$$

thus the simultaneous recursion becomes a simple limited recursion

$$F(0, \vec{y}) = [\![h_1(x, \vec{y}), \ldots, h_k(x, \vec{y})]\!]^{(k)}$$

$$F(x+1, \vec{y}) = [\![\ldots, g_i(x, \vec{y}, \Pi_1^k F(x, \vec{y}), \ldots, \Pi_k^k F(x, \vec{y})), \ldots]\!]^{(k)}$$

$$F(x, \vec{y}) \leq [\![\ldots, b_i(x, \vec{y}), \ldots]\!]^{(k)}$$

where the inequality is valid since the J is increasing with respect to both variables and composing it a finite number of times with itself does not change this fact.

By the placement of the $[\![\ldots]\!]^{(k)}$ and the Π_i^k (all in $\mathcal{E}^2$), it is $F \in \mathcal{E}^n$, $n \geq 2$. Hence so is $f_i = \Pi_i^k F$.   □

---

[10] It is quite common to write Kz for $K(z)$ and $\Pi_i^k z$ for $\Pi_i^k(z)$.

We apply the above to the simulation via simultaneous primitive recursion of URM computations (cf. Lemma 15.2.1 and Theorem 2.4.20).

15.2.3 Lemma. *The simulating functions for a given URM, i.e., $\lambda y \vec{x}.X_i(y, \vec{x})$ and $\lambda y \vec{x}.IC(y, \vec{x})$, are all in $\mathcal{E}^2$.*

Proof It suffices to prove that the simultaneous recursion we introduced in Theorem 2.4.20 is *limited*, with bounds in $\mathcal{E}^2$, while the part-functions and predicates in the definition by cases in the iterator of said proof are all in K_1^{sim} and $K_{1,*}^{sim}$ hence trivially also in $\mathcal{E}^2$ and $\mathcal{E}_*^2$, respectively.

As for the bounding functions, we note that, for all variables of the under simulation URM M, we have

$$X_i(y, \vec{x}) \leq \max(\vec{x}) + \max\{a : a \text{ occurs in an } X \leftarrow a \text{ instruction of } M\} + y$$

The estimate above is so since in *each* of y steps the most we can add to any variable is 1. Also, $IC(y, \vec{x}) \leq k$, where k labels **stop** in M.

Thus the bounding functions are all in $\mathcal{E}^1 \subseteq \mathcal{E}^2$.   □

15.2.4 Lemma. *For $n \geq 2$, $\mathcal{E}^n$ is closed under $\sum_{i \leq z}$.*

Proof Let $f \in \mathcal{E}^n$, $n \geq 2$. We will show that

$$g = \lambda z \vec{y}. \sum_{i \leq z} f(i, \vec{y}) \in \mathcal{E}^n \tag{1}$$

Indeed,

$$g(0, \vec{y}) = f(0, \vec{y})$$
$$g(z + 1, \vec{y}) = f(z + 1, \vec{y}) + g(z, \vec{y})$$

The iterator $f(z + 1, \vec{y}) + w$ being in $\mathcal{E}^n$ we next look for a bound for g.

- First, the $n = 2$ case:

$$g(z, \vec{y}) \leq \sum_{i \leq z} f(i, \vec{y}) \leq \sum_{i \leq z} \left(c \Big(\max(i, \vec{y}) \Big)^r + l \right) \text{ (cf. Exercise 15.4.13)}$$

$$\leq c(z + 1) \Big(\big(\max(z, \vec{y}) \big)^r + l \Big)$$

The bound $\lambda z \vec{y}.c(z+1)\Big(\big(\max(z, \vec{y}) \big)^r + l \Big)$ is in $\mathcal{E}^2$ by the fact that $\lambda z w.zw \in \mathcal{E}^2$.

- The $n > 2$ case: As A_{n-1}^k majorises *every* function in $\mathcal{E}^n$, for a k that depends on the function, we have

$$g(z, \vec{y}) \le \sum_{i \le z} f(i, \vec{y}) \le \sum_{i \le z} A_{n-1}^k \left(\max(i, \vec{y}) \right) \le (z+1) A_{n-1}^k \left(\max(z, \vec{y}) \right)$$

(1)

Since $\lambda xy.xy \in \mathcal{E}^n$, for $n \ge 2$, we have that the bound in (1) is in $\mathcal{E}^n$.

$\square$

15.2.5 Lemma. *For $n \ge 2$, if $\lambda \vec{x}_k . f(\vec{x}_k) \in \mathcal{E}^n$, then there is a URM M —such that $f = M_{X_1}^{\vec{X}_k}$— that runs within time $\lambda \vec{x}_k . t(\vec{x}_k) \in \mathcal{E}^n$, that is, for every input $\vec{x}_k$, $M_{X_1}^{\vec{X}_k}$ halts (i.e., reaches* **stop***) in $\le t(\vec{x}_k)$ steps.*

Proof As noted earlier (section preamble before Lemma 15.2.2), it suffices to carry out our programming on the loop programs formalism. The proof is by induction on $n \ge 2$. For each n there is another induction on the closure that the set $\mathcal{E}^n$ is.

n-Basis: $n = 2$. Induction on $\mathcal{E}^2$ requires to verify the contention for the initial functions $\lambda x.x$, $\lambda x.x + 1$ and $\lambda xy.xy$ first. The first two need one-line (loop) programs whose run time is 1 —a constant. But $\mathcal{E}^2$ contains all constant functions.

As for $\lambda xy.xy$, the loop program

$$P : \begin{cases} \quad \textbf{Loop } X \\ \qquad \textbf{Loop } Y \\ \qquad\quad Z \leftarrow Z + 1 \\ \qquad \textbf{end} \\ \quad \textbf{end} \end{cases}$$

satisfies $P_Z^{XY} = \lambda xy.xy$. Its run time is —since each execution of **end** counts for two steps and of **Loop** counts for one (except *upon first entry*, where an extra one step is charged)— $\le ((4y+1)+3)x + 1$. Clearly, $\lambda xy.((4y+1)+3)x + 1 \in \mathcal{E}^2$.

We can also observe simply that ignoring constant overhead we can estimate the run time quickly seeing how many times we go around the loop and say at once that the run time is $O(xy)$ (cf. Sect. 0.4.2).

We postpone the induction over $\mathcal{E}^2$ as we prefer to do all these simultaneously for $n \ge 2$ (the reasoning is identical, for each n).

So we next bound from above the run time when computing the *initial* functions of $\mathcal{E}^n$, $n > 2$, on appropriate loop programs (as proxies for appropriate URMs).

We will find a run time *upper bound*, T, of a well chosen loop program that computes A_{n-1}, $n > 2$, and show that $T \in \mathcal{E}^n$.

Now, $A_{n-1} \in K_{n-1}^{sim} = \mathcal{L}_{n-1}$.

Let then M be a loop program in L_{n-1} (Definition 15.1.19) such that

$$A_{n-1} = M_Y^X$$

Modify M into a new $\widetilde{M}$ that is still in L_{n-1} and still computes A_{n-1}, namely,

$$A_{n-1} = \widetilde{M}_Y^X$$

The modification is easy: We choose a variable Z that *does not occur* in M and strategically place several instructions $Z \leftarrow Z + 1$ in it to obtain $\widetilde{M}$.

These are placed, *one copy* before *each* non-**Loop**, non-**end** instruction of M. For instructions **Loop** or **end** we act according to our simulation protocol of loop programs by URMs, so we place *one copy* of $Z \leftarrow Z + 1$ *before* a **Loop** instruction and *one after*, while we put *two* copies *before* an **end** instruction.

See below.

$$\vdots$$

$Z \leftarrow Z + 1$ **Comment**. *Entry of loop count*

Loop X

$Z \leftarrow Z + 1$ **Comment**. *Loop's second count*

$$\vdots$$

$Z \leftarrow Z + 1$ **Comment**. *Count for next M-instruction*

An original $M-$**instruction** : $U \leftarrow 0$ *or* $U \leftarrow U + 1$ *or* $U \leftarrow W$

$$\vdots$$

$Z \leftarrow Z + 1$ **Comment**. *Execution of* **end** *count one*

$Z \leftarrow Z + 1$ **Comment**. *Execution of* **end** *count two*

end

$$\vdots$$

Clearly $\widetilde{M}_Y^X = A_{n-1}$ and $\widetilde{M}_Z^X$, in $\mathcal{L}_{n-1}$ is *the run time* for A_{n-1} *according to* M. Since $\widetilde{M}_Z^X \in K_{n-1}^{sim} = \mathcal{L}_{n-1}$, for some r, this function is majorised by $A_{n-1}^r(x)$, for all inputs x that are "read" in X. A_{n-1}^r is the "T" we promised:

1. It majorises the run time of $A_{n-1} = M_Y^X$ on some URM M.
2. Since $A_{n-1} \in \mathcal{E}^n$, and $\mathcal{E}^n$ is closed under substitution, this run-time majorant, $T = A_{n-1}^r$, is also in $\mathcal{E}^n$.

We finally are ready to do induction over the closure $\mathcal{E}^n$, $n \geq 2$, to conclude, that if $f \in \mathcal{E}^n$, then, for some URM (loop program will do by earlier remarks) M, $f = M_Y^{\vec{X}_k}$ and the run time is majorised by a function in $\mathcal{E}^n$.

We have established the Basis, that all *initial* functions of $\mathcal{E}^n$ have the property. We note

- The property propagates with substitution. So let h and g have the property, and consider $f = h(g(\vec{x}), \vec{y})$. We write t_f, etc., for the run time majorising function on some appropriate URM. Programming in the obvious way (superposition) we have

$$t_f(\vec{x}, \vec{y}) = t_h(g(\vec{x}), \vec{y}) + t_g(\vec{x})$$

As our $\mathcal{E}^n$ is closed under substitution, contains $\lambda xy.x + y$ and h and g have the property, we have $t_f \in \mathcal{E}^n$. The other substitution cases are trivial.
- The property propagates with limited recursion. So let f be given by the following schema, from h, g, b in $\mathcal{E}^n$:

$$f(0, \vec{y}) = h(\vec{y})$$
$$f(x + 1, \vec{y}) = g(x, \vec{y}, f(x, \vec{y}))$$
$$f(x, \vec{y}) \leq b(x, \vec{y})$$

The recursion is implemented on a loop program as

$$i \leftarrow 0$$
$$z \leftarrow h(\vec{y})$$

Loop X

$$z \leftarrow g(i, \vec{y}, z)$$
$$i \leftarrow i + 1$$

end

We next estimate —recalling how **Loop-end** works— an upper bound $t_f(x, \vec{y})$ for the run time of the above.

$$t_f(x, \vec{y}) = t_h(\vec{y}) + 2 + \sum_{i < x}(4 + t_g(i, \vec{y}, f(i, \vec{y})))$$

By the I.H. $t_g \in \mathcal{E}^n$ and so is f. Thus, by Lemma 15.2.4 and the I.H. on t_h we have that $t_f \in \mathcal{E}^n$. $\qquad \square$

15.2.6 Theorem. (The Ritchie -Cobham Property of $\mathcal{E}^n$, $n \geq 2$) $f \in \mathcal{E}^n$, for $n \geq 2$, iff f runs on some URM M —that is, $f = M_Y^{\vec{X}_k}$— within time $t_f \in \mathcal{E}^n$.

Proof The *only if* is Lemma 15.2.5. For the *if*, say $f = M_Y^{\vec{X}_k}$ and t_f on that M is in $\mathcal{E}^n$, $n \geq 2$. By Lemma 15.2.3, $\lambda y\vec{x}_k.Y(y, \vec{x}_k) \in \mathcal{E}^2$. But $f(\vec{x}_k) = Y(t_f(\vec{x}_k), \vec{x}_k)$ for all $\vec{x}_k$, and clearly $f \in \mathcal{E}^n$ by assumption, via substitution. $\qquad \square$

15.2.7 Corollary. $\mathcal{L}_n = \mathcal{E}^{n+1}$, for $n \geq 2$.

Proof A function $\lambda\vec{x}.f(\vec{x})$ in $\mathcal{L}_n$ is computable by a loop program M —i.e., $f = M_Y^{\vec{X}}$— whose run time is bounded by A_n^k, for some k. The latter function being in $\mathcal{E}^{n+1}$ implies via Theorem 15.2.6 (*if*) that $f \in \mathcal{E}^{n+1}$.

Conversely, say $f \in \mathcal{E}^{n+1}$. Let a URM M compute f —that is $f = M_Y^{\vec{X}_k}$— in time $t_f \in \mathcal{E}^n$. Then for some r, $t_f(\vec{x}_k) \leq A_n^r(\max(\vec{x}_k))$.

Therefore, $f(\vec{x}_k) = Y\left(A_n^r(\max(\vec{x}_k)), \vec{x}_k\right)$, for all $\vec{x}_k$. It follows that $f \in K_n^{sim}$, since A_n^r is and so is the simulating function Y (in fact the latter is in K_2^{sim} (Lemma 15.2.1)). But $K_n^{sim} = \mathcal{L}_n$. □

15.2.8 Corollary. $K_n^{sim} = \mathcal{E}^{n+1}$, *for* $n \geq 2$.

15.2.9 Corollary. $K_n \subseteq \mathcal{E}^{n+1}$, *for* $n \geq 2$.

Proof $K_n \subseteq K_n^{sim}$, $n \geq 0$. Now apply Corollary 15.2.8. □

15.2.10 Corollary. $K_n = \mathcal{E}^{n+1}$, *for* $n \geq 4$.

Proof We prove $\mathcal{E}^{n+1} \subseteq K_n$, for $n \geq 4$.

We work as in Lemma 15.2.2 converting the simultaneous recursion of Theorem 2.4.20 into a single recursion. Of course, no K_n, or even K_n^{sim}, is closed under primitive recursion straight off the Definition (15.1.1). (It turns out that from $n \geq 4^{11}$ —the present result— the K_n *are* closed under (single) *limited* recursion.)

We start with the quadratic pairing function $J = \lambda xy.(x+y)^2 + x \in K_2$. Thus $I_k \overset{Def}{=} \lambda\vec{x}_k.[\![\,\vec{x}_k\,]\!]^{(k)} \in K_2$, for any fixed k (cf. Proposition 15.1.11).

In Example 2.4.32 we derived $Kz = z \dotminus \lfloor\sqrt{z}\rfloor^2$ and $Lz = \lfloor\sqrt{z}\rfloor \dotminus Kz$. Since $\lambda z.\lfloor\sqrt{z}\rfloor \in K_3$ (Exercise 15.4.33), so are K and L and thus the Π_i^k as well.

Therefore, imitating the proof of Lemma 15.2.2 (*not* using a bounding function here), we set $F(y,\vec{x}) = I_k(f_1(y,\vec{x}), \ldots, f_k(y,\vec{x}))$ and thus the iteration equation for F is

$$F(y+1,\vec{x}) = [\![\,\ldots, g_i(y,\vec{x}, \Pi_1^k F(y,\vec{x}), \ldots, \Pi_k^k F(y,\vec{x})), \ldots\,]\!]^{(k)} \tag{1}$$

This (since the $\Pi_i^k \in K_3$) shows that $F \in K_4$ and so $f_i = \lambda y\vec{x}.\Pi_i^k F(y,\vec{x}) \in K_4$ as well.

We apply the above to Theorem 2.4.20.

So let $n \geq 4$, $\mathcal{E}^{n+1} \ni f = M_{X_1}^{\vec{X}_r}$, M chosen so (Lemma 15.2.5) that its run time is bounded by a function in $\mathcal{E}^{n+1}$.

Let X_i, $1 \leq i \leq m(\geq r)$ be all the variables of M. Thus in this context we set

$$F(y,\vec{x}_r) = I_{m+1}(X_1(y,\vec{x}_r), \ldots, X_m(y,\vec{x}_r), IC(y,\vec{x}_r))$$

---

[11] Or from $n \geq 2$ (Muller 1973) or from $n \geq 3$ (Schwichtenberg 1969).

Now, for some l, the run time of M is $\leq A_n^l(\max(\vec{x}_r))$, for all $\vec{x}_r$. Thus, $f = \lambda \vec{x}_r . \Pi_1^{m+1} F(A_n^l(\max(\vec{x}_r)), \vec{x}_r) \in K_n, n \geq 4$.   □

The following refinement of Corollary 15.2.10 is within the techniques developed in this chapter, benefiting from a more careful exploration of the properties of the pairing function from Example 2.4.34.

15.2.11 Proposition. (Schwichtenberg (1969)) $K_n = \mathcal{E}^{n+1}$, *for* $n \geq 3$.

Proof We abandon the quadratic pairing in favour of the pairing in Example 2.4.34. In fact, we do not need to iterate J, but rather we can define "I_m" directly, for any fixed m:

$$I_m \overset{Def}{=} 2^{x_1+x_2+\cdots+x_m+m} + 2^{x_2+\cdots+x_m+m-1} + 2^{x_3+\cdots+x_m+m-2} + \cdots + 2^{x_m+1} + 1$$

By an extension of the informal argument in Example 2.4.34, I_m is 1-1.

For $1 \leq i \leq m+1$ define by induction on i the functions

$$t_i(z) \overset{Def}{=} 2^{\left\lfloor \log_2(z \dotminus \sum_{1 \leq j < i} t_j(z)) \right\rfloor}$$

Setting $z = I_m(\vec{x}_m)$ we can prove that $t_i(z) = 2^{x_i+x_{i+1}+\cdots+x_m+m-i+1}$ (cf. Exercise 15.4.34). But then, trivially, for $1 \leq i \leq m$,

$$\Pi_i^m(z) = \left\lfloor \log_2 t_i(z) \right\rfloor \dotminus \left\lfloor \log_2 t_{i+1}(z) \right\rfloor \dotminus 1 \qquad (\dagger)$$

Now this seems like hardly being an improvement (over Corollary 15.2.10), given that $\lambda x.2^x \in K_2$ and the projections Π_i^m are given in terms of $\lambda x. \lfloor \log_2 x \rfloor$, which is in K_3 (cf. Proposition 15.1.11). Thus the iteration equation (1) in the previous proof will push things into $K_n, n \geq 4$ once again.

The ingenious idea in Schwichtenberg (1969) is *not to* first project F —to get $f_i(y, \vec{x}) = \Pi_i^k F(y, \vec{x})$ and *then* to manipulate f_i by g_j (of Lemma 15.2.2)— but rather operate, for the simple operations needed (such as $+1$ and $\dotminus 1$), by applying two K_2-functions (other than the Π_i^m) to $I_m(\ldots, x_i, \ldots)$ to effect a $+1$ or $\dotminus 1$ change to x_i.

These functions are a_i^m ("a" for "add") and s_i^m ("s" for "subtract") below. They are chosen so that

(A) $a_i^m \in K_2$ and $s_i^m \in K_2$ (cf. definitions below, and Proposition 15.1.11).
(B) $a_i^m I_m(x_1, \ldots x_i, \ldots, x_m) = I_m(x_1, \ldots x_i + 1, \ldots, x_m)$.
(C) $s_i^m I_m(x_1, \ldots x_i, \ldots, x_m) = I_m(x_1, \ldots x_i \dotminus 1, \ldots, x_m)$.

They are given by

$$a_i^m(z) \overset{Def}{=} 2\left(t_1(z) + \cdots + t_i(z)\right) + t_{i+1}(z) + \cdots + t_{m+1}(z)$$

and

$$s_i^m(z) \overset{Def}{=} \text{if } \Pi_i^m(z) = 0 \text{ then } z \text{ else } \left\lfloor \frac{t_1(z)}{2} \right\rfloor + \cdots + \left\lfloor \frac{t_i(z)}{2} \right\rfloor +$$

$$t_{i+1}(z) + \cdots + t_{m+1}(z)$$

both definitions for $i = 1, \ldots, n$. The reader should be able to verify that what the a_i^m and s_i^m do is respectively what (B) and (C) above indicate (Exercise 15.4.34).

One last detail for the reader to verify: In Theorem 2.4.20 we encounter conditions such as $IC(y, \vec{x}) = L$ and $X_i(y, \vec{a}_n) > 0$. Let us translate the first: Assume I_m codes all the X_i functions and lastly the IC as the m-th function, that is, $F(y, \vec{x}) = I_m(X_1(y, \vec{x}), \ldots, X_{m-1}(y, \vec{x}), IC(y, \vec{x}))$. Thus $IC(y, \vec{x}) = L$ appears in the single recursion as $\Pi_m^m F(y, x) = L$. We claim that $\lambda z.\Pi_i^m z = L$ is in $K_{2,*}$, for *all* $1 \le i \le m$. Indeed

$$\Pi_i^m z = L \equiv t_i(z) = 2^{L+1} t_{i+1}(z)$$

which is in $K_{2,*}$. Similarly, testing $\Pi_m^m F(y, \vec{x}) = 0$. is inexpensive: The iterator tests $\Pi_i^m(z) = 0 \equiv t_i(z) = 2t_{i+1}(z)$, which is in $K_{2,*}$.

All told, Schwichtenberg's coding of the simultaneous recursion iterates on *at most K_2 functions and relations*, so $I_m \in K_3$.   □

15.2.12 Remark. The Ritchie-Cobham property does two things:

1. Connects the definitional with the computation complexity of primitive recursive functions.
2. Is a powerful tool to compare primitive recursive hierarchies.
   Note however that the correspondence between definitional and computational complexity might be overstated sometimes, inflating the computation complexity a priori estimate. For example the program P below

Loop X

    Loop Y

        Loop Z

            $W \leftarrow W + 1$

        end

    end

end

computes $P_W^{XYZ} = \lambda xyz.xyz$, a function in $\mathcal{L}_2$ running in $O(xyz)$ time. The predictions of our theory are (correct but) pessimistic: P_W^{XYZ} runs in time $O(A_3^m(\max(x, y, z)))$, for some m.

The realisation that $\lambda xyz.xyz$ is computable in $O(xyz)$ time on some URM (that simulates our loop program) means that this function is in $\mathcal{E}^3 = \mathcal{L}_2$. The question of placing an arbitrary function given by a program in the lowest level possible of the hierarchies is recursively unsolvable (Meyer and Ritchie 1967). See also Exercise 15.4.35.   □

15.3   Cobham's Class of Feasibly Computable Functions

This section looks into the area of *analysis of algorithms*, but with a strong computability flavour. It retells a result of Cobham (1964) (also retold in Tourlakis 1984) that mathematically characterises the functions that are *deterministic*[12] *polynomial time computable*.

Analysis of algorithms as the title suggests studies the complexity of specific algorithms, devises lower bounds of complexity —that is, run-time functions for selected *problems* that, provably, no algorithm can do faster— and also devises strategies for building *efficient* (invariably meaning "*fast*") algorithms.

In our dealings with complexity (Chaps. 14 and 15) we took the *systemic*, *hierarchy approach* rather than focusing on this or that problem. Most importantly, our complexity (run time) functions were in terms of the *numerical value* of the algorithm's *inputs*. In the area of analysis of algorithms the literature opts on run times that are invariably expressed in terms of the *length* of the *input* (if the input is a *number*, then *length* would refer to the *notation* of said number, be it base-10, or base-2, or some other notational scheme). One influencing factor for this choice is that analysis of algorithms has in its repertoire a significant number of non-numerical algorithms, such as algorithms for *sorting arrays, traversing directed graphs*, checking a *Boolean formula* for *satisfiability*, etc. It would be awkward and rather roundabout —and unnatural— to state the run time of an algorithm that traverses a binary tree by first converting the (naturally) non numerical coding of the tree *into a number* just for the sake of quoting the run time!

 15.3.1 Remark. This section will adopt the approach of quoting algorithm run time in terms of the *input length* of the algorithm.

Nevertheless, *we will continue in our analyses to consider inputs as being numerical* and our programming formalism, to be used whenever we want to write a program, will be an augmented URM that will define shortly.

Needless to say that in quoting *run time*, we will do so in terms of the *notation length* of the numerical input.

As we remarked many times (cf. in particular Chaps. 7, 8 and 12), every set of *strings* over an alphabet Σ of m distinct symbols can be viewed as a set of numbers,

---

[12] "Deterministic" refers to the property of URMs according to which each instruction determines a *unique* next instruction in the computation process.

written in m-adic notation (cf. Definition 7.0.22). Let us estimate the length of a number x so written down by the logarithm function:

The length of a string x over Σ —viewed as a number base-m— is essentially "Big-O" (Sect. 0.4.2) the logarithm base-m of x. To verify, consider

$$x = a_n m^n + a_{n-1} m^{n-1} + \cdots + a_0$$

Then $\overline{\langle 1 \ldots 1 \rangle}_m \leq x \leq \overline{\langle m \ldots m \rangle}_m$, that is,

$$\frac{m^{n+1} - 1}{m - 1} \leq x \leq m \frac{m^{n+1} - 1}{m - 1}$$

or $m^{n+1} - 1 \leq (m - 1)x \leq m(m^{n+1} - 1) < m^{n+2} - 1$ hence

$$m^{n+1} \leq (m - 1)x + 1 < m^{n+2} \tag{1}$$

Now, $n + 1 = |x|$ where $|x|$ here is m-adic length of x. By (1),

$$|x| = n + 1 = \left\lfloor \log_m \big((m - 1)x + 1 \big) \right\rfloor = O(\log_m x) \tag{2}$$

Since

$$\log_b z = \frac{\log_c z}{\log_c b}$$

we can simply state $\log_b z = O(\log_c z)$. Symmetrically, $\log_c z = O(\log_b z)$.

 This observation is at the root of quoting logarithmic bounds as "$O(\log z)$", *omitting the logarithm base*. In particular, we write (2) as

$$|x| = O(\log x) \tag{3}$$

Often we take our time bound functions $t(n)$ from some class of functions, e.g., the class of polynomials, or the class of all of $\mathcal{PR}$. Obviously, if a URM runs within time $O(n^c)$ for some constant c (a so-called *polynomial run time*) it makes a significant difference if we are quoting run time as a function of the *input value x* or of the *input length $|x|$*, when the input alphabet has $m \geq 2$ symbols: This is because x —the input *value*— is $O(m^{|x|})$ by the preceding analysis, and thus a URM, which runs within time $O(x^c)$ *with respect to input value* runs within time $O\big((m^{|x|})^c\big)$,

that is, $O\big(m^{c|x|}\big)$, that is, an exponential bound *with respect to input length*.

For time bounds from the class of polynomials we *invariably quote run time with respect to input length*.

On the other hand, if $t(n)$ comes from a class that contains exponentials, e.g., t is from $\mathcal{PR}$ or even from the class $\mathcal{L}_2$ that contains $\lambda x.2^x$, then it is irrelevant how we quote run time, whether with respect to input value or input size, since $\lambda x.m^{t(x)} \in \mathcal{PR}$ or in $\mathcal{L}_2$ as well. □

15.3.2 Definition. (Feasible Computations) Cobham (1964) A URM computation is called, *technically*,[13] *feasible* iff its run time is bounded by a polynomial, *the bound being quoted with respect to input length.* □

15.3.3 Long Remark.

(A) Thus, *unfeasible*, or also *intractable* —these are all technical terms— computations are those that require exponential run time or more. The distinction is fairly natural and leads to a nice theory. However, from a *practical* point of view, a URM that runs within time $n^{1,000,000}$ has no computations that are any more "feasible" than one that runs within time 2^n. The former is as bad with an input of length 2 as the latter is with an input of length 1,000,000.

Cobham (1964) has given a *machine-independent characterization* of the *class* of *feasibly computable* functions that we present in this section.

(B) "*Machine-independent*" refers to two observations.

   a. His class $\mathcal{L}$ —the context will fend for itself to avoid confusion with the *other* $\mathcal{L} = \mathcal{PR}$— is given as a *closure* of two functions under two operations, with no reference to programs or computations.

   b. Whether one uses URMs (as we do) or Turing Machines for programming, it is well known that each model can simulate the other within polynomial time, meaning, if URM M runs in time $O(t(n))$ then a *simulating* Turing machine for M can be constructed that runs in time $O(t(n)^k)$, for some k where n is the length of the input. And conversely, if Turing Machine M runs in time $O(T(n))$ then a *simulating* URM for M can be constructed that runs in time $O(T(n)^r)$, for some r where n is the length of the input.[14] Therefore, for those cases where we need to quote computation times (with respect to input length) it is immaterial if we do so for the URM or the Turing Machine (which is what was used in Cobham 1964).

The modified URM for this section is this:

15.3.4 Definition. (Modified URM) The modified URM will still be working *on number theoretic functions, however* its *instructions* will also emulate string operations (see below) thus the model is suitable also for computations of functions on *strings*, $f : \Sigma^* \to \Sigma^*$, for some fixed alphabet of Σ of $m > 1$

---

[13] The term is used *officially*!.

[14] Cf. Minsky (1967), Tourlakis (1984). Notably, Papadimitriou (1994), shows that a TM can simulate in polynomial time even an "*augmented*" URM that can access an unbounded number of registers via indirect addressing. Such a URM is called a *random access machine* in the literature, a *RAM*.

symbols. It will *not* have the instructions $Z \leftarrow Z + 1$, $Z \leftarrow Z \div 1$, $Z \leftarrow a$, or if $X = 0$ goto L else goto L' as *primitives* (cf. Sect. 1.1). However these *can* be *emulated* (they are "*macros*").

We may as well call the model, *temporarily*, the m-URM reflecting that its string manipulation is over an alphabet of $m > 1$ symbols.

Each instruction (a)–(d) below is "internally"[15] implemented in *one* step (as an operation on strings).

The *primitive* instructions of the modified URM are (a)–(d). The instructions (1)–(2) are *macros* that are useful when programming (e.g., Exercises 15.4.41 and 15.4.42). Their run time is $O(|X|)$:

(a) if $|X| = 0$ goto L else goto L'.
(b) (Right successor) $X \leftarrow X * d$, for any $d \in \Sigma$ —the concatenation is m-adic.
(c) $X \leftarrow \lfloor X/m \rfloor$ —in m-adic notation, i.e., remainders r are *not* Euclidean $(0 \le r < m)$ but are m-adic: $1 \le r \le m$. This operation assigns to X the string obtained by removing the rightmost m-adic digit of X, assigns 0 (zero) to X if X was zero.
      In the notation of Lemma 15.3.9 below this is known as $X \leftarrow init(X)$.
(d) $X \leftarrow \mod(X, m)$. This operation assigns to X its rightmost digit; assigns 0 (zero) to X if $X = 0$. In Lemma 15.3.9 we write "$last(X)$" for the operation on the right hand side of "$\leftarrow$".

Macro instructions

(1) (Left successor) $X \leftarrow d * X$, for any $d \in \Sigma$ —the concatenation is m-adic.
(2) $X \leftarrow$ the leftmost digit of X, 0 if $X = 0$. In Lemma 15.3.9 we write "$first(X)$" for the operation on the right hand side of "$\leftarrow$".

I used the synonym of "rem", "$\mod$", above to notationally make a distinction. Let us agree that $\mod$ is in the m-adic sense, that is, possible remainders r satisfy $1 \le r \le m$. rem applies in the original Euclidean sense, that is, remainders r satisfy $0 \le r < m$.   □

Looking ahead, let us at this stage —for the sake of convenience— fix the alphabet $\Sigma = \{1, 2\}$. That is, $m = 2$.

One can program an m-URM, M, so that $M_Z^{XY} = \lambda xy.x + y$. One uses the "school method" for addition, that is, proceeding from right (least significant position) to left, adding same-position digits of x and y, *digit by digit*, and *generating* and *propagating* the "carry". Note that

Adding the digits d and d' and a carry of c we get:

- (Case $d = 1$, $d' = 1$ and $c = 1$): We get $3 = 1.2 + 1$, thus, resulting new digit $D = 1$ and new carry $c = 1$.
- (Case $d = 1$, $d' = 2$ and $c = 1$): We get $4 = 1.2 + 2$, thus, resulting new digit $D = 2$ and new carry $c = 1$.

---

[15] In an imaginary compiler.

- (Case $d = 2$, $d' = 2$ and $c = 1$): We get $5 = 2.2 + 1$, thus, resulting new digit $D = 1$ and new carry $c = 2$.
- (Case $d = 1$, $d' = 1$ and $c = 2$): We get $4 = 1.2 + 2$, thus, resulting new digit $D = 2$ and new carry $c = 1$.
- (Case $d = 1$, $d' = 2$ and $c = 2$): We get $5 = 2.2 + 1$, thus, resulting new digit $D = 1$ and new carry $c = 2$.
- (Case $d = 2$, $d' = 2$ and $c = 2$): We get $6 = 2.2 + 2$, thus, resulting new digit $D = 2$ and new carry $c = 2$.

In the same manner (modulo a lot of detail!) —that is, using the "school method"— one can also program $\lambda xy.xy$ on a 2-URM. These "school method" programs will run within $O(|x| + |y|)$ and $O(|x| \cdot |y|)$ times respectively.

See Exercises 15.4.41 and 15.4.42.

We continue our long remark:

(C) **15.3.5 Definition. (The $\mathcal{L}(\Sigma)$)** Cobham's class $\mathcal{L}(\Sigma)$ of the *feasibly computable functions*, where Σ has m symbols, is defined to be $\{f : \mathbb{N}^k \to \mathbb{N} : k \geq 1 \wedge f = M_Y^{\vec{X}_k},$ where M is an m-URM, which for all inputs x_i runs in time $O((\sum_{i=1}^{k} |x_i|)^r)$, for some $r > 0$, where $|x_i|$ indicates m-adic length$\}$. □

(D) Since we can convert a number from m-adic to n-adic ($m > 1$ and $n > 1$) notation in polynomial time with respect to the length of x (cf. Exercise 15.4.40) **Pause**. Which length? m-adic or n-adic? See (3) on p.592◄

it follows that $\mathcal{L}(\Sigma)$ is independent of Σ, and henceforth we drop Σ from the notation and write just $\mathcal{L}$ for Cobham's class of the *feasibly computable functions*.

(E) Cobham's main theorem is that the class of functions, $\mathcal{L}(\Sigma)$, that can be computed in polynomial time with respect to input length is the closure of the set below,

$$\{\lambda x.x * d \text{ (for all } d \in \Sigma), \lambda xy.x^{|y|}\} \tag{1}$$

under *substitutions* (Exercise 2.1.10) and *bounded (right) recursion on m-adic notation*, the latter meaning the schema

$$f(0, \vec{y}_n) = h(\vec{y}_n)$$

$$f(x * d, \vec{y}_n) = g_d(x, \vec{y}_n, f(x, \vec{y}_n)), \text{ for all } d \in \Sigma$$

$$|f(x, \vec{y}_n)| \leq |B(x, \vec{y}_n)|$$

where the functions h, $(g_d)_{d \in \Sigma}$, and B are given and $*$ is m-adic concatenation, m being the cardinality of Σ.

The closure just described we denote by $\mathcal{L}'(\Sigma)$.

We will prove that $\mathcal{L} = \mathcal{L}'(\Sigma)$. Thus, Σ that crept in once again in order to provide a context to speak of "$x * d$" and "$|y|$", once again proves to be immaterial (as long as $m > 1$).

The notation in (1) and in the bounded recursion schema above should be read *number-theoretically*. Thus, $x^{|y|}$ in (1) is ordinary exponentiation but the exponent is the m-adic *length* of y.

We call $x * d = mx + d$, for each $d \in \Sigma$, a *right successor* of x. Its length is, trivially, $|x| + 1$. Note that we identify the empty string λ with 0, because m-adic notation applies to $x \geq 1$ (cf. Chap. 7).

Cobham's theorem is *independent of the choice of alphabet* Σ just as $\mathcal{L}(\Sigma)$ is.

In the proof that the number theoretic definition $\mathcal{L}'(\Sigma)$ and $\mathcal{L}$ coincide *we will need to "fix" a Σ of cardinality $m > 1$ and proceed on this basis.* □

15.3.6 Lemma. $\mathcal{L}'(\Sigma) \subseteq \mathcal{L}$.

Proof Induction over the closure, $\mathcal{L}'(\Sigma)$, and some programming:
For the initial functions of $\mathcal{L}'(\Sigma)$ note

- $R_d = \lambda x.x * d$. Consider the URM M below:
  1 : $X \leftarrow X * d$
  2 : **stop**
  Clearly, $R_d = M_X^X$ and the run time is $O(1)$.
- The case for $\lambda xy.x^{|y|}$. Consider the URM N below
  1 : $E \leftarrow 1$
  $\begin{cases} 2: & \text{if } |y| = 0 \text{ goto } 6 \text{ else goto } 3 \\ 3: & E \leftarrow E \cdot x \text{ (multiplication)} \\ 4: & y \leftarrow \lfloor y/m \rfloor \\ 5: & \text{goto } 2 \end{cases}$
  6 : **stop**

Pause. Our m-URMs do not have the instruction $E \leftarrow 1$ as primitive. Is this an insurmountable problem? (cf. Exercise 15.4.43)◄

In instruction 3 above we used the "macro instruction $E \leftarrow E \cdot x$" justified in doing so by Exercise 15.4.42.

Clearly, $N_E^{xy} = \lambda xy.x^{|y|}$.

Using Remark 15.3.1(3) to simplify working with lengths, we estimate (upper bound of) the run time of N as $O(|x|^2|y|^2)$:

Indeed, the "running product" E *length* is bounded by $O(\log(x^{|y|})) = O(|y| \log x) = O(|y| |x|)$. Thus an upper bound of *each* execution of instruction 3 is $O(|y| |x|^2)$.

Each other instruction (1, 2, 4, 5) has charge 1 (each invocation). So overall, the latter *together* cost an additional $O(|y|)$ *after* $|y|$ *cycles around the loop*.

Instruction 3, $|y|$ times around the loop, thus costs an upper bound of $O(|x|^2|y|^2)$. This is also the overall for N since it dominates $O(|y|)$.

The claim propagates with

1. *Substitution.* We look at the "interesting" case, $\lambda \vec{x}\vec{y}.g(h(\vec{x}), \vec{y})$. By the I.H. on the closure we have timing functions t_g and t_h such that $t_g(z, \vec{y}) =$

$O(\max(|z|, \ldots, |y_i|, \ldots)^n)$, for some n, and $t_h(\vec{x}) = O(\max(\ldots, |x_i|, \ldots)^r)$, for some r.

Thus, $t_f(\vec{x}, \vec{y}) = t_h(\vec{x}) + t_g(h(\vec{x}), \vec{y})$. Equipped with the observation that $|h(\vec{x})| = O(\max(\ldots, |x_i|, \ldots)^r)$ (the length can grow by no more than 1 for each computation step of h) we get —dropping vector notation, i.e., all vectors have length=1, for clarity

$$t_f(x, y) = O((|x|)^r) + O\left(\max(|h(x)|, |y|)^n\right)$$

$$= O((|x|)^r) + O\left(\max(O((|x|)^r), |y|)^n\right)$$

$$= O\left(\max(|x|, |y|)^{n+nr}\right)$$

2. *Bounded recursion on notation*. See Exercise 15.4.44.

□

The converse is straightforward but tedious. The aim is to use the analog of Theorem 2.4.20. But that needs coding in order to handle *simultaneous primitive recursion on notation*. And coding does not come cheap.

15.3.7 Lemma.   *The following are in $\mathcal{L}'(\Sigma)$.*

(1) $\lambda x.x$
(2) $\lambda x.0$
(3) $L_d = \lambda x.d * x;$ *a left successor, for all $d \in \Sigma$.*
(4) $\lambda x.rev(x)$ *(rev(x) has the same sequence of m-adic digits as x, but in* reverse *order).*

Proof

(1) $\lambda x.x$. Grzegorczyk substitution $(y \leftarrow 1)$ in $\lambda xy.x^{|y|}$.
(2) $\lambda x.0$. Grzegorczyk substitution $(x \leftarrow 0)$ in $\lambda x.x$.
(3) $L_d = \lambda x.d * x$. Consider

$$L_d(0) = d$$

$$L_d(x * r) = \left(L_d(x)\right) * r, \text{ for all } r \in \Sigma$$

$$|L_d(x)| \leq |x * 1|$$

In equation two we used the initial function $\lambda z.z * r$ as iterator, for each $r \in \Sigma$. Bounding function is R_1, an initial function.

(4) $\lambda x.rev(x)$. Consider

$$rev(0) = 0$$

$$rev(x * r) = \left(r * rev(x)\right), \text{ for all } r \in \Sigma$$

$$|rev(x)| \leq |x|$$

Equation two can also be written as $rev(x * r) = L_r\big(rev(x)\big)$. The bounding
function is $\lambda x.x \in \mathcal{L}'(\Sigma)$ by (1).

$\square$

15.3.8 Corollary. $\mathcal{L}'(\Sigma)$ *is closed under limited* left *recursion on notation.*

Proof Consider the limited *left* recursion below and assume that h, $(g_d)_{d\in\Sigma}$ and b
are all in $\mathcal{L}'(\Sigma)$.

$$\begin{cases} f(0, \vec{y}) & = h(\vec{y}) \\ f(d * x, \vec{y}) & = g_d(x, \vec{y}, f(x, \vec{y})), \text{ for all } d \in \Sigma \\ |f(x, \vec{y})| & \leq |b(x, \vec{y})| \end{cases} \tag{1}$$

Define a new function $\widetilde{f}$ by a right *recursion* on notation:

$$\begin{cases} \widetilde{f}(0, \vec{y}) & = h(\vec{y}) \\ \widetilde{f}(x * d, \vec{y}) & = g_d(rev(x), \vec{y}, \widetilde{f}(x, \vec{y})), \text{ for all } d \in \Sigma \\ |\widetilde{f}(x, \vec{y})| & \leq |b(rev(x), \vec{y})| \end{cases} \tag{2}$$

First note that

$$\widetilde{f}(x, \vec{y}) = f(rev(x), \vec{y}) \tag{†}$$

for all x, $\vec{y}$ (induction on $|x|$ or equivalently on the formation of x).

Indeed, for $|x| = 0$, we have $x = 0$, $rev(0) = 0$ and thus $\widetilde{f}(0, \vec{y}) = h(\vec{y}) = f(0, \vec{y}) = f(rev(0), \vec{y})$.

Fix $|x|$ and assume (†) —the I.H.

Then for $|x| + 1$ we work with $x * d$, for each $d \in \Sigma$:

$$\widetilde{f}(x * d, \vec{y}) \stackrel{(2)}{=} g_d(rev(x), \vec{y}, \widetilde{f}(x, \vec{y}))$$

$$\stackrel{I.H.}{=} g_d(rev(x), \vec{y}, f(rev(x), \vec{y}))$$

$$\stackrel{(1)}{=} f(d * rev(x), \vec{y})$$

$$= f\big(rev(x * d), \vec{y}\big)$$

Thus, from the established (†), it is also $f(x, \vec{y}) = \widetilde{f}(rev(x), \vec{y})$, and thus $f \in \mathcal{L}'(\Sigma)$ if we validate the bound of $\widetilde{f}$ and hence its membership in $\mathcal{L}'(\Sigma)$.

Indeed, the bounding function in (2) is in $\mathcal{L}'(\Sigma)$ since b and rev are, and
the bounding inequality in (2) is *correct* since $|\widetilde{f}(x, \vec{y})| = |f(rev(x), \vec{y})| \leq |b(rev(x), \vec{y})|$.

$\square$

15.3.9 Lemma.   *The following functions are in $\mathcal{L}'(\Sigma)$.*

(1) $init = \lambda x.$"*what remains after removing the rightmost digit*". *By definition,*
     $init(0) = 0.$
(2) $first = \lambda x.$"*the leftmost digit*". *By definition, $first(0) = 0.$*
(3) $last = \lambda x.$"*the rightmost digit*". *By definition, $last(0) = 0.$*
(4) $tail = \lambda x.$"*what remains after removing the leftmost digit*". *By definition,*
     $tail(0) = 0.$
(5) $\lambda xyz.if\ x = 0\ then\ y\ else\ z.$
(6) $\lambda x.1 \dotminus x.$
(7) $\lambda x.x + 1.$
(8) $\lambda x.|x|.$
(9) $\lambda x.x \dotminus 1.$
(10) $ones = \lambda x.if\ x = 0\ then\ 0\ else\ 1^{|x|},$ *where* "$1^{|x|}$" *denotes a string of $|x|$ ones.*
(11) $sub = if\ |x| \le |y|\ then\ 0\ else\ 1^{|x|-|y|}.$
(12) $\lambda xy.x * y.$

Proof   We sample a few and leave the rest to the reader (Exercise 15.4.45).

(1) $init.$

$$init(0) = 0$$
$$init(x * d) = x, \text{ for all } d \in \Sigma$$
$$|init(x)| \le |x|$$

(2) $first.$

$$first(0) = 0$$
$$first(d * x) = d, \text{ for all } d \in \Sigma$$
$$|first(x)| \le |x * 1|$$

(5) Let us call the "if-then-else" sw (switch) as before. Thus

$$sw(0) = y$$
$$sw(x * d) = z, \text{ for all } d \in \Sigma$$
$$|sw(x, y, z)| \le |(y * 1)^{|z*1|}|$$

(7) $A = \lambda x.x + 1.$

$$A(0) = 1$$
$$A(x * d) = x * (d + 1), \text{ for all } m > d \in \Sigma$$
$$A(x * m) = A(x) * 1$$
$$|A(x)| \le |x * 1|$$

The penultimate line in the schema above says that if there is a carry of 1, then add it to $x = init(x * m)$ and flip m to 1. The 2nd equation uses "$d + 1$". This expression just indicates *which* digit out of *finitely many* we use as the *new* rightmost digit. It does *not* indicate an *operation* as in $x + 1$.

(8) $\lambda x.|x|$.

$$|0| = 0$$

$$|x * d| = |x| + 1, \text{ for all } d \in \Sigma$$

$$||x|| \leq |x|$$

(10) *ones* $= \lambda x.$if $x = 0$ then 0 else $1^{|x|}$.

$$ones(0) = 0$$

$$ones(x * d) = ones(x) * 1, \text{ for all } d \in \Sigma$$

$$|ones(x)| \leq |x|$$

(11) $sub = $ if $|x| \leq |y|$ then 0 else $1^{|x|-|y|}$.

$$sub(x, 0) = ones(x)$$

$$sub(x, y * d) = init(sub(x, y)), \text{ for all } d \in \Sigma$$

$$|sub(x, y)| \leq |x|$$

(12) $\lambda xy.x * y$.

Hint. Left recursion with x as the recursion variable, or right recursion with y as the recursion variable. See Exercise 15.4.45.

□

15.3.10 Remark. We define as before (Definition 15.1.12) the class of $\mathcal{L}'(\Sigma)_*$ predicates by

$$\mathcal{L}'(\Sigma)_* \overset{Def}{=} \{f(\vec{x}) = 0 : f \in \mathcal{L}'(\Sigma)\}$$

That is, a predicate $P(\vec{x})$ is in is $\mathcal{L}'(\Sigma)_*$ iff for some $f \in \mathcal{L}'(\Sigma)$ it is equivalent to the predicate "$f(\vec{x}) = 0$".

In view of Lemma 15.3.9(6), the above definition says the same thing as "$P(\vec{x})$ is in is $\mathcal{L}'(\Sigma)_*$ iff $c_P \in \mathcal{L}'(\Sigma)$".

Recall the notation xBy, xEy and xPy (defined in the Lemma 7.0.31). Correspondingly, we will also make use here of *part of* quantification introduced by Definition 7.0.32. □

15.3.11 Lemma. $\mathcal{L}'(\Sigma)$ *is closed under Boolean operations, and also under* $(\exists x)_{Bz}$ *and* $(\exists x)_{Ez}$ *(also under* $(\forall x)_{Bz}$ *and* $(\forall x)_{Ez}$ *by the Boolean part).*

Proof

- (Boolean operations.) Closure under $\neg$ is due to the $\lambda x.1 \mathbin{\dot-} x$ being in $\mathcal{L}'(\Sigma)$ while the membership of "switch" sw in this set accounts for closure under $\vee$. There is nothing new to these two results.
- (Closure under part of ("B" and "E") quantification.) Let $P(y, \vec{x}) \equiv p(y, \vec{x}) = 0$ and $p \in \mathcal{L}'(\Sigma)$. Let $c_\exists$ be the characteristic function of $(\exists y)_{Bz} P(y, \vec{x})$. Then

$$\begin{cases} c_\exists(0, \vec{x}) & = p(0, \vec{x}) \\ c_\exists(y * d, \vec{x}) & = \text{if } p(y * d, \vec{x}) = 0 \text{ then } 0 \text{ else } c_\exists(y, \vec{x}) \\ |c_\exists(y, \vec{x})| & \leq |p(y, \vec{x})| \end{cases}$$

For $(\exists y)_{Ez} P(y, \vec{x})$ we argue as above but with left limited recursion on notation. □

15.3.12 Lemma. $\lambda x.x = d$, for all $d \in \mathcal{L}'(\Sigma)$, is in $\mathcal{L}'(\Sigma)_*$.

Proof Define f by

$$f(0) = 1, \text{ false}$$

$$f(x * d') = 1, \text{ for all } d' \neq d\text{: false}$$

$$f(x * d) = \text{if } x = 0 \text{ then } 0 \text{ else } 1$$

$$|f(x)| \leq |x * 1|$$

Thus $f \in \mathcal{L}'(\Sigma)$ and $x = d \equiv f(x) = 0$. □

How about $\lambda x.x = 0$? Well, $x = 0 \equiv Id(x) = 0$ where $Id = \lambda x.x$, which is in $\mathcal{L}'(\Sigma)$ (Lemma 15.3.7).

15.3.13 Lemma. *The following predicates are in $\mathcal{L}'(\Sigma)_*$.*

$$\overset{\text{all } d}{}$$

(1) $tally_d(x)$ *meaning* $0 \neq x = \overbrace{dd \cdots d}^{\text{all } d}$.
(2) $\lambda xy.|x| \leq |y|,\ \lambda xy.|x| < |y|,\ \lambda xy.|x| = |y|,\ \lambda xy.|x| \neq |y|$.
(3) $\lambda xy.x = y,\ \lambda xy.x \neq y$.
(4) $\lambda xy.x\,By,\ \lambda xy.x\,Ey$.
(5) $\lambda xyz.z = x * y$.
(6) $\lambda xy.x\,Py$.

Proof

(1) $tally_d(x)$. We are placing the characteristic function, $c_{tally\text{-}d}$, of this predicate in $\mathcal{L}'(\Sigma)$:

$$c_{tally\text{-}d}(0) = 1, \text{ false}$$

$$c_{tally\text{-}d}(x * d') = 1, \text{ for all } d' \neq d\text{: false}$$

$$c_{tally\text{-}d}(x * d) = \text{if } x = 0 \text{ then } 0$$
$$\text{else if } c_{tally\text{-}d}(x) = 0 \text{ then } 0 \text{ else } 1$$
$$|c_{tally\text{-}d}(x)| \leq |x * 1|$$

(2) $\lambda xy.|x| \leq |y|, \lambda xy.|x| < |y|, \lambda xy.|x| = |y|, \lambda xy.|x| \neq |y|$.

- We note that $|x| \leq |y| \equiv sub(x, y) = 0$.
- Next, $|x| < |y| \equiv \neg |y| \leq |x|$,
- $|x| = |y| \equiv |x| \leq |y| \wedge |y| \leq |x|$,
- and $|x| \neq |y| \equiv \neg |y| = |x|$.

(3) $\lambda xy.x = y, \lambda xy.x \neq y$.

$$x = y \equiv (\forall w)_{Bx}(\exists u)_{By}(|w| = |u| \wedge last(w) = last(u))$$

That is, w and u mark equal distances from the beginning of *each* of x and y and thus also mark their *last digits*. These last digits are to be the same (iff condition for $x = y$) for *all choices* of w (and u). As for "$last(w) = last(u)$" this is shorthand for

$$last(w) = 1 \wedge last(u) = 1 \vee \ldots \vee last(w) = m \wedge last(u) = m$$

With $x = y$ settled, we just note $x \neq y \equiv \neg x = y$.

(4) $\lambda xy.xBy, \lambda xy.xEy$.

$$xBy \equiv (\exists w)_{B}x = w$$

$$xEy \equiv (\exists w)_{E}x = w$$

(5) $\lambda xyz.z = x * y$. By (3), let $f \in \mathcal{L}'(\Sigma)$ such that $z = w \equiv f(z, w) = 0$. Then $\lambda zxy.f(z, x * y) \in \mathcal{L}'(\Sigma)$ and $z = x * y \equiv f(z, x * y) = 0$.

(6) $\lambda xy.xPy$. $xPy \equiv (\exists w)_{By}(\exists u)_{Ey}(y = w * u \wedge xEw)$.

$\square$

15.3.14 Corollary. $\mathcal{L}'(\Sigma)_*$ *is closed under* $(\exists y)_{Pz}$ *and* $(\forall y)_{Pz}$.

Proof Let $P(y, \vec{x})$ be in $\mathcal{L}'(\Sigma)_*$.

$$(\exists y)_{Pz}P(y, \vec{x}) \equiv (\exists w)_{Bz}(\exists u)_{Ez}\Big(z = w * u \wedge (\exists y)_{Ew}P(y, \vec{x})\Big) \qquad \square$$

15.3.15 Definition. (The Bounded max Operator) For any predicate $P(y, \vec{x})$ we define

$$(\max y)_{Bz}P(y, \vec{x}) = \begin{cases} \max\{y : yBz \wedge P(y, \vec{x})\} \\ 0 \text{ if the maximum does not exist} \end{cases}$$

Similarly,

$$(\max y)_{Ez} P(y, \vec{x}) = \begin{cases} \max\{y : yEz \wedge P(y, \vec{x})\} \\ 0 \text{ if the maximum does not exist} \end{cases}$$

□

15.3.16 Lemma. *If $P(y, \vec{x}) \in \mathcal{L}'(\Sigma)_*$, then both max-operators defined in Definition 15.3.15 produce a function in $\mathcal{L}'(\Sigma)$.*

Proof The proof is similar to that for closure of $\mathcal{PR}$ under $(\mu y)_{\leq z}$. Let $c_\exists$ be the characteristic function of $(\exists y)_{Bz} P(y, \vec{x})$, and p that of P. Both are in $\mathcal{L}'(\Sigma)$, the latter by assumption and the former by Lemma 15.3.11. Now set $g = \lambda z \vec{x}.(\max y)_{Bz} P(y, \vec{x})$.

$$g(0, \vec{x}) = 0$$

For all $d \in \Sigma$

$$g(z * d, \vec{x}) = \text{if } c_\exists(z * d, \vec{x}) = 0 \text{ then } \begin{cases} \text{if } p(z * d, \vec{x}) = 0 \\ \quad \text{then } z * d \\ \quad \text{else } g(z, \vec{x}) \end{cases} \text{ else } 0$$

$$|g(z, \vec{x})| \leq |z|$$

Thus $g \in \mathcal{L}'(\Sigma)$.

The case for $(\max y)_{Ez}$ is entirely analogous, but instead one should use left recursion (Exercise 15.4.46). □

We next look back to Lemma 7.0.35 where $maxtal_1$ was essentially defined, where however $\Sigma = \{1, 2\}$, that is, $m = 2$. This is not a problem, and while we sailed ahead to this point with an *unspecified* fixed $m > 1$ *we hereby fix it to* 2. We will use the Quine-Smullyan-Bennett coding (Quine 1946; Smullyan 1961; Bennett 1962).

On the programming side we have the freedom —as we have noted— to work with *any* Σ without changing the URM-generated class $\mathcal{L}(\Sigma)$ (note no prime!), which led to naming it just $\mathcal{L}$. Thus, we opt to work with $\Sigma = \{1, 2\}$ for the balance of the section.

As the purpose here is to work in $\mathcal{L}'(\Sigma)$, *not* in $\mathcal{R}_{alt}$ —with the Σ finally disclosed— we will repurpose Lemma 7.0.35 and restate and reprove the parts we need here.

Notation follows the conventions of the current section.

15.3.17 Lemma. *The following predicates and functions are in $\mathcal{L}'(\Sigma)_*$ or $\mathcal{L}'(\Sigma)$.*

(1) $tally_1(x) \equiv^{Def} x$ is a tally of ones
(2) $y = maxtal_1(x)$; means that y is the maximum-length tally of ones that is part of x. Convention: If $1Px$ is false, then $maxtal_1(x) = 0$.

(3) $\lambda x.maxtal_1(x)$.

(4) $J(x, y)$, which means that $z = J(x, y)$ is dyadic (1-1) code for the pair (x, y) of numbers (see proof below for the exact coding used).

Proof

(1) This is proved in Lemma 15.3.13 with the definition requiring that $x \neq 0$. In Chap. 7 we did not so require, but this does not present a technical problem in either direction.

(2) $y = maxtal_1(x) \equiv tally_1(y) \wedge yPx \wedge \neg 1yPx \vee \neg 1Px \wedge y = 0.$[16]

(3) $\lambda x.maxtal_1(x) \in \mathcal{L}'(\Sigma)$ follows from (2) above and Lemma 15.3.16 noting that $maxtal_1(x) = \max(y)_{B\ ones(x)}(y = maxtal_1(x))$.

(4) $J(x, y)$ codes (x, y) via dyadic concatenation as $x * \# * y$ where the separator "#" depends on x and y as was the case with coding arbitrary sequences in Lemma 7.0.35. Thus we define

$$J(x, y) = x * \# * y \quad \text{Quine (1946), Smullyan (1961)} \qquad (\dagger)$$

where $\# = 2 * maxtal_1(x * 2 * y) * 1 * 2.$[17] Thus, $J \in \mathcal{L}'(\Sigma)$.

We next obtain K and L. So let $z = J(x, y)$. Define,

$$g(z) = (\max w)_{Bz} maxtal_1(z)Ew$$

Hence $g(z) = Kz * 2 * maxtal_1(z)$, and $z = Kz * 2 * maxtal_1(z) * 2 * Lz$. Now "solving" for K: $Kz = (\max w)_{Bz}z = w * 2 * maxtal_1(z)$. Clearly $K \in \mathcal{L}'(\Sigma)$ and the proof for L is entirely similar.

□

We will not be using here the coding detailed in the discussion immediately following Lemma 7.0.34. Instead, we will use the coding/decoding *for pairs only* — J, K, L defined above— to define (fixed length) sequence coding using the method of Lemma 15.2.2.

We can now state without proof an obvious lemma, following Definition 2.4.35 with the J, K, L of Lemma 15.3.17:

15.3.18 Lemma. (Sequence Coding in $\mathcal{L}'(\Sigma)$) For each $n \geq 1$, the $I_n = \lambda \vec{x}_n.[\![\,\vec{x}_n\,]\!]^{(n)}$ and Π_i^n, $i = 1, 2, \ldots, n$ defined in Definition 2.4.35 —using the J, K, L from Lemma 15.3.17— are in in $\mathcal{L}'(\Sigma)$.

15.3.19 Proposition. $\mathcal{L}'(\Sigma)$ is closed under limited simultaneous recursion on notation (both left and right).

---

[16] As in Chap. 7 we will often omit the concatenation operator $*$ and write xy for $x * y$ where the context will not allow the confusion of xy with "implied multiplication" $x \times y$. Incidentally, it should be clear that "$\neg 1yPx$" is an abbreviation of "$\neg\big((1y)Px\big)$". Also note that I could have written "$\neg y1Px$" instead of "$\neg 1yPx$".

[17] The "2" in $x * 2 * y$ protects against a super long $tally_1$ that ends x and begins y.

Proof Let h, g_i^d and b_i be in $\mathcal{L}'(\Sigma)$. Then so are the f_i defined below.

$$\text{for } i = 1, \ldots, k \begin{cases} f_i(0, \vec{y}) & = h_i(\vec{y}) \\ f_i(x * d, \vec{y}) & = g_i^d(x, \vec{y}, f_1(x, \vec{y}), \ldots, f_k(x, \vec{y})), \ d = 1, 2 \\ |f_i(x, \vec{y})| & \leq |b_i(x, \vec{y})| \end{cases}$$

(1)

We convert the simultaneous recursion to a simple (single) limited recursion using the coding I_k and Π_i^k of Lemma 15.3.18. The projections are expressed as in Example 2.4.37, as substitutions using only K and L, the latter defined in Lemma 15.3.17(4) (see also Exercise 2.5.9).

So, following the process of Theorem 2.4.17, we set

$$F(x, \vec{y}) \overset{Def}{=} [\![f_1(x, \vec{y}), \ldots, f_k(x, \vec{y})]\!]^{(k)}$$

thus the simultaneous recursion becomes a simple limited recursion

$$F(0, \vec{y}) = [\![h_1(x, \vec{y}), \ldots, h_k(x, \vec{y})]\!]^{(k)}$$

For $d = 1, 2$, $F(x * d, \vec{y}) = [\![\ldots, g_i^d(x, \vec{y}, \Pi_1^k F(x, \vec{y}), \ldots, \Pi_k^k F(x, \vec{y})), \ldots]\!]^{(k)}$

$$|F(x, \vec{y})| \leq [\![\ldots, |b_i(x, \vec{y})|, \ldots]\!]^{(k)}$$

where the inequality is valid since the J is clearly (review Lemma 15.3.17(4)) increasing with respect to both variables and composing it a finite number of times with itself does not change this fact.

By the placement of the $[\![\ldots]\!]^{(k)}$ and the Π_i^k (all in $\mathcal{L}'(\Sigma)$), it is $F \in \mathcal{L}'(\Sigma)$. Hence so is $f_i = \Pi_i^k F$.

The case for *left* limited simultaneous recursion on notation is left to the reader (Exercise 15.4.48). □

15.3.20 Proposition. *Let M be a 2-URM with variables $x_1, x_2, \ldots x_{n+1}, x_{n+2}, \ldots, x_m$, of which x_i, for $i = 1, \ldots, n$, are input variables while x_1 is also the output variable.*

With reference to Definition 15.3.4 and the above discussion, the simulating functions $\lambda y \vec{a}_n . X_i(y, \vec{a}_n)$, for $i = 1, \ldots m$, and $\lambda y \vec{a}_n . IC(y, \vec{a}_n)$ are in $\mathcal{L}'(\Sigma)$.

Proof (outline) The *simulating recurrence equations* for a given 2-URM, M, can be practically copied from those for the URM (Theorem 2.4.20).

We have the following simultaneous recursion that defines the simulating functions *using only primary instructions* (a)–(d) in Definition 15.3.4:

$$X_i(0, \vec{a}_n) = a_i, \text{ for } i = 1, \ldots, n$$

$$X_i(0, \vec{a}_n) = 0, \text{ for } i = n + 1, \ldots, m$$

$$IC(0, \vec{a}_n) = 1$$

For $y \geq 0$ and $i = 1, \ldots, m$, and $d = 1, 2$,

$$X_i(y * d, \vec{a}_n) = \begin{cases} X_i(y, \vec{a}_n) * d & \text{if } IC(y, \vec{a}_n) = L \text{ and "}L : \mathbf{x}_i \leftarrow \mathbf{x}_i * d\text{" } \in M \\ init\big(X_i(y, \vec{a}_n)\big) & \text{if } IC(y, \vec{a}_n) = L \text{ and "}L : \mathbf{x}_i \leftarrow init(\mathbf{x}_i)\text{" } \in M \\ last\big(X_i(y, \vec{a}_n)\big) & \text{if } IC(y, \vec{a}_n) = L \text{ and "}L : \mathbf{x}_i \leftarrow last(\mathbf{x}_i)\text{" } \in M \\ X_i(y, \vec{a}_n) & \text{othw} \end{cases}$$

In general, for each $\mathbf{x}_i$, we scan the program M from top to bottom and write a case for each instruction of types (a)–(d) that involves $\mathbf{x}_i$ (Definition 15.3.4):

The recurrence for IC is

$$IC(y * d, \vec{a}_n) = \begin{cases} L' & \text{if } IC(y, \vec{a}_n) = L \text{ and "}L : \text{ if } |\mathbf{x}_i| = 0 \text{ goto } L' \text{ else} \\ & \text{goto } L''\text{" is in } M \text{ and } |X_i(y, \vec{a}_n)| = 0 \\ L'' & \text{if } IC(y, \vec{a}_n) = L \text{ and "}L : \text{ if } |\mathbf{x}_i| = 0 \text{ goto } L' \text{ else} \\ & \text{goto } L''\text{" is in } M \text{ and } |X_i(y, \vec{a}_n)| > 0 \\ k & \text{if } IC(y, \vec{a}_n) = k \text{ and "}k : \textbf{stop}\text{" is in } M \\ IC(y, \vec{a}_n) + 1 & \text{othw} \end{cases}$$

Since the iterator functions only utilize the functions $\lambda z.z * d$, $\lambda z.init(z)$, $\lambda z.last(z)$, $\lambda z.z + 1$, $\lambda z.z$, and the iterator predicates use only $\lambda z.|z| = 0$, and $\lambda z.|z| > 0$ —all in $\mathcal{L}'(\Sigma)$ and $\mathcal{L}'(\Sigma)_*$— it will follow that all the simulating functions are in $\mathcal{L}'(\Sigma)$, once we show that all the functions on the right hand side are bounded by some functions b_i in $\mathcal{L}'(\Sigma)$.

To this end note

1. $IC(y, \vec{a}_n) \leq k$, where k is the label of **stop**. Thus $|IC(y, \vec{a}_n)| \leq |k|$.
   Note that $\lambda y\vec{a}_n.|k| \in \mathcal{L}'(\Sigma)$. (Why?)
2. As the y is being built by adding digits to its right (top iterator only contributes to growth of length) we note that, at "step" y, $|X_i(y, \vec{a}_n)|$ must be bounded *at most* by the length of the expression equal to the *maximum input* — $\max_{i \leq n}(\ldots, a_i, \ldots)$— concatenated with y:

$$|X_i(y, \vec{a}_n)| \leq |\max_{i \leq n}(\ldots, a_i, \ldots) * y|$$

But $\lambda y\vec{a}_n. \max_{i \leq n}(\ldots, a_i, \ldots) * y \in \mathcal{L}'(\Sigma)$ (Why?).

□

The step counting variable y indicates elapsed time via its length $|y|$, that is, for each i, $X_i(y, \vec{a}_n)$ is the value of $\mathbf{x}_i$ at step $|y|$ of the computation.

We can now prove

15.3.21 Theorem. (Cobham (1964)) $\mathcal{L} \subseteq \mathcal{L}'(\Sigma)$.

Proof Let $\lambda \vec{x}_n . f(\vec{x}_n) \in \mathcal{L}$. That is,

$$f = M_{x_1}^{\vec{x}_n} \tag{†}$$

and suppose that the 2-URM M runs within

$$O(|x_1 * x_2 * \ldots * x_n|^r)$$

time.

That is, for some constant C', the run time, $t_f(\vec{x}_n)$, is *bounded, for all* $\vec{x}_n$, by

$$\underbrace{\left| (x_1 * x_2 * \ldots * x_n * 1) * \ldots * (x_1 * x_2 * \ldots * x_n * 1) \right|}_{C' \text{ occurrences of } (x_1 * x_2 * \ldots * x_n * 1)}^{r}$$

where the addition of "$*1$" along with the C' contributes the "C and L" discussed in Definition 0.4.29 towards obtaining the display $(1')$ on p.69.

Now $B = \lambda \vec{x}_n . (x_1 * x_2 * \ldots * x_n * 1) * \ldots * (x_1 * x_2 * \ldots * x_n * 1) \in \mathcal{L}'(\Sigma)$ (Exercise 15.4.49).

How do we legitimise (in $\mathcal{L}'(\Sigma)$) the bound $|B(\vec{x}_n)|^r$, for a fixed r, via a function $\mathcal{B} \in \mathcal{L}'(\Sigma)$ that satisfies $|B(\vec{x}_n)|^r \le |\mathcal{B}(\vec{x}_n)|$, for all $\vec{x}_n$?

Define a sequence of functions, $g_1(y) = y$, $g_2(y) = (g_1(y))^{|y|}$, and in general

$$g_{k+1}(y) = (g_k(y))^{|y|}$$

Trivially (substitution) all the g_i (for any fixed i) are in $\mathcal{L}'(\Sigma)$.

Next, by induction on k we can prove $|g_k(y)| = O(|y|^k)$. Indeed, $|g_1(y)| = |y|$. Then

$$|g_{k+1}(y)| = \left| (g_k(y))^{|y|} \right| \stackrel{I.H. \text{ and } \text{Remark } 15.3.1(3)}{=} O(|y| \times |y|^k) = O(|y|^{k+1})$$

Thus, for a properly chosen constant C'',

$$|B(\vec{x}_n)|^r \le \overbrace{|g_r(B(\vec{x}_n) * 1) * \ldots * g_r(B(\vec{x}_n) * 1)|}^{C'' \text{ occurrences of } g_r(B(\vec{x}_n) * 1)}$$

Clearly, if we set

$$\mathcal{B}(\vec{x}_n) = \overbrace{g_r(B(\vec{x}_n) * 1) * \ldots * g_r(B(\vec{x}_n) * 1)}^{C'' \text{ occurrences of } g_r(B(\vec{x}_n) * 1)}$$

we have $\mathcal{B} \in \mathcal{L}'(\Sigma)$, and also $f = \lambda \vec{x}_n . X_1(\mathcal{B}(\vec{x}_n), \vec{x}_n) \in \mathcal{L}'(\Sigma)$. □

15.3.22 Remark. Cobham's theorem also establishes the independence from Σ of Definition (E) that appears within Remark 15.3.3.

Cobham (1964) does not contain a proof of his Theorem 15.3.21 and it appears that Cobham never published a proof. Earlier proofs, all with Turing machines as the programming language, appeared in Von Henke et al. (1973) — with no reference to Cobham (1964) and using a special TM model which is not straightforwardly provable as being equivalent to the "standard TM— also in Dowd (1976) and Tourlakis (1984). (Odifreddi 1999, VIII.2) contains extensive coverage on polynomial time bounded computation and derives Cobham's theorem from a later result: Buss (1986); the programming vehicle again being the Turing machine.

The class of $\mathcal{L}$-relations, $\mathcal{L}_*$, is nowadays denoted usually by the symbol "P" —the class of *deterministic* polynomial time *decidable* relations. Along with the corresponding class "NP" of *nondeterministic* polynomial time *verifiable* relations these two notations participate in the statement of one of the major open problems of complexity theory, "is $P=NP$?".

Note the distinction of *decidable* vs. *verifiable*. Procedurally, a nondeterministic machine —whether a TM or a URM— allows the user to exercise "guessing" in choosing the next instruction to execute during a computation, much as a user exercises choice when engaged in writing a (logical) *proof*. That is, a nondeterministic programming model has instructions that have *no unique* successor instruction.

Mathematically, a *nondeterministic* process, independently of any programming formalism, *verifies* an *existential statement*, $(\exists y)P(y, \vec{x})$ —for a given input $\vec{x}$— if the statement is true for said input. The process does not care about $\neg(\exists y)P(y, \vec{x})$, so it does not care to *decide*. This is totally consistent with how we view semi-recursive relations.

Now, in the complexity domain, one wants to do their verifications *efficiently*. Thus, we may view NP, that is, the *nondeterministic version* of $\mathcal{L}_*$ of Cobham's, as the set of *polynomial time bounded* verifiable problems.

Hence one may *define* NP —or $\mathcal{L}_*^{ND}$ ("ND" for nondeterministic)— as the set of relations of the form

$$(\exists y)_{\leq f(\vec{x}_n)} P(y, \vec{x}_n), \text{ for some } k \text{ and } f \in \mathcal{L}, \text{ with } |f(\vec{x})| = O\left(\left(\sum_{1 \leq i \leq n} |x_i|\right)^k\right)$$

where $P(y, \vec{x}_n) \in \mathcal{L}_* = P$.

With this mathematical definition of "*polynomial time verifiable relation*" the following becomes straightforward (Exercise 15.4.49):

15.3.23 Theorem. $P = NP$ (that is, $\mathcal{L}_* = \mathcal{L}_*^{ND}$) iff $\mathcal{L}_*$ is closed under $(\exists y)_{\leq z}$.

The above theorem was originally proved by Cook using the encoding of nondeterministic Turing machine computations by Boolean formulas that he introduced in his proof of the NP-completeness of the problem "SAT": "is the Boolean formula A satisfiable?"

Cook circulated his proof of Theorem 15.3.23 to his graduate complexity class (University of Toronto, 1971). The author who was a student in this class gave a

different proof at the time, based on the Quine-Smullyan arithmetisation of nondeterministic TM computations, and constructed a Kleene predicate $T(z, x, y)$ in $\mathcal{L}_*$
(this is retold in Tourlakis 1984). Thus an input x is accepted by the nondeterministic
TM of code z by a computation of length $O(|x|^k)$ iff $(\exists y)_{\leq \mathscr{B}(x)} T(z, x, y)$ where
$\mathscr{B} \in \mathcal{L}$ is a version of the "$\mathscr{B}$" constructed in the proof of Theorem 15.3.21 above
and satisfies $|\mathscr{B}(x)| = O(|x|^k)$. □

15.4 Exercises

1. Prove Proposition 15.1.2.
2. Prove the majorising lemma for $\mathcal{K}^{sim}$ (Lemma 15.1.8).
3. Prove Proposition 15.1.9.
4. Complete the proof of Proposition 15.1.11, including the details missing from
   the parts of the proof substantially given in the text.
5. Prove Lemma 15.1.13.
6. Complete the missing details in the proof of Theorem 15.1.17, namely, why is
   $K_0 \subseteq \mathcal{C}$ and why is $ifte \in \mathcal{C}$?
7. Prove Theorem 15.1.22.
8. Prove that the E^n, $n \geq 0$, are closed under $(\mu y)_{<z}$ and $(\mu y)_{\leq z}$.
9. Prove that if f of n arguments $\vec{x}$ is in $\mathcal{E}^{-1}$, then there is some $i \leq n$ (depending
   on f) such that, for all $\vec{x}$, $f(\vec{x}) \leq x_i$, or there is a k such that, for all $\vec{x}$, $f(\vec{x}) \leq$
   k.
10. Prove that if f of n arguments $\vec{x}$ is in $\mathcal{E}^{-1}$, then there is some $i \leq n$ (depending
    on f) such that $f(\vec{x}) \leq x_i$ a.e.
11. Prove that if f of n arguments $\vec{x}$ is in $\mathcal{E}^0$, then for some $i \leq n$ and k (depending
    on f) we have, for all $\vec{x}$, $f(\vec{x}) \leq x_i + k$.
12. Prove that if f of n arguments $\vec{x}$ is in $\mathcal{E}^1$, then for some k and l (depending on
    f) we have, for all $\vec{x}$, $f(\vec{x}) \leq k \max(\vec{x}) + l$.
13. Prove that if f of n arguments $\vec{x}$ is in $\mathcal{E}^2$, then for some c, l and k (depending
    on f) we have, for all $\vec{x}$, $f(\vec{x}) \leq c(\max(\vec{x}))^k + l$.
14. Prove $\mathcal{E}^{-1} \subseteq \mathcal{E}^0$. Is this a proper inclusion? Why?
15. Prove that if-then-else is not in $\mathcal{E}^0$.
16. Prove that $\lambda xy. \max(x, y)$ is not in $\mathcal{E}^0$.
17. Prove that $\lambda xy. \min(x, y)$ *is* in $\mathcal{E}^{-1}$.
18. Compete the proof of Theorem 15.1.37.
19. Prove that $\mathcal{E}^0$ is closed under *restricted summation*, that is, if f is in $\mathcal{E}^0$, then
    so is $\lambda z\vec{x}. \sum_{i \leq x} (1 \dotminus f(i, \vec{x}))$.
20. Prove that $\lambda xyz.z = g_n(x, y) \in \mathcal{E}_*^{-1}$, where g_n is Grzegorczyk's Ackermann
    function.
21. Use Kalmár's definition of elementary functions $\mathcal{E}$ and prove that $\lambda xy.xy$ and
    $\lambda xy.x^y$ belong to this class.

22. Add to Kalmár's definition of elementary functions the initial functions $\lambda xy.xy$ and $\lambda xy.x^y$ and drop the operation $\prod_{\leq x}$ from the given outright list.
  Show that this new closure (under substitution and $\sum_{\leq x}$ only) is still $\mathcal{E}$.

23. Using 21. proceed to prove that $\mathcal{E}$ contains if-then-else and is closed under $(\overset{\circ}{\mu}y)_{\leq x}$, and moreover $\mathcal{E}_*$ is closed under Boolean operations and $(\exists y)_{\leq x}$.

24. Define $pow(n, x) = (\overset{\circ}{\mu}y)_{\leq x}(\neg n^{y+1}|x)$.
  Show this function is in $\mathcal{E}$ (Grzegorczyk 1953).

25. Set $next(x, y)$ to mean "x and y are primes and y is the least prime after x".
  Show that $next(x, y) \in \mathcal{E}_*$ (Grzegorczyk 1953).

26. Prove that the projections of the pairing function $J(x, y) = (x + y)^2 + x$ are in $\mathcal{E}^{-1}$. Find other pairing functions with projections in $\mathcal{E}^{-1}$

27. Show that $\mathcal{E}$ is closed under limited recursion (prime power coding; using 24. and 25. help).

28. Prove that the Kleene T-predicate (for functionals with no type-1 inputs) introduced as $T^{(n,0)}$ in Theorem 13.4.4 —as defined (ibid), *with no changes*— is in $\mathcal{E}_*^{-1}$.

29. Prove that every semi-recursive set is the range of an $\mathcal{E}^{-1}$-function (Grzegorczyk 1953 shows so for $\mathcal{E}^0$ which was in the original Grzegorczyk hierarchy —$\mathcal{E}^{-1}$ was not).

30. Prove that the equivalence problem of functions in any $\mathcal{E}^n$, $n \geq -1$, is not semi-recursive.

31. Grzegorczyk (1953) Show that $\mathcal{E} = \mathcal{E}^3$.

32. Show that that

$$\left. \begin{matrix} 2^{\cdot^{\cdot^{\cdot^{2^x}}}} \\ 2^{\cdot} \end{matrix} \right\} x \text{ 2's}$$

is *not* in $\mathcal{E}$. Conclude that $\mathcal{E} \subsetneq \mathcal{PR}$.

33. Prove that $\lambda z.\lfloor \sqrt{z} \rfloor \in K_3$.

34. Prove all claims made with incomplete or no proofs in the proof of Proposition 15.2.11. Incidentally, what is t_{m+1}?

35. Prove the problem of finding the smallest n such that a given $f \in \mathcal{L}_k$ (some k) is in $\mathcal{L}_n$ is unsolvable (Meyer and Ritchie 1967).

36. Show that if the input alphabet has only one symbol, then quoting run time as a function of the value of the input or as a function of the length of the input makes no difference.

37. Prove that $\mathcal{PR}$ has the Ritchie-Cobham property.

38. Prove that $\mathcal{E}_*$ —the *elementary* predicates of Kalmár— is the closure of $\{\lambda xyz.z = xy\}$ under *explicit transformations*, Boolean operations, bounded quantification and substitution of the call 2^x into variables.

By *explicit transformations* we mean the Grzegorczyk substitution operations but with the substitution of an arbitrary function call for a variable *not* included.[18]

Hint. Let $\mathcal{E}'_*$ be the closure defined above. Let

$$\mathcal{E}'' \overset{Def}{=} \left\{ f : \lambda y \vec{x}.y = f(\vec{x}) \in \mathcal{E}'_* \wedge f(\vec{x}) \le 2^{\cdot^{\cdot^{2^{\max(\vec{x})}}}} \left.\right\} \begin{matrix} k \ 2s \\ \text{a.e. for some } k \end{matrix} \right\}$$

Show that $\mathcal{E}'' = \mathcal{E}$. Conclude that $\mathcal{E}_* = \mathcal{E}''_* = \mathcal{E}'_*$.

39. What class of predicates do we obtain if in 38 we change "substitution of a call 2^x into a variable" by "substitution of a call to a polynomial into a variable", where the class of *polynomials* is defined to be the closure of $\{\lambda xy.x + y, \lambda xy.xy, \lambda xy.x \dot{-} y\}$ under *substitution*.

   By *explicit transformations* we mean the Grzegorczyk substitution operations but with the substitution of an arbitrary function call for a variable not included.[19]

   Hint. Explore the claim that we obtain the class of predicates that is *alternatively defined* as the closure of $\{\lambda xyz.z = x + y, \lambda xyz.z = x \times y\}$ under *explicit transformations*, Boolean operations, and bounded quantification ($(\exists y)_{\le z}$ and $(\forall y)_{\le z}$). The latter definition is said to define the class of *Constructive Arithmetic* predicates.

40. Prove that conversion from m-adic notation of x to n-adic ($m > 1$ and $n > 1$) of x can be done in time $O(|x|^k)$ for some small k where $|\ldots|$ is m-adic length.

41. Program addition $\lambda xy.x + y$ on a 2-URM so that the program provably it runs within time $O(|x| + |y|)$.

42. Program multiplication $\lambda xy.xy$ on a 2-URM so that the program provably runs within time $O(|x| \cdot |y|)$.

43. Show that we can program the "macro" $X \leftarrow a$, for any constant a, in an m-URM.

44. Complete the proof of Lemma 15.3.6.

45. Conclude the proof of Lemma 15.3.9.

46. Conclude the proof of Lemma 15.3.16.

47. Prove Lemma 15.3.18.

48. Complete the proof of Proposition 15.3.19 (case of limited left recursion on notation).

49. Complete the proof of Theorem 15.3.21 settling any unproved claims therein.

50. Prove that for any $d \in \Sigma$, $maxtal_d \in L'(\Sigma)$.

51. Prove Theorem 15.3.23.

---

[18] This does not disallow the special case of "2^x" that is explicitly postulated.

[19] This does not disallow the special case of a call to a polynomial that is explicitly postulated.

References

P. Axt, Iteration of primitive recursion. Z. Math. Logik **11**, 253–255 (1965)

J. Barwise, *Admissible Sets and Structures* (Springer, New York, 1975)

J. Barwise (ed.), *Handbook of Mathematical Logic* (North-Holland, Amsterdam, 1978)

J. Bennett, *On Spectra*. Ph.D. Thesis, Princeton University, (1962)

R. Bird, *Programs and Machines* (Wiley, Hoboken, 1976)

M. Blum, A machine-independent theory of complexity of recursive functions. J. ACM **14**, 322–336 (1967)

G. Boolos, *The Unprovability of Consistency; An Essay in Modal Logic* (Cambridge University Press, Cambridge, 1979)

G.S. Boolos, J.P. Burgess, R.C. Jeffrey, *Computability and Logic* (Cambridge University Press, 2003)

A. Borodin, Complexity classes of recursive functions and the existence of complexity gaps, in *Proc. ACM Symposium on Theory of Computing* (1969), pp. 67–78

A. Borodin, Computational complexity and the existence of complexity gaps. Technical Report 19, Computer Science, University of Toronto, 1970

N. Bourbaki, *Éléments de Mathématique; Théorie des Ensembles* (Hermann, Paris, 1966)

S.R. Buss, *Bounded Arithmetic* (Bibliopolis, Napoli, 1986)

R. Carnap, *Logische Syntax der Sprache* (Springer, Berlin, 1934)

A. Church, A note on the Entscheidungsproblem. J. Symbolic Logic **1**, 40–41, 101–102 (1936a)

A. Church, An unsolvable problem of elementary number theory. Am. J. Math. **58**, 345–363 (1936b) (Also in M. Davis, *The Undecidable* (Raven Press, Hewlett, 1965), pp. 89–107)

A. Cobham, The intrinsic computational difficulty of functions, in *International Congress for Logic, Methodology and Philosophy of Science*, ed. by Y. Bar-Hillel (North-Holland, Amsterdam, 1964), pp. 24–30

N.J. Cutland, *Computability; An Introduction to Recursive Function Theory* (Cambridge University Press, 1980)

M. Davis, *Computability and Unsolvability* (McGraw-Hill, New York, 1958a)

M. Davis, *The Undecidable* (Raven Press, Hewlett, 1965)

M. Davis, *Computability and Unsolvability* (McGraw-Hill, New York, 1958b).

R. Dedekind, *Was sind und was sollen die Zahlen?* (Vieweg, Braunschweig, 1888). (In English translation by W.W. Beman (R. Dedekind, *Essays on the Theory of Numbers*, Dover Publications, New York, 1963))

R. Dedekind, *Essays on the Theory of Numbers* (Dover Publications, New York, 1963). (First English edition translated by W.W. Beman and published by Open Court Publishing, 1901)

J.C.E. Dekker, Productive sets. Trans. Am. Math. Soc. **78**, 129–149 (1955)

E.W. Dijkstra, Go to statement considered harmful. Commun. ACM **11**(3), 147–148 (1968)

© Springer Nature Switzerland AG 2022
G. Tourlakis, *Computability*, https://doi.org/10.1007/978-3-030-83202-5

M.J. Dowd, Primitive recursive arithmetic with recursion on notation and boundedness. Technical Report TR No. 88, Dept. of Computer Science, University of Toronto, Toronto, 1976

H.B. Enderton, *A Mathematical Introduction to Logic* (Academic Press, New York, 1972)

J.E. Fenstad, *General Recursion Theory; An Axiomatic Approach* (Springer, New York, 1980)

R.M. Friedberg, Two recursively enumerable sets of incomparable degrees of unsolvability. Proc. Nat. Acad. Sci. U.S.A. **43**, 236–238 (1957)

F.W. Von Henke, K. Indemark, K. Weihrauch, Hierararchies of primitive recursive word functions and transactions defined by automata, in *Automata, Languages and Programming (Proc. Symp. Rocquencourt, 1972)* (North-Holland, Amsterdam, 1973), pp. 549–561

K. Gödel, Über formal unentscheidbare Sätze der Principia Mathematica und verwandter Systeme I. Monatshefte für Math. Phys. **38**, 173–198 (1931) (Also in English in M. Davis, *The Undecidable* (Raven Press, Hewlett, 1965), pp. 5–38)

D. Gries, F.B. Schneider, *A Logical Approach to Discrete Math* (Springer, New York, 1994)

A. Grzegorczyk, Some classes of recursive functions. Rozprawy Matematyczne **4**, 1–45 (1953)

P. Halmos, *Naive Set Theory* (Van Nostrand, New York, 1960)

W. Heinermann, *Untersuchungungen über die Rekursionzahlen rekursiver Funktionen*. Ph.D. Thesis, Münster Univers., 1961

D. Hilbert, P. Bernays, *Grundlagen der Mathematik I and II* (Springer, New York, 1968)

P.G. Hinman, *Recursion-Theoretic Hierarchies* (Springer, New York, 1978)

S.T. Hu, *Elements of General Topology* (Holden-Day, San Francisco, London, Amsterdam, 1964)

L. Kalmár, A Simple Example of an Undecidable Arithmetical Problem (Hungarian with German abstract). Mate. Fizikai Lapok **50**, 1–23 (1943)

L. Kalmár, An argument against the plausibility of Church's thesis, in *Constructivity in Mathematics Proc. of the Colloquium* (Amsterdam, 1957), pp. 72–80

E. Kamke, *Theory of Sets* (Dover Publications, New York, 1950). Translated from the 2nd German edition by F. Bagemihl

J.L. Kelley, *General Topology* (Van Nostrand, Princeton, 1955)

S.C. Kleene, General recursive functions of natural numbers. Math. Ann. **112**, 727–742 (1936)

S.C. Kleene, Recursive predicates and quantifiers. Trans. Am. Math. Soc. **53**, 41–73 (1943). (Also in M. Davis, *The Undecidable* (Raven Press, Hewlett, 1965), pp. 255–287)

S.C. Kleene, *Introduction to Metamathematics* (North-Holland, Amsterdam, 1952)

S.C. Kleene, Recursive functionals and quantifiers of finite types, I. Trans. Am. Math. Soc. **91**, 1–52 (1959)

D.E. Knuth, *The Art of Computer Programming*, volume 1. Fundamental Algorithms (Addison-Wesley, Reading, 1973)

D.L. Kreider, R.W. Ritchie, *Notes on Recursive Function Theory*. Lecture Notes for Mathematics 89 (Seminar in Logic), Dartmouth College, Winter Term 1965

K. Kunen, Combinatorics, in (J. Barwise (ed.), *Handbook of Mathematical Logic* (North-Holland, Amsterdam, 1978)), chapter B.3, pp. 371–401

W.J. LeVeque, *Topics in Number Theory*, volumes I and II (Addison-Wesley, Reading, 1956)

A. Levy, *Basic Set Theory* (Springer, New York, 1979)

Z. Manna, *Mathematical Theory of Computation* (McGraw-Hill, New York, 1974)

A.A. Markov, Theory of algorithms. Trans. Am. Math. Soc. **2**(15) (1960)

E. Mendelson, *Introduction to Mathematical Logic*, 3rd edn. (Wadsworth & Brooks, Monterey, 1987)

A.R. Meyer, D.M. Ritchie, Computational complexity and program structure. Technical Report RC-1817, IBM, 1967

M.L. Minsky, *Computation: Finite and Infinite Machines* (Prentice-Hall, Englewood Cliffs, 1967)

J. Moldestad, *Computations in Higher Types* (Springer, New York, 1977)

J.H. Moris, Lambda-Calculus Models of Programming Languages. Technical Report MAC-TR-57, MIT, Cambridge, 1968

Y.N. Moschovakis, Abstract first-order computability. Trans. Am. Math. Soc. **138**, 427–464; 465–504 (1969)

Y.N. Moschovakis, Axioms for computation theories—first draft, in *Logic Colloquium '69*, ed. by R.O. Gandy, C.E.M. Yates (North-Holland, Amsterdam, 1971), pp. 199–255

Y.N. Moschovakis, Kleene's amazing second recursion theorem. Bull. Symb. Logic **16**(2), 189–239 (2010)

A. Mostowski, On definable sets of positive integers. Fund. Math. **34**, 81–112 (1947)

A.A. Muchnik, Negative answer to the problem of reducibility of the theory of algorithms (Russian). Dokl. Akad. Nauk SSSR **108**, 194–197 (1956)

A.A. Muchnik, Solution of Post's reduction problem and of certain other problems in the theory of algorithms. I (Russian). Tr. Mosk. Mat. Obs. **7**, 391–405 (1958)

H. Muller, Characterisation of the elementary functions in terms of depth of nesting of primitive recursive functions. Recursive Funct. Theory Newslett. **5**, 14–15 (1973)

J. Myhill, Creative sets. Z. Math. Logik **1**, 97–108 (1955)

J. Myhill, J.C. Shepherdson, Effective operations on partial recursive functions. Z. Math. Logik **1**, 310–317 (1955)

A. Nerode, General topology and partial recursive functionals, in *Talks in Cornell Summer Inst. Symb. Logic* (1957), pp. 247–251

P. Odifreddi, *Classical Recursion Theory, Volume I*. Number 125 in Studies in Logic and the Foundations of Mathematics (Elsevier, Amsterdam, 1989)

P. Odifreddi, *Classical Recursion Theory, Volume II*. Number 143 in Studies in Logic and the Foundations of Mathematics (Elsevier, Amsterdam, 1999)

C.H. Papadimitriou, *Computational Complexity* (Addison-Wesley, Reading, 1994)

R. Péter, *Recursive Functions* (Academic Press, New York, 1967)

E.L. Post, Finite combinatory processes. J. Symb. Logic **1**, 103–105 (1936)

E.L. Post, Recursively enumerable sets of positive integers and their decision problems. Bull. Am. Math. Soc. **50**, 284–316 (1944)

W.V. Quine, Concatenation as a basis for arithmetic. J. Symb. Logic **11**, 105–114 (1946)

H.G. Rice, On completely recursively enumerable classes and their key arrays. J. Symb. Logic **21**(3), 304–308 (1956)

D.M. Ritchie, *Complexity Classification of Primitive Recursive Functions by their Machine Programs*. Term paper for Applied Mathematics, vol. 230 (Harvard University, Cambridge, 1965)

R.W. Ritchie, Classes of predictably computable functions. Trans. Am. Math. Soc. **106**, 139–173 (1963)

R.M. Robinson, An essentially undecidable axiom system, in *Proc. of the International Congress of Mathematicians*, vol. 1 (1950), pp. 729–730. (Abstract)

H. Rogers, *Theory of Recursive Functions and Effective Computability* (McGraw-Hill, New York, 1967)

J. Barkley Rosser, Extensions of some theorems of Gödel and Church. J. Symb. Logic **1**, 87–91 (1936)

G.E. Sacks, On degrees less than $\mathbf{0}'$. Ann. Math. **77**, 211–231 (1963)

L.P. Sasso, *Degrees of unsolvability of partial functions*. Ph.D. Thesis, University of California, Berkeley, 1971

H. Schwichtenberg, Rekursionszahlen und die Grzegorczyk-Hierarchie. Arch. Math. Logik **12**, 85–97 (1969)

D. Scott, Outline of a mathematical theory of computation, In *Proc. 4th Annual Princeton Conf. on Information Sciences and Systems* (Princeton University, Princeton, 1970), pp. 169–176

J.C. Shepherdson, H.E. Sturgis, Computability of recursive functions. J. ACM **10**, 217–255 (1963)

J.R. Shoenfield, *Mathematical Logic* (Addison-Wesley, Reading, 1967)

J.R. Shoenfield, *Degrees of Unsolvability*. Mathematics Studies, vol. 2 (Elsevier, New York, 1971)

C. Smoryński, The incompleteness theorems, in (J. Barwise (ed.), *Handbook of Mathematical Logic* (North-Holland, Amsterdam, 1978)), chapter D.1, pp. 821–865

R.M. Smullyan, *Theory of Formal Systems*. Number 47 in Annals of Math. Studies (Princeton University Press, Princeton, 1961)

R.M. Smullyan, *Gödel's Incompleteness Theorems* (Oxford University Press, Oxford, 1992)

R.I. Soare, The infinite injury priority method. J. Symb. Logic **41**, 513–530 (1976)

R.I. Soare, *Recursively Enumerable Sets and Degrees*. Perspectives in Mathematical Logic, Omega Series (Springer, Berlin, 1987)

A.L. Tarski, General principles of induction and recursion; the notion of rank in axiomatic set theory and some of its applications. Bull. Am. Math. Soc. **61**, 442–443 (1955). (2 abstracts)

G. Tourlakis, *Computability* (Reston Publishing, Reston, 1984)

G. Tourlakis, Some reflections on the foundations of ordinary recursion theory, and a new proposal. Z. Math. Logik **32**(6), 503–515 (1986)

G. Tourlakis, Recursion in partial type-1 objects with well-behaved oracles. Math. Logic Q. **42**, 449–460 (1996)

G. Tourlakis, *Lectures in Logic and Set Theory, Volume 1: Mathematical Logic* (Cambridge University Press, Cambridge, 2003a)

G. Tourlakis, *Lectures in Logic and Set Theory, Volume 2: Set Theory* (Cambridge University Press, Cambridge, 2003b)

G. Tourlakis, *Mathematical Logic* (Wiley, Hoboken, 2008)

G. Tourlakis, *Theory of Computation* (Wiley, Hoboken, 2012)

B.A. Trakhtenbrot, *Complexity of Algorithms and Computations*. University of Novosibirsk, USSR (course notes), 1967

D. Tsichritzis, P. Weiner, *Some Unsolvable Problems and Partial Solutions*. Technical Report 69, Dept. of Electrical Eng. Comp. Sciences Lab, Princeton University, Princeton, 1968

A.M. Turing, On computable numbers, with an application to the Entscheidungsproblem. Proc. Lond. Math Soc. **2**(42, 43), 230–265, 544–546, (1936, 1937). (Also in M. Davis, *The Undecidable* (Raven Press, Hewlett, 1965), pp. 115–154)

V.A. Uspenskii, On enumeration operators. Dokl. Akad. Nauk SSSR **103**, 773–776 (1955)

O. Veblen, J.W. Young, *Projective Geometry*, vol. I (Ginn and Company, Boston, 1916)

J.C. Warkentin, *Small Classes of Recursive Functions and Relations*. Technical Report CSRR 2052, Dept. of Appl. Analysis and Comp. Science Research Report, University of Waterloo, 1971

R.L. Wilder, *Introduction to the Foundations of Mathematics* (Wiley, New York, 1963)

Notation Index

Chapter 0

S_a, 75

$((\exists x)A)$, 18

$((\forall x)A)$, 15

$(A, <)$, 69

$(A \equiv B)$, 18

$(A \wedge B)$, 18

$(A \vee B)$, 15

$(A \to B)$, 18

$(\neg A)$, 15

$(\vec{x}_n)$, 61

1_A, 66

2^A, 61

$A(\ldots, s, \ldots)$, 22

$A(x, y, \ldots, z)$, 17

$A[s]$, 22

$A[x := s]$, 22

$A[x]$, 17

$A \xrightarrow{f} B \xrightarrow{g} C$, 65

$A \subset B$, 56

$A \subseteq B$, 56

$A \subsetneqq B$, 56

$A \times B$, 61

A^n, 61

B^A, 75

$R \circ S$, 65

R^{-1}, 64

$[x]_R$, 68

$\Gamma \vdash A$, 26

$\Gamma \vdash_{\mathcal{T}} A$, 26

$\bigcap A$, 58

$\bigcup A$, 58

$\emptyset$, 12

$\dfrac{A, A \to B}{B}$, 24

$\dfrac{A \to B}{A \to (\forall x)B}$, 24

$\mathring{S}_a$, 75

$\mathscr{P}(A)$, 61

$\text{Cl}(\mathcal{I}, \mathcal{O})$, 88

$\text{dom}(f)$, 62

$\text{ran}(f)$, 62

$\vdash A$, 26

$\vdash_\Gamma A$, 26

$\vdash_{\mathcal{T}, \Gamma} A$, 26

$\vec{x}_n$, 61

$\{n \in \mathbb{N} : Q(n)\}$, 5

$a R b$, 64

c_S, 87

$f(a) \downarrow$, 61

$f(a) \simeq b$, 61

$f(a) \uparrow$, 61

$f(x) = O(g(x))$, 69

$f(x) = o(g(x))$, 70

$f(x) \sim g(x)$, 70

$f(x) \simeq g(y)$, 63

$f : A \to B$, 62

$f = O(g)$, 69

© Springer Nature Switzerland AG 2022
G. Tourlakis, *Computability*, https://doi.org/10.1007/978-3-030-83202-5

Chapter 1

Chapter 2

Chapter 3

Chapter 4

Chapter 5

Chapter 6

Chapter 7

Chapter 8

Chapter 9

Chapter 14

Chapter 15

Index

© Springer Nature Switzerland AG 2022
G. Tourlakis, *Computability*, https://doi.org/10.1007/978-3-030-83202-5

Printed in the United States
by Baker & Taylor Publisher Services